T0313883

HANDBOOK OF ROAD ECOLOGY

HANDBOOK OF ROAD ECOLOGY

Edited by Rodney van der Ree, Daniel J. Smith and Clara Grilo

WILEY Blackwell

Library of Congress Cataloging-in-Publication Data

Handbook of road ecology / edited by Rodney van der Ree, Daniel J. Smith, and Clara Grilo.
    pages cm
  Includes bibliographical references and index.
  ISBN 978-1-118-56818-7 (cloth)
1. Roadside ecology. 2. Roads–Design and construction–Environmental aspects. 3. Ecological assessment (Biology)
4. Wildlife conservation. I. Van der Ree, Rodney, editor. II. Smith, Daniel J. (Daniel Joseph), 1957– editor.
III. Grilo, Clara, editor.
  QH541.5.R62H36 2015
  577.5'5–dc23
                            2014046643

A catalogue record for this book is available from the British Library.

Cover image: Flap-neck Chameleon photo by and reproduced with permission of Wendy Collinson.
"Bonville Bypass Land Bridge", Pacific Highway, NSW, Australia, photo by Rodney van der Ree

Set in 9/11pt Photina by SPi Publisher Services, Pondicherry, India

CONTENTS

NOTES ON CONTRIBUTORS

Dr Isobel M. Abbott completed her PhD in 2012 on the effects of major roads on bats, at University College Cork, Ireland. She works as a freelance ecologist with broad interests in nature conservation and research, specialising in bat ecology. Recent projects include ecological impact assessments of roads and windfarms on bats.

Dr Tamar Achiron-Frumkin (DPhil, Oxford) is an ecologist and environmental advisor. Since the early 2000s, she has worked on various biodiversity conservation projects, gradually focusing on habitat connectivity and traffic-related issues. She was involved in the design of a few ecoducts and in monitoring usage. She produced a local version of COST 341, and initiated a project investigating the use and prioritization of existing road structures as animal passages to minimise fragmentation for the Israel National Transportation Infrastructure Company.

Graham Alexander is a professor at the University of the Witwatersrand, South Africa, where he heads a research laboratory that focuses on ecology, physiology, biogeography, and conservation of reptiles. He has a particular interest in elucidating causality of range limitation in reptiles, and using this information for conservation purposes.

John Altringham is Professor of Animal Ecology and Conservation at the University of Leeds in the UK. He has published three books on bats and numerous scientific papers and popular articles, many of them on bats and bat conservation. He works with a wide variety of conservation organisations.

Dr Kimberly M. Andrews has a joint position between Jekyll Island State Park Authority and University of Georgia (UGA) Savannah River Ecology Laboratory and is graduate faculty at the UGA Odum School of Ecology. Her lab focuses on small vertebrates and conducts field research on wildlife spatial ecology, land-use effects on habitat quality, and wildlife-human interactions. These field data serve as the basis for habitat management, land-use planning and infrastructure development, conflict resolution, and public education practices.

Fernando Ascensão holds a PhD in conservation biology. He studies animal–road interactions, including its effects on animal movement, landscape connectivity, gene flow, and population persistence. He undertakes his research in Portugal and Brazil, where he is currently a postdoc fellow in Brazilian Road Ecology Research Group (Lavras University). His goal is contributing to the development of sustainable transportation networks worldwide.

Alex Bager is the Coordinator of the Brazilian Road Ecology Centre (CBEE), and Professor at the Universidade Federal de Lavras, Brazil. Since 1995, he has been working on animal–vehicle collisions and barrier effects of roads. He is also coordinator of Urubu (Vulture) System, a social networking application to reduce the environmental impacts of roads and railroads on biodiversity.

Niko Balkenhol is Professor of Wildlife Management at the Faculty of Forestry and Forest Ecology of the University of Göttingen, Germany. His teaching and research focus on spatial and molecular approaches for wildlife ecology, conservation and management.

Christophe Barbraud is a researcher at Centre d'Études Biologiques de Chizé (CEBC) – Centre National de la Recherche Scientifique (CNRS) since 2001. He is mainly interested in modelling population dynamics of vertebrates in relation to climate variability. He uses long-term individual monitoring data (demography, biometrics, ecology at sea) of birds and marine mammals.

Hans Bekker was an eco-engineer with Rijkswaterstaat of the Dutch Ministry of Transport, Public Works and Water Management. He was a program leader with environmental issues, mainly wildlife, roads/rails and traffic, and acted as a bridge between civil engineers and ecologists, policy and projects, scientists and practitioners. Hans co-initiated the Infra Eco Network Europe (IENE), in 1995, was a member of steering committee of ICOET (USA), chaired COST 341 and ran the Dutch Long-Term De-fragmentation Program. Hans retired in 2015.

Dr Roberta Bencini is an associate professor in the Wildlife Research Group, School of Animal Biology, at the University of Western Australia. Since 2005, she has been investigating methods to mitigate the negative effects of roads and other developments on wildlife, including underpasses and rope bridges.

Dr Anna Berthinussen completed her PhD in 2013 on the effects of roads on bats and the effectiveness of current mitigation practice, at the University of Leeds, UK. She is currently working on a Defra-funded study of the interactions between bats and roads, and has plans to continue her career in research and wildlife conservation.

Bradley F. Blackwell, PhD, serves as a research wildlife biologist for the NWRC. His research focuses on exploiting animal sensory ecology and antipredator behavior in the development of technology, for example vehicle lighting systems, and habitat-management guidelines, to reduce animal–vehicle collisions, particularly wildlife–aircraft collisions.

Matt Blank, PhD, is an assistant research professor at the Western Transportation Institute and the Department of Civil Engineering at Montana State University. As part of the Road Ecology program at WTI, Dr Blank performs research focused on fish passage, fish swimming abilities, dam removal and river hydraulics. He has over 20 years of experience in the academic and practical engineering world. He also teaches applied fluid mechanics at MSU, and does water resource consulting with Environmental Resources Management (ERM).

Martijn Boonman is a consultant at Bureau Waardenburg, the Netherlands. He is involved in projects considering monitoring, environmental impact assessments (EIA), ecological infrastructure and the effectiveness of fauna crossings under motorways. An important part of the EIA's consists of studies on bats in windfarms.

Dr Carlos E. Borghi is the Director of the Centro de Investigaciones de la Geósfera y la Biósfera (CIGEOBIO) – Universidad Nacional de San Juan and Consejo Nacional de Investigaciones Científicas y Técnicas (Argentina). His research focuses on animal ecology and animal–plant interaction in deserts and the effect of human perturbations on wildlife.

Stan Boutin is a professor in the Department of Biological Sciences, University of Alberta, and Alberta Biodiversity Conservation Chair. His research interests include forestry–wildlife interactions, cumulative effects, integrated landscape management and population dynamics of boreal vertebrates.

Dr Brian K. Chambers is an assistant professor in the Wildlife Research Group, School of Animal Biology, at the University of Western Australia. The focus of his research revolves around the issue of urbanisation and its impact on native Australian mammals and reptiles through habitat modification, fragmentation and the construction of linear infrastructure such as roads.

Dr Yung En Chee is a quantitative ecologist at The University of Melbourne with research experience in statistical, spatial and ecological modelling. Often working in multi- and interdisciplinary teams, her research focuses on applying ecological and decision analysis theory, models and methods to provide data-driven, practical decision support for conservation and ecosystem management problems. She has authored a reference resource of tools designed to guide and enhance the rigour of Strategic Environmental Assessments (Chee et al. 2011, http://www.academia.edu/3412596/Methodologies_and_Tools_for_Strategic_Assessments_under_the_EPBC_Act_1999).

Professor Jiding Chen received his MSc degree in ecology from Peking University in 1992 and since then works for China Academy of Transportation Sciences (CATS). Now he is the vice president of CATS. His research area includes road ecology, green transportation, scenic byway planning and management, and environmental impact assessment.

Fernando Colchero is an assistant professor in the Department of Mathematics and Computer Science and in the Max-Planck Odense Center for the Biodemography of Aging, University of Southern Denmark. He is a member of the Scientific Board of Jaguar Conservancy, A.C. He is an ecologist by training, and his work focuses on developing statistical methods to understand demographic and spatial dynamics of wild animal populations.

Wendy Collinson is the project executant of the Endangered Wildlife Trust's Wildlife and Roads Project. She has an MSc (Rhodes University, South Africa), which examined the impacts of roads on South African wildlife. She is currently driving initiatives that address the now-recognised threat of roads in South Africa. In addition, she is creating a national network to raise awareness and further quantify road ecology issues through proactive mitigation measures such as a Roadkill Sensitivity Map and best practice guidelines for road development.

Dalia A. Conde is an Assistant Professor in the Institute of Biology and in the Max-Planck Odense Center for the Biodemography of Aging, University of Southern Denmark. She is a member of the Scientific Board of Jaguar Conservancy, A.C. She did her PhD in ecology at the Nicholas School of the Environment, Duke University. Her dissertation focused on the impact of roads on biodiversity.

Patricia Cramer is a research assistant professor at Utah State University in the United States, and an independent wildlife researcher. An expert on wildlife, roads and crossing structures, she works with Departments of Transportation and Wildlife Agencies to research wildlife near roads in Utah, Montana, Idaho, Oregon, Washington and other states. She received the Denver Zoo's 2010 Conservationist of the Year award, and the 2013 US Federal Highways Environmental Excellence Award in Research for her work in Utah.

Dr David B. Croft is a visiting fellow in the School of Biological, Earth and Environmental Sciences at UNSW. With a PhD from the University of Cambridge, David taught vertebrate biology, animal behaviour and ecology, and natural resource management in the arid lands. He has published research on invertebrates, marsupials, sheep, marine mammals and primates. His specialty is the behavioural ecology of kangaroos with a recent focus on interactions with people in livestock enterprises, on roads and in wildlife tourism.

Pamela Cunneyworth is a Director of Colobus Conservation overseeing the primate research and conservation activities of the organisation. She has worked in Africa since 1992 in advocacy related to the international conventions of biodiversity and desertification as well as in testing and implementing solutions addressing primate and forest conservation issues. Currently, she is working to develop best practice guidelines for human–primate conflicts for south-east Kenya.

Gino D'Angelo is the Deer Research Project Leader for the Farmland Wildlife Populations and Research Group of Minnesota Department of Natural Resources in Madelia, MN, USA. Gino's research has focused on the evaluation and development of strategies to minimize deer–vehicle collisions, physiological capabilities of white-tailed deer, deer movement ecology, and management of wildlife damage.

Dr Harriet Davies-Mostert is Head of Conservation at the Endangered Wildlife Trust, one of South Africa's largest conservation NGOs. She provides strategic scientific oversight to conservation projects across southern Africa, promoting practical evidence-based research as the basis for effective strategies to conserve southern Africa's rich biodiversity heritage. President of the South African Wildlife Management Association, Harriet is also a member of the Cat, Canid and Conservation Breeding specialist groups of the IUCN's Species Survival Commission.

Dr W. Richard J. Dean is an ornithologist and research associate of the Percy FitzPatrick Institute for African Ornithology, University of Cape Town. After retiring from academia, he started an indigenous nursery and ecological consulting and restoration business in the arid Karoo region of South Africa with Sue Milton. See http://www.renu-karoo.co.za.

Travis L. DeVault, PhD, serves as a Research Wildlife Biologist and Project Leader for the NWRC. His research centres on wildlife ecology and behaviour, with an emphasis on resolution of human–wildlife conflicts. He is particularly interested in how land-use practices on and near airports can be modified to reduce wildlife–aircraft collisions, while increasing revenue potential and renewable energy production for airports.

Andrea Donaldson is the Conservation Manager at the Kenyan based Colobus Conservation and a PhD student affiliated to Durham University in the United Kingdom. She co-ordinates interdisciplinary research and monitoring projects relating to primate and forest conservation, including human-primate conflicts.

Benjamin Dorsey is a research assistant at Montana State University, where he also obtained his MSc on wildlife mortality along the Canadian Pacific Railway. He has also worked for several years on the Trans-Canada Highway Wildlife Crossings project. When not working he enjoys travelling the world by rail, and rock climbing.

Clinton W. Epps studies connectivity, gene flow, animal movement and wildlife conservation in Tanzania, southern Africa and the United States. He is an associate professor in the Department of Fisheries and Wildlife at Oregon State University. He is interested in documenting, modelling and conserving connectivity of animal populations in fragmented landscapes.

Lenore Fahrig is professor of biology at Carleton University, Ottawa, Canada. The overall goal of her research is to understand how landscape structure – for example spatial patterning of roads, forestry and agricultural regions – affects the abundance, distribution and persistence of organisms. A particular focus is on the effects of roads and traffic on wildlife populations, using a combination of spatial simulation modelling and field studies. She is currently a board member of the Ontario Road Ecology Group.

Philip M. Fearnside is a research professor at Brazil's National Institute for Research in Amazonia (INPA) in Manaus. He lived with settlers on Brazil's Transamazon Highway (BR-230) for 2 years and has also studied the BR-364, BR-163 and BR-319 Highways. He has over 500 publications on developments such as roads and their impacts (see http://philip.inpa.gov.br). Recipient of numerous awards, in 2006 Thompson ISI identified him as the world's second most highly cited scientist in the area of global warming.

Michelle E. Gadd works at the US Fish and Wildlife Service where she oversees the African elephant and African rhino conservation programs. Her research interests include conservation outside of parks and the effects of barriers on African mammals.

Jeffrey W. Gagnon has worked for AZGFD since 1997, currently as a statewide, research biologist, focusing for the past decade on wildlife–highway interactions throughout Arizona, including State Route 260 and US Highway 93 wildlife crossing projects. Jeff works

closely with Arizona Department of Transportation on numerous projects to ensure wildlife concerns are properly addressed. Jeff received his MS from Northern Arizona University where he studied the effects of traffic volumes on elk movements associated with highways and wildlife underpasses.

Dr T. Ganesh is a senior fellow at ATREE. For over three decades, he has worked and advised students on various ecological aspects primarily focussing on plant–animal interaction; bird and primate ecology; ecological restoration and long-term monitoring of forests. He was pivotal in establishing nature clubs and conducting outreach activities in various schools in Western Ghats and also authored a bilingual multi-taxa field guide. He is an avid bird watcher and enjoys travelling to natural landscapes.

Miriam Goosem is Principal Research Fellow in the Centre for Tropical Environmental Sustainability Science at James Cook University, Cairns, Australia. Her research in the field of rainforest road ecology spans 25 years including a variety of vegetation and wildlife fragmentation impacts. She was involved, together with colleagues, in the implementation in rainforest of the first purpose-built underpasses and rope canopy bridges between 1995 and 2005 and continues to monitor their effectiveness.

Clara Grilo is a postdoctoral researcher of the Department of Biology and CESAM, at the University of Aveiro, Portugal. Her primary interest is applied ecological research in support of active conservation projects. Currently, much of her research is focused on the impact of anthropogenic changes to the landscape and effects on wildlife.

Sanjay Gubbi is a wildlife biologist who works on understanding tigers, leopards and their interactions with development and other aspects. He is currently leading research on leopard distribution, density estimation and understanding leopard–human conflict in protected areas, multiple use forests and human dominated landscapes. He is keenly interested in applied conservation activities that have resulted in various on-ground conservation successes in the Western Ghats, southern India. He works with the government to reduce impacts of roads in ecologically sensitive areas.

Éric Guinard is a civil engineer, doctor in Ecology in the Centre d'Études et d'expertise sur les Risques, l'Environnement, la Mobilité et l'Aménagement – Direction Territoriale du Sud-Ouest (CEREMA – DTerSO) near Bordeaux since 2005. He is in charge of expertise and management assistance of ecological studies on road and motorway projects. He also participates in the development of methods and conducts applied research projects, mainly concerning interactions between transportation infrastructure or urban extension and natural habitats.

Kari Gunson has worked for 15 years informing road–wildlife mitigation projects throughout North America. She lives in Ontario, Canada, and works for Eco-Kare International, translating road ecology science into practical mitigation solutions. She has provided expertise for design, placement and monitoring of mitigation measures for a variety of animals. Her work has contributed to 14 peer-reviewed published articles in the fields of road ecology and geographic information science.

Dr Andrew J. Hamer is an ecologist at the Australian Research Centre for Urban Ecology, a division of the Royal Botanic Gardens Melbourne and located at the University of Melbourne. His research is directed towards understanding the drivers underpinning how amphibians and freshwater turtles respond to urbanisation. He is currently involved in a research project investigating the behaviour of Australian frogs at under-road tunnels. He is also researching broad-scale trends in amphibian and turtle populations in the face of increasing urbanisation.

PD Dr-Ing. Heinrich Reck studied agricultural biology and landscape conservation at Hohenheim and Stuttgart Universities and obtained his post-doctoral lecturing qualification (Habilitation) in landscape ecology at Kiel University. He works as a senior researcher and lecturer on the interface between spatial environmental planning and animal ecology and is a member of the state planning council of Schleswig-Holstein, Germany. He has worked on road ecology and application-oriented research on impact mitigation and compensation work since 1990.

Marcel P. Huijser received his MS in population ecology (1992) and his PhD in road ecology (2000) at Wageningen University, the Netherlands. He studied plant–herbivore interactions in wetlands (1992–1995), hedgehog traffic victims and mitigation strategies (1995–1999), and multifunctional land use issues (1999–2002) in the Netherlands. Marcel has been conducting road ecology research for the Western Transportation Institute at Montana State University

(USA) since 2002, and he is currently a visiting professor at the University of São Paulo, Brazil (ESALQ, Piracicaba campus).

Pierre L. Ibisch, Professor for Nature Conservation with Eberswalde University for Sustainable Development, Germany. He holds a research professorship on 'Biodiversity and natural resource management under global change' and is Co-director of the Centre for Econics and Ecosystem Management. He has special interests in adaptation to global change and integration of risk management in adaptive biodiversity conservation management, functionality of ecosystems and conservation priority setting, spatial planning, and protected area management.

Sandra Jacobson is a wildlife biologist for USDA Forest Service, Pacific Southwest Research Station specializing in transportation ecology. She designs mitigation for highway impacts to species ranging from elephants to butterflies internationally. Her projects and graduates have received numerous awards, including from the USA FHWA. She is a Steering Committee member of ICOET, a charter member of the Transportation Research Board's Committee on Ecology and Transportation and a Steering Committee member of the ARC Design Forum for wildlife crossing structures.

Jochen A. G. Jaeger is an associate professor in the Department of Geography, Planning and Environment at Concordia University in Montreal, Canada. He received his PhD in Environmental Sciences from the Swiss Federal Institute of Technology (ETH) in Zurich in 2000. His research is in the fields of landscape ecology with a focus on landscape fragmentation and urban sprawl, road ecology, ecological modelling, environmental indicators, environmental impact assessment and novel concepts of problem-oriented trans-disciplinary research.

Darryl Jones is an Professor at Griffith University, in Brisbane, Australia, and Deputy Director of the Environmental Futures Research Institute at that university. He has been actively engaged in urban ecology since the early 1980s and in road ecology research for over 10 years.

Dr Nina Klar is working at the federal administration of Hamburg, Germany, being responsible for native species conservation. She is especially interested in wildlife species living in human-dominated landscapes. After her research on wildcats and road ecology, she is now conducting conservation projects for urban wildlife.

Angela Kociolek is a Research Scientist at the Western Transportation Institute, Montana State University-Bozeman, Bozeman, USA, where she conducts road ecology research and outreach to transportation professionals. Angela is currently the Technology Transfer Initiative Leader for ARC, a partnership seeking to make wildlife crossing structures a standard practice across North America.

Yaping Kong is a Professor who received her MS degree in ecology from Beijing Normal University in 2002, and since then has worked for the China Academy of Transportation Sciences (CATS). Now she is the vice-director of the Research Centre for Environmental Protection and Transportation Safety. Her research area includes vegetation restoration, water resource protection, road geological disaster control, ecological highway planning and management, transportation policy making, EIA and road ecology.

Stefan Kreft is a researcher with the Centre for Econics and Ecosystem Management, Eberswalde University for Sustainable Development, Germany. Under the impression of rapid land-use changes in South America, his research priorities have gradually shifted away from species conservation to ecosystem-based conservation approaches, addressing adaptation to climate change in particular. Besides a current focus on Europe, developing and transitional countries remain of great interest to him. He is member of the Roadless Areas Initiative of the Society for Conservation Biology.

Dr Tom A. Langen is Professor of biology, Clarkson University. He conducts road-related environmental research including winter road management, predictive modelling of road mortality hotspots, design of wildlife barriers and passageways for turtles, and the impact of highways on habitat connectivity in Costa Rican National Parks. He leads workshops in Latin America and North America on the environmental impact of roads and other infrastructure.

Thomas E. S. Langton is an International Consultant Ecologist based in Suffolk, UK, specialising in the conservation of herpetofauna and their communities and habitats. He has a wide range of experience working for government, industry and the non-profit sectors including linear transport developments and mitigation and applied road ecology solutions. His main activities include practical aspects of habitat and species surveys, habitat restoration and construction, and

species and habitat management. He also works on wildlife law implementation.

Dr Scott LaPoint is a wildlife ecologist at the Max Planck Institute for Ornithology in Radolfzell, Germany. He has investigated mammalian responses to roads as an undergraduate and throughout his graduate studies, including his dissertation where he investigated urban landscape connectivity via movement data collected on free-ranging carnivores.

A. David M. Latham is a wildlife ecologist with Landcare Research, New Zealand. His research interests include vertebrate pest research; predator–prey ecology; spatial ecology; large mammal ecology, conservation and management; and human disturbance–wildlife interactions.

William F. Laurance is a distinguished research professor and Australian Laureate at James Cook University in Cairns, Australia, and also holds the Prince Bernhard Chair in International Nature Conservation at Utrecht University, the Netherlands. He studies the ecology and conservation of tropical forests throughout the world, and to date has authored seven books and over 400 scientific and popular articles. He is a fellow of the American Association for the Advancement of Science and former president of the Association for Tropical Biology and Conservation. He is also director of the Centre for Tropical Environmental and Sustainability Science at James Cook University as well as founder and director of the leading international scientific organisation ALERT—the Alliance of Leading Environmental Researchers and Thinkers.

Dr Enhua Lee is a senior ecologist at the environmental consulting company, Eco Logical Australia. She has prepared numerous biodiversity strategies, biodiversity and natural resource management plans, and environmental impact assessments. Enhua conducted her PhD at UNSW on the ecological impacts of roads in arid ecosystems, investigating impacts on soil, vegetation, kangaroo, small mammal and lizard distributions and abundance, and kangaroo behaviour and mortality.

Dr David Lesbarrères is an associate professor at the Centre for Evolutionary Ecology and Ethical Conservation, Laurentian University in Sudbury, Canada. His main interests are focused on theoretical and applied questions about the evolution and ecology of amphibian species and communities. His research program is currently centred on population genetics in human dominated landscapes, road ecology and emerging infectious diseases, ultimately integrating all these aspects to understand the declines of amphibian populations.

Dr Juan E. Malo is an associate professor and researcher at the Terrestrial Ecology Group of Universidad Autónoma de Madrid. His research interests include ecological interactions and the effects of human activities on wildlife populations, with a special focus to environmental impact assessment of infrastructures and fragmentation.

Carlos Manterola is the General Director of Grupo Anima Efferus A.C. and the Director of Conservation of Jaguar Conservancy, A.C., in Mexico. He was General Director of the conservation NGO Unidos para la Conservación. He has led numerous conservation projects including the establishment of Protected Areas in Mexico, the protection and recovery of the pronghorn antelope in Mexico, management of desert bighorn sheep on Tiburon Island and the conservation of jaguars and their habitat in Mexico and Central America.

Dr Cristina Mata is a postdoctoral researcher at the Terrestrial Ecology Group of Universidad Autónoma de Madrid (Spain). Her main research is focused on monitoring and assessment of mitigation measures aimed at the reduction of habitat fragmentation by roads and railways.

Dr Vinod B. Mathur is the Director, Wildlife Institute of India. He obtained his doctoral degree in wildlife ecology from the University of Oxford in 1991. He is Regional Vice-Chair of the IUCN-World Commission on Protected Areas (WCPA-South Asia). He is a member of UN-IPBES Multidisciplinary Expert Panel (MEP). His areas of interest are Impact Assessment and Road Ecology.

Dr Markus Melber studies the impact of roads on bats as well as the effectiveness of mitigation projects for bats along a heavy-traffic motorway but also the ecology of forest-living bats. Besides working as a research associate at the University of Greifswald, Germany, he has also worked for several German federal agencies. He often acts as an advisor for public agencies on mitigation projects and on conservational topics. His work has resulted in several scientific publications, book chapters and reports.

Dr Suzanne J. Milton is a plant ecologist and research associate of the Percy FitzPatrick Institute, University of Cape Town. After retiring from academia, she started an indigenous nursery and ecological consulting and restoration business in the arid Karoo region of South Africa with Richard Dean. See http://renu-karoo.co.za/. Sue Milton and Richard Dean also founded the Wolwekraal Conservation and Research Organisation.

Mike Misso has been the Manager of Christmas Island and Pulu Keeling National Parks since late 2010. Prior to moving to Christmas Island, Mike worked as a Natural Resource Management facilitator, and prior to this in a range of national park management roles at Kakadu and Uluru Kata Tjuta National Parks in Australia, including as a Planning Officer, Chief Ranger and Natural Resource Manager.

Christa Mosler-Berger is a wildlife biologist and co-manager of the non-profit association WILDTIER SCHWEIZ and responsible for the Swiss Wildlife Information Service. She has been involved in the evaluation of animal detection systems (ADS) since they were first installed in 1993 in Switzerland.

Rob Muller has worked as the Chief Ranger of Christmas Island National Park since mid 2010. One of Rob's key responsibilities is, with other Ranger staff, to coordinate the road management activities for conserving red crabs during their annual breeding migration. Prior to moving to Christmas Island, Rob worked as a Ranger (including as a Chief Ranger), at Kakadu National Park in Australia for over 20 years.

Benezeth Mutayoba is an awardee of 2014 National Geographic/Buffett Award in 'Leadership in African Conservation' and works on wildlife movements, road kill dynamics, connectivity and gene flow in isolated wildlife populations as well as on wildlife health and forensics. He is a professor in the Department of Veterinary Physiology, Biochemistry, Pharmacology and Toxicology, Faculty of Veterinary Medicine, Sokoine University of Agriculture, Tanzania.

Katarzyna Nowak has studied primates and elephants in flooded and montane forests in Tanzania and South Africa. She is currently a junior research fellow at Durham University, UK, and a research associate at the University of the Free State, Qwaqwa, South Africa. She is interested in how flexibility in behavior affects species' capacity for persistence in human-dominated landscapes. She is currently researching samango monkeys' landscape of fear.

Kirk A. Olson has been promoting conservation of migratory ungulates and grazing ecosystems in Mongolia and Central Asian region since 1998. Kirk completed his PhD at the University of Massachusetts, Amherst, and his dissertation focused on the ecology and conservation of Mongolian gazelles. Kirk is a Research Associate at the Smithsonian Conservation Biology Institute and most recently worked with Fauna and Flora International's saiga conservation program.

Mattias Olsson has a PhD in biology and is working at EnviroPlanning AB and part time at SLU (Swedish University of Agricultural Sciences) in the Triekol research program. His research and enquiries are about wildlife and infrastructure, and he regularly works with civil engineers and landscape architects in order to mitigate the negative effects of highways and railroads. When he is not working, he spends time with the family and as a coach for a girl's handball team and a boy's soccer team.

Fabrice Ottburg, BSc, is a research scientist involved in applied and multi-disciplinary research, consultancy and acquisition for various projects in ecology (fundamental ecological research) and habitat fragmentation. He has extensive experience in ecological impact assessments in landscape areas and mitigation/compensation/monitoring studies for large-scale projects. He is also qualified in studies on nature development, ecological nature and juridical development and animal ecology (fishes, amphibians and reptiles).

Eugenia Pallares is General Director of the Mexican conservation NGO Jaguar Conservancy. She has collaborated and coordinated various projects on the conservation of jaguars and their habitat in Mexico, mitigation of the impact of roads on biodiversity in the Mayan Forest, and projects involving environmental policies. She has worked on editorial boards where a number of books, calendars, brochures and other materials have been produced. She is also a member of the Board of the Council for Sustainable Development in Mexico.

Dr Dan Parker is a wildlife biologist, based at Rhodes University in Grahamstown, South Africa. He supervises a large and vibrant post-graduate research school and is particularly interested in the biology and conservation of Africa's large carnivores.

Dr Kirsten M. Parris is a Senior Lecturer in the School of Ecosystem and Forest Sciences, The University of Melbourne. Her research interests include the ecology of urban systems, ecology and conservation biology of amphibians, bioacoustics, field survey methods and ecological ethics.

Ms Claire Patterson-Abrolat runs the Endangered Wildlife Trust's Special Projects Programme which covers a range of projects dealing with the development of innovative, economically viable alternatives to address harmful impacts to the benefit of people and biodiversity.

Sarah E. Perkins is a Lecturer in Ecology at Cardiff University. Sarah established and runs 'Project Splatter' a UK-wide citizen science initiative to collate wildlife roadkill using social media. Sarah is a strong supporter of the value of crowd-sourced data to both scientists and citizens. Away from roads her research focuses on the ecology of wildlife diseases.

H.C. Poornesha works on conservation of wildlife habitats in the Western Ghats of India through GIS analysis and conservation planning. He has also contributed largely to applied conservation issues in the landscape (see http://ncf-india.org/people/h-c-poornesha for further details).

Roger Prodon is a professor at the École Pratique des Hautes Études (EPHE) where he led for 12 years a research team working on vertebrate ecology in Mediterranean and mountain areas. He is mainly interested in bird community dynamics following disturbance (e.g. after fire), long-term monitoring, bird elevational gradients and island ecology.

Dr Asha Rajvanshi heads the EIA Cell of the Wildlife Institute of India (WII). She works in the area of road ecology and has developed a range of best practice guidance manuals for mainstreaming biodiversity in impact assessment in different economic sectors including roads. She has been part of several global EIA initiatives and is a member of IAIA.

Dr Lisa J. Rew is an associate professor at Montana State University. Her research concentrates on the dispersal, distribution and dynamics of weedy plant species, and how best to manage them at a local scale. She is involved with this project due to her interest in how seeds are dispersed by vehicles, and how that could impact wildlife. When she isn't working she can often be found playing in the mountains.

Kevin Roberts is currently the Section Leader – Environment with consulting firm Cardno. From 2007 until 2014, he was the Senior Environmental Specialist (Biodiversity) for the NSW Roads and Maritime Services, Australia. Kevin's responsibilities were developing policy and procedures for managing biodiversity across the organisation. Prior to working for RMS, Kevin has held a range of senior roles in the NSW agencies responsible for regulating and planning for biodiversity conservation.

Dr Carme Rosell is a senior consultant at Minuartia and is part of a research group at the University of Barcelona. She has led numerous projects to design and monitor wildlife passages in roads and high speed railways. Her recent projects are focused on reducing animal-vehicle collisions and improving road maintenance practices. She has co-authored guidelines including the COST341 handbook *Wildlife and Traffic*. Carme is a member of the Infra Eco Network Europe Steering Committee.

Trina Rytwinski is currently working as a post-doc in the Geomatics and Landscape Ecology Research Lab, at Carleton University, Ottawa, Canada. Her research focuses on understanding the circumstances in which roads affect population persistence, specifically looking at species traits and behavioural effects of roads, and ways to mitigate road effects.

Thomas W. Seamans, MS, serves as a supervisory wildlife biologist for the NWRC. His primary research focus is the development and evaluation of wildlife repellents and methods intended to reduce human–wildlife conflicts.

Helio Secco is biologist who graduated from the State University of Northern Rio de Janeiro (UENF), and obtained his MSc in Applied Ecology at Federal University of Lavras (UFLA). In recent years, he participated in several projects at the Brazilian Center for Research in Road Ecology. Helio is currently interested in research areas related to the assessment of environmental impacts of anthropogenic structures on tropical wildlife.

Dr Andreas Seiler received his PhD in wildlife biology in 2003 from the Swedish University of Agricultural Sciences. Since 1994, he has been working on traffic and wildlife related issues, mainly research on animal–vehicle collisions and traffic-related mortality and barrier effects, and broader landscape fragmentation issues. He has been active in COST-341 action and is a member of the Steering Committee and Secretariat of IENE (Infra Eco Network Europe) with a special responsibility for the IENE international conferences.

Dr Nuria Selva is an associate professor at the Institute of Nature Conservation in Krakow, Polish Academy of Sciences. Her research within animal ecology is broad, including large carnivores and scavengers, and conservation biology. She has recently focused on brown bears in the Carpathians, as well as the effects of supplementary feeding and global change on this bear population. She also focuses on conservation policies at European and international levels to protect ecological processes and wilderness, including roadless areas.

K. S. Seshadri is pursuing his PhD in biology at the National University of Singapore. He has varied interests spanning birds, herpeto-fauna and canopy science. He is a recipient of the 'Future Conservationist' award and is actively involved in conservation, education and outreach activities. Though he primarily studies amphibians, he has studied the impact of roads on fauna in south India. He is passionate about bird watching and nature photography.

Fraser Shilling is the Co-Director of the Road Ecology Center and research scientist in the Department of Environmental Science and Policy, University of California, Davis. He obtained his ecology-focused Ph.D. from the university of Southern California. He is a member of several Transportation Research Board committees and leads road ecology research for state and national transportation agencies. He is the lead scientist for wildlifeobserver.net and wildlifecrossing.net, both crowd-sourced datasets for wildlife observation. He also leads research in intermediate-scale monitoring of sea level rise and infrastructural adaptation.

Leonard E. Sielecki is the Wildlife and Environmental Specialist for the British Columbia Ministry of Transportation and Infrastructure. Since 1996, Leonard has been the Province of British Columbia's subject matter expert on wildlife accident monitoring and mitigation. He serves on committees of the National Academies of Sciences, the Transportation Research Board, and the International Conference on Ecology and Transportation (ICOET). Leonard is completing his PhD at the University of Victoria where he developed the Wildlife Hazard Rating System® for motorists.

Anders Sjölund is the National Biodiversity Coordinator for the Swedish Transport Administration. He is also Chair of the nature and cultural heritage group at The Nordic Road Association (NVF), Chair of the Steering Committee for the Infra Eco Network Europe (IENE), member of the Swedish Wildlife Accident Council and member of the Steering Committee for the International Conference on Transport and Ecology (ICOET).

Dr Daniel J. Smith is a research associate and member of the graduate faculty in the Department of Biology at the University of Central Florida and a member of the National Academies Transportation Research Board Subcommittee on Ecology and Transportation. He has over 20 years of experience in the fields of ecology and environmental planning. His primary focus is studying movement patterns and habitat use of terrestrial vertebrates and integrating conservation, transportation and land-use planning.

Kylie Soanes is a PhD candidate at the University of Melbourne, Australia, and is part of the Australian Research Centre for Urban Ecology and the Australian Research Council Centre for Excellence in Environmental Decisions. Her PhD project evaluates the effectiveness of wildlife crossing structures for a gliding marsupial over a major highway. Kylie is interested in evaluating the success of conservation management and restoration projects and designing effective monitoring programs.

Josie Stokes is the Senior Biodiversity Specialist (Environmental Policy) at the NSW Roads and Maritime Services (RMS). Her role is to develop operational environmental policy to assist the RMS in minimising its impact on the environment, review environmental impact assessments and provide expert technical advice to project teams. She has also been an ecologist for the Australian Museum and Parsons Brinckerhoff. She has over 17 years of experience in assessing the impacts of development, particularly of linear infrastructure, on biodiversity across Australia.

Dr Emma Stone is a Research Associate in the Bat Ecology and Bioacoustics Lab at the University of Bristol, UK. She conducts experimental research on the impacts of roost exclusions and the effectiveness of mitigation for bats. Her PhD was on the impact of street lighting on bats and the effectiveness of mitigation legislation for bats. Emma is now conducting applied research on the conservation of bats and carnivores in Malawi and has established the charity Conservation Research Africa to assist.

Martin Strein is a biologist with the German Federal state of Baden-Württemberg who is advising on the implementation of a statewide biotope network. When focusing on wildlife mitigation measures, he uses a broader ecological perspective, rather than a species-specific solution, to support important ecological functions and biodiversity. He is also skilled in the management of large protected areas and has spent many years working for and evaluating national parks, mainly in Africa.

Richard P. J. H. Struijk is a herpetologist at RAVON Foundation (Reptile Amphibian and Fish Conservation, the Netherlands) and is graduate faculty at the Wageningen University and Research Centre. Coordinating several monitoring projects on the use of crossing structures by herpetofauna, he is involved in infrastructural planning and evaluation of mitigation measures. Privately he is working on the conservation and captive propagation of endangered Asian box turtles (*Cuora* sp.).

Paul Sunnucks is a researcher and educator in the School of Biological Sciences at Monash University, Australia. His research interests focus on population biology of animals in natural habitat and those altered by human activities, working with stakeholders to manage landscapes and ecological processes. He has a particular fondness for all ecosystems and life forms.

Adam Switalski is Principal Ecologist for the environmental consulting company, Inroads Consulting LLC. He specializes in the management of forest roads and is an expert in road restoration science and practice. His research is focused on the impact of restoring roads on fish and wildlife habitat. He is working to establish more cost-effective and ecologically sustainable transportation systems in the US Northern Rockies.

Stephen Tonjes has worked 28 years in environmental compliance for the Florida Department of Transportation, and now consults part-time. Before FDOT, he served in the US Coast Guard, taught marine science in the Florida Keys, and monitored compliance for the Coast Guard bridge permit program in Juneau, Alaska, and for the US Fish and Wildlife Service in Washington, DC. He has a special interest in communicating wildlife ecology to transportation professionals and transportation development to wildlife ecologists.

Marguerite Trocmé has been responsible for setting the environmental standards for the Swiss highways since 2008 at the Federal road office. She began working on roads and environmental issues in 1989 as an environmental project reviewer at the Swiss federal office for the environment. She was vice-chairman of the European COST 341 project on habitat fragmentation due to transport infrastructure and is currently president of the VSS commission on traffic and wildlife and has initiated a number of research projects in the field.

Edgar A. van der Grift is a senior research scientist in the Environmental Science Group at Alterra, part of Wageningen University and Research Centre. His research focuses on the impacts of habitat fragmentation on wildlife and the effectiveness of measures that aim to restore habitat connectivity across roads and railroads. He also consults to policy makers, road planners and conservation groups during the preparation and implementation phase of projects that aim for the establishment of effective ecological networks and environmental friendly transport systems.

Dr Rodney van der Ree is an Associate Professor and the Deputy Director of the Australian Research Centre for Urban Ecology, a division of the Royal Botanic Gardens Melbourne, based at the University of Melbourne. His research broadly focuses on quantifying and mitigating the impacts of human activities, such as roads and cities, on the natural environment. He is currently leading research projects on the effectiveness of mitigation techniques for wildlife in south-east Australia and is interested in road ecology issues in developing countries.

Paul J. Wagner is a wildlife ecologist with the Washington State Department of Transportation, Washington, USA. Active with Road Ecology for over 20 years, he serves on research committees of the National Academies of Sciences, the Transportation Research Board Committee on Ecology and Transportation and the Infra-Eco Network Europe (IENE). Paul is a founding member and past Chair of the International Conference on Ecology and Transportation (ICOET).

Dr Yun Wang is an associate professor at the China Academy of Transportation Sciences (CATS). He obtained his PhD from the China Academy of Sciences in road, landscape and ecological protection in 2007. In 2005, he translated *Road Ecology: Science and Solution* by Richard Forman into Chinese and in 2009, he co-wrote *Road Ecology in China*. His research now focuses on the interactions of roads and wildlife, landscape fragmentation and road ecology.

Susie Weeks has been the Executive Officer of the Mount Kenya Trust since 2001. She and her team have managed a number of successful private–public conservation partnerships to protect the integrity of Mount Kenya's forests and wildlife. The Mount Kenya Trust spearheaded the pioneering Mount Kenya Elephant Corridor project alongside the project's partner organisations. Susie is a gazetted Kenya Wildlife Service Honorary Warden.

Cameron Weller is an environmental manager with Jacobs and has over 7 years experience, primarily in the delivery of large infrastructure projects in Australia. He also has experience in working on large multi-disciplinary design teams as the environmental design lead. His work involves designing and managing the installation of fauna mitigation measures, writing environmental management plans and ensuring environmental compliance.

Patricia White began the US Habitat and Highways Campaign in 2000 to address impacts of highways on wildlife and encourage transportation planning that incorporates conservation. Her first report, *Second Nature: Improving Transportation without Putting Nature Second* was awarded the 2004 NRCA Award of Achievement for best publication. Patricia was a founding member of the International Conference on Ecology and Transportation (ICOET) Steering Committee, a founding member of the TRB Committee on Ecology and Transportation and proud founder of the TransWild Alliance.

Brendan Whittington-Jones is currently based in Oman and authoring a book on African wild dog conservation in South Africa. During his seven years working at the Endangered Wildlife Trust he coordinated the KwaZulu-Natal Wild Dog Advisory Group and the National Wild Dog Metapopulation Project. His MSc focused on the conservation and conflict implications of wild dogs ranging outside of protected areas in KwaZulu-Natal province, South Africa.

Fernanda Zimmermann Teixeira is a biologist interested in conservation biology, applied ecology and EIA. She is a PhD student in ecology at Federal University of the Rio Grande do Sul State (UFRGS) in Brazil, studying spatial patterns of wildlife–vehicle collision and impacts of road networks on the landscape. During her Master's research, she studied the similarity of road-kill hotspots among different groups and the influence of carcass removal and detectability on road-kill estimates.

FOREWORD

Roads smoothly and efficiently move us from place to place, and, by concentrating movement in somewhat straight strips, limit the big footprint of impacts on nature. But most roads were built before the rise and spread of ecology through society. As a consequence in part, roads with traffic cause significant and widely permeating effects on natural systems. Mitigation of today's surface transportation system therefore stands as a primary challenge of society and transportation. Furthermore in rapidly developing areas worldwide new roads proliferate, which now can be built with solid ecological foundations.

Nature within the strip of road and roadside is, of course, degraded. Mitigation reduces that effect, but especially minimizes the outward-rippling degradation across the land. What nature is affected, or natural systems disrupted? Three dimensions are central: (1) habitat and plants, (2) water quantity and quality and (3) wildlife. Roads and wildlife are the highlight of this book, though valuable insights on the other two dimensions appear.

The pages in your hand are a *tour-de-force*, a gem, indeed a treasure chest. I find it readable, interesting, practical, useful and ambitious. The remarkable cast of authors has uncovered a goldmine for us. The editors catalysed extra rigor and consistency, thus encouraging comparisons and usability. Virtually, every chapter begins with several succinct topic statements, which pinpoint the essence and also provide an overview. These statements are then analysed as the sections of text. Mitigation is the focus, though new road construction in developing nations is included. Wildlife, including different faunal groups and different regions, is emphasised. An international perspective thoroughly permeates the presentation.

Policy, planning and practice are highlighted alongside research and state-of-the-science results. I gained insight into every chapter perused.

Building on this accomplishment, analogous books highlighting roads and vehicles relative to vegetation and water would be valuable. Habitat, vegetation and plants are emasculated by roadside cutting and mowing. Fortunately, converting most (though not all) roadside area from grassy to woody vegetation is consistent with traffic safety and cost efficiency. Consider the numerous ecological and societal benefits. New habitat created, and existing adjacent woody habitat enhanced. Wildlife populations increased, probably well exceeding any increase in roadkills. Road crossing facilitated, thus reducing the habitat fragmentation and barrier-to-movement effect against wildlife and pollinators. There was reduced spread of airborne chemical pollutants from roadway and vehicles. Rare plants, animals and habitats enhanced on roadsides, especially important where scarce in agricultural and urban landscapes.

Water in varied forms poses endless problems, both familiar and as surprises, for transportation. Think of road-closure flooding, washouts/roadbed failures, wet driving surfaces, drainage-ditch filling, eroded roadsides, mudslides/landslips, frost cracks and potholes, snow-and-ice surfaces, blowing snow and too much snow. Water quantity-and-quality problems for nature are also severe. The soil water table is widely altered (raised or lowered) by roads. Where the water table is close to ground surface, wetlands are altered (drained or expanded). Fortunately, 'eco-piping' or permeating the roadbed with pipes crossing beneath a road maintains more natural water tables and wetlands. With permeated roadbeds, floodwaters seldom reach road surfaces and rarely wash out roads. The hydrologic connectivity through roadbeds supports more natural fish movements, and happy anglers. The same pipes connect the land for many small terrestrial animals. Drilling and inserting horizontal pipes is a routine, and in view of this array of benefits, cost-effective technology.

Water-quality pollution benefits follow suit. Most vehicle- and road/roadside-generated chemicals are readily 'treated' near roads in elongated mitigation structures (depressions, wetlands, ponds). Soil and microbes mainly clean the water. Polluted heated ditch-water entering nearby water bodies is largely eliminated using familiar stream features (convoluting, step-damming) plus tall vegetation (wind-and-sun evapotranspiration pumping). Again these manifold water quantity and quality benefits are consistent with safety and efficiency, cost effectiveness, and engineering design creativity.

A decade ago, four transportation leaders, a leading hydrologist, and nine ecology-research scholars co-wrote the book, *Road Ecology: Science and Solutions*. This synthesized a scattered literature and articulated principles linking roads/vehicles, soil/water/air and plants/animals. One of our dreams was the highly useful compendium now in your hand.

The scientist in me inexorably jumps from this treasure chest of insight to pregnant and important research frontiers awaiting us. How do our current ecological science results apply to the diverse types of roads and traffic levels criss-crossing the land? The ecology of road segments and especially road networks in a landscape cries out for study. Where is the ecology of different truck, car, tire, even road surface types? What is the (ecology and cost) optimum distance between road-crossing structures for different wildlife types? How can the ubiquitous utility poles along roads be used in mitigation solutions? To understand roads and wildlife populations, the non-roadkill dimensions now need much greater emphasis. As suggested earlier, habitat/plant and water quantity/quality dimensions of road ecology are lurking giants, awaiting a few prescient researchers and leaders.

My government-and-citizen-side hones in on the need and opportunity to accelerate solutions now for transportation, the land and us. Every roadbed, bridge and culvert repair/replacement is the cost-effective moment to concurrently address other goals of society, such as walking/biking paths, reduced flooding, enhanced fish movement, reconnected split communities and so forth. Roadsides represent a massive little-used resource (for nature and us) at our doorstep. Roadside food production, trail networks, stormwater and pollution mitigation, history-and-nature education effectively create variegated roadsides, bulging with useful solutions for society. Light, noise, vibration and wind can be dispersed or concentrated, as well as decreased or increased. Eco-piping or pipe-perforated roadbeds provide lots of benefits quickly. The 'road-effect zone' provides a ready framework for ecologically planning, engineering and mitigating roads. In parks, towns and sprawl areas, curvy, slightly bumpy and seemingly narrow roads slow traffic and reduce effects on wildlife. In every jurisdiction, remove a road segment or two to create continuous ecologically valuable, large natural-habitat patches. By lowering (e.g. 2–3 m) short stretches of roads in good-drainage areas, inexpensive green-bridges (with some 10 cm of sandy soil) will help re-establish semi-natural wildlife movement patterns across the land. And just on the horizon, a transportation system slightly above or below ground level, using lightweight renewable-energy automated pods, effectively recovers an extensive area of road/roadside-covered terrain. Furthermore this 'netway system' reconnects today's fragmented land for nature and us. Indeed, on an exhilarating netway ride at London's airport I experienced the future.

Road ecology and this book's impressive synthesis highlight a great opportunity for planners, engineers and ecologists to collaborate for new successes, and receive important accolades together. History will record that transportation, land-and-water, and society are the big beneficiaries.

Richard T. T. Forman
Harvard University

PREFACE

This book brings together some of the leading researchers, academics, practitioners and transportation agency personnel from around the world to focus on the challenge of improving the ecological sustainability of the linear infrastructure – primarily road, rail and utility easements – that dissects and fragments most landscapes around the world. Where possible, we aimed to have co-authors from different continents on every chapter – and indeed, many authors are collaborating together for the first time on this book.

When authors were invited to contribute, we gave them this initial challenge: 'Imagine you are in charge of your professional world for a day, and could change anything to improve the ecological sustainability of roads (or other linear infrastructure) and traffic: what six to eight things would you change or want people to learn and do differently?' Conversely, a second challenge posed to the authors was slightly more pessimistic: 'Identify the six to eight mistakes that you regularly see or experience in your area of practise and write about those and how to avoid them'. This approach appeared to stimulate our authors and provided a tangible grounding for their writing – but the *real* challenge came when we tried to impose an average word limit for each chapter of 3,000 words! In hindsight, the word limit was probably too restrictive for some topics, but it forced authors to be concise and succinct – which we hope you, the reader, appreciate!

Chapters are written as a series of lessons, insights or principles (hereafter referred to only as lessons) that forced authors to be very specific about their key points. Many struggled with this style – but our hope is that it allows you to quickly identify the pertinent information to help you in your day to day tasks. We realised

that time is precious – and for most of you – time is money (yours or your bosses!) and we have designed the book so you can quickly and efficiently find the answers to your questions and get back to the planning, designing, building, maintaining or granting approvals to build roads or other transportation infrastructure. And in the likely event that this book does not answer all your questions, the further readings and up-to-date reference lists for each chapter should point you to the extra information you need.

The chapters span the project continuum – starting with planning and design, through construction and into maintenance and management. Research and monitoring is such an important aspect that it sits like an umbrella, encompassing all phases of a transportation project. Rigorous monitoring and evaluation of the impacts of a road or effectiveness of mitigation often requires the collection of data before the road or mitigation is built – hence the chapters on monitoring, evaluation and maintenance come before the impacts and mitigation are described. A significant proportion of the book focuses on impacts and solutions for species groups and specific regions. The rate of major road construction in the United States, Australia and Western Europe has slowed, while developing countries are expanding their road and rail networks at an incredibly rapid rate. This book highlights some of the unique regional challenges with case studies from Asia, South America and Africa.

Chapters are designed to be stand-alone – you do not need to read the book from cover to cover, or even from front to back, to be able to use its contents. We envisage that readers will come to our book when facing a challenge – or rather an opportunity – and they can dive into the relevant chapter to improve their understanding of the major problems and the array of current possible solutions. Nevertheless, we have endeavoured to ensure that chapters build upon and complement each other – so reading (or even skimming) it from cover to cover won't be a waste of time. Extensive cross-referencing among chapters directs the reader to relevant material elsewhere in the book.

We should point out what this book is not: it is not a series of standards for the design of roads or mitigation measures. These standards and guidelines already exist in many countries, states or regions and we did not want to repeat them here. If they don't exist in your region, there are enough around to borrow from in order to develop your own. And because the optimal design and placement of, for example, crossing structures, fences or wildlife detection systems should evolve as our understanding and technology improves, such specific information would be quickly out of date. All the authors in this book have strived to identify the greatest challenges and opportunities and write about them in a way that is timeless.

Our sincere hope is that this book improves the way roads and other linear infrastructure are planned, designed, approved, built, maintained and studied.

Rodney van der Ree
Daniel J. Smith
Clara Grilo
September, 2014

ACKNOWLEDGEMENTS

Edited books such as the *Handbook of Road Ecology* are a combined effort of many people – not the least of which are the 115 people from 25 countries that contributed to the 62 chapters. It was a privilege for us to combine your individual expertise into a product that truly exceeded the sum of its parts! Zoe Metherell and Scott Watson generously illustrated numerous figures, and many others assisted in developing the concept, providing figures and editing the final product – including Alan Crowden, Ward Cooper and Kelvin Mathews at Wiley, Radjan LourdeSelvanadin at SPi Global, Lee Harrison, Cindy van der Ree and Marcel Huijser.

Road ecology is most definitely a collaboration between industry and academia, researchers and engineers, government agencies and road construction companies. This handbook is no different. Numerous companies, government agencies, not-for-profit conservation groups and university/research centres have contributed financially to the production and distribution of this book. Funds provided by these generous supporters have allowed us to provide over 200 copies of this handbook to practitioners in developing countries. Further details are available at www.handbookofroadecology.net. We sincerely thank these organisations for their support: ACO Polycrete Pty Ltd, Animex – Animal Exclusion Solutions, the Australian Research Centre for Urban Ecology at the Royal Botanic Gardens Melbourne, Chinese Academy of Transportation Sciences, Eco-Kare International, Florida Wildlife Federation, Melbourne Sustainable Society Institute, School of Ecosystem and Forest Sciences at the University of Melbourne and the Federal Road Office of Switzerland of the Department of Environment, Transport, Energy and Communications.

Rodney: I thank the many ecologists, road practitioners and government regulators who have shared time and experiences discussing, planning, designing and building better roads and other linear infrastructure. I am grateful to Andrew Bennett for guiding me through the perilous days of designing and completing a PhD in what was the nascent days of 'road ecology'. Mark McDonnell encouraged me to write and edit this volume, and he and the Baker Foundation supported this undertaking. To my co-editors – thanks for bringing complementary skills to the editing table and for sharing the vision of this book! I am particularly appreciative of Cindy's continuing love and support, who once again allowed me to disappear from family life to complete this project. To Ethan and Ezra – thanks for tolerating my absences and may you forever contemplate the solutions to roadkill and barrier effects as you travel life's roads.

Dan: I extend a special thank you to four individuals that encouraged, mentored and guided me toward a career in the disciplines of landscape ecology, road ecology and conservation planning – Larry Harris, Gary Evink, Leroy Irwin and Richard Forman. I'd also like to thank Reed Noss for his support and collaboration, which has helped me sharpen my research design and analytical skills and furthered my success as a scientist. My sincere appreciation goes to my co-editors, even though each of us ended up spending many long nights and days on this collaboration, the spirited camaraderie made it an enjoyable learning experience.

Last but not least, thank you to family, friends and colleagues that have inspired me and kept me steadfast towards making a positive difference in the world.

Clara: I thank my parents for their support and enthusiasm for my research and John A. Bissonette for inspiring me to work on road and landscape ecology. Thanks to Rodney for inviting me to participate in this book project. A very special thanks to IENE, ICOET and ICCB conference organisers that allowed me to meet researchers from all parts of the world. Social events at these conferences were priceless to meet most of the authors who contributed to this book, who also share a passion for road ecology as well as the love of good beer and hilarious moments.

ABOUT THE COMPANION WEBSITE

This book is accompanied by a companion website:

www.wiley.com\go\vanderree\roadecology

The website includes:
- Powerpoints of all figures from the book for downloading
- Pdfs of all tables from the book for downloading

Chapter 1

THE ECOLOGICAL EFFECTS OF LINEAR INFRASTRUCTURE AND TRAFFIC: CHALLENGES AND OPPORTUNITIES OF RAPID GLOBAL GROWTH

Rodney van der Ree[1], Daniel J. Smith[2] and Clara Grilo[3]

[1]Australian Research Centre for Urban Ecology, Royal Botanic Gardens Melbourne, and School of BioSciences, The University of Melbourne, Melbourne, Victoria, Australia
[2]Department of Biology, University of Central Florida, Orlando, FL, USA
[3]Departamento de Biologia & CESAM, Universidade de Aveiro, Aveiro, Portugal

SUMMARY

Roads, railways and utility easements are integral components of human society, allowing for the safe and efficient transport of people and goods. There are few places on earth that are not currently traversed or impacted by the vast networks of linear infrastructure. The ecological impacts of linear infrastructure and vehicles are numerous, diverse and, in most cases, deleterious. Recognition and amelioration of these impacts is becoming widespread around the world, and new roads and other linear infrastructure are increasingly planned to avoid high-quality areas and designed to minimise or mitigate the deleterious effects. Importantly, the negative effects of the existing infrastructure are also being reduced during routine maintenance and upgrade projects, as well as targeted retrofits to fix specific problem areas.

Handbook of Road Ecology, First Edition. Edited by Rodney van der Ree, Daniel J. Smith and Clara Grilo.
© 2015 John Wiley & Sons, Ltd. Published 2015 by John Wiley & Sons, Ltd.
Companion website: www.wiley.com\go\vanderree\roadecology

1.1 Global road length, number of vehicles and rate of per capita travel are high and predicted to increase significantly over the next few decades.

1.2 The 'road-effect zone' is a useful conceptual framework to quantify the negative ecological and environmental impacts of roads and traffic.

1.3 The effects of roads and traffic on wildlife are numerous, varied and typically deleterious.

1.4 The density and configuration of road networks are important considerations in road planning.

1.5 The costs to society of wildlife-vehicle collisions can be high.

1.6 The strategies of avoidance, minimisation, mitigation and offsetting are increasingly being adopted around the world – but it must be recognised that some impacts are unavoidable and unmitigable.

1.7 Road ecology is an applied science which underpins the quantification and mitigation of road impacts.

The global rates of road construction and private vehicle ownership as well as travel demand will continue to rise for the foreseeable future, including at a rapid rate in many developing countries. The challenge currently facing society is to build a more efficient transportation system that facilitates economic growth and development, reduces environmental impacts and protects biodiversity and ecosystem functions. The legacy of the decisions we make today and the roads and railways we construct tomorrow will be with us for many years to come.

INTRODUCTION

Since ancient times, trails and roads have connected settlements and facilitated the movement of goods and people around the world. The Appian Way (over 500 km long), built in the second and third centuries BC in Italy for military and trade purposes, was one of the first improved (hard-surfaced) highways. Portions of this road still remain today, a testament to the high-quality engineering and construction practices of the Roman Empire and the importance of roads to human society. Up until the early 1900s, the majority of the roads linking cities and towns were mostly unimproved, and paving with brick, concrete or asphalt only became common when mass production of vehicles began and the demand for better quality roads and more efficient routes increased. Depression-era public work programs designed to provide employment opportunities and stimulate economies also facilitated a significant increase in paved roads. Today, road construction is still an important driver of economic growth, both during construction and for its long-term effects. Roads are now conspicuous components of almost all landscapes globally, and set to expand even further into the future (Lesson 1.1).

Transportation infrastructure and roads, in particular, are pivotal to economic and social development by providing access to markets, places of employment, businesses, health and family care, leisure activities and education. Governments and international development banks see the construction of new roads and improvement of existing roads as priorities to improve livelihoods. However, the benefits of improved access vary regionally and by road type (e.g. Fan & Chan-Kang 2005), and not all rural road projects result equally in increased agricultural productivity and/or poverty reduction (Laurance et al. 2014; Chapter 2), and in some cases the costs outweigh the benefits. Once built, roads are nearly permanent elements in the landscape, and the wrong road (e.g. motorway/expressway vs. unpaved road) in the wrong place (e.g. roadless wilderness vs. agricultural landscape) can have long-term consequences for both society and the environment. Planning and impact assessment processes must properly account for all the costs, benefits and environmental impacts to ensure that the future road network is as sustainable as possible, particularly in regions where the rate of road construction is currently high or set to increase (see Chapter 5).

The broad aim of this chapter is to provide the necessary background and context for the many topics covered in this book. While primarily focused on roads and vehicles, the lessons in this chapter and book can be applied to all types of linear infrastructure.

LESSONS

1.1 Global road length, number of vehicles and rate of per capita travel are high and predicted to increase significantly over the next few decades

The total length of paved and unpaved roads on earth currently exceeds 64 million km; enough for 83 round-trips to the moon (CIA 2013). Roads dominate most landscapes worldwide – for example, 83% of the continental United States is now within 1 km of the nearest road of any type (Riitters & Wickham 2003). There is approximately 5 million km of road across the 27 countries of the European Union (EFR 2011). The emerging economies of China, India and Brazil are already among the top five countries in road length (4.1, 4.7 and 1.6 millions km, respectively) (CIA 2013) and they have ambitious plans to further increase the capacity of their transportation networks (Chapters 50, 52 and 57). Globally, an additional 25 million lane-kilometre of paved road are to be built by 2050, 90% of which will be in non-Organisation for Economic Co-operation and Development (OECD) countries (Dulac 2013). The 870 million vehicles around the world in 2009 are expected to more than double by 2050 to between 1.7 and 2.8 billion (WEC 2011; Meyer et al. 2012). The majority of these cars will still be in developed countries (with a 33% increase from 2000 to 2050), even though non-OECD countries will have a five-fold increase in vehicles by 2050 (Fulton & Eads 2004). In 2000, the total vehicular travel worldwide was estimated at 32 trillion passenger kilometre per year (up from 2.8 trillion in 1950), and by 2050 is predicted to be 105 trillion passenger kilometre per year, of which about 42% will be by car, the remainder by bus, rail and air (Schafer & Victor 2000).

The predictions of growth in road length, per capita travel and car ownership are based on models with a range of assumptions and will ultimately be influenced by fuel availability and pricing, climate change limits, a desire for increased mobility and other technological, economic, environmental and social priorities and constraints. While the magnitude of the predictions may be debated, all models predict a massive increase in the number of vehicles, road length and travel distances. The challenge for society is to acknowledge this potential rate of growth and decide (i) if it is necessary or desired; (ii) where it should occur; (iii) the preferred mode of transport (e.g. cars,

high-speed trains or air travel); and (iv) the design and management of the transport network (e.g. road design and type of mitigation). Importantly, the impacts and solutions proposed in this book and the wider road ecology literature are based on the scale and extent of the current road network. The predictions of growth, even if only partially correct, require urgent and effective actions now.

1.2 The 'road-effect zone' is a useful conceptual framework to quantify the negative ecological and environmental impacts of roads and traffic

The 'road-effect zone' is defined as the area over which the ecological effects of roads and traffic extend into the adjacent landscape (Forman & Deblinger 2000), including noise, light and chemical pollution; disturbance effects; and habitat modification (Fig. 1.1). The size of the road-effect zone is determined by the characteristics of the (i) road (width, surface type, elevation relative to adjacent landscape); (ii) traffic (volume, speed); (iii) adjacent landscape (topography, hydrography, vegetation type, habitat quality); (iv) prevailing wind speed and direction; and (v) species traits and their sensitivity to the impact. Road effects have been observed many hundreds to thousands of metres from the road itself (Reijnen et al. 1995; Forman & Deblinger 2000; Boarman & Sazaki 2006; Eigenbrod et al. 2009; Benítez-López et al. 2010; Shanley & Pyare 2011). The impacts are usually greatest closer to the road and either diminish gradually with increasing distance from the road or exhibit thresholds with steep changes in responses (Eigenbrod et al. 2009). The road-effect zone is a useful approach to quantify and mitigate the negative effects of roads and traffic because it helps regional planners calculate the extent of the area impacted by existing roads (e.g. 15–22% of continental United States) (Forman 2000) or likely to be impacted by proposed roads (e.g. Williams et al. 2001).

1.3 The effects of roads and traffic on wildlife are numerous, varied and typically deleterious

Roads and traffic can significantly affect individual wildlife, populations and communities, and landscapes (Figs 1.1 and 1.2). These impacts can begin during construction and may continue as long as the road

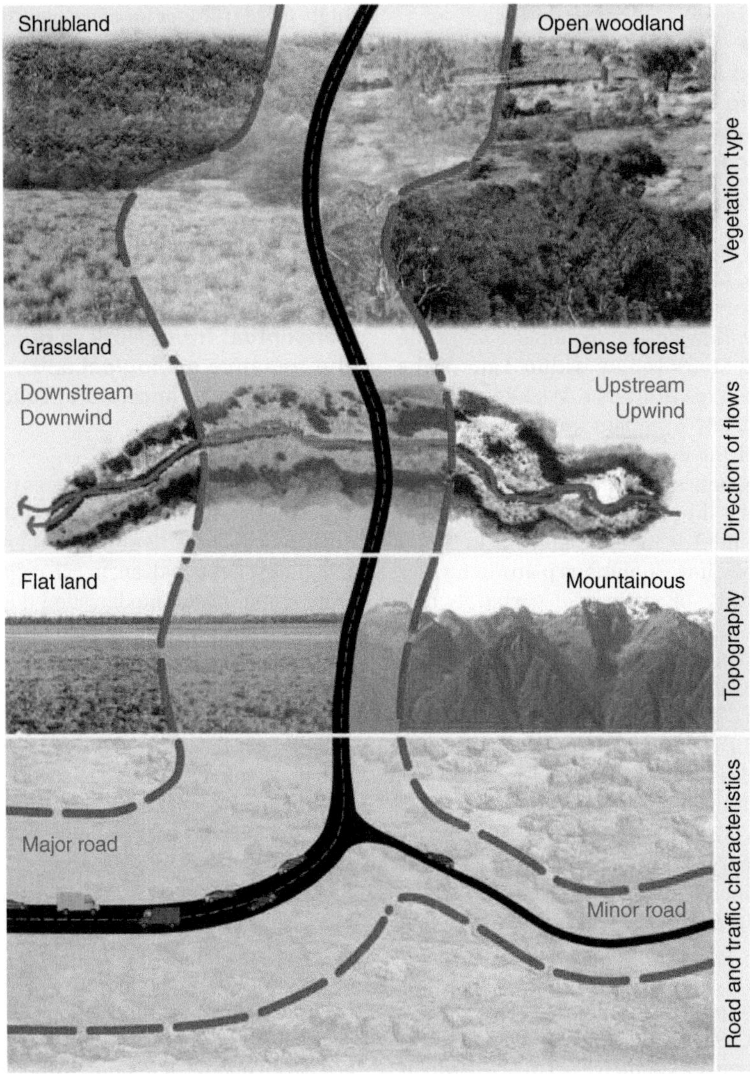

Figure 1.1 The road-effect zone, showing the area over which the ecological impacts of roads and traffic extend. The size of the road-effect zone is affected by a range of parameters – here we show four: (1) vegetation type; (2) direction of flows such as wind and water; (3) topography; and (4) road and traffic characteristics. The relative size of the road-effect zone for each parameter is illustrative only and not indicative; for example, the road-effect zone is not necessarily three times larger in flat than mountainous terrain. Source: Photograph by Zoe Metherell. Reproduced with permission of Zoe Metherell.

remains operational or until the impacts are mitigated. The majority of impacts are typically deleterious, and if severe enough, can reduce the size of populations of wildlife, with a concomitant increase in the risk of local extinction. These impacts are summarised here, and expanded on in subsequent chapters:

• **Habitat loss**: The construction and expansion of transportation corridors results in the clearing of vegetation and a loss of habitat at and adjacent to the road (Figs 1.2 and 2.1). Roads attract people and encourage further development, often resulting in further clearing of vegetation after road construction. Indirect loss of habitat also occurs through degradation, and this can exceed the amount of habitat directly cleared for the road.

• **Habitat degradation**: Due to a range of interacting biotic and abiotic effects, habitat quality often declines

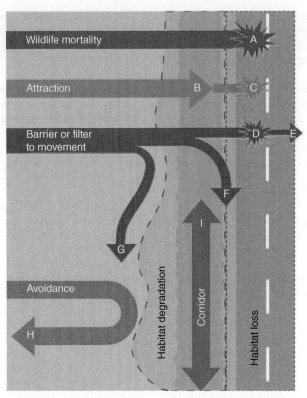

Figure 1.2 Impacts of roads on individual wildlife, populations and ecosystems. Habitat is lost to build the road and habitat adjacent to the road is degraded. The most obvious impact of roads and traffic on wildlife is mortality due to Wildlife-vehicle collisions WVC (A). Some species are attracted to resources (e.g. carrion, spilled grain or heat for basking) on the road or roadside (B) which, depending on the animals ability to avoid traffic, may result in death due to WVC (C). The barrier or filter effect reduces the movement of animals across the road and a proportion of individuals that attempt to cross are killed due to WVC (D) and some make it across (E), while others are deterred from crossing by the road (F) or degraded roadside habitat (G). Other species actively avoid the road or degraded habitat (H). By contrast, some species use the roadside vegetation as habitat and/or as a corridor for movement (I). Source: Illustration by Zoe Metherell. Reproduced with permission of Zoe Metherell.

adjacent to linear infrastructure. For example, the abrupt edges along linear clearings modify microclimatic conditions and encourage weed invasion, and specialist 'habitat interior' species of plants and animals are often outcompeted by 'edge-adapted' generalist species. Edge effects are particularly pronounced in tropical ecosystems (Chapter 49).

• **Barrier or filter to movement**: The creation of gaps in habitat can prevent or restrict the movement of wildlife that avoid clearings, and the noise, light, and chemical pollution and disturbance from vehicles will exacerbate these effects. Road width, whether it is paved or unpaved, and traffic volume affect the severity of the barrier effect (Riley et al. 2006) and species-specific thresholds exist. The type of movement affected varies, including (i) individuals' daily access

to important resources; (ii) seasonal migrations of entire populations; and (iii) once-in-a-lifetime dispersal events, all of which can have significant consequences for individual survival, gene flow and population persistence.

• **Wildlife mortality** due to wildlife-vehicle collisions or WVC: Animals that attempt to cross roads or are attracted to the road surface have an increased risk of being involved in WVC and being killed or injured (e.g. Figs 26.2A, 32.2, 32.3, 33.1, 35.1, 38.2).

• **Avoidance**: Some species of wildlife avoid the road-effect zone due to traffic disturbance and/or habitat degradation, resulting in a reduction of habitat or a barrier to movement.

• **Attraction**: Roads and roadsides can attract some species by providing resources or enhanced opportunities. For example, reptiles may bask on the warm surface of the road, herbivores may forage on the enhanced plant growth on roadsides and scavengers can be attracted to feed on roadkill (e.g. Figs 26.2B, 26.3A, 26.4, 46.6).

• **Habitat and/or corridor for movement**: In some highly modified landscapes, roadside strips can provide the majority of habitat for wildlife (e.g. Fig. 46.3). Many adaptable species of wildlife, including invasive species (Seabrook & Dettmann 1996), use the cleared roadways and railways to efficiently move around the landscape (Fig. 26.3B).

The nature and severity of these effects vary among species because of their different morphological, ecological and behavioural traits. Importantly, most effects rarely operate in isolation (e.g. Farji-Brener & Ghermadi 2008), and many act synergistically. For example, animals that avoid roads have low rates of mortality due to WVC because they rarely attempt to cross, but barrier to movement effects may be high, potentially subdividing the population into smaller sub-populations. This arrangement is often called a metapopulation – a set of discrete populations of the same species occurring within the same area that exchange individuals through dispersal, migration or human-assisted movement (after Hanski & Simberloff 1997). The persistence of the metapopulation depends on the number and size of the sub-populations and the level of connectivity among them, and the risk of extinction increases as sub-populations become fewer, smaller and/or less connected. Species that are attracted to roads may suffer high rates of mortality due to WVC if they are unable to avoid oncoming vehicles, or conversely, low rates of mortality if they avoid oncoming vehicles (e.g. low-mobility species such as amphibians versus high-mobility species such as scavenging carnivores).

A recent review demonstrated that roads and traffic have had detectable population-level effects by reducing the size or density of populations near roads for many species (Fahrig & Rytwinski 2009; Chapter 28). These included frogs and toads (Fahrig et al. 1995; Hels & Buchwald 2001), salamanders (Gibbs & Shriver 2005), turtles (Steen & Gibbs 2004), birds (Erritzoe et al. 2003), European hares (Roedenbeck & Voser 2008), badgers (Clarke et al. 1998), bobcats and coyotes (Riley et al. 2006), Iberian lynx (Ferreras et al. 1992) and bighorn sheep (Epps et al. 2005). Roads and traffic can also alter population structure by affecting specific groups of animals, resulting in populations with skewed age or sex ratios (e.g. Aresco 2005; Nafus et al. 2013). These impacts are of particular concern when

roads pass through protected areas or ranges of rare and threatened species or sever access to important breeding areas.

1.4 The density and configuration of road networks are important considerations in road planning

The density and configuration of the road network across the landscape are important drivers of the scale and intensity of road impacts on wildlife. Road density is a measure of the abundance of roads within a region, and is measured as the length of road per unit area. Thresholds in road density have been identified for populations of a number of species, including gray wolves in the Great Lakes region, USA which generally avoided landscapes when road density exceeded approximately 0.6 km per km^2 (Thiel 1985). The configuration of the network describes how roads and other linear infrastructure are arranged – such as bundled together or spread out across the landscape. Road networks are typically (i) rectangular/block/grid patterns that decrease in density from urban to rural areas; (ii) radial spokes and concentric rings that form around a city or other central feature; or (iii) linear configuration typically following natural features in the landscape. Road configuration has an enormous bearing on the scale of road impacts across the landscape, and bundling them together and having fewer roads with higher traffic volume is almost always preferred to having them spread out (Jaeger et al. 2006; Rhodes et al. 2014; Chapter 3).

1.5 The costs to society of wildlife-vehicle collisions can be high

The cost to society of WVC with large animals is high, primarily from human injury and loss of life, as well as costs associated with damage and repair of vehicles. There are approximately two million WVC with large mammals in the United States every year, injuring 29,000 people and killing 200 more (Conover et al. 1995), and there were an estimated 500,000 WVC with ungulates in Europe during 1995 (Groot Bruinderink & Hazebroek 1996). The likely rates of collisions are undoubtedly much higher because (i) collisions resulting in minor or negligible damage remain unreported, and (ii) the cause of single-vehicle collisions with roadside objects (e.g. trees) that result in human death may be due to swerving to avoid collisions with wildlife, which remain unreported. The death of wildlife due to WVC will also reduce the size of animal populations, which in some regions are an important source of food for people or income via tourism or hunting. Reduced populations of other species due to WVC may also impact people if such species are important pollinators or perform other critical ecosystem services (e.g. insectivorous bats and birds that help control populations of mosquitoes and other flying insect pests).

1.6 The strategies of avoidance, minimisation, mitigation and offsetting are increasingly being adopted around the world – but it must be recognised that some impacts are unavoidable and unmitigable

The impacts of roads and traffic have been recognised globally as significant threats to the persistence of species and functioning of healthy ecosystems. The principles of the hierarchy of avoiding, minimising, mitigating and offsetting these impacts have also been widely adopted and increasingly practised (Chapter 7). Many governments and communities around the world have accepted the challenge and additional cost of building an efficient transportation network that is safe for wildlife and people. In some regions, priority has shifted to retrofitting the existing network to reduce its impacts on biodiversity. The global proliferation of numerous professional networks (Chapter 61) and non-government organisations with the intent to improve best-practice road mitigation and the membership that includes planners, designers, regulators, ecologists and engineering/construction firms is a testament to this.

However, not all impacts can be fully mitigated, and not all mitigation measures are equally effective. For example, it is difficult and likely impossible in some locations to control the effects of human activities after roads are built, such as increased land clearing and development, the migration and movement of people, and increased hunting or poaching (Chapters 2 and 51). Similarly, the inclusion of mitigation measures in a proposed road project does not automatically mean that all effects have been mitigated and the project should proceed. For example, the likelihood of crossing structures effectively permitting the annual migration of hundreds of thousands of mammals in the Serengeti is extremely low (Chapter 56). Therefore, it is essential to include a 'no-road' option when ranking different route options during the planning of new roads or expansion of existing roads in remote and/or intact ecosystems (Selva et al. 2011; Chapter 3).

1.7 Road ecology is an applied science which underpins the quantification and mitigation of road impacts

The accurate quantification and effective mitigation of road impacts relies on scientifically rigorous research and monitoring (Chapter 10). The first published road ecology studies reported rates of WVC, the most visible ecological effect of roads and traffic (e.g. Stoner 1925; DeVos 1949; Fitch 1949). As road networks expanded and traffic volumes increased in the latter half of the 20th century, research began to focus on quantifying and reducing rates of WVC with large herbivores to save human lives and reduce societal costs. More recently, attention has expanded to include smaller species and encompass a range of biological and ecological parameters such as species distribution, abundance, reproductive rate, behaviour and dispersal (e.g. Legagneux & Ducatez 2013). There have also been recent calls to understand effects at larger spatial and temporal scales and to focus on populations, communities of species and ecosystems (van der Ree et al. 2011). However, quantifying the full breadth of impacts and the effectiveness of mitigation measures as well as reporting practical issues associated with road planning and management are still scarce in research findings (Roedenbeck et al. 2007). Consequently, a large proportion of published road ecology studies appear to have little influence on road planning and design. In moving forward, road agencies should recognise and support good-quality research, scientists and practitioners should collaborate more effectively and researchers should ask applied questions that provide relevant information which road agencies need (Chapter 10).

CONCLUSIONS

The global network of roads, railways, artificial waterways, trails and utility easements is extensive in its length and spread. The total number of vehicles in use is escalating and already difficult to comprehend, and the total distances travelled annually even more so. However, these statistics are to be dwarfed over the next 20–40 years, even if the predictions in growth of road length, number of vehicles and travel distances are only partially met. The impacts of linear infrastructure and vehicles on many species and ecosystems are sufficiently well known to allow the development of effective strategies to avoid, minimise, mitigate and offset most negative effects. The challenge facing society is to identify and retrofit the worst parts of the existing network and build and manage a network for tomorrow that is as good for biodiversity as it is for people.

FURTHER READING

Beckman et al. (2010): An edited volume focussing on North America that aims to collate and integrate information and approaches from various disciplines, as well as a series of case studies that demonstrate effective innovations in planning and mitigation.

Benítez-López et al. (2010): A meta-analysis of almost 50 studies, demonstrating that populations of many species of wildlife declined in close proximity to infrastructure, including up to about 1 km for birds and 5 km for mammals.

Forman et al. (2003): A seminal and comprehensive review and introduction to the field of road ecology, encompassing ecological concepts, planning, wildlife and vegetation, and pollution.

van der Ree et al. (2011): The introduction to a special issue of the open access journal, *Ecology and Society*, which contains 17 articles focussed on the 'Effects of roads and traffic on wildlife populations and landscape function' (http://www.ecologyandsociety.org/issues/view.php/feature/41).

REFERENCES

Aresco, M. J. 2005. The effect of sex-specific terrestrial movements and roads on the sex ratio of freshwater turtles. Biological Conservation **123**:37–44.

Beckman, J. P., A. P. Clevenger, M. P. Huijser and J. A. Hilty, editors. 2010. Safe passages: highways, wildlife and habitat connectivity. Island Press, Washington, DC.

Benítez-López, A., R. Alkemade and P. A. Verweij. 2010. The impacts of roads and other infrastructure on mammal and bird populations: a meta-analysis. Biological Conservation **143**:1307–1316.

Boarman, W. I. and M. Sazaki. 2006. A highway's road-effect zone for desert tortoises (*Gopherus agassizii*). Journal of Arid Environments **65**:94–101.

Central Intelligence Agency (CIA). 2013. The world fact book 2013–14. CIA, Washington, DC. Available from https://www.cia.gov/library/publications/the-world-factbook/fields/2085.html#xx (accessed 17 September 2014).

Clarke, G. P., P. C. L. White and S. Harris. 1998. Effects of roads on badger *Meles meles* populations in south-west England. Biological Conservation **86**:117–124.

Conover, M. R., W. C. Pitt, K. K. Kessler, T. J. DuBow and W. A. Sanborn. 1995. Review of human injuries, illnesses, and economic losses caused by wildlife in the United States. Wildlife Society Bulletin **23**:407–414.

DeVos, A. 1949. Timber wolves (*Canis Lupus Lycaon*) killed by cars on Ontario highways. Journal of Mammalogy **30**:197.

Dulac, J. 2013. Global land transport infrastructure requirements: estimating road and railway infrastructure capacity and costs to 2050. International Energy Agency, Paris.

Eigenbrod, F., S. Hecnar and L. Fahrig. 2009. Quantifying the road-effect zone: threshold effects of a motorway on anuran populations in Ontario, Canada. Ecology and Society **14**:24.

Epps, C. W., P. J. Palsboll, J. D. Wehausen, G. K. Roderick, R. R. Ramey and D. R. McCullough. 2005. Highways block gene flow and cause a rapid decline in genetic diversity of desert bighorn sheep. Ecology Letters **8**:1029–1038.

Erritzoe, J., T. D. Mazgajski and Ł. Rejt. 2003. Bird casualties on European roads – a review. Acta Ornithologica **38**:77–93.

European Union Road Federation (EFR). 2011. European road statistics handbook 2011. ERF, Brussels.

Fahrig, L., J. H. Pedlar, S. E. Pope, P. D. Taylor and J. F. Wegner. 1995. Effect of road traffic on amphibian density. Biological Conservation **73**:177–182.

Fahrig, L. and T. Rytwinski. 2009. Effects of roads on animal abundance: an empirical review and synthesis. Ecology and Society **14**:21.

Fan, S. and C. Chan-Kang. 2005. Road development, economic growth, and poverty reduction in China. Research report 138. International Food Policy Research Institute, Washington, DC.

Farji-Brener, A. G. and L. Ghermadi. 2008. Leaf-cutting ant nests near roads increase fitness of exotic plant species in natural protected areas. Proceedings of the Royal Society B **275**:1431–1440.

Ferreras, P., J. J. Aldama, J. F. Beltrán and M. Delibes. 1992. Rates and causes of mortality in a fragmented population of Iberian lynx *Felis pardina* Temminck, 1824. Biological Conservation **61**:197–202.

Fitch, H. S. 1949. Road counts of snakes in western Louisiana. Herpetologica **5**:87–90.

Forman, R. T. T. 2000. Estimate of the area affected ecologically by the road system in the United States. Conservation Biology **14**:31–35.

Forman, R. T. T. and R. D. Deblinger. 2000. The ecological road-effect zone of a Massachusetts (USA) suburban highway. Conservation Biology **14**:36–46.

Forman, R. T. T., D. Sperling, J. A. Bissonette, A. P. Clevenger, C. D. Cutshall, V. H. Dale, L. Fahrig, R. France, C. R. Goldman, K. Heanue, J. A. Jones, F. J. Swanson, T. Turrentine, and T. C. Winter. 2003. Road ecology. Science and solutions. Island Press, Washington, DC.

Fulton, L. and G. Eads. 2004. IEA/SMP model documentation and reference case projections. World Business Council for Sustainable Development, Geneva.

Gibbs, J. P. and W. G. Shriver. 2005. Can road mortality limit populations of pool-breeding amphibians? Wetlands Ecology and Management **13**:281–289.

Groot Bruinderink, G. W. T. A. and E. Hazebroek. 1996. Ungulate traffic collisions in Europe. Conservation Biology **10**:1059–1067.

Hanski, I. and D. Simberloff. 1997. The metapopulation approach. Pages 5–26. In I. Hanski and M. Gilpin, editors. Metapopulation biology: ecology, genetics and evolution. Academic Press, San Diego, CA.

Hels, T. and E. Buchwald. 2001. The effect of road kills on amphibian populations. Biological Conservation **99**:331–340.

Jaeger, J. A. G., L. Fahrig and K. Ewald. 2006. Does the configuration of road networks influence the degree to which roads affect wildlife populations? Pages 151–163. In C. L. Irwin, P. Garrett and K. P. McDermott, editors. 2005 International Conference on Ecology and Transportation (ICOET). Center for Transportation and the Environment, North Carolina State University, Raleigh, NC.

Laurance, W. F., G. R. Clements, S. Sloan, C. O'Connell, N. D. Mueller, M. Goosem, O. Venter, D. P. Edwards, B. Phalan, A. Balmford, R. van der Ree and I. B. Arrea. 2014. A global strategy for road building. Nature **513**:229–232.

Legagneux, P. and S. Ducatez. 2013. European birds adjust their flight initiation distance to road speed limits. Biology Letters **9**:20130417.

Meyer, I., S. Kaniovski and J. Scheffran. 2012. Scenarios for regional passenger car fleets and their CO_2 emissions. Energy Policy **41**:66–74.

Nafus, M. G., T. D. Tuberville, K. A. Buhlmann and B. D. Todd. 2013. Relative abundance and demographic structure of Agassiz's desert tortoise (*Gopherus agassizii*) along roads of varying size and traffic volume. Biological Conservation **162**:100–106.

Reijnen, R., R. Foppen, C. Terbraak and J. Thissen. 1995. The effects of car traffic on breeding bird populations in woodland III. Reduction of density in relation to the proximity of main roads. Journal of Applied Ecology **32**:187–202.

Rhodes, J. R., D. Lunney, J. Callaghan and C. A. McAlpine. 2014. A few large roads or many small ones? How to accommodate growth in vehicle numbers to minimise impacts on wildlife. PLoS One **9**:e91093.

Riitters, K. H. and J. D. Wickham. 2003. How far to the nearest road? Frontiers in Ecology and the Environment **1**:125–129.

Riley, S. P. D., J. P. Pollinger, R. M. Sauvajot, E. C. York, C. Bromley, T. K. Fuller and R. K. Wayne. 2006. A southern California freeway is a physical and social barrier to gene flow in carnivores. Molecular Ecology **15**:1733–1741.

Roedenbeck, I. A., L. Fahrig, C. S. Findlay, J. E. Houlahan, J. A. G. Jaeger, N. Klar, S. Kramer-Schadt and E. A. van der Grift. 2007. The Rauischholzhausen agenda for road ecology. Ecology and Society **12**:11. Available from http://www.ecologyandsociety.org/vol12/iss11/art11/ (accessed 17 September 2014).

Roedenbeck, I. A. and P. Voser. 2008. Effects of roads on spatial distribution, abundance and mortality of brown hare (*Lepus europaeus*) in Switzerland. European Journal of Wildlife Research **54**:425–437.

Schafer, A. and D. G. Victor. 2000. The future mobility of the world population. Transportation Research Part A: Policy and Practice **34**:171–205.

Seabrook, W. A. and E. B. Dettmann. 1996. Roads as activity corridors for cane toads in Australia. Journal of Wildlife Management **60**:363–368.

Selva, N., S. Kreft, V. Kati, M. Schulck, B. Jonsson, B. Mihok, H. Okarma and P. Ibisch. 2011. Roadless and low-traffic areas as conservation targets in Europe. Environmental Management **48**:865–877.

Shanley, C. S. and S. Pyare. 2011. Evaluating the road-effect zone on wildlife distribution in a rural landscape. Ecosphere **2**:art 16.

Steen, D. A. and J. P. Gibbs. 2004. Effects of roads on the structure of freshwater turtle populations. Conservation Biology **18**:1143–1148.

Stoner, D. 1925. The toll of the automobile. Science **61**:56–57.

Thiel, R. P. 1985. Relationship between road densities and wolf habitat suitability in Wisconsin. American Midland Naturalist **113**:404–407.

van der Ree, R., J. A. G. Jaeger, E. A. van der Grift and A. P. Clevenger. 2011. Effects of roads and traffic on wildlife populations and landscape function: road ecology is moving toward larger scales. Ecology and Society **16**:1–9.

Williams, N. S. G., E. J. Leary, K. M. Parris, and M. J. McDonnell. 2001. The potential impacts of freeways on native grassland. The Victorian Naturalist **118**:4–15.

World Energy Council (WEC). 2011. Global transport scenarios 2050. WEC, London.

Chapter 2

BAD ROADS, GOOD ROADS

William F. Laurance

Centre for Tropical Environmental and Sustainability Science (TESS), College of Marine and Environmental Science, James Cook University, Cairns, Queensland, Australia

SUMMARY

Roads greatly influence the footprint of human activity, but they are often constructed with little consideration of their environmental impacts, especially in developing nations. Here, differences between environmentally 'good' and 'bad' roads are highlighted, and it is argued that a proactive road-zoning system is direly needed at international and national scales. Such a zoning system could identify areas where the environmental costs of roads are likely to be high and their socioeconomic benefits low, as well as areas where road improvements could have modest environmental costs and large societal benefits.

2.1 Land-use pressures will rise sharply this century and will be strongly influenced by roads.
2.2 Agricultural yield increases alone will not spare nature – land-use zoning is crucial too.
2.3 Roads in pristine areas are environmentally dangerous – the first cut is critical.
2.4 Paved highways have especially large-scale impacts.
2.5 Roads can be environmentally beneficial in certain contexts.
2.6 Roads are amenable to policy modification.
2.7 A recently proposed global road-mapping scheme could serve as a potential model for these efforts.

This road-planning scheme could be an important tool for prioritising road investments and for underscoring the transformative role of roads in determining environmental change. An overriding priority is to proactively zone roads at a range of spatial scales while highlighting their critical role in provoking environmental change. Keeping roads out of surviving irreplaceable natural areas is among the most tractable and cost-effective ways to protect crucial ecosystems and the vital services they provide, whereas roads in the right places can facilitate increases in agricultural productivity and efficiency.

Handbook of Road Ecology, First Edition. Edited by Rodney van der Ree, Daniel J. Smith and Clara Grilo.
© 2015 John Wiley & Sons, Ltd. Published 2015 by John Wiley & Sons, Ltd.
Companion website: www.wiley.com\go\vanderree\roadecology

INTRODUCTION

Many would be surprised to learn that the Amazon, the world's greatest rainforest, now has over 260,000 km of legal and illegal roads (Barber et al. 2014) – enough to encircle the Earth more than six times. This is not an isolated example. Even in formerly remote corners of the world – from the Congo to Borneo and Siberia to Namibia – roads and transportation networks are expanding apace.

The global road rush is being driven by escalating demands for minerals, fossil fuels, timber and arable land, and by the needs of developing nations to improve their transportation and energy infrastructures (Laurance et al. 2009). Road expansion is favoured by many economists and international donors and lenders (e.g. Jacoby 2000), who see it as a cost-effective way to promote regional integration and spur economic growth.

Scientists, however, often see roads in a negative light because they can open a Pandora's box of environmental problems. In the Amazon, for instance, new roads in forested areas often promote illegal colonisation, mining, hunting and land speculation (Laurance et al. 2001, 2002; Fearnside & Graça 2006). As a result, nearly 95% of the deforestation (Fig. 2.1), fires and atmospheric carbon emissions in Amazonia occur within 5 km of roads (Barber et al. 2014). In Equatorial Africa, road expansion and associated hunting are driving major declines of forest elephants (Laurance et al. 2006; Blake et al. 2007) and other vulnerable wildlife (Fig. 2.2).

Here, I argue that roads can either benefit or harm nature, depending on their location and design. Understanding how roads affect land-use dynamics will be vital for balancing future development needs and the environment.

LESSONS

2.1   Land-use pressures will rise sharply this century and will be strongly influenced by roads

The 21st century will witness profound changes in land use, many of which are necessary and unavoidable. Meeting the needs of a projected 11 billion people for food, fibre and biofuels will require a major increase in the footprint of agriculture. According to projections of strong, consistent relationships between economic growth and food consumption, food production alone will need to increase 100–110% by the middle of this century (Tilman et al. 2011). Based on current trends in farming practices this would require about 1 billion ha of additional farming and grazing land (Tilman et al. 2011), an area larger than Canada.

The tsunami-like changes in land use this century will be strongly influenced by patterns of road development. This follows from massive road building in the past; by the year 2000, roads totalled over 28 million km in length globally (CIA 2008). Roads are sometimes

Figure 2.1 Forest clearing along roads in Rondônia, Brazil, 1989. Source: Google Earth (Imagery date 7 August 1989, 10°02'43.59"S, 63°10'03.82W).

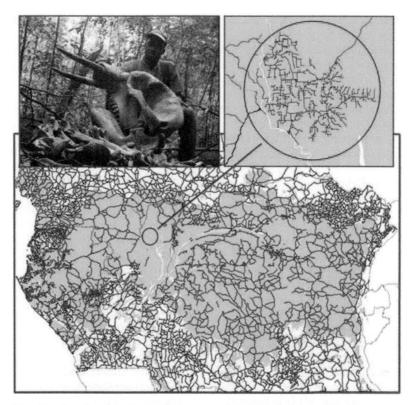

Figure 2.2 A rapid proliferation of roads is allowing hunters to penetrate into the heart of the Congo Basin, imperilling wildlife such as the forest elephant. Insets: gunshot elephant in Gabon, and smaller logging roads not shown in the larger image. Source: Photograph by Ralph Buij. Reproduced with permission of Ralph Buij; Small and large road images by Stephen Blake. Reproduced with permission of Stephen Blake and the World Resources Institute.

built specifically to promote agricultural expansion but, often, agriculture follows roads created for other purposes, such as mining or logging (Laurance et al. 2009). This can result in farms and ranches expanding into places with marginal soils or climates, or that are too far from markets to be cost-effective (Chapter 51; Fearnside 1986).

2.2   Agricultural yield increases alone will not spare nature – land-use zoning is crucial too

Given the escalating demands for food and biofuel, many environmental scientists and agronomists have highlighted a need to improve agriculture – using modern crop varieties, fertilisers, pest control and improved transportation to raise yields while limiting the footprint of agriculture and thereby 'sparing' lands for nature conservation (Green et al. 2005; Edwards et al. 2010; Phalan et al. 2011). Unfortunately, improving yields alone is unlikely to conserve nature. If it increases farming profitability, yield increases can actually do the opposite – encourage conversion of vast areas of land for production (Angelsen & Kaimowitz 2001). This is occurring today with the rapid expansion of lucrative oil palm plantations across the tropics, often at the expense of biodiversity-rich rainforests (Koh & Wilcove 2008; Butler & Laurance 2009).

Increasing agricultural yields will only benefit nature if it is coupled with effective land-use planning (Balmford et al. 2012). A key element of such planning is roads, which profoundly influence the footprint of human activities.

2.3 Roads in pristine areas are environmentally dangerous – the first cut is critical

While many factors influence road planning, a few key principles can help guide their siting and design. The environmentally most dangerous roads are those that penetrate into relatively pristine regions, such as a large forest tract (Laurance et al. 2001, 2002, 2009; Chapter 3). Deforestation is highly contagious spatially, such that the probability that a land parcel will be cleared rises dramatically if it is adjacent to an area that has already been cleared (Boakes et al. 2010). For this reason the first cut into a forest is the critical one; if it occurs, then other cuts are likely to follow.

2.4 Paved highways have especially large-scale impacts

Paved highways typically have much larger-scale environmental impacts than do unpaved roads (Laurance et al. 2002; Kirby et al. 2006; Barber et al. 2014). In wetter environments, paved roads provide year-round access to natural resources such as timber, minerals or agricultural land, whereas unpaved roads can become seasonally impassable (Fig. 51.4). Paved roads are also typically wider and have more traffic that is faster-moving than is the case for unpaved roads, and thereby are a greater danger and movement-barrier to wildlife (Laurance et al. 2009).

Disentangling the specific contributions of paved and unpaved roads to environmental damage is challenging because paved roads tend to spawn networks of secondary, unpaved roads (Laurance et al. 2009). Nevertheless, paved roads are much stronger predictors of deforestation than are unpaved roads, and their effects extend for considerably larger distances away from roads (Laurance et al. 2002; Kirby et al. 2006). For instance, the paved Belém-Brasília Highway, completed in the early 1970s, has today evolved into a 400-km-wide slash of forest destruction and secondary roads across the eastern Brazilian Amazon (Laurance et al. 2009). In the wrong place, a paved road can provoke an environmental disaster.

2.5 Roads can be environmentally beneficial in certain contexts

Although many roads promote environmental damage, paving and other road improvements can be socially and environmentally beneficial in certain contexts. In areas well-suited for agricultural development, road improvements can act as 'magnets', attracting migrants away from vulnerable frontier areas (Andersen et al. 2002; Weinhold & Reis 2008; Rudel et al. 2009). Concentrating people in carefully defined areas is beneficial because the relationship between deforestation and human population density is nonlinear, such that later migrants into an area clear much less forest on average than do those who arrive initially (Laurance et al. 2002). Better transportation infrastructure also increases access to markets, cutting waste and improving farmers' profits.

As a result, building high-quality roads in places where farming is already widespread, where there is little intact habitat, and where sizeable gaps between current and potential farm yields exist can help increase agricultural production (Weinhold & Reis 2008). This can enhance rural livelihoods and limit the negative environmental impacts of farming, by raising production efficiency and helping to keep farming more contained and localised. The global road-mapping scheme described in Lesson 2.7 and in Laurance et al. (2014) highlights a strategy for advancing these aims.

2.6 Roads are amenable to policy modification

It is notable that roads are much more amenable to policy modification than are socially complex problems such as human population growth and overconsumption. Roads can be re-routed, projects cancelled or construction delayed. Many large road projects are funded by taxpayers, investors or international donors that are responsive to environmental concerns. If publicly named and shamed, corporations that build environmentally bad roads can lose customers and shareholders. For instance, a Malaysian logging corporation, Concord Pacific, was publicly vilified for bulldozing a 180-km-long road into the highlands of Papua New Guinea – ostensibly to aid local communities. After the company took more than US$60 million in illegal timber, it was fined $97 million by the national court of Papua New Guinea (Greenpeace 2002).

2.7 A recently proposed global road-mapping scheme could serve as a potential model for these efforts

Given the environmentally transformative roles of roads, it has recently been argued that a global zoning exercise is needed to identify areas that should ideally remain road-free as well as those where transportation

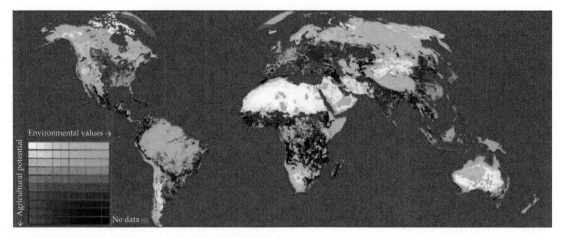

Figure 2.3 A global roadmap that attempts to estimate the relative risks and rewards of road building. Green-shaded areas are where road building would have high environmental costs, whereas red-shaded areas are where new or improved roads could help to promote increased agricultural production. Dark-shaded areas are 'conflict zones' where environmental costs and potential road-building benefits are both high. Light-shaded areas are lower priorities for environmental values and road building. Source: From Laurance et al. (2014).

improvements are a priority (Laurance & Balmford 2013; Laurance et al. 2014). This strategy integrated spatial data on remaining intact habitats and wilderness areas, existing transport infrastructure, agricultural yields and losses, biodiversity indicators, carbon storage and other relevant attributes (Fig. 2.3).

The key goal of this road-zoning effort is to promote roads and road improvements in areas that contain existing rural development and increase agricultural yields while at the same time limiting roads where the prospects for environmental damage are great. Some examples of the latter include the proposed Serengeti Highway that could disrupt one of the world's great remaining wildlife migrations (Chapter 56; Pimm 2010); Brazil's Manaus-Porto Velho Highway, which when completed will link major population centres to the heart of the Amazon (Chapter 51; Fearnside & Graça 2006); and the proposed Ladia Galaska road network, which threatens the largest surviving block of forest in northern Sumatra, Indonesia (Gaveau et al. 2009).

Beyond reducing overall habitat destruction, road zoning would also focus on safeguarding rare environments and areas with many endemic species, such as remaining intact habitats within biodiversity hotspots (Myers et al. 2000). In regions where transportation projects are unavoidable but environmental costs are high, alternatives such as railroads or river transport might be effective compromises (Laurance et al. 2009). Such projects can move people and products while stopping only at specific places, limiting their human footprint.

CONCLUSIONS

An overriding priority is to zone roads proactively on varying spatial scales while highlighting their critical role in provoking environmental change. Keeping roads out of surviving irreplaceable natural areas is among the most tractable and cost-effective ways to protect crucial ecosystems and the vital services they provide, whereas roads in the right places can facilitate increases in agricultural productivity and efficiency. In a world struggling to conserve nature and support human well being as land-use pressures intensify, managing transportation networks is where the rubber meets the road.

FURTHER READING

Laurance (2009): A hard-hitting essay on the high environmental costs of many roads.

Laurance and Balmford (2013): Highlights a global roadmapping scheme designed to maximize the social and economic benefits of roads while minimizing their environmental costs.

Laurance et al. (2009): A balanced overview of the diverse impacts of roads on tropical forests and their biodiversity.

Laurance et al. (2014): Presents a global scheme for prioritising road building based on their relative environmental costs and their potential societal benefits, particularly for promoting increased food production.

REFERENCES

Andersen, L. E., C. W. J. Granger, E. J. Reis, D. Weinhold and S. Wunder. 2002. The dynamics of deforestation and economic growth in the Brazilian Amazon. Cambridge University Press, Cambridge.

Angelsen, A. and D. Kaimowitz, Editors. 2001. Agricultural technologies and tropical deforestation. CABI Publishing, Wallingford.

Balmford, A., R. Green and B. Phalan. 2012. What conservationists need to know about farming. Proceedings of the Royal Society B **279**:2714–2724.

Barber, C. P., M. A. Cochrane, C. M. Souza Jr and W. F. Laurance. 2014. Roads, deforestation, and the mitigating effect of protected areas in the Amazon. Biological Conservation **177**:203–209.

Blake, S., S. Strindberg, P. Boudjan, C. Makombo, I. Bila-Isia, O. Ilambu, F. Grossmann, L. Bene-Bene, B. de Semboli, V. Mbenzo, D. S'hwa, R. Bayogo, L. Williamson, M. Fay, J. Hart and F. Maisels. 2007. Forest elephant crisis in the Congo Basin. PLoS Biology **5**(4):e111, doi:10.1371/journal.pbio.0050111.

Boakes, E. H., G. M. Mace, P. McGowan and R. A. Fuller. 2010. Extreme contagion in global habitat clearance. Proceedings of the Royal Society B **277**:1081–1085.

Butler, R. A. and W. F. Laurance. 2009. Is oil palm the next emerging threat to the Amazon? Tropical Conservation Science **2**:1–10.

Central Intelligence Agency (CIA). 2008. The World Factbook Transportation statistics. The US Government Printing Office, Washington, DC. Available: http://www.nationmaster.com/graph/tra_hig_tot-transportation-highways-total (accessed 1 June 2013).

Edwards, D. P., J. A. Hodgson, K. C. Hamer, S. Mitchell, A. Ahmad, S. Cornell and D. S. Wilcove. 2010. Wildlife-friendly oil palm plantations fail to protect biodiversity effectively. Conservation Letters **3**:236–242.

Fearnside, P. M. 1986. Human carrying capacity of the Brazilian rainforest. Columbia University Press, New York.

Fearnside, P. M. and P. Graça. 2006. BR-319: Brazil's Manaus-Porto Velho Highway and the potential impact of linking the arc of deforestation to central Amazonia. Environmental Management **38**:705–716.

Gaveau, D. L. A., S. Wich, J. Epting, D. Juhn, M. Kanninen and N. Leader-Williams. 2009. The future of forests and orangutans (*Pongo abelii*) in Sumatra: predicting impacts of oil palm plantations, road construction, and mechanisms for reducing carbon emissions from deforestation. Environmental Research Letters **4**:034013, doi:10.1088/1748-9326/4/3/034013.

Green, R. E., S. J. Cornell, J. P. W. Scharlemann and A. Balmford. 2005. Farming and the fate of wild nature. Science **307**:550–555.

Greenpeace. 2002. Partners in crime: Malaysian loggers, timber markets and the politics of self-interest in Papua New Guinea. Greenpeace International, Amsterdam. Available: http://www.greenpeace.org/international/Global/international/planet-2/report/2002/4/partners-in-crime-malaysian-l.pdf (accessed 1 June 2013).

Jacoby, H. C. 2000. Access to markets and the benefits of rural roads. Economic Journal **110**:713–737.

Kirby, K. R., W. F. Laurance, A. K. Albernaz, G. Schroth, P. M. Fearnside, S. Bergen, E. M. Venticinque and C. da Costa. 2006. The future of deforestation in the Brazilian Amazon. Futures **38**:432–453.

Koh, L. P. and D. S. Wilcove. 2008. Is oil palm agriculture really destroying tropical biodiversity? Conservation Letters **1**:60–64.

Laurance, W. F. 2009. Roads to ruin. New Scientist, 29 August, pp. 24–25.

Laurance, W. F. and A. Balmford. 2013. A global map for road building. Nature **495**:308–309.

Laurance, W. F., M. A. Cochrane, S. Bergen, P. M. Fearnside, P. Delamonica, C. Barber, S. D'Angelo and T. Fernandes. 2001. The future of the Brazilian Amazon. Science **291**:438–439.

Laurance, W. F., A. K. M. Albernaz, G. Schroth, P. M. Fearnside, E. Venticinque and C. Da Costa. 2002. Predictors of deforestation in the Brazilian Amazon. Journal of Biogeography **29**:737–748.

Laurance, W. F., B. M. Croes, L. Tchignoumba, S. A. Lahm, A. Alonso, M. Lee, P. Campbell and C. Ondzeano. 2006. Impacts of roads and hunting on central-African rainforest mammals. Conservation Biology **20**:1251–1261.

Laurance, W. F., M. Goosem and S. G. Laurance. 2009. Impacts of roads and linear clearings on tropical forests. Trends in Ecology and Evolution **24**:659–669.

Laurance, W. F., G. R. Clements, S. Sloan, C. O'Connell, N. D. Mueller, M. Goosem, O. Venter, D. P. Edwards, B. Phalan, A. Balmford, R. van der Ree and I. Burgues Arrea. 2014. A global strategy for road building. Nature **513**:229–232.

Myers, N., R. A. Mittermeier, C. G. Mittermeier, G. A. B. da Fonseca and J. Kent. 2000. Biodiversity hotspots for conservation priorities. Nature **403**:853–858.

Phalan, B., M. Onial, A. Balmford and R. E. Green. 2011. Reconciling food production and biodiversity conservation: Land sharing and land sparing compared. Science **333**:1289–1291.

Pimm, S. L. 2010. The Serengeti road to disaster. National Geographic Online, Washington, DC. Available: http://newswatch.nationalgeographic.com/2010/06/18/serengeti_road/ (accessed 4 June 2013).

Rudel, T. K., R. DeFries, G. P. Asner and W. F. Laurance. 2009. Changing drivers of tropical deforestation create new challenges and opportunities for conservation. Conservation Biology **23**:1396–1405.

Tilman, D., C. Balzer, J. Hill and B. L. Befort. 2011. Global food demand and the sustainable intensification of agriculture. Proceedings of the National Academy of Sciences USA **108**, 20260–20264.

Weinhold, D. and E. Reis. 2008. Transportation costs and the spatial distribution of land use in the Brazilian Amazon. Global Environmental Change **18**:54–68.

Chapter 3

WHY KEEP AREAS ROAD-FREE? THE IMPORTANCE OF ROADLESS AREAS

Nuria Selva[1], Adam Switalski[2], Stefan Kreft[3] and Pierre L. Ibisch[3]

[1]Institute of Nature Conservation, Polish Academy of Sciences, Krakow, Poland
[2]InRoads Consulting, LLC, Missoula, MT, USA
[3]Centre for Econics and Ecosystem Management, Eberswalde University for Sustainable Development, Eberswalde, Germany

SUMMARY

Roadless and low-traffic areas are typically large, natural or semi-natural areas that have no roads or few roads with low-traffic volume. They are relatively unaffected by roads and subsequent developments, and therefore, represent relatively undisturbed ecosystems, which provide important benefits for biodiversity and human societies. Roadless areas are rapidly becoming rare across the globe due to construction of road networks that serve widely expanding human activity. With a few exceptions, roadless and low-traffic areas are not considered in national or international legislation; and consequently, they have been widely neglected in transport planning.

3.1 Roadless areas contribute significantly to the preservation of biodiversity and ecosystem services.

3.2 Planning of new transport routes should identify existing roadless areas and avoid them.

3.3 Subsequent ('contagious') development effects of road construction should be avoided in roadless and low-traffic areas.

3.4 Unnecessary and ecologically damaging roads should be reclaimed to enlarge roadless areas and restore landscape-level processes.

3.5 It is crucial to systematically evaluate the need for and location of proposed roads and implement the principle of 'no net loss' of unfragmented lands when there is no alternative.

An important question during planning is whether the proposed road is really needed, and if so, where should it be placed. When the dissection of a roadless area is absolutely unavoidable, measures to prevent contagious development should be implemented, as well as compensation measures to restore the same amount of unfragmented habitat.

Handbook of Road Ecology, First Edition. Edited by Rodney van der Ree, Daniel J. Smith and Clara Grilo.
© 2015 John Wiley & Sons, Ltd. Published 2015 by John Wiley & Sons, Ltd.
Companion website: www.wiley.com\go\vanderree\roadecology

INTRODUCTION

With more than 64 million km of roads worldwide (CIA 2013), road networks play a primary role in shaping the environment. Approximately 90% of the world's land surface can be reached within 48 hours of travel by road or rail from the nearest city (Williams 2009). The ecological effects of roads extend far beyond the edge of the road itself; and despite the efforts to minimise road impacts in the past decades, a large portion of the planet is affected by roads (e.g. about one-fifth of the continental United States, Forman 2000). Among the numerous impacts of roads, probably the most important is what we have termed 'contagious' development: roads provide access to previously remote areas, thus opening them up for more roads and developments, and triggering land-use changes, resource extraction and human disturbance (Fig. 2.1, Chapter 51). In this context, the importance of keeping the remaining large unfragmented lands road-free becomes an urgent task.

Roadless and low-traffic areas either have no roads or few roads with low-traffic volumes (see Lesson 3.2 for definitions). They have become a rare element of the landscape; only 3% of the conterminous United States is more than 5 km away from a road (Riitters & Wickham 2003). Consideration of unfragmented lands is typically neglected in road planning and biodiversity conservation. The aims of this chapter are to highlight the value of roadless and low-traffic areas, the need to consider them in sustainable transport planning and the importance of road removal to restore them.

LESSONS

3.1 Roadless areas contribute significantly to the preservation of biodiversity and ecosystem services

Lands without roads have not been altered by road effects such as traffic, noise pollution or wildlife mortality due to collision with vehicles. Roadless areas contain natural and semi-natural habitats with a low level of human disturbance, where wide arrays of ecological processes are preserved. Habitats that are more intact provide greater benefits for biodiversity and human societies than degraded habitats (see reviews in DellaSala and Strittholt (2003) and Selva et al. (2011)).

Roadless areas are biodiversity reservoirs. They are important for wildlife and have the potential to conserve sensitive and endangered species (Loucks et al. 2003). They are crucial for species that move across large tracts of habitat, such as brown bears, wolves or elephants (e.g. Blake et al. 2008). Even large unfragmented areas which have been moderately modified (e.g. for agriculture) can still provide landscape connectivity. Roadless areas are known strongholds for salmonids and other fish species (Quigley & Arbelbide 1997), and a significant refuge for native wildlife and plants (Gelbard & Harrison 2003). They also serve as a barrier against invasive and exotic species, and diseases of wildlife, livestock and humans. For instance, the risk of humans contracting Lyme disease is reduced in larger patches of unfragmented forest, where the diversity of vertebrate hosts is higher (Allan et al. 2003).

Roadless and low-traffic areas perform numerous ecosystem services that are vital for humans. These include the maintenance of healthy soil, clean air and clean and reliable supply of water (DellaSala & Strittholt 2003). While some managers suggest that roads are needed to manage fire and pests, roadless areas are generally characterized by lower fire risk and lower frequency of insect outbreaks than roaded areas (DellaSala & Frost 2001). The social and economic benefits of roadless areas, such as non-motorised outdoor recreation, education and scientific values, are large and well documented (e.g. Loomis & Richardson 2000). As human population increases, the demand for undisturbed land and for wilderness experiences will likewise increase.

Roadless and low-traffic areas are important in the context of climate change (Selva et al. 2011). Undisturbed and mature ecosystems provide buffering capacity, moderate weather extremes (e.g. by retaining water) and help to stabilize local climates (e.g. Norris et al. 2012), thereby protecting against the impacts of storm events, like flooding or landslides. Roadless and low-traffic areas of mature forest and peatland are significant in the sequestration of carbon. Roadless areas accommodate adaptations and range shift responses by plants and animals to climate change by providing important landscape connections and moderating the rate of change of local environmental conditions.

With the current rate of road encroachment, biodiversity crisis and global change processes such as climate change, roadless and low-traffic areas may far exceed roaded areas for their benefits provided to human societies (Selva et al. 2011). Therefore, it seems sensible that sustainable transport policies retain and re-establish unroaded lands in order to conserve biodiversity and maintain the health of ecosystems on which we depend (Textbox 3.1).

3.2 Planning of new transport routes should identify existing roadless areas and avoid them

While roadless and low-traffic areas can be broadly defined as natural and semi-natural areas without roads or with few roads of low-traffic intensity, respectively, there are different legal descriptions and criteria used around the globe to identify them. Although road-free areas and areas with low road density or low traffic volumes are not automatically considered in conservation and transport planning, there are two basic approaches to incorporate roads in spatial planning. The first approach identifies road-free areas of a minimum size (e.g. Wilderness and Inventoried Roadless Areas in the United States) or areas with traffic volume below a specified threshold (e.g. Unfragmented Areas by Traffic in Germany, see Textbox 3.1), and the second approach identifies areas with high conservation status. Under this approach (e.g. Last of the Wild global program or areas of good conservation status in the Chiquitano dry forest, Bolivia), roads and their impacts are combined with other indicators, such as human population density, deforestation or cattle grazing, in order to prioritize areas for biodiversity conservation (Table 3.1).

Roadlessness typically correlates with relatively good conservation status. Therefore, indices that assess the environmental impact of roads by identifying roadless and low-traffic areas should be applied during spatial planning (e.g. SPROADI, Freudenberger et al. 2013). The definition of thresholds to identify such areas, such as the minimum size of roadless areas or the maximum tolerable traffic volume, depends on the landscape context. For example, the dissection of relatively small roadless areas (e.g. Fig. 3.2) is a conservation issue in highly populated regions like central Europe, while large road-free areas are a priority in relatively pristine and unfragmented regions, like the Amazon or Siberia.

Textbox 3.1 Recognition and protection of roadless and low-traffic areas in the world.

Wilderness and roadless area protection in the United States

In the United States, many roadless areas were first protected when the Wilderness Act (1964) was passed. Wilderness was defined as 'an area where the earth and its community of life are untrammeled by man, where man himself is a visitor and does not remain'. Wilderness areas in the United States do not allow permanent improvements or human habitation and were originally required to be larger than 2024 ha (Table 3.1). The National Wilderness Preservation System in the United States has grown to more than 40 million ha today. In 2001, the US Forest Service protected an additional 24 million ha of road-free areas larger than 405 ha under the 'Roadless Conservation Rule'. These inventoried areas are protected from building new roads, although they still allow for motorized use, such as all-terrain vehicles, helicopter logging and other uses that are prohibited in wilderness. Walking trails are common in both Wilderness and Inventoried Roadless Areas.

Low-traffic and unfragmented areas in Europe

Large roadless areas are rare in Europe, and, instead, definitions referring to low-traffic areas have been developed. The concept of unfragmented areas by traffic (UAT) was developed by the German Federal Agency for Nature Conservation as a landscape assessment tool (Table 3.1). The UATs are greater than 10,000 ha and not dissected by roads with more than 1000 vehicles/day, by railway lines (twin-track and single-track electrified lines) or by human settlements, airports or channels. The 2008 inventory identified about 9 million ha of UATs in Germany, of which a quarter are protected under European Directives. The eastern part of Germany contains more UATs than western Germany (Fig. 3.1), which may be illustrative of the different degree of fragmentation between eastern and western Europe.

Global roadless areas

A prototype map of roadless areas in the world was developed in 2012 by Google Earth, the Society for Conservation Biology – Europe Section and Members of the European Parliament (http://earthengine. google.org/). Here, roadless areas were defined by using buffers of different distances (from 1 to 10 km) from the nearest road (including dirt roads), rail or navigable waterway (Table 3.1). This map was presented in 2012 at the Rio + 20 Conference in Brazil and at the eleventh meeting of the Conference of the Parties to the Convention on Biological Diversity in India to demonstrate that roadlessness is the most cost-efficient and effective way to protect biodiversity.

Table 3.1 Examples of initiatives across the world which have identified roadless or low-traffic areas for biodiversity conservation.

Initiative	Framework	Region	Criteria	Indicators	Legal enforcement
Wilderness areas	Wilderness Act, 1964	USA	Size Relative naturalness Roads	≥2024 ha — Presence/absence	Yes
Last frontier forests	World Resources Institute, 1997	Global	'Wilderness areas'[a] Forest cover[b] Size[b]	Sierra Club's global 'wilderness areas' map Dominance Ability to support viable populations of large, wide-ranging animals for a century	No
Inventoried Roadless Areas	Roadless Conservation Rule, 2001	USA	Naturalness of structure and composition: various indicators[b] Size	Dominance ≥405 ha	Yes
Areas of good conservation status	Chiquitano dry forest conservation and sustainable development plan, 2002	Bolivia	Population/km^2 Roads, railways, pipelines Navigable rivers and lakes Deforestation Cattle grazing Forest concessions, forests accessible by roads	Five classes ranging from 0–4.4 to ≥95.3 Eight classes ranging from 'principal road' to 'pipeline crossing wet savannah' Presence/absence Two classes: values > and ≤30% Four classes ranging from 'semi-intensive cattle-grazing in open and semi-open areas' to 'no cattle grazing' Presence/absence	No
Last of the wild	Wildlife Conservation Society and Center for International Earth Science Information Network, 2002–2005	Global	Size Population/km^2 Railroads Major roads Navigable rivers Coastlines Night-time stable lights values[a] Urban polygons Land cover categories	>500 ha 10 classes ranging from 0–0.5 to >9.5 Beyond 2 km buffer Beyond 2 and 15 km buffer, respectively Beyond 15 km buffer Beyond 15 km buffer Four classes ranging from 0 to ≥89 Presence/absence Four classes ranging from 'urban' to 'forests, tundra and deserts'	No
Unfragmented areas by traffic (UAT)	German Federal Agency for Nature Conservation, 2008	Germany	Size Heavily used roads, railway lines, channels, urban centres Traffic intensity	10,000 ha Presence/absence <1000 vehicles/day	No
Roadless areas	Google Earth and Society for Conservation Biology, 2012	Global	Roads, rails or navigable waterways	Beyond 1 and 10 km buffers, respectively	No

[a]McCloskey and Spalding (1989).
[b]Assessed by qualified expert opinion.

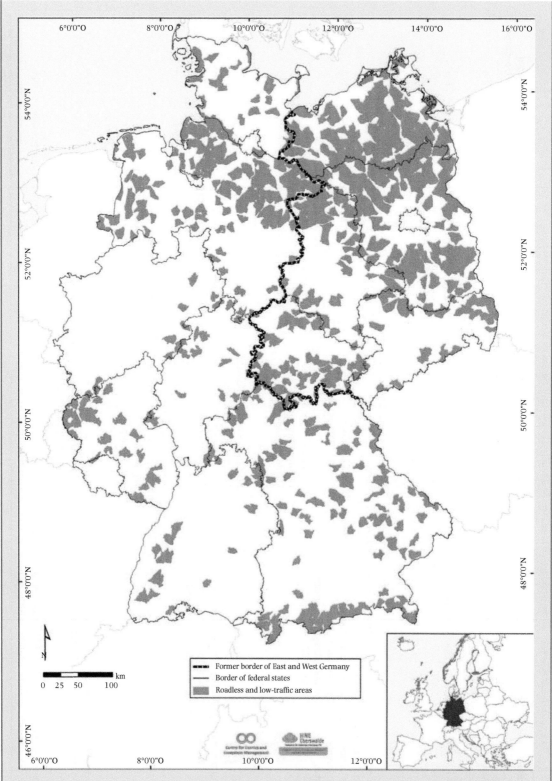

Figure 3.1 Low-traffic areas in Germany (Unfragmented Areas by Traffic, UAT). UATs cover 45% of the new federal states (eastern Germany) versus 18% of the old federal states (western Germany). Source: Adapted from Selva et al. (2011).

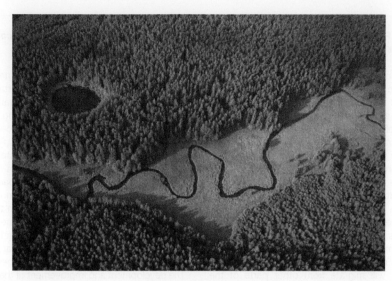

Figure 3.2 The Rospuda valley in northeastern Poland (6.3 km²) is the last pristine percolating fen (or active peatland) of the European temperate zone. Its untouched hydrological system guarantees the stability of the ecosystem (no succession) and the presence of endangered and relict species. In 2007, a road project dissecting this peatland was stopped due to legal infringements of the EU nature directives, after more than 10 years of campaign. The road was finally re-routed through agricultural fields, something that could have been done with proper transport planning years before, thus avoiding high social conflicts and economic costs. Source: Photograph by Piotr Małczewski. Reproduced with permission of Piotr Małczewski.

The roads that cause the greatest environmental damage are those dissecting pristine and unfragmented ecosystems. Even in human-dominated landscapes, the construction of new roads may add additional disturbances to those related to land uses. In this sense, an agricultural landscape without roads still might provide better habitat quality (e.g. connectivity for wildlife) than the same farmland with numerous roads. Given the intensification of land-use pressures across the globe, influencing the patterns of road development to keep roads out of natural areas is the most tractable way to conserve nature (Laurance & Balmford 2013; Laurance et al. 2014; Chapter 2).

3.3 Subsequent ('contagious') development effects of road construction should be avoided in roadless and low-traffic areas

Roads are one of the main drivers of ecosystem change. By facilitating access to previously remote areas, new roads trigger a cascade of land-use changes and habitat degradation (Chapter 51). Roads are almost inevitably followed by urban and agricultural development, and they promote mining, hunting, fishing and logging (Wilkie et al. 2000; Southworth et al. 2011).

In Central Africa, logging roads, which represent 38% of all road length, boost unsustainable hunting and the massive loss of wildlife; for example, wildlife densities decreased by 25% 3 weeks after logging roads were opened in Congo (Laporte et al. 2007; Wilkie et al. 2011). The role of roads in deforestation is undisputable and the most rapid rate of forest clearing occurs within 10 km of the road, especially if paved. As demonstrated in the Amazon, greater than 95% of deforestation, fires and atmospheric carbon emissions occur within 50 km of roads (Laurance et al. 2001; Southworth et al. 2011).

Roads also accelerate human migration to the area and subsequent illegal colonization and land speculation (Chapter 51). Road paving, demand for agriculture and cattle ranching areas and ambiguous land tenure systems promote new settlements in undisturbed areas (Southworth et al. 2011). New roads, as well as road improvements in low-traffic areas, have important economic and social impacts, mainly derived from facilitated market access. These collateral or contagious development effects of roads are often more destructive than the direct impacts of the road itself. Sensible transport and land-use planning should carefully regulate contagious development and be supported by appropriate law enforcement (Textbox 3.2).

Textbox 3.2 Roads in developing countries. The case of conservation planning and 'contagious' development in Bolivia.

Road development is often used as an indicator of socioeconomic development. Roads improve mobility of people, but also catalyse the extraction of natural resources and subsequent degradation of ecosystems, and cause profound changes in local socioeconomic systems (Chapter 2).

Bolivia is a socioeconomically poor and biologically rich country that still has a significant portion of its territory covered by natural ecosystems; the latter partly due to a poorly developed road infrastructure (Fig. 3.3, Ibisch & Mérida 2004). However, as in most developing countries, the pressure on ecosystems is increasing rapidly, making the contagious development effect of roads particularly troubling. Whenever new roads provide access to formerly remote areas, people will migrate from other parts of the country and establish (often illegally) new settlements (Chapter 2).

Recent landscape-scale planning in Bolivia used roads as indicators of biodiversity degradation (e.g. Araujo et al. 2010). Roadlessness was taken as a proxy for functional and intact ecosystems and used as a criterion for identifying important areas for conservation. However, the implementation of conservation measures (e.g. land-use planning, including the creation of protected areas) has not been enough to safeguard the high-priority regions. In 2002, an internationally financed road was constructed through the Chiquitano dry forest ecoregion in southeastern Bolivia. A decade later, the indirect impacts of the road (namely forest clearing and expansion of agriculture) have exceeded those outlined in even the most pessimistic environmental impact assessment (S. Reichle, personal communication). The fear that the impacts of new roads cannot be effectively mitigated by accompanying conservation measures has been confirmed. The development and improvement of the road network across Bolivia has continuously accelerated deforestation and other forms of biodiversity degradation. This highlights the importance of keeping unfragmented and natural habitats free of roads as the most effective way to conserve them.

Figure 3.3 Especially in forests, even small and unpaved roads give access for land use such as agriculture or settlement, which may ultimately replace the original ecosystem. Porongo, Santa Cruz, Bolivia. Source: Photograph by Pierre L. Ibisch.

3.4 Unnecessary and ecologically damaging roads should be reclaimed to enlarge roadless areas and restore landscape-level processes

Land managers are restoring roaded areas by closing and reclaiming unneeded or ecologically damaging roads (Fig. 3.4). Many of these roads are historical legacies, but new roads built to support resource extraction should be restored once the activity ceases. There are various treatments possible, ranging from simply blocking the road entrance to full removal and recontouring of the roadbed which allows hydrological and ecological processes and properties to return (Switalski et al. 2004). Increased infiltration and revegetation reduces fine sediment erosion from roads into streams, improving habitat quality for fish and other aquatic species (McCaffery et al. 2007).

Reclaimed roads improve wildlife habitat quality primarily through limitation of motorised access and the restoration of vegetation providing food and shelter for wildlife. Black bears were found to use recontoured roads at much higher rates than roads open to traffic, but also at greater rates than roads closed to traffic with a gate or other barrier (Switalski & Nelson 2011). Similarly, grizzly bears expanded their distribution in Montana, USA, following extensive road reclamation (Summerfield et al. 2004), and moose populations increased following road removal in Nova Scotia, Canada (Crichton et al. 2004). Removing roads at a large scale such as is occurring in the United States has increased the size of core wildlife habitat and has the potential to restore landscape-level connectivity.

Road reclamation efforts and the expansion of roadless areas increase the resilience of ecosystems and help mitigate climate change. For example, as larger storms become more common in the face of climate change, more culverts catastrophically fail during high flows, releasing large amounts of sediment into streams. Removing culverts and restoring stream crossings eliminates this risk and associated negative impacts on aquatic habitats (Chapters 44 and 45). Additionally, when roads are decompacted during reclamation, vegetation and soils can develop more rapidly and sequester large amounts of carbon. Total soil carbon storage increased 6-fold to 65 metric tons C/km (to 25 cm depth) in the northwestern United States compared with untreated abandoned roads (Lloyd et al. 2013). With more than 100,000 km of roads slated for reclamation in the United States alone in the coming decades, road reclamation has the potential to sequester large amounts of carbon.

Figure 3.4 After treatment, vegetation recolonises reclaimed roads reducing erosion and providing food and cover for animals. This photo was taken 10 years after road reclamation on the Clearwater National Forest in the northwestern USA. Source: Photograph by Adam Switalski.

3.5 It is crucial to systematically evaluate the need for and location of proposed roads and implement the principle of 'no-net-loss' of unfragmented lands when there is no alternative

It is important to systematically evaluate whether a road is *really* needed; and if so, explore alternative route options before dissecting and eliminating roadless areas or increasing traffic volumes in low-traffic areas (Fig. 3.5). Infrastructure development and, particularly, road construction should avoid dissecting roadless areas. Road-free areas of natural and semi-natural habitats should be maintained by concentrating traffic on existing highly travelled roads and bundling infrastructure close together (Chapter 5). When this is not possible, it is crucial to protect the remaining area by avoiding contagious development and to apply compensation policies of no net loss to unfragmented lands (Chapter 7). Measures such as road reclamation, promotion of railroads or speed and traffic limitation should also be considered. The implementation of sustainable development schemes at large spatial scales should help prevent the degradation of roadless and low-traffic areas (Fig. 3.5)

CONCLUSIONS

Roadless and low-traffic areas have become scarce, indicating a reduction in well-preserved and functioning ecosystems worldwide. The maintenance of roadless areas is more cost-effective than measures to mitigate or minimise road impacts, or even road reclamation. In this context, a vital task is to identify, map and describe the remaining roadless and low-traffic areas, and to promote their maintenance and protection. Developed countries are removing unnecessary roads and restoring landscape processes to enlarge roadless areas. This exemplifies the need for rewilding in a human-dominated planet. Roadless and low-traffic areas are a timely tool to preserve intact functioning ecosystems at local and global scales in the face of climate change. Their rarity and the services they provide to society call for systematically considering them in modern land-use and road planning.

ACKNOWLEDGEMENTS

This is part of the 'Roadless Areas Initiative' of the Policy Committee of the Society for Conservation Biology – Europe Section. The contribution by NS was

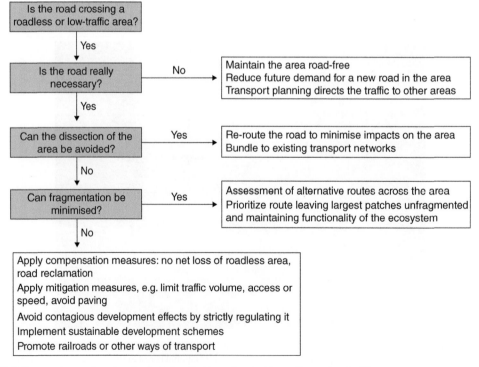

Figure 3.5 Four main questions to ask when planning a road project in roadless or low-traffic areas.

supported by the Institute of Nature Conservation. PLI acknowledges the awarding of the research professorship 'Biodiversity and natural resource management under global change' by Eberswalde University for Sustainable Development. Research by SK was partially funded by the Federal Ministry for Education and Research (INKA BB project).

FURTHER READING

DellaSala and Strittholt (2003): Review of the ecological, social and economic benefits of roadless areas conservation in the USA, with special focus on the conservation assessments of two case studies of roadless areas.

Selva et al. (2011): Identifies the importance of roadless and low-traffic areas for biodiversity conservation and ecosystem services to society, and urges for their inventory and inclusion in urban and transport planning. It includes a legal analysis of roadless areas in Europe and their overlap with the Natura 2000 network, using Germany as a case study.

Switalski et al. (2004): Summary of the current understanding in the science and practice of road reclamation. Taking a multi-disciplinary approach, the article reviews how road reclamation benefits and impacts different natural resources and identifies knowledge gaps.

http://earthengine.google.org/: This is the Google platform for environmental data at a planet scale. It includes a prototype map of global roadless areas.

http://roadlessland.org/: This is an interactive website that shows the inventoried roadless areas in the US and has a number of maps and scientific resources for roadless areas.

REFERENCES

Allan, B. F., F. Keesing and R. S. Ostfeld. 2003. Effects of forest fragmentation on Lyme disease risk. Conservation Biology **17**: 267–272.

Araujo, N., R. Müller, C. Nowicki and P. L. Ibisch (eds). 2010. Prioridades de conservación de la biodiversidad de Bolivia. SERNAP, FAN, TROPICO, CEP, NORDECO, GEF II, CI, ZNC, WCS, Universidad de Eberswalde. Editorial FAN, Santa Cruz, Bolivia.

Blake, S., S. L. Deem, S. Strindberg, F. Maisels, L. Momont, I. B. Isia, I. Douglas-Hamilton, W. B. Karesh and M. D. Kock. 2008. Roadless wilderness area determines forest elephant movements in the Congo Basin. PLoS One **3**: e3546.

Central Intelligence Agency (CIA). 2013. The world fact book. CIA, Washington, DC. Available from https://www.cia.gov/library/publications/the-world-factbook/geos/html#xx (accessed 24 September 2014).

Crichton, V., T. Barker and D. Schindler. 2004. Response of a wintering moose population to access management and no hunting. A Manitoba experiment. Alces **40**: 87–94.

DellaSala, D. A. and E. Frost. 2001. An ecologically based strategy for fire and fuels management in national forest roadless areas. Fire Management Today **61**: 12–23.

DellaSala, D. A. and J. Strittholt. 2003. Scientific basis for roadless area conservation. Report prepared by the World Wildlife Fund and Conservation Biology Institute.

Forman, R. T. T. 2000. Estimate of the area affected ecologically by the road system in the United States. Conservation Biology **14**: 31–35.

Freudenberger, L., P. R. Hobson, S. Rupic, G. Péer, M. Schluck, J. Sauermann, S. Kreft, N. Selva and P. L. Ibisch. 2013. Spatial Road Disturbance Index (SPROADI) for conservation planning: a novel landscape index, demonstrated for the State of Brandenburg, Germany. Landscape Ecology **28**: 1353–1369.

Gelbard, J. L. and S. Harrison. 2003. Roadless habitats as refuges for native grasslands: interactions with soil, aspect, and grazing. Ecological Applications **13**: 404–415.

Ibisch, P. L. and G. Mérida. 2004. Biodiversity: the richness of Bolivia. State of knowledge and conservation. Ministerio de Desarrollo Sostenible y Planificación, Editorial FAN, Santa Cruz, Bolivia.

Laporte, N. T., J. A. Stabach, R. Grosch, T. S. Lin and S. J. Goetz. 2007. Expansion of industrial logging in Central Africa. Science **316**: 1451.

Laurance, W. F. and A. Balmford. 2013. A global map for road building. Nature **495**: 308–309.

Laurance, W. F., G. R. Clements, S. Sloan, C. S. O'Connell, N. D. Mueller, M. Goosem, O. Venter, D. P. Edwards, B. Phalan, A. Balmford, R. van der Ree and I. B. Arrea. 2014 A global strategy for road building. Nature **513**: 229–232.

Laurance, W. F., M. A. Cochrane, S. Bergen, P. M. Fearnside, P. Delamonica, C. Barber, S. D'Angelo and T. Fernandes. 2001. The future of the Brazilian Amazon. Science **291**: 438–439.

Lloyd, R. A., K. A. Lohse and T. P. A. Ferré. 2013. Influence of road reclamation techniques on forest ecosystem recovery. Frontiers in Ecology and the Environment **11**: 75–81.

Loomis, J. B. and R. Richardson. 2000. Economic values of protecting roadless areas in the United States. The Wilderness Society, Washington, DC.

Loucks, C., N. Brown, A. Loucks and K. Cesareo. 2003. USDA Forest Service roadless areas: potential biodiversity conservation reserves. Conservation Ecology **7**: 5.

McCaffery, M., T. A. Switalski and L. Eby. 2007. Effects of road decommissioning on stream habitat characteristics in the South Fork Flathead River, Montana. Transactions of the American Fisheries Society **136**: 553–561.

McCloskey, J. M. and H. Spalding. 1989. A reconnaissance level inventory of the amount of wilderness remaining in the world. Ambio **18**: 221–227.

Norris, C., P. Hobson and P. L. Ibisch. 2012. Microclimate and vegetation function as indicators of forest thermodynamic efficiency. Journal of Applied Ecology **49**: 562–570.

Quigley, T. M. and S. J. Arbelbide (eds). 1997. An assessment of ecosystem components in the interior Columbia Basin and portions of the Klamath and Great Basins. Volume **3**.

USDA Forest Service and USDI Bureau of Land Management, Portland, OR.

Riitters, K. H. and J. D. Wickham. 2003. How far to the nearest road? Frontiers in Ecology and the Environment **1**: 125–129.

Selva, N., S. Kreft, V. Kati, M. Schluck, B. G. Jonsson, B. Mihok, H. Okarma and P. L. Ibisch. 2011. Roadless and low-traffic areas as conservation targets in Europe. Environmental Management **48**: 865–877.

Southworth, J., M. Marsik, Y. Qiu, S. Perz, G. Cumming, F. Stevens, K. Rocha, A. Duchelle and G. Barnes. 2011. Roads as drivers of change: trajectories across the tri-national frontier in MAP, the Southwestern Amazon. Remote Sensing **3**: 1047–1066.

Summerfield, B., W. Johnson and D. Roberts. 2004. Trends in road development and access management in the Cabinet-Yaak and Selkirk Grizzly Bear Recovery Zones. Ursus **15**(Workshop Supplement): 115–122.

Switalski, T. A., J. A. Bissonette, T. H. DeLuca, C. H. Luce and M. A. Madej. 2004. Benefits and impacts of road removal. Frontiers in Ecology and the Environment **2**: 21–28.

Switalski, T. A. and C. R. Nelson. 2011. Efficacy of road removal for restoring wildlife habitat: black bear in the Northern Rocky Mountains. Biological Conservation **114**: 2666–2673.

Wilkie, D. S., E. L. Bennett, C. A. Peres and A. A. Cunningham. 2011. The empty forest revisited. Annals of the New York Academy of Sciences **1223**: 120–128.

Wilkie, D., E. Shaw, F. Rotberg, G. Morelli and P. Auzel. 2000. Roads, development, and conservation in the Congo basin. Conservation Biology **14**: 1614–1622.

Williams, C. 2009. Where's the remotest place on Earth? New Scientist **2704**: 40–43.

Chapter 4

INCORPORATING BIODIVERSITY ISSUES INTO ROAD DESIGN: THE ROAD AGENCY PERSPECTIVE

Kevin Roberts[1] and Anders Sjölund[2]

[1]Cardno, St Leonards, New South Wales, Australia
[2]The Swedish Transport Administration, Borlänge, Sweden

SUMMARY

Road agencies have a responsibility to design, build and operate roads in an environmentally sensitive manner, which includes addressing ecological issues. Agencies that manage other linear infrastructure, such as railways and utility easements, have similar responsibilities. All major infrastructure projects follow similar stages and processes from inception through planning, design, construction, operation and maintenance. Within this process, there are limited and specific opportunities to most effectively implement ecologically sensitive planning and design.

4.1 Road planning, design, construction and operation are complex challenges that attempt to balance environmental, economic and social demands.

4.2 Road projects have a typical series of stages that begins with strategic planning and ends with operation.

4.3 Appropriate ecological input into a road project should occur in every stage.

4.4 Standards and guidelines are critical to ensure a consistent and high-quality approach to roads and road mitigation.

Road agencies around the world are responding to the changes that society is demanding by including greater consideration of ecological issues when planning, building and managing the road network. This is an important challenge for road agencies because their traditional role as managers of the transportation network is expanding and becoming more complicated. It is imperative that road agencies successfully adapt to these changes to ensure the future road network is as environmentally friendly as possible.

Handbook of Road Ecology, First Edition. Edited by Rodney van der Ree, Daniel J. Smith and Clara Grilo.
© 2015 John Wiley & Sons, Ltd. Published 2015 by John Wiley & Sons, Ltd.
Companion website: www.wiley.com\go\vanderree\roadecology

INTRODUCTION

Most countries have government agencies that are responsible for the planning, construction and maintenance of road networks. Roads are important drivers of economic and social development (Chapter 2), and road agencies are focussed on building bigger, better, safer and more efficient roads to cater for growing demand for vehicle movement (Chapter 1). Increasingly, road agencies are being challenged to respond to community and government expectations to protect and preserve the environment, often through the requirements of environmental legislation. Most governments also have agencies for other linear infrastructure, such as railways and utility easements; and whilst this chapter (and book) focuses on roads and traffic, these other agencies face similar expectations and processes to balance competing demands.

Planning and managing the road network is a complex interaction among various levels of government, private infrastructure companies and the community, each with different responsibilities and expectations. The policy and legislation of planning and regulatory agencies, funding arrangements, government and political priorities, historical legacies, economic circumstances, changing technology, road safety expectations and competing transport priorities all influence how roads are developed, built and managed (Chapter 8). All road projects require a broad coalition of public and government support, which usually entails properly considering and adequately addressing the impacts of the project on communities and the environment. Above all, the project needs to deliver value to the community.

The reasons to initiate road projects are mostly political, strongly supported by prevailing economic models, and cost-benefit analysis which are often influenced by community demands. These models and demands rarely consider environmental costs or benefits, especially ecological ones. However, the best outcome for the environment typically occurs when ecological thinking influences road planning early in the project development cycle. The aims of this chapter are to illustrate the processes that road agencies typically follow when designing, building and managing the road network, and highlight some of the key challenges facing road agencies and the scientific community when incorporating ecological and environmental safeguards into road development.

LESSONS

4.1 Road planning, design, construction and operation are complex challenges that attempt to balance environmental, economic and social demands

Road agencies must trade off a range of competing demands when planning and designing new roads and/or upgrading existing roads. Using information from a diverse range of specialists, road planners must consider matters such as safe and sustainable road design, reducing the impact on property and business, constructability, construction techniques and materials, traffic management as well as the environment. Planners and designers also receive input from the community at several stages in the process. This public response may focus the planner/designer on resolving localised (but very real) impacts on a community rather than the broader and more technical environmental issues. When considering environmental impacts, the planning and design team address a range of issues, including impacts of noise on neighbours during construction and operation; impacts on cultural heritage; air quality and greenhouse gas emissions; flooding and impacts on water quality and aquatic and groundwater-dependent ecosystems; and minimising direct and indirect impacts on habitat and species. There is rarely one ideal solution and trade-offs among environmental impacts are usually required because the ecological relationships within an ecosystem are diverse and complex. For example, the construction of a wildlife overpass may require greater clearing of habitat during construction than a modified culvert (Chapter 21), but in the long-run will provide connectivity for a wider array of species. Similarly, a road on a viaduct will have lower barrier effects, but it may have greater noise pollution issues than a road at ground level. Development of the road is an iterative process between designers, engineers, planners, regulators, scientists and the community. As the environmental assessment proceeds, new impacts are examined and mitigation or design measures proposed. The final design reflects this decision-making process.

Road agencies will apply a 'value' test to a road proposal. This can be formally applied or, in many cases, is a concept that underpins how decision-making on road development occurs. The question still remains, 'Is it feasible or reasonable to provide a mitigation measure for a particular species or ecological value?'

The feasibility of a design or mitigation measure is whether it can be constructed or installed. Can it be constructed without compromising the objectives of the road or creating other unacceptable impacts? The reasonableness of a proposed design or mitigation measure is difficult to define, but assesses if the cost of the measure is acceptable compared to the nature and extent of the mitigation benefit. Put simply, is the measure of good value? For example, it is technically feasible to construct roads as viaducts or tunnels, but the construction and operational challenges will substantially increase the cost, potentially to the point where it is no longer viable and thus unreasonable. But the same mitigation may be reasonable on another project or at another time.

Whilst the road planning and design team make every effort to address impacts, many projects will have impacts that cannot be avoided, minimised or mitigated (Lesson 1.6, Chapter 7).

4.2 Road projects have a typical series of stages that begins with strategic planning and ends with operation

Imagine that you have been given the task by government to improve the capacity and safety of a highway connecting two towns. The government has promised to open the improved road by a set date, and funding has been allocated for planning and design with a forward commitment for construction. As with all road projects, you would develop and implement the project in a series of stages, each with different issues to address (Fig. 4.1).

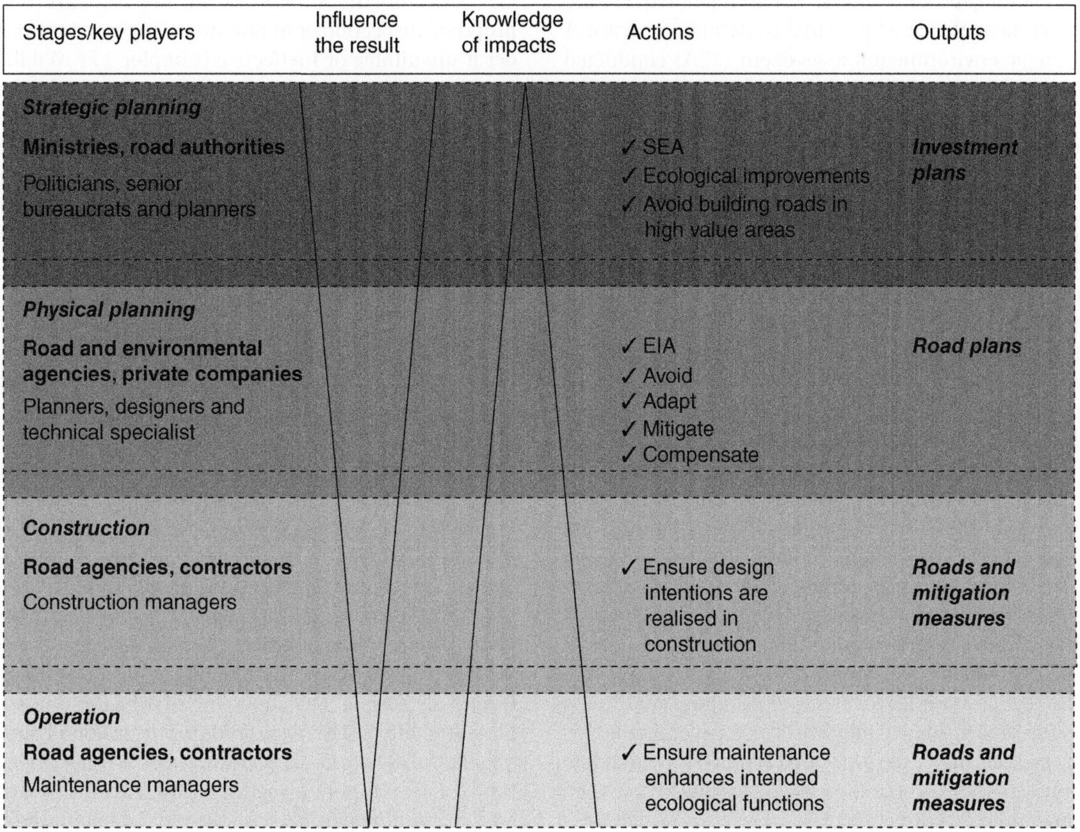

Stages/key players	Influence the result	Knowledge of impacts	Actions	Outputs
Strategic planning **Ministries, road authorities** Politicians, senior bureaucrats and planners			✓ SEA ✓ Ecological improvements ✓ Avoid building roads in high value areas	**Investment plans**
Physical planning **Road and environmental agencies, private companies** Planners, designers and technical specialist			✓ EIA ✓ Avoid ✓ Adapt ✓ Mitigate ✓ Compensate	**Road plans**
Construction **Road agencies, contractors** Construction managers			✓ Ensure design intentions are realised in construction	**Roads and mitigation measures**
Operation **Road agencies, contractors** Maintenance managers			✓ Ensure maintenance enhances intended ecological functions	**Roads and mitigation measures**

Figure 4.1 The typical stages, key players, actions and outputs of most road projects. *Note that not all projects or jurisdictions follow all stages or utilize techniques such as SEA.* The width of the two triangles represents the extent to which the final result can be modified or the amount of knowledge of potential impacts at each stage. EIA, environmental impact assessment; SEA, strategic environmental assessment.

Most projects start with a strategic planning stage (Fig. 4.1), where route options are developed and key project goals established. Politicians and senior bureaucrats are often influential at this stage, and planning is typically based on economic models and calculations. There may be one logical route option or many, each with different negative and positive impacts. The road may need to be designed for large trucks, high speeds, or high traffic volumes. The project may be a new road, a widening of an existing road, a series of general improvements or targeted works at safety hot spots. The strategic planning stage usually concludes with the preparation of an investment plan for government treasuries over a 3-, 5- or 10-year time period. Investment plans usually identify targets for cost, traffic access, safety and design, but rarely provide targets for environmental or ecological issues. This is a significant handicap for ecological matters through the remainder of the planning process. At best, environmental issues are sometimes considered potential constraints or risks. A strategic environmental assessment (SEA) conducted during the strategic planning stage has the potential to improve decision-making by taking a broader perspective and increasing transparency, although they are not yet routinely undertaken (Chapter 5).

The second stage is the physical planning of the road project, where the location is decided, the concept design developed and the environmental impact assessment (EIA) performed. This stage concludes when some form of project or planning approval is given to the project. Whilst every jurisdiction has its own form of EIA and approval processes, which also varies with project size and likely impact (Chapter 6), there are some common elements. By the time a project reaches this stage, there are often only very limited opportunities to *avoid* environmental impact, partially because EIAs rarely account for all potential impacts (Chapter 5). Consideration therefore is on avoiding the environmental impact of the road by minor alterations to the location of the road, minimising through design and mitigation and in some jurisdictions, provision of compensatory (or offset) measures (Chapter 7). The regulatory authority may also impose conditions on the road agency to improve ecological outcomes for the project, such as to provide a certain level of wildlife connectivity or to monitor impact or effectiveness of mitigation on a target species.

The third stage is the construction of the road project. At this point, there is considerable risk that the good intentions of the detailed planning can be lost through misinterpretation or re-interpretation by construction personnel (Chapter 9). It is also the stage where the engineers and ecologists involved in planning and design have usually moved to new projects, potentially leaving a vacuum of knowledge and understanding. Changes are often made during construction to simplify the process or reduce costs, and these may undermine the ecological values of the project. Typical examples include the use of areas set aside as ecological buffers as temporary roads or soil storage areas, or design changes to crossing structures that reduce their effectiveness. These mistakes or design changes are more likely to occur if the ecological measures are unclear in the project designs, drawings or specifications (Chapter 8). The key to avoiding such mistakes is to provide clear information and control processes, such as project briefings and requirements for ecological approval for activities to commence.

The final stage of a project is operation and maintenance, which continues indefinitely. This is a critical stage affecting the success of mitigation because inappropriate inspection or maintenance regimes may render it unsuitable or ineffective (Chapter 17). Wildlife fencing, crossing structures, revegetation works and other mitigation measures require maintenance to remain effective, and they should be designed to facilitate maintenance (Chapters 17, 20 and 21). This is particularly important where the measure (such as planting a wildlife corridor) is designed to develop or evolve and deliver its maximum ecological outcome many years into the future.

4.3   Appropriate ecological input into a road project should occur in every stage

Ecological advice should be sought and considered at all stages of the project to ensure that key ecological issues are identified as early as possible and to allow maximum opportunity for cost-effective solutions. The nature and extent of the ecological information required at each stage of a project will vary – the point is that it is required at every stage (Table 4.1). At present, ecologists are primarily engaged as consultants to provide specific advice for the EIA in the physical planning and design stage, and they occasionally participate in the strategic planning, despite the importance of such input during this stage. The EIA process is often running parallel with the physical planning in order to expedite the process, with variable levels of integration between the two. This means that trade-offs that must be made during both planning stages may be based on incomplete knowledge. The ability to incorporate new information and modify the design

Table 4.1 Type and detail of ecological input required in each stage of a road project. Project stages are explained further in Figure 4.1.

Stage in road project	Type and detail of ecological input required
Strategic planning	Focus on options that avoid or improve ecological outcomes based on strategic environmental assessment. Examples of key questions include: Can the impact on important wildlife migration routes be avoided? Can the project enhance wildlife connectivity by restoring connections? Can areas without roads be avoided?
Physical planning	Focus on road designs that minimise, mitigate or offset impacts based on detailed ecological analysis. Examples of key questions include: Where should fauna crossings be located? Can the road design be modified to minimise impact on important habitat?
Construction	Ensure that ecologically sensitive designs are easily translated to construction. Examples of key questions include: Has the design of wildlife crossing structures met the required standards for the target species? Has the detailed drainage design considered the impact on adjacent important habitat?
Operation	Ongoing ecological management, maintenance of mitigation measures, review and adaptive management. Examples of key questions include: Is there a plan for monitoring and maintenance in place to ensure crossing structures remain effective over time? Are areas of important habitat adjacent to the road project being managed to ensure that they are not degraded by indirect impacts of the operation of the road?

will depend on the feasibility and reasonableness of the proposed measures.

The greatest opportunity to influence the location of a road is during the strategic planning stage. Therefore, information provided during this stage (possibly from a formal SEA process, such as the identification of ecological corridors or populations of endangered species) can be influential in reducing the ecological impacts of the project. In reality, however, knowledge about ecological impacts is often low during strategic planning and may be limited to protected areas or location of some species.

The best outcomes for a project can be achieved if (i) accurate ecological knowledge is available early; (ii) ecological requirements are stated so as to fit the actual planning or design questions considered during the planning process; (iii) maintenance requirements of mitigation measures are considered in the design; and (iv) the accepted starting position in the road planning process is to try to improve ecological functions.

4.4 Standards and guidelines are critical to ensure a consistent and high-quality approach to roads and road mitigation

Standards and guidelines provide specifications on a range of road designs and measures to ensure agreed practices are being uniformly adopted across a region (Chapter 59). Planners and engineers require

standards and guidelines for all aspects of roads, including mitigation measures for wildlife. Importantly, these standards and guidelines should not stifle innovation or experimentation, because there is still much uncertainty about the design of many features for some species, habitats and landscapes. Standard designs for mitigation measures should provide clear guidance on what is known to work, what may not or does not work and where further innovation or experimentation is required. Chapter 59 provides examples of standards and best practice guidelines that help provide clear direction for improved design outcomes, as well as key issues that need to be considered when developing such manuals.

CONCLUSION

Road agencies, in partnership with planning and environmental agencies and road ecology experts have made substantial progress over the past 20–30 years in developing a base of knowledge and applying techniques to minimise impacts of roads on wildlife. Key challenges for the future are the building of effective partnerships among road planners, designers, engineers, managers and ecological experts to establish and standardise effective approaches and designs, and to provide a framework for ongoing innovation and improvement through testing and monitoring of existing and new approaches.

Chapter 5

IMPROVING ENVIRONMENTAL IMPACT ASSESSMENT AND ROAD PLANNING AT THE LANDSCAPE SCALE

Jochen A. G. Jaeger

Department of Geography, Planning and Environment, Concordia University Montreal, Montréal, Québec, Canada

SUMMARY

There is increasing concern about insufficient consideration of potential ecological effects of roads in project-specific environmental impact assessment (EIA) and other, more advanced types of EIA that exist in various countries. Local impacts are often treated superficially, and landscape-scale effects are usually neglected.

5.1 EIAs of road projects are generally poor.

5.2 Landscape-scale effects of road networks are neglected in EIAs.

5.3 There is a lack of knowledge of thresholds in the cumulative effects of landscape fragmentation and habitat loss on the viability of wildlife populations.

5.4 Wildlife populations may have long response times to increases in landscape fragmentation ('extinction debt').

5.5 There are large uncertainties about many potential ecological effects of roads; they need explicit consideration in EIA, and decision-makers should more rigorously apply the precautionary principle.

5.6 Landscape fragmentation should be monitored because it is a threat to biodiversity and a relevant pressure indicator.

5.7 Maintaining ecological corridor networks is less costly than paying for their restoration at a later date.

5.8 Limits to control landscape fragmentation are needed.

5.9 Caring about the quality of the entire landscape is essential, not just protected areas and wildlife corridors.

Handbook of Road Ecology, First Edition. Edited by Rodney van der Ree, Daniel J. Smith and Clara Grilo.
© 2015 John Wiley & Sons, Ltd. Published 2015 by John Wiley & Sons, Ltd.
Companion website: www.wiley.com\go\vanderree\roadecology

Major efforts are necessary to improve the quality of project-specific EIAs, landscape-scale cumulative effect assessment (CEA), strategic environmental assessment (SEA), road planning, and land-use planning. Given that road networks subdivide wildlife populations into a patchwork of sub-populations, future studies should directly address ecological effects at the landscape scale.

INTRODUCTION

Most wildlife populations are at higher risk of decline and extinction when their habitats are fragmented by roads. So how are such threats from road construction considered in environmental impact assessment (EIA) and road planning? The purpose of EIA is to systematically identify and evaluate the potential impacts of proposed projects to ensure environmental protection and sustainable development. Proponents are required to describe and assess all potential direct and indirect effects of their projects on living organisms, soil, water, air, climate and the landscape, the interaction between these factors, material assets and the cultural heritage. In Europe, for example, the assessment needs to include 'direct effects and any indirect, secondary, cumulative, short, medium and long term, permanent and temporary, positive and negative effects' (CEC 1997, p. 15). The Canadian Environmental Assessment Act (1995, section 16(1)) explicitly requires a cumulative effects assessment (CEA) that considers 'any cumulative effects that are likely to result from the project in combination with other projects or activities that have been or will be carried out'. However, it has long been recognised that the focus of EIA on individual projects makes CEA difficult. This is problematic because many effects of roads occur at landscape scales (as discussed throughout this chapter and book) and are seldom sufficiently covered in project-specific EIAs (Duinker & Greig 2006). In addition, the planning of road networks is not covered by the project-level EIAs; and therefore, an assessment of the environmental effects of network plans is also needed. This is the subject of strategic environmental assessment (SEA): to assess the environmental effects of a proposed plan, policy or program. CEA and SEA are of particular interest for landscape-scale effects of roads and road networks. However in many countries, CEA and SEA are not (or not yet) a requirement.

The aims of this chapter are to identify the typical deficiencies of EIAs for road projects, discuss the implications of these inadequacies and suggest solutions.

LESSONS

5.1 EIAs of road projects are generally poor

Recent reviews of EIAs from Europe, the United Kingdom and the United States (e.g. Atkinson et al. 2000; Byron et al. 2000; Söderman 2005; Gontier et al. 2006; Tennøy et al. 2006; Karlson et al. 2014) have concluded that many were deficient in the following areas:
- The degree to which designated sites would be affected could rarely be ascertained.
- It was generally unclear whether reasonable searches had been carried out to detect rare or protected species.
- To address biodiversity issues, there was a significant gap between current EIA practice and the state of the art in GIS-based modelling.
- Fragmentation and barrier effects were seldom considered.
- The impact assessments were often just descriptive rather than analytical and predictive.
- Indirect impacts were rarely considered.
- The focus on the local scale did not allow prediction and assessment of ecological effects of habitat fragmentation, nor the consideration of scales of ecological processes.
- Information on biodiversity was often absent from the landscape sections.

Thus, the assessment of biodiversity-related impacts is still far from meeting its goals (see Textbox 5.1), and the development and implementation of new methods appear necessary to meet regulations and recommendations on the consideration of biodiversity in EIA and SEA (Gontier et al. 2006; Karlson et al. 2014). For example, the width of corridors investigated in EIAs is often only a few hundred meters. Thus, they will inevitably miss the wider-ranging effects of roads, since it is known that declines in species abundances range between 40 and 2800 m from the road for birds, between 250 and 1000 m (and possibly more) for amphibians, and up to 17 km for mammals (Benítez-López et al. 2010).

Textbox 5.1 Example of the neglect of biodiversity issues in a current EIA of a road project.

The Transportation Ministry of Quebec recently proposed to widen a 6.5km section of Highway 5 (Transport Canada et al. 2010). The EIA report was strongly criticized for its numerous deficiencies (Findlay et al. 2011), including the following:
- "There is no mention in the report of wildlife mortality or road kill, even though there is accumulating evidence that traffic mortality negatively affects population viability." (p. 9)
- "The report states that 'The portion of the trail to be preserved under the highway along La Pêche River will facilitate wildlife movement from one side of the highway to the other in this area' (p. 41). No evidence is adduced to support this prediction. The effectiveness of crossing structures depends on a number of factors, including habitat/landscape context, the particular wildlife species in question, use by humans and a number of specific design/construction attributes. None of these factors are considered in the report. Hence the conclusion is completely unsubstantiated." (p. 9)
- "The report goes on to conclude that wildlife are 'not interested' in crossing in this area. 'Considering the low interest among wildlife species to cross from one side of the right-of-way to the other and their abundance on its western side, the environmental effect is currently considered to be of low intensity. Its duration will be permanent and its scope, local. Therefore, the residual environmental effect is considered not significant' (p. 42). As no data were collected that can inform patterns of wildlife movement through or adjacent to the A5 right-of-way, inferences about wildlife's motivation to do so – or not – are complete speculation. The conclusion that these effects will be insignificant is therefore completely unsubstantiated." (p. 9)
- "Summary and conclusions: The final report is seriously deficient in at least four important respects:
  i. It fails to consider a wide range of potential environmental impacts of highways.
  ii. It has failed to incorporate obvious sources of existing data.
  iii. Most of the conclusions are based on little or no evidence, are completely unsubstantiated, or conflict directly with the current state of scientific knowledge.
  iv. A cumulative effect assessment is completely lacking. There is, therefore, no evidence to support the report's conclusions that the cumulative effects will not be significant.
We conclude the proponents have not adequately discharged their responsibilities under the Canadian Environmental Assessment Act. The next phase of the project ought not to proceed until these responsibilities have been adequately discharged." (p. 14).

There was no response to the submission of the criticism.

Cumulative effects, that is potential effects resulting from the combination of several projects or activities together, including earlier and likely future projects, deserve particular attention because they constitute the most relevant effects worth assessing in most EIAs (Duinker & Greig 2006). They would require that the *total effects* of all human stresses on valued ecosystem components be kept within tolerable and acceptable levels. However, cumulative effects are rarely assessed properly: 'The promise and the practice of CEA are so far apart that continuing the kinds and qualities of CEA currently undertaken in Canada is doing more damage than good' (Duinker & Greig 2006, p. 153). Guidance available in the EIA literature has been largely ignored in the domain of actual EIA practice, and CEA in particular has largely failed to deliver on its promises. Six serious problems in CEA are: (i) major difficulties in applying CEA in project-level EIAs; (ii) a focus of EIA on project approval instead of environmental sustainability; (iii) a general lack of understanding of ecologic impact thresholds;

(iv) inappropriate separation of cumulative effects from project-specific impacts; (v) weak interpretations of cumulative effects by practitioners and analysts; and (vi) inappropriate handling of potential future developments, for example due to narrowly focused scenarios (Duinker & Greig 2006). Fundamental improvements are required, for example through regional environmental assessments in combination with regional land-use planning, in addition to more rigorous CEA analysis in project EIAs. Such improvements are necessary since all effects are cumulative, and thus the aggregate stresses acting on valued ecosystem components need to be assessed (Duinker & Greig 2006).

All these findings demonstrate that (i) most EIAs are too vague or make unsubstantiated predictions; (ii) most EIAs do not consider the landscape scale; and (iii) almost none use state-of-the art modelling methods to predict likely effects. As a consequence, we almost never see an EIA that concludes that the road should not be built. This is usually not because the mitigation measures were likely to be so successful that the road will have no significant negative impact, but because many effects were not sufficiently covered in the EIA, in particular landscape-scale and cumulative effects. The poor quality of EIAs poses a significant concern considering that various specific guidelines on biodiversity/ecological assessment issues have been available for two decades in the United States, Canada, and parts of Europe (e.g. CEQ 1993; DoT 1993; CEAA 1996; DIREN 2002), and more recently in Asia (Chapter 53). These guidelines are not effectively applied, probably because many EIAs are prepared by consultants who depend on continued support from their clients. Concluding that there are significant environmental effects might result in being cut off from the preparation of EIAs in the future, which is not in their interest. This is a structural flaw of the current EIA system that needs to be fixed, for example through independent peer review and through the publication of good textbooks providing detailed instructions.

5.2 Landscape-scale effects of road networks are neglected in EIAs

Even though landscape-scale effects are known to be highly important for wildlife populations, they have not yet been studied very well in road ecology (van der Ree et al. 2011), and accordingly, they are poorly covered in EIAs. Since the design of conclusive landscape-scale studies is more difficult than small-scale studies, their results will inevitably also be less certain:

> For research questions concerned with landscape-scale ecological effects and long-term consequences, the inferential strength of any feasible study will always be comparatively low. Consequently, it is inevitable that the uncertainty associated with any conclusion will necessarily be high. It is a cruel irony in road ecology that, the more important the question, the more uncertainty is associated with the answers. (Roedenbeck et al. 2007, p. 17)

For example, long-distance dispersal of animals is rare but is ecologically important for re-colonizing empty habitats (e.g. in meta-population dynamics), allowing range shifts of populations in response to climate change, and gene flow. However, data on long-distance movements are difficult to collect, and studying

populations across multiple sites requires longer time scales and greater investments than studies at individual sites. Landscape-scale effects may not be detectable at the local scale, but landscape effects are real and can have a large impact on the success of mitigation. In addition, various cumulative effects and consequences on ecological communities, such as predator-prey dynamics, changes in the food chain, source-sink dynamics, and cascading effects are still unknown.

5.3 There is a lack of knowledge of thresholds in the cumulative effects of landscape fragmentation and habitat loss on the viability of wildlife populations

There are thresholds in the effects of increasing road density and habitat loss on the viability of wildlife populations (Fig. 5.1). When roads are added to a landscape, population viability does not decrease linearly, but usually exhibits a threshold after which there is a dramatic decline. For example, road density was negatively related to the species richness of amphibians and reptiles at ponds in Ontario, Canada (Findlay & Houlahan 1997) and to the presence of European tree frogs at ponds in Europe (Pellet et al. 2004). The detrimental effect of landscape fragmentation by roads is a primary cause of the decline of endangered brown hare populations in Switzerland. High road densities have made the hare populations – once one of the most abundant mammal species in Switzerland – much more vulnerable to unfavourable weather, to the intensification of agricultural practices, and habitat loss (Roedenbeck & Voser 2008).

Several empirical studies reported values in road density above which certain species do not occur any more, for example wolves in Ontario, Canada, and Minnesota, USA (Jensen et al. 1986; Mech et al. 1988), and grizzly bears in Montana, USA (Mace et al. 1996). However, little information is available about the thresholds of decline in population viability (Robinson et al. 2010).

What do these thresholds and the lack of information about them mean for transportation planning? They imply that nobody knows how close the wildlife populations already are to their thresholds, and the decline of wildlife populations may come as a surprise. If populations have so far survived all road construction in a landscape, this does not mean that the populations will survive further road construction. The next new road may push the population across the threshold and cause extinction. Even worse, when

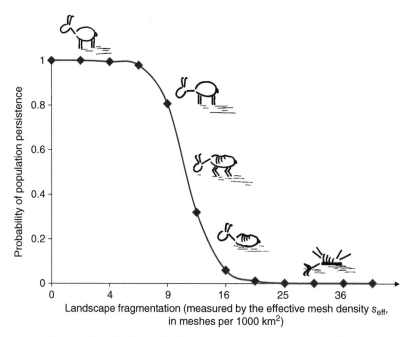

Figure 5.1 Illustration of thresholds in the effect of landscape fragmentation caused by roads on the viability of wildlife populations, determined from computer simulations. The specific values of the thresholds depend on the particular species, traffic volumes on the roads, and the amount and quality of available habitat. Once the threshold has been passed and the so-called 'point of no return' has been crossed, it may become impossible to rescue a declining population. Effective mesh density is a metric to quantify the degree of landscape fragmentation (see Lesson 5.6 and Textbox 5.2). Source: Jaeger and Holderegger (2005), reprinted with permission from *GAIA: Ecological Perspectives for Science and Society*.

the 'point of no return' has been crossed and the population is already in decline, it will likely be impossible to reverse the trend and rescue the population even if relatively drastic protection measures were taken. These thresholds are likely to depend on the species and the landscape, and therefore it is unlikely that they will be known any time soon. Long-term studies would be required to elucidate these thresholds, including species that are not (or not yet) endangered. As a consequence of the current practice of considering only endangered species in EIAs, many species that are declining but not (yet) endangered are pushed closer and closer to their thresholds.

5.4 Wildlife populations may have long response times to increases in landscape fragmentation ('extinction debt')

Wildlife populations react to the fragmentation of their habitats with variable response times. The response may take several decades (e.g. Findlay & Bourdages 2000), indicated by the time lag in Fig. 5.2.

The response times to the main four mechanisms affecting a population may differ: The effect of (i) habitat loss is almost immediate, (ii) reduced habitat quality and (iii) traffic mortality may take longer and (iv) reduced connectivity even longer still. After this time lag, the population is smaller and more vulnerable to extinction. The response times for most species are not known, and this realisation is important for EIA because it implies that the decline and loss of populations will continue for several decades after road construction. The term 'extinction debt' is used to denote the number of populations that will go extinct because of changes that have already occurred (Tilman et al. 1994).

Population persistence is influenced by all past and present land uses that contribute to habitat loss and fragmentation. Thus, EIA and landscape conservation planning should take into account the effects of all land uses on animal survival and movement and the associated response times. Research approaches to investigate the response times and the resulting extinction debt have only recently been suggested (Kuussaari et al. 2009).

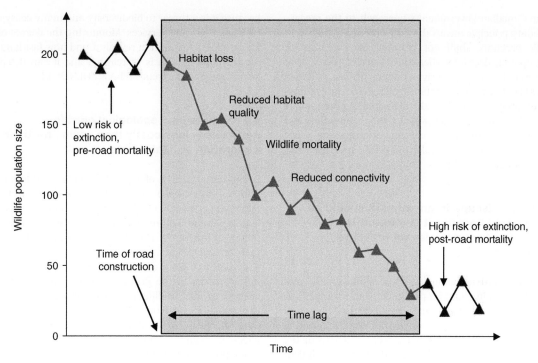

Figure 5.2 Four major ecological impacts of roads and traffic on animal populations and the time lag for their cumulative effect. After the time lag (often in the order of decades, shown in grey), population size is smaller, exhibits greater relative fluctuations over time, is more vulnerable, and the risk of extinction is high. Source: Adapted from Forman et al. (2003).

5.5 There are large uncertainties about many potential ecological effects of roads; they need explicit consideration in EIA, and decision-makers should more rigorously apply the precautionary principle

There are many uncertainties about the potential landscape-scale ecological effects of roads, for example influences of configuration of the road network on wildlife populations. The bundling of transportation infrastructure to leave other parts of the landscape unfragmented decreases the impact of the road network (Jaeger et al. 2006; Chapters 2 and 3). Even though the barrier effect of a bundle of transport routes will be higher than the barrier effect of a single transport route, bundling is preferable because more core habitat remains unaffected by edge effects. In addition, wildlife crossing structures could then traverse all the transport infrastructure in one go; however, there may be limits to the length of crossing structures that different species of wildlife will use. Similarly, the upgrading of existing highways will usually be less detrimental than the construction of new highways

elsewhere (Jaeger et al. 2006). However, research about the role of road network configuration is lacking, even though it is urgently needed to inform EIA and landscape-scale road planning.

As we do not know the thresholds in road density, the response times of wildlife populations to new roads, or the influence of road network configuration, **these uncertainties need to be explicitly incorporated into decision-making.** We know thresholds exist, but we cannot wait another 30 or 40 years for research to identify thresholds and response times before they are considered in EIA. This requires a shift from a reactive to a proactive mode of mitigation and more rigorous application of the precautionary principle (EEA 2001) and the concept of environmental threat (Jaeger 2002). This shift to more proactive decision-making is supported by the insight that the failure of detecting environmental impacts that exist (Type II error) usually has more detrimental consequences than the erroneous detection of impacts that do not exist (Type I error) (Kriebel et al. 2001). The precautionary principle is promoted in environmental policy in Europe, and to some degree in other countries as well. For example, the Canadian government's approach to the precautionary principle means that 'the absence of full scientific certainty shall not be used as a reason for postponing decisions where there is a risk of serious or irreversible harm' (Privy Council Office, Canada 2003). Wise policy therefore avoids an increase of fragmentation from the start. In addition, EIA practitioners should be more explicit about their assumptions and knowledge gaps, disclosing uncertainties such that decision-makers can make more informed decisions (Tennøy et al. 2006).

5.6 Landscape fragmentation should be monitored because it is a threat to biodiversity and a relevant pressure indicator

Many countries monitor their biodiversity, and Switzerland include one parameter that measures the pressure on the landscape caused by fragmentation due to transportation infrastructure and urbanisation (Jaeger et al. 2008). This metric is calculated using the method of effective mesh size and effective mesh density (Fig. 5.3, Textbox 5.2). Further increases in the level of landscape fragmentation need to be avoided

because it is a threat to biodiversity and many ecosystem functions and services. Monitoring the degree of landscape fragmentation reveals if and how fast landscape fragmentation is increasing, and it can detect any changes in the trends (EEA & FOEN 2011).

5.7 Maintaining ecological corridor networks is less costly than paying for their restoration at a later date

In Switzerland, 218 of the 303 wildlife corridors of national importance were disturbed or disconnected (Holzgang et al. 2001). Their restoration has required a large amount of money and will need additional money in the future. Therefore, it is a good strategy to map ecological corridors and keep them sufficiently wide and free from development and transportation infrastructure in the first place. It is also more cost effective to build wildlife crossing structures during the upgrading or construction of unavoidable new roads than retrofitting existing roads. The Netherlands have allocated about €410 million to a national defragmentation program that aims to retrofit crossing structures to existing infrastructure (van der Grift 2005). This is an important lesson because countries can save a lot of

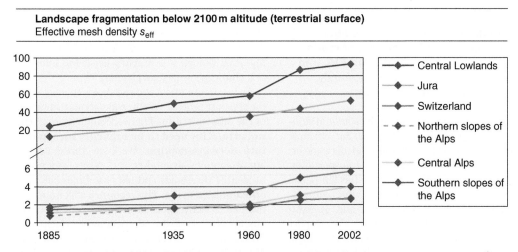

Landscape fragmentation below 2100 m altitude (terrestrial surface)
Effective mesh density s_{eff}

Central Lowlands
Jura
Switzerland
Northern slopes of the Alps
Central Alps
Southern slopes of the Alps

Aid to understanding: The effective mesh density s_{eff} (i.e. the effective number of meshes per 1000 km^2) indicates the probability of two randomly chosen points within an area being divided by barriers (e.g. a road or a built-up area). The higher the s_{eff}, the greater the degree of landscape fragmentation.

Figure 5.3 Presentation of the increase of the indicator 'landscape fragmentation' in Switzerland (and its five ecoregions) since 1885 by the Swiss Government in *Swiss Environmental Statistics: A Brief Guide 2007* (FSO & FOEN 2007). The Swiss Government presented this information to the general public in this publication, together with a number of other indicators. The level of fragmentation is quantified by the effective mesh density (s_{eff}), see Textbox 5.2. Source: FSO & FOEN (2007).

Textbox 5.2 Definition of effective mesh size m_{eff} and effective mesh density s_{eff}.

The effective mesh size m_{eff} is based on the probability that two randomly located points (or animals) in an area are connected (or in the same patch) and are not separated by a barrier (e.g. roads and urban area) (Fig. 5.4). Thus, it indicates the ability of animals to move freely in the landscape and is ecologically more meaningful than measuring road density. The second part (multiplication by the size of the region) converts this probability into a measure of area. This area is the 'mesh size' of a regular grid pattern showing an equal degree of fragmentation and can be directly compared with other regions. The smaller the effective mesh size, the more fragmented the landscape.

The effective mesh density gives the effective number of meshes per square kilometre, that is the density of the meshes. The effective mesh density value rises when fragmentation increases. The two measures contain the same information about the landscape, but the effective mesh density is more suitable for detecting trends and changes in trends. Both have been used to quantify the degree of landscape fragmentation for environmental monitoring (Jaeger et al. 2008; EEA & FOEN 2011). A detailed description of both metrics can be found in Jaeger (2002) and Jaeger et al. (2008).

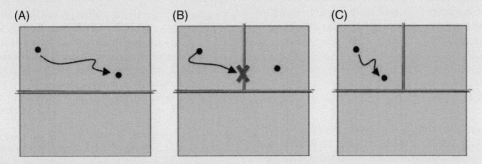

Figure 5.4 Illustration of the effective mesh size metric: Two randomly chosen points in the landscape may be connected (A) or separated by a barrier (B). The more barriers in the landscape, the fewer points are connected and the lower the effective mesh size. For a landscape size of 16 km², the effective mesh size is 8 km² in (A) and 6 km² in (B) and (C). The effective mesh size is an expression of the probability of any two randomly chosen points in the landscape being connected. This corresponds to the definition of landscape connectivity as 'the degree to which the landscape facilitates or impedes movement among resource patches' (Taylor et al. 1993). Source: EEA & FOEN (2011, p. 22).

money by addressing the issue of landscape fragmentation now rather than ignoring the need for these measures during road construction and having to deal with the increased costs of adding them later. Many countries worldwide, including developing countries with a rapid rate of road construction (e.g. Chapters 50, 52 and 57), are in a good position to avoid the mistakes that various countries in Europe have committed in the past.

5.8 Limits to control landscape fragmentation are needed

In 1985, the German Federal Government declared the goal to 'reverse the trend in land consumption and landscape fragmentation' (BdI 1985). There is also an explicit intention to preserve large, un-fragmented spaces with little traffic (Chapter 3) which is a central principle of regional and landscape planning in Germany. However, landscape fragmentation has increased unabatedly since 1985, and there is an ongoing discussion about introducing quantitative limits to the degree of fragmentation. The German Environmental Agency recently proposed to establish limits to the rate of increase of landscape fragmentation based on the effective mesh size (Penn-Bressel 2005): The remaining large unfragmented areas in Germany are to be preserved, and enlarged where possible; and in already highly fragmented areas, the trend is to be slowed. This lesson is important because it provides the option of introducing targets and limits that can be evaluated to assess whether or not they have been achieved. Thus, it provides a regulatory ground for administrative action for curtailing fragmentation when the targets are exceeded.

5.9 Caring about the quality of the entire landscape is essential, not just protected areas and wildlife corridors

Many wildlife species suffer from high mortality when moving around in the landscape outside of protected areas. This implies that we should **always** be concerned about the ecological effects of roads and improve the ecological quality of the landscape – inside and outside of protected areas – and it is essential to keep enough habitat (Fahrig 2001). Otherwise, if there is not sufficient habitat left, wildlife corridors and crossing structures will be useless. The location of crossing structures must be considered in relation to the location, quality and amount of habitat within adjacent landscapes, as well as in relation to future cumulative impacts of human activities during landscape planning to ensure the effectiveness of mitigation.

CONCLUSION

It is dangerous to think that roads can be built anywhere if they have wildlife crossing structures and fences – while wildlife habitat is subdivided. This danger is not visible in the EIAs of single projects, in particular when their cumulative effects are inadequately evaluated. We must acknowledge that crossing structures and fences mitigate only **some** of the effects of roads, but not all (Lesson 1.6).

Unfortunately, EIAs rarely study the ecological effects of new roads sufficiently. Studies often neglect the many cumulative effects and uncertainties about the ecological effects. A central database of road EIAs should be established to enable learning from previous studies and share experiences in a more systematic way. Accountability should be increased for unanticipated ecological damages and effects with long time lags, for example through the introduction of an obligatory insurance. Decision-makers must embrace better approaches for judgment under uncertainty, such as the precautionary principle and the establishment of quantitative limits or objectives to limit road density or the degree of landscape fragmentation (Penn-Bressel 2005; Roedenbeck et al. 2007). If these improvements are not implemented, we will swiftly move further away from the goal of sustainability. The conduct of regionally focused CEA in a transparent process should become the responsibility of governments in the context of land-use planning or integrated resource management planning (Duinker & Greig 2006). Regional

CEA should explicitly document the assumptions made regarding threshold levels of stress, and should implement a rigorous monitoring program. However, if it turns out that CEA cannot be done adequately, then the proposed projects should not be permitted to proceed.

FURTHER READING

EEA (2001): provides 12 late lessons from early warnings highlighting the consequences of neglecting uncertainties based on a set of 15 case studies from 1896 to 2000 and stresses strategies such as the precautionary principle for addressing uncertainties to avoid repeating mistakes.

EEA and FOEN (2011): a report on the degree of landscape fragmentation in 28 European countries using the effective mesh density and explaining how this information can be used for monitoring landscape fragmentation, for the establishment of objectives and limits, and for EIA through regional environmental assessment and CEA.

Duinker and Greig (2006): discuss six problems of current EIA practice and conclude that the CEA in particular has largely failed to deliver on its promises.

Gontier et al. (2006): a review of EIAs for road and railway projects in Europe that demonstrates the existing gap between research in GIS-based ecological modelling and current practice in biodiversity assessment in environmental assessment.

Karlson et al. (2014): a recent review of EIA and CEA for road and railway projects, highlighting a number of persistent inadequacies and areas for improvement.

REFERENCES

Atkinson, S. F., Bhatia, S., Schoolmaster, F. S., Waller, W. T. 2000. Treatment of biodiversity impacts in a sample of US environmental impact statements. Impact Assessment and Project Appraisal **18**: 271–282.

Benítez-López, A., Alkemade, R., Verweij, P. A. 2010. The impacts of roads and other infrastructure on mammal and bird populations: A meta-analysis. Biological Conservation **143**: 1307–1316.

Bundesminister des Innern (BdI). 1985. Bodenschutzkonzeption der Bundesregierung, Bundestags-Drucksache 10/2977 vom 7. März 1985. Kohlhammer, Stuttgart.

Byron, H. J., Treweek, J. R., Sheate, W. R., Thompson, S. 2000. Road developments in the UK: An analysis of ecological assessment in environmental impact statements produced between 1993 and 1997. Journal of Environmental Planning and Management **43**: 71–97.

Canadian Environmental Assessment Agency (CEAA). 1996. A guide on biodiversity and environmental assessment. Ministry of Supply and Services, Ottawa. 15 pp.

Commission of the European Communities (CEC). 1997. Council directive 1997/11/EEC of 3 March 1997, amending directive 1985/337/EEC on the Assessment of the Effects of Certain Public and Private Projects on the Environment. Official Journal of the European Communities **L073**: 5–15.

Council on Environmental Quality (CEQ). 1993. Incorporating biodiversity considerations into environmental impact analysis under the national environmental policy act. CEQ, Washington, DC. 29 pp.

Department of Transport (DoT). 1993. Design manual for roads and bridges-DMRB. Environmental assessment, Vol. **11**. HMSO, London.

Direction régionale de l'environnement de Midi-Pyrénées (DIREN). 2002. Guide sur la prise en compte des milieux naturels dans les études d'impact. 75 pp. Published by DIREN Midi-Pyrénées, Toulouse, France. Available at http://www.environnement.gouv.fr/midi-pyrenees/.

Duinker, P. N., Greig, L. A. 2006. The impotence of cumulative effects assessment in Canada: Ailments and ideas for redeployment. Environmental Management **37**: 153–161.

European Environment Agency (EEA). 2001. Late lessons from early warnings: The precautionary principle 1896–2000. Environmental Issue Report No 22. European Environment Agency, Copenhagen.

European Environment Agency & Swiss Federal Office for the Environment (EEA & FOEN); Jaeger, J. A. G., Soukup, T., Madriñán, L. F., Schwick, C., Kienast, F. 2011. Landscape fragmentation in Europe. Joint EEA-FOEN report. EEA Report No. 2/2011. Publications Office of the European Union, Luxembourg. 87 pp. doi:10.2800/78322.

Fahrig, L. 2001. How much habitat is enough? Biological Conservation **100**: 65–74.

Findlay, C. S., Houlahan, J. 1997. Anthropogenic correlates of species richness in southeastern Ontario wetlands. Conservation Biology **11**: 1000–1009.

Findlay, C. S., Bourdages, J. 2000. Response time of wetland biodiversity to road construction on adjacent lands. Conservation Biology **14**: 86–94.

Findlay, C. S., Callaghan, C., Chapleau, F., Fahrig, L., Jaeger, J. A. G., Morin, A., Woodley, S. 2011. Comments concerning the final screening report of the proposed highway A5 extension (CEAR Reference 08-01-3981). Submitted to the Federal Environmental Assessment Coordinator at Transport Canada, Ottawa, on 12 January 2011. 19 pp.

Forman, R. T. T., Sperling, D., Bissonette, J. A., Clevenger, A. P., Cutshall, C. D., Dale, V. H., Fahrig, L., France, R., Goldman, C. R., Heanue, K., Jones, J. A., Swanson, F. J., Turrentine, T., Winter, T. C. 2003. Road ecology. Science and solutions. Island Press, Washington, DC.

Gontier, M., Balfors, B., Mörtberg, U. 2006. Biodiversity in environmental assessment – current practice and tools for prediction. Environmental Impact Assessment Review **26**: 268–286.

Holzgang, O., Pfister, H. P., Heynen, D., Blant, M., Righetti, A., Berthoud, G., Marchesi, P., Maddalena, T., Müri, H., Wendelspiess, M., Dändliker, G., Mollet, P., Bornhauser-Sieber, U. 2001. Korridore für Wildtiere in der Schweiz – Grundlagen zur überregionalen Vernetzung von Lebensräumen, BUWAL, SGW und Vogelwarte Sempach, Schriftenreihe Umwelt Nr. 326, Bern, 116 pp. (also available in French).

Jaeger, J. A. G. 2002. Landscape fragmentation – a transdisciplinary study according to the concept of environmental threat (in German: Landschaftszerschneidung – Eine transdisziplinäre Studie gemäß dem Konzept der Umweltgefährdung). Verlag Eugen Ulmer, Stuttgart, 447 pp.

Jaeger, J. A. G., Holderegger, R. 2005. Thresholds of landscape fragmentation (in German; Schwellenwerte der Landschaftszerschneidung). GAIA **14**: 113–118.

Jaeger, J. A. G., Fahrig, L., Ewald, K. 2006. Does the configuration of road networks influence the degree to which roads affect wildlife populations? In: Irwin, C. L., Garrett, P., McDermott, K. P. (eds). Proceedings of the 2005 International Conference on Ecology and Transportation (ICOET). Center for Transportation and the Environment, North Carolina State University, Raleigh, NC, pp. 151–163.

Jaeger, J. A. G., Bertiller, R., Schwick, C., Müller, K., Steinmeier, C., Ewald, K. C., Ghazoul, J. 2008. Implementing landscape fragmentation as an indicator in the Swiss Monitoring System of Sustainable Development (MONET). Journal of Environmental Management **88**: 737–751.

Jensen, W. F., Fuller, T. K., Robinson, W. O. 1986. Wolf (*Canis lupus*) distribution on the Ontario–Michigan border near Sault Ste. Marie. Canadian Field-Naturalist **100**: 363–366.

Karlson, M., Mörtberg, U., Balfors, B. 2014. Road ecology in environmental impact assessment. Environmental Impact Assessment Review **48**: 10–19.

Kriebel, D., Tickner, J., Epstein, P., Lemon, J., Levins, R., Loechler, E. L., Quinn, M., Rudel, R., Schettler, T., Stoto, M. 2001. The precautionary principle in environmental science. Environmental Health Perspectives **109**: 871–876.

Kuussaari, M., Bommarco, R., Heikkinen, R. K., Helm, A., Krauss, J., Lindborg, R., Öckinger, E., Pärtel, M., Pino, J., Rodà, F., Stefanescu, C., Teder, T., Zobel, M., Steffan-Dewenter, I. 2009. Extinction debt: A challenge for biodiversity conservation. Trends in Ecology and Evolution **24**: 564–571.

Mace, R. D., Waller, J. S., Manley, T. L., Lyon, L. J., Zuuring, H. 1996. Relationships among grizzly bears, roads and habitat in the Swan Mountains, Montana. Journal of Applied Ecology **33**: 1395–1404.

Mech, L. D., Fritts, S. H., Radde, G., Paul, W. J. 1988. Wolf distribution and road density in Minnesota. Wildlife Society Bulletin **16**: 85–87.

Pellet, J., Guisan, A., Perrin, N. 2004. A concentric analysis of the impact of urbanization on the threatened European tree frog in an agricultural landscape. Conservation Biology **18**: 1599–1606.

Penn-Bressel, G. 2005. Begrenzung der Landschaftszerschneidung bei der Planung von Verkehrswegen. GAIA **14**: 130–134.

Privy Council Office, Canada. 2003. A framework for the application of precaution in science-based decision making about risk. 12 pp. http://www.pco-bcp.gc.ca/docs/information/publications/precaution/Precaution-eng.pdf. Accessed on 18 October 2013.

Robinson, C., Duinker, P. N., Beazley, K. F. 2010. A conceptual framework for understanding, assessing, and mitigating ecological effects of forest roads. Environmental Review **18**: 61–86.

Roedenbeck, I. A., Voser, P. 2008. Effects of roads on spatial distribution, abundance and mortality of brown hare (*Lepus europaeus*) in Switzerland. European Journal of Wildlife Research **54**: 425–437.

Roedenbeck, I. A., Fahrig, L., Findlay, C. S., Houlahan, J., Jaeger, J. A. G., Klar, N., Kramer-Schadt, S., van der Grift, E. A. 2007. The Rauischholzhausen-agenda for road ecology. Ecology and Society **12**: 11. http://www.ecologyandsociety.org/vol12/iss1/art11/. Accessed on 15 September 2014.

Söderman, T. 2005. Treatment of biodiversity issues in Finnish environmental impact assessment. Impact Assessment and Project Appraisal **22**: 87–99.

Swiss Federal Statistical Office and the Swiss Federal Office for the Environment (FSO & FOEN). 2007. Swiss environmental statistics: A brief guide 2007. Swiss Confederation, Neuchatel/Berne, 34 pp.

Taylor, P. D., Fahrig, L., Henein, K., Merriam, G. 1993. Connectivity is a vital element of landscape structure. Oikos **68**: 571–573.

Tennøy, A., Kværner, J., Gjerstad, K. I. 2006. Uncertainty in environmental impact assessment predictions: The need for better communication and more transparency. Impact Assessment and Project Appraisal **24**: 45–56.

Tilman, D., May, R. M., Lehman, C. L., Nowak, M. A. 1994. Habitat destruction and the extinction debt. Nature **371**: 65–66.

Transport Canada, Fisheries and Oceans Canada, National Capital Commission. 2010. Extension of Highway 5 Project between Farm Point and the Connection to Road 366. Screening Report. Final Version. Ottawa, Canada, December 2010. 65 pp.

Van der Grift, E. A. 2005. Defragmentation in the Netherlands: A success story? GAIA **14**: 144–147.

Van der Ree, R., Jaeger, J. A. G., van der Grift, E., Clevenger, A. P. 2011. Guest editorial: Effects of roads and traffic on wildlife populations and landscape function: Road ecology is moving towards larger scales. Ecology and Society **16**: 48. http://www.ecologyandsociety.org/vol16/iss1/art48/. Accessed on 15 September 2014.

Chapter 6

WHAT TRANSPORTATION AGENCIES NEED IN ENVIRONMENTAL IMPACT ASSESSMENTS AND OTHER REPORTS TO MINIMISE ECOLOGICAL IMPACTS

Josie Stokes

NSW Roads and Maritime Services, New South Wales, North Sydney, Australia

SUMMARY

The construction and maintenance of roads often impacts adversely upon biodiversity values. To minimise these impacts, road agencies rely on technical information and expert advice provided by ecologists. This information is provided in a variety of documents ranging from formal environmental impact assessment (EIA) documents to small-scale ecological assessment reports. Road agencies expect the following from a high quality EIA:

6.1 The consultant must have a thorough understanding of the scope for the EIA.

6.2 The EIA should include accurate and expert technical advice.

6.3 Adequate methods are used to conduct EIA surveys and analyse the results.

6.4 The EIA should be easy to read and comprehend.

6.5 The EIA must adequately assess the potential impacts of the project or action on biodiversity.

6.6 The EIA should follow the mitigation hierarchy (i.e. avoid, minimise, mitigate and lastly offset) and recommend realistic measures to protect the environment.

This chapter provides best practice guidance for delivering high quality EIA and is relevant to road agencies as well as environmental consultants. The recommendations in this chapter apply equally to other projects or actions that do not require a formal EIA, and yet have the potential to have significant environmental impacts.

Handbook of Road Ecology, First Edition. Edited by Rodney van der Ree, Daniel J. Smith and Clara Grilo.
© 2015 John Wiley & Sons, Ltd. Published 2015 by John Wiley & Sons, Ltd.
Companion website: www.wiley.com\go\vanderree\roadecology

INTRODUCTION

The construction and operation of roads and other linear infrastructure can impact adversely upon the environment and biodiversity. Road agencies have a legal, social and environmental obligation to minimise their ecological and environmental impacts. In order to adequately assess the potential impacts of a road project (or action) on biodiversity, an environmental impact assessment (EIA) is often required (Chapter 5). It is usually only major road projects that are required to follow the formal EIA process because of the potentially large-scale significant impacts they may have. For small-scale road projects or actions, there is still a legislative requirement to consider potential environmental impacts, but these are usually addressed through smaller or informal ecological assessments. Agencies may also request surveys of biodiversity or roadkill for a range of reasons not related to specific projects (e.g. road safety). In all these situations, the assessments and reports must address the same six lessons as outlined in this chapter for formal EIA.

Road agencies rely on the technical information and advice provided by environmental specialists (private consultants or in-house) to assist them to make decisions that avoid, minimise and mitigate potential impacts. Unfortunately, the EIAs (and other reports) received from consultants vary in quality (e.g. Karlson et al. 2014), which can result in delays to projects, additional costs to the agency and, ultimately, further impact on the environment.

Environmental consultants must deliver reports to clients on time and try to come in under budget so that their company can make a profit. This can lead to 'cutting corners', including the use of inadequate survey methods, employing a less-experienced team member to write the report and of course, the dreaded 'cut and paste' from previous unrelated reports. The aims of this chapter are to highlight the expectations of a road agency for EIAs (and other surveys and assessments) and outline the requirements for a high-quality EIA.

LESSONS

6.1   The consultant must have a thorough understanding of the scope for the EIA

When preparing an EIA, the consultant must have a thorough understanding of the scope in order to deliver a high-quality report that adequately assesses the potential impacts of a project (or action) upon biodiversity. It is therefore critical that the road agency knows what questions need to be answered so that they can provide the consultant with a clear, concise scope for the EIA, including the following:
• A description of the ecological characteristics of the study area including identifying protected and threatened terrestrial and aquatic flora and fauna species, populations and ecological communities and their habitats.
• An identification of the direct and indirect impacts of the proposed activity on terrestrial and aquatic flora and fauna species, populations, ecological communities and critical habitat.
• An assessment of the nature, extent, frequency, duration and timing of potential impacts.
• An evaluation of the extent to which the proposed activity contributes to processes threatening the survival of biota on the site.
• An analysis of the significance of the potential impact of the proposed activity on species, ecological communities and populations listed under relevant legislation.
• A section that proposes measures to avoid, minimise, mitigate and, if necessary, offset impacts.

6.2   The EIA should include accurate and expert technical advice

The advice and technical information provided by consultants guides the avoidance, minimisation and mitigation of potential impacts. It is important that the information provided in the EIA is thorough and accurate because road agencies rely on the technical information in order to comply with policies and legal mandates regarding the potential impacts of a proposal on the environment.

For example, by assuming that threatened species are absent from an area (through lack of survey effort, inexperienced surveyors, incorrect timing of surveys, lack of consultation with researchers/species experts) when they actually do occur, or have a high likelihood of occurring, presents a high risk. This is especially challenging for agencies when a construction contract for a project has been awarded and a threatened species that was dismissed as being absent from the site is suddenly discovered in the project area (Textbox 6.1).

6.3   Adequate methods are used to conduct EIA surveys and analyse the results

It is widely recognised that one of the limitations of ecological surveys conducted for EIA is that they are often just 'snapshots in time'. It is therefore important

---

Textbox 6.1 Endangered frogs stop work on Australian highway upgrade.

Surveys for the endangered green and golden bell frog (Fig. 6.1) along a section of the Princes Highway upgrade project in eastern Australia were undertaken in 2009, coinciding with a major drought. The consultant did not use adequate survey methods or follow guidelines for this species, and consequently it was dismissed as being 'unlikely to occur'. When construction for the highway upgrade started 2 years later (and after the drought had broken), this species appeared in high numbers across the project site. Construction was suspended for 5 months while the relevant approvals were obtained. This 'unexpected' threatened species discovery cost the road agency approximately AUD 4 million in delays and additional mitigation measures.

Figure 6.1 The endangered green and golden bell frog. Source: Photograph by Josie Stokes.

methods must also be adequately described in the EIA to ensure subsequent surveys conducted by someone else are as consistent as possible.

If previous ecological studies and field work have been undertaken in the same location as the proposal (i.e. for corridor or route selection studies), they should be used as background information and the data included in the EIA. Previous reports should be identified and include a summary of survey techniques and effort. All raw data should be submitted as part of the EIA in a useable and retrievable format and carefully stored by the road agency. Approaches and techniques to analyse and synthesise data are continually improving. Unfortunately, methods used in many EIAs are often outdated and do not represent best practice in the field (Lesson 5.1). Consequently, the amount of information being obtained from existing data sets is not being maximised, which likely results in an inadequate assessment of the potential impacts of the proposed project. EIAs and other reports should use the most current analysis and modelling techniques to analyse data and predict likely impacts.

6.4 The EIA should be easy to read and comprehend

EIAs are usually available to the public, and overly technical language should be avoided to ensure that its content is understandable by a wide and often lay audience. Similarly, every consultant has a different way of presenting information, and sometimes it can be repetitive, illogically structured and even worse, may contain irrelevant information cut and pasted from a previous, unrelated report.

EIA reports can be very large, and extend to multiple volumes that focus on a range of environmental matters, particularly for large or complicated projects. Therefore, they should be logically organised to allow the reader to easily find relevant information. By following a standard structure (e.g. Table 6.1), the road agency can devote more time to reviewing the technical content, rather than the structure of the report.

6.5 The EIA must adequately assess the potential impacts of the project or action on biodiversity

EIAs that fail to identify all of the potential impacts of a project can have serious consequences during the project approval process and for biodiversity during

that adequate methods are employed by the consultant in order to provide a detailed and accurate description of the site. Regulatory agencies are important stakeholders that review EIAs, and their comments are often focused on the inadequacy of the field survey techniques to detect a particular threatened species or accurately describe the amount of potential habitat for threatened species (Chapter 5).

Numerous government policy documents and species recovery plans describe survey methods for different scales of development, regions and threatened species (see references for Australian examples). The road agency should review the proposed methods provided by the consultant against methods that are recommended by regulatory agencies. Where available, the field survey techniques and survey effort used should be supported by scientific literature to ensure a high-quality contribution to the EIA. Field survey

Table 6.1 Example of a logical structure for the ecological component of an environmental impact assessment. Note that the format and/or order of contents may vary regionally and on the size and/or potential impact of the proposed project or action.

Section heading	Section sub-heading	Contents of section
Table of contents	List of tables List of figures List of appendices	
Executive summary		The executive summary should concisely summarise the key findings of the report in non-technical terms.
Introduction	Background	Briefly summarise the background of the project, including a map where relevant.
	Description of the project Legislative (planning) context	Present a concise description of the project and any key design elements. Identify the planning context for biodiversity issues.
Methods	Background research Nomenclature Field survey Survey site selection	Provide details of the literature reviewed and databases searched. Describe the nomenclature followed for species surveyed. Describe the field techniques employed, including for targeted surveys. Explain the rationale behind the selection of field techniques and survey areas/sites selected.
	Survey effort Likelihood of occurrence	Summarise the field survey effort for each field technique. Describe the criteria used to assess the likelihood of occurrence of threatened species.
	Vegetation condition Personnel	Describe the vegetation condition categories applied, e.g. low, medium, high. Summarise the key personnel, their role in the investigation/report and their qualifications.
	Limitations	Discuss any limitations to the field survey effort or reporting.
Existing environment	Landscape context	Identify the bioregional context, catchments and other relevant aspects of the landscape.
	Land use Vegetation communities and habitat	Identify notable past and current land uses and disturbance history. Assess the condition of the study area, including factors that contribute to its existing condition. Include all relevant information from existing data sources and field surveys. Describe terrestrial, aquatic and riparian vegetation communities and habitats. Sub-headings and tables may be useful for presenting this information.
	threatened ecological communities, species, endangered populations and migratory/marine species.	Determine the presence and extent of endangered ecological communities, threatened flora and fauna species, endangered populations, migratory/marine species and their habitats in the survey area. Assess if any threatened biota identified in the background research as being likely to occur, were not detected during field surveys.

	Fauna movement corridors Critical habitat	Analyse the local and regional significance of fauna movement corridors. Determine if critical habitat is present within the study area.

Potential impacts	1. Loss of vegetation/habitat (including aquatic habitat, threatened ecological communities and critical habitat).
	2. Habitat fragmentation, edge and barrier effects
	3. Aquatic habitat impacts
	4. Injury and mortality
	5. Weeds
	6. Pests and pathogens
	7. Changed hydrology
	8. Impacts on groundwater dependent ecosystems
	9. Noise, vibration and light
	10. Cumulative impacts
		– The impact assessment chapter should commence with an introductory paragraph on the range of potential impacts likely to result from the proposed activity.
		– The impact description section of the EIA must identify and discuss each of the potential impacts of the proposed activity.
		– The EIA must clearly distinguish between impacts anticipated during construction and impacts that may arise from the operational phase of the project.
		– The quality of the discussion on the potential impacts of the project must:
		  • Identify both direct and indirect impacts.
		  • Characterise the frequency, duration and intensity of the impact.
		  • Differentiate between construction and operational impacts.
		  • Consider the local and regional scale of the impact.
		  • Considers the impacts relative to existing pressures and impacts.
		  • Discusses the significance of the impact, supported by the conclusions of any assessments of significance or 'effect determinations' (see next section).

Significance of impacts on threatened species, populations and ecological communities		In Australia, assessments of the significance of impacts of a proposed activity on a threatened species, population or ecological community are required under State and Commonwealth legislation (see DECC 2007; DoE 2013). In the USA, a similar assessment of impacts is required but is called 'effect determinations'. Effect determinations are needed for listed species and designated critical habitat (see WDOT 2012). Assessments of significance or effect determinations for each threatened species or ecological community must be summarised in this section and included in full as an appendix. Assessments of significance or effect determinations should only be grouped where species share a similar life history and habitat requirements, e.g. threatened woodland birds, cave roosting bats, large forest owls. Each assessment of significance or effect determination must end with a conclusion on the significance of the impact.

(Continued)

Table 6.1 (*Continued*)

Section heading	Section sub-heading	Contents of section
Proposed safeguards/mitigation measures		Proposed environmental safeguards/mitigation measures must: • Respond to potential impacts identified in the assessment of project impacts. • Be structured so that there is no repetition. • Safeguards/mitigation measures should be developed in consultation with the road agency and the relevant regulatory agency. • Where practical, mitigation and safeguards should be supported by the road agency as reasonable and feasible.
Conclusions		The conclusions chapter should summarise (without introducing new information or ideas), the following information: • Key ecological and environmental features of the study area. • Key findings of the report, including key impacts. • Whether the proposal is likely to significantly affect any species, populations, communities or their habitats. • Key mitigation actions.
Appendices	Species lists (field survey) Likelihood of occurrence tables	Provide tables of all flora and fauna species recorded during field surveys. Provide tables for threatened flora, fauna or ecological communities considered likely to occur based on background research. Include habitat requirements and life cycle attributes and whether the species or community has a low, moderate or high likelihood of occurring.
	Assessments of significance or effect determinations	Provide complete assessments of significance or effect determinations for those threatened species or communities with a moderate to high likelihood of occurring within the study area that may potentially be impacted by the project.
	Raw data sheets and raw data in electronic format Copies of permits, licences and other relevant documents.	Examples of raw data useful for road agencies include microbat call sonograms and vegetation plot/quadrat data. Provide scanned copies of relevant ethics approvals, scientific licence numbers and accredited assessor numbers/details.

or after construction (Chapter 5). The recommended minimum information for adequately assessing potential ecological impacts in an EIA includes the following:

• Identification of direct and indirect impacts.
• Characterising the frequency, duration and intensity of the impact.
• Differentiating between construction and operational impacts.
• Consideration of the scale of the impact from a local and regional perspective.
• Consideration of the impacts resulting from the proposal relative to existing pressures and impacts.
• Discussion of the significance of the impact, supported by the conclusions of any assessments of significance (or effect determination) as appropriate.

6.6 The EIA should follow the mitigation hierarchy (i.e. avoid, minimise, mitigate and lastly offset) and recommend realistic measures to protect the environment

EIAs are highly variable in their proposed safeguards, and sometimes none are proposed. Other EIAs don't provide any correlation between the potential impact and how the proposed mitigation may ameliorate that impact.

The Roads and Maritime Services of New South Wales, Australia, has developed a process to identify and evaluate potential mitigation measures (NSW RMS 2012). Mitigation measures and the discussion in the EIA should:

• Respond specifically to the potential impacts identified in the EIA.
• Be structured to avoid repetition.
• Be developed in consultation with the road agency. In some instances, it will be appropriate to involve regulatory agencies in these discussions if they have expertise in the likely effectiveness of the proposed mitigation.
• Be supported by the road agency as feasible and reasonable (Chapter 4).

Examples of guidelines and standards for best practice mitigation measures are given in Chapter 59.

CONCLUSIONS

Inadequate EIAs can result in delays to projects, additional costs to the agency (and hence society) and often significant impacts to biodiversity. There are numerous lessons for road agencies, environmental consultants and regulatory agencies to learn in order to raise the quality of EIAs. EIA reports must be logically structured, use sound and reliable survey and analysis techniques, include comprehensive and thorough technical advice, and accurately and honestly assess the potential impacts of the proposed project or action on biodiversity.

ACKNOWLEDGEMENTS

Thanks to NSW Roads and Maritime Services for allowing me to submit material, and most importantly to the Australian environment which constantly keeps me inspired to help save what we have left of this beautiful and unique biota.

FURTHER READING

DoE (2013): This document provides guidance for assessing the likely significance of impacts of a project (or action) on federally-listed species and ecological communities in Australia.

NSW DEC (2004): A guidance document for decision makers when considering a project (or action). It also provides information and assistance to individuals or organisations that may be required to consider the effect of a project (or action) on threatened biodiversity.

NSW RMS (2011): Guidelines providing best practice advice to road agencies, construction workers and ecological consultants to minimise impacts during construction of roads; also applicable to other types of linear infrastructure.

NSW RMS (2012): This document provides advice and guidance on how to undertake best practice environmental impact assessments. It was developed to assist ecologists and road agency staff in NSW, Australia.

UK EA (2002): A handbook to guide EIA scoping activities and explain the role and importance of scoping in EIA in the United Kingdom. It also provides guidance on key issues common to many project types.

WSDOT (2012): This document provides a template and tools to write better biological assessments when analysing the potential impacts of actions on the environment in the United States.

REFERENCES

Department of the Environment (DoE). 2013. Matters of national environmental significance – assessment of significance guidelines. DoE, Canberra. Available from http://www.environment.gov.au/epbc/publications/significant-impact-guidelines-11-matters-national-environmental-significance (accessed 28 January 2015).

Karlson, M., Mörtberg, U., & Balfors, B. 2014. Road ecology in environmental impact assessment. Environmental Impact Assessment Review **48**: 10–19.

New South Wales Department of Environment and Climate Change (NSW DECC). 2007. Threatened species assessment guidelines – assessment of significance. NSW DECC, Sydney. Available from http://www.environment.nsw.gov.au/resources/threatenedspecies/tsaguide07393.pdf (accessed 28 January 2015).

New South Wales Department of Environment and Conservation (NSW DEC). 2004. Threatened biodiversity survey and assessment: guidelines for developments and activities (working draft). NSW DEC, New South Wales.

New South Wales Roads and Maritime Services (NSW RMS). 2011. Biodiversity guidelines – protecting and managing biodiversity on RTA projects. NSW RMS, Sydney. Available from http://www.rms.nsw.gov.au/documents/ about/environment/biodiversity_guidelines.pdf (accessed 28 January 2015).

New South Wales Roads and Maritime Services (NSW RMS). 2012. Environmental impact assessment practice note – biodiversity assessment. NSW RMS, New South Wales.

United Kingdom Environment Agency (UK EA). 2002. Environmental impact assessment: a handbook for scoping projects. UK EA, Reading. Available from https://www.gov.uk/government/publications/handbook-for-scoping-projects-environmental-impact-assessment (accessed 28 January 2015).

Washington State Department of Transportation (WSDOT). 2012. Washington State Department of Transportation biological assessment guide. WSDOT, Olympia, WA. Available from http://www.wsdot.wa.gov/environment/biology/ba/baguidance.htm (accessed 28 January 2015).

Chapter 7

PRINCIPLES UNDERPINNING BIODIVERSITY OFFSETS AND GUIDANCE ON THEIR USE

Yung En Chee

School of BioSciences, The University of Melbourne, Parkville, Victoria, Australia

SUMMARY

Biodiversity offsets (also known as compensatory measures) are increasingly being applied as a tool for balancing development and conservation. Offsets may be appropriate compensation for residual impacts after genuine attempts have been made to avoid, minimize and mitigate impacts from development activities. Offsetting activities usually involve the management of habitat/land at locations separate from the impact site. To be considered successful, offsetting must at a minimum, produce 'no net loss', by delivering adequate conservation gains to balance the losses imposed by developmental impacts.

7.1 Adhering to the mitigation hierarchy is essential for maintaining the legitimacy of offsets and compensatory measures.

7.2 Early identification and understanding the limits of what can be offset is an essential step in offset mitigation planning.

7.3 The concept of offsets requires an understanding of what is required to achieve 'no net loss'.

7.4 Uncertainties and risks that might affect delivery of 'no net loss' are foreseeable and should be anticipated and accounted for in planning for offsets.

7.5 Effective management and governance is imperative for achieving 'no net loss'.

The concept of biodiversity offsets is simple. However, they are complex to design, implement and deliver in practice, particularly given the attendant risks and uncertainties. Although their use has expanded worldwide, evidence regarding actual conservation outcomes is still limited.

Handbook of Road Ecology, First Edition. Edited by Rodney van der Ree, Daniel J. Smith and Clara Grilo.
© 2015 John Wiley & Sons, Ltd. Published 2015 by John Wiley & Sons, Ltd.
Companion website: www.wiley.com\go\vanderree\roadecology

INTRODUCTION

Despite the best intentions, many road development projects will entail biodiversity impacts such as increased mortality of species that are attracted to roads, or disruption of habitat connectivity because topography or terrain prohibits re-connection, or suitable crossing structures cannot be built for the species in question. Biodiversity offsets are a relatively new tool for balancing development and conservation. Although their use is still in its infancy, they are now part of the statutory framework mandated by legislation in a growing number of countries including Australia, USA, Canada, Brazil, Columbia, New Zealand, South Africa and some in Europe (ten Kate et al. 2004; Gordon et al. 2011). They are also used in a voluntary capacity in projects around the world (Doswald et al. 2012).

What exactly are biodiversity offsets? Although different definitions exist, the definition developed by the Business Biodiversity Offsets Programme (BBOP, Textbox 7.1) encompasses the key principles regarding the use of offsets, namely, the primacy of adhering to the mitigation hierarchy (i.e. avoid, minimise, mitigate, then finally, offset), appropriate quantification of biodiversity losses and conservation gains, and the concept of 'no net loss'. In the statutory frameworks of different jurisdictions, the policy objective ranges from 'net gain' to 'no net loss' to general statements about addressing adverse impacts (McKenney & Kiesecker 2010). However, the most common goal is 'no net loss and preferably a net gain' (BBOP 2009).

This chapter explains the key concepts of offsets as compensation for residual project impacts and describes the principles required to ensure that offsets

have the best prospect of delivering 'no net loss' of biodiversity under real-world uncertainties that occur in practice.

LESSONS

7.1 Adhering to the mitigation hierarchy is essential for maintaining the legitimacy of offsets and compensatory measures

Biodiversity offsets are controversial because of the perception that they could encourage regulators to approve projects with severe biodiversity impacts as long as offsets are provided as compensation. Statutory and voluntary frameworks for offsetting address this concern by requiring adherence to the mitigation hierarchy (e.g. BBOP 2012a; DSEWPaC 2012). In theory, this designates offsetting as the least favoured measure, to be used as a 'last resort' to compensate for *residual impacts* after all reasonable measures have been taken to avoid, minimise and mitigate biodiversity impacts *in situ* (Fig. 7.1). Lending institutions (particularly Equator Principles Financial Institutions) are increasingly incorporating the mitigation hierarchy (or elements of it) into their policies for assessing and managing environmental, social and reputational risk that might impact on their social license to operate (UNEP-WCMC 2011; Doswald et al. 2012).

7.2 Early identification and understanding the limits of what can be offset is an essential step in offset mitigation planning

Biodiversity offset actions can produce gains in biodiversity value at an offset site in three ways: through enhancement, averted loss and creation (Textbox 7.2).

It is important to note that offset actions to generate gains in biodiversity from enhancement or averted loss must be over and above any duty of care that is already required at the offset site. Under a policy of 'no net loss', only gains that would not have occurred without the offset actions are counted. This is the principle of **additionality**.

In practice, proponents of developments should undertake a risk assessment at the earliest possible stage in the planning process to identify if impacts that cannot be offset are likely (BBOP 2012a). Doing this at the early planning stages provides data to inform early decisions concerning project siting, design and risks. A more detailed analysis of impacts that may not be

---

Textbox 7.1 Definition of biodiversity offsets.

Biodiversity offsets are '*measureable conservation outcomes* of actions designed to compensate for significant *residual* adverse biodiversity impacts arising from project development *after appropriate prevention and mitigation measures have been taken*. The goal of biodiversity offsets is to achieve *no net loss* and preferably a net gain of biodiversity on the ground with respect to species composition, habitat structure, ecosystem function and people's use and cultural values associated with biodiversity'. (BBOP 2009, italics mine).

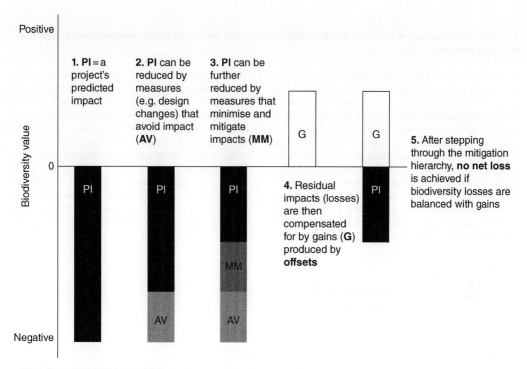

Figure 7.1 The sequential steps of applying the mitigation hierarchy (left to right) and the role of offsets in ensuring 'no net loss'. Source: Adapted from BBOP (2009) and Quétier and Lavorel (2012).

possible to offset should also form part of the formal EIA. Some biodiversity impacts cannot be offset because (BBOP 2012b; Maron et al. 2012; Pilgrim et al. 2013):
• The affected entity (e.g. species/community/ecosystem) occurs only at a few sites or populations and is effectively **irreplaceable** because there are too few viable offset sites outside the area affected by the development.
• The affected entity is endangered and close to the threshold of minimum viability (e.g. because the majority of the original population(s) and/or habitat has been lost).
• The affected entity is in good to excellent condition; and there are few, if any, opportunities to make gains at offset sites via **enhancement** (Textbox 7.2).
• The background rates of loss for the affected entity are low; and there are few, if any, opportunities to obtain gains through **averted loss** (Textbox 7.2).
• Lack of knowledge or effective restoration techniques mean it is unclear if ecologically equivalent gains can be made at the offset site within an acceptable timeframe and level of certainty (e.g. attempting to restore 'full' floristic diversity or 'old-growth' habitat);
• The resources required to generate gains at offset sites is prohibitive.

7.3   The concept of offsets requires an understanding of what is required to achieve 'no net loss'

'No net loss' is central to the concept of offsetting. It seeks to ensure that biodiversity losses due to direct and indirect impacts of a project are balanced by biodiversity gains at offset sites such that there is no net reduction overall in the type, amount and quality/ condition of biodiversity over some defined space and timeframe (BBOP 2012c).

An explicit spatial and temporal scale must be specified to provide the context for:
• Estimating the extent, severity and duration of direct and indirect impacts due to the project; this is important if project activities have cumulative or cascading effects on biodiversity;
• Considering landscape-scale processes that might be important for achieving gains at offset sites (e.g. distance and functional connectivity between the offset site and source populations of affected entities) and understanding the timeframe for delivering and maintaining biodiversity gains from offset actions.

Textbox 7.2 Three ways in which biodiversity actions can generate gains: enhancement, averted loss and creation.

Solid lines in Fig. 7.2 show the projected trajectory of biodiversity value at three hypothetical offset sites under existing land use, management regime or threatening processes (e.g. logging, grazing, weed or predator invasion). Dotted lines indicate the corresponding trajectories of biodiversity value under an offsets action regime at each site. At the topmost site, which is in relatively good condition and not subject to degrading forces, rehabilitation or restoration actions (e.g. weed control, reinstatement of woody debris) have the potential to produce gains in biodiversity value through enhancement. Note however, that offset sites that are in excellent or very good condition may have limited potential for further improvement. The middle-placed site is of poorer quality than the topmost site to begin with and is subject to strong degrading forces. To produce a gain from averted loss, this site will require offset actions that can effectively mitigate threatening processes and reverse degradation to bring about improved biodiversity value. This might require enacting a set of several actions such as fencing, predator control and re-vegetation, concurrently. Finally, gains from creation require offset actions that can effect measureable increases in biodiversity value from a site where they were initially absent.

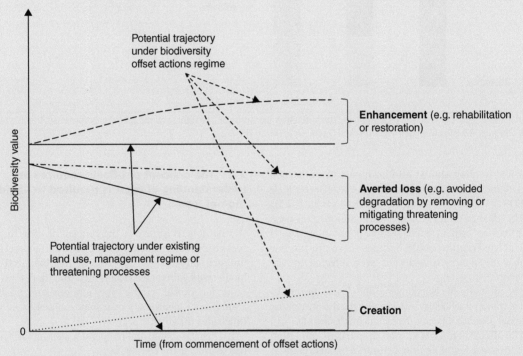

Figure 7.2 Schematic representation of how enhancement, averted loss and creation can theoretically generate biodiversity gains. Source: Yung En Chee.

Road agencies frequently purchase land along highway easements for road building and then sell off unused parcels after the project is completed. However, parcels close to the highway that have native vegetation on them may be retained to serve as biodiversity offsets. Does this satisfy a 'no net loss' test? The answer depends on the biodiversity components under consideration. Research has shown that apparent survival rates for squirrel gliders next to a highway in south-east Australia is approximately 70% less than that of gliders greater than 5 km away (McCall et al. 2010). So sites along a highway that have elevated mortality rates are inadequate for achieving 'no net loss', and appropriate offset sites must be sought outside the road-effect zone. Similarly, stormwater retention ponds that treat runoff from roads and other impervious surfaces are unlikely to be an appropriate offset for amphibians if levels of pollutants exceed tolerable thresholds (Chapter 31).

Because the concept of biodiversity encompasses such complexity, methods for quantifying biodiversity losses and gains are inevitably an imperfect simplification based on surrogates for the full complement. In practice, all operational approaches to offsetting focus on a subset of biodiversity components. To limit misunderstandings that could arise from this inherent limitation, it is vital to clearly document and justify the offset design elements outlined in Table 7.1.

7.4 Uncertainties and risks that might affect delivery of 'no net loss' are foreseeable and should be anticipated and accounted for in planning for offsets

Uncertainty arises at every stage of the offsetting process (Textbox 7.3). Many of these risks or uncertainties are foreseeable, and should be anticipated, identified and accounted for in the offsetting process. Except in cases where offset 'credits' must be paid up front before any project impacts are permitted, most offsetting situations involve immediate and certain biodiversity losses in exchange for uncertain, future gains. This is problematic because the expectation that mitigation, rehabilitation and restoration techniques can generate the required gains is not supported by the available evidence (Maron et al. 2012; Curran et al. 2014).

When there is high uncertainty about whether biodiversity components can be successfully restored and/or the time lag required is unacceptably long, proponents should revisit the feasibility of offsetting the impacts. In all other cases, the risk of failure should be underwritten (Burgin 2008). A common approach for incorporating this risk is to use multipliers or offset ratios that are established to guarantee a favourable outcome (Moilanen et al. 2009). The method for setting multipliers or offset ratios should be clearly documented and justified. A recent simulation analysis showed that a comprehensive accounting of uncertainty can result in very high offset ratios (in the order of 10–100s) to be robust (Moilanen et al.

2009); in practice, these are unlikely to be politically or economically acceptable.

7.5 Effective management and governance is imperative for achieving 'no net loss'

The enterprise of delivering 'no net loss' of biodiversity from offsetting requires effective management and oversight of the issues related to offset design, uncertainties in loss–gain quantification, and onground implementation. An adaptive management approach is required that includes monitoring and evaluation, supported by a structured, iterative process for using the resulting data to update the knowledge base for improved decision-making under uncertainty. Monitoring should not only track whether offsets were actually implemented on the ground but also their effectiveness at achieving the gains predicted based on theory and design under conditions of real-world uncertainties. Adaptive management (Nichols & Williams 2006; Duncan & Wintle 2008) and management strategy evaluation (Smith et al. 1999; Sainsbury et al. 2000) are examples of useful frameworks for guiding purposeful monitoring, evaluation, learning and uncertainty reduction.

Transparent governance arrangements should address several key issues:
• Roles and responsibilities: specify exactly who is responsible for each stage of the offsetting process, what needs to be done at each stage, what minimum standards will be applied and how those standards will be enforced. Offsetting should be integrated into the road planning process (Chapters 4 and 9).
• Resourcing: financing provisions for all stages of the offsetting process up to implementation, monitoring and finalisation must be adequate, sustainable and sufficiently robust to withstand changes in economic conditions (Burgin 2008).
• Reporting and auditing regime: periodic reporting of monitoring and auditing results and the rationale for implementation adjustments is important for demonstrating accountability and providing assurance that required gains will be delivered (even if temporary setbacks occur). Lessons learnt will be important for designing future offsets.
In addition to their central role in delivering 'no net loss' of biodiversity, effective management and governance underpins responsible risk management and contributes towards the management of

Table 7.1 Offset design elements and some key considerations and guidance on their use.

Design element	Key considerations, guidance and examples
Choice of biodiversity components and their measures	Not everything can be measured, so what are the key components or features that need to be explicitly captured? These can be specific to affected entities such as threatened species and their particular life history requirements, relate to biodiversity patterns more generally, or some combination of both.
	Potential measures for:
	• Target species include number of breeding pairs, number of young produced per year, survival rates, abundance or amount of high-quality habitat
	• Biodiversity include composition, structure or ecological processes. For instance, native plant species richness, basal area and density of overstorey, mid-storey and understorey vegetation, and site spatial characteristics such as size, shape, configuration, juxtaposition with other habitat types in the region, connectivity value and irreplaceability. Some measure of irreplaceability is important to avoid assigning low value to what may be an instance of a degraded but highly threatened and irreplaceable habitat (Bekessy et al. 2010)
Methods/protocols for estimating these measures	Methods for estimating the selected measures should be fit-for-purpose, rigorous and clearly described for correct and consistent application by assessors. In some jurisdictions, regulatory authorities have specified standards for minimum desktop analysis requirements, minimum field survey effort, survey techniques for target taxa, accredited site assessment methods and modelling and mapping of species and/or habitats (e.g. DSE 2010).
How is the 'currency' constructed?	Once selected, biodiversity components and their associated measures are integrated into a biodiversity **currency** or **metric** that represents biodiversity value. The currency needs to be devised with care because it forms the basis for what is meant by losses and gains. Importantly:
	• Characteristics not adequately incorporated into the currency will only be protected by chance and risk being lost in the exchange (Walker et al. 2009; BBOP 2012c); and
	• Care must be taken to ensure that construction of the currency reflects the underlying ecological rationale/intent.
	For instance, an 'area × condition' index is a common form of currency where the condition is based on scoring a set of attributes, weighting and then calculating the sum (e.g. Habitat Hectares, see Parkes et al. (2003); and BioMetric, see Gibbons et al. (2009)). The use of additive scoring implies that the attributes are independent and substitutable (Burgman 2001). If attributes are not actually independent, then the use of additive scoring implies a degree of 'double-counting' for correlated attributes. Substitutability also implies that a low score in one attribute can be perfectly compensated for by a higher score in another attribute. This can lead to perverse scenarios where negative changes in an attribute, for instance, the felling of standing trees, can be perfectly compensated for by say, an increase in the abundance of logs (McCarthy et al. 2004). A multiplicative index would prevent this problem (McCarthy et al. 2004).
	The 'area × condition' form may also be problematic if it allows a small, high biodiversity value site to be compensated for by a much larger but low quality offset site. In practice, 'exchange rules' (see below) are used to prevent this by stipulating the degree of exchangeability between area and condition.
	Continuous improvement through careful testing and user feedback is important for ensuring that a currency is robust and ecologically meaningful.

Table 7.1 (_Continued_)

Design element	Key considerations, guidance and examples
The 'accounting model'	The accounting model specifies the rules for estimating net balance with regard to type, amount and quality/condition of biodiversity over some defined space and timeframe (BBOP 2012c). Demonstrating ecological equivalence across type, space and time is challenging because no two sets of biodiversity, separated in space and time are going to be identical. To preserve equivalent or near-equivalent exchanges, 'exchange rules' are used to limit out-of-kind exchanges that might undermine the delivery of 'no net loss'. So for instance, rules can be set to: • Prohibit exchange of biodiversity components of high irreplaceability for components of lower conservation/threat status. • Limit exchanges to the same species, communities or ecosystem types (i.e. impacts to a given species cannot be compensated for by improvements to another species). • Disallow exchange of area for condition (i.e. cannot exchange a small, high-quality site for a larger but lower quality offset site) • Require exchanges within the same watershed or biogeographic region. • Stipulate the permitted time lag (if any) between loss and gain.
The explicit calculation of losses and gains at matched impact and offset sites	In cases where selected biodiversity components include (i) a mix of specific entities, (ii) more general habitat attributes, and (iii) site characteristics with landscape-context measures, multiple loss–gain assessments may be required to transparently account for the different components and/or currencies (BBOP 2012c).

Textbox 7.3 Types and examples of uncertainty in offsetting.

Uncertainties in quantifying losses and gains in offsetting can arise from the following:
• measurement error of biodiversity components due to inaccuracies in measurement relating to equipment, observer technique and instrument/operator error;
• systematic error due to biases in survey or sampling methods or excluding an overlooked biodiversity component;
• natural variation in dynamic systems that undergo cycles of disturbance and recovery (e.g. fire- or flood-prone ecosystems), thereby confounding attempts to obtain 'representative' measurements from a single time period (McCarthy et al. 2004);
• models of uncertainty, that is working conceptions and representations of the system of interest. This could lead to error in projections of predicted impacts and/or expected gains from offset actions (e.g. misdiagnosis of threatening processes);
• subjective judgement as a result of data interpretation and/or use of expert judgements for estimates when data are scarce and/or error prone.

Uncertainties that may cause offsets to fail are many and varied, and include for example:
• Environmental stochasticity (e.g. storms, fires, landslides or floods);
• Insecurity of tenure at offset sites (e.g. changes in planning regulations);
• Insufficient funds to resource the full package of offset actions;
• Partial or total failure of threat mitigation, rehabilitation or restoration techniques relied on to generate enhancement, averted loss and creation gains.

---

reputational risk and the social license to operate. These factors are also important to the investment community and lending institutions, who may require demonstrated impact management and environmental and social risk management as a condition of finance and continued access to resources (Doswald et al. 2012).

CONCLUSIONS

The concept of biodiversity offsets is simple. However, they are complex to design, implement and deliver in practice. Although their use has expanded worldwide, evidence regarding actual conservation outcomes is still limited. The lack of rigorous post-implementation monitoring and evaluation makes it unclear whether offsets are a reliable tool for delivering 'no net loss' of biodiversity. This reinforces the importance of managing impacts via the mitigation hierarchy until further research can provide evidence-based guidance on the performance of different types of offset actions under real-world conditions.

ACKNOWLEDGEMENTS

I thank Ascelin Gordon (RMIT University) and participants of the 'Change and Uncertainty in Biodiversity Offsets' workshop (November 2011) hosted by the Australian Research Council Centre of Excellence for Environmental Decisions. In particular, I benefited greatly from discussions with Theo Stephens, Susan Walker and Bill Langford. I also acknowledge the support of ARC Linkage LP110100304 and the Centre of Excellence for Biosecurity Risk Analysis.

FURTHER READING

BBOP (2012a): Accessible compendium of relevant issues, methodologies and tools for offset designers. Explains requirements at each step of the offset design process and outlines useful tools.

Moilanen et al. (2009): Influential analysis demonstrating the importance and impact of uncertainty on the prospects of achieving no net loss using biodiversity offsets.

Walker et al. (2009): Detailed critique of the very premise of biodiversity offsets, strongly argued from ecological, political, regulatory and administrative perspectives.

REFERENCES

Bekessy, S.A., B.A. Wintle, D.B. Lindenmayer, M.A. McCarthy, M. Colyvan, M.A. Burgman and H.P. Possingham. 2010. The biodiversity bank cannot be a lending bank. Conservation Letters 3:151–158.

Burgin, S. 2008. BioBanking: an environmental scientist's view of the role of biodiversity banking offsets in conservation. Biodiversity and Conservation 17:807–816.

Burgman, M.A. 2001. Flaws in subjective assessments of ecological risks and means for correcting them. Australian Journal of Environmental Management 8:219–226.

Business Biodiversity Offsets Programme (BBOP). 2009. Business, biodiversity offsets and BBOP: an overview. BBOP, Washington, DC.

Business Biodiversity Offsets Programme (BBOP). 2012a. Biodiversity offset design handbook-updated. BBOP, Washington, DC.

Business Biodiversity Offsets Programme (BBOP). 2012b. Resource paper: limits to what can be offset. BBOP, Washington, DC.

Business Biodiversity Offsets Programme (BBOP). 2012c. Resource paper: no net loss and loss-gain calculations in biodiversity offsets. BBOP, Washington, DC.

Curran, M., S. Hellweg and J. Beck. 2014. Is there any empirical support for biodiversity offset policy? Ecological Applications 24: 617–632.

Department of Sustainability and Environment (DSE). 2010. Biodiversity precinct structure planning kit. DSE, Melbourne, Victoria.

Department of Sustainability, Environment, Water, Populations and Communities (DSEWPaC). 2012. Environment Protection and Biodiversity Conservation Act of 1999: environmental offsets policy. DSEWPaC, Canberra, Australian Capital Territory.

Doswald, N., M. Barcellos Harris, M. Jones, E. Pilla and I. Mulder. 2012. Biodiversity offsets: voluntary and compliance regimes. A review of existing schemes, initiatives and guidance for financial institutions. United Nations Environment Programme-World Conservation Monitoring Centre (UNEP-WCMC), Cambridge and United Nations Environment Programme-Finance Initiative (UNEP-FI), Geneva.

Duncan, D.H. and B.A. Wintle. 2008. Towards adaptive management of native vegetation in regional landscapes. Pages 159–182 in Landscape analysis and visualisation: spatial models for natural resource management and planning. C. Pettit, W. Cartwright, I. Bishop, K. Lowell, D. Pullar and D. Duncan, editors. Springer, Berlin.

Gibbons, P., S.V.A. Briggs, D. Ayers, J. Seddon, S. Doyle, P. Cosier, C. McElhinny, V. Pelly and K. Roberts. 2009. An operational method to assess impacts of land clearing on terrestrial biodiversity. Ecological Indicators 9:26–40.

Gordon, A., W.T. Langford, J.A. Todd, M.D. White, D.W. Mullerworth and S.A. Bekessy. 2011. Assessing the impacts of biodiversity offset policies. Environmental Modelling & Software 26:1481–1488.

Maron, M., R.J. Hobbs, A. Moilanen, J.W. Matthews, K. Christie, T.A. Gardner, D.A. Keith, D.B. Lindenmayer and C.A. McAlpine. 2012. Faustian bargains? Restoration realities in the context of biodiversity offset policies. Biological Conservation 155:141–148.

McCall, S., M.A. McCarthy, R. van der Ree, M.J. Harper, S. Cesarini and K. Soanes. 2010. Evidence that a highway reduces apparent survival rates of squirrel gliders. Ecology and Society 15(3):27.

McCarthy, M.A., K.M. Parris, R. Van Der Ree, M.J. McDonnell, M.A. Burgman, N.S.G. Williams, N. McLean, M.J. Harper, R. Meyer, A. Hahs and T. Coates. 2004. The habitat hectares approach to vegetation assessment: an evaluation and suggestions for improvement. Ecological Management and Restoration 5:24–27.

McKenney, B.A. and J.M. Kiesecker. 2010. Policy development for biodiversity offsets: a review of offset frameworks. Environmental Management 45:165–176.

Moilanen, A., A.J.A. van Teeffelen, Y. Ben-Haim and S. Ferrier. 2009. How much compensation is enough? A framework for incorporating uncertainty and time discounting when calculating offset ratios for impacted habitat. Restoration Ecology **17**:470–478.

Nichols, J.D. and B.K. Williams. 2006. Monitoring for conservation. Trends in Ecology and Evolution **21**:668–673.

Parkes, D., G. Newell and D. Cheal. 2003. Assessing the quality of native vegetation: the 'habitat hectares' approach. Ecological Management & Restoration **4**:S29–S38.

Pilgrim, J.D., S. Brownlie, J.M.M. Ekstrom, T.A. Gardner, A. von Hase, A., K. ten Kate, C.E. Savy, R.T.T. Stephens, H.J. Temple, J. Treweek, G.T. Ussher, and G. Ward. 2013. A process for assessing the offsetability of biodiversity impacts. Conservation Letters 6: 376–384.

Quétier, F. and S. Lavorel. 2012. Assessing ecological equivalence in biodiversity offset schemes: key issues and solutions. Biological Conservation **144**:2991–2999.

Sainsbury, K.J., A.E. Punt and A.D.M. Smith. 2000. Design of operational management strategies for achieving fishery ecosystem objectives. ICES Journal of Marine Science **57**:731–741.

Smith, A.D.M., K.J. Sainsbury and R.A. Stevens. 1999. Implementing effective fisheries-management systems – management strategy evaluation and the Australian partnership approach. ICES Journal of Marine Science **56**:967–979.

ten Kate, K., J. Bishop and R. Bayon. 2004. Biodiversity offsets: views, experience and the business case. International Union for Conservation of Nature and Natural Resources (IUCN), Gland and Cambridge and Insight Investment, London.

United Nations Environment Programme-World Conservation Monitoring Centre (UNEP-WCMC). 2011. Review of the biodiversity requirements of standards and certification schemes: a snapshot of current practices. UNEP-WCMC and CBD, Montréal.

Walker, S., A.L. Brower, R.T.T. Stephens and W.G. Lee. 2009. Why bartering biodiversity fails. Conservation Letters **2**:149–157.

Chapter 8

CONSTRUCTION OF ROADS AND WILDLIFE MITIGATION MEASURES: PITFALLS AND OPPORTUNITIES

Cameron Weller

Environmental Management and Planning, Jacobs Group Pty Ltd, St Leonards, New South Wales, Australia

SUMMARY

The construction stage of a road project is when years of planning and design are realised. The construction stage relies on a team of people with a diverse range of skills and expertise to deliver a product according to a set of detailed designs. The construction period is also when any mistakes made during planning and design become evident and when errors are actually made. Mistakes range from incorrectly designed or constructed mitigation measures to missed opportunities for improvement. The ease and cost to fix such mistakes vary significantly, as well as the consequences for biodiversity and reputations if they remain unaddressed.

8.1 Pre-construction planning and dedicated environmental staff are essential to identify opportunities and avoid mistakes.

8.2 A pre-construction review of road and mitigation designs is important to assess constructability and identify opportunities for improvement.

8.3 Clearing of vegetation must be carefully planned and strictly monitored.

8.4 Early installation and regular maintenance of fauna exclusion fences can help to reduce wildlife mortality during construction.

8.5 Early construction of fauna mitigation measures can minimise impacts and allow adaptation of designs if required.

8.6 Effectiveness of mitigation can be reduced if the quality of the finishing is inadequate.

8.7 Appropriate education targeted at the needs of different construction personnel can help to achieve the best ecological outcomes.

Handbook of Road Ecology, First Edition. Edited by Rodney van der Ree, Daniel J. Smith and Clara Grilo.
© 2015 John Wiley & Sons, Ltd. Published 2015 by John Wiley & Sons, Ltd.
Companion website: www.wiley.com\go\vanderree\roadecology

The quality of the completed project will vary depending on the level of environmental monitoring or supervision, the engagement of the contractors and the degree to which the construction team have been educated during the project. Every effort should be made to identify potential mistakes before they happen and to redress any that occur, especially if the potential consequences are high.

INTRODUCTION

The construction and completion stages of a road project are usually the most visible and anticipated part of the process. Despite a long period of detailed planning and careful design, construction rarely occurs without mishap, design modifications or periods where there is some uncertainty about 'will this actually work?' Roads and mitigation measures are built to highly detailed designs and specifications that have been reviewed, agreed upon and funded. Mistakes occur when aspects are not built to the specified standard, or if the design was incorrect. Missed opportunities for improvements to the design are also potentially a problem. Mistakes and missed opportunities can be small or large, and cheap or prohibitively expensive to fix.

It is impossible to specify a step-by-step approach that eliminates all risk of mistakes because each project is different. A major difference among projects is the delivery method, which includes construct only, design and construct, or various levels of partnership between government and private industry (referred to in this chapter as an alliance). These different approaches influence cost, contractual responsibility for different aspects of a project and the ability to identify and deal with mistakes and missed opportunities. A 'construct-only' contract is where the client (i.e. road or infrastructure agency) engages a private company to build a road from a fully developed design. Consequently, the construction company is not responsible for mistakes which become evident during construction but which occurred during the planning or design, and they are unlikely to rectify such issues unless paid to do so. Therefore, it is critical that the road agency has thoroughly planned and designed the project from the outset. A 'design and construct' approach is where the client engages a company or consortium to design the project according to a scope of works and then, after approval, to build it. This arrangement has more interaction between the client and the construction and design teams than a construct-only approach and potential errors and missed opportunities are also more easily identified and resolved. However, any mistakes or missed opportunities that are

a consequence of a poorly prepared scope of works is the responsibility of the client, not the design or construction teams. Private companies are unlikely to take responsibility to fix such mistakes, unless they can recoup costs. An alliance is where the client partners with one or more private design and construction companies to develop the design and then build the project together. This approach often achieves optimal outcomes for biodiversity because the road agency is involved in all stages of the project and can ensure high standards. It is the collective responsibility of all parties to ensure the project is constructed on time, within budget and as designed.

The diversity of factors that influence the delivery of a project complicates the ability to present a checklist approach to all the possible mistakes or missed opportunities. Rather, in this chapter, I have focused on a number of common issues that occur relatively frequently. The lessons are ordered to correspond with the typical stages in large infrastructure projects (Chapters 4 and 9).

LESSONS

8.1 Pre-construction planning and dedicated environmental staff are essential to identify opportunities and avoid mistakes

Chapter 4 and Figure 4.1 show the major stages in a road project. An additional planning stage usually occurs after a contract has been awarded and is often a key milestone in gaining approval to commence construction. This pre-construction planning is usually focussed on details of the construction program including detailed construction plans to ensure that the project runs efficiently and is completed on schedule and within budget. Environmental issues should also be considered during this pre-construction planning, including any outstanding flora and fauna surveys required to be undertaken prior to clearing of vegetation, preparation of detailed plans for vegetation removal, strategies to reduce wildlife mortality during construction and what to do if threatened species are found during construction. Inadequate pre-construction planning will inevitably lead to costly mistakes and missed opportunities to improve the ecological outcome of a road project.

All major projects should employ dedicated environmental staff to be involved in all aspects of the project. Environmental staff may be employed directly by the contractor or seconded from elsewhere, but they must maintain high standards and be open to inspection and review. The environmental staff should educate the planning and construction personnel as early as possible about the ecological issues and why environmental protection is required. It is the responsibility of the public authority/client to develop ways to ensure the contractor is considering the environment on projects without environmental staff. One approach is through contractual measures or key performance indicators, which should be considered even on projects with environmental staff. Contractual conditions may include a requirement to document all construction planning sessions that must consider environmental constraints and issues; or enforce the requirement of a vegetation clearing limit that cannot be exceeded. There could also be a profit bonus scheme to reward the contractor for good performance; such as for clearing less vegetation than the contractual limit.

8.2   A pre-construction review of road and mitigation designs is important to assess constructability and identify opportunities for improvement

A review of designs by the construction team during pre-construction planning to assess constructability is an important review over and above those conducted during planning and design. If a mitigation measure cannot be constructed (e.g. materials not available or too expensive) or if an opportunity for improvement (e.g. modified design or materials to be used, inclusion of additional mitigation measures) is missed, the outcome may be more expensive and/or less effective. In all design modifications, the species expert should be consulted to ensure the new design is still suitable for the target species.

This pre-construction review should be undertaken by a team with experience in construction and who also understand the intent of the design, such as an environmental manager or species expert. The intent of traditional design elements, such as bridges, drainage culverts or pavements is readily understood by engineers; however, they may not understand the intent of a wildlife crossing structure, and just build it as it is designed. For example, the construction of a canopy bridge (Chapters 40 and 41) is relatively easy, and the supervising engineer and construction team just need to follow the detailed plans. However, they may not understand the importance of ensuring that the ends of the canopy bridge need to be in close proximity to trees for it to be effective. Similarly, a low-flow channel in the base of a culvert for fish passage is easily built (Chapters 44 and 45); however, the engineer may not be aware of the critical importance of the scour protection at the inlet and outlet at facilitating or impeding fish passage.

Cross-discipline checks are also required to ensure that there are no clashes among different design elements, which may reduce effectiveness of the mitigation measure or compromise another aspect. For example, the location and entrances of wildlife crossing structures should not conflict with other infrastructure such as fences, drains, retention ponds, utilities, adjacent roads, as well as satisfy road safety requirements.

8.3   Clearing of vegetation must be carefully planned and strictly monitored

Most road projects invariably remove vegetation, rock and other features that provide habitat for wildlife. Pre-clearing vegetation surveys occur immediately (i.e. weeks or months) before clearing, and can help identify important habitat features, such as large hollow-bearing trees, coarse woody debris or snags and riffle pools in waterways. These assets can then be protected *in situ* or collected and stored to be reinstated after construction. Pre-clearing fauna surveys should occur days (or even months) before clearing to ensure wildlife have vacated the area and are not killed during clearing. These surveys will also inform whether wildlife rescue personnel or translocations are required to protect individuals. For example, tree clearing could follow a two-stage process, where trees without hollows are cleared on day 1, and trees with hollows the following day, on the basis that hollow-dependent fauna will move after the disturbance of the initial clearing. For some species, the only viable option is to trap and translocate because any clearing of habitat will result in mortality.

Monitoring of clearing methods and limits is important because excessive clearing or removal of critical elements may reduce the effectiveness of mitigation. For example, tall trees may encourage bats to fly above traffic, and the removal of these may result in increased rates of mortality (Chapter 34). Similarly, careless clearing may destroy hollow logs and other habitat features that were to be stored and used in the landscape rehabilitation after construction. Consequently, effectiveness of mitigation may be reduced while vegetation grows and habitat components may need to be imported to the project or elements of the road or mitigation may need to be redesigned.

8.4 Early installation and regular maintenance of fauna exclusion fences can help to reduce wildlife mortality during construction

Mortality of wildlife during construction does occur, although reliable records are rarely kept because the focus is on construction, not wildlife. While rates of mortality are probably generally low because of the disturbance and removal of vegetation, any additional mortality of rare or endangered species could be significant and should be avoided.

Consideration should be given to installing permanent fencing as early as possible in the project to exclude animals from the construction zone (Chapter 20). Temporary fencing may be useful on projects without permanent exclusion fencing or if the exclusion fencing can't be built until later. It is important to ensure the road design is sufficiently finalised before installing permanent fences to avoid costly removal and re-installation or relocation of the fence. Exclusion fencing may also need to be installed on adjacent roads if animals are displaced from the main construction site. Targeted fencing around specific habitat elements, such as wetlands, may be necessary if these occur within the right of way. Regular inspection and repair of exclusion fencing during construction is essential (Chapter 20).

Existing structures such as bridges and culverts that are to be removed or impacted by the construction process should be carefully inspected for birds and bats and other species that may be roosting in or on the structure (Chapters 33 and 34). Approaches to deal with these situations must be developed well in advance of the structure being impacted and be accounted for in any contractual milestones. Methods to minimise impacts include conducting works outside the period of use by the species (e.g. for migratory species or during breeding) or excluding access by wildlife to the structure during construction. However, the impacts of permanently or temporarily restricting access by wildlife to a roost must also be considered.

8.5 Early construction of fauna mitigation measures can minimise impacts and allow adaptation of designs if required

Where possible, always install and finalise permanent wildlife crossing structures as early as possible. Underpasses and overpasses for wildlife are usually installed during the bulk earth works stage of a project but are rarely properly finished until the landscaping stage at the conclusion of a project. Depending on project size, this may result in an ineffective crossing structure for 12–24 months during construction, plus as long as it takes for animals to habituate and use the structure (Chapter 21). Greater effort should be made to landscape and finish the crossing structure more quickly, giving wildlife the opportunity to find and use the structure before the road is opened to traffic. If exclusion fencing has also been installed, the crossing structure may be immediately effective for some species.

Early installation of crossing structures and other mitigation measures also allows opportunity to rapidly and cost-efficiently repair or modify structures during construction. For example, disturbance from construction within or around waterways can affect the in-stream environment through erosion and increased turbidity (Chapter 44). Similarly, recently constructed crossings and fish habitat such as log riffles, riffle pools and plunge pools are also susceptible to in-stream erosion, potentially killing fish and invertebrates and creating barriers to movement (Chapter 45). Early installation means that if erosion does occur, repairs and remediation work is more easily undertaken because the necessary earth-moving machinery is still on-site. Modifications to the design of mitigation measures made during construction will be cheaper and easier, while machinery and personnel are still nearby than after the project has been completed.

8.6 Effectiveness of mitigation can be reduced if the quality of the finishing is inadequate

The effectiveness of crossing structures and other mitigation measures is often dependent on the 'attention to detail' displayed by the construction team when completing a project. Because the finishing of mitigation measures usually occurs at the end of a project, construction teams are often already disbanding and moving onto new jobs. In these circumstances, the momentum and enthusiasm to finish the project to a high standard may have waned (especially if the infrastructure has already been officially opened), and it may not be completed as well as originally envisaged. This is particularly problematic if contractual arrangements did not specify the minutiae of the logs, rocks, branches and plantings to be installed within and around crossing structures, because it is difficult to hold construction teams accountable. The relocation of machinery and personnel to other jobs also complicates the repair of major landscaping mistakes (Lesson 8.5). However, construction firms are often contractually obliged to repair structural defects in the road for a period of time after construction, and similar conditions should be built into contracts for mitigation measures.

On-going monitoring and supervision of the construction of mitigation measures is essential to identify and resolve any issues (or opportunities for improvements) as quickly as possible. In addition, a final inspection is necessary to ensure it is built as designed and intended (Chapter 9) and finished to a high standard.

8.7 Appropriate education targeted at the needs of different construction personnel can help to achieve the best ecological outcomes

Education and engagement of the planning, design and construction teams is critical to the success of a project and the innovation and support for future projects. While it is not critical that every construction team member understands the specifics about why an underpass for a frog is a certain design or that the approach ramp on a land bridge can't exceed a certain angle, it is helpful for them to have a general understanding. Education on a construction site can be challenging because of the range of different types of people with varying levels of education, personalities and opinions. Education does not need to be extensive or intensive, and the project or site induction is a good starting point. However, the often significant amount of safety information that is presented during site inductions can limit the amount of environmental information that can be effectively delivered. Consequently, other approaches,

such as the use of targeted information sessions or 'toolbox' meetings for specific groups of personnel can be used to deliver key messages. The use of information posters in the lunch room or site toilets is an effective approach, provided they are well designed and changed periodically (similar to road signs, Chapter 24). Educational material should be based on interesting, factual information and avoid negative phrasing (i.e. avoid 'Don't do this' or 'You will be fined or sent to jail if you do...'). In the hyper-masculine environment of many construction sites, the use of negative language to reinforce a point can have the opposite effect and create an apathetic or adversarial (e.g. 'You can't tell me what to do') response.

Mitigation measures are often perceived as a waste of money by many construction team members, and education should include examples of the mitigation measures being used by the targeted species on other projects. Similarly, demonstrating that certain mitigation measures are cost-effective and that society can save money through reductions in the rate of wildlife-vehicle collision and human death, injury and property repair can be a powerful message. This can often negate the concerns of whether the mitigation measures work, and can assist in building the commitment of the construction team in installing the mitigation measures as per the design, and to ensure their effectiveness.

CONCLUSION

The construction stage of a road project is typically the final opportunity to build major structural elements on a road, at least until the next major upgrade or repair. Well-designed and properly built roads may not require any significant construction for 20–50 or more years. Therefore, mitigation measures should be built and finished to the highest possible standard to ensure long-term effectiveness. The construction stage is also often the final opportunity to cost-effectively identify opportunities and make any improvements. Appropriate education targeted at the needs of different construction personnel is an essential step in ensuring the mitigation is built to the highest possible standard and to engender a process of innovation for future projects.

Chapter 9

ENSURING THE COMPLETED ROAD PROJECT IS DESIGNED, BUILT AND OPERATED AS INTENDED

Rodney van der Ree[1], Stephen Tonjes[2] and Cameron Weller[3]

[1]Australian Research Centre for Urban Ecology, Royal Botanic Gardens Melbourne, and School of BioSciences, The University of Melbourne, Melbourne, Victoria, Australia
[2]District Five Environmental Management Office, Florida Department of Transportation (retired) and private environmental consultant, DeLand, FL, USA
[3]Environmental Management and Planning, Jacobs Group Pty Ltd, St Leonards, New South Wales, Australia

SUMMARY

Roads are built by a team of people from a range of disciplines who must collaborate to ensure the road is built within budget, on time and to the highest possible standards with the least environmental impact. Ecologists, and especially those with expertise on the species of concern likely to be impacted by the proposed road, must be involved in the early planning stages of the project and throughout the process.

9.1 Road planning, design, construction and operation is a truly collaborative process.

9.2 Engage ecologists and biologists with expertise on the ecosystems or species of concern at the earliest planning stages to ensure the best outcome for biodiversity.

9.3 Large-scale or expensive mitigation measures need to be identified during the route selection process so that costs and benefits can be properly evaluated.

9.4 Clearly define the ecological goals of the mitigation.

9.5 Mitigation measures need to be identified during the planning or early design stages to prevent unnecessary costs.

9.6 Misinterpretation of concepts and designs can (and often do) occur at each stage in a road project.

Handbook of Road Ecology, First Edition. Edited by Rodney van der Ree, Daniel J. Smith and Clara Grilo.
© 2015 John Wiley & Sons, Ltd. Published 2015 by John Wiley & Sons, Ltd.
Companion website: www.wiley.com\go\vanderree\roadecology

9.7 Ensure that species or ecosystem experts continue to be included in each design and construction stage of a project to ensure the effectiveness of mitigation measures.

Genuine collaboration among road planners, engineers, designers, construction teams and ecologists is essential. Without it, roads may be built in poor locations, and mitigation measures are likely to be ineffective, financially wasteful and fail to protect species from local extinction.

INTRODUCTION

All road projects have an environmental or ecological impact and can affect wildlife. Road projects can be lengthy and involve a convoluted process (Chapter 4), with planning and design stages usually taking longer than the construction itself. It is important to assess the potential ecological impacts as early as possible to ensure effective outcomes. There are numerous approaches to this assessment (Chapters 5, 6 and 13), ranging from quick, desk-based analyses to in-depth, long-term field assessments. The important point is that the potential impacts on wildlife need to be assessed at an early stage to ensure the road project is as ecologically sustainable as possible.

The aim of this chapter is to bring insights from the perspective of the ecologist to the planning and design teams to help ensure that roads and mitigation measures are designed as intended, built as designed and operated effectively.

LESSONS

9.1   Road planning, design, construction and operation is a truly collaborative process

Many individuals and teams are involved in the delivery of a road project. While it may appear adversarial, most team members are keen to see the best outcome for the road, the community and the environment. Engineers and planners are willing to be creative with designs and solutions, but ecologists must recognise that supervisors and decision-makers need to give the designers permission to spend time and money developing specifications outside the standard menus. Accommodating wildlife and habitat is not likely to be considered unless mandated by regulation or formal agreement which is a result of public sentiment and government policy.

Ecologists must communicate the needs of wildlife to engineers and planners, but should be mindful of the constraints that engineers and planners work under.

Similarly, planners and engineers should remember that ecologists may be unable to talk in absolutes (e.g. specifying standard dimensions for crossing structures) because the environment is highly variable, and there is still much to learn about the impacts of roads and the effectiveness of mitigation on different wildlife.

It is the responsibility of all involved in the project to recognise where and when expert input is required. One opportunity often missed is the inclusion of maintenance engineers during design (Chapter 17) to develop crossing structures that can be maintained in an efficient and cost-effective manner. Consultant planners and designers need to alert their clients (typically road agencies) of the need to bring in ecologists and species experts at the earliest possible stage and retain them throughout the design process. Attempting to lower expenses by engaging ecologists who are not experts in the ecosystems or species of concern will typically result in incorrect assessments of ecological impacts and ineffective mitigation measures, as illustrated by the examples given in this chapter. Genuine collaboration will help to ensure that the roads we build today are as ecologically sustainable as possible.

9.2   Engage ecologists and biologists with expertise on the ecosystems or species of concern at the earliest planning stages to ensure the best outcome for biodiversity

The best outcomes are usually achieved when ecologists and biologists with expertise in the ecosystems or species of concern are brought into the project at the earliest possible stages (Fig. 4.1). Species and ecosystem experts often identify issues earlier than general ecologists, including identifying specific impacts, key habitats to be avoided and the methods of mitigation most likely to be effective (e.g. the number, type, size and location of crossing structures).

The ability to add or modify a mitigation measure (or modify the location or design of the road) decreases as the project approaches delivery (Fig. 4.1). Mitigation strategies added or modified later in the process attract additional costs and may be less effective than those included earlier (Lesson 9.3). While early input from species experts is easier, cheaper and more effective, road agencies should remain willing to incorporate design modifications at any stage where evidence of a significant benefit exists.

9.3 Large-scale or expensive mitigation measures need to be identified during the route selection process so that costs and benefits can be properly evaluated

It is very difficult to convince a road agency to include an expensive mitigation option that is not in the original budget, after the funding has been approved and obtained. The planning and early design stages of a road usually include an assessment of alternative road alignments. The scope of this 'route-selection' varies depending on the size of the project and the sensitivity of the areas the road traverses. An important element of this decision is the extent and expense of mitigation measures required for each option. Preliminary environmental analysis should be used during the route selection phase to estimate the extent and need for mitigation measures and their indicative costs, rather than leaving this until the detailed design stage when the footprint of the road has been precisely established. The early use of this information provides better outcomes at a lower cost.

9.4 Clearly define the ecological goals of the mitigation

Mitigation measures are installed because (i) an existing road negatively impacts wildlife or (ii) an assessment for a proposed road has identified that it will likely have a negative impact on wildlife. It is important to clearly define the ecological goals involved in this process, and Textbox 16.1 explains the specific, measurable, achievable, realistic and time-framed (SMART) approach to goal setting. General goals, such as 'restore connectivity' or 'reduce mortality' are suitable for high-level, strategic plans or possibly the early planning stages of a proposed road, but not the EIA or detailed planning documents for specific projects. Sufficient support to include mitigation measures in projects is more likely when the specifics of the problem and detailed goals for mitigation are given.

The planning concepts will also be improved when the target species and goals for mitigation are clearly articulated. This will also provide a benchmark against which proposals made during construction to alter the design of the road or mitigation can be assessed.

9.5 Mitigation measures need to be identified during the planning or early design stages to prevent unnecessary costs

The plans for a road project become more detailed as the project progresses, with every feature carefully designed and positioned. For example, requesting a culvert be moved 100 m from its proposed location to a better location after the design stage has finished or construction has commenced, is likely to be very expensive (Chapter 4). Thousands of cubic metres of fill may be required to raise the road for hundreds of meters either side of the new location for the culvert. If construction has already commenced, the extra fill may need to be sourced from outside the project because the cut and fill balance for the highway has already been met. Nevertheless, any concerns about any aspect of the design of the road or mitigation should be raised at any stage of the project. The decision-making process will include some form of cost-benefit analysis and if possible, a positive outcome can be achieved.

9.6 Misinterpretation of concepts and designs can (and often do) occur at each stage in a road project

Well-managed projects will encourage input by experts from a range of disciplines (e.g. ecologists, hydrologists, landscape architects and engineers) at the relevant and appropriate stages. Textbox 9.1 illustrates the potential for miscommunication associated with the number of people and steps typically involved in the process of constructing a new road or reconstructing an existing road.

Unfortunately, this is not an isolated or exaggerated example (Figs 9.1 and 9.2). At each stage in the project, different people are involved in designing, refining and providing input to the project and while most changes are intended as improvements, at times no thought is given to the ecological consequences. Compounding matters, the costs to retrofit or modify mitigation measures are so prohibitive that funds are seldom available to correct such errors.

Textbox 9.1 The errors and misunderstandings that led to the construction of an ineffective land bridge.

In the early planning stages of the process, a species expert, planner and an engineer may spend a few hours together brainstorming about design ideas for an overpass to facilitate safe crossings by an endangered species. The ecologist admits that they know all about the movement patterns of the species within intact forest, but knows little about how the species may behave near major highways, and more importantly, whether they will even use a vegetated land bridge. Nevertheless, the ecologist suggests some important design principles (based on the biology and ecology of the target species) to the engineer, who then creates a conceptual design for the overpass. The initial concept design then goes to the detailed-design team, who incorporate the concept into the road design and determines the best place (from a construction perspective, rather than an ecological one) where the overpass should be located. A landscape architect may also get involved deciding aesthetic issues, including the selection of shrubs (which may or may not encourage use by the target species) along the highway verge that seamlessly integrates with plantings on the overpass. Following construction, the species expert visits the much anticipated land bridge, only to find that the overpass looks nothing like what they envisioned when they spoke with the planners and engineers a couple of years earlier. The overpass, poorly located and inappropriately vegetated for ecological purposes is largely ineffective and the threatened species has become locally extinct.

9.7   Ensure that species or ecosystem experts continue to be included in each design and construction stage of a project to ensure the effectiveness of mitigation measures

Transportation professionals who are untrained in wildlife ecology cannot be expected to account for all of the wildlife-related issues that arise during the design and construction of mitigation measures.

Numerous manuals on designing wildlife mitigation measures have been published in recent years (Chapter 59), but non-ecologists should not be expected to apply them appropriately to a project any more than drainage engineers, for example, should be expected to apply structural or pavement standards. Most of these wildlife manuals are not standard-based, but rather guidance-based from a relatively recent and rapidly developing body of literature, research and experience. They are also

Figure 9.1 The concrete shelf and elevated poles designed to facilitate crossings by small to medium-sized and arboreal mammals in this wildlife underpass are less effective (and possibly ineffective) because they have not been connected to the surrounding habitat. Source: Photograph by R. van der Ree.

Figure 9.2 Due to relatively minor modifications, mistakes or misunderstandings during design and construction, the three drainage culverts remain dry during rainfall events and the wildlife underpass is flooded. Source: Photograph by and reproduced with permission of Scott Watson.

usually written from the perspective of a certain region.

Just as species and ecosystem experts need to be included in the earliest planning stages, they also need to be included in the earliest design stages. A common procedure in design is for successive iterations of plans to be circulated among the various engineering disciplines for review. Whilst it is important to have species and ecosystem experts participate in these reviews, a formal review–comment–response procedure is usually too cumbersome to allow the adjusting of standard engineering designs to the unfamiliar needs of wildlife. For example, in a standard review procedure, the species or ecosystem expert might identify a problem in one discipline, which is corrected in the next round of plans. In the subsequent review, a different discipline may identify a new issue or suggest changes that in turn creates a different problem (Lesson 8.2). As highlighted in Textbox 9.1, there are many opportunities for seemingly small or insignificant changes in designs to compound in the construction stage and result in an ineffective mitigation measure.

In the state of Florida, USA, a task team approach was used to put a project back on track after some design issues around mitigation measures could not be resolved during the review of plans. The 'Wildlife Crossing Design Team' included roadway, structures and drainage engineers from the design department, engineers from the maintenance, construction and conceptual design teams, and an experienced wildlife ecologist. All the members of the team met together with the design project manager to discuss early concept sketches before any detailed plans were drafted. With all the specialists in the same room, unintended consequences of design alternatives and suggested changes from any of the specialists were identified and addressed immediately. Issues that had stalled in formal reviews of plans over months were resolved in one or two meetings and a few e-mail exchanges.

Another opportunity for error exists in the formal review of a project by a team of engineers which has *not* been involved in the project design (known in the United States as 'value engineering'). This 'fresh look' often results in significant cost savings and improved designs, but it also requires careful scrutiny by species and ecosystem experts because wildlife mitigation measures are usually the least understood aspects of a road project, and they may be targeted for inappropriate reductions or alterations. Similarly, construction teams may suggest cost-saving measures and may respond to unexpected site conditions by modifying the design or slightly re-positioning structures. Any single large modification or numerous small ones, not approved by the species or ecosystem expert, may render a mitigation measure ineffective.

Experienced transportation ecologists should be consulted when a design brief or scope of works to outsource any stage of the design process is being written by the transportation agency. Contract writers and estimators at the agency may misinterpret the details or requirements of the EIA or other planning documents. If time and money is not budgeted for collaboration with species and ecosystem experts, the designers are already starting on the wrong foot, potentially leading to contractual arguments and ineffective mitigation measures. One solution may be to include a caveat in the scope of works or design brief to allow the adaptive design of mitigation measures, specifying that the road agency will pay for the design of fauna mitigation measures using a different model (e.g. a schedule of rates, rather than a lump sum).

CONCLUSIONS

Relative to the operational life of a road, there is only a small window of time to influence its location and design. In addition, the high costs associated with retrofitting existing roads mean that the most cost-effective time to install a mitigation structure is when the road is being built or upgraded. Species and ecosystem experts need to be involved with road projects at the earliest possible stages of the planning process and continue to be involved throughout the development of detailed construction plans and until project completion. This will ensure the potential impacts of the new infrastructure on the ecosystem or species of concern are properly identified and quantified and that (i) the road avoids areas of high-quality habitat or populations of threatened species; (ii) the road includes the most effective mitigation techniques placed in optimal locations; (iii) the mitigation measures are built as originally envisaged; and (iv) the completed road has the lowest possible impact on the ecosystem or species of concern.

ACKNOWLEDGEMENTS

Thanks to the many planners, designers, engineers and construction personnel that we have collaborated with on projects in Australia and the United States. Thanks for teaching us some of your trade and allowing us to share a little of ours.

Chapter 10

GOOD SCIENCE AND EXPERIMENTATION ARE NEEDED IN ROAD ECOLOGY

Rodney van der Ree[1], Jochen A. G. Jaeger[2], Trina Rytwinski[3] and Edgar A. van der Grift[4]

[1]Australian Research Centre for Urban Ecology, Royal Botanic Gardens Melbourne, and School of BioSciences, The University of Melbourne, Melbourne, Victoria, Australia

[2]Department of Geography, Planning and Environment, Concordia University Montreal, Montréal, Québec, Canada

[3]Geomatics and Landscape Ecology Research Laboratory, Department of Biology, Carleton University, Ottawa, Ontario, Canada

[4]Alterra, Wageningen UR, Environmental Science Group, Wageningen, Netherlands

SUMMARY

Scientifically rigorous research that produces accurate information is required to identify and mitigate the negative impacts of roads and traffic on wildlife, communities and ecosystems. The current approach to road planning and construction is not conducive to doing good science or incorporating explicit learning in the road development process. This typically results in inadequate information about road impacts, and poor-quality monitoring that rarely answers relevant questions about mitigation effectiveness, often leads to equivocal outcomes and does not improve 'best practice' as it potentially could and should. With some improvements and more experiments with roads and mitigation measures, the planning, design, construction and management of roads and road impacts could be significantly enhanced.

10.1 Rigorous science is essential to assess, avoid, minimise, mitigate and offset the impacts of roads and traffic.

10.2 Effective monitoring is an essential tool in road ecology.

10.3 Getting the question right is a critical first step in research and monitoring.

10.4 Study design matters.

10.5 Monitoring should be seen as an integral and valuable part of road projects.

Handbook of Road Ecology, First Edition. Edited by Rodney van der Ree, Daniel J. Smith and Clara Grilo.
© 2015 John Wiley & Sons, Ltd. Published 2015 by John Wiley & Sons, Ltd.
Companion website: www.wiley.com\go\vanderree\roadecology

10.6 Experiments investigating road impacts and mitigation effectiveness are an important way forward in road ecology and better management of roads.

10.7 Research and monitoring should be strategically planned and coordinated across jurisdictional boundaries.

10.8 The data and findings need to be accessible to relevant user groups, e.g. scientists, planner, and decision-makers, ideally also the public.

A diverse suite of people are involved in the planning, design, construction, and maintenance of roads and other linear infrastructure, and they all require reliable and accurate information to make decisions – not just about a particular location, but knowledge about ecological patterns and ecological relationships. Rigorous research, experiments and meaningful monitoring are urgently required to ensure linear infrastructure is ecologically sensitive and mitigation is effective ecologically and in terms of costs.

INTRODUCTION

Decisions in environmental management should be based on the best available scientific evidence and adopt ethical precautionary principles when dealing with uncertainties that remain. In road ecology, the scientific method has been used since 1925 (Stoner 1925) to document the impacts of roads and traffic on wildlife, and it is clear that roads have significant negative effects on species, communities and ecosystems (e.g. Forman et al. 2003; other chapters in this book). Options to avoid or mitigate these negative effects are numerous and have been widely implemented around the world, especially in Europe and North America. Over the past 25 years, studies have demonstrated that wildlife will use crossing structures to safely traverse roads (Chapter 21) and that fences can reduce animal mortality due to wildlife–vehicle collisions (WVC) and funnel wildlife towards crossing structures (Chapter 20). Evidence of the severity of the impacts and efficacy of mitigation originates from numerous sources with widely varying levels of scientific rigour, from well-designed and replicated observational studies and experiments (e.g. Clevenger & Waltho 2000; McClure et al. 2013) to groups of local citizens collecting incidental roadkill data (Chapters 60 and 62). All types of data from a range of sources have value, and our current knowledge of impacts and mitigation originates from these contributions.

Good decisions require good information, and poorly designed or executed research and monitoring run the risk of providing incomplete or incorrect information (Legg & Nagy 2006). In this chapter, we highlight why research and monitoring must be scientifically rigorous in order to fully understand the impacts of roads and traffic and achieve effective mitigation. We propose a series of changes that will significantly improve the quality and breadth of our understanding and also promote a mindset of continual improvement in road-impact assessment and mitigation.

The terms 'research' and 'monitoring' are often used interchangeably, but in this chapter, we attribute specific meanings. Research is broadly defined as the systematic collection and analysis of information to increase our understanding of a topic or issue. Research can be conducted over a range of temporal and spatial scales using a variety of techniques. Monitoring is a specific form of research and involves the repeated measuring of certain variables, usually over an extended period of time (e.g. years or decades). Experiments, which are used in both research and monitoring, use manipulation and testing under controlled conditions to understand the causal relationships between two or more variables. The types of questions posed in research and monitoring are often different. Research could be used to quantify the impacts of street lighting on wildlife or to decide how many crossing structures are required along a proposed road, while monitoring would be used to assess the effectiveness of those crossing structures over time. The collection of data to determine the frequency and locations of WVC and the variables that influence this rate and the locations could be research or monitoring, depending on how the study is carried out. Monitoring is further explained in Lesson 10.2 and experiments in Lesson 10.6.

LESSONS

10.1 Rigorous science is essential to assess, avoid, minimise, mitigate and offset the impacts of roads and traffic

Road ecology research has three main foci that are fundamental to the day-to-day needs of transportation agencies: (i) what are the (likely) impacts of the (proposed) road and/or traffic on an individual, population, community or ecosystem; (ii) what are the best strategies to avoid those impacts; and (iii) what is the effectiveness of mitigation and offsets (Chapters 7 and 16)? Well-designed and executed research and monitoring have already provided answers to many important questions in road ecology, but many more remain unanswered. Many of these questions that require further research have been identified in other chapters of this book.

High-quality research can be expensive and logistically challenging to achieve, and this partially explains why many studies of mitigation efficacy were poorly designed (van der Ree et al. 2007). Nevertheless, the benefits of investing in high-quality research programmes are numerous and include:
- Findings are more likely to be correct, complete and reliable.
- Results can be more confidently extrapolated to other locations, species or ecosystems.
- The impacts of roads and traffic can be properly quantified, and effective strategies to avoid, mitigate and offset impacts can be comprehensively identified.
- Based on good data, we can gain general insights, discover regularities and develop theories that can be tested by other studies.
- Mitigation can be modified to ensure it is ecologically effective and cost-efficient.

Therefore, if we use poor-quality research to identify impacts and to develop and assess methods to avoid, mitigate and offset impacts, we may draw incorrect or incomplete conclusions resulting in a decline in biodiversity and inappropriate use of public investments.

10.2 Effective monitoring is an essential tool in road ecology

Monitoring is the periodic 'recording of the condition of a feature of interest to detect or measure compliance with a predetermined standard' (Hellawell 1991). In road ecology, most monitoring is commissioned by road agencies to assess the ecological impacts of a road (e.g. wildlife mortality or barrier effects) or the efficacy of mitigation. If we already knew the impacts of a proposed construction project or the effectiveness of mitigation, such conditions of approval would not be imposed. Unfortunately, and contrary to the original expectations and intentions, it turns out that it has often been a waste of time and money because it does not provide reliable or broadly useful information (Legg & Nagy 2006; Lindenmayer & Likens 2010). This is the case for much monitoring in conservation biology, because (Legg & Nagy 2006; Lindenmayer & Likens 2010):
- The goals or questions are poorly conceived and/or articulated.
- There is no hypothesis or theory specified to be tested.
- The survey design is inadequate, for example, the necessary comparisons cannot be made to make relevant inferences.
- The wrong data is collected, or the data is of poor quality because it is inconsistently or incorrectly collected.
- The study has insufficient statistical power to detect a relevant change.
- For a diverse suite of reasons, the monitoring programme is not finished properly.
- Insufficient funds or support is provided by the commissioning client.

However, these issues can be overcome when monitoring is valued and executed properly. There are three different types of monitoring, albeit with some overlap: (i) curiosity or passive monitoring (monitoring devoid of specific questions, lacking an experimental design and often done out of inquisitiveness); (ii) mandated monitoring (monitoring that has been stipulated by government legislation or directive); and (iii) question-driven monitoring (monitoring guided by a conceptual model and rigorous experimental design) (Lindenmayer & Likens 2010). All types of monitoring are undertaken in road ecology; however, the majority is mandated by the permitting agency as a condition of approval for the project. Unfortunately, mandated monitoring is often under-resourced and not taken seriously such that the opportunity to maximise the information learned from each monitoring programme is lost. To prevent such failure, good 'monitoring needs questions, an experimental design, a conceptual framework, and data integrity through repeatable application of appropriate field protocols' (Lindenmayer & Likens 2010: p. 3).

10.3 Getting the question right is a critical first step in research and monitoring

It is beyond the scope of this chapter to fully identify and detail all the steps to undertake scientifically robust research and monitoring (but see further readings for additional information). An important first step is to formulate the question(s) to be addressed by the study. Road ecology is an applied discipline and questions must be based on the needs of road planners, decision-makers and managers. However, there is a fine balance between questions that are so directly applied (to a single project) that they have little application outside the specific problem, and questions that are so general that they fail to provide any practical or applied information. Road agencies typically invest in research and monitoring that are limited to the very specific needs of a particular project (e.g. Fig. 10.1A). Where possible, research and monitoring should strive to be broadly applicable and generalisable (i.e. Fig. 10.1C). In other words, studies should be designed to answer bigger questions than simply: 'Does organism X cross the road at location Y?' or 'What is the rate of wildlife mortality due to WVC at location Z?' These simple questions are not wrong, but with a little extra thought and resources, the same research could provide information relevant to different species, landscapes, roads and road projects, that is, to reveal patterns or regularities such as relationships between wildlife response and characteristics of mitigation structures (Fig. 10.1C). Explicitly stating the questions helps to ensure that the study design and the data collected will be sufficient to answer the questions. For example, the inclusion of multiple roads in a study of WVC will allow the identification of factors influencing rates of roadkill, which can then be applied to non-studied roads. Similarly, assessing rates of crossing by wildlife through various types of culverts along different roads or landscapes will identify patterns in rates of crossing and some of the factors influencing those patterns, for example, different designs and placement of culverts. In both examples, road agencies will collect the specific information they need, as well as help them to gain a more complete understanding of the impacts of roads and the effectiveness of mitigation. This may reduce the need to 're-do' the same study elsewhere, thereby saving significant amounts of money.

Doing more monitoring than the mandated minimum will be more expensive on a per-project basis, but in the medium- and long-term, the cost savings could be significant because the same monitoring does not need to be undertaken at every project. For example, including extra study sites (i.e. crossing structures and/or roads) in the monitoring will provide more reliable data on rates of use as well as the factors that influence use, which can then potentially be extrapolated to other sites (i.e. Fig. 10.1C). If an expanded monitoring programme demonstrated effectiveness (see Chapter 16 on how to define SMART goals to measure effectiveness) of a certain mitigation measure, it may not need to be monitored again. The alternative method, which would typically be the current approach, would be to conduct separate (and potentially inadequate) monitoring programmes on

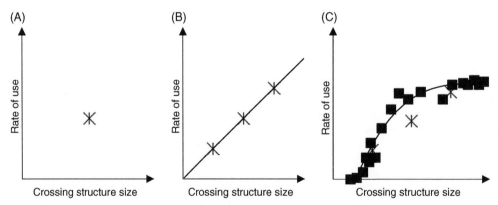

Figure 10.1 The relationship between the number of data points and the depth of understanding. In this hypothetical example, the rate of use of a crossing structure by wildlife has been monitored over time. In (A), the question was, 'How often does species X use the structure?' (a typical road agency question), and a single data point has been collected which cannot be extrapolated or generalised to other locations. In (B), the question is slightly broader, and three crossing structures have been monitored, but because the sample size is still insufficient, one might conclude that very large crossing structures will always perform better than smaller ones. With sufficient replication and a sufficient range of the independent variable covered (C), the data show that there is a threshold in the benefit of increasing the size of a crossing structure, beyond which the higher cost of larger structures would not be justified. Such a study has rarely been done, but a noteworthy example is Pfister et al. (1999). Source: Reproduced with permission from Rytwinski et al. (2015).

each road project (i.e. Fig. 10.1A and B), with the overall cost to the agency multiplied by the number of projects. Regulatory agencies should also become involved by increasing the standard of their mandated monitoring.

While study questions can be simple (e.g. How often/ at what rate does species X use this underpass?), we recommend that they be broadened and made more useful by testing a hypothesised regularity or theory. This does not need to be an 'all-encompassing, grand theory' of road ecology – but could be quite simple. If we think that size, substrate type and amount of vegetation at the entrances of crossing structures are important variables influencing the rate of use, we should explicitly attempt to incorporate these factors into the study question. Other predictions about the effect of characteristics of species or road and landscape conditions on the vulnerability of species to road impacts (e.g. Jaeger et al. 2005; Fahrig & Rytwinski 2009) could also be explicitly tested as part of the monitoring programme. By identifying and testing a theory, even simple theories or hypothesised relationships, the results of the research and monitoring can be more readily used to inform environmental impact assessments (EIAs) and road placement and design elsewhere.

10.4   Study design matters

The inability of science to strongly influence the location and design of new roads in the last decades has been attributed, in part, to a lack of research that directly answers practical questions posed by road agencies and because much research that has been done has low inferential strength (Roedenbeck et al. 2007). Assuming the question is right (Lesson 10.3), the next major challenge is getting the study design right.

There are two main considerations to study design: study class and type (Roedenbeck et al. 2007). Study class relates to whether the study is manipulative or non-manipulative. In manipulative studies, the researcher has control over the variable of interest and the response to manipulating it is measured. Non-manipulative studies (sometimes called observational or natural experiments) occur when the researcher takes advantage of changes that have happened (by using existing data) or are about to happen (by taking measurements, i.e. making observations) to understand its effect. Most studies in road ecology are non-manipulative because they measure a response to the construction of a new road or mitigation, but

the researcher has very little, if any, control over any of the variables of interest, such as the size or placement of a crossing structure. While it may be argued that the construction of a road or mitigation is a manipulation (by the road agency), and therefore an experiment in some sense, the monitoring is technically non-manipulative because the researcher had no control over the design. The engineers and planners decided on the design of the road or crossing structure for reasons unrelated to study design considerations, and the researcher must monitor what was built. These are also sometimes called pseudo-experiments. In contrast, if the researcher was engaged at the design stage of a road project, they could experimentally modify a design parameter to investigate its effect and thus perform a manipulative experiment.

The strongest type of study is one in which data are collected **B**efore and **A**fter the Impact (e.g. road construction or installation of mitigation), at sites where the **I**mpact has occurred and at **C**ontrol sites which have not been affected by the impact (Fig. 10.2). This study design is referred to as a BACI design and provides the highest level of inferential strength, which is a measure of the ability of the study to fully answer the research question. If well conducted (e.g. adequate replication, sufficient before data collected), the information from BACI studies can also be used to inform road planning and design decisions.

Depending on the question, the impacts of construction activities can be assessed by also taking measurements **D**uring construction (i.e. BDACI) (Fig. 10.2). Other study types include CI, where measurements are taken after the road has been constructed (or a manipulation has occurred) at Control and Impact sites, or BA, where measurements are taken Before and After at impact sites only. The definition of a control site will vary depending on the question and may include road-free areas, areas with narrow or low-traffic-volume roads, unmitigated roads or unmodified mitigation measures (e.g. see first point in Textbox 10.1).

In all types of study designs, an important consideration is the need for replication of sites. The natural environment is inherently variable (see scatter of plots in Fig. 10.1C), and having multiple control and impact sites will help to account for random variation among sites, that is, to distinguish between the natural variation and the effect of the variable of interest, for example, size of the wildlife passage (Chapter 16). When selecting study sites, it is important to avoid introducing bias into the design. There are a number of approaches to

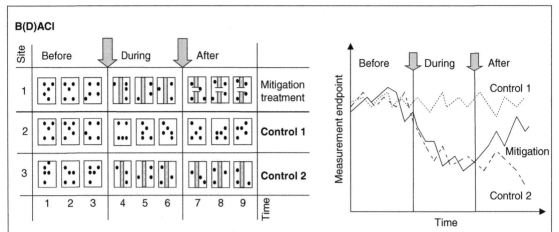

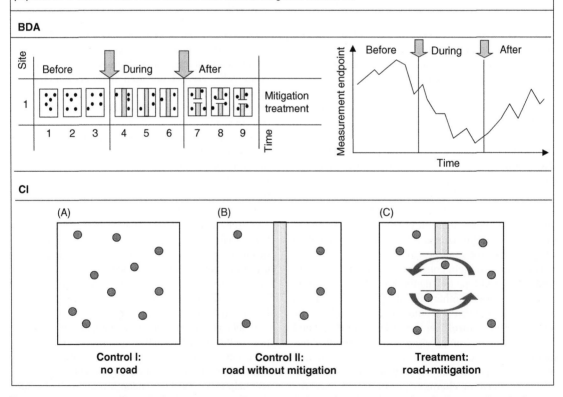

Figure 10.2 Three possible study designs to answer the question 'What is the effectiveness of road mitigation?' Study class could be manipulative or non-manipulative. Study types are before(–during)–after–control–impact (BDACI), before(–during)–after (BDA) and control–impact (CI). The dots and arrows symbolise animals moving in the landscape and across the overpass, respectively. Source: Roedenbeck et al. 2007.

Textbox 10.1 Incorporating manipulative experiments in road ecology.

The range of potential experiments in road ecology are numerous and encompass quantifying impacts, as well as developing strategies to avoid, mitigate and offset impacts (see also Rytwinski et al. 2015). Hypothetical examples are as follows:

- Problem: 20 underpasses were installed under a highway, and use by small mammals is too low.

  Typical solution: Place logs in all underpasses to provide shelter and encourage movement.

  Experimental solution: Hypothesis is that small mammals need cover. Test this hypothesis by placing logs which provide cover in 10 randomly selected underpasses, and leave the remaining 10 structures unmodified as controls. Compare the rate of use by small mammals between the two treatments, as well as before and after the placement of logs.

  Result: Rate of use in the 10 treated underpasses quadruples and does not change in untreated underpasses.

  Conclusion: Logs are critical to the movement of small mammals and should be included in all future mitigation programmes.

- Problem: Bat activity near a 15 km-long section of road is low.

  Typical solution: Provide offset habitat away from the road.

  Experimental solution: Hypothesis is that street lights are the primary cause of reduced bat activity, not increased mortality because traffic volume at night is low. Test hypothesis by modifying lighting (e.g. changing globes, turning off, modifying housing) in sections of road 2 km in length, with identical unmodified sections as controls.

  Result: No difference in bat activity before or after the treatment or between treated sections.

  Conclusion: Lighting is not the primary cause of the decline near this road.

- Problem: An endangered species of bird does not cross a road through a protected area.

  Typical solution: Build crossing structure or provide offset habitat.

  Experimental solution: Hypothesis is that traffic volume is the cause. Test hypothesis by periodically diverting traffic (e.g. 1 week open, 1 week closed) onto an alternative road outside the protected area, and measure rate of crossing by birds. Compare rates of crossing before and after at both the treated road and other nearby roads and in roadless areas using a design similar to that used by McClure et al. (2013) (Fig. 10.3).

  Result: Rate of crossing by birds at closed road is equal to rate of crossing the same distance in roadless area (i.e. a no-road control), and crossing ceases when road is open to traffic.

  Conclusion: Traffic, and not the clearing or paved surface, is responsible for the barrier effect in this species of bird. To facilitate connectivity to allow for dispersal and gene flow, periodically close road to allow crossing, especially during dispersal periods.

---

achieve this, including selecting sites at random (e.g. simple, cluster, stratified) or systematically. For example, a study evaluating if land bridges are more effective than culverts at facilitating wildlife movement will automatically be biased because land bridges are typically built where the road is in a cutting or at ground level, while culverts are more frequently installed where the road is already elevated (e.g. over drainage lines) and because a different suite of species of wildlife will occupy drainage versus upslope areas. One solution to this problem is to use experiments that explicitly identify and attempt to control this bias (Lesson 10.6, Textbox 10.1).

Well-replicated BACI designs may not always be possible (see limitations discussed in Roedenbeck et al. (2007)). However, regulatory and road agencies should always strive to ensure the studies and monitoring programmes they demand are conducted to the highest possible scientific standard.

10.5 Monitoring should be seen as an integral and valuable part of road projects

Unfortunately, monitoring typically happens at the end of a project, and it is often underfunded, undersupervised and poorly designed and therefore is unable to achieve what was specified in the conditions of approval. Instead, monitoring has been undertaken and is checked off on the list of duties, but the outcomes for improved understanding and knowledge are unclear. Rather, monitoring should be a collaboration between road agencies and researchers and be integrated into all stages of a road project, not just the last task before the project is finalised. Specifically, the need for post-construction monitoring and the questions to be answered should be identified at the earliest stages of a project to develop useful and important questions, establish an appropriate study design, secure sufficient funding and collect any necessary before data (Chapters 9, 15 and 16). Depending on the question, this may require several years of data collection before construction of the road or mitigation begins. Obviously, once construction begins, it is impossible to collect before data (see Chapter 14 for possible exceptions).

To improve research and monitoring, road agencies should engage staff with scientific training and understanding to supervise and collaborate on research projects. It is unrealistic to expect all regional or project-based staff to understand the intricacies of study design and statistical analysis and to be able to prepare sensible project briefs and adequately supervise research projects. Simple common mistakes include (i) awarding contracts to the cheapest bidder (often with inadequate sample size and insufficient statistical power of their study) and/or those with an inappropriate study design or techniques; and (ii)awarding the 'before' and 'after' phases of monitoring to different contractors which means that any change observed (or not) could be a result of the project or the change in personnel (Ginevan & Splitstone 2004). While it is preferable that these staff be employed by the road agency, they can also be engaged from elsewhere (e.g. natural resource agencies, universities). Similar limitations also hamper EIA for other types of projects and could be significantly improved (Ginevan & Splitstone 2004; Chapter 5).

10.6 Experiments investigating road impacts and mitigation effectiveness are an important way forward in road ecology and better management of roads

An experiment is a systematic approach to understanding how something works. It is systematic because we formulate hypotheses (or develop an understanding) of how something operates and then take measurements to determine which hypothesis was correct. Experiments can be opportunistic-testing the effect of natural or unpredictable disturbances (e.g. fire, flood), or designed-assessing the impact of a known or predictable disturbance (e.g. road construction, mitigation, tidal inundation). As explained in Lesson 10.4, research and monitoring can be manipulative or non-manipulative. For example, a simple manipulative experiment recently tested if the avoidance of roads by birds was due to traffic noise by propagating roadway sounds through speakers within a roadless area used as a stopover during migration (McClure et al. 2013; Fig. 10.3). Other confounding variables were either eliminated (i.e. visual disturbance of vehicles, pollution, habitat changes due to proximity of the road) or held constant (i.e. the timing, duration and intensity of roadway sound) at their impact site. The study concluded that many migratory birds avoided the stopover area due to noise. The conclusion from this experiment is

(A)   (B)

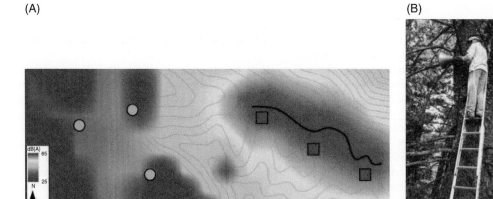

Figure 10.3 (A) Experimental set-up and background noise levels during the 'noise-on' phase of a manipulative experiment to test the effect of road noise on birds. The solid black line is a 500 m-long 'phantom road' (no paving, no traffic, just the loudspeakers) in a roadless area where migrating birds stop over in Idaho, United States. Road noise (higher db denoted in red, lower in blue) was propagated through 15 pairs of speakers (B, not shown on map) in 4-day intervals from 04:30 to 21:00, and the occurrence of birds was measured in the treatment areas (red squares) and control plots (green circles). Bird abundance was, on average, 22% lower at treatment plots and 12% higher at control plots, between noise-on and noise-off periods. Source: (A) Reproduced with permission of Christopher McClure and (B) Photograph by and reproduced with permission of Christopher McClure.

groundbreaking – for the first time, there is unequivocal data that noise is responsible (but not necessarily solely so) for the road-effect zone for birds.

However, most studies in road ecology are non-manipulative, where we collect data to investigate the effect of a disturbance that has happened or we know is about to happen, without modifying any of the parameters for the sake of improving the experiment.

The best monitoring programmes also involve an experiment, where confounding variables are removed or held constant while the parameter of interest is varied. For example, planners often ask about the minimum size of crossing structures to achieve a certain level of connectivity. If the experimental monitoring was incorporated into the mitigation, the size of the structure could be varied to see how the rate of use is affected. For example, underpass size could be experimentally varied by adding and removing a false wall or ceiling to existing structures. Depending on the question and the scale of the experiment, it may be necessary to explicitly incorporate the experiment into the design of the road or mitigation. For example, if 10 waterways of approximately equal size need to be crossed, and both a culvert and extended bridge would satisfy hydrological requirements, then half could be crossed with a bridge and the remainder with a culvert (Fig. 10.4) (or a bridge but its width would be temporarily reduced by the researchers). This example is 'experimenting with mitigation', because the mitigation explicitly sets out to test how the structure type affects the rate of use. The study would randomly allocate treatments to the pool of available sites and collect data before and after the treatment, as well as at control and impact sites. Control sites in this example are bridges whose sizes are not experimentally reduced.

10.7 Research and monitoring should be strategically planned and coordinated across jurisdictional boundaries

It is not possible to conduct well-designed BACI experiments or monitoring programmes on every project. Therefore, road agencies need to take a regional and coordinated approach to research and monitoring (e.g. Fraser et al. 2013). Road agencies usually operate within state or provincial boundaries and with smaller planning divisions, complicating the ability to communicate effectively across jurisdictional boundaries. If research and monitoring were centrally managed, it would be easier to coordinate research projects to increase the quality of the study design and increase sample size (and avoid duplication of research projects that cannot be compared to each other because their study design differs too much). With this approach, road agencies could more easily pool money for research across a number of projects and plan more comprehensive monitoring programmes that achieve far-superior outcomes. If each region or project continues to manage their own projects independently, they are not taking advantage of the potential benefits of pooling funds and study sites to achieve a more robust study design. The employment of a 'research and monitoring coordinator' or group that operates at a state, national or international level will allow road agencies to combine multiple projects that span across jurisdictional boundaries and infrastructure types (e.g. road and rail) or are completed at different times into a single study by ensuring the same methods are adopted. This approach is increasingly being employed, even on a global scale, called 'coordinated distributed

(A)

(B)

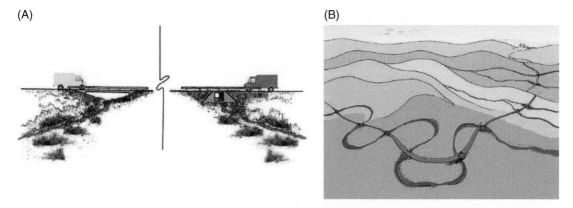

Figure 10.4 Experimenting with mitigation – a controlled, manipulative experiment to compare the use by wildlife of bridges (indicated with yellow trucks) versus culverts (red trucks) as detailed in (A). On a new road project, culverts and bridges are installed alternately (B), and the rate of use by wildlife can be measured and compared. Source: Reproduced with permission of Zoe Metherell.

experiments' (Fraser et al. 2013). With this approach, studies are conducted across the world, and because they all use standardised and controlled protocols, the meta-analyses (Lesson 10.8) are much more powerful and lead to greater insights.

10.8 The data and findings need to be accessible to relevant user groups e.g. scientists, planners and decision-makers, ideally also the public

Large-scale and cross-jurisdictional studies are not always feasible. Nevertheless, the data and a description of the theory being tested, the conceptual framework and methods must be accessible to planners, designers and researchers to facilitate meta-analyses and systematic reviews of existing data (Arnqvist & Wooster 1995; Pullin & Stewart 2006) to identify trends and develop theories (e.g. Fahrig & Rytwinski 2009). Data from earlier studies can also be added to current projects to increase the number of replicates or data points. For this to occur, road agencies must stipulate as a contractual agreement that the raw data, with the necessary metadata to describe it, be made available to relevant user groups within a certain period of time after the project has been completed. The number of publications in peer-reviewed scientific journals is a significant measure of a researcher's productivity, and researchers should be given sufficient time to publish their work, without unduly limiting the ability of others to use the data in a meta-analysis.

Unfortunately, many studies and EIAs remain on shelves or on computer hard drives and, as time progresses, become increasingly difficult to find and access. Consequently, the opportunity to learn from previous experiences is not capitalised. National or international databases with information about impacts, mitigation measures, monitoring programmes and results would be invaluable when designing future studies (to identify appropriate sites) and when conducting meta-analyses. These databases could be managed by the existing national or international road ecology networks (Chapter 61).

CONCLUSIONS

Road ecology is an applied discipline focusing on quantifying and mitigating the negative ecological effects of roads and traffic and other linear infrastructure. Research and monitoring have often been perceived by transport agencies as a hindrance to the core business of building and managing the infrastructure, but this perception should be corrected to realise the potential gains in knowledge, performance and cost savings from coordinated experimental research projects. The future of road ecology requires a greater level of genuine collaboration among those planning, designing, building and maintaining the infrastructure and those who conduct research and monitoring. For this to succeed, transportation agencies must adopt the mantra of incorporating best-practice science, including experiments, into best-practice mitigation. In addition, researchers must become more engaged with transportation agencies to design and undertake high-quality research and monitoring that provide practical and applied information that has relevance beyond the immediate question or problem.

ACKNOWLEDGEMENTS

Rodney van der Ree is supported by the Baker Foundation. We thank Jeff Houlahan, Lenore Fahrig and Scott Findlay for discussions that lead to some of the ideas in this chapter and Kylie Soanes for comments on an early draft.

FURTHER READING

Fraser et al. (2013): A review of the increasing role for coordinated distributed experiments in ecology, where standardised methods and controlled protocols allow for much stronger meta-analyses and hence insights into the impacts of roads and traffic and the effectiveness of mitigation.

van der Grift et al. (2013): Guidelines to measure the effectiveness of wildlife crossing structures and other forms of road mitigation for wildlife.

Lindenmayer and Likens (2010): A detailed and readable book about the challenges of conducting meaningful monitoring and how to ensure that monitoring projects deliver the most reliable information and do not become a waste of time and money.

Roedenbeck et al. (2007): A detailed description of the types of questions relevant to road ecology and the study designs that are available to answer those questions.

REFERENCES

Arnqvist, G. and D. Wooster. 1995. Meta-analysis: synthesizing research findings in ecology and evolution. Trends in Ecology and Evolution **10**:236–240.

Clevenger, A. P. and N. Waltho. 2000. Factors influencing the effectiveness of wildlife underpasses in Banff National Park, Alberta, Canada. Conservation Biology **14**:47–56.

Fahrig, L. and T. Rytwinski. 2009. Effects of roads on animal abundance: an empirical review and synthesis. Ecology and Society **14**:21. http://www.ecologyandsociety.org/vol14/iss1/art21/ (accessed 18 September 2014).

Forman, R. T. T., D. Sperling, J. A. Bissonette, A. P. Clevenger, C. D. Cutshall, V. H. Dale, L. Fahrig, R. France, C. R. Goldman, K. Heanue, J. A. Jones, F. J. Swanson, T. Turrentine and T. C. Winter. 2003. Road ecology. Science and solutions. Island Press, Washington, DC USA.

Fraser, L. H., H. A. Henry, C. N. Carlyle, S. R. White, C. Beierkuhnlein, J. F. Cahill Jr., B. B. Casper, E. Cleland, S. L. Collins, J. S. Dukes, A. K. Knapp, E. Lind, R. Long, Y. Luo, P. B. Reich, M. D. Smith, M. Sternberg and R. Turkington. 2013. Coordinated distributed experiments: an emerging tool for testing global hypotheses in ecology and environmental science. Frontiers in Ecology and the Environment **11**:147–155.

Ginevan, M. E. and D. E. Splitstone. 2004. Statistical tools for environmental quality measurement. Chapman & Hall/CRC, Boca Raton, FL USA.

Hellawell, J. M. 1991. Development of a rationale for monitoring. Pages 1–14 in F. B. Goldsmith, editor. Monitoring for conservation and ecology. Chapman & Hall, London UK.

Jaeger, J. A. G., J. Bowman, J. Brennan, L. Fahrig, D. Bert, J. Bouchard, N. Charbonneau, K. Frank, B. Gruber and K. T. von Toschanowitz. 2005. Predicting when animal populations are at risk from roads: an interactive model of road avoidance behavior. Ecological Modelling **185**:329–348.

Legg, C. J. and L. Nagy. 2006. Why most monitoring is, but need not be, a waste of time. Journal of Environmental Management **78**:194–199.

Lindenmayer, D. B. and G. E. Likens 2010. Effective ecological monitoring. CSIRO Publishing, Collingwood, Australia.

McClure, C. J. W., H. E. Ware, J. Carlisle, G. Kaltenacker and J. R. Barber. 2013. An experimental investigation into the effects of traffic noise on distributions of birds: avoiding the phantom road. Proceedings of the Royal Society of London – Series B: Biological Sciences **280**:20132290.

Pfister, H. P., D. Heynen, V. Keller, B. Georgii and F. Von Lerber. 1999. Häufigkeit und Verhalten ausgewählter Wildsäuger auf unterschiedlich breiten Wildtierbrücken (Grünbrücken). [Translation: Frequency and behavior of selected wild mammals on wildlife bridges (green bridges) of differing widths.] Schweizerische Vogelwarte Sempach, Sempach, Switzerland, 48 pages and Appendix.

Pullin, A. S. and G. B. Stewart. 2006. Guidelines for systematic review in conservation and environmental management. Conservation Biology **20**:1647–1656.

Roedenbeck, I. A., L. Fahrig, C. S. Findlay, J. E. Houlahan, J. A. G. Jaeger, N. Klar, S. Kramer-Schadt and E. A. van der Grift. 2007. The Rauischholzhausen agenda for road ecology. Ecology and Society **12**:11. http://www.ecologyandsociety.org/vol12/iss1/art11/ (accessed 18 September 2014).

Rytwinski, T., R. van der Ree, G. M. Cunnington, L. Fahrig, C. S. Findlay, J. Houlahan, J. A. G. Jaeger, K. Soanes, and E. A. van der Grift. 2015. Experimental study designs to improve the evaluation of road mitigation measures for wildlife. Journal of Environmental Management **154**:48–64.

Stoner, D. 1925. The toll of the automobile. Science **61**:56–57.

van der Grift, E. A., R. van der Ree, L. Fahrig, S. Findlay, J. Houlahan, J. A. G. Jaeger, N. Klar, L. F. Madriñan and L. Olson. 2013. Evaluating the effectiveness of road mitigation measures. Biodiversity and Conservation **22**:425–448.

Van der Ree, R., E. A. van der Grift, C. Mata and F. Suarez. 2007. Overcoming the barrier effect of roads – how effective are mitigation strategies? An international review of the effectiveness of underpasses and overpasses designed to increase the permeability of roads for wildlife. Pages 423–431 in C. L. Irwin, D. Nelson and K. P. McDermott, editors. International conference on ecology and transportation center for transportation and the environment. North Carolina State University, Raleigh, NC USA.

Chapter 11

FIELD METHODS TO EVALUATE THE IMPACTS OF ROADS ON WILDLIFE

Daniel J. Smith[1] and Rodney van der Ree[2]

[1]Department of Biology, University of Central Florida, Orlando, FL, USA
[2]Australian Research Centre for Urban Ecology, Royal Botanic Gardens Melbourne, and School of BioSciences, The University of Melbourne, Melbourne, Victoria, Australia

SUMMARY

Our understanding of the ecological impacts of roads and traffic, and indeed other linear infrastructure such as railways and utility easements, has burgeoned in the past two decades. These ecological effects are numerous and diverse and can extend for many kilometres beyond the road itself. The suite of survey techniques and study designs to quantify these effects is broad, and there are a number of important steps or key points to help ensure the results of surveys are reliable, collected in a cost-efficient manner, explanatory and inform management.

11.1 Formulating and articulating the research and monitoring questions is essential to designing relevant field surveys.

11.2 Locate and use existing studies and data whenever possible.

11.3 Study parameters are influenced by a number of interrelated and potentially competing demands.

11.4 Local- and landscape-level data are typically both necessary to comprehensively evaluate road impacts on wildlife.

11.5 There are many survey techniques available, and each has inherent biases, strengths and weaknesses.

11.6 Ensure high standards for the collection, management, analysis and reporting of data.

11.7 Several housekeeping issues are important for a successful study including personnel and resource management, funding and budgets, obtaining the necessary permits and legal obligations.

Despite increased knowledge, there are still many ecosystems and species for which we know little about their specific or general responses to roads and/or mitigation measures. Hence, there remains an urgent need for high-quality studies that tackle relevant questions and knowledge gaps. Making use of existing data and identifying the best and most appropriate methods for the collection of new data are essential to this endeavour.

Handbook of Road Ecology, First Edition. Edited by Rodney van der Ree, Daniel J. Smith and Clara Grilo.
© 2015 John Wiley & Sons, Ltd. Published 2015 by John Wiley & Sons, Ltd.
Companion website: www.wiley.com\go\vanderree\roadecology

INTRODUCTION

Monitoring and data collection are essential to avoiding, minimising, mitigating and offsetting the negative environmental effects of roads and traffic. The scale of studies is broad, and can range from short to long term, single or multiple roads, and may occur on old, new or proposed roads. In all cases, success is enhanced when a structured process is followed: (i) formulate appropriate research questions and study objectives; (ii) locate and use valid and reliable existing data; (iii) construct an effective study design and identify target species; (iv) select best methods for data collection; (v) analyse data using appropriate techniques; and (vi) publish the results and data.

Environmental impact assessments (EIAs; Chapters 6, 7 and 10) for many road projects typically involve data collection during the pre-construction phase to assess the likely ecological impacts of the proposed road. In contrast, most research or monitoring projects that quantify the impacts of existing roads or the success of mitigation are usually mandated as a condition of approval for a project or are initiated and conducted more independently, such as by academics, citizen scientists or governments, but outside the approval process for a major project. Impact studies encompass a range of methods and approaches, including (i) roadkill surveys; (ii) analyses of animal movements, such as road avoidance, barrier effect and increased overlapping of home ranges; (iii) population-level studies that assess population size or density, survival rates, sex ratios and reproductive output; and (iv) species occurrence and distribution. These studies occur at a range of spatial scales, from a single section of road to many roads. Most typically address effects on individual animals and occasionally populations, and recent reviews have recommended that studies need to be broadened to encompass communities, ecosystems and ecosystem processes (van der Ree et al. 2011; Chapter 10). The aims of this chapter are to highlight some important concepts and steps associated with field surveys and to identify and describe the range of field methods, including their pros and cons, to help road agencies and practitioners select the best field methods for the job. This chapter should be read in conjunction with Chapters 10 and 12–16, which provide more specific and critical information about study design considerations and applying research methods to evaluate the success of mitigation.

LESSONS

11.1 Formulating and articulating the research and monitoring questions is essential to designing relevant field surveys

The process of formulating the questions to be answered in research and monitoring is critically important. Lesson 10.3 explains how questions need to be explicitly stated to ensure the design and methods used to answer it are appropriate. It also highlighted that the questions posed by road agencies are often very species or location centric and that many benefits and cost savings would arise if questions were made broader to test ecological theories or hypotheses. We reiterate aspects of Lesson 10.3 and Chapter 10 here because they are fundamental to doing field-based research and monitoring. The two broad categories of field research in road ecology are related to (ii) actual or proposed road projects and (ii) general ecological research on the impacts of roads and traffic or effectiveness of mitigation – although the distinction between these two categories is often blurred. The research conducted for actual or proposed road projects is typically focused on:
• Identifying the species likely to be impacted by a road project and the type and severity of those impacts, often as part of an EIA (Chapters 5 and 6);
• Designing the mitigation strategy, including the type and location of mitigation required;
• Quantifying the success of mitigation (Chapters 15 and 16);
• Species that are rare or threatened, high profile or large enough to threaten human life if involved in wildlife-vehicle collision (WVC);
General ecological research is often broader than that associated with road projects and includes:
• A broad range of species, not necessarily rare, threatened or high profile;
• Multiple sites spread over large geographic areas;
• A greater focus on testing ecological theories and hypothesised regularities;
• Sometimes manipulative/experimental (see Chapter 10).
Identifying the key research questions is necessary to develop a set of objectives and choose appropriate survey methods. Identifying the knowns and unknowns within the study area narrows the focus of the questions and simplifies the data collection and analysis. To effectively evaluate the unknowns, the questions should be phrased as measurable or testable objectives or hypotheses.

11.2   Locate and use existing studies and data whenever possible

Before collecting new information, it is important to obtain any pre-existing studies or data on the study area and actual (or similar) species of interest and assess its usefulness. As explained in Lesson 10.8, the data and findings from research and monitoring need to be accessible, and with a bit of luck, there may be sufficient existing data to fully answer the question and no further fieldwork is required, thereby avoiding a duplication of effort. However, the most likely scenario is that the existing data will need (i) extra synthesis or analysis; (ii) can be used to inform the design of the new study; or (iii) provide additional data points to improve the conclusions that can be drawn (i.e. Fig. 10.1). The benefits of this approach are numerous, not the least of which are the potential cost savings and the improved reliability and accuracy of findings.

Useful studies and data are often difficult to find, principally because many road agencies do not require data or findings to be published. Their primary focus is to complete the impact study and keep the road project moving forward. Even when a relevant study is found, it may only contain summary information and the actual data may be hard to access or use (often due to incompatible data formats, copyright or intellectual property constraints or user restrictions). Relevant local data on wildlife movement and roadkill data are typically more difficult to find than regional- or landscape-scale data (e.g. biodiversity atlas data, land use, roads, hydrology, topography), and such Geographic Information System (GIS) data is often available online. Even so, site- and species-specific studies are increasingly more accessible through road agency websites, proceedings of road ecology conferences (Chapter 61), email and online discussion groups and internet searches of scientific literature.

11.3   Study parameters are influenced by a number of interrelated and potentially competing demands

If fieldwork is required, the next step is to design the study. Clearly beyond the scope of this single lesson, key decisions focus on (i) study class (manipulative or non-manipulative; Lesson 10.4); (ii) study type (i.e. BACI, CI, BA; Lesson 10.4); (iii) survey methods (Lesson 11.5); (iv) study duration; and (v) target species (Lesson 16.1). It is important to reconsider the original

question or hypothesis as each decision is being made to ensure the study remains relevant. Road agencies should collaborate with ecologists to ensure their questions and hypotheses address the specific problem or issue while remaining as broad and generally applicable as possible.

Key decisions on study design affect the scope of the project and breadth of species examined. The scope is defined by its spatial and temporal limits, which affects the amount of resources required to undertake the study. Spatial considerations include (i) the length and width of the existing (or proposed) road; (ii) landscape context; and (iii) variability in the amount, type and quality of adjacent habitat and the size of the road-effect zone (Fig. 1.1). Determining the relative value of different land use/vegetation types can narrow the focus of the study, eliminating the need to evaluate areas with little to no value as habitat for the target species. The timing of surveys should take into account the likely movements of the target species to maximise detectability. These movements may be predictable for species that migrate seasonally such as some amphibians (Chapter 31) or large ungulates (Chapters 42 and 56) or are active at certain times of the day (e.g. nocturnal species). The movements of other species are less predictable, such as nomadic species whose movements are triggered by rainfall or food availability. Selection of the species for study can be affected by its legal status, profile and risk to motorists (Chapters 15 and 16) and may include multiple species. Ideally, the species should represent those most impacted by the road, but this is not always possible (Lesson 16.1). The species selected for study will have a significant bearing on the types of data sought and methods of collection (Lesson 15.5).

11.4   Local- and landscape-level data are typically both necessary to comprehensively evaluate road impacts on wildlife

The suitability of an area as habitat for a species is influenced by local- and landscape-level variables, and data from both spatial scales are usually required to evaluate the direct and indirect effects of roads and traffic. Relevant local-scale factors include (i) biological data, for example, vegetation type and extent, habitat quality, species–habitat associations and predator–prey effects; (ii) physical data, for example, terrain, soil characteristics and water quality and quantity; and (iii) environmental attributes, for example, microclimate, lighting and noise levels. Landscape-scale data includes land cover/use, road-network configuration, soils, geology, topography and hydrology, and can often be sourced from maps and other publications, with digital data relatively easy to manipulate and analyse with GIS and mathematical models (Chapter 13). Roadway features include road width, surface type and number of lanes, as well as right-of-way characteristics, traffic speed and volume including seasonal and daily cycles, and any mitigation measures (Chapters 20 and 21).

11.5 There are many survey techniques available, and each has inherent biases, strengths and weaknesses

A wide range of methods have been used to assess the impacts of roads and traffic on wildlife (Table 11.1), and their suitability for different species and situations varies (Table 15.1). Numerous publications detail the application of these methods (e.g. NRC 2005; Silvy 2012; 'Further Reading'). Each method has its limitations and biases, and using multiple techniques will increase the precision and quality of information collected. In this lesson, we summarise the methods most commonly used to assess road impacts.

Roadkill surveys record the number, location and species of wildlife that are killed on the road due to wildlife-vehicle collisions (WVC) and are frequently used to identify locations of unsuccessful crossing attempts by wildlife, especially for those species that are easily detected after WVC (Fig. 11.1). If the landscape, road and traffic data are also collected at each WVC, the influence of these factors on collision rates can be identified and predictive roadkill hotspot models built (e.g. Santos et al. 2013; Chapter 13). Factors to consider include mode of survey (i.e. driving or walking), timing and frequency of surveys, study duration and observer safety (Clevenger et al. 2003; Chapter 12). Driving allows surveyors to cover more road length, but faster travel speeds will reduce the detectability of small animals (Lesson 54.3). Walking allows for more thorough detection but reduces the length of road that can reasonably be surveyed. Safety issues for both the observers and motorists must be considered, such as traffic speed and volume at the time of the survey, observer visibility to oncoming vehicles and verge width. Detecting seasonal differences would require sampling across multiple seasons. If only a limited survey period is practical, then surveys should be conducted when the target species are most active and likely to encounter roads, such as during migration, breeding or dispersal. Chapters 12

and 54 discuss specific considerations of roadkill surveys such as the removal of roadkill by scavengers, survey timing, frequency and duration, and Chapter 13 discusses techniques to analyse roadkill data. However, roadkill surveys do not identify locations of successful road crossings (see Textbox 35.1) nor where crossings are not being attempted (Seiler 2004; Coffin 2007). Also, roadkill data is not useful (except for predictive modelling; Chapter 13) in cases involving the siting of a new road corridor.

Surveying animal tracks on the roadside is a simple method that provides data on the location and frequency of successful and aborted attempts by wildlife to cross roads (e.g. Alexander et al. 2005), the direction of their movements and the location of sites for mitigation (e.g. Manley et al. 2004; Alexander 2008; Smith 2012). It can also be complementary to roadkill surveys on the same stretch of road and is particularly applicable when significant lengths of road (i.e. many kilometres; Smith 2006) are being evaluated. Track imprints, usually footprints, enable quantification of crossing frequency at the species or species group level but are not useful in identifying particular individuals (e.g. Smith 2006). Track beds, made up of naturally occurring substrates (Fig. 11.2A) or specially laid tracking plots (Fig. 15.1), are set up parallel to and as close to the road as possible (Smith 2006; Hardy & Huijser 2007). Suitable materials for recording track imprints include snow, sand, gypsum or other soft, smooth substrates; the best soils are loamy (i.e. a blend of mostly sand and silt and less clay) that hold imprints under wet and dry conditions. Local environmental conditions affect imprint quality and persistence, and rain, floods, snow and wind can erase tracks (Fig. 15.3B) – frequent inspections may be necessary. For short-term assessments (e.g. a few months or less), three or more times per week may be warranted, while inspections once or twice weekly may be adequate for long-term (one or more years) studies. Training is necessary to identify tracks and the method can be labour intensive. Vegetation regrowth in the plot can be a maintenance problem and may require periodic tilling or herbicide. If many kilometres of road are being monitored, it can be costly to prepare and maintain.

Camera traps automatically detect and record wildlife and are deployed for variable periods of time in the field, usually weeks to months (Fig. 11.3). Most cameras use sensors that are triggered by animal movement or body heat and record photographs or short videos. When used in combination with track and roadkill surveys, camera traps can be used to identify the location and frequency of successful and

Table 11.1 Generalised comparison of the characteristics of the main survey methods to evaluate road impacts on wildlife.

Method	Data type	Animal handling	Spatial extent	Resolution/ scale	Complexity	Effort	Cost	Reference
Roadkill surveys	Point	No	Small–large	Fine	Low	Moderate	Low	Chapters 12 and 13
Animal tracks[a]	Point, line	No	Small–large	Fine	Low	Moderate	Low–moderate	Silvy (2012)
Camera traps	Point	No	Small–medium	Medium	Low	Low	Low	Silvy (2012)
Wildlife census: observational	Point, line, area	No	Small–medium	Medium	Low	Moderate	Low	Sutherland (2006)
Wildlife census: interventional	Point	Yes	Small	Medium	Moderate	High	High	Silvy (2012), Willson and Gibbons (2009)
Animal tracking[a]	Point, line, area	Yes	Small–medium	Medium	High	Low–high	Low–high	Silvy (2012)
Genetics	Point	Yes/no	Small–large	Medium	High	Moderate	Moderate	Chapter 14
Landscape/GIS models	Point, line, area	No	Large	Coarse	High	Low	Low	Chapters 13 and 36

[a]Surveys of animal tracks are the reading and recording of animal footprints. Tracking surveys are where the movement path of an animal is followed.

Figure 11.1 Roadkill black bear being recorded, including its sex, age, GPS coordinates, location on road, lane direction, date and time. Source: Photograph by D.J. Smith.

(A)   (B)

Figure 11.2 (A) Preparation of a track bed parallel to the road using naturally occurring soil. The track bed should be wide enough to ensure the target species will leave several footprints when crossing. (B) Red wolf imprint. Source: Photographs by D.J. Smith.

unsuccessful road crossings (Smith 2012). Camera traps are commonly used to monitor wildlife crossing structures and other mitigation (Chapter 15). They are also useful in determining population density of large mammals (Carbone et al. 2001; Rowcliffe et al. 2008). In India, camera traps were used to show that chital, gaur and elephants avoided roads with high traffic density (Gubbi et al. 2012). Further, video cameras can reveal behavioural response of individual animals when approaching or crossing roads (Clevenger & Huijser 2011). Unique markings can help identify

specific individuals (Mendoza et al. 2011; O'Connell et al. 2011) or sex/age classes (Soanes et al. in review); otherwise, photographs only provide data to the species or genus level. Cameras can now be deployed for long periods of time without checking or maintaining them, especially if solar powered and when using a mobile phone network to transmit images to an office. In comparison to track surveys, camera traps were more precise in species identification, whereas track surveys were more beneficial for quick surveys on a limited budget (Lyra-Jorge et al. 2008). Limitations of this

(A)

(B)

Figure 11.3 (A) Trail camera attached to guardrail and (B) Florida panther photographed on a camera trap crossing the road at night. Source: Photographs by D.J. Smith.

technology include the short functional range of the sensor (maximum 30 m) and limited ability to detect animals that are small and slow moving or whose external body temperature is similar to ambient; however, camera trap technology is continually improving. Given the limited range of the sensors, camera trap surveys can be costly when applied over many kilometres of roads because many cameras are needed. Large numbers of images, including false triggers, may be generated, and the cost of staff time to download and review images needs to be considered – however, volunteers or automated systems may reduce these costs. Theft and vandalism can also be a significant problem and may require special precautions at additional expense.

Wildlife census is a general term describing a variety of techniques used to enumerate populations, and numerous books have been written about this (e.g. Sutherland 2006; Silvy 2012). A census can be **observational**, where researchers are able to identify species and enumerate the population with minimal interference to the wildlife (e.g. bird counts, spotlighting, camera trapping, aerial surveys, hair tubing), or **interventional**, where researchers physically capture and handle or otherwise disturb the wildlife during the process (e.g. most trapping techniques). For species that are readily visible or vocal and easily recognised (e.g. birds), observational methods are applicable. Interventional methods (Fig. 11.4) are more applicable for species that are cryptic and difficult to detect (e.g. amphibians, small mammals) or if additional information is required.

In road-impact studies, these methods are primarily used to census populations (e.g. species abundance and

diversity) and evaluate road avoidance and crossing frequency. For instance, observational surveys showed that the occurrence of particular species of breeding birds was lower near busy roads, which was attributed to impacts of traffic noise (Reijnen & Foppen 2006). The arrangement of survey points along the road and at control plots is determined by the specific question or aims of the study, and survey points may be distributed randomly, stratified or uniformly along a transect or grid (Sutherland 2006; Smith et al. 2015a).

While observational techniques reduce the amount of stress on individual animals, they typically provide less data on the age, sex and health of individual animals than trapping techniques. Capture–mark–recapture involves marking individuals for later recognition if recaptured or resighted, which is valuable for estimating population size or density, recruitment, survival and potentially relocation (Sutherland 2006; Silvy 2012). Markings can be temporary or permanent (e.g. leg bands, ear tags, tattoos or microchips), and the choice of method depends on the target species and duration of the study. In trapping studies, special consideration should be given to animal care and welfare, marking techniques, capture efficiency of different trap types and frequency of trap inspections (Willson & Gibbons 2009; Silvy 2012). Baiting is often used in trapping surveys but may introduce bias into the data set by providing attractive food sources (Sutherland 2006). Trapping is usually more labour intensive and costly than observational techniques.

Animal tracking includes a range of sophisticated and simple techniques to detect animal movements and use of space. Radio-tracking involves fitting

Figure 11.4 Examples of common interventional methods for surveying wildlife: (A) drift fence and trapping array for capturing small vertebrates, (B) funnel trap with eastern diamondback rattlesnake, (C) pitfall trap with barking tree frog, (D) bobcat in live cage trap.

individual animals with a radio transmitter, often on a collar around their neck, glued to their skin or shell or surgically implanted (Figs. 11.5 B, C, D and 38.1) and then tracked using a radio receiver. Tracking can be done on foot (Fig. 11.5A), from a vehicle or aircraft or via stationary towers, and animals can be located via triangulation or by homing in (White & Garrott 1990). Radio-tracking is used to describe (i) home range size and shape; (ii) habitat preferences; (iii) movement paths; and (iv) distances and routes travelled for foraging, dispersal and migration (Manley et al. 2004; Silvy 2012). Depending on which individuals within a population are tracked, this information can refer to the whole population or specific age or sex classes (Silvy

2012). In road-impact studies, they can also locate potential and actual road crossings as well as avoidance behaviour (Chapter 36; Rouse et al. 2011). For example, telemetry studies demonstrated one species of snake and two turtle species crossed roads less often than expected by chance (Shepard et al. 2008). Numerous telemetry studies have examined movement patterns of animals in relation to roads (e.g. Rondinini & Doncaster 2002; Tigas et al. 2002; Dickson et al. 2005; Chapter 36). This information is valuable in not only assessing impacts but also for identifying locations for mitigation.

The two most common tracking devices are very-high-frequency (VHF) radio transmitters and Global Positioning System (GPS) satellite transmitters (Fig. 11.5).

(E)

(F)

(G)

(H)

Figure 11.4 (*Continued*) (E) harp trap to catch insectivorous bats, (F) insectivorous bat in harp trap, (G) hair tube to collect hair samples and (H) Elliott trap on a bracket to catch small arboreal mammals. Source: (A, B, C and D) Photographs by D.J. Smith and (E, F, G and H) Photographs by R. van der Ree.

VHF equipment is relatively inexpensive but requires intensive field monitoring. GPS equipment is more costly upfront but automated and programmable and can employ remote data collection options (Guthrie 2012). The quality of the data from both systems is affected by the strength of the transmitter, precision of the antenna and receiver, frequency at which locations or fixes are collected and user error (Silvy 2012). To estimate movement pathways or road-crossing locations, fixes must be collected frequently, which may be at 15 minute intervals or less, depending on the speed of the target species (Dickson et al. 2005). When describing animal home ranges, fixes can be collected less frequently, such as daily or weekly. Transmitter lifespan is dependent on battery size and signal pulse rate (i.e. the number of pulses per minute). Transmitters for small bats often have a battery life of just 10 days, compared to many

Figure 11.5 Examples of animal tracking methods and equipment: (A) field technician using a radio-telemetry receiver and antenna to locate an animal's transmitter signal, (B) gopher tortoise with glue-on radio transmitter, (C) common brushtail possum under sedation being fitted with VHF transmitter and (D) black bear with GPS collar recovering from anaesthesia at release. Source: (A, B) Photographs by D.J. Smith, (C) Photograph by R. van der Ree and (D) Photograph by and used with permission of Mike Orlando..

years for larger animals. This limits the long-term study potential on smaller animals without frequent recaptures, which is difficult for some species and more stressful for those that can be easily recaptured. Therefore, users must specify the pulse rate (and hence transmitter life) that maximises the collection of data necessary to answer the research questions. Tracking technology is continually developing – becoming smaller and more powerful and with longer battery life. For example, data loggers can be automated to detect and record transmitters that move within a certain range, which is useful when recording crossings through an underpass. The duty cycle of some transmitters can be programmed to change when different behaviours are expected (e.g. resting or active, at night or during migration) to save battery life.

Other tracking methods can be useful to identify movements and habitat use when radio or satellite tracking is not a suitable option. These methods include (i) dusting animals with fluorescent powder and following the powder trail they leave (e.g. McDonald & St. Clair 2004; Graeter & Rothermel 2007); (ii) releasing an animal with a spool and line attached and following the thread which unravels (e.g. Boonstra & Crane 1986; Dodd 2002); and (iii) snow tracking (e.g. Bellis 2008). These techniques provide data on movement paths and habitat use but are labour intensive and generally limited to short durations and movement distances by individuals (Furman et al. 2011).

Genetics is a relatively new frontier as a method for detecting road impacts on wildlife and is comprehensively explained in Chapter 14. Major advances in DNA analysis techniques and the emergence of landscape genetics have proven it a valuable approach to assessing population-level effects of roads (Storfer et al. 2007),

(A)   (B)   (C)

Figure 11.6 A barbed-wire hair snare adjacent to the road can identify highway crossing locations by black bears: (A) double strand attached to a row of trees, (B) attached to the top of a guardrail and (C) tuft of black bear hair caught on a barb. Source: Photographs by D.J. Smith.

and we recommend incorporating it into all future studies. As an example, barbed-wire strung parallel to roads in Florida, North Carolina and Virginia, United States, was used to obtain hair from black bears for DNA analysis (Fig. 11.6); results identified locations and frequency of road crossings by specific individuals (Wills 2008; Vaughan et al. 2011 Smith et al. 2015b). DNA identification can reveal locations, number of individuals and sex ratio of animals crossing roads (Sawaya & Clevenger 2010) and evaluate barrier effects and fragmentation of populations (Proctor et al. 2005; Textbox 14.2). For example, microsatellites were used to evaluate the effect of road size on increased genetic differentiation in red-backed salamanders (Marsh et al. 2008). Low-cost methods exist for genetic sampling, including from road-kills, hair traps, scats and shed skin from reptiles, and can be coupled with other already occurring trapping surveys and marking techniques (e.g. ear notching; Fig. 14.4). Laboratory work (e.g. DNA processing) can appear expensive but is comparable and often more cost-efficient than many field techniques, and it is important to use care in collection and storage to preserve sample quality for subsequent processing in the laboratory (Chapter 14).

11.6 Ensure high standards for the collection, management, analysis and reporting of data

Important parameters to consider regarding data collection include (i) identifying the appropriate number of treatment and control sites needed and their location; (ii) setting applicable timelines and data collection efforts to obtain sufficient data to perform statistical analysis; and (iii) minimising collection bias by employing standardised field methods.

Three different scenarios exist for road-impact studies (see Chapters 15 and 16 for study designs that evaluate mitigation measures): single site, multiple sites or a continuous stretch of road (Smith et al. 2015a). For a single location, the effort is limited to one treatment. Selecting two or three control sites similar to the treatment site allows for a comparative analysis. For multiple impact sites, a similar number of control sites should be selected. While evaluations of a single site only describe what is occurring at that location, studies on multiple sites measure the size of the impact and variance across all sites. For appropriate comparison, control sites should be similar in character to treatment sites (e.g. road and habitat features, topography). Finally, for long stretches of a single road, subsampling (e.g. 100 m stretches repeated every 2 km) is appropriate. Randomly selected sample sites should represent all habitat types and significant natural features (e.g. stream corridors) associated with the target species. This provides the ability to make inferences regarding variability in impacts to wildlife across habitat types or other significant natural features (see Chapter 10 for more detail on study design, sampling and replication).

The sampling intensity and duration should increase as the level of variability and complexity in the study area increases (Smith et al. 2015a). More survey effort is necessary to detect road impacts and change over time when results are highly variable. Pilot surveys that estimate spatial and temporal variation in wildlife activity should occur before the monitoring is designed. These surveys can yield long-term benefits in efficiency by identifying appropriate sampling frequency and duration. For situations with moderate to high variation, power analysis (e.g. Gibbs et al. 1998) or computer simulation (e.g. Rhodes & Jonzén 2011) is useful in identifying the ideal number and seasonal timing of surveys. In practice, actual frequency of data collection may depend as much on logistics (i.e. available time and resources to perform surveys) as on statistical power. It is critical that studies have sufficient statistical power to detect an impact to ensure the impacts of the road are comprehensively identified and effective mitigation is developed.

Comprehensive road-impact studies may take place over multiple years, and consequently, multiple staff are likely to be involved in surveys, data analysis and reporting. Standardised protocols should be adopted to ensure consistent data collection over time because multiple observers can be a major source of variation (e.g. Sauer et al. 1994) due to differences in detection and identification abilities, interpretation of written protocols and decision-making. Standard forms with easy-to-follow data keys and instructions that outline specific information to be collected and how it should be recorded are required (see Heyer 1994; Wilson et al. 1996; McDiarmid et al. 2012).

11.7 Several housekeeping issues are important for a successful study including personnel and resource management, funding and budgets, obtaining the necessary permits and legal obligations

The majority of this chapter has focused on the field-based considerations of conducting road-impact studies. There are numerous 'housekeeping' issues that must also be considered and addressed to ensure that field studies are successful. First, the project scope must be comprehensive to ensure sufficient resources including staff, equipment and funding are available. The amount of funding will affect the choice of methods and the frequency and duration of data collection. Importantly, the road project should include sufficient funds to adequately assess the impacts – if insufficient, it may be necessary to scale back the scope to match the available budget. However, a threshold in funding required to undertake the study exists, below which the quality and reliability of the research are so compromised that it is no longer worth attempting.

Second, researchers will need permits from relevant wildlife and/or environmental agencies if the study involves any handling of animals or occurs in protected areas. Approval from an animal ethics committee may also be required, ensuring animals are handled caringly (Silvy 2012). Finally, the parameters of the study may be defined by legal requirements, mutual agreements or land use or transportation plans. Most major road projects require review and approval from relevant levels of government that environmental impacts have been adequately avoided, minimised or mitigated.

CONCLUSIONS

Commonly used field methods to quantify the impact of roads include roadkill surveys, population censuses, tracking of animal movements and analyses of gene flow. Time, effort and expense can be saved by using existing data and the best and most appropriate methods for new data collection to identify specific impacts on wildlife and to design mitigation. Several steps are important to ensure the data is efficiently collected and the results are reliable, explanatory and inform management. The methods and concepts discussed in this chapter are integrally connected to several sections in this book, most notably research monitoring and maintenance, impacts and mitigation, fauna and landscape issues and regional issues. These sections contain chapters which provide additional information about study design and research methods, including specific examples and applications.

ACKNOWLEDGEMENTS

Rodney van der Ree is supported by the Baker Foundation.

FURTHER READING

Dodd (2009): A manual of amphibian ecology and conservation that includes a practical review of field monitoring techniques used to survey and study amphibians.

Graeter et al. (2013): Describes and summarises sampling techniques for censusing and monitoring reptile and amphibian populations.

McDiarmid et al. (2012): A comprehensive guide to survey techniques to study reptiles, including a detailed discussion of each method and data analysis techniques.

Sutherland (2006): A detailed yet practical book describing how to plan, implement and analyse the results of field surveys for plants, invertebrates, reptiles, mammals, fish, birds and amphibians.

Sutherland et al. (2004): A manual of the ecology and conservation of birds that includes a practical review of field monitoring techniques.

Wilson et al. (1996): A complete volume on approaches to investigations of mammalian populations. It discusses study design, survey planning, field methods and statistical techniques.

REFERENCES

Alexander, S.M. 2008. Snow-tracking and GIS: Using multiple species-environment models to determine optimal wildlife crossing sites and evaluate highway mitigation plans on the Trans-Canada Highway. The Canadian Geographer **52**:169–187.

Alexander, S.M., N.M. Waters and P.C. Pacquet. 2005. Traffic volume and highway permeability for a mammalian community in the Canadian Rocky Mountains. The Canadian Geographer **49**:321–331.

Bellis, M.A. 2008. Evaluating the effectiveness of wildlife crossing structures in southern Vermont. Masters Thesis. University of Massachusetts, Amherst.

Boonstra, R. and I.T.M. Crane. 1986. Natal nest location and small mammal tracking with a spool and line technique. Canadian Journal of Zoology **64**:1034–1036.

Carbone, C., S. Christie, K. Conforti, et al. 2001. The use of photographic rates to estimate densities of tigers and other cryptic mammals. Animal Conservation **4**:75–79.

Clevenger, A.P. and M.P. Huijser. 2011. Wildlife crossing structure handbook: Design and evaluation in North America. Report No. FHWA-CFL/TD-11-003. Federal Highway Administration, Washington, DC.

Clevenger, A.P., B. Chruszcz and K.E. Gunson. 2003. Spatial patterns and factors influencing small vertebrate fauna roadkill aggregations. Biological Conservation **109**:15–26.

Coffin, A.W. 2007. From roadkill to road ecology: A review of the ecological effects of roads. Journal of Transport Geography **15**:396–406.

Dickson, B.G., J.S. Jenness and P. Beier. 2005. Influence of vegetation, topography, and roads on cougar movement in southern California. Journal of Wildlife Management **69**:264–276.

Dodd, Jr., C.K. 2002. North American box turtles: A natural history. University of Oklahoma Press, Norman, OK.

Dodd, Jr., C.K., ed. 2009. Amphibian ecology and conservation: A handbook of techniques. Oxford University Press, Oxford, 464 pp.

Furman, B.L.S., B.R. Scheffers and C.A. Paszkowski. 2011. Use of fluorescent powdered pigments as a tracking technique for snakes. Herpetological Conservation and Biology **6**:473–478.

Gibbs, J.P., S. Droege and P. Eagle. 1998. Monitoring populations of plants and animals. BioScience **48**:935–940.

Graeter, G.J. and B.B. Rothermel. 2007. The effectiveness of fluorescent powdered pigments as a tracking technique for amphibians. Herpetological Review **38**:162–165.

Gubbi, S., H.C. Poomesha and M.D. Madhusudan. 2012. Impact of vehicular traffic on the use of highway edges by large mammals in a South Indian wildlife reserve. Current Science **102**:1047–1051.

Graeter, G.J., K.A. Buhlmann, L.R. Wilkinson and J.W. Gibbons, eds. 2013. Inventory and monitoring: Recommended techniques for reptiles and amphibians. Partners in Amphibian and Reptile Conservation, Technical Publication IM-1, Birmingham, AL.

Guthrie, J.M. 2012. Modeling movement behavior and road crossing in the black bear of south central Florida. Masters Thesis. University of Kentucky, Lexington.

Hardy, A.R. and M.P. Huijser. 2007. US 93 preconstruction wildlife monitoring field methods handbook. Final Report, Project FHWA/MT-06-008/1744-2. Federal Highways Administration, Washington, DC and Montana Department of Transportation, Helena. Available from: http://www.mdt.mt.gov/other/research/external/docs/research_proj/wildlife_crossing/field_handbook.pdf (accessed 5 May 2014).

Heyer, W.R. 1994. Measuring and monitoring biological diversity: Standard methods for amphibians. Smithsonian Institution Press, Washington, DC.

Lyra-Jorge, M.C., G. Ciocheti, V.R. Pivello and S.T. Meirelles. 2008. Comparing methods for sampling large- and medium-sized mammals: Camera traps and track plots. European Journal of Wildlife Research **54**:739–744.

Manley, P.N., W.J. Zielinski, M.D. Schlesinger and S.R. Mori. 2004. Evaluation of a multiple-species approach to monitoring species at the ecoregional scale. Ecological Applications **14**:296–310.

Marsh, D.M., R.B. Page, T.J. Hanlon, R. Corritone, E.C. Little, D.E. Seifert and P.R. Cabe. 2008. Effects of roads on patterns of genetic differentiation in red-backed salamanders, *Plethodon cinereus*. Conservation Genetics **9**:603–613.

McDiarmid, R.W., M.S. Foster, C. Guyer, J.W. Gibbons and N. Chernoff. 2012. Reptile diversity: Standard methods for inventory and monitoring. University of California Press, Berkeley, CA.

McDonald, W. and C.C. St. Clair. 2004. Elements that promote highway crossing structure use by small mammals in Banff National Park. Journal of Applied Ecology **41**:82–93.

Mendoza, E., P.R. Martineau, E. Brenner and R. Dirzo. 2011. A novel method to improve individual animal identification based on camera-trapping data. Journal of Wildlife Management **75**:973–979.

National Research Council (NRC). 2005. Assessing and managing the ecological impacts of paved roads. National Academies Press, Washington, DC.

O'Connell, A.F., J.D. Nichols and K.U. Karanth. 2011. Camera traps in animal ecology: Methods and analysis. Springer, New York.

Proctor, M.F., B.N. McLellan, C. Strobeck and R.M.R. Barclay. 2005. Genetic analysis reveals demographic fragmentation of grizzly bears yielding vulnerably of small populations. Proceedings of the Royal Society B **272**:2409–2416.

Reijnen, R. and R. Foppen. 2006. Impact of road traffic on breeding bird populations. In The ecology of transportation: Managing mobility for the environment, J. Davenport and J.L. Davenport, eds. Springer, Dordrecht, pp. 255–274.

Rhodes, J.R. and N. Jonzén. 2011. Monitoring temporal trends in spatially-structured populations: How should sampling effort be allocated between space and time? Ecography **34**:1040–1048.

Rondinini, C. and C.P. Doncaster. 2002. Roads as barriers to movement for hedgehogs. Functional Ecology **16**:504–509.

Rouse, J.D., R.J. Willson, R. Black and R.J. Brooks. 2011. Movement and spatial dispersion of *Sistrurus catenatus* and *Heterodon platirhinos*: Implications for interactions with roads. Copeia **2011**(3):443–456.

Rowcliffe, J.M., J. Field, S.T. Turvey and C. Carbone. 2008. Estimating animal density using camera traps without the need for individual recognition. Journal of Applied Ecology **45**:1228–1236.

Santos, S.M., R. Lourenco, A. Mira and P. Beja. 2013. Relative effects of road risk, habitat suitability, and connectivity on wildlife roadkills: The case of tawny owls (*Strix aluco*). PLoS One **8**:1–11.

Sauer, J.R., B.G. Peterjohn and W.A. Link. 1994. Observer differences in the North American Breeding Bird Survey. The Auk **111**:50–62.

Sawaya, M.A. and A.P. Clevenger. 2010. Using non-invasive genetic sampling methods to assess the value of wildlife crossings for black and grizzly bear populations in Banff National Park. Pages 42–55 in The Proceedings of the 2009 International Conference on Ecology and Transportation, P.J. Wagner, D. Nelson and E. Murray, eds. Center for Transportation and Environment, North Carolina State University, Raleigh, NC.

Seiler, A. 2004. Trends and spatial patterns in ungulate-vehicle collisions in Sweden. Wildlife Biology **10**:301–313.

Shepard, D.B., A.R. Kuhns, M.J. Dreslik and C.A. Philips. 2008. Roads as barriers to animal movement in fragmented landscapes. Animal Conservation **11**:288–296.

Silvy, N.J. 2012. The wildlife techniques manual: Volume 1: Research. Volume 2: Management, 7th edition. Johns-Hopkins University Press, Baltimore, MD, 1136 pp.

Smith, D.J. 2006. Ecological impacts of SR 200 on the Ross Prairie ecosystem. Pages 380–396 in Proceedings of the 2005 International Conference on Ecology and Transportation, C.L. Irwin, P. Garrett and K.P. McDermott, eds. Center for Transportation and the Environment, North Carolina State University, Raleigh, NC.

Smith, D.J. 2012. Determining location and design of cost effective wildlife crossing structures along US 64 in North Carolina. Transportation Research Record **2270**:31–38.

Smith, D.J., K.E. Gunson, D.M. Marsh and S. Tonjes. 2015a. Monitoring road effects and mitigation measures, and using adaptive management approaches. Ch. 12 In Roads and ecological infrastructure: Concepts and applications for small animals, K.M. Andrews, P. Nanjappa and S.P.D. Riley, eds. Johns-Hopkins University Press, Baltimore, MD.

Smith, D.J., M.K. Grace, H.R. Chasez and M.J.W. Noss. 2015b. State Road 40 Pre-Construction Wildlife Movement Monitoring: Areas A, B and F. Final Report, Contract No. BDK78, TWO #501-3. Florida Department of Transportation, District Five, Deland, FL. 90 pp. + appendices.

Soanes, K., P. Vesk and R. van der Ree. In review. Monitoring the use of road-crossing structures by arboreal marsupials: Insights gained from motion-triggered cameras and passive integrated transponder (PIT) tags. Wildlife Research.

Storfer, A., M.A. Murphy, J.S. Evans, C.S. Goldberg, S. Robinson, S.F. Spear, R. Dezzani, E. Delmelle, L. Vierling and L.P. Waits. 2007. Putting 'landscape' in landscape genetics. Heredity **98**:128–142.

Sutherland, W.J. 2006. Ecological census techniques: A handbook, 2nd edition. Cambridge University Press, Cambridge, 446 pp.

Sutherland, W.J., I. Newton and R. Green. 2004. Bird ecology and conservation: A handbook of techniques. Oxford University Press, Oxford, 408 pp.

Tigas, L.A., D.H. Van Vuren and R.M. Sauvajot. 2002. Behavioral responses of bobcats and coyotes to habitat fragmentation and corridors in an urban environment. Biological Conservation **108**:299–306.

van der Ree, R., J. A. G. Jaeger, E. A. van der Grift and A. P. Clevenger. 2011. Effects of roads and traffic on wildlife populations and landscape function: Road ecology is moving toward larger scales. Ecology & Society **16**:1–9.

Vaughan, M.R., M.J. Kelly, and J.A. Trent. 2011. Evaluating potential effects of widening US 64 on the Black Bear Population of Alligator River National Wildlife Refuge, Dare County, North Carolina. Final Report. VT-NCDOT Contract No. MA-2009-02. 82 pp.

White, G.C. and R.A. Garrott. 1990. Analysis of wildlife radio-tracking data. Academic Press, San Diego, CA.

Wills, J. 2008. DNA-based hair sampling to identify road crossings and estimate population size of black bears in Great Dismal Swamp National Wildlife Refuge, Virginia. MS Thesis. Virginia Tech, Blacksburg, VA.

Willson, J.D. and J.W. Gibbons. 2009. Drift fences, coverboards, and other traps. Pages 229–245 in Amphibian ecology and conservation: A handbook of techniques, C.K. Dodd, Jr., ed. Oxford University Press, Oxford.

Wilson, D.E., F.R. Cole, J.D. Nichols, R. Rudran and M.S. Foster. 1996. Measuring and monitoring biological diversity: Standard methods for mammals. Smithsonian Institution Press, Washington, DC.

Chapter 12

CASE STUDY: A ROBUST METHOD TO OBTAIN DEFENDABLE DATA ON WILDLIFE MORTALITY

Éric Guinard[1], Roger Prodon[2] and Christophe Barbraud[3]

[1]Cerema-DTer Sud-Ouest/DAIT/GBMN, Saint-Médard-en-Jalles, France
[2]Ecologie et Biogéographie des Vertébrés (E.P.H.E.), Centre d'Ecologie Fonctionnelle & Evolutive, Montpellier, France
[3]Centre d'Etudes Biologiques de Chizé, CNRS, Institut Ecologie et Environnement, Villiers-en-Bois, France

INTRODUCTION

The impact of road mortality on animal populations is an important conservation issue (Chapters 1 and 28). Detecting the key factors involved in wildlife mortality due to wildlife-vehicle collision (WVC) – a prerequisite to propose effective mitigation measures – requires robust and unbiased estimates of the number of animals killed. Various methods are currently used (e.g. Lesson 54.3), making it difficult to compare the results from different studies (Erritzøe et al. 2003). However, few studies take into account biases that influence mortality estimates.

Counts of wildlife roadkill from a slow-moving vehicle enable long sections of road to be surveyed. However, such surveys have lower detection rates than those by foot (Erritzøe et al. 2003), especially on roadside verges (Guinard et al. 2012).

Three parameters need to be quantified to reduce bias in surveys of roadkill due to WVC: (i) The persistence probability of a carcass (the probability that the carcass was not removed from the road between two consecutive counts); (ii) the entry probability of a carcass (the probability that a new carcass appears on the road between two consecutive counts); and (iii) the probability of detecting the carcass. The persistence rate is mainly affected by the activity of scavengers and by the destruction of carcasses by vehicles (Fig. 12.1). Other factors that also affect persistence and entry probabilities include species characteristics, carcass age, position on the road (i.e. on median, traffic lane or verge), traffic volume and scavenger abundance (Ponce et al. 2010; Guinard et al. 2012). The detectability of a carcass is affected by several factors, including the ability of fieldworkers, carcass characteristics, survey method, weather conditions and traffic volume.

Handbook of Road Ecology, First Edition. Edited by Rodney van der Ree, Daniel J. Smith and Clara Grilo.
© 2015 John Wiley & Sons, Ltd. Published 2015 by John Wiley & Sons, Ltd.
Companion website: www.wiley.com\go\vanderree\roadecology

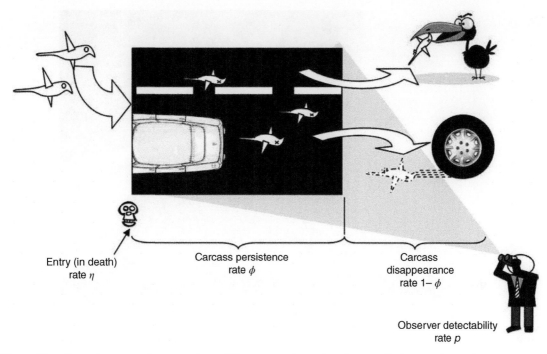

Figure 12.1 Successive steps in the status of a wildlife carcass on the road, and its possible detection by an observer. An animal enters the population of carcasses with a probability of entry η when it is hit and killed by a vehicle. The carcass persists between successive counts within a session with a probability of persistence Φ. If the carcass persists, it may be detected by an observer with probability p. The carcass may disappear with a probability $1 - \Phi$. Source: E. Guinard.

TESTING THE METHOD

Four sections of motorway in southwest France, totalling 169 km, were surveyed for 2.5 days each season for 4 years (i.e. 10 days surveyed annually). All vertebrate carcasses detected along the roads were recorded and data analysed, but only birds are reported here in this chapter. Each section of motorway had relatively similar climatic and landscape conditions and shared a similar avifauna, but had different traffic volumes (two sections had average annual daily traffic volumes of 35,000 vehicles per day and the remaining two with 7000–8000 vehicles per day). Each survey session consisted of five successive counts (two per day and one on the morning of the third day) to estimate carcass detectability, entry and persistence rates. Counts were made from a car travelling at 40–50 km per hour on the emergency lane (Fig. 12.2), with a driver and always with the same observer.

During the last afternoon of each 2.5-day session, a count by foot was made by the same observer walking on verges, on randomly selected sub-sections about

10 km in length. The position of each carcass on the road, verge or median was recorded.

During each count we recorded all carcasses, identified them to species where possible and individually marked them by painting a mark on the pavement (Figs. 12.3 and 12.4), allowing us to distinguish between new deaths and those that remained from the previous sessions. We also recorded the location (± 10 m) and age (fresh (<3 days) or old (>3 days; Figs. 12.3 and 12.4) of each carcass.

To estimate carcass persistence and the entry and detection probabilities, we analysed the detection histories of 512 carcasses using the Cormack–Jolly–Seber model (Lebreton et al. 1992) and POPAN data type (Schwarz & Arnason 1996) in Program MARK 5.1 (White & Burnham 1999). Such analyses can also be done with the program E-SURGE (Choquet et al. 2009). Only fresh carcasses were taken into account when calculating probability of entry.

Comparing the number of carcasses counted from the vehicle (the less accurate method) and by foot (the reference method) allowed us to calculate the detectability of the carcasses.

Figure 12.2 Barn owl found on motorway emergency lane in southwest France. Source: Photograph by E. Guinard.

Figure 12.3 Fresh carcass of blackcap on motorway in southwest France. Source: Photograph by E. Guinard.

Figure 12.4 Old carcass of Phasianidae marked on the pavement of a motorway in southwest France. Source: Photograph by E. Guinard.

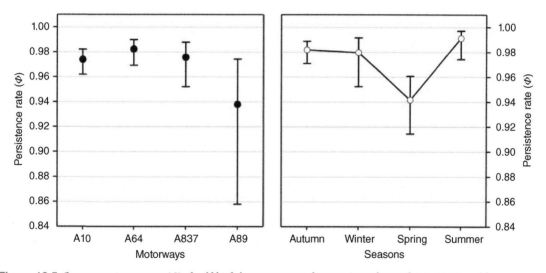

Figure 12.5 Carcass persistence rate (Φ) of wild birds by motorway and season (error bars indicate ±95% confidence interval). Source: Guinard et al. (2012). Reproduced with permission of Elsevier.

THE IMPORTANCE OF INCORPORATING CARCASS DETECTABILITY AND PERSISTENCE INTO ESTIMATES OF ROADKILL

Counts by car and by foot provided similar estimates of carcass numbers, except on the verges where the numbers of carcasses recorded from a car represent only 6% of the counts by foot (Guinard et al. 2012). Carcass persistence differed among motorways and among seasons, being higher during summer and lower during spring (Fig. 12.5), mainly due to variation in scavenger activity. Old and large carcasses such as owls persisted longer compared to fresh and small carcasses such as passerines. Carcass detectability was constant over time. Taking into account persistence and detection probabilities, the uncorrected carcass numbers recorded during a 2.5-day session were underestimated by 10% for owls and 30% for passerines. Overall, surveys by vehicles underestimated mortality for all birds by 33%, compared to walking surveys.

Average barn owl mortality rates estimated from the uncorrected counts of carcasses recorded from a vehicle were 15% lower than when the entry, persistence and detection rates were accounted for (1.88 ± 0.44 ($\pm$s.e.) vs 2.24 ± 0.06 individuals per kilometre per year). The underestimate was 36% for the European robin (Guinard 2013).

CONCLUSIONS

This study showed that not taking into account the detection and persistence rates of the carcasses leads to an underestimate of the rate of mortality due to WVC. The persistence of carcasses is lower on the road during breeding season (spring), probably due to higher activity of scavengers. Carcass persistence is higher for old carcasses presumably because they are less desirable to scavengers.

To overcome the inconsistencies in estimates of roadkill that result from the different census methods currently used, a standardised method should be followed to minimise bias and make data more comparable. Standardised survey analyses must take into account carcass characteristics, detectability, persistence and entry rates to obtain unbiased estimates of roadkills. These rates can be estimated by performing several counts by car within a short period of time, with trained observers working in good weather conditions. We recommend that surveys occur monthly in 2-day sessions with at least four counts per session. Counts from vehicles can be made on long sections (>20 km) at the landscape scale, but shorter foot-based surveys are necessary to correct the bias in vehicle surveys. Comparing robust mortality estimates with population abundance estimates in neighbouring areas and relating them to habitat variables (road profile, proximity to wildlife crossing structures, vegetation structure and traffic measurements) can help explain, predict and effectively mitigate the mortality of wildlife on roads.

ACKNOWLEDGEMENTS

We thank DGITM/DIT/GRN/ARN5 – a Department of Transportation of the French Ministry of Ecology, Sustainable Development and Energy – for funding.

REFERENCES

Choquet, R., L. Rouan, and R. Pradel. 2009. Program E-SURGE: a software application for fitting multi-event models. In: Modeling demographic processes in marked populations. Springer USA. pp. 845–865.

Erritzøe, J., Mazgajski, T. D. and L. Rejt. 2003. Bird casualties on European roads-a review. Acta Ornithologica **38**(2): 77–93.

Guinard, É. 2013. Infrastructures de transport autoroutières et avifaune: les facteurs influençant la mortalité par collision (PhD thesis). École Pratique des Hautes Études, Paris.

Guinard, É., Julliard, R. and C. Barbraud. 2012. Motorways and bird traffic casualties: carcasses surveys and scavenging bias. Biological Conservation **147**: 40–51.

Lebreton, J. D., Burnham, K. P., Clobert, J. and D. R. Anderson. 1992. Modelling survival and testing biological hypotheses using marked animals: a unified approach with case studies. Ecological Monographs **62**: 67–118.

Ponce, C., Alonso, J. C., Argandoña, G., Garcia Fernández, A. and M. Carrasco. 2010. Carcass removal by scavengers and search accuracy affect bird mortality estimates at power lines. Animal Conservation **13**(6): 603–612.

Schwarz, C. J. and A. N. Arnason. 1996. A general methodology for the analysis of open-model capture recapture experiments. Biometrics **52**: 860–873.

White, G.C. and K. P. Burnham. 1999. Program MARK: survival estimation from populations of marked animals. Bird Study **46** Supplement: 120–138. http://warnercnr.colostate.edu/~gwhite/mark/mark.htm (accessed 19 September 2014).

Chapter 13

ROAD–WILDLIFE MITIGATION PLANNING CAN BE IMPROVED BY IDENTIFYING THE PATTERNS AND PROCESSES ASSOCIATED WITH WILDLIFE-VEHICLE COLLISIONS

Kari Gunson[1] and Fernanda Zimmermann Teixeira[2]

[1]Eco-Kare International, Peterborough, Ontario, Canada
[2]Post Graduate Program in Ecology, Bioscience Institute, Federal University of Rio Grande do Sul, Porto Alegre, Brazil

SUMMARY

Collisions between vehicles and wildlife impact human safety and wildlife conservation. Transportation planners are increasingly involved in planning and implementing road-wildlife mitigation measures to lessen the risk of wildlife-vehicle collision (WVC) as well as provide connectivity opportunities for safe wildlife movement. An understanding of where, when and why WVC occur is essential to avoid high-risk areas and design effective mitigation measures.

13.1 Information about when, where and why WVC occur along roads can be used to inform where mitigation would be most effectively placed to reduce WVC.

13.2 Global Positioning Systems are essential for the rapid and accurate collection of large volumes of WVC data for use in mitigation planning.

Handbook of Road Ecology, First Edition. Edited by Rodney van der Ree, Daniel J. Smith and Clara Grilo.
© 2015 John Wiley & Sons, Ltd. Published 2015 by John Wiley & Sons, Ltd.
Companion website: www.wiley.com\go\vanderree\roadecology

13.3 There are numerous methods available to identify where and when WVC hotspots and hot moments are located along roads that can instruct mitigation planners.

13.4 When WVC data is not available, models can be used to predict WVC hotspots and hot moments; however, more rigorous study designs are required for application to mitigation planning.

13.5 There are several inexpensive and accessible tools that have been developed to measure when, where and why WVC occur.

There are many tools available to assist transportation planners and decision-makers in determining the location of mitigation measures for wildlife. These tools use empirical data to calculate hotspots and hot moments of WVC along roads. When empirical data is not available, predictive models can be applied to roads that have similar road and landscape conditions as the modelled site.

INTRODUCTION

Animals move through the landscape for a variety of reasons and often interact with roads, traffic and other linear infrastructure. There is a risk of a collision with a vehicle if the animal attempts to cross the road, potentially resulting in injury or death (roadkill) to the animals and/or occupants of the vehicle. The rate of wildlife-vehicle collisions (WVC) has been increasing globally, and the number of WVC with deer in the United States that resulted in fatalities of motorists has increased from 131 in 1994 to 223 in 2007 (www.deercrash.com). The loss of wildlife from WVC is substantial (Chapter 28) and is one of the main human-caused sources of wildlife mortality (Forman & Alexander 1998). Furthermore, WVC are expensive, costing Americans an estimated US$8 billion annually in property damage and health-care costs (Huijser et al. 2008; Chapter 42).

The location and timing of WVC are influenced by the location of the road in the landscape, traffic volume and vehicle speed (see review in Gunson et al. (2011)). Identifying spatial (hotspots) and temporal (hot moments; see Beaudry et al. (2010)) patterns of WVC and understanding the factors that influence their occurrence are essential to avoiding high-risk areas and designing effective mitigation measures. In this chapter, we discuss methods that are often used to measure where, when and why WVC occur along roads as well as the application of these methods to mitigation planning. It is important to distinguish between the use of WVC and roadkill data as opposed to where wildlife successfully cross roads in mitigation planning, because sometimes different factors, such as traffic volume, influence whether animals cross a road safely or not (Fig. 13.1) (Clevenger & Ford 2010; Neumann et al. 2012).

LESSONS

13.1 Information about when, where and why WVC occur along roads can be used to inform where mitigation would be most effectively placed to reduce WVC

Transportation planners aim to lessen the impacts of WVC on wildlife populations and increase road safety for motorists. An understanding of where and when WVC occur will inform the placement and design of mitigation measures on existing roads, and being able to predict the location of WVC along proposed roads will allow planners to avoid high-risk areas. For example, large ungulates may cross roads anywhere along a 1–2 km section of road, requiring mitigation that spans the entire section of road to prevent roadkill (e.g. moose; Krisp & Durot 2007; see also Chapter 56). In contrast, WVC may occur within several hotspots, each 100 m in length, for species with low mobility or specialised habitat requirements (e.g. some species of turtles), requiring localised mitigation at each hotspot (e.g. Langen et al. 2012). These patterns inform where mitigation measures, including the number of crossing structures and associated length of fencing, would be most effectively placed.

Often, temporary mitigation measures, such as intermittent road closures, speed reductions, wildlife warning signs and awareness campaigns, can be used to decrease the rate of WVC. Understanding when WVC occur will inform the timing and duration of when this type of mitigation will be most effective. Warning signs strategically placed when wildlife are crossing roads during seasonal movements are more effective than permanent signs, because motorists are less likely to habituate to their message (Sullivan et al. 2004; Gunson & Schueler 2012; Chapter 24). Furthermore, because road closures and speed reductions are not desirable for

(A)

(B)

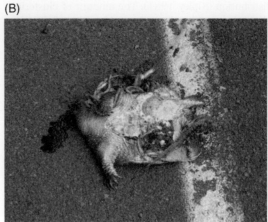

Figure 13.1 (A) Not all crossings of roads by wildlife end in collision and mortality. However, turtles, in this case common snapping turtle (B), often have high rates of roadkill. Source: (A) Photograph by and reproduced with permission of Paul L. Clark and (B) Photograph by Kari Gunson.

motorists or transportation planners (Chapters 37 and 52), these measures are typically only implemented for limited time periods.

13.2 Global positioning systems are essential for the rapid and accurate collection of large volumes of WVC data for use in mitigation planning

The location accuracy of WVC data is extremely important because it determines the confidence in its use to place mitigation measures along roads (Gunson et al. 2009). Some studies in North America and Europe have used over 30 years of WVC data collected by natural resource, police and transportation agencies to detect patterns in WVC occurrence (e.g. Nielsen et al. 2003; Seiler 2005). Frequent limitations in the data are that WVC location is typically only collected for large animals and its spatial error can vary from 800 to 6500 m, when locations are referenced to the closest road distance marker or landmark, respectively (Gunson et al. 2009).

The timing and location of WVC can now be accurately recorded with Global Positioning System (GPS) technology included with cellular phones and digital cameras. Location information can be uploaded to a centralised database as part of a citizen science awareness or research project or entered into specific online databases (e.g. Textbox 50.1, Chapter 62). Other studies have integrated personal digital devices with GPS technology for road

maintenance crews (Ament et al. 2011) and truckers (Hesse et al. 2010) to collect WVC data.

An advantage of these initiatives is the potential to collect accurate and abundant WVC data for a broader range of species. However, the ease of using a GPS device has led to an explosion of uncoordinated efforts among academic and citizen science projects. Data is collected to meet the needs of each project; however, it is rarely integrated into centralised databases for use by transportation agencies for mitigation planning. Furthermore, reliability and accuracy of the data need to be established before it can be used for mitigation planning (Chapter 12, Chapter 62). Coordinated programmes accompanied with education and awareness campaigns have the potential to maximise the accuracy and amount of WVC data collected for large and small species to inform mitigation planning.

13.3 There are numerous methods available to identify where and when WVC hotspots and hot moments are located along roads that can instruct mitigation planners

There are several methods used in road ecology to quantify the aggregation (or clustering) and distribution of WVC along roads. First, the Ripley's K technique measures the aggregation or clustering of roadkill along a road and whether it is statistically significant, that is, differs spatially from a random distribution (Ripley 1981). The amount of clustering along the road can be measured in units of distance (e.g. meter) and is often expressed as a peak distance to indicate the scale at which clustering occurs (Fig. 13.2). For example, Langen et al. (2012) found turtle-vehicle collisions were most clustered along 250 m-long sections of highway.

Once that clustering has been found to be statistically significant, a logical next step is to determine where this clustering occurs. Kernel density estimation (detailed in Bailey and Gatrell (1995)) is the analysis most often used, and it measures where along the road an aggregation of WVC occurs. The estimation requires a user-defined search distance to calculate the density of WVC along a specific road segment (e.g. Ramp et al. 2005; Krisp & Durot 2007; Mountrakis & Gunson 2009). Defining the search distance (length of road) to measure density can be guided by the objectives of the study (Krisp & Durot 2007) and by the biological movement scale of the target species (Ramp et al. 2005). Additionally, by conducting the Ripley's K analysis first, a significant clustering distance can be used to inform the search distance.

A disadvantage of kernel density estimation is that it usually does not include a measure of statistical significance in the available software. A HotSpot Identification analysis recently developed by Coelho et al. (2014) compares the density of observed WVC along the road with a simulated Monte Carlo random distribution. Confidence intervals (similar to the Ripley's K analysis) are used to determine significance, and clustering is considered significant when the density of observed WVC is above the upper confidence interval, thereby indicating where mitigation should be prioritised (Fig. 13.3).

In addition to being clustered in space, WVC may also be clustered in time (e.g. during seasonal migrations) and can be referred to as 'hot moments' (Beaudry et al. 2010). The same methods described earlier in this lesson to determine the spatial distribution of WVC can also be used to evaluate *when* WVC are aggregated (Mountrakis & Gunson 2009). The difference between measuring hot moments and hotspots is that WVC are plotted along a defined timeline (period) rather than along a length of road. To further illustrate this, in the kernel density analysis, the search distance is defined as a period of time relevant to when WVC occur (e.g. season). Space and time both influence the occurrence of WVC independently but can also interact to increase the risk of a collision (Mountrakis & Gunson 2009).

13.4 When WVC data is not available, models can be used to predict WVC hotspots and hot moments; however, more rigorous study designs are required for application to mitigation planning

WVC models are advantageous because they can predict likely roadkill patterns and the need for mitigation planning on proposed roads or on roads without data on WVC or wildlife movement. WVC-based models are typically applied to a species or group of species impacted by a road or road network. Relevant landscape- and road-related factors are grouped as independent variables into multivariate models, and the dependent variable is typically the number or presence/absence of WVC that have occurred along a road segment (Gunson et al. 2011). Landscape factors include anthropogenic land use, wildlife habitat and terrain, which influence animal distribution, abundance and movement patterns (e.g. Malo et al. 2004). Road factors such as traffic volume, road alignment, motorist visibility and road grades also influence the risk of WVC (e.g. Seiler 2005).

Unfortunately, there are few published examples that have applied the results of WVC models to mitigation planning on new or existing roads. One reason is that model validation is often conducted with WVC data collected from the same study area where the model was developed (but see exception in Seiler (2005)). These validation techniques are data driven and of limited application outside the study area because the reliability and scalability of model results are unknown.

Another reason that lessens model applicability for mitigation planning is that the more intuitive factors such as species-specific habitat are routinely modelled in different landscapes to explain where WVC occur. To build on what is already expected, more integral spatial relationships such as type, shape, size or configuration of a species preferred habitat with respect to roads should be included in models (Gunson et al. 2011). Selection of factors relevant to specific road–wildlife mitigation projects can be improved with preliminary consultation among transportation planners, engineers and ecologists before model development.

Statistically significant models that include confounding and interacting variables provide mixed results and render interpretation and application to mitigation planning difficult. For example, it is difficult for a transportation planner to know whether clearing roadside vegetation will decrease WVC because motorist visibility is increased or be counterproductive

Textbox 13.1 Using free software to conduct spatial analyses of WVC along roads.

Coelho et al. (2012) explored the spatial patterns of frog and toad roadkill along a 4.4 km section of a two-lane highway in southern Brazil. This section of road neighbours a peri-urban reserve, the Itapeva State Park in the Atlantic Forest Biosphere Reserve. This protected area has high ecosystem diversity and high species richness of frogs and toads (28 species). A total of 1333 frogs and toads from 13 species and 6 families were found dead on the road during 18 months of road surveys conducted by foot (Coelho et al. (2012)). The data is summarised using a combination of Siriema software (Coelho et al. 2014; Lesson 13.5) and a spatial analysis tool for ArcGIS software (SANET; http://sanet.csis.u-tokyo.ac.jp; Okabe et al. 2006) to determine the spatial distribution and density of frog and toad roadkill.

A Ripley's K analysis (a plot of the L statistic) shows that WVC are clustered more than expected by chance when the black line is above the upper confidence limit (grey line) (Fig. 13.2). In this case, the rate of frog and toad roadkill was greater than expected between 0 and 4.24 km along the road and less than expected between 4.25 and 4.41 km, and peak clustering occurred at approximately 1.8 km. In other words, the roadkill is aggregated on road segments ranging from 0 to

4.24 km in size. This information can be used to both inform the search distance in an analysis to identify the location of hotspots along the road and plan the scale at which mitigation is required along the road length.

The result of kernel density analysis performed with the SANET tool is plotted in Figure 13.3 and shows highest-density aggregations of roadkill in red and lowest density in yellow. A search distance (bandwidth) of 50 m was chosen for this kernel analysis because (i) the Ripley's K analysis identified that roadkill was significantly clustered at a spatial scale of 50 m, (ii) 50 m is relevant to the movement scale of frogs and toads and (iii) 50 m is an appropriate scale for implementing mitigation measures. The results of the kernel density analysis are plotted on the road with a land-use layer to aid in interpretation (Fig. 13.3).

Last, we present the results from Coelho et al. (2012) who conducted a HotSpot Identification analysis to supplement the kernel density estimation used (Fig. 13.4). When the intensity of roadkill (black line) is above the upper confidence limit (grey line), then that road segment has more collisions than expected by chance. When using a search distance of 50 m, they found the highest frequency of frog and toad deaths occurred at approximately 1.5 and 3.2 km along the

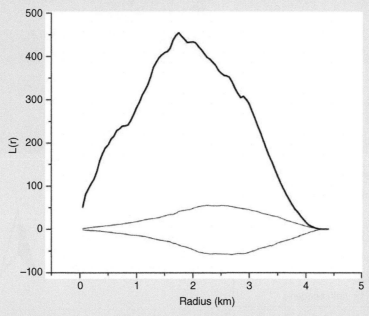

Figure 13.2 L statistic (K observed − K simulated mean) as a function of scale distance (radius) and 90% confidence limits (grey lines) for frog and toad roadkills along a 4.4 km road section in southern Brazil. Source: Coelho et al. (2012). Reproduced with permission of Elsevier.

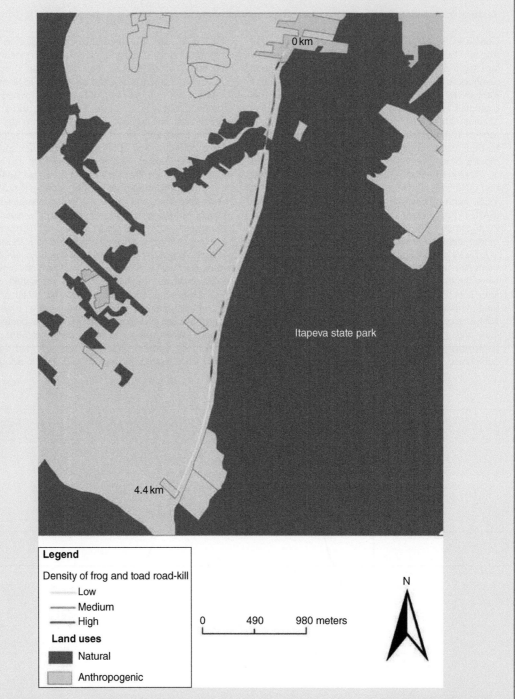

Figure 13.3 Kernel density estimates of frog and toad roadkill along a highway adjacent to Itapeva State Park in the Atlantic Forest Biosphere Reserve, Brazil. Light grey shading is human-modified areas and darker areas are more natural habitat. Source: Fernanda Zimmermann Teixeira.

road. Similar results were also found in the SANET analysis (Fig. 13.2), and the HotSpot Identification analysis showed that these aggregations were significant. These peaks are obvious locations to prioritise more effective, localised and spatially explicit mitigation measures.

Collectively, these results can guide the scale of mitigation planning as well as where to place mitigation measures along a road. The majority of the road length has more roadkills than expected, and because the highest densities occurred at approximately 1.5 and 3.2 km, mitigation measures, such as crossing structures, could be focused at these locations. Other practical suggestions along the road length include a reduction in speed limits and temporary road closures during high crossing events that can be predicted by seasonal weather patterns.

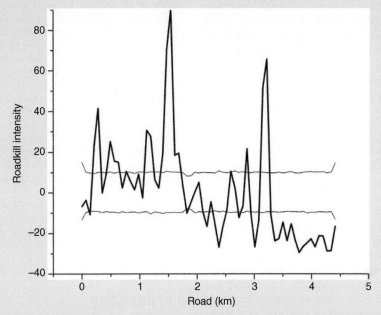

Figure 13.4 Anuran roadkill intensity (black line) and 99% confidence limits (grey lines) along a 4.4 km road section in southern Brazil. Source: Coelho et al. (2012). Reproduced with permission of Elsevier.

because ungulates are now attracted to roadsides for foraging (Gunson et al. 2011; Chapters 42 and 46). Study designs that control for interacting variables within models need to be more widely used (Chapter 10), especially when the objectives are for application to road mitigation planning projects.

13.5 There are several inexpensive and accessible tools that have been developed to measure when, where and why WVC occur

Once WVC data is collected, it is essential to plot, visualise and explore where they occur in relation to the landscape (Chapter 62). Google Earth (http://www.google.com/earth/index.html) is a free and useful tool that can be used to plot and visualise WVC on aerial imagery. These data can also be imported into a Geographic Information System (GIS) to conduct more sophisticated analyses. A commonly used commercial software is ArcGIS (ESRI 2011); however, a free alternative is Quantum GIS (QGIS; http://www.qgis.org/).

Free software tools that perform spatial and temporal analyses are continually evolving and improving. In 2001, researchers used software developed to analyse spatial patterns of crime (CrimeStat; Levine 2000) (http://www.icpsr.umich.edu/CrimeStat/download.html) to evaluate spatial hotspots of WVC along roads in Canada (Clevenger et al. 2001). Soon after, more sophisticated spatial analysis toolkits, such as Ripley's K statistics, were developed with use in a GIS (e.g. SANET; Okabe et al. 2006) to identify spatial patterns of plants and animal occurrence along roads (Clevenger et al. 2003; Spooner et al. 2004; Ramp et al. 2005). More recently, researchers have programmed spatial and temporal analysis tools, for example, Ripley's K in programming software such as Matlab 7.1 that determine spatiotemporal patterns of WVC along roads (Mountrakis & Gunson 2009).

Of particular note is Siriema, a free software package developed with a user-friendly interface that evaluates spatial distribution of WVC along roads (www.ufrgs.br/siriema). Analyses can be conducted along a road by first straightening it or by using the road as is and considering its sinuosity (Coelho et al. 2014). The software performs two options: a Ripley's K function (see Coelho et al. 2008) and also a HotSpot Identification analysis (Lesson 13.3 and Textbox 13.1).

CONCLUSIONS

An understanding of when, where and why WVC occur is essential to inform management, retrofit mitigation on existing roads, avoid high-risk areas when building new roads and install mitigation on new or proposed roads. Wildlife mitigation research on roads has come a long way over the past 30 years and is benefitting from the rapid development of new tools and techniques. User-friendly tools such as Siriema software can substantially improve and facilitate integration of science into practical mitigation solutions for wildlife on roads.

The challenge ahead lies in improving the collection of WVC data and developing models that can be applied to mitigation planning. It is not practical or feasible to assume that accurate and systematic WVC data can be collected on all existing and newly planned roads, especially when wildlife populations are declining. As a result, there is a need for more rigorous models with predictive capabilities that can be applied to new roads. The integration of research with transportation planning requires a multidisciplinary approach that includes transportation planners, biologists and engineers through all planning stages of transportation projects.

ACKNOWLEDGEMENTS

We acknowledge Andreas Kindel, Igor P. Coelho and Jochen G. Jaeger for earlier discussion on concepts and Artur V.P. Coelho and Giorgos Mountrakis for assistance with programming the spatial analysis tools presented here. We thank Coordenação de Aperfeiçoamento de Ensino Superior (CAPES) and Eco-Kare International for funding FZT and KG, respectively, to write this chapter.

FURTHER READING

Coelho et al. (2012): This paper presents a more detailed description of the methods and results used for the Ripley's K and the HotSpot analysis presented in the case study.

Coelho et al. (2014): This is the user's guide for Siriema, a freely available software that outlines the Ripley's K test and the HotSpot Identification analysis presented in the case study.

Gunson et al. (2011): Provides a thorough review of research studies that have created WVC hotspot models with a discussion of improved methodologies to meet the aims and objectives of mitigation planning.

Mountrakis and Gunson (2009): An in-depth study that combined both space and time to look at patterns and distributions of wildlife-vehicle collisions temporally and along roads.

Okabe et al. (2006): A user's guide for the ArcGIS extension Spatial Analysis along Networks (SANET), which includes a set of analysis to evaluate spatial patterns along networks, including network K functions and kernel density estimation.

REFERENCES

Ament, R., D. Galarus, D. Richter, K. Bateman, M. Huijser and J. Begley. 2011. Roadkill Observation Collection System (ROCS): Phase III Development. A report prepared for the US Highway Administration and the Deer Vehicle Crash Information and Research Center.

Bailey, T.C. and A.C. Gatrell. 1995. Interactive Spatial Data Analysis. Longman Scientific and Technical, Essex.

Beaudry, F., P.G. Demaynadier and M.L. Hunter Jr. 2010. Identifying hot moments in road-mortality risk for freshwater turtles. Journal Wildlife Management **74**: 152–159.

Clevenger, A.P. and A.T. Ford. 2010. Wildlife crossing structures, fencing and other highway design considerations. Pages 17–49 *in* J.P. Beckmann, A.P. Clevenger, M.P. Huijser and J.A. Hilty, editors. Safe Passages: Highways, Wildlife and Habitat Connectivity. Island Press, Washington, DC.

Clevenger, A.P., B. Chruszcz and K.E. Gunson. 2001. Highway mitigation fencing reduces wildlife-vehicle collisions. Wildlife Society Bulletin **29**: 646–653.

Clevenger, A.P., B. Chruszcz and K.E. Gunson. 2003. Spatial patterns and factors influencing small vertebrate fauna road-kill aggregations. Biological Conservation **109**: 15–26.

Coelho, I.P., A. Kindel and A.V.P. Coelho. 2008. Road-kills of vertebrate species on two highways through the Atlantic Forest Biosphere Reserve, southern Brazil. European Journal of Wildlife Research **54**: 689–699.

Coelho, A.V.P., I.P. Coelho, F.Z. Teixeira and A. Kindel. 2014. Siriema: Road Mortality Software. User's Guide V.2.0. NERRF, UFRGS, Porto Alegre. Available at www.ufrgs.br/siriema (accessed 20 September 2014).

Coelho, I.P., F.Z. Teixeira, P. Colombo, A.V.P. Coelho and A. Kindel. 2012. Anuran road-kills neighboring a peri-urban reserve in the Atlantic Forest, Brazil. Journal of Environmental Management **112**: 17–26.

Environmental Systems Research Institute (ESRI). 2011. ArcGIS Desktop: Release 10. ESRI, Redlands, CA.

Forman, R.T.T. and L.E. Alexander. 1998. Roads and their major ecological effects. Annual Review on Ecology and Systematics **29**: 207–231.

Gunson, K.E. and F.W. Schueler. 2012. Effective placement of road mitigation using lessons learned from turtle crossing signs in Ontario. Ecological Restoration **30**: 329–334.

Gunson, K.E., A.P. Clevenger, A. Ford, J.A. Bissonette and A. Hardy. 2009. A comparison of data sets varying in spatial accuracy used to predict the occurrence of wildlife-vehicle collisions. Environmental Management **44**: 268–277.

Gunson, K.E., G. Mountrakis and L. Quackenbush. 2011. Spatial wildlife-vehicle collision models: a review of current work and its application to transportation mitigation projects. Journal of Environmental Management **92**: 1074–1082.

Hesse, G., Rea, R.V., Klassen, N. Emmons, S., and Dickson, D. 2010. Evaluating the potential of the Otto® Wildlife GPS device to record roadside moose and deer locations for use in wildlife vehicle collision mitigation planning. Wildlife Biology Practice **6**: 1–13.

Huijser, M.P., P. McGowen, J. Fuller, A. Hardy, A. Kociolek, A.P. Clevenger, D. Smith and R. Ament. 2008. Wildlife-Vehicle Collision Reduction Study: Report to Congress. Final Report. Western Transportation Institute, Montana State University, Bozeman, MT. 254p.

Krisp, J.M., S. Durot. 2007. Segmentation of lines based on point densities – an optimisation of wildlife warning sign placement in southern Finland. Accident Analysis & Prevention **39**: 38–46.

Langen, T., K.E. Gunson, C.A. Scheiner and J.T. Boulerice. 2012. Road mortality in freshwater turtles: identifying causes of spatial patterns to optimize road planning and mitigation. Biodiversity and Conservation **21**: 3017–3034. doi:10.1007/s10531-012-0352-9.

Levine, N. 2000. CrimeStat: A Spatial Statistics Program for the Analysis of Crime Incident Locations. Ned Levine & Associates, Annandale, VA and National Institute of Justice, Washington, DC.

Malo, J., F. Suárez and A. Díez. 2004. Can we mitigate animal–vehicle accidents using predictive models? Journal of Applied Ecology **41**: 701–710.

Mountrakis, G. and K.E. Gunson. 2009. Multi-scale spatiotemporal analyses of moose-vehicle collisions: a case study in northern Vermont. International Journal of Geographic Information Science **23**: 1389–1312.

Neumann, W., G. Ericsson, H. Dettki, N. Bunnefeld, N. Keuler, D. Helmers and V. Radeloff. 2012. Difference in spatiotemporal patterns of wildlife road-crossing and wildlife-vehicle collisions. Biological Conservation **135**: 70–78.

Nielsen, C.K., R.G. Anderson and M.D. Grund. 2003. Landscape influences on deer vehicle accident areas in an urban environment. Journal of Wildlife Management **67**: 46–51.

Okabe, A., K. Okunuki and S. Shiode. 2006. The SANET toolbox: new methods for network spatial analysis. Transactions in GIS **10**: 535–550.

Ramp, D.J., J. Caldwell, K.A. Edwards, D. Warton and D.B. Croft. 2005. Modelling of wildlife fatality hotspots along the snowy mountain highway in New South Wales, Australia. Biological Conservation **126**: 474–490.

Ripley, B.D. 1981. Spatial Statistics. John Wiley & Sons, Inc., New York.

Seiler, A. 2005. Predicting locations of moose–vehicle collisions in Sweden. Journal of Applied Ecology **42**: 371–382.

Spooner, P.G., I.D. Lunt, A. Okabe and S. Shiode. 2004. Spatial analysis of roadside *Acacia* populations on a road network using the network K-function. Landscape Ecology **19**: 491–499.

Sullivan, T.L., A.E. Williams, T.A. Messmer, L.A. Hellinga and S.Y. Kyrychenko. 2004. Effectiveness of temporary warning signs in reducing deer vehicle collisions during mule deer migrations. Wildlife Society Bulletin **32**: 907–915.

Chapter 14

INCORPORATING LANDSCAPE GENETICS INTO ROAD ECOLOGY

Paul Sunnucks[1] and Niko Balkenhol[2]

[1]School of Biological Sciences, Monash University, Melbourne, Victoria, Australia
[2]Department of Wildlife Sciences, Georg-August-University Göttingen, Göttingen, Germany

SUMMARY

Many investigative challenges in road ecology can be addressed by analyses that incorporate genetic data. However, genetic approaches in road ecology are underutilised, partially due to insufficient communication between researchers and stakeholders about the strong applicability, efficiency and cost effectiveness of genetic data for addressing key issues in ecological management. Here, we outline some of the strengths of genetics and summarise important data types and analytical methods. We use the term 'landscape genetics' as a catch-all for the application of genetic techniques in road ecology.

14.1 Landscape genetics is effective in evaluating the barrier effects of roads and their influence on population persistence.

14.2 Patterns of genetic variation in individuals and populations can be used to estimate biological processes that are highly relevant for road ecologists.

14.3 The power of landscape genetics in road-related research can be optimised by choice of study design and sampling protocol.

14.4 Landscape genetics is cost effective and available by collaboration with suitable providers.

Through these four points, we seek to encourage much greater consideration of genetic approaches for understanding and mitigating the ecological impacts of transportation infrastructure. Genetic approaches have repeatedly been shown to be valuable in these contexts, and they are constantly improving. Thus, their appropriate application should lead to substantial benefits for practitioners.

Handbook of Road Ecology, First Edition. Edited by Rodney van der Ree, Daniel J. Smith and Clara Grilo.
© 2015 John Wiley & Sons, Ltd. Published 2015 by John Wiley & Sons, Ltd.
Companion website: www.wiley.com\go\vanderree\roadecology

INTRODUCTION

Road ecology generally seeks to understand the interactions of plants and animals with transportation infrastructure. This includes day-to-day and long-term movements of individuals, as well as gene flow, demography and persistence of populations and communities; but these are challenging to measure.

For more than two decades, genetic techniques have provided efficient solutions to the challenges of measuring dispersal, gene flow and other biological variables. Individuals of a species can be tested for differences in their DNA, which offspring inherit from their parents. This variability in genes is important for road ecologists in two distinct ways. First, adaptive genes have certain ecological functions and directly impact individual fitness (e.g. health, survival, reproductive success). Thus, changes in adaptive genetic variability caused by roads could directly impact population viability and the potential of species to adapt to environmental change. Second, selectively neutral DNA does not directly impact individual fitness, but can be used to evaluate many ecological processes that affect neutral genetic variation, such as population abundance, reproduction and dispersal. Using landscape genetics to analyse the distribution of neutral genetic variation across space has greatly increased our understanding of underlying ecological processes and has tremendous potential for road ecology.

However, despite its efficacy, use of landscape genetics by road ecologists and agencies responsible for land and resource management and transportation planning has been limited. It can be challenging for potential end users to get an overview of the utility, cost effectiveness and availability of genetic techniques (see Sunnucks and Taylor (2008), Balkenhol and Waits (2009) and Simmons et al. (2010) for examples of how useful it can be). Several misconceptions may inhibit an increased application of genetic approaches in road ecology. One common misunderstanding is that genetic signatures of reduced dispersal caused by roads take many generations to be detectable. In fact, modern genetic methods can detect effects of roads, including inhibited movement of individuals, almost immediately. Genetic approaches can be a very effective path to evaluate the influence of roads on movement, gene flow and population trajectories at timescales of interest to ecological managers.

The objectives of this chapter are to (i) provide a general understanding of the principles and value of landscape genetics in road ecology; (ii) motivate transportation planners, scientists, land managers, decision makers and other stakeholders to consider landscape genetics as part of their toolbox; and (iii) provide a straightforward explanation of how to collect, use, analyse and interpret genetic data to answer questions related to ecological road effects.

LESSONS

14.1 Landscape genetics is effective in evaluating the barrier effects of roads and their influence on population persistence

Several extensive reviews have detailed the strengths of landscape genetics for assessing gene flow and genetic structure (ones targeted for ecological managers include DeYoung and Honeycutt (2005), Sunnucks and Taylor (2008) and Balkenhol and Waits (2009)). Some of the most important and directly relevant advantages of genetic applications in road ecology are:
• Genetic data can show the extent to which dispersal leads to gene flow, a critical contributor to thriving, persistent populations. Movement of individuals (or their gametes, such as plant pollen) is necessary but not sufficient to cause gene flow: individuals that move must also successfully reproduce. For example, while some carnivores crossed a freeway in California, gene flow across the freeway was very low, indicating that many of the individuals that crossed did not subsequently breed. Thus, few individual road crossings were actually relevant to future population persistence, and genetic approaches were necessary to demonstrate this (Riley et al. 2006).
• Larger populations are less likely to go extinct than smaller ones. Specifically, a measure of population size based on evolutionary genetics, effective population size (N_e), should be related to population health. There are many measures of N_e, but they can be thought of as reflecting the size of the pool of individuals with a fair chance of contributing genes to the next generation. N_e will be reduced by impacts of roads that lead to fewer breeders or increased inequality in breeding success. Downward trends in N_e would usually indicate increased extinction risk. While estimating effective population size from field-based biological data is tortuous, genetic estimators can accomplish this task quite well.
• Reduced dispersal and gene flow induced by roads can be detected genetically before serious demographic harm occurs. Thus, a genetic 'early warning' can allow for effective intervention while situations are still relatively salvageable.

• Genetic sampling does not require animal recaptures, which saves time and money. Non-invasive sampling, such as collection of faeces, skin layers, feathers and plucked or shed hair (Fig. 14.5), can create great efficiencies for sampling endangered or dangerous species. These sampling benefits facilitate surveys at large scales and intensities, opening up entirely new possibilities. For example, much larger sample sizes are possible with genetic analysis than with telemetry or mark–recapture studies, making it possible to infer about processes such as mobility or gene flow over small and large temporal and spatial scales. Genetic techniques also increase the chance of detecting rare, long-distance and sex-biased dispersal, which may be key drivers of ecological systems. 'Wildlife forensic' applications provide the ability to quantify rates of roadkill by using genetic techniques to identify otherwise-unrecognisable traces of tissue or blood sampled from roads and motor vehicles.

• Because all organisms have genetic material, there are virtually no limits to the taxa that can be studied. The true drivers of ecosystems, such as plants, fungi and microorganisms, are difficult to study by traditional ecological methods but are amenable to genetic approaches.

• It is often possible to obtain DNA from museum collections and other stored material to estimate past population processes. Obvious applications include time series analyses that examine changing levels and distribution of genetic variation in landscapes changed by transportation networks.

The value of landscape genetics has been demonstrated by over 30 studies that used genetic data to investigate road impacts and assess the effectiveness of mitigation measures (reviewed in Balkenhol and Waits (2009); Holderegger and Di Giulio (2010); Jackson and Fahrig (2011)). Overall, genetic approaches have great potential to address the five issues considered to be most critical to understanding the ecological effects of roads and traffic (Roedenbeck et al. 2007).

14.2   Patterns of genetic variation in individuals and populations can be used to estimate biological processes that are highly relevant for road ecologists

Every individual of every species has a unique genetic signature because some changes in DNA occur in transmission from parents to offspring. To measure this genetic variation, laboratory techniques are used that detect differences in the DNA among samples.

Quantifying genetic variation provides the basis for estimating the genetic relationships among samples of individuals and populations. To do so, genetic markers are normally used to detect differences in specific comparable sections of DNA among samples. Examples of typical genetic markers relevant for road ecology include microsatellites and mitochondrial DNA (mtDNA) sequences. Microsatellites are usually inherited from both parents and because they evolve rapidly they enable us to detect genetic differences among samples over relatively short time periods. In contrast, mtDNA is inherited only from mothers in most animals. Microsatellites and mtDNA markers can also be used together to compare recent to current levels of genetic variation or to detect sex-specific road effects on dispersal. Many other types of genetic markers exist, each having specific advantages and limitations. Also, emerging genomic and next generation sequencing approaches can provide much greater power and resolution by scanning the entire genome for DNA differences. The principles of genomic approaches are nonetheless similar to those of established landscape genetics.

Once molecular genetic data have been collected, two different components of genetic variation are commonly estimated: (i) the amount of genetic variation ('genetic diversity') and (ii) the spatial distribution of genetic variation ('genetic structure' or 'genetic differentiation') (Fig. 14.1). The latter is often most interesting for road ecologists, because the distribution of genetic variation provides information on levels of

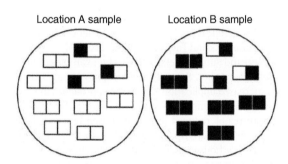

Figure 14.1 Schematic to illustrate the distinction between genetic diversity and genetic differentiation. Consider two comparable genetic variants, 'white' and 'black', within individuals depicted as a pair of squares. Sample A and sample B have identical amounts of genetic diversity: both have 3 copies of 1 genetic type and 17 of the other. However, genetic differentiation between the two population samples is large: one is predominantly black and the other white. Source: Paul Sunnucks.

genetic exchange occurring between different locations (e.g. on either side of a road).

Many key questions of direct relevance to road ecology can be addressed by landscape genetics (Table 14.1). A detailed coverage of the numerous analyses of genetic data relevant to ecological management that focus on transportation is beyond the scope of this chapter but can be found in sources cited earlier. Here, we emphasise two main points: first, the possibilities for road-related, ecological research that utilises genetic data are extensive. For example, genetic data can characterise individuals, identify their parents (parentage analysis), delineate genetic populations (genotypic clustering), identify migrants into populations (assignment tests), quantify environmental influences on gene flow and genetic structure and estimate census and effective population sizes. Second, the analytical (i.e. statistical) methods used for these different investigations have advanced substantially in recent years, providing greater reliability and improved ecological meaningfulness. A key area of advance, as explained in Textbox 14.1, has been improvement in 'direct' methods for quantifying genetic population structure and gene flow as distinct from longer-established 'indirect' methods.

Table 14.1 Some key tasks, questions and analyses central to road ecology that are readily addressed with molecular population genetics.

Topic and relevant questions	Genetic approach
Definition and mapping of populations	
How many populations are there and where are they located?	Genotypic clustering and spatially explicit landscape genetics
How long have they been there?	Comparisons of outcomes of genotypic, frequency-based and DNA sequence based analyses
Between-population processes	
How are populations structured and how different are they?	Spatial autocorrelation, Mantel tests. Frequency differentiation tests of genotypes, gene frequencies and DNA sequences
What are the rates, patterns of dispersal and gene flow?	Assignment/parentage tests for contemporary estimates, medium-term estimates from gene frequencies, long-term estimates from DNA sequences and coalescent analyses
What kinds of individuals disperse with what probability?	
What proportions of individuals are from different sources and how mixed is the ancestry of individuals?	Assignment tests and mixed stock analyses
Within-population processes	
What is the effective population size (N_e)?	Patterns of genetic variation, particularly in microsatellites and DNA sequences
Has the effective population size changed recently?	Tests based on loss of genetic variation and coalescence
Have fundamental population processes (e.g. mating systems, kin interactions) changed?	Assignment, parentage and kinship tests → local dispersal, social/mating systems and kin structure
Monitoring and mitigation	
Species identification of roadkill	DNA sequence comparisons with databases
Causes of mortality	Forensic applications, for example, DNA sampled from classes of motor vehicle or roads
Censusing, births, deaths, reproductive success, migration, sex ratio, space use including road crossing and use of mitigation structures	Non-invasive sample collection, genotype matching, genetic capture–mark–recapture analysis
Properties of large-scale systems	
Relationship between road networks and functional connectivity, barrier/filter effects of natural and built landscape features	Connectivity modelling, isolation by resistance, partial Mantel tests and emerging improvements
Relationship between road impacts and population persistence	Demogenetic modelling

Indirect genetic approaches to estimating population processes

Classic approaches to identifying population structure are based on differences in frequencies of genetic variants among samples from different locations. These data are summarised into statistics such as the commonly used measure of genetic differentiation among populations called F_{ST}, which is closer to zero when genetic variation is similar among populations (which could be caused by high levels of gene flow) and closer to one when genetic variation is dissimilar (more likely under low levels of gene flow). These are referred to as 'indirect' approaches because they do not actually identify immigrant individuals, but instead, by making assumptions and applying theoretical relationships, use data from population samples to estimate population subdivision and infer levels of gene flow. Despite their widespread use, indirect approaches have significant limitations, including that nature frequently violates the assumptions of indirect methods. This is particularly so in settings of recent change such as surrounding transport infrastructure. Indirect measures such as F_{ST} also can respond quite slowly to reductions in gene flow and so may not detect recent impacts of road construction. Thus, direct genetic measures may often be more useful to ecological managers.

Direct genetic approaches to estimating population processes

Recent direct genetic approaches focus on unique genetic variation of individuals and provide information at the finest spatial and shortest temporal scales. This additional capacity is beneficial in a number of important respects. Perhaps the most profound is that the genome of each individual bears highly resolving information (not detected by indirect approaches) that can be used to infer events in the life of that individual, including dispersal (Fig. 14.2). Additionally, the focus of direct approaches on individuals avoids many of the stringent (and often unlikely) assumptions of indirect methods; for example, F_{ST} assumes gene flow to be equally likely among all habitat patches in a system. Direct methods for analysing genetic data provide extraordinary opportunities for addressing road-related ecological questions in highly efficient ways.

Through their focus on genetic signatures of individuals, direct approaches can be used for sophisticated analyses of movement and reproduction across landscape features, including roads. A key example is the application of assignment tests. In their simplest application, these tests can be used to estimate which population an individual was born into (Fig. 14.3). With sufficient sampling and power, it is possible to characterise dispersal events for large sets of individuals and examine which sex and ages of individuals disperse and what landscape features promote or hinder mobility. Such approaches are effective in detecting even subtle reductions in movement and gene flow across roads.

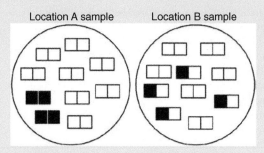

Location A sample   Location B sample

Figure 14.2 Schematic contrasting individual and population measures of genetic difference. Individual genetic variation can be distinctive even when population frequencies are identical. The representation of individuals and genetic variation follows that of Figure 14.1. Samples A and B have the same genetic frequencies (each is $4/20 = 20\%$ black) and so do not show genetic differentiation by indirect measures. However, the genetic compositions of individuals in the two populations are different: location A has two individuals that are 'black', whereas the black characters in sample B are not concentrated into individuals. Such distinguishing of within- and among-sample variation is useful – in this case, the black individuals in A could be recent migrants that have not yet bred with residents. With sufficiently powerful genetic tests, such migration is detectable as soon as an individual arrives. Source: Paul Sunnucks.

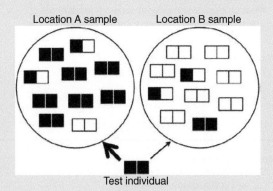

Location A sample   Location B sample

Test individual

Figure 14.3 Schematic of the principle of assignment tests. The most likely birth population of an individual is calculated on the basis of the location in which its genotype has the highest probability of being derived. The representation of individuals and genetic variation follows that of Figure 14.1. In this example, a test individual has two copies of the 'black' genetic variant (one from its mother, one from its father). Black is four times more common in location A than B, so on these data it is 16 times more likely (symbolised by size of the arrows) that the individual was born in location A than B. Source: Paul Sunnucks.

14.3 The power of landscape genetics in road-related research can be optimised by choice of study design and sampling protocol

Sound planning and good study design are necessary for strong inferences in research applications. The most robust study design is BACI (Chapter 10), where data are gathered before and after roads or after mitigation measures are constructed, and from areas with (impact) and without (control) transportation infrastructures. When BACI is not feasible, less powerful but less demanding alternatives may be useful to apply. One good approach is space-for-time substitution, in which reference sites are identified that can be taken to represent an impacted area before impact. For example, an investigation of the impacts of a major road on crossing by a species of gliding marsupial used reference sites away from the road but with similar physical and habitat conditions mirroring that of road sites (van der Ree et al. 2010). The case study in Textbox 14.2 presents an example of a control–impact (CI) design using space-for-time substitution.

Design of genetic sampling is not only important but complex and requires expert input. Appropriate numbers and spatial distribution of samples depend on the age of the road, the spatial scale and the research questions (Chapter 15). Computer simulations are often beneficial for scoping major design issues (e.g. Hoban et al. 2013). It can be very beneficial to conduct feasibility and pilot studies to determine optimal sampling schemes.

Traditional field research programmes often do not include genetic sampling, despite its relatively minor additional cost and effort (Sunnucks 2011). Routine collection of genetic samples would maximise the ability of future research to apply a strong study design, notably BACI. A typical genetic sample might be a few cubic millimetres of tissue (such as 2 mm diameter ear clips obtained during marking of small mammals (Fig. 14.4), a tail tip of a lizard, 10–50 µl of bird blood or some legs of an invertebrate such as a beetle). The essential issue with preservation is that DNA should be dehydrated rapidly and effectively (e.g. storing in >95% ethanol) and kept cool.

Some rare, elusive or sensitive species may require non-invasive sampling. Non-invasive material such as faeces, shed skin, hairs (Fig. 14.5) or feathers is often somewhat more difficult and expensive to work with in the laboratory than are samples obtained from captured animals, but this may be more than compensated for by benefits in the sampling programme. To optimise preservation approaches, pilot studies may be required (e.g. Sloane et al. 2000).

14.4 Landscape genetics is cost effective and available by collaboration with suitable providers

Ecological managers rarely have the expertise and infrastructure to carry out genetics research, in which case they have two main options: (i) engaging commercial genetics laboratories or (ii) collaborating with universities or other research institutions. The laboratories must have expertise in data interpretation and would ideally be able to design tailored genetic tests. The ability of genetics laboratories to increase their involvement in management-oriented research would

---

Textbox 14.2 Using genetics to evaluate the efficiency of road-crossing structures for wildlife.

In an interesting application of molecular techniques for road ecology, Kuehn et al. (2007) collected genetic samples from 222 roe deer along a 28-km stretch of a fenced motorway in Switzerland. They used a control–impact (CI) sampling design (Chapter 10) that included four paired sampling locations on both sides of the approximately 20-year-old motorway ('impact site'), as well as three sampling locations not separated by the road ('control site'). Wildlife crossing structures (over- and/or underpasses) were present at the four impact sites along the motorway. Using 12 microsatellite markers, Kuehn et al. then assessed whether genetic differentiation among the impacted sites was significantly higher than among the control sites. Consistent with the hypothesised barrier effect of the highway, sites separated by the road generally exhibited significantly higher genetic differentiation (i.e. lower levels of gene flow). However, one pair of sites did not show elevated genetic differentiation, principally because the overpass connecting these two sites was the only crossing structure that was embedded into forest patches, the primary habitat of roe deer. Thus, the results suggest that the motorway has reduced genetic exchange of roe deer and that only one of the crossing structures in the area is functioning as intended. The study illustrates that genetic data make it possible to detect barrier effects of roads on a timescale relevant to decision makers and managers and that it can be used to efficiently evaluate the functionality of wildlife crossing structures.

Figure 14.4 DNA collected via ear notching – visible on the right ear of this Mitchell's hopping mouse. Source: Photograph by and reproduced with permission of Marie Lochman/Lochman Transparencies.

Figure 14.5 Non-invasive genetic sampling of American black bear (pictured here) and brown bear populations using barbed-wire hair trap, Banff National Park, Alberta, Canada. Source: Photograph by and reproduced with permission of Michael Sawaya.

be promoted by a culture of collaboration throughout: during resource acquisition, project design, data interpretation and management recommendations. Ultimately, fruitful interactions between management agencies and laboratories with good-quality, peer-reviewed research publications will be greatly facilitated by training of management staff in principles of landscape genetics.

Genetic approaches are cost effective if properly applied and conducted – operating costs are typically around 20% of payroll costs, which is comparable with other disciplines. Cost effectiveness is continuously being improved by technological advances. Deriving sets of microsatellites for a 'new' species typically took several weeks and cost approximately US$8000 (2012 equivalent) in materials, whereas next generation sequencing has transformed that to 1 or 2 days work and US$1500. Ample genetic markers are now available for most organisms, and this catalogue continues to grow (Gardner et al. 2011).

An excellent illustration of the improvements made possible through genetics is the non-invasive census of the northern hairy-nosed wombat, an endangered species in Queensland, Australia. Environmental managers are now estimating demography and investigating biology of the population with a non-invasive genetic approach (collecting hairs at burrow entrances, for genotyping) that is considerably cheaper and less invasive than traditional field surveys and yields additional management-relevant information (Banks et al. 2003).

CONCLUSIONS

In summary, genetic approaches hold great promise for road ecology. The utility and applicability of molecular ecology will further improve through ongoing developments in the field that will make data acquisition and analysis cheaper, more reliable and more meaningful. These improvements will help identify road effects on all levels of biodiversity, particularly when combined with field-based methods and simulation modelling. Together, these research approaches can focus on many questions that are otherwise difficult to address including (i) behavioural changes due to traffic noise (e.g. bird calls and related mating success across roads); (ii) identification of optimal routes for conservation corridors or wildlife crossing structures; or (iii) road effects on species diversity at broad spatial scales (e.g. facilitated through e-DNA). Moving towards such 'molecular road ecology' will provide research findings that can significantly improve decision making with regard to construction and management of transportation infrastructure that avoids, minimises and mitigates negative ecological impacts when possible. This will be facilitated by a culture of collaboration between researchers, end users and other stakeholders, and by improving the capacity to deliver cost-effective services.

ACKNOWLEDGEMENTS

PS thanks Rodney van der Ree for the introduction to road ecology in 2006. Funding for roads research in his group has come from the Australian Research Council, the VicRoads, the NSW Roads and Maritime Services and the Holsworth Wildlife Fund. NB thanks the members of the Washington Wildlife Habitat Connectivity Working Group that first motivated him to think about landscape genetic applications in road ecology.

FURTHER READING

Balkenhol and Waits (2009), Holderegger and Di Giulio (2010) and Simmons et al. (2010): Three reviews of landscape genetics in road ecology.

Frankham et al. (2005): A very concise summary of important population genetic concepts and of their relevance for conservation.

Allendorf et al. (2013): An in-depth treatment of population genetic concepts.

REFERENCES

Allendorf, F. W., G. H. Luikart and S. N. Aitken. 2013. Conservation and the Genetics of Populations. Wiley-Blackwell, Oxford.

Balkenhol, N. and L. P. Waits. 2009. Molecular road ecology: exploring the potential of genetics for investigating transportation impacts on wildlife. Molecular Ecology **18**:4151–4164.

Banks, S. C., S. D. Hoyle, A. Horsup, P. Sunnucks and A. C. Taylor. 2003. Demographic monitoring of an entire species (the Northern hairy-nosed wombat, *Lasiorhinus krefftii*) by genetic analysis of non-invasively collected material. Animal Conservation **6**:101–107.

DeYoung, R. W. and R. L. Honeycutt. 2005. The molecular toolbox: genetic techniques in wildlife ecology and management. Journal of Wildlife Management **69**: 1362–1384.

Frankham, R., J. D. Ballou and D. A. Briscoe. 2005. A primer of Conservation Genetics. Cambridge University Press, Cambridge.

Gardner, M. G., A. J. Fitch, T. Bertozzi and A. J. Lowe. 2011. Rise of the machines – recommendations for ecologists when using next generation sequencing for microsatellite development. Molecular Ecology Resources **11**:1093–1101.

Hoban, S., O. Gaggiotti, ConGRESS Consortium, G. Bertorelle. 2013. Sample Planning Optimization Tool for conservation and population Genetics (SPOTG): a software for choosing the appropriate number of markers and samples. Methods in Ecology and Evolution **4**:299–303.

Holderegger, R. and M. Di Giulio. 2010. The genetic effects of roads: a review of empirical evidence. Basic and Applied Ecology **11**:522–531.

Jackson, N. D. and L. Fahrig. 2011. Relative effects of road mortality and decreased connectivity on population genetic diversity. Biological Conservation **144**:3143–3148.

Kuehn, R., K. E. Hindenlang, O. Holzgang, J. Senn, B. Stoeckle and C. Sperisen. 2007. Genetic effect of transportation infrastructure on roe deer populations (*Capreolus capreolus*). Journal of Heredity **98**:13–22.

Riley, S. P. D., J. P. Pollinger, R. M. Sauvajot, E. C. York, C. Bromley, T. K. Fuller and R. K. Wayne. 2006. A southern California freeway is a physical and social barrier to gene flow in carnivores. Molecular Ecology **15**:1733–1741.

Roedenbeck, I. A., L. Fahrig, C. S. Findlay, J. E. Houlahan, J. A. G. Jaeger, N. Klar, S. Kramer-Schadt and E. A. van der Grift. 2007. The Rauischholzhausen agenda for road ecology. Ecology and Society **12**:11. http://www.ecologyandsociety.org/vol12/iss1/art11/ (accessed 9 January 2015).

Simmons, J. M., P. Sunnucks, A. C. Taylor and R. van der Ree. 2010. Beyond roadkill, radiotracking, recapture and F_{ST} – a review of some genetic methods to improve understanding of the influence of roads on wildlife. Ecology and Society **15**:9. http://www.ecologyandsociety.org/vol15/iss1/art9/ (accessed 9 January 2015).

Sloane, M. A., P. Sunnucks, D. Alpers, L. B. Beheregaray and A. C. Taylor. 2000. Highly reliable genetic identification of individual northern hairy-nosed wombats from single remotely collected hairs: a feasible censusing method. Molecular Ecology **9**:1233–1240.

Sunnucks, P. 2011. Towards modelling persistence of woodland birds: the role of genetics. Emu **111**:19–39.

Sunnucks, P. and A. C. Taylor. 2008. The application of genetic markers to landscape management. Pages 211–234 in C. Pettit, W. Cartwright, I. Bishop, K. Lowell, D. Pullar and D. Duncan, editors. Landscape Analysis and Visualisation: Spatial Models for Natural Resource Management and Planning. Springer, Berlin.

van der Ree, R., S. Cesarini, P. Sunnucks, J. L. Moore and A. Taylor. 2010. Large gaps in canopy reduce road crossing by a gliding mammal. Ecology and Society **15**:35. http://www.ecologyandsociety.org/vol15/iss4/art35/ (accessed 9 January 2015).

Chapter 15

GUIDELINES FOR EVALUATING USE OF WILDLIFE CROSSING STRUCTURES

Edgar A. van der Grift[1] and Rodney van der Ree[2]

[1]Alterra, Wageningen UR, Environmental Science Group, Wageningen, Netherlands
[2]Australian Research Centre for Urban Ecology, Royal Botanic Gardens Melbourne, and School of BioSciences, The University of Melbourne, Melbourne, Victoria, Australia

SUMMARY

Wildlife crossing structures help animals cross safely under or over roads or other linear infrastructure and hence play an important role in the conservation of biodiversity. Measuring the rate of use by wildlife is an important first step in almost every evaluation of wildlife crossing structures. Unfortunately, the majority of studies of the use of crossing structures by wildlife lack a proper study design which limits the quality or reliability of the findings. The design and methods of each study to evaluate the use of crossing structures must be tailor-made because of differences among structures in their design, goals, target species, landscape and road conditions.

15.1 Identify and describe the target species for the wildlife crossing structure being evaluated.

15.2 For each target species, define the intended type and frequency of use.

15.3 Design the study to enable a comparison of actual rate of use and minimum expected rate of use.

15.4 Use data from control plots to estimate the minimum expected rate of use of a crossing structure.

15.5 Select survey methods that monitor multiple species simultaneously and use more than one survey method for each species.

15.6 The timing, frequency and duration of the monitoring should allow for rigorous estimates of crossing structure use.

15.7 Measure explanatory variables to enable a comprehensive analysis of the monitoring data and comparison of crossing structure functioning.

15.8 Thorough analysis, reporting and sharing of data are critical.

Taken individually, each study of the use of crossing structures by wildlife provides an important but basic understanding of their function. Adopting the guidelines presented in this chapter will improve the quality of each monitoring programme as well as permit robust meta-analyses to optimise design, placement and management of wildlife crossing structures at much broader spatial scales.

Handbook of Road Ecology, First Edition. Edited by Rodney van der Ree, Daniel J. Smith and Clara Grilo.
© 2015 John Wiley & Sons, Ltd. Published 2015 by John Wiley & Sons, Ltd.
Companion website: www.wiley.com\go\vanderree\roadecology

INTRODUCTION

Wildlife crossing structures, encompassing a broad range of underpasses and overpasses, help animals to cross roads safely and hence play an important role in the conservation of biodiversity (Chapter 21). Monitoring to investigate if the target species are using the structures is an important first step, although it does not provide a complete insight into the effectiveness of the structure (van der Grift et al. 2013; Chapter 16). Well-designed studies of use can identify the rate of crossing by different species, enable the identification of design and landscape features that influence crossing rates (e.g. Clevenger & Waltho 2000, 2005; Ng et al. 2004; Ascensão & Mira 2007; Grilo et al. 2008) and provide an insight into interactions among species (e.g. predation (Chapter 23), territoriality) and human activities that may affect crossing rates (e.g. Doncaster 1999; Ford & Clevenger 2010; van der Grift et al. 2012; Chapter 22).

Unfortunately, the majority of studies that evaluate the use of crossing structures lack a proper study design (see Chapter 10 for discussion on study design). In many studies, monitoring is limited to registering passing animals at the crossing structures, without measurements at control sites. Essential variables are often not measured, which hinders the interpretation of the data and makes it impossible to compare the functioning of multiple crossing structures. Not surprisingly, most monitoring studies are published in research reports or journals without peer review (see van der Ree et al. (2007) for some exceptions). These limitations hinder our ability to make full use of study results and increase our general understanding of the main drivers behind wildlife crossing structure use. In this chapter, we present guidelines to improve study design to enable the accurate evaluation of the use of wildlife crossing structures by wildlife. The overall aim is to improve the quality of future studies on crossing structure use and permit robust meta-analyses (Chapter 10) to optimise the design, placement and management of future crossing structures.

LESSONS

15.1   Identify and describe the target species for the wildlife crossing structure being evaluated

A critical first step in evaluating the use of a crossing structure by wildlife is to make a list of all the target species that should or could use the crossing structure(s) being evaluated. This list is important because it

influences all subsequent decisions on study design, sampling scheme and survey methods. For example, track beds are suitable for recording large carnivores or ungulates but are unsuitable for most small mammal or reptile species. Ensure the list is specific and identifies each species and not species groups, such as 'small mammals' or 'frogs', because species within the same group may also require different sampling methods.

15.2   For each target species, define the intended type and frequency of use

To assess if a crossing structure has achieved its goals, the intended type and frequency of use should be clearly described for each target species. In general, three types of use can be distinguished, each with its own corresponding crossing frequency: (i) daily use, to access resources within an individual's home range on a daily or almost daily basis, for example, to access foraging grounds, water sources or roosts; (ii) seasonal use, to access seasonal habitats, for example, annual migration to access breeding ponds, wintering grounds or calving areas; and (iii) occasional use, where animals, often juveniles, make long-distance movements to establish their own territory, that is, dispersal movements, which are important for demographic and genetic exchange between existing populations and recolonisation of vacant habitat patches.

Clear insight into the expected type and frequency of use is essential information to make decisions about study design, sampling scheme and survey methods. For example, the sampling technique at structures intended to assist recolonisation of empty habitat patches by dispersing individuals should be capable of recognising dispersing individuals. Or if a crossing structure is to allow home range movements of neighbouring social groups, the sampling technique must be able to identify members of the different groups.

15.3   Design the study to enable a comparison of actual rate of use and minimum expected rate of use

Simply registering the number of crossings per species does not provide a solid basis to make conclusions about the rate of use of crossing structures or their effectiveness (Chapter 16). The rate of crossing is largely dependent on population density and is usually greater in areas with high population densities than in areas where a species is rare. Therefore, the study design should allow for comparing actual use (or observed rate of crossing) with estimates of minimum expected use. Use must be expressed in terms of *crossing rate*, that is, the number of crossings of a species per unit of time. In the case of animals that use crossing structures as habitat or can only slowly move across a structure, use should be expressed as *relative abundance*, that is, the number of animals of a species per unit of area. Measuring actual and minimum expected crossing rates or relative abundances permits calculation of species performance ratios (Lesson 15.4). If multiple crossing structures are monitored or data from different studies are being compared, such performance ratios provide a solid basis for further analyses of what factors influence structure use.

15.4 Use data from control plots to estimate the minimum expected rate of use of a crossing structure

Preferences of species towards certain types of crossing structures can only be identified if baseline information on the presence/absence and/or abundance of the target species from the surrounding area is available. Ideally, this baseline information should be collected at multiple nearby locations, that is, control plots, simultaneously with measurements of use at the structure itself. Measurements at control plots are used to calculate the minimum expected rate of use of the crossing structure, provided the survey methods, such as plot size, survey technique and sampling scheme (Textbox 15.1), at control plots are similar to those at

---

Textbox 15.1 Estimating the minimum expected rate of use of a crossing structure.

In this textbox, we illustrate how to estimate the minimum expected rate of use of a crossing structure using data from a wildlife overpass in the Netherlands (van der Grift et al. 2012). The rate of crossing by medium-sized mammals was surveyed using two 15 m-long track beds that spanned the full width of the overpass (Fig. 15.1A). Simultaneously, 24 control plots were established within suitable habitat for the target species in the immediate vicinity of the overpass (Fig. 15.1B). Each control plot was randomly positioned and consisted of one 15 m-long track bed. Track beds at the overpass and in the surroundings were surveyed three to four times per week over a 1-year period. The minimum expected crossing rate of the species at the overpass is equal to the mean crossing rate at the control plots. If the crossing rate at the overpass is higher than this mean, the species

seems to actively select a route across the overpass (funnelling effect). If the crossing rate at the overpass is lower than this mean, the species seems to actively avoid the overpass. At the studied overpass, three mammal species – badger, polecat and rabbit – crossed more often than the estimated minimum crossing rate for the species. Species performance ratios for these species were, respectively, 11, 12 and 1.5. This means that, for example, badger crossed the track beds on the overpass 11 times more often than the track beds at the control plots. Five species – roe deer, red fox, European hare and red squirrel – crossed less frequently than expected. These species seem to avoid the overpass. Comparison with species performance ratios at other overpasses indicated that overpass width and design were likely the main factors that affected use of the structure by these target species.

(A)

(B)

Figure 15.1 Track bed at the wildlife overpass (A) and at a randomly positioned control plot within suitable habitat in the vicinity of the overpass (B). Source: Photographs by and reproduced with permission of Fabrice Ottburg.

the crossing structure. Measurements at control plots should be expressed in terms of crossing rate or relative abundance, similar to those at the crossing structure (Lesson 15.3).

Control plots should be positioned close to the crossing structure being evaluated in order to assess populations that may potentially benefit from the crossing structure. However, control plots should be placed outside the road-effect zone (Forman & Deblinger 2000; Lesson 1.2) of the species being monitored because species presence and abundance is often reduced within this zone (Benítez-López et al. 2010). Control plots outside the road-effect zone will more accurately measure the 'pre-road' conditions than plots within the zone and allow a stronger and more reliable comparison of structure effectiveness at achieving the goal of maintaining pre-road

connectivity levels. This approach is not perfect, but it provides a consistent approach that can be applied across a range of studies. The road-effect zone is usually asymmetric, as some road effects only occur at specific locations, or road and landscape features result in different effect distances along the road length (Forman et al. 2003; Fig. 1.1). The size of the road-effect zone can be determined by estimating the abundance of each target species at increasing distances from the road.

The control plots should be positioned within a circular area around the crossing structure (Fig. 15.2). The radius of this area should be minimised to decrease travel time during surveys but large enough to position all control plots outside the road-effect zone. Control plots should be randomly placed only within suitable habitat for the target species, and not in rarely or never

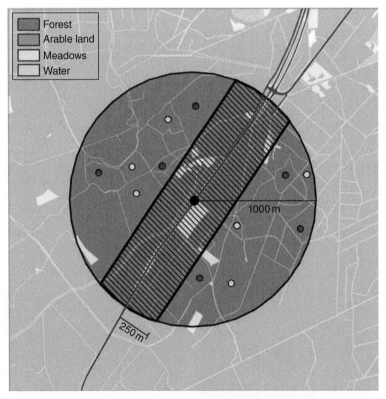

Figure 15.2 The distribution of control plots in a hypothetical study of the use of a wildlife crossing structure (black dot) by a woodland dependant species. Control plots are randomly placed within deciduous (red dots) and coniferous (yellow dots) forest used by the species but not in arable land, meadows and water, which the species avoids. The radius of the area in which control plots are placed is 1000 m. The road-effect zone (denoted by stripes) is 250 m, in which no control plots are placed. Power analyses determined that a minimum of 6 control plots were needed per suitable habitat type, which results in a total of 12 control plots. Source: Adapted from van der Grift (2010).

used habitats (Fig. 15.2). More control plots generally lead to increased precision, and the minimum number should be identified before the study commences using a power analysis, preferably based on the results of a pilot study. The pilot study should determine the variation in the studied variable and therefore how many control plots are required to allow for rigorous estimates of expected use.

15.5 Select survey methods that monitor multiple species simultaneously and use more than one survey method for each species

A range of survey methods exist to monitor crossing structure use, but not all are equally effective or informative (Table 15.1, Fig. 15.3A, B, Chapter 11). For example, camera traps work well for large mammals but are usually ineffective for most birds and invertebrates. Track beds provide data on species, direction of movement and gait, and cameras provide this as well as the time of crossing, some weather conditions and, depending on the species, sex, age class, behaviour and unique markings. Cost-effectiveness also differs among methods (Ford et al. 2009). Survey methods that monitor multiple species simultaneously are recommended because they provide more information for similar effort and cost. In addition, the use of multiple survey methods for each species can decrease bias and provide better estimates of the parameter of interest (Textbox 15.2). Consistent use of the same methods and personnel over time, both at the crossing structures and at the control sites, is important to provide comparable results.

15.6 The timing, frequency and duration of the monitoring should allow for rigorous estimates of crossing structure use

The timing of monitoring depends primarily on the objectives of the mitigation and objectives of the monitoring. If the aim is to evaluate the extent to which a crossing structure restores access to seasonal habitats (e.g. spring migration of amphibians to breeding ponds), the monitoring can be limited to the period in which those migrations occur. Furthermore, the life cycle of the species will influence the timing of monitoring if predictable periods of presence/absence (e.g. migratory species) or inactivity (e.g. hibernation during winter) can be

identified. If the aim is to evaluate a species' daily movements in an area that encompasses a crossing structure, monitoring throughout the year – excluding periods in which the species is inactive or does not occur – is recommended. Most species show different activity and movement patterns throughout the year (Fig. 15.5); hence, measuring use of crossing structures for only a few weeks or months may not provide a comprehensive estimate of use.

Monitoring frequency employed in past studies is highly variable, ranging from daily to once per month, as shown by a review of 121 studies (van der Ree et al. 2007). The frequency of monitoring is closely related to the timing of the monitoring. For example, weekly surveys may be required if the structure is to provide year-round home range movements, while daily surveys are likely required for structures that provide seasonal connectivity for a 1-month period. The choice of survey method is critical, as these differ considerably in the amount of effort required to collect data. For example, camera traps can collect data 24 hours per day, 7 days per week with little effort, while track beds demand frequent visits by field personnel often resulting in monitoring intervals of once per week or less.

The duration of monitoring in published studies also varies considerably, ranging from 4 nights to 8 years (van der Ree et al. 2007). Monitoring for multiple years is recommended, as crossing structure use may vary considerably across years. For example, the initial rate of use of structures after installation is often low, increasing after months or years when animals become accustomed to its presence. The need for monitoring across multiple years increases if population size of the target species is known to vary from year to year or if the trend in the rate of crossing structure use is to be identified. It may be important to survey over many years because the vegetation at or around crossing structures will mature over time, potentially affecting rate of use.

15.7 Measure explanatory variables to enable a comprehensive analysis of the monitoring data and comparison of crossing structure functioning

In addition to recording use by animals, a range of other variables should be also measured during the monitoring to improve interpretation of the results. These explanatory variables allow for better comparisons among study sites and stronger inferences concerning the causes of

Table 15.1 The suitability of commonly used survey methods for each species group.

Survey method	Species group										
	Large mammals	Medium-sized mammals	Small mammals	Bats	Non-flying birds	Flying birds	Reptiles	Amphibians	Non-flying insects	Flying insects	Other invertebrates
Track bed (coarse sands)	**	**	0/–[a]	–	**	–	0/–[a]	–	–	–	–
Track bed (fine sands)	**	**	0	–	**	–	0	0	–	–	–
Track plate	–	**	0	–	–	–	0	0	–	–	–
Snow tracking	*	*	–	–	–	–	–	0	–	–	–
Photo/video camera	**	**	*/–[b]	–	**	?	?	*[c]	–	–	–
Infrared trail monitor	0	0	0	0	0	–	–	–	–	–	–
Artificial shelters	–	–	*	–	–	–	**	**	*	–	*
Acoustic monitoring (e.g. bat detectors, birdsong recorders)	–	–	–	**	*/–[b]	*/–[b]	–	–	–	–	–
Survey of animals by direct observations	–	–	–	*	–	**	**	**	–	**	–
Survey of animal signs (e.g. feeding sign, dropping)	*	*	*	–	–	–	–	–	–	–	–
Hair trap – hair identification	*	*	*	–	–	–	–	–	–	–	–
Hair trap – DNA analysis	*	*	*	–	–	–	–	–	–	–	–
Capture–mark–recapture	–	*/–[b]	**	–	*/–[b]	*/–[b]	**	**	**	**	*

Capture–mark–monitor (e.g. PIT tag, ear tag, leg banding)	*	*	–	*	*	*/–[b]	*/–[b]	*/–[b]
Capture–tracking (e.g. radio tracking, GPS/satellite tracking)	*	–	–	*/–[b]	*/–[b]	*/–[b]	*/–[b]	*/–[b]
Capture–release (e.g. live trap, pitfall trap, mist net)	–	**	*	*	**	**	**	**
Capture–kill (e.g. pitfall trap)	–	–	–	–	*	*	–	*

*, suitable; **, highly suitable; 0, registration of crossing, but not able to identify species; –, not suitable; ?, unknown.

[a]Registration, but not at species level, for only some species within this species group.

[b]Suitable for only some species within this species group.

[c]If used in small wildlife underpasses.

(A)

(B)

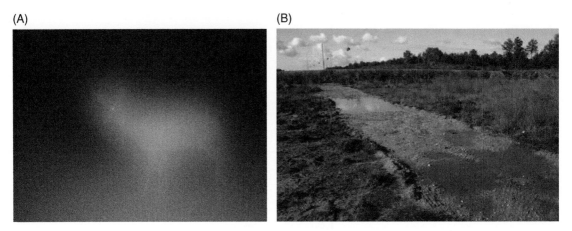

Figure 15.3 Every survey method has its weaknesses. For example, fog may prevent accurate identification of passing animals with camera traps (A) or heavy rainfall may obscure animal footprints on track beds (B). Source: Photographs by Edgar van der Grift.

Textbox 15.2 Using multiple techniques to assess rates of crossing of canopy bridges and glider poles.

There is rarely a single survey method that is without error or bias. In Australia, crossing structures installed for arboreal mammals include canopy bridges and glider poles, as well as underpasses for those species that come to the ground (Figs 40.1 and 40.5). A recent study (Soanes et al. 2013) measured the crossing rate by squirrel gliders using radiotracking, motion-triggered cameras and microchip (also called passive integrated transponder (PIT)) scanners (Fig. 15.4). The nightly rate of glider crossings at one canopy bridge was 2.66 using cameras, 0.40 using radiotracking and 1.37 using microchip scanners. The advantage of radiotracking and microchip scanners is that the identity of the individual can be identified, which cameras are unable to do. However, individuals that have not been fitted with a transmitter or microchip will be missed, while cameras register both tagged and non-tagged individuals. Radiotracking is extremely labour intensive but is able to detect crossings that occurred without the use of crossing structures, such as via canopy connectivity (Fig. 40.1A) or at unmonitored crossing structures. Trapping is similarly labour intensive and relies on recapturing marked animals on both sides of the road, which rarely occurred. In addition to differences in detectability and effort, the use of multiple techniques builds in some redundancy in case of equipment malfunction.

(A)   (B)   (C)

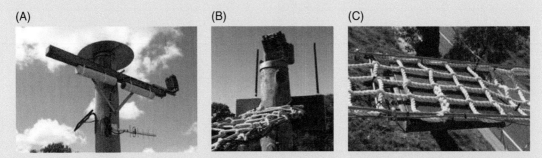

Figure 15.4 Cameras on (A) glider poles and (B) both ends of canopy bridges were used to record crossings by squirrel gliders in south-east Australia. The (C) microchip scanners on rope bridges record the identity of individuals as they cross over, assuming they have been previously captured and fitted with microchips. The combination of microchip scanner and camera provides some redundancy in case one method fails. Source: Photographs by Rodney van der Ree.

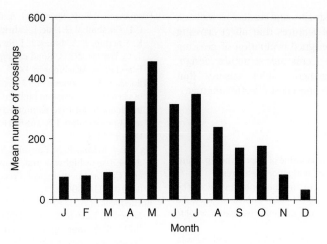

Figure 15.5 The average number of successful crossings per month of roe deer at a wildlife overpass in the Netherlands. Use of the overpass varies seasonally, and if monitoring was only conducted during winter (D, J, F), the rate of annual use would be underestimated. In contrast, annual estimates derived only from summer data (J, J, A) would be an overestimate. Source: Adapted from van der Grift et al. (2009).

observed differences in crossing rates. These should include documenting the spatial and temporal variability in (i) road design and traffic; (ii) crossing structure design, management and co-use by people, domestic animals or livestock; (iii) structural features of the surrounding landscape; and (iv) weather conditions.

Road design variables include road width, whether the road is in a cut or elevated on fill, and presence and type of pavement, street lights, fences, noise screens, median strip and road verges. Traffic volume and speed should be documented at several temporal scales, for example, daily, seasonally or annually. Crossing structure variables include the size and characteristics of the crossing structure, the distance between crossing structures, the presence and design of wildlife fences, the type and frequency of management and the presence and frequency of use by non-target species, humans and domestic animals. Information on the duration of the construction period and the date that the road/crossing structure was ready for use may also be important. Important landscape variables include altitude, topography, land use, type and amount of vegetation and the occurrence of important landscape elements, such as hedgerows or ponds. The landscape features should be mapped and monitored over a circular area around the crossing structure, with a radius that is similar to the one recommended for positioning control plots (Lesson 15.4). Finally, weather conditions during monitoring should be documented, including, where relevant, temperature, cloudiness, precipitation, snow cover depth and wind speed.

15.8   Thorough analysis, reporting and sharing of data are critical

Monitoring programmes are usually considered completed when the data has been analysed and the report has been submitted to the transport agency. A major limitation is that most reports are not peer reviewed and neither the report nor raw data are made widely available. Peer review of reports and publication in scientific journals is recommended to improve the quality and rigour of the scientific methods as well as improve access to the findings. This will help to ensure that future studies and mitigation programmes can build on existing knowledge. The raw data should be published to allow for meta-analyses, where multiple data sets from a range of sources are pooled and analysed together (Lesson 10.8). One of the aims of promoting the consistent approach and improved quality of monitoring is to provide a greater number of studies that can be used in this way.

CONCLUSIONS

Proper evaluation of the use of wildlife crossing structures requires a tailor-made monitoring plan. The monitoring approach – including study design, sampling scheme and survey technique – depends on the target species for mitigation and the objectives of the study. The set of guidelines presented here allows for improved comparisons among study sites and studies, as well as a better understanding of features that affect crossing structure use. A well-designed evaluation of crossing structure use optimises crossing structure design, placement and management, which ensures that resources are spent in the most cost-effective manner.

FURTHER READING

van der Grift et al. (2013): An outline of approaches to evaluate the effectiveness of road mitigation measures.

van der Ree et al. (2007): An international review of the study design and methods of 121 published scientific papers and reports on the use and effectiveness of wildlife crossing structures.

Roedenbeck et al. (2007): A framework for a research agenda for road ecology and recommendations for experimental designs that maximise inferential strength, given existing constraints.

REFERENCES

Ascensão, F. and A. Mira. 2007. Factors affecting culvert use by vertebrates along two stretches of road in southern Portugal. Ecological Research **22**: 57–66. doi:10.1007/s11284-006-0004-1.

Benítez-López, A., R. Alkemade and P. A. Verweij. 2010. The impacts of roads and other infrastructure on mammal and bird populations: a meta-analysis. Biological Conservation **143**: 1307–1316.

Clevenger, A. P. and N. Waltho. 2000. Factors influencing the effectiveness of wildlife underpasses in Banff National Park, Alberta, Canada. Conservation Biology **14**: 47–56.

Clevenger, A. P. and N. Waltho. 2005. Performance indices to identify attributes of highway crossing structures facilitating movement of large mammals. Biological Conservation **121**: 453–464. doi:10.1016/j.biocon.2004.04.025.

Doncaster, C. P. 1999. Can badgers affect the use of tunnels by hedgehogs? A review of the literature. Lutra **42**: 59–64.

Ford, A. and A. P. Clevenger. 2010. Validity of the prey-trap hypothesis for carnivore-ungulate interactions at wildlife-crossing structures. Conservation Biology **24**: 1679–1685. doi:10.1111/j.1523-1739.2010.01564.x.

Ford, A., A. P. Clevenger and A. Bennett. 2009. Comparison of methods of monitoring wildlife crossing-structures on highways. Journal of Wildlife Management **73**: 1213–1222. doi:10.2193/2008-387.

Forman, R. T. T. and R. D. Deblinger. 2000. The ecological road-effect zone of a Massachusetts (USA) suburban highway. Conservation Biology **14**: 36–46.

Forman, R. T. T., D. Sperling, J. A. Bissonette, A. P. Clevenger, C. D. Cutshall, V. H. Dale, L. Fahrig, R. France, C. R. Goldman, K. Haenue, J. A. Jones, F. J. Swanson, T. Turrentine and T. C. Winter. 2003. Road ecology – science and solutions. Island Press, Washington, DC.

Grilo, C., J. A. Bissonette and M. Santos-Reis. 2008. Response of carnivores to existing highway culverts and underpasses: implications for road planning and mitigation. Biodiversity and Conservation **17**: 1685–1699. doi:10.1007/s10531-008-9374-8.

Ng, S. J., J. W. Dole, R. M. Sauvajot, S. P. D. Riley and T. J. Valone. 2004. Use of highway undercrossings by wildlife in southern California. Biological Conservation **115**: 499–507. doi:10.1016/S0006-3207(03)00166-6.

Roedenbeck, I. A., L. Fahrig, C. S. Findlay, J. Houlahan, J. A. G. Jaeger, N. Klar, S. Kramer-Schadt and E. A. van der Grift. 2007. The Rauischholzhausen agenda for road ecology. Ecology and Society **12**: 11. http://www.ecologyandsociety.org/vol12/iss1/art11/ (Accessed 16 January 2015).

Soanes, K., M. Carmody Lobo, P. A. Vesk, M. A. McCarthy, J. L. Moore and R. van der Ree. 2013. Movement re-established but not restored: inferring the effectiveness of road crossing mitigation for a gliding mammal by monitoring use. Biological Conservation **159**: 434–441.

van der Grift, E. A. 2010. Guidelines for measuring the use of wildlife crossing structures. Alterra, Wageningen. http://www.mjpo.nl/publicaties/monitoringsrapporten/ [in Dutch] (accessed 16 January 2015).

van der Grift, E. A., F. G. W. A. Ottburg and J. Dirksen. 2009. Use of wildlife overpass Zanderij Crailoo by people and wildlife. Alterra-report 1906. Alterra, Wageningen [in Dutch].

van der Grift, E. A., F. Ottburg, R. Pouwels and J. Dirksen. 2012. Multiuse overpasses: does human use impact the use by wildlife? In: P. J. Wagner, D. Nelson and E. Murray (eds.). Proceedings of the 2011 International Conference on Ecology and Transportation: 115–123. Center for Transportation and the Environment, North Carolina State University, Raleigh, NC.

van der Grift, E. A., R. van der Ree, L. Fahrig, S. Findlay, J. Houlahan, J. A. G. Jaeger, N. Klar, L. F. Madriñan and L. Olson. 2013. Evaluating the effectiveness of road mitigation measures. Biodiversity and Conservation **22**: 425–448. doi:10.1007/s10531-012-0421-0.

van der Ree, R., E. A. van der Grift, N. Gulle, K. Holland, C. Mata and F. Suarez. 2007. Overcoming the barrier effect of roads – how effective are mitigation strategies? An international review of the use and effectiveness of underpasses and overpasses designed to increase the permeability of roads for wildlife. In: C. L. Irwin, D. Nelson and K. P. McDermott (eds.). Proceedings of the International Conference on Ecology and Transportation: 423–431. Center for Transportation and the Environment, North Carolina State University, Raleigh, NC.

Chapter 16

GUIDELINES FOR EVALUATING THE EFFECTIVENESS OF ROAD MITIGATION MEASURES

Edgar A. van der Grift[1], Rodney van der Ree[2] and Jochen A. G. Jaeger[3]

[1]Alterra, Wageningen UR, Environmental Science Group, Wageningen, Netherlands
[2]Australian Research Centre for Urban Ecology, Royal Botanic Gardens Melbourne, and School of BioSciences, The University of Melbourne, Melbourne, Victoria, Australia
[3]Department of Geography, Planning and Environment, Concordia University Montreal, Montréal, Québec, Canada

SUMMARY

Wildlife crossing structures – underpasses and overpasses – have been constructed around the world and are used by many species of wildlife to safely cross roads and other linear infrastructures. However, there is still much to learn about their effectiveness at contributing to the preservation of biodiversity. How many and what kinds of structures do we need to reach the goals of mitigation? Without clear insights into the effectiveness of wildlife crossing structures, we run the risk of losing wildlife populations (or even species) and wasting money. The evaluation of the effectiveness of mitigation requires a good experimental design and should be incorporated into road planning.

16.1 Identify and describe the target species and goals of mitigation.
16.2 Monitor target species that are likely to demonstrate statistically significant effects with comparatively little sampling effort in space and/or time.
16.3 Select parameters of interest that are most closely related to the outcome of real concern.
16.4 Adopt a study design that allows for rigorous conclusions.
16.5 Use model simulations to determine the best sampling scheme.
16.6 Select mitigation sites to be monitored based on the objective(s) of the evaluation.
16.7 Choose control sites based on the goals of mitigation.
16.8 Measure explanatory variables that provide the best possible estimates of mitigation effectiveness.
16.9 Utilise survey methods that monitor multiple species simultaneously.

16.10 Ensure that the evaluation of mitigation has sufficient resources, is integrated into the planning and construction of roads and is coordinated across boundaries.

A comprehensive evaluation of the extent to which road mitigation measures reduce the risk of decline and extinction of wildlife populations is essential to ensure that conservation funds are being allocated in the most cost-effective manner. Researchers need to be involved in the design of the evaluation programmes from the earliest stages of the road construction or road mitigation project. Although the set of guidelines presented here is ambitious, we are convinced that they are necessary to improve our understanding of the effectiveness of road mitigation measures.

This chapter is a summary of a paper published in *Biodiversity and Conservation* (van der Grift et al. 2013), in which more details on these guidelines are presented.

INTRODUCTION

The construction of wildlife crossing structures has become standard practice in many parts of the world to help restore wildlife movement across roads and contribute to the conservation of biodiversity. Numerous studies have shown that, if well designed and placed, crossing structures are used by many species, but little is known about their effectiveness at achieving specific goals (Fig. 16.1). Do crossing structures do what we expect them to do? To what extent do they contribute to the preservation of wildlife populations and species? How many and what kinds of wildlife crossing structures do we need to reach the goals of mitigation? Finding the right answers to these questions is essential for the development of efficient road mitigation strategies. Without clear insights into the effectiveness of crossing structures, we run the risk of losing wildlife populations and species, despite all our efforts to prevent just that. A lack of such insights may also result in an inefficient use of financial resources. If we underestimate their effectiveness, we may build more than are needed. If we overestimate their effectiveness, we may build

Figure 16.1 Roe deer using a wildlife overpass in the Netherlands. Evidence showing that the target species uses the crossing structure is essential, but not the full story, because use of a wildlife crossing structure does not necessarily equate to its effectiveness. Source: Photograph by Edgar van der Grift.

too few, also resulting in the loss of our investments as biodiversity continues to decline. If we do not compare the effectiveness of different kinds of crossing structures, we may build an inferior type. And if we do not compare the effectiveness of crossing structures against other mitigation strategies, we are unable to identify the most effective strategy. In this chapter, we outline the essential elements of a good experimental design to evaluate the effectiveness of wildlife crossing structures and recommend how to incorporate such evaluations into road planning. While the focus is on crossing structures, similar principles apply for other types of mitigation.

LESSONS

16.1 Identify and describe the target species and goals of mitigation

A comprehensive evaluation of the effectiveness of wildlife crossing structures requires a clear definition of success. Effectiveness is defined as the extent to which the goals of mitigation are reached. In practice, the goal of mitigation is often a list of species that should use the crossing structure. Such lists are important, but not enough. A clear and testable description of the goal for each species and structure is needed. We recommend the SMART approach, that is, goals that are specific, measurable, achievable, realistic and time framed (Textbox 16.1). The goals should (i) specify which road impact is to be addressed; (ii) quantify the reduction in impacts aimed for; (iii) preferably be agreed upon by all stakeholders; (iv) match available resources; and (v) specify the time span over which the reduction is to be achieved.

16.2 Monitor target species that are likely to demonstrate statistically significant effects with comparatively little sampling effort in space and/or time

Studies of the use of wildlife crossing structures usually focus on species that are rare and/or in decline or those large enough to cause significant damage to vehicles and humans if involved in wildlife-vehicle collisions (WVC). If limited resources prevent an evaluation of all threatened species, one should choose those that are most likely to demonstrate statistically significant effects with comparatively little sampling effort in space and/or time. The following criteria will assist in

---

Textbox 16.1 SMART mitigation goals.

If a wildlife crossing structure is meant to reduce the number of roadkills of a particular species, a SMART goal (Lesson 16.1) may be: *In year X after mitigation, all (or Z% of) traffic-related mortality of species Y is prevented.* Goals that relate to the permeability of the road for wildlife may be expressed as: *In year X after mitigation, the number of (i) between-population movements, (ii) home range movements or (iii) seasonal migrations across the road of species Y is similar to (or at least Z% of) pre-road-construction movement numbers.* Similarly, goals that relate to the desired size and eventually the viability of the connected populations may be expressed as: *The post-mitigation population density of species Y at time X is similar to (or at least Z% of) the pre-road-construction population density.* Or if the pre-road-construction situation is unknown: *In year X post-mitigation, population density of species Y is similar to (or at least Z% of) population density in control population(s) Q.* Empirical studies and model simulations may help formulate proper goals, for example, to set a realistic time frame or to assess what is needed to maintain viable wildlife populations.

---

selecting species for sampling: (i) strong responses to roads and traffic, for example, high roadkill rates, or unable/reluctant to cross roads; (ii) short response times to road mitigation, for example, quickly habituate to wildlife crossing structures; (iii) relatively widespread, as this will increase opportunities to find replication and control sites; (iv) low natural variability in population densities over time, as high variability will decrease the statistical power to detect effects; and (v) readily and easily surveyed. If the list of species for evaluation still exceeds available resources, other criteria, such as even representation of different species groups, habitats and/or trophic levels, can be used.

16.3 Select parameters of interest that are most closely related to the outcome of real concern

The main drivers behind road mitigation initiatives are human safety, animal welfare and wildlife conservation. When human safety drives the mitigation efforts, the primary objective is to reduce human casualties and property damage. In this context, the most informative parameter is the number of humans killed or injured due to WVC or unsuccessful collision avoidance. Less informative but still useful parameters include (i) insurance money spent on damage to vehicles due to WVC; (ii) number of human hospitalisations due to WVC; and (iii) number of WVC in total.

When animal welfare drives the mitigation efforts, the primary objective is to prevent wildlife mortality due to WVC. The most direct measures of success are (i) the number of animals killed or injured on roads and (ii) the number of animals killed or with ill health (e.g. inbreeding depression, increased competition due to overcrowding, reduced access to food resources; Textbox 39.1) due to isolation through the barrier effect of roads. These parameters complement each other as each addresses a different mechanism through which wildlife can be positively affected by wildlife crossing structures, that is, through fewer roadkill or through increased road permeability and hence increased access to resources.

When wildlife conservation drives the mitigation efforts, the primary objective is to maintain or restore population viability. Since population viability cannot be directly measured in the field, we need to measure attributes of the population that reflect or influence the likelihood of population persistence. The most informative attribute in this respect is the trend over time in the size – or density – of the local population. For example, if existing roads are having population-level effects and crossing structures are successful in mitigating those effects, we would expect to see increases in population size after the structures are installed. If the crossing structures are installed along with the construction of a new road, successful mitigation would be indicated by no change in the size of the wildlife population. When it is not possible to estimate population size or trend, less indicative attributes may be measured, such as the number of roadkill, reproductive success, age structure, sex ratio, between-population movements, genetic differentiation or genetic variability within the population (Chapter 14). However, conclusions about the effectiveness of mitigation will be harder to make as these attributes are less closely tied to population viability.

16.4   Adopt a study design that allows for rigorous conclusions

An appropriate study design is critical for determining the effectiveness of mitigation. The optimal study design includes collection of data before and after road construction, at sites where mitigation is installed (impact sites or treatment sites) and at sites without mitigation (control sites) (i.e. BACI design, Chapter 10). Collecting data before and after road construction – or before and after installing mitigation measures at an existing road – allows the 'before' situation to be used as a reference. Collecting data at control sites ensures that measured changes can be attributed to the mitigation. For example, mitigation may reduce roadkill, but an observed reduction in roadkill could also be caused by other factors, such as a decrease in population density, increased road avoidance behaviour or changes in traffic volume. An important assumption is that impact and control sites are similar in all relevant respects. As this assumption is rarely met, replication of both impact and control sites is recommended (Fig. 16.2).

16.5   Use model simulations to determine the best sampling scheme

The sampling scheme includes (i) the duration of monitoring before and after mitigation; (ii) the frequency of monitoring; and (iii) the number of replicate sites. As these parameters are unlikely to be independent, model simulations can help elucidate the optimal sampling scheme by exploring the relationship between the probability of detecting an effect of mitigation, the duration of monitoring and the number of replicate sites (Fig. 16.3). Figure 16.3 shows that to achieve an 80% probability of detecting an effect of road mitigation that is 80% effective, we should measure the parameter of interest for about 26 years, assuming we monitor at only one study site. However, monitoring duration can be significantly shortened to 12.5 years if 10 study sites are monitored simultaneously. Increased replication will reduce the uncertainty in effect size, allowing a reliable conclusion about road mitigation effectiveness to be reached sooner. The ability to detect a significant effect also depends on the accuracy of the measurements – and the more accurate they are, the shorter the monitoring period. In the event of insufficient funds and/or sites for adequate replication, novel sampling designs and the pooling of data across studies have the potential to achieve reliable outcomes (Chapter 10). Similar graphs can be produced for other relationships between sampling variables such as the duration of monitoring and sampling frequency, or the number of replicate sites and sampling frequency.

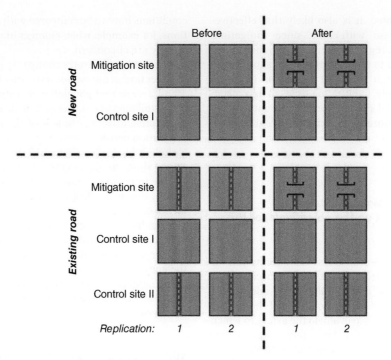

Figure 16.2 The optimal study design to evaluate the effectiveness of wildlife crossing structures when installed at a new or existing road. Two types of control sites are possible: (i) no road and (ii) no mitigation. Preferably, both mitigation and control sites are replicated. Note that the actual number of replicates required will likely exceed two, and it depends on a range of factors. Source: E. van der Grift.

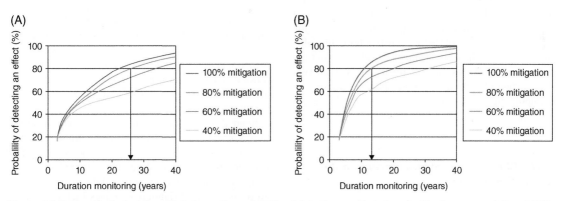

Figure 16.3 Hypothetical relationship between the probability of detecting an effect of road mitigation on population viability, the duration of monitoring after the mitigation measures are put in place and the number of study sites: 1 study site (A) or 10 study sites (B). The four scenarios illustrate variations in the expected effectiveness of mitigation, for example, road mitigation is expected to reduce the road effect by 100, 80, 60 or 40%. Source: van der Grift et al. (2009).

There is no 'one sampling scheme fits all' approach. The sampling scheme is closely related to the chosen parameter of interest and the characteristics of the studied species. For example, a species with a low reproductive rate and long lifespan would require monitoring over a longer period to assess a change in population density, compared to a species with a high reproductive rate and short lifespan. Because the rate of use of crossing structures often increases over time, and if use and effectiveness are positively correlated, it is also likely that effectiveness will increase with time since mitigation. Therefore, sampling should not begin before an effect is expected to have occurred and should continue long enough to detect lagged or transient effects, for example, habituation times to new structures. Using model simulations prior to data collection can help ensure the monitoring programme is cost-effective and achievable and occurs at the appropriate time.

16.6   Select mitigation sites to be monitored based on the objective(s) of the evaluation

If the evaluation is to assess the effectiveness of multiple crossing structures, it may be necessary to sample a subset of those available, the selection of which should be guided by the objective of the evaluation. If the objective is to evaluate the extent to which the mitigation is effective for a target species, one should choose a random sample of mitigation sites from the total available. If the objective is to evaluate whether a specific type of crossing structure mitigates road impacts for a target species, one should choose sites that are most likely to demonstrate statistically significant effects. In that case, select sites where (i) the road effect is known or expected to be high; (ii) the timing of mitigation allows for sufficient time for repeated measurements before construction; (iii) sufficient replicate sites are available; and (iv) multiple mitigation measures are planned for a relatively long section of road as this may allow for phasing or manipulating mitigation in an experimental design (Chapter 10).

16.7   Choose control sites based on the goals of mitigation

Control sites must be carefully selected to ensure the comparison between the mitigation and control is valid. The goals for mitigation determine which types of control sites are needed: control sites with an unmitigated road, control sites with no road or both (Fig. 16.2). The former option applies when post-mitigation conditions have to be compared with pre-mitigation conditions, for example, when comparing movements before and after mitigation. Control sites without roads are required when post-mitigation conditions have to be compared with pre-road conditions, for example, when changes in population size/density are of concern.

Figure 16.4 illustrates changes in population density over time at mitigation and control sites. Scenarios 1 and 2 show that population density increased with the installation of road mitigation measures. However, a comprehensive assessment of the extent to which population density improves can only be made if we include some 'no-road' control sites. The other scenarios show no improvement (Scenario 3) or even a decline in population density (Scenario 4) after mitigation, due to ineffective mitigation. Proper assessments of the extent to which declines in population density have been mitigated can only be made if we include 'road without mitigation' control sites. Similar scenarios can be developed for cases where the construction of a new road and mitigation take place simultaneously, except that the trajectories would all start at the level of the no-road control at time zero.

16.8   Measure explanatory variables that provide the best possible estimates of mitigation effectiveness

Variables other than the parameters of interest should also be measured to improve interpretation of the results, provide better comparisons among study sites and allow for stronger inferences concerning the causes of observed differences. We recommend documenting spatial (among sites) and/or temporal (over time) variability in (i) road design and traffic; (ii) crossing structure design and use; and (iii) features of the surrounding landscape.

Road design variables include road width, elevation, pavement, lighting, fences, noise screens, median strip and verge widths and vegetation. Traffic volume and speed should be documented at several temporal scales, for example, daily, seasonally or annually. Road mitigation variables include characteristics of the mitigation measure, passage through the crossing structure by the target and non-target species and presence and frequency of co-use by humans and domestic animals. Information on the duration of the construction period and the date that the mitigation measures were ready for use may also be important. Finally, landscape variables include altitude, topography, land use, type and amount of vegetation and the occurrence of characteristic landscape elements.

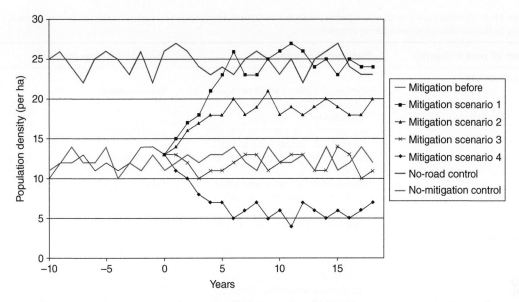

Figure 16.4 Hypothetical result when evaluating the effectiveness of road mitigation measures at an existing road. Mitigation measures are installed at time zero. In addition to the mitigation site, measurements are carried out – before and after mitigation – at a no-mitigation control site and a no-road control site. Generally, there are four possible scenarios: (1) the road mitigation measures are 100% effective, and population density increases to the level of the no-road control site; (2) the road mitigation measures are only partly effective, and population density increases compared to the unmitigated situation but does not reach the level of the no-road control site; (3) the road mitigation measures are not effective, and population density remains similar to the density in the unmitigated situation; and (4) the road mitigation measures are not effective and even worsen the situation by reducing population density compared to the unmitigated situation. Source: van der Grift et al. (2013).

16.9 Utilise survey methods that monitor multiple species simultaneously

Survey methods that monitor multiple species simultaneously should be used because they provide more information for similar effort (Table 16.1, Chapter 11). Where possible, more than one survey method should be used for each species to decrease bias and improve estimates of the parameter of interest. Consistent use of the same methods and personnel is essential across control/mitigation sites and important over time to provide comparable results (Lesson 10.5).

16.10 Ensure that the evaluation of mitigation has sufficient resources, is integrated into the planning and construction of roads and is coordinated across boundaries

Discussion about the evaluation of mitigation often begins after construction has finished. It is, however, essential that the evaluation of the mitigation be integrated into all other aspects of the road project (Chapter 10). Indeed, the monitoring plan should be a condition of approval, and minimum standards set to ensure it is fully planned and funded before construction begins. This is essential to ensure that sufficient 'pre-construction' (i.e. before) data can be collected and that there are sufficient resources to complete the monitoring and determine effectiveness of mitigation. A feasibility assessment prior to commencing a study may result in the inclusion of additional sites to allow for some that may become unsuitable during the study. If there is insufficient money to adequately complete a project, we recommend conducting fewer scientifically rigorous studies that are more likely to contribute new knowledge than numerous poorly designed studies. One approach to achieve adequate funding and better study design is to pool resources (including funding, study sites) across projects and jurisdictional boundaries (Lesson 10.7). The data should be analysed and reported in a timely manner to ensure existing structures can be modified within an adaptive framework and the design of future mitigation measures is improved.

Table 16.1 Potential survey method(s) for each parameter of interest. The list provides some examples of frequently used survey methods and is not aimed at being complete.

Driver of road mitigation	Parameter of interest	Potential survey method
Human safety		
	Number of humans killed or injured due to WVC or to collision avoidance	Questionnaire
	Insurance money spent on material/immaterial damage due to WVC	Questionnaire
	Number of hospitalisations due to WVC	Questionnaire
	Number of WVC with species that potentially impact human safety, regardless of whether they resulted in human injury or death	Road surveys
Animal welfare		
	Number of animals killed or injured while crossing roads	Road surveys
	Number of animals killed or with ill health due to isolation from needed resources through the barrier effect of roads	Field surveys
Wildlife conservation		
	Trend in population size/density	Capture–mark–recapture; point/transect counts or calling surveys; pellet counts; nest/den counts; tracking arrays, for example, photo/video cameras, track pads
	Number of animals killed	Road surveys
	Reproductive success	Counts of eggs/young
	Age structure	Capture, direct observation
	Sex ratio	Capture, direct observation
	Between-population movements	Capture–mark–recapture, radio-tracking, direct observation, tracking arrays
	Genetic differentiation	Invasive DNA sampling after capture; non-invasive DNA sampling, for example, through hair traps, scat collection, antler/skin collection
	Genetic variability	Invasive DNA sampling after capture, non-invasive DNA sampling

Source: van der Grift et al. (2013).

CONCLUSIONS

A comprehensive evaluation of the extent to which road mitigation measures reduce the risk of decline and extinction of wildlife populations is essential to ensure that funds are being cost-effectively allocated. To achieve this, transportation and permitting agencies need to move beyond asking consultants or researchers to simply record the use or measure the rate of crossing by fauna, rather to insist that evaluations determine the extent to which wildlife crossing structures have mitigated the negative effects of the road. Researchers must be involved in the design of the evaluation programmes from the earliest stages of the project and need to inform the road agency of the essential components of good study design. The importance and benefits of road mitigation measures should be better communicated to all stakeholders. The guidelines presented here are ambitious, but they are necessary to improve our understanding of the effectiveness of road mitigation measures.

FURTHER READING

Clevenger and Sawaya (2010): A pilot study to assess potential population-level benefits of wildlife crossing structures through non-invasive genetic sampling.

Fahrig and Rytwinski (2009): A review of the empirical literature on the effects of roads and traffic on animal abundance and distribution and the development of a set of predictions of the conditions that lead to negative or positive effects or no effect of roads on animal abundance.

van der Ree et al. (2009): A modelling study in which population viability analysis was used to assess the effectiveness of tunnels for the endangered mountain pygmy possum in Australia.

Roedenbeck et al. (2007): A research agenda for road ecology and recommendations for experimental designs that maximise inferential strength.

REFERENCES

Clevenger, A. P. and M. A. Sawaya. 2010. Piloting a non-invasive genetic sampling method for evaluating population-level benefits of wildlife crossing structures. Ecology and Society **15**: 7. http://www.ecologyandsociety.org/vol15/iss1/art7 (accessed 22 September 2014).

Fahrig, L. and T. Rytwinski. 2009. Effects of roads on animal abundance: An empirical review and synthesis. Ecology and Society **14**: 21. http://www.ecologyandsociety.org/vol14/iss1/art21 (accessed 22 September 2014).

Roedenbeck, I. A., L. Fahrig, C. S. Findlay, J. E. Houlahan, J. A. G. Jaeger, N. Klar, S. Kramer-Schadt and E. A. van der Grift. 2007. The Rauischholzhausen agenda for road ecology. Ecology and Society **12**: 11. http://www.ecologyandsociety.org/vol12/iss1/art11 (accessed 22 September 2014).

van der Grift, E. A., H. A. H. Jansman, H. P. Koelewijn, P. Schippers and J. Verboom. 2009. Effectiveness of wildlife passages in transport corridors – guidelines for the set-up of a monitoring plan. Alterra-report 1942. Alterra, Wageningen.

van der Grift, E. A., R. van der Ree, L. Fahrig, S. Findlay, J. Houlahan, J. A. G. Jaeger, N. Klar, L. F. Madriñan and L. Olson. 2013. Evaluating the effectiveness of road mitigation measures. Biodiversity and Conservation **22**: 425–448. doi:10.1007/s10531-012-0421-0.

van der Ree, R., M. A. McCarthy, D. Heinze and I. M. Mansergh. 2009. Wildlife tunnel enhances population viability. Ecology and Society **14**: 7. http://www.ecologyandsociety.org/vol14/iss2/art7 (accessed 22 September 2014).

Chapter 17

HOW TO MAINTAIN SAFE AND EFFECTIVE MITIGATION MEASURES

Rodney van der Ree[1] and Stephen Tonjes[2]

[1]Australian Research Centre for Urban Ecology, Royal Botanic Gardens Melbourne, School of BioSciences, The University of Melbourne, Melbourne, Victoria, Australia
[2]District Five Environmental Management Office, Florida Department of Transportation (retired) and private environmental consultant, DeLand, FL, USA

SUMMARY

Roads and their associated infrastructure require regular inspection and maintenance to detect and repair faults before they pose a hazard to motorists and while the cost of repair remains relatively low. Mitigation measures for wildlife (e.g. crossing structures, wildlife detection systems, fencing) also need to be inspected and maintained to ensure they remain structurally sound and functional. Monitoring equipment must also be adequately maintained to ensure the monitoring programme is not disrupted. However, maintenance is often overlooked or ignored, resulting in ineffective mitigation measures that ultimately waste money and endanger wildlife and road users.

17.1 Routine roadside maintenance must not reduce the effectiveness of a mitigation measure.

17.2 Inspection and maintenance of wildlife mitigation measures must address both structural and functional integrity.

17.3 Ecological experts and maintenance engineers should develop and review maintenance programmes together.

17.4 Develop specific systems, procedures and funding for maintenance of mitigation measures, and incorporate them into existing programmes.

17.5 Maintenance engineers must be involved in the design stages of the project.

17.6 The maintenance programme should facilitate research and monitoring.

The development and implementation of schedules and procedures to maintain mitigation measures is critical, and transportation agencies must incorporate the maintenance of mitigation measures into everyday practice.

Handbook of Road Ecology, First Edition. Edited by Rodney van der Ree, Daniel J. Smith and Clara Grilo.
© 2015 John Wiley & Sons, Ltd. Published 2015 by John Wiley & Sons, Ltd.
Companion website: www.wiley.com\go\vanderree\roadecology

INTRODUCTION

New roads usually take years to plan, design and construct and are often eagerly awaited to improve safety, congestion or access. Road maintenance, on the other hand, is like housework – nobody notices unless it *doesn't* get done. Maintenance of mitigation measures is often neglected because it is far removed from the planning and design stages of a road project. Planning and design usually receive much more attention from wildlife experts and the general public than routine maintenance regimes. This is extremely problematic because even minor deteriorations or small maintenance errors (like the small design errors discussed in Chapter 9) can seriously compromise the proper functioning of mitigation measures.

Maintenance programmes must be developed and implemented for all mitigation measures to ensure they continue to achieve their goals over their full lifespan. A maintenance programme consists of four key elements: (i) an inventory of the asset; (ii) an inspection schedule; (iii) routine upkeep or repairs triggered by deterioration of condition; and (iv) an adaptive response to new knowledge or understanding about maintenance standards or techniques. Transportation agencies must include mitigation measures in their routine maintenance programmes (as they do with other features of the road network) to ensure the financial investment is not wasted, human safety is not compromised and wildlife are not placed at further risk of endangerment due to poorly maintained structures. The aim of this chapter is to highlight why the maintenance of wildlife mitigation measures must occur, how to ensure it occurs and what needs to be done. Maintenance of roadsides and roadside vegetation is primarily addressed in Chapter 46.

LESSONS

17.1   Routine roadside maintenance must not reduce the effectiveness of a mitigation measure

The standards and processes for routine road maintenance programmes have developed over many years, usually without consideration of wildlife. Therefore, it is not surprising that some of these practices are incompatible with the needs of wildlife. Standard roadside maintenance practices are frequently applied to mitigation measures or the areas around them, often with disastrous consequences. For example, wildlife culverts with a sandy substrate have been 'cleaned out' to bare concrete, which is a standard maintenance procedure for drainage culverts. Mowing or pruning of roadside vegetation around structures can result in unsuitable access to crossing structures for some species of wildlife. Depending on the specific needs of the target species, the vegetation at entrances to crossing structures may need to be left longer when cutting, mowed less frequently or not mowed at all. Mitigation measures are rarely cheap, so inappropriate maintenance of structures or adjacent areas can make them ineffective and financially wasteful.

One challenge is to ensure that maintenance crews know the specific requirements of each mitigation measure, and this can be achieved through careful mapping, physical demarcation of sites with signs, regular induction meetings and site inspections. This challenge is exacerbated by the relatively frequent turnover of maintenance crews and contractors.

17.2   Inspection and maintenance of wildlife mitigation measures must address both structural and functional integrity

For each mitigation measure, maintenance programmes need to cover structural integrity (e.g. will the bridge, underpass or fence collapse; is the infrared beam on the wildlife detection system operational?) and functional integrity (e.g. is the exclusion fence, overpass or noise wall achieving its purpose?).

The inspection and maintenance of structural integrity will be similar to those for standard road features. In contrast, the inspection and maintenance of functional integrity is more complicated and is likely to be outside the current scope of standard maintenance activities. For example:
• Underpasses usually function better as wildlife passages if they have a natural (rather than concrete) floor, and road agencies must ensure that the right amount of soil and leaf litter remains on the floor of the underpass.
• Combined wildlife underpasses and drainage culverts (Lesson 21.3) may need periodic cleaning to keep them open while retaining sufficient substrate material, especially open-topped tunnels for amphibians, which may accumulate toxic de-icing chemicals in the winter.
• Some species may need 'fauna furniture' (e.g. rocks or brush; Figs 39.1 and 39.2) in underpasses for

shelter; however, excess furniture can accumulate after floods, impeding access. Sufficient cleaning is required to restore 'openness' without excessive removal of cover.

• The density of shrubs and trees on wildlife overpasses and at the entrances to underpasses influences the rate of use by different species, and it is important to maintain the vegetation density and mix of species such that it favours the target species.

• Vegetation, fencing and/or soil berms used as screening between people and wildlife on co-use crossing structures (Chapter 22) must be sufficiently dense to minimise human disturbance and encourage use by wildlife.

Although individual mitigation measures may need different maintenance treatment, these activities should be incorporated into the same maintenance programme employed for most other road features. These maintenance (or asset management) programmes should include (i) an inventory of each asset; (ii) an inspection schedule; (iii) what to inspect; (iv) what thresholds in condition will trigger a response; and (v) the details of the response or specific maintenance procedures needed (i.e. what maintenance or remedial actions are required and who is responsible).

The maintenance programme for mitigation measures should also provide information on the goal of the mitigation measure, the target species and some relevant biological and ecological information. This is to ensure that any subsequent maintenance decisions are beneficial for the species and to allow new knowledge or understanding about maintenance standards to be incorporated adaptively.

17.3   Ecological experts and maintenance engineers should develop and review maintenance programmes together

A big challenge facing managers is that the specifics of a maintenance programme for a mitigation measure are often unknown. For example, a timber pole needs to be replaced when it fails a torque resistance test (a measure of wood rot). In contrast, we may not know the optimum depth of substrate on a culvert floor nor the threshold in depth of water pooling at an underpass entrance that will prevent wildlife from using it. While these factors might not reduce the efficacy of the mitigation measure for some species, it may be critical for others.

A common-sense approach is probably sufficient to develop an initial cost-effective maintenance programme. For example, the preferred habitat of most target species is reasonably well understood, and maintenance should aim to reproduce those conditions in and around the mitigation measure. Barrier fencing (to prevent wildlife from entering the roadway; Chapter 20) for a species that migrates annually across a road (such as a frog; Chapter 31) should be inspected and repaired shortly before migration commences and periodically throughout the migration (Lesson 31.7). If there is any doubt about the specifics, an experimental approach (Chapter 10) should be used to refine the maintenance programme in response to measured outcomes. An adaptive approach and sharing of experiences among agencies (perhaps through the regional road ecology networks; Chapter 61) would also help to refine maintenance programmes to ensure structural and functional integrity.

17.4   Develop specific systems, procedures and funding for maintenance of mitigation measures, and incorporate them into existing programmes

There is a tendency around the world to build mitigation measures and then 'forget' about them. Agencies responsible for maintenance need ecological assistance and funding to develop and implement the maintenance programme. Road agencies often have insufficient funds for standard road maintenance programmes, let alone the maintenance of mitigation measures as well. The maintenance budget is also one of the first to be cut because it is typically the largest expenditure of road agencies and is perceived politically as the least exciting. The amount of funding for maintenance should be accounted for at the beginning of the project and provided for in the long-term planning of transportation agency budgets.

As described in Lesson 17.2, maintenance of mitigation programmes must be incorporated into established road maintenance programmes. Caution should be exercised if mitigation maintenance is not fully integrated with existing programmes because this exacerbates the risk of substandard or inappropriate maintenance. Programmes should also ensure a failsafe or double-check system to ensure the maintenance has occurred. The incorporation of maintenance regimes into best-practice manuals, guidelines and standards is essential (Chapter 59).

17.5   Maintenance engineers must be involved in the design stages of the project

Mitigation measures may be unnecessarily difficult and expensive to maintain if they are designed without input from maintenance experts. Maintenance engineers should be included in the development and review of project concepts and design plans, preferably as part of a design task team (Chapter 9). Including maintenance considerations at this early stage will increase the ease and efficiency of maintenance after construction. Furthermore, wildlife experts and maintenance engineers should collaboratively develop protocols for maintaining each individual feature and review and refine these over time. However, ease of maintenance should not be the primary influence on the design – an easily maintained but inappropriate structure is not an optimal solution!

17.6   The maintenance programme should facilitate research and monitoring

Research and monitoring is frequently conducted to assess the use and effectiveness of mitigation measures (Chapter 10). Maintenance such as mowing grass, pruning trees or clearing obstructions in a culvert may affect the rate of use of a crossing structure by wildlife or modify the effectiveness of a wildlife exclusion fence.

If the research and monitoring team is unaware of these maintenance activities, they may incorrectly conclude the structure is more or less effective than it truly is. Ideally, maintenance activities would be experimentally incorporated into the research and monitoring programme to improve the maintenance programme adaptively (Lesson 17.3).

Mitigation measures can also be designed to include features to facilitate monitoring. A simple example is the inclusion of gates in exclusion fencing to allow staff to access the crossing structure. Locating glider poles outside the 'clear zone' or behind guard rails will eliminate the need to divert or slow traffic during maintenance (Fig. 17.1). The provision of access tracks can facilitate maintenance of crossing structures, and especially exclusion fences. In some regions, workers require specialist training to enter enclosed spaces, such as small underpasses, and the installation of larger structures may eliminate this extra maintenance cost. Of course, it is critical to assess if such modifications are feasible within the constraints of the project and whether they would reduce effectiveness for the target species.

Equipment to monitor the effectiveness of mitigation will also inevitably require maintenance, and these maintenance needs should be incorporated into the design of the road/mitigation measure. For example, routine annual maintenance of cameras on glider poles (Fig. 17.1, Chapter 40) can be scheduled for times

(A)   (B)

Figure 17.1 Placing glider poles behind safety rails with enough clearance for truck access and solid ground (A) allows year-round access without the added cost of traffic control (B). Source: Photographs by R. van der Ree.

of year when the ground is dry enough to drive a travel tower on site. Emergency repairs during winter may need to be delayed until the ground dries out or could involve expensive earthworks if needed urgently. Earthworks to improve access to a mitigation measure (e.g. improving drainage and laying a stable base, avoiding steep access tracks, filling in of ditches) that are completed during the construction process will be more cost-effective than trying to retrofit to a completed project.

CONCLUSIONS

Transportation agencies have standard maintenance programmes for road assets that must be modified to include the specialised needs of wildlife mitigation measures. Due to the wide variety of mitigation measures deployed around the world and their diverse range of specific goals, it is impossible to develop a one-size-fits-all maintenance programme within the limits of this chapter. We advocate a common-sense approach based on the needs of the target species initially, with regular review and adaptation over time to refine the programme. Importantly, we recommend a greater level of cooperation and sharing of knowledge among agencies to ensure functional integrity of mitigation measures and avoid costly mistakes.

ACKNOWLEDGEMENTS

We thank Kylie Soanes for helpful comments on an early draft. RvdR is supported by the Baker Foundation.

Chapter 18

UNDERSTANDING AND MITIGATING THE NEGATIVE EFFECTS OF ROAD LIGHTING ON ECOSYSTEMS

Bradley F. Blackwell, Travis L. DeVault and Thomas W. Seamans

USDA/APHIS/WS National Wildlife Research Center, Ohio Field Station, Sandusky, OH, USA

SUMMARY

Natural light plays an integral role in biological systems, one that can be disrupted by the intrusion of other light sources. Specifically, artificial lighting, including road lighting, poses negative effects on plant and animal physiology, animal behaviour and predation rates. These effects are cumulative as multiple, artificial light sources contribute.

18.1 Light functions as a natural stimulus.
18.2 Metrics used to quantify artificially produced light are generally not biologically relevant.
18.3 Species response to artificial light varies by visual system.
18.4 Light emitted varies relative to the type of lighting technology.
18.5 Planning for road lighting must include zoning relative to light levels and light-fixture placement.
18.6 Mitigating the negative effects of road lighting requires research collaboration.

Negative effects of artificial lighting, including road lighting, are manageable. By better understanding the ecosystems through which roads pass and how light affects resident organisms, we can adapt lighting fixtures, fixture design and zoning to minimise site-specific effects, as well as contributions to cumulative light pollution.

Handbook of Road Ecology, First Edition. Edited by Rodney van der Ree, Daniel J. Smith and Clara Grilo.
© 2015 John Wiley & Sons, Ltd. Published 2015 by John Wiley & Sons, Ltd.
Companion website: www.wiley.com\go\vanderree\roadecology

INTRODUCTION

A critical aspect of road planning involves driver and pedestrian safety, and road lighting is a key component (IDA/IES 2011). However, decisions on how, where and when to use artificial lighting have immediate implications for the well-being of ecosystems through which roads pass. Specifically, light is a natural stimulus that affects the physiology, behaviour and movements of all organisms. Artificial lighting alters the length of natural photoperiod (duration of daily exposure to light) and contrasts in intensity and spectrum with natural, ambient light, thus unavoidably affecting the sensory ecology of organisms. Further, artificial light poses cumulative effects on ecosystems because multiple light sources are often present in a given area (Fig. 18.1). Cumulative effects are expressed differentially across species, because not all light sources are equal in their effects on physiology or behaviour.

To mitigate negative effects to natural systems by artificial lighting used on roads, planners must first consider whether lighting is necessary. If so, they must consider not only the varying sensitivity of the human eye to different light wavelengths relative to driver and pedestrian safety but also the biological relevance of lighting to the resident organisms. Our goal is to provide road practitioners, engineers and ecologists with a concise review of resources available to aid in the reduction of the negative effects of road lighting on ecosystems.

LESSONS

18.1   Light functions as a natural stimulus

Light exists as particles (photons) and waves and is described relative to wavelength (Fig. 18.2). Natural light plays a significant role in the sensory ecology of animals, particularly with regard to photoperiod, which stimulates (i) circadian rhythms important to the basic health and development of plants and animals (e.g. growth, reproduction and disease resistance) and (ii) daily and seasonal physiology and behaviour of animals (e.g. foraging, breeding, dispersal and migration). In addition, animals use light cues in predator detection, habitat selection and vehicle avoidance (Gaston et al. 2012, 2013).

18.2   Metrics used to quantify artificially produced light are generally not biologically relevant

Consideration given to design of light fixtures and emission spectra (i.e. the distribution of wavelengths emitted by a lamp; Fig. 18.2) generally fails to consider the

Figure 18.1 Multiple light sources, including road lighting, from Dubai, UAE, contributing to cumulative artificial light pollution. Photograph credit: Expedition 30 Crew to the International Space Station for the Earth Observations Experiment and Image Science & Analysis Laboratory, Johnson Space Center; U.S. National Aeronautics and Space Administration (http://earthobservatory.nasa.gov/IOTD/view.php?id=77360). Source: Photograph from Earth Observatory, NASA.

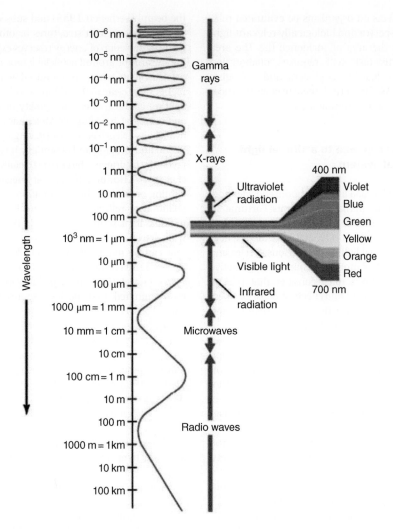

Figure 18.2 The electromagnetic spectrum and the portion of the spectrum visible to most animals, represented in nanometres (nm or 1×10^{-9} m) from 400 to 700 nm. Notably, many non-primate species of animals have the capability to detect wavelengths in the near ultraviolet (300–400 nm). Figure credit: U.S. National Aeronautics and Space Administration; http://science-edu.larc.nasa.gov/EDDOCS/Wavelengths_for_Colors.html. Source: NASA.

biological relevance of the light stimulus. For example, light emitted from artificial sources is typically not quantified relative to wavelength, but in lumens (i.e. the luminous flux or power from a light source) and illuminance (the total luminous flux incident on a surface per unit area). However, we cannot effectively understand animal response to light stimuli if the measurements (metrics) of fixture design and performance are in units of power.

Plants and animals respond directly to the intensity or number of photons per wavelength striking photoreceptors in their eyes (Endler 1990; Rich & Longcore 2006; Gaston et al. 2013). For example, 1 W of light at 400 nm (Fig. 18.2) has only 57% of the photon flux as 1 W of light at 700 nm (Endler 1990). In other words, the total energy reported is 1 W at both wavelengths, but the biologically relevant metric, photon flux by wavelength, differs by greater than 50%. As such, the lumen and luminous flux are inaccurate metrics for discerning biological effects because they do not take into account the density of photons striking photoreceptors. We suggest that light fixtures and potential effects on organisms be evaluated relative to emission spectra and biologically relevant light intensity within the area of incidence (i.e. the area illuminated). This task will require collaboration among planners, lighting engineers and ecologists (Lesson 18.5). As for actual measurements, these should be taken via spectroradiometer.

18.3 Species response to artificial light varies by visual system

Effective planning for road lighting should consider how light affects organisms in roadside habitats. Fortunately, recent research (e.g. Rich & Longcore 2006; Horváth et al. 2009; IDA 2010; Gaston et al. 2013) details the negative effects of artificial lighting on various species and ecosystems. In short, planning for road lighting relative to potential biotic effects must consider that relative brightness of artificial light and effects of emission spectra on organisms vary with the sensory (plants) and visual physiology of the animals affected.

For example, human vision is trichromatic, meaning that we possess three independent channels for detecting and processing colour. However, many non-primate animals perceive the world in a much different way. Birds are generally tetrachromatic and capable of detecting wavelengths within the ultraviolet portion of the electromagnetic spectrum (Hart 2001; Fig. 18.2), whereas few bird species rely on scotopic or rod-dominated vision (i.e. rod photoreceptors are primarily sensitive to light intensity, such as under dim-light conditions). Further, the ability to perceive colour is dependent on the number of different visual pigments present in cone photoreceptors.

The influence of natural light is evident with changes in photoperiod that influence the timing of seasonal events in birds (e.g. effects on breeding physiology) and even mate selection (Dawson et al. 2001; de Molenaar et al. 2006). The addition of artificial light can interfere with this natural stimulus (de Molenaar et al. 2006). Also, a light-sensitive 'magnetic compass' aids orientation during night-time migration (especially when cloud cover prevents the use of stars as visual cues); this innate navigational ability can be confounded by specific wavelengths from artificial lighting (e.g. >500 nm; Poot et al. 2008).

Perhaps the most well-known effect of artificial light on birds is the attraction to, and disorientation by, high-intensity glare from warning beacons on communication towers, offshore oil platforms and other structures (Gauthreaux & Belser 2006). Birds migrating at night and attracted to such lighting can become 'trapped by

the beam' (Verheijen 1985) and subsequently die from direct collisions with structures or other birds or indirectly by depletion of energy reserves expended while flying towards or around artificial lights. Bird attraction to artificial lights is more pronounced on cloudy and misty nights than clear nights (Montevecchi 2006). Artificial lighting can also affect the quality of breeding habitat and timing of breeding (de Molenaar et al. 2006), prey availability (Negro et al. 2000), singing patterns (Miller 2006) and foraging and potentially increase exposure to predators by drawing birds to artificially lit areas (Santos et al. 2010). However, the primary negative effect of road lighting on birds is the contribution to cumulative light pollution of reflected or escaping light skywards from multiple light sources (Fig. 18.1) (light that interferes with detection of celestial migration cues), a problem that can be managed by fixture design (Lesson 18.4; Fig. 18.3) and zoning (Lesson 18.5).

In contrast, the visual capability of bats is primarily rod dominated, and species response to road lighting varies by level of illumination and area affected (Lesson 18.5). Foraging opportunities for bats can be enhanced due to insect attraction to light (Eisenbeis 2006; Lesson 34.3), but increased competition with other bat species and avoidance of lighting can also pose negative effects (Rydell 2006; Zurcher et al. 2010; Stone et al. 2012). Bats attracted to road lighting are also susceptible to vehicle collisions (Zurcher et al. 2010; Chapter 34). For the most part, however, effects of road lighting on bat species are manageable via attention to light-fixture location, lamp illuminance and shielding (Fig. 18.3; Lessons 18.4 and 18.5).

Similarly, other terrestrial mammal species (e.g. rodents) are also susceptible to disruption in photoperiod and migration, as well as enhanced predation associated with artificial lighting. As with bats, light-fixture location, lamp illuminance and shielding (Lessons 18.4 and 18.5) can be adapted to the particular species affected by road lighting (see Rich and Longcore (2006) for detailed discussion of artificial lighting effects on terrestrial mammals).

Few studies have examined the effects of road lighting on amphibians and reptiles or reported biologically relevant metrics of light intensity for these species or other taxa (Perry et al. 2008). An exception is the well-documented negative effect of artificial lighting on sea turtles (Salmon 2006). Also, as with birds, the magnetic compass in amphibians is affected by light wavelengths greater than 500 nm (Diego-Rasilla et al. 2010), a spectral range falling within that of sodium-vapour lamps often used along roads (Rydell 1992). An effective management approach to reduce negative effects of road

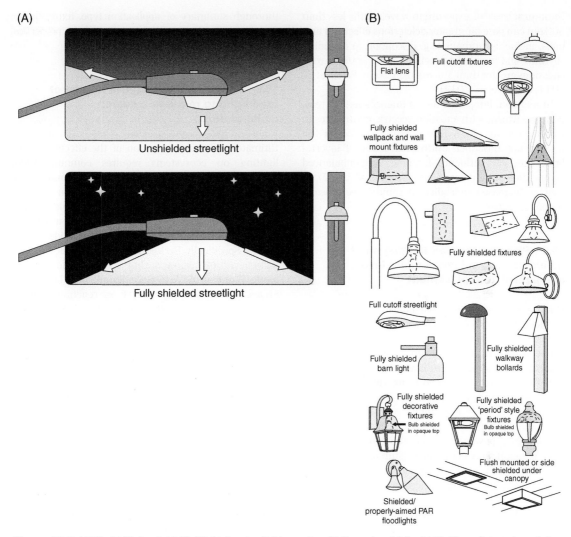

Figure 18.3 (A) Unshielded and shielded light showing light escaping. (B) Examples of fully shielded lamp fixtures intended for structures as well as road applications. Source: Reproduced with permission of R. Crelin (www.BobCrelin.com).

lighting on amphibians and reptiles includes species-specific considerations relative to light-fixture location, position above or in the road (i.e. road-embedded lighting), emission spectra and intensity, shielding (Fig. 18.3) and on/off schedules (Lessons 18.4, 18.5 and 18.6).

18.4 Light emitted varies relative to the type of lighting technology

Current options for selection of road lighting technology include standard high-intensity discharge sources (a lamp technology with emission from 550 to 650 nm;

Rich & Longcore 2006; Fig. 18.2) and the more recently introduced solid-state light-emitting diodes (LEDs), often marketed as 'cool white' LEDs. Despite the name (associated with how humans perceive light from these devices), energy emitted by these LEDs commonly include wavelengths from 450 to 460 nm, thus falling into the blue range of the electromagnetic spectrum (IDA 2010; see also Gaston et al. 2012; Fig. 18.2). Advocates of these devices contend that they afford lower illumination levels because of the sensitivity of human rod cells to shorter wavelengths (IDA 2010; Falchi et al. 2011). However, caution is recommended when considering widespread use of this lighting.

Unnatural levels of exposure to wavelengths less than 500 nm can pose far greater deleterious effects on animals, including humans (e.g. disruption in circadian rhythms and metabolic function), than sources with emissions greater than 500 nm (IDA 2010; Falchi et al. 2011; Gaston et al. 2012).

In addition, lamp type also influences fixture temperature which, with emission spectra, contributes to insect attraction, as well as energy required for full illumination (Eisenbeis 2006). Attraction of invertebrate prey can influence foraging and imbalanced competition among bat species (Stone et al. 2012), as well as increased mortality to some insect species (Eisenbeis 2006).

18.5 Planning for road lighting must include zoning relative to light levels and light-fixture placement

Questions that should be asked during road planning include: What level of illumination is required, if any? How would planned lighting contribute to cumulative artificial light pollution within an ecosystem? What emission spectra would pose fewer direct negative consequences to species exposed to lighting? How might lighting indirectly affect animals by attracting and concentrating prey? How might light-fixture design, zoning and placement help reduce negative effects on organisms? Typically, recommended light-fixture type, area of effect and cumulative illumination by road lighting vary by human population density, level of human activity and the interspersion of protected natural areas.

The IDA/IES (2011) provides zoning guidance to balance illumination relative to the needs of people and ecosystems adjacent to the road, though guidance is not specific to biological light intensity. Within specified zones, and considering the type of site (e.g. road through residential or non-residential area), the IDA/IES recommends Total Initial Luminaire (TIL; lumens per site) and Maximum Allowable Backlight, Uplight and Glare (i.e. 'BUG') ratings. Essentially, each zone and associated TIL/BUG rating represents a broad approach to mitigating effects of light pollution. In addition, fixture orientation and shielding (also affecting the TIL/BUG rating) should limit upward incidental reflection or direct emission so as to reduce light escaping skywards (Fig. 18.3), which contributes to skyglow and attraction of insects or migrating birds (Eisenbeis 2006; Salmon 2006; Luginbuhl et al. 2009; Falchi et al. 2011; IDA/IES 2011). The IDA/IES (2011), in particular, provides a

thorough summary of application type, fixture/lamp designs, associated metrics describing light properties and guidance on zoning and BUG ratings.

18.6 Mitigating the negative effects of road lighting requires research collaboration

Ultimately, effective mitigation of the effects of road lighting on ecosystems requires communication among road planners, lighting engineers and ecologists. An example of such collaboration is an advance in lighting technology that allows for complete elimination of traditional overhead road lighting where the intent is for driver orientation and not roadside illumination. Specifically, Bertolotti and Salmon (2005) and Salmon (2006) showed that road-embedded LEDs along Highway A1A in Boca Raton, Florida, United States, prevented stray light from reaching nearby beaches, thus reducing the nocturnal disorientation of dispersing sea turtle hatchlings. In addition, we suggest that future research in the development of lighting technology and application consider (i) light-fixture performance measured in terms of biologically relevant light intensity; (ii) lamp designs that are easily adaptable to wavelength and intensity requirements; and (iii) daily and seasonal scheduling for operation relative to the ecosystem affected.

CONCLUSIONS

Depending upon concerns for driver or pedestrian safety, an obvious solution to managing negative effects of road lighting in conservation areas is to avoid the use of road lighting altogether. However, where lighting is deemed necessary, it is also important to recognise that a 'one-size-fits-all' approach to road lighting will not minimise negative effects to ecosystems. Collaboration among planners, lighting engineers and ecologists will allow for the tailoring of lighting technology that maximises driver and pedestrian safety while reducing or eliminating the effects of artificial light on ecosystems. Where data on wavelength sensitivity of affected taxa are unavailable, we suggest that a conservative approach is to reference findings from taxonomically related species. These findings might include behavioural responses to biologically relevant measures of emission spectra or to light measured at levels of luminous flux (see Gaston et al. (2013)). Another option is to make conservative decisions on lighting (e.g. avoiding emission spectra <500 nm; see Lesson 18.4). We also concur with Falchi et al. (2011) that where artificial lighting is necessary, these sources should (i) not release light directly at and above the horizontal; (ii) limit downward emission outside the area to which lighting is required; (iii) limit emission of short-wavelength spectra; (iv) be zoned and spaced to minimise unnecessary lighting; and (v) be operated via on/off scheduling where appropriate.

ACKNOWLEDGEMENTS

During the preparation of this chapter, the authors were supported by the U.S. Department of Agriculture, Animal and Plant Health Inspection Service, Wildlife Services, National Wildlife Research Center (NWRC). We thank E. Poggiali (NWRC) for assistance with manuscript preparation.

FURTHER READING

Endler (1990): Suggested a quantitative approach to measure colour reflected from animals and their visual backgrounds relative to the conditions of ambient lighting, an approach distinguished from use of measures of energy flux.

Fahrig and Rytwinski (2009): Review of the empirical literature on effects of roads (including effects such as road lighting) and traffic on animal abundance and distribution.

Forman et al. (2003): The first detailed and wide-ranging book on road ecology.

Gaston et al. (2013): Proposed a framework for consideration of how artificial lighting alters natural light regimens and influences biological systems.

Rich and Longcore (2006): Published the first detailed assessment of the negative consequences of artificial night lighting on ecosystems.

REFERENCES

Bertolotti, L. and M. Salmon. 2005. Do embedded roadway lights protect sea turtles? Environmental Management **36**:702–710.

Dawson, A., V. M. King, G. E. Bentley and G. F. Ball. 2001. Photoperiodic control of seasonality in birds. Journal of Biological Rhythms **16**:365–380.

de Molenaar, J. G., M. E. Sanders and D. A. Jonkers. 2006. Road lighting and grasslands birds: local influence of road lighting on a black-tailed godwit population. Pages 114–136 in C. Rich and T. Longcore, editors. Ecological consequences of artificial night lighting. Island Press, Washington, DC.

Diego-Rasilla, F. J., R. M. Luengo and J. B. Phillips. 2010. Light-dependent magnetic compass in Iberian green frog tadpoles. Naturwissenschaften **97**:1077–1088.

Eisenbeis, G. 2006. Artificial night lighting and insects: attraction of insects to streetlamps in a rural setting in Germany. Pages 281–304 in C. Rich and T. Longcore, editors. Ecological consequences of artificial night lighting. Island Press, Washington, DC.

Endler, J. A. 1990. On the measurement and classification of colour in studies of animal colour patterns. Biological Journal of the Linnean Society **41**:315–352.

Fahrig, L. and T. Rytwinski. 2009. Effects of roads on animal abundance: an empirical review and synthesis. Ecology and Society **14**:21. http://www.ecologyandsociety.org/vol14/iss1/art21/ (accessed 2 October 2014).

Falchi, F., P. Cinzano, C. D. Elvidge, D. M. Keith and A. Haim. 2011. Limiting the impact of light pollution on human health, environment and stellar visibility. Journal of Environmental Management **92**:2714–2722.

Forman, R. T. T., D. Sperling, J. A. Bissonette, A. P. Clevenger, C. D. Cutshall, V. H. Dale, L. Fahrig, R. France, C. R. Goldman, K. Heanue, J. A. Jones, F. J. Swanson, T. Turrentine and T. C. Winter, editors. 2003. Road ecology. Science and solutions. Island Press, Washington, DC.

Gaston, K. J., J. Bennie, T. W. Davies and J. Hopkins. 2013. The ecological impacts of nighttime light pollution: a mechanistic appraisal. Biological Reviews **88**:912–927.

Gaston, K. J., T. W. Davies, J. Bennie and J. Hopkins. 2012. Reducing the ecological consequences of night-time light pollution: options and developments. Journal of Applied Ecology **49**:1256–1266.

Gauthreaux, S. A., Jr. and C. G. Belser. 2006. Effects of artificial night lighting on migrating birds. Pages 67–93 in C. Rich and T. Longcore, editors. Ecological consequences of artificial night lighting. Island Press, Washington, DC.

Hart, N. S. 2001. The visual ecology of avian photoreceptors. Progress in Retinal Eye Research **20**:675–703.

Horváth, G., G. Kriska, P. Malik and B. Robertson. 2009. Polarized light pollution: a new kind of ecological photopollution. Frontiers in Ecology and the Environment **7**:317–325.

International Dark-Sky Association (IDA). 2010. Visibility, environmental, and astronomical issues associated with blue-rich white outdoor lighting. IDA, Tucson, AZ. http://www.darksky.org/assets/documents/Reports/IDA-Blue-Rich-Light-White-Paper.pdf (accessed 2 September 2014).

International Dark-Sky Association–Illuminating Engineering Society (IDA/IES). 2011. Joint IDA-IES model lighting ordinance (MLO)-2011, with user's guide. http://www.darksky.org/outdoorlighting/mlo (accessed 2 September 2014).

Luginbuhl, C. B., G. W. Lockwood, D. R. Davis, K. Pick and J. Selders. 2009. From the ground up I: light pollution sources in Flagstaff, Arizona. Publications of the Astronomical Society of the Pacific **121**:185–203.

Miller, M. W. 2006. Apparent effects of light pollution on singing behavior of American robins. Condor **108**:130–139.

Montevecchi, W. A. 2006. Influences of artificial light on marine birds. Pages 94–113 in C. Rich and T. Longcore, editors. Ecological consequences of artificial night lighting. Island Press, Washington, DC.

Negro, J. J., J. Bustamante, C. Melguizo, J. L. Ruiz and J. M. Grande. 2000. Nocturnal activity of lesser kestrels under artificial lighting conditions in Seville, Spain. Journal of Raptor Research **34**:327–329.

Perry, G., B. W. Buchanan, R. N. Fisher, M. Salmon and S. E. Wise. 2008. Effects of artificial night lighting on amphibians and reptiles in urban environments. Pages 239–256 in J. C. Mitchell, R. E. Jung Brown and B. Bartholomew, editors. Urban herpetology. Herpetological conservation, Vol. 3. Society for the Study of Amphibians and Reptiles, Salt Lake City, UT.

Poot, H., B. J. Ens, H. de Vries, M. A. H. Donners, M. R. Wernand and J. M. Marquenie. 2008. Green light for nocturnally migrating birds. Ecology and Society **13**:47. http://www.ecologyandsociety.org/vol13/iss2/art47/ (accessed 23 September 2014).

Rich, C. and T. Longcore, editors. 2006. Ecological consequences of artificial night lighting. Island Press, Washington, DC.

Rydell, J. 1992. Exploitation of insects around streetlamps by bats in Sweden. Functional Ecology **6**:744–750.

Rydell, J. 2006. Bats and their insect prey at streetlights. Pages 43–60 in C. Rich and T. Longcore, editors. Ecological consequences of artificial night lighting. Island Press, Washington, DC.

Salmon, M. 2006. Protecting sea turtles from artificial night lighting at Florida's oceanic beaches. Pages 141–168 in C. Rich and T. Longcore, editors. Ecological consequences of artificial night lighting. Island Press, Washington, DC.

Santos, C. D., A. C. Miranda, J. P. Granadeiro, P. M. Lourenco, S. Saraiva and J. M. Palmeirim. 2010. Effects of artificial illumination on the nocturnal foraging of waders. Acta Oecologica **36**:166–172.

Stone, E. L., G. Jones and S. Harris. 2012. Conserving energy at a cost to biodiversity? Impacts of LED lighting on bats. Global Change Biology **18**:2458–2465.

Verheijen, F. J. 1985. Photopollution: artificial light optic spatial control systems fail to cope with. Incidents, causations, remedies. Experimental Biology **44**:1–18.

Zurcher, A. A., D. W. Sparks and V. J. Bennett. 2010. Why the bat did not cross the road? Acta Chiropterologica **12**:337–340.

Chapter 19

ECOLOGICAL IMPACTS OF ROAD NOISE AND OPTIONS FOR MITIGATION

Kirsten M. Parris

School of Ecosystem and Forest Sciences, The University of Melbourne, Burnley, Victoria, Australia

SUMMARY

Roads and traffic alter the physical environment of species and ecological communities. They also change their acoustic environment through the introduction of noise, both during construction and when a road is open to traffic. Road-construction noise, such as that produced during earthworks, pile driving and road surfacing, can be of high intensity but usually of limited duration. In contrast, road-traffic noise is often persistent over time – busy highways can carry substantial traffic for many hours per day, day after day and year after year. Road noise has a number of ecological impacts on wildlife living in nearby habitats.

19.1 Road noise may be stressful for animals.

19.2 Road noise makes it harder for animals to hear each other, their predators and their prey.

19.3 Road noise may cause temporary or permanent hearing loss in animals.

19.4 High levels of road-construction noise may cause other injuries to animals in aquatic habitats.

19.5 Animals and their acoustic environment may need protection from road noise.

19.6 Mitigation of road noise to protect animals and their acoustic environment should be considered prior to road construction.

19.7 There is an urgent need for more research into the various effects of road noise on animals and their ecological communities.

Road noise has a variety of ecological impacts, including effects on the physiology, behaviour, communication, reproduction and survival of animals that live in or move through the noise-affected areas. While further research is required to better understand the ecological consequences of introducing road-construction and road-traffic noise to a given habitat, measures to mitigate the known impacts of road noise on wildlife should be implemented as part of new road projects. Further research will also help to improve the effectiveness of measures to protect animals and their acoustic environment from road noise.

Handbook of Road Ecology, First Edition. Edited by Rodney van der Ree, Daniel J. Smith and Clara Grilo.
© 2015 John Wiley & Sons, Ltd. Published 2015 by John Wiley & Sons, Ltd.
Companion website: www.wiley.com\go\vanderree\roadecology

INTRODUCTION

Sound is a radiant energy, transmitted as waves of pressure through a material medium such as air, water or soil. The pressure of a simple sound cycles up and down in a regular pattern over space and time. The frequency (or pitch) of a sound, measured in hertz (Hz), is the number of cycles it completes per second. The amplitude (also known as volume or loudness) of a sound can be measured in pressure or intensity, and both are expressed in decibels (dB). The decibel level of a sound is expressed relative to a reference pressure, commonly $20\,\mu Pa$ in air and $1\,\mu Pa$ in water. As the decibel scale is a logarithmic scale, an increase of 10 dB corresponds to a 10-fold increase in amplitude. A-weighted decibels, abbreviated to dB(A) or dBA, describe the relative loudness of sounds in air as perceived by humans.

Noise can be defined as unwanted sound or as the background sound found in any environment. Road noise – including the noise generated during road construction (road-construction noise) and the noise caused by vehicles travelling on existing roads (road-traffic noise) – differs from natural noise in a number of ways. During road construction, heavy machinery is used to clear vegetation, move rocks and soil and prepare the road surface. In addition, blasting and pile driving may be used during construction of tunnels and bridges. Road-construction noise can be of high intensity but usually of limited duration, ranging from weeks to months at a given location. In contrast, road-traffic noise is usually of a lower intensity than road-construction noise but more persistent over time. Road-traffic noise is generally louder and lower pitched than natural noise, with much of its energy occurring in the lower-frequency bands below 2000 Hz (Patricelli & Blickley 2006). Depending on the volume and speed of traffic, local topography, road surface and prevailing weather conditions, the noise of vehicles travelling on a road can extend more than 2 km across the landscape on either side (also known as the road-effect zone; Lesson 1.2). Road noise has a range of ecological impacts, and it is important to consider these when planning new roads or when attempting to reduce the impact of noise from existing roads on wildlife.

The aim of this chapter is to summarise the important effects of road noise (including road-construction noise and road-traffic noise) on ecological communities and to discuss noise-mitigation options that may help to protect animals and their acoustic environment. Just as an animal lives in, and may be adapted to, a particular physical environment (e.g. a forest, a desert or a seagrass meadow), it also lives in, and may

be adapted to, a particular acoustic (sound) environment (Morton 1975). Road noise can disrupt the acoustic environment of animals, with a number of important ecological consequences (Barber et al. 2010; Slabbekoorn et al. 2010).

LESSONS

19.1 Road noise may be stressful for animals

The noise of road construction and road traffic can startle nearby animals, causing a physiological stress response (Kight & Swaddle 2011). As a consequence, animals may move away from the noise-affected area, either temporarily or permanently. Permanent avoidance of areas affected by road noise will lead to a permanent decrease in the amount of habitat available for noise-sensitive species. In places with a high density of roads, such as parts of Europe and North America, this decrease can be quite dramatic (Forman & Deblinger 2000; Barber et al. 2010). On the other hand, noise-sensitive animals that do not or cannot move away from areas close to busy roads will experience ongoing exposure to road-traffic noise, which may lead to chronic physiological stress (McEwen & Wingfield 2003; Blickley et al. 2012b). This, in turn, can lead to secondary effects such as weakened immune function and reduced breeding success (Wikelski & Cooke 2006; Kight & Swaddle 2011; Blickley et al. 2012b).

An experimental study of greater sage-grouse in North America found that male birds avoided lekking sites (places they call from to attract females for mating) where speakers played recorded traffic noise (Blickley et al. 2012a). Over three breeding seasons, the abundance of males was 73% lower at lekking sites with road noise than at quiet control sites. In addition, male birds that did use the noisy lekking sites had higher levels of stress hormones (glucocorticoids) than males at quiet sites (Blickley et al. 2012b). These results show that chronic road noise can cause sage-grouse to avoid otherwise suitable habitat and can increase stress levels in the male grouse who remain in noisy areas. In marine habitats, some species of seals, porpoises and dolphins have been observed to move away from areas affected by pile-driving noise for the duration of construction works, although in at least one case, Indo-Pacific humpback dolphins returned when construction had finished (reviewed in Jefferson et al. (2009)).

A recent study of great tits in the Netherlands found that breeding success was lower in noisy areas close to a busy motorway, with female birds laying smaller clutches of eggs and fewer chicks fledging than in quieter areas nearby (Halfwerk et al. 2011). Earlier research in the Netherlands found reduced bird densities next to noisy roads (Reijnen et al. 1995, 1996) and that territories near busy roads were more likely to be occupied by inexperienced male birds who then struggled to attract a mate (Reijnen & Foppen 1994). Similarly, a range of large mammals are known to be less abundant in habitats near roads, including caribou, African elephant, zebra and blue wildebeest (Newmark et al. 1996; Benítez-López et al. 2010). In many of these examples, it is unclear whether the observed effects of roads are due to physiological stress caused by road noise, difficulty communicating in noise (Lesson 19.3) or the nearness of vehicles travelling on the road. The design of these studies means that different effects of roads such as noise, visual disturbance or mortality caused by collision with vehicles cannot easily be separated.

19.2 Road noise makes it harder for animals to hear each other, their predators and their prey

Many animal groups – including insects, fish, frogs, birds and mammals – communicate using sound. For example, birds use songs and calls to attract mates; defend territories from rivals; keep in contact with their mate, parent or chicks; beg for food; and warn other birds of danger from potential predators. In addition, many animals rely on hearing to detect the sound of approaching predators or the location of potential prey. Acoustic interference or masking occurs when background noise reduces the distance over which a sound can be detected. While animals have a number of strategies to make themselves heard in a background of natural noise (e.g. Brumm & Slabbekoorn 2005), those that live in habitats near roads must also contend with noise from road construction and road traffic.

Acoustic interference from road noise can disrupt important social interactions among animals and may have significant consequences for both individuals and populations. These include difficulty attracting and maintaining a mate, reduced breeding success, population declines and changes in the composition of ecological communities in areas affected by road noise (Patricelli & Blickley 2006; Kight et al. 2012; Proppe et al. 2013). While some animals are known to sing or call differently in road-traffic noise to increase the distance over which they can be heard (e.g. singing or calling more loudly, at

a higher pitch or at quieter times of the day to avoid peak periods of road-traffic noise; Barber et al. 2010), these changes are not large enough to regain all the communication distance that is lost (Parris et al. 2009; Parris & McCarthy 2013; Textbox 19.1).

Acoustic interference from road noise may also increase the vulnerability of animals to predators or decrease foraging success by animals that rely on sound to detect their prey (Barber et al. 2010). High levels of background noise increase watchfulness (also known as vigilance behaviour) in animals; because they cannot hear predators approaching, animals spend more time watching out for them and less time foraging for food (Barber et al. 2010). There is also evidence that animals such as bats avoid foraging near noisy highways where it would be difficult to hear their prey (Schaub et al. 2008). A recent behavioural experiment using simulated highway noise found that the foraging efficiency of the greater mouse-eared bat declined dramatically as it moved closer to the simulated road (Siemers & Schaub 2011; Chapter 34). These kinds of noise effects may have further consequences for predator–prey interactions and food webs in ecological communities (Siemers & Schaub 2011).

19.3 Road noise may cause temporary or permanent hearing loss in animals

The noise of road construction or road traffic may cause temporary or permanent hearing loss in animals in nearby habitats. The hearing threshold of an animal refers to the point at which a sound is just loud enough to be heard – the higher the threshold, the louder the sound must be to be detected. High levels of noise can damage the cochlea in the inner ear, leading to a temporary or permanent increase in the hearing threshold of affected animals (Kight & Swaddle 2011). These kinds of auditory injuries can result from a single, extreme noise event or from chronic exposure to high levels of noise. In general, the higher the noise level and the longer it continues, the greater the change in hearing threshold, the longer it will take until normal hearing is recovered, and the greater the chance of permanent hearing loss (Kight & Swaddle 2011).

An early laboratory study of the effects of vehicle noise on species accustomed to a quiet, desert environment exposed the Mojave fringe-toed lizard and the desert kangaroo rat to 10 minutes of intermittent dune-buggy noise (Brattstrom & Bondello 1983). Both species suffered hearing loss for a number of weeks and during this time were unable to detect and respond

Textbox 19.1 **Effects of road-traffic noise on the grey shrike-thrush**

The grey shrike-thrush is a common, sedentary Australian songbird (Fig. 19.1) found in a variety of habitats including forests and woodlands. The song of the grey shrike-thrush is melodious and varied, containing pure tones, trills and whistles. Males sing to attract a mate, while both males and females sing to defend territories from neighbouring birds (Higgins & Peter 2002). The frequency (pitch) of the grey shrike-thrush song overlaps the frequency of road-traffic noise, so we would expect these birds to have difficulty hearing each other in areas with high levels of road-traffic noise.

Parris and Schneider (2009) studied the effects of road-traffic noise and daily traffic volume on the grey shrike-thrush at 58 roadside sites on the Mornington Peninsula in south-eastern Australia. The roads ranged in size from narrow, unsealed roads with very little traffic to multi-lane freeways carrying 32,000 vehicles per day. However, each road had a narrow strip of native vegetation on each verge, providing habitat for birds and other animals (Fig. 19.2). This study design reduced other habitat differences between quiet and noisy sites that may have influenced the birds.

As the level of traffic noise at a study site increased, the grey shrike-thrush sang at a higher frequency (Fig. 19.3). While this change would help it to be heard above the noise of the traffic, it is not large enough to overcome the masking effect of the noise entirely – the bird would only get back around 10% of the communication distance lost in noise (Parris & McCarthy 2013). In addition, the chance of finding one or more grey shrike-thrushes on a visit to a site decreased from around 85% at the quietest sites to 15% at the noisiest sites (Fig. 19.4), suggesting that the birds are less likely to be present at noisier sites. Acoustic interference from road-traffic noise may be making it more difficult for these birds to establish and maintain territories, attract mates and maintain pair bonds, possibly leading to reduced breeding success in noisy roadside habitats (Parris & Schneider 2009). Given the narrowness of the roadside verges and the active nature of these birds, we would expect to detect the birds if they were present, even in very noisy conditions.

Figure 19.1 The grey shrike-thrush. Source: Photograph by and reproduced with permission of Rob Drummond/Lochman Transparencies.

Figure 19.2 A roadside study site on the Mornington Peninsula in south-eastern Australia, showing the narrow strip of *Eucalyptus* woodland on each side of the road. Source: Photograph by Kirsten Parris.

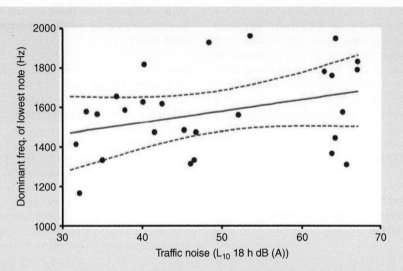

Figure 19.3 The grey shrike-thrush increases the frequency of its song (as measured by the dominant frequency of the lowest note) as the level of traffic noise at a site increases. The sound-pressure level of the traffic noise is expressed as L_{10} 18 h dB (A), which is the 90th percentile of the distribution of traffic noise experienced in the 18 hours between 6 am and midnight. The solid green line shows the predicted relationship, the dotted green lines the 95% credible intervals, and the purple circles the data points, with one point for each site where the species was recorded. Source: Adapted from Parris and Schneider (2009). © Kirsten Parris.

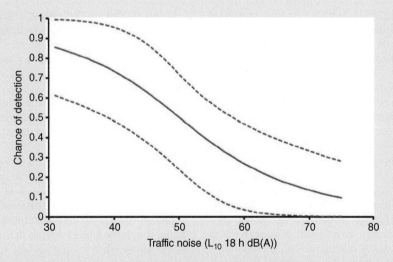

Figure 19.4 The grey shrike-thrush was less likely to be found at sites with high levels of traffic noise, with the chance of detection decreasing as the level of traffic noise increased. The sound-pressure level of the traffic noise is expressed as L_{10} 18 h dB (A), which is the 90th percentile of the distribution of traffic noise experienced in the 18 hours between 6 am and midnight. The solid green line shows the predicted relationship, and the dotted green lines the 95% credible intervals. Source: Adapted from Parris and Schneider (2009). © Kirsten Parris.

to recorded calls of their predators. More recent research has focused on acoustic injury in fish caused by road-construction noise in aquatic habitats, such as impulsive noise from pile driving (reviewed in Popper and Hastings (2009) and Slabbekoorn et al. (2010)). While more research is needed on the effects of road noise on hearing in birds, continuous noise above 93 dB(A) may cause a temporary increase in the hearing threshold of birds, while impulsive noise above 125 dB(A) may cause permanent hearing damage (Dooling & Popper 2007).

19.4   High levels of road-construction noise may cause other injuries to animals in aquatic habitats

Construction of bridges or causeways across shallow bodies of water such as bays and estuaries often involves the driving of supporting piles into the marine substrate. Depending on the substrate, depth and the size of a project, pile driving can continue for many days or even months (Jefferson et al. 2009). Pile driving can produce such high levels of impulsive sound that the pressure waves of sound moving through the water cause internal injuries to fish. This type of injury is known as barotrauma and results from rapid changes in the volume of gases within the body of the fish and in the solubility of gas in its blood and tissues (Halvorsen et al. 2012). Barotrauma injuries caused by underwater sound waves include emboli (gas bubbles that form in the blood and tissues when gas leaves solution), resulting tissue damage, and the rapid expansion of gas-filled organs such as the swim bladder. In more extreme cases, the swim bladder may rupture, or gas bubbles in the gills or heart may kill the fish instantly (Halvorsen et al. 2012).

A controlled laboratory study identified the threshold for the onset of injury to juvenile Chinook salmon as a cumulative sound exposure level of 210 dB re 1 µPa^2.s (Halvorsen et al. 2012). This exposure level could be achieved, for example, from 960 pile-driving strikes at a sound exposure level of 180 dB re 1 µPa^2.s each (the typical number of strikes needed to drive a single pile). If this threshold were to be exceeded, it would be likely to affect the survival of the exposed fish. However, in a later study, this level of sound exposure caused substantial barotrauma in another species of fish, the hybrid striped bass (Casper et al. 2013). Impulsive sound from pile driving may occasionally be intense enough to cause barotrauma in small marine mammals such as dolphins and porpoises (Jefferson et al. 2009).

19.5   Animals and their acoustic environment may need protection from road noise

Given the variety and potential seriousness of its ecological impacts, animals and their acoustic environment may need protection from road noise. This will particularly apply in areas supporting populations of threatened species or where levels of road-construction or road-traffic noise are expected to be high. Measures that can be used to mitigate road noise vary depending on the type of noise (road-construction or road-traffic noise), the size of the road and expected traffic volumes and the type of habitats that may be affected by noise (terrestrial or marine). However, one widely applicable approach is to reduce or exclude the noise-generating activity at times when and places where animals of concern are known to be present or are expected to be particularly vulnerable to the effects of noise. In terrestrial habitats, this could include ceasing road construction, closing roads or reducing traffic speeds (and thus traffic noise) during the breeding season of animals that communicate using sound, such as birds and frogs (Parris et al. 2009; Halfwerk et al. 2011). In marine habitats, pile driving could stop during the main calving season of mammals, as it does in Hong Kong to protect mothers and calves of the finless porpoise and Indo-Pacific humpback dolphin (Jefferson et al. 2009). However, this approach requires information on the distribution and breeding activities of the animals in question.

A number of other measures have been trialled to reduce the impact of pile-driving noise on fish and mammals in marine habitats, such as ramping up and the use of air-bubble curtains or jackets (Jefferson et al. 2009; Popper & Hastings 2009). Ramping up, in which the intensity of pile driving starts at a low level and then increases over time, is intended to warn fish and other marine animals and give them a chance to leave the area before sound-pressure levels are high enough to be dangerous. More research is needed on the best way to use this strategy for different groups of animals (Jefferson et al. 2009). Curtains of air bubbles can also be used to reduce the transmission of underwater sound. A small-scale experimental study in Denmark found that noise levels from pile driving were reduced substantially when such a curtain was in operation, and avoidance behaviour seen in nearby, captive harbour porpoises before the curtain was installed was no longer observed (Lucke et al. 2011).

Sound barriers have been used for decades to protect humans from high levels of road-traffic noise. However, they are rarely used specifically to protect non-human animals in this way. Sound barriers are usually constructed of solid materials such as earth, concrete, wood, steel or glass and can reach from the ground up to 5 m or more in height (Figs 33.3 and 33.4). While these kinds of structures are very effective at reducing levels of road-traffic noise in areas near busy roads, they also form a barrier to the movement of many animals that walk or hop along the ground, such as mammals, reptiles, amphibians and invertebrates. In addition, flying birds can collide with sound barriers made of glass unless they are patterned or coloured to make them more conspicuous. While it may be difficult for many ground-dwelling animals to cross a busy road safely, it becomes impossible where solid sound barriers are installed (Chapter 20). Alternatives to solid sound barriers include barriers made of dense vegetation or barriers that have a small gap at the bottom to allow animals to pass underneath them. However, these kinds of barriers may be less effective at reducing road noise than solid barriers. The noise of vehicles travelling on a road can also be reduced through the use of quiet paving materials and by improvements in engine, muffler and tyre design.

19.6 Mitigation of road noise to protect animals and their acoustic environment should be considered prior to road construction

Given that mitigation of both road-construction noise and road-traffic noise may require substantial planning and investment, it should be considered in the early stages of a road project, well before construction begins (Chapters 4 and 9). Suitable preparation will include collection of ecological information such as the distribution and seasonal activities of different groups or particular species of animals in areas to be affected by the noise of the road. Planners will then need to consider the expected levels of noise during construction and when the road is opened to traffic, and choose or develop suitable mitigation options. This issue currently receives very little attention, particularly in terrestrial habitats.

19.7 There is an urgent need for more research into the various effects of road noise on animals and their ecological communities

While a range of ecological impacts of road noise have been identified, many questions about these impacts remain unanswered. Key areas for future research include the short- and long-term consequences of road noise for social interactions, foraging behaviour, stress levels, survival, breeding success and abundance of animals such as frogs, fish, birds, bats, invertebrates and marine mammals. In addition, much more research is needed on the design and effectiveness of strategies to mitigate the expected ecological effects of road-construction noise and road-traffic noise. The best way to improve our understanding of the ecological impacts of road noise (and ways to mitigate them) will be to set up well-designed experiments (Chapter 10; Fig. 10.3) to address one or more of these questions while constructing or operating new roads.

CONCLUSIONS

Road-construction noise and road-traffic noise can have a multitude of effects on animals and their acoustic environment, in both terrestrial and aquatic habitats. At the level of an individual animal, these effects may include behavioural changes, increased physiological stress, injury or death. At the level of a population, they may include a lower probability of survival and reduced breeding success in habitats affected by road noise. As a consequence, animals and their acoustic environment may need protection from high levels of noise during road construction and/or operation, and suitable noise-mitigation strategies should be considered early in the road planning process.

FURTHER READING

Barber et al. (2010): Reviews the costs of chronic noise exposure for terrestrial animals.
Kight and Swaddle (2011): Reviews the physiological effects of noise on animals.
Slabbekoorn et al. (2010): Reviews the impacts of increasing underwater noise on fish.
Warren et al. (2006): Reviews the effects of urban noise on animal communication and animal behaviour.

REFERENCES

Barber, J. R., K. R. Crooks and K. M. Fristrup. 2010. The costs of chronic noise exposure for terrestrial organisms. Trends in Ecology & Evolution **25**:180–189.
Benítez-López, A., R. Alkemade and P. A. Verweij. 2010. The impacts of roads and other infrastructure on mammal and bird populations: a meta-analysis. Biological Conservation **143**:1307–1316.

Blickley, J. L., D. L. Blackwood and G. L. Patricelli. 2012a. Experimental evidence for the effects of chronic anthropogenic noise on abundance of greater sage-grouse at leks. Conservation Biology **26**:461–471.

Blickley, J. L., K. Word, A. H. Krakauer, J. L. Phillips, S. Sells, C. C. Taff, J. C. Wingfield and G. L. Patricelli. 2012b. The effect of experimental exposure to chronic noise on fecal corticosteroid metabolites in lekking male greater sage-grouse (*Centrocercus urophasianus*). PLoS One **7**:e50462.

Brattstrom, B. H. and M. C. Bondello. 1983. Effects of off-road vehicle noise on desert vertebrates. Pages 167–206 in R. H. Webb and H. H. Wilshire, editors. Environmental effects of off-road vehicles. Springer-Verlag, New York.

Brumm, H. and H. Slabbekoorn. 2005. Acoustic communication in noise. Advances in the Study of Behavior **35**:151–209.

Casper, B. M., M. E. Smith, M. B. Halvorsen, H. Sun, T. J. Carlson and A. N. Popper. 2013. Effects of exposure to pile driving sounds on fish inner ear tissues. Comparative Biochemistry and Physiology Part A **166**:352–360.

Dooling, R. J. and A. N. Popper. 2007. The effect of highway noise on birds. California Department of Transportation, Sacramento, CA.

Forman, R. T. T. and R. D. Deblinger. 2000. The ecological road-effect zone of a Massachusetts (U.S.A.) suburban highway. Conservation Biology **14**:36–46.

Halfwerk, W., L. J. M. Holleman, C. M. Lessells and H. Slabbekoorn. 2011. Negative impact of traffic noise on avian reproductive success. Journal of Applied Ecology **48**:210–219.

Halvorsen, M. B., B. M. Casper, C. M. Woodley, T. J. Carlson and A. N. Popper. 2012. Threshold for onset of injury in Chinook salmon from exposure to impulsive pile driving sounds. PLoS One **7**(6):e38968.

Higgins, P. J. and J. M. Peter (eds). 2002. Handbook of Australian, New Zealand and Antarctic birds. Volume 6: pardalotes to Shrikethrushes. Oxford University Press, Melbourne, Victoria.

Jefferson, T. A., S. K. Hung and B. Würsig. 2009. Protecting small cetaceans from coastal development: impact assessment and mitigation experience in Hong Kong. Marine Policy **33**:305–311.

Kight, C. R. and J. P. Swaddle. 2011. How and why environmental noise impacts animals: an integrative, mechanistic review. Ecology Letters **14**:1052–1061.

Kight, C. R., M. S. Saha and J. P. Swaddle. 2012. Anthropogenic noise is associated with reductions in the productivity of breeding Eastern Bluebirds (*Sialia sialis*). Ecological Applications **22**:1989–1996.

Lucke, K., P. A. Lepper, M. Blanchet and U. Siebert. 2011. The use of an air bubble curtain to reduce the received sound levels for harbor porpoises (*Phocoena phocoena*). Journal of the Acoustical Society of America **130**:3406–3412.

McEwen, B. S. and J. C. Wingfield. 2003. The concept of allostasis in biology and biomedicine. Hormones and Behavior **43**:2–15.

Morton, E. S. 1975. Ecological sources of selection on avian sounds. The American Naturalist **109**:17–34.

Newmark, W. D., J. I. Boshe, H. I. Sariko and G. K. Makumbule. 1996. Effects of a highway on large mammals in Mikumi National Park, Tanzania. African Journal of Ecology **34**:15–31.

Parris, K. M. and M. A. McCarthy. 2013. Predicting the effect of urban noise on the active space of avian vocal signals. The American Naturalist **182**:452–464.

Parris, K. M. and A. Schneider. 2009. Impacts of traffic noise and traffic volume on birds of roadside habitats. Ecology and Society **14**:29. http://www.ecologyandsociety.org/vol14/iss1/art29/ (accessed 23 September 2014).

Parris, K. M., M. Velik-Lord and J. M. A. North. 2009. Frogs call at a higher pitch in traffic noise. Ecology and Society **14**:25. http://www.ecologyandsociety.org/vol14/iss1/art25/ (accessed 23 September 2014).

Patricelli, G. L. and J. L. Blickley. 2006. Avian communication in urban noise: causes and consequences of vocal adjustment. The Auk **123**:639–649.

Popper, A. N. and M. C. Hastings. 2009. The effects of anthropogenic sources of sound on fishes. Journal of Fish Biology **75**:455–489.

Proppe, D. S., C. B. Sturdy and C. Cassady St Clair. 2013. Anthropogenic noise decreases urban songbird diversity and may contribute to homogenization. Global Change Biology **19**:1075–1084.

Reijnen, R. and R. Foppen. 1994. The effects of car traffic on breeding bird populations in Woodland. I. Evidence of reduced habitat quality for willow warblers (*Phylloscopus trochilus*) breeding close to a highway. Journal of Applied Ecology **31**:85–94.

Reijnen, R., R. Foppen, C. ter Braak and J. Thissen. 1995. The effects of car traffic on breeding bird populations in woodland. III. Reduction of density in relation to the proximity of main roads. Journal of Applied Ecology **32**:187–202.

Reijnen, R., R. Foppen and H. Meeuwsen. 1996. The effects of traffic on the density of breeding birds in Dutch agricultural grasslands. Biological Conservation **75**:255–260.

Schaub, A., J. Ostwald and B. M. Siemers. 2008. Foraging bats avoid noise. Journal of Experimental Biology **211**:3174–3180.

Siemers, B. M. and A. Schaub. 2011. Hunting at the highway: traffic noise reduces foraging efficiency in acoustic predators. Proceedings of the Royal Society B **278**:1646–1652.

Slabbekoorn, H., N. Bouton, I. van Opzeeland, A. Coers, C. ten Cate and A. N. Popper. 2010. A noisy spring: the impact of globally rising underwater sound levels on fish. Trends in Ecology and Evolution **25**:419–427.

Warren, P. S., M. Katti, M. Ermann and A. Brazel. 2006. Urban bioacoustics: it's not just noise. Animal Behaviour **71**:491–502.

Wikelski, M. and S. J. Cooke. 2006. Conservation physiology. Trends in Ecology and Evolution **21**:38–46.

Chapter 20

FENCING: A VALUABLE TOOL FOR REDUCING WILDLIFE-VEHICLE COLLISIONS AND FUNNELLING FAUNA TO CROSSING STRUCTURES

Rodney van der Ree[1], Jeffrey W. Gagnon[2] and Daniel J. Smith[3]

[1] Australian Research Centre for Urban Ecology, Royal Botanic Gardens Melbourne, and School of BioSciences, The University of Melbourne, Melbourne, Victoria, Australia
[2] Arizona Game and Fish Department, Phoenix, AZ, USA
[3] Department of Biology, University of Central Florida, Orlando, FL, USA

SUMMARY

Fences prevent animals from accessing roads, thereby reducing the rate of wildlife-vehicle collisions (WVC). Fences also funnel animals towards crossing structures, making them an essential component of the success of this form of mitigation. Fencing can be used for a variety of species, ranging from frogs and turtles to deer and bears. Consequently, fence designs are almost as varied as the species they target.

20.1 Fencing is an essential component of mitigation and must be comprehensively integrated into the mitigation programme for it to be effective.

20.2 Fencing must be designed for the target species.

20.3 Consider alternatives to traditional fences.

20.4 Animals inevitably breach fences, and when they do, they must be able to exit the roadway.

Handbook of Road Ecology, First Edition. Edited by Rodney van der Ree, Daniel J. Smith and Clara Grilo.
© 2015 John Wiley & Sons, Ltd. Published 2015 by John Wiley & Sons, Ltd.
Companion website: www.wiley.com\go\vanderree\roadecology

20.5 Fence ends and planned breaks in fences must be designed to reduce the rate of WVC.

20.6 Fences need to be maintained forever.

Appropriately designed fences and crossing structures can cost-effectively reduce or eliminate WVC. The potential negative effects of fences must also be considered, including increasing the barrier effect when installed without crossing structures and mortality of wildlife at fence ends and if poorly designed or maintained. Careful consideration of a small number of design and maintenance parameters is essential to achieve and maintain effectiveness.

INTRODUCTION

Fencing is an integral component of mitigation, and its primary purposes are to prevent animals from accessing the road (also known as exclusion or barrier fencing), thereby reducing the rate of wildlife-vehicle collisions (WVC), and/or to funnel animals towards crossing structures (hereafter funnel fencing). Funnel fencing can be shaped like a funnel or be parallel to the road, and in both cases, it has the effect of directing animals towards crossing structures. Fencing and its intended function (i.e. barrier or funnel) must be defined relative to the target species. For example, frogs can pass directly through the large mesh of a deer fence. Similarly, a fence that directs long-ranging carnivores towards underpasses that are a few kilometre apart (i.e. funnel fencing) may act as a barrier fence for smaller animals if the crossing structures are beyond their movement capabilities. In this book, all fencing designed to influence the movement of wild animals is called wildlife fencing. This chapter does not focus on fencing that is used to demarcate ownership or management responsibilities of different parcels of land (boundary fencing).

The first wildlife fences along roads were intended to prevent large animals from entering the roadway and colliding with vehicles (e.g. Puglisi et al. 1974; Falk et al. 1978). Fencing can prevent WVC and is particularly effective when combined with properly designed and located crossing structures that allow faunal movement across roads (Dodd Jr. et al. 2004; Huijser et al. 2007; Dodd et al. 2012; Sawyer et al. 2012). Fence designs and standards differ among regions and species, and for some species, alternatives to traditional fences are available. The aims of this chapter are to highlight the key considerations necessary for an effective fencing strategy.

LESSONS

20.1 Fencing is an essential component of mitigation and must be comprehensively integrated into the mitigation programme for it to be effective

The primary role of fences is to reduce the number of animals that move onto the road, thereby reducing the frequency of WVC that result in human injuries and fatalities, property damage and wildlife injury and mortality. Effective fences are barriers to movement of the target species, and when installed without crossing structures can fragment habitat and increase the risk of population decline and extinction (Jaeger & Fahrig 2004). For example, reduced connectivity caused by roads, fences and human settlements has resulted in wildlife population crashes and local extinction of numerous migratory species (Chapters 42, 56 and 58). Therefore, it is essential to consider animal movements when determining the spacing of crossing structures installed with fencing – placing them too far apart and beyond distances a given species will travel to find them will reduce connectivity. This is particularly problematic when the same fence acts as a funnel for a highly mobile species but acts as a barrier for a less mobile species. In other words, fencing should never be installed without considering the movements of both target and non-target species (Glista et al. 2009).

There is clear evidence from around the world that the effectiveness of wildlife crossing structures is improved when animals are funnelled towards them, and fencing is the most effective way to achieve this for most species (excluding birds, bats and some arboreal animals) (Jackson & Tyning 1989; Gagnon et al. 2010a; McCollister & Van Manen 2010; Sawyer et al. 2012; Textbox 20.1). Crossing structures and fencing that are designed and installed simultaneously are usually more effective than when installed separately, and this approach is also more cost-effective. A tight join between a fence and structure is critical to prevent breaching by wildlife (Fig. 20.1), and this is more easily and cost-effectively achieved when both structures are designed and installed as a single unit. Similarly, using landscaping, rocks or other material to reduce gaps under fences is best achieved when the crossing structure is being constructed because earth-moving machinery is still on site. Any works that are not part of the original construction project (i.e. retrofits to improve functionality, repairs, unscheduled maintenance) will be logistically more difficult and costly than if conducted and allowed for during road construction.

Although fences and crossing structures should be installed simultaneously, there are many occasions where retrofits are justified (Fig. 20.2). Installing fences in the absence of other works should be considered when (i) rates of WVC are unacceptably high; (ii) wildlife are not being adequately funnelled to crossing structures; or (iii) there are no short- to medium-term plans to rebuild a stretch of road. If existing crossing or drainage structures are adequate for wildlife, fencing retrofits are less expensive than a highway upgrade (Aresco 2005; Gagnon et al. 2010a). There are numerous examples from around the world which demonstrate that fencing retrofits can have positive outcomes. For example, the number of elk–vehicle collisions along a 48 km

Figure 20.1 Depending on the species, the join between this bridge abutment and fauna fence may be too large, allowing animals to breach the fence and access the roadway. Source: Photograph by R. van der Ree.

(A)

(B)

Figure 20.2 (A) This cost-effective temporary retrofit of a 1.2 m-high livestock fence to a height of 2.4 m along Interstate 17 in Arizona reduced the rate of elk–vehicle collision by 100%. Note the coloured tape on the top strand to warn birds of the new fence. (B) Simple wire mesh added to existing guard rail to reduce WVC with snapping turtle, New York. In all mitigation and especially retrofits, it is essential to ensure that a cheap option is also ecologically effective. Source: (A) Photograph by J.W. Gagnon and (B) Photograph by and reproduced with permission from Tom Langen.

stretch of Interstate 17 (I-17) in Arizona, United States, exceeded 80 per year. To reduce the number of collisions, the standard livestock fencing along a 9.6 km section of I-17 with the highest occurrence of elk–vehicle collisions was simply raised to 2.4 m. This retrofit connected four crossing structures (two open-span bridges and two modified highway interchanges), and within the first 2 years, collisions with elk were reduced by 100% and use of the two bridges increased by 100%. Given the cost and timelines of reconstruction for I-17, this relatively inexpensive project was a short-term solution until appropriate wildlife crossings and fencing could be completed (Gagnon et al. 2013). Temporary fencing adjacent to construction projects may be necessary to prevent wildlife from accessing construction zones and to prevent wildlife that flee construction zones from moving onto adjacent roads and being involved in WVC (Chapter 8). While temporary fences do not need to be as solidly built as permanent fencing, they still need to be effective and designed for the target species (Lesson 20.2).

20.2 Fencing must be designed for the target species

Fences must be designed with respect to the size, behaviour and jumping or climbing ability of the target species (Woltz et al. 2008; Clevenger & Huijser 2011; Grandmaison 2011). Deer and kangaroos typically require fences that are at least 2–2.4 m in height, while fences 0.2–1.0 m in height may be suitable for small animals (Fig. 32.4) (Grandmaison 2011; Gulsby et al. 2011; Andrews et al. 2015; Chapters 31, 32 and 39). For species that climb, a smooth fence that provides

no purchase, or a 'floppy top' or 'overhanging lip' (Fig. 20.4) may be necessary to prevent them from climbing over (e.g. Klar et al. 2009).

Other species are powerful and determined, and fences may need to be fortified with heavier gauge wire or steel posts to prevent them from breaking the wire or pushing the fence down. In some instances, electrified fencing may be required, such as for larger ungulates or to reduce the potential for larger animals, such as bears, from climbing over (Seamans & VerCauteran 2006; Leblond et al. 2007; Gagnon et al. 2010a; Chapters 43 and 54). Mesh size also matters and must be small enough to prevent juveniles from passing through without their parent. Inappropriate mesh size may also cause entanglement (Fig. 58.2), and some gliding and flying species may become impaled on barbed wire (van der Ree 1999). The base of fences may need to be buried or include a skirt to prevent burrowing animals from breaching the fence. The colour and opacity of fences may also influence effectiveness. Turtles and amphibians have been recorded moving faster along an opaque fence than a translucent or transparent one (e.g. desert tortoises, Ruby et al. 1994). This is probably important because wildlife spend less time and energy attempting to find a way through a fence if they perceive it as a barrier, which is more likely if opaque than translucent or transparent. Many types of fencing materials for reptiles and amphibians have been experimented with, including aluminium flashing, fibreglass and plastics, woven and bonded screen and synthetic cloth, and different-sized wire mesh (e.g. Ruby et al. 1994; Smith & Noss 2011; Chapters 31 and 32).

Fence designs for many common and large-bodied species have been developed and standardised in North America, Europe and some other parts of the world

---

Textbox 20.1 'If you build it they won't come … without fencing'.

Seven open-span bridges, each with 300 m of wildlife fencing on each side (i.e. a total of 600 m per entrance), were constructed in 2003 along an 8 km stretch of State Route (SR) 260 in central Arizona, United States, during an upgrade from a two- to four-lane divided highway (Fig. 20.3). After construction, researchers monitored WVC, elk and deer use of the underpasses via video surveillance, and the rate and location of road crossing by 110 elk fitted with GPS collars. During 2004, 54 elk–vehicle collisions were reported, and only 12% of elk and deer crossed SR260 via the underpasses, while the remainder

avoided the underpasses and crossed the road at grade or passed under the bridge of one carriageway and entered the roadway via the median. Data from the GPS-collared elk was used to identify locations where they crossed the highway to determine where to add fencing to intercept at-grade crossings. In the year following fencing, collisions were reduced by 87%, and successful use of the underpasses increased dramatically (Fig. 20.3). Further, once elk were guided to the underpasses, permeability of the highway returned to nearly pre-construction levels within the first year (Dodd et al. 2007).

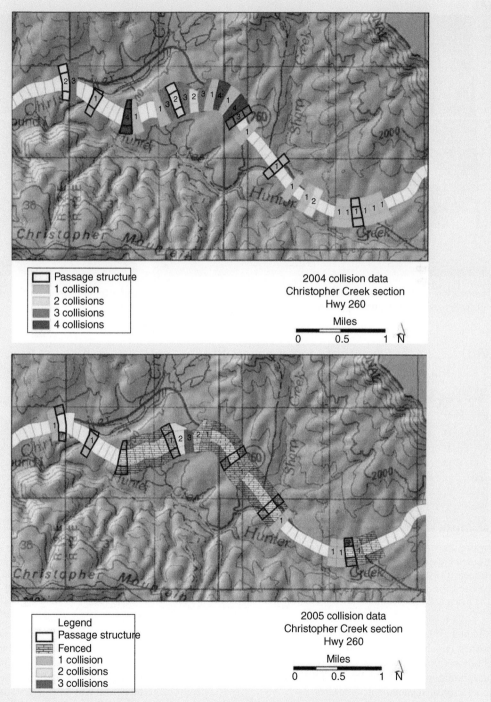

Figure 20.3 Seven open-span bridges for wildlife (shown in black outline) were constructed along an 8 km stretch of State Route 260 in Arizona. Prior to fencing (upper figure), 54 collisions with elk occurred in 1 year (number of elk collisions shown in each 0.1 mile segment). Following fencing, collisions were reduced to 8 in the following year (lower figure), pointing to the necessity of fencing in the success of wildlife crossing structures. Source: Dodd et al. 2007.

(A)
(B)

Figure 20.4 (A) Floppy top fence for koalas and (B) curved fence for desert tortoise in Arizona. Source: (A) Photograph by R. van der Ree and (B) Photograph by and reproduced with permission from Scott Sprague.

(A)
(B)
(C)

Figure 20.5 Examples of combination fencing for multiple species. (A) A 2.4 m-high wire mesh fence with sheet metal below to prevent arboreal species and kangaroos from climbing or jumping over, Victoria, Australia; (B) fencing for deer and tortoise in Arizona; (C) a 0.635 cm$_2$ mesh fence for reptiles and amphibians with 1.2 m aboveground and 30 cm belowground, with a stock fence to 1.5 m and ElectroBraid to 2.4 m, Florida. Source: (A) Photograph by R. van der Ree, (B) Photograph by J.W. Gagnon and (C) Photograph by D.J. Smith.

(Chapter 59). Information on fence designs for new species can be adapted from existing guidelines or obtained from zoos, conservation reserves and farms and should be experimentally tested (Chapter 10) prior to widespread deployment.

Fencing for a single target species is rarely recommended because it is often cost-effective to install a single fence with different panels or mesh sizes (Figs. 20.5A, B and C) that benefit a wide range of species for a relatively small extra cost. For example, reptiles and amphibians are stopped by the small-sized mesh at the base of the fence and large ungulates are stopped by the larger mesh and strands reaching a height of greater than 2 m (Figs. 20.5B and C). Some arboreal species are prevented from climbing fences that are fitted with sheet metal (Fig. 20.5A). Fences can also be solid walls, such as the 1.1 m-tall concrete walls with overhanging lip at Paynes Prairie State

Preserve, Florida, United States (Fig. 20.6). Known for its high species diversity of wildlife and high rates of roadkill, the multi-species barrier wall and eight culverts at Paynes Prairie reduced mortality for all species (excluding tree frogs) by 93.5% and 51 species used the crossings (Dodd Jr. et al. 2004).

20.3 Consider alternatives to traditional fences

Walls and soil berms are frequently used to reduce traffic noise and light for humans (Chapter 48, Figs. 33.3 and 33.4), and with minor modifications, they may also function as wildlife fencing. Land

bridges must include fences to protect vehicles underneath from falling objects, and the fence should be combined with barriers to reduce noise and light. These include concrete walls, artistic features and berms planted with dense vegetation (Fig. 22.2); some designs can almost eliminate all vehicle noise and light. Concrete Jersey barriers are effective fences for some species, but the small drainage gaps (known as scuppers) at the base of some may allow movement of small-bodied species and should be filled. However, Jersey barriers placed in the median should retain these gaps, and potentially include large openings, to prevent animals that do make it onto the road from becoming trapped and forced back into oncoming traffic (Fig. 20.7). While not ideal, this provision may

(A)   (B)

Figure 20.6 Solid wall at Paynes Prairie, Florida, immediately after construction (A) and a few years later (B). Wall height is 1.1 m, culvert is 1.8 m × 1.8 m, and the overhanging lip protrudes by about 15 cm. Source: Photographs by D.J. Smith.

(A)   (B)

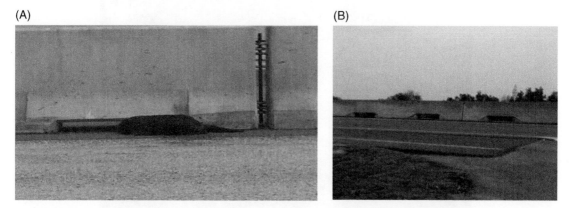

Figure 20.7 Jersey median barriers with drainage holes can trap animals, such as this river otter, and/or force them back into traffic (A), and while not ideal, animals that encounter such barriers in the centre median with larger holes (B) at least have a chance to make it across the road. Source: Photographs by D.J. Smith.

at least give these animals a chance of making it safely to the other side.

Dense plantings and other elements that provide food or shelter (e.g. a line of tree stumps, logs or artificial structures) can be used to direct animals towards crossing structures (Chapter 21). Riprap (Fig. 20.8, Chapter 44) can also be used, because many species, including ungulates, avoid walking on sharp, uneven or unstable surfaces. However, its effectiveness may diminish over time as the riprap will settle and dirt and debris may fill the spaces.

Noise and light walls (Figs. 33.3, 33.4) may act as wildlife fencing if their design takes into account the needs of the target species (i.e. gap size, height, etc.). They can force flying animals to fly up and over them, potentially above traffic height, although further research on the efficacy of this approach is required (Chapters 33 and 34). A similar potentially effective approach is rows of poles on bridges that encourage birds to fly above them. This method was applied for royal terns migrating along the Atlantic Coast that were being struck by vehicles when flying low over a bridge in Florida, United States. To address this problem, 122 silver-coloured metal poles 3 m tall and 21 cm in diameter were attached to the bridge at 3.7 m intervals, resulting in a 64% reduction in bird strikes with vehicles (Bard et al. 2002). This was a relatively unobtrusive and inexpensive solution, costing a total of US$5900 in 1994.

20.4 Animals inevitably breach fences, and when they do, they must be able to exit the roadway

Fences can be breached by wildlife through breaks in the fence (e.g. caused by accidents, storms, vandalism or natural wear and tear) and by digging under the fence, climbing over or passing through the fence, or entering at the ends of fences (Puglisi et al. 1974; Falk et al. 1978; Clevenger et al. 2001; Gulsby et al. 2011; Clevenger 2013). Frequent options to exit the roadway must be provided, especially if the fences extend for long distances, relative to the movement ability of the species.

Right-of-way (ROW) escape methods include one-way gates, jump-outs (Fig. 20.9) (also referred to as earthen or escape ramps) and climb-outs, all of which must be designed specifically for the target species, preferably in collaboration with species experts. One-way gates are intuitively appealing but are plagued by design faults and issues that have yet to be fully resolved (e.g. Reed et al. 1974a; Sielecki 2007). One-way gates frequently jam open (allowing wildlife into the ROW), or jam shut (preventing animals from exiting the ROW), and many animals appear unwilling to push into them to force them open. In some instances, individuals have learned to open the gates backwards and access the roadway. Wildlife can also impale themselves on the tines of improperly designed one-way gates. Jump-outs (Textbox 20.2 and Fig. 20.9) allow animals to escape the

Figure 20.8 This riprap was installed to funnel deer and elk towards crossing structures. While initially successful, deer and elk managed to negotiate their way through and fencing had to be retrofitted. Source: Photograph by J.W. Gagnon.

(A)  (B)  (C)

Figure 20.9 Jump-outs allow animals trapped inside a fenced roadway to escape. In these examples, ungulates walk along the fence until they reach the jump-out, and (A, B) they go up the ramp and jump down and out, or (C) the jump-out is built where the road is elevated on fill, and the animals jump down and out of the road reserve. Short lengths of guide fence perpendicular to the jump-out in (C) and in the middle of the entrance help to direct animals to the opening. The jump-out in (B) has electrified strands of wire to prevent animals from climbing up and into the roadway. Source: (A and C) Photographs by J.W. Gagnon and (B) Photograph by D.J. Smith.

Textbox 20.2 Adaptively designing right-of-way jump-outs for the target species.

Numerous crossing structures and wildlife fencing were installed along 27 km of US Highway 93 in Arizona, United States. Jump-outs were installed for desert bighorn sheep based on the design of successful jump-outs for elk. The jump-outs were monitored (Gagnon et al. 2014), and the sheep accessed the roadway by jumping or climbing them in the wrong direction (Fig. 20.10A). Plain fencing wire was installed across the opening and sheep used the ramp in the proper direction by going over or under the wire, while sheep jumping in the wrong direction collided with the wire. To prevent collisions with the wire, visibility was improved by encasing it in a plastic pipe and the height was adjusted (Fig. 20.10B). Eventually, the optimal height of the pipe was identified that maximised use in the proper direction while eliminating reversed use. All jump-outs were then modified with metal pipes providing a 0.4 m opening which allowed larger males to jump over the bars and ewes and lambs to pass underneath. Further monitoring documented that 96% of sheep approaching the jump-out from the proper direction used them to exit the roadway and 0% of sheep approaching from the bottom gained access to the roadway.

(A)  (B)

Figure 20.10 Desert bighorn sheep entering the roadway via a jump-out without modification (A) and a sheep successfully exiting the roadway via a jump-out with plastic pipe to provide a visible obstruction to deter access from outside the roadway while continuing to allow proper use (B). Eventually, the plastic pipe was replaced with metal bars at a height of 0.4 m. Source: Photographs by J.W. Gagnon.

roadway by 'jumping out and down' and outside of the fence and are most useful for ungulates, including elk, deer and bighorn sheep (Gagnon et al. 2014; Siemers et al. 2014). Jump-outs need to be high enough to allow animals to jump down and outside of the roadway but not back up and inside the roadway. The backside should be smooth enough to prevent agile species from climbing up. Electric fencing can also be used on the backside to discourage climbing. Jump-outs can either be built where the road is elevated on fill or the ramp can be built up, so the inside of the ramp cannot be seen from the outside (so animals are not apt to climb into the ROW). Guide wings can be included to direct animals into the jump-out, rather than walk straight past it (Lesson 20.5). Climb-outs work in the same way as jump-outs, except they are vertical poles against fences that allow climbing animals to exit, but not enter, the roadway. While simple concepts, the specifications of all escape mechanisms must be thoroughly tested prior to widespread deployment (Textbox 20.2).

20.5 Fence ends and planned breaks in fences must be designed to reduce the rate of WVC

Sometimes, the entire lengths of roads through conservation areas are fenced, but this is rarely the norm. Fencing is usually focused at high-risk localities and crossing structures and often includes breaks to allow traffic from side roads to enter the fenced road. Gates are the most effective approach at containing larger fauna; however, they are impractical when traffic volume on the lateral road is high and gates can be unintentionally left open (Sawyer et al. 2012). When gates are not an option, wildlife guards and electrified mats

(Fig. 20.11) can be effective (Reed et al. 1974b; Seamans & Helon 2008; Allen et al. 2013). The effectiveness of these approaches depends on the gap between the steel bars, the depth of the pit below the guards (to discourage animals from placing their feet between the bars) and the width of the guard or mat (to prevent jumping over) and should be thoroughly tested for new conditions before widespread deployment. Other approaches include (i) installing animal detection systems at fence ends (Chapter 24); (ii) angling fence ends away from the road, so animals that reach the 'wing' are directed away from the road and turn around; or (iii) having fence ends terminate in unsuitable habitat, so animals turn around to avoid it. Aerial electrified wires suspended above the main road have been used to deter elephants crossing at grade from entering the roadway (Chapter 43).

The location and length of fencing is ideally based on rates of WVC or known wildlife movements (Chapter 13). Tracking data (Chapter 11) can identify the locations with the highest rates of crossings, and fencing placed in these locations can funnel animals towards the crossing structures (Gagnon et al. 2010b; Dodd et al. 2012). It should be noted that wildlife movement paths may vary over time (e.g. Chapter 56) and fences should extend some distance beyond the current preferred crossing location in anticipation of this. When data on WVC or wildlife movement are not available, fence length can be based on the extent of habitat or other features important to the target species. For example, wetlands are critical for many amphibians and reptiles, and fencing (with crossing structures) should extend beyond known breeding sites to allow movement and prevent road-related mortality (Chapters 31 and 32). Similarly, many species of wildlife use riparian corridors as habitat and for movement,

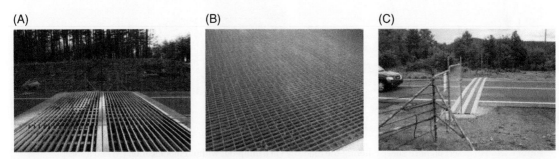

Figure 20.11 Strategies to prevent wildlife from accessing a fenced roadway where designed breaks in fencing occur. Wildlife guards: rows of steel tube (A) or steel mesh (B) prevent many ungulates from accessing the fenced roadway. (C) Electrified mats across the road is an option to consider for species that are capable of crossing wildlife guards. In both cases, the technique must be designed and tested on the target species to ensure effectiveness. Source: Photographs by J.W. Gagnon.

and fencing on either side of a bridge or culvert should encompass the riparian habitat to prevent roadkill and facilitate connectivity. Fencing in wetlands must be positioned to ensure that wildlife that require terrestrial habitats are able to access dry land during flooding events (e.g. Bager & Fontoura 2013).

20.6 Fences need to be maintained forever

Fences must be regularly maintained to maximise effectiveness and should be designed to facilitate maintenance (Chapter 17). Fences should be placed in accessible locations to allow repairs and vegetation removal, designed to facilitate inspection and built with robust materials. Fence design and materials are important considerations in areas where sand or snow drifts collect against fences, where trees can fall on fences, or if subject to repeated inundation, especially by saline water. Un-maintained fences allow more frequent breaching by wildlife, increasing the number of wildlife accessing the roadway. For example, growth of vegetation on fences or barrier walls (Fig. 20.6B) provides the means for small mammals, reptiles and amphibians to climb over and access the roadway. Inspection and maintenance of fencing for migratory species should be scheduled just before the migration usually occurs, which for some species can be quite accurately predicted. Inspection regimes need to assess structural integrity (e.g. are the posts rotting or the mesh intact?), breaches (e.g. errant vehicles, trespassers cutting fence) and natural breaches (e.g. overhanging or fallen trees, vines, grass or shrubs, piled-up sand or snow, washouts).

CONCLUSIONS

Fencing is an integral component of most mitigation programmes. Appropriately located, designed and installed fencing can reduce rates of WVC to almost zero and cost-effectively improve human safety, reduce property damage and help conserve wildlife populations. Fencing should always be integrated with properly spaced and located wildlife crossing structures or else the barrier effect of the road will be amplified. While fencing can dramatically reduce rates of wildlife mortality, the negative effects of reduced connectivity for wildlife from poorly designed fencing can be significant, resulting in reduced migration, dispersal and gene flow, further endangering wildlife. Maintenance is an ongoing requirement, and fences designed and installed with maintenance in mind will be cheaper and easier to maintain, resulting in longer service and increased benefits.

FURTHER READING

Clevenger and Huijser (2011): Focusing on North America, this handbook features extensive species-specific guidance on the design of mitigation measures, including fencing, escape ramps and wildlife guards.

Iuell et al. (2003): A broad-ranging summary and description of mitigation measures, including a range of general and specific recommendations on the use and design of fences in Europe.

Pepper et al. (2006): A detailed guide, focusing on the United Kingdom, on the design and construction of wildlife fences.

REFERENCES

Allen, T. D. H., M. P. Huijser and D. W. Wiley. 2013. Effectiveness of wildlife guards at access roads. Wildlife Society Bulletin **37**: 402–408.

Andrews, K. M., P. Nanjappa and S. P. D. Riley (Eds). 2015. Roads and Ecological Infrastructure: Concepts and Applications for Small Animals. Johns Hopkins University Press, Baltimore, MD.

Aresco, M. J. 2005. Mitigation measures to reduce highway mortality of turtles and other herpetofauna at a north Florida lake. Journal of Wildlife Management **69**: 549–560.

Bager, A. and V. Fontoura. 2013. Evaluation of the effectiveness of a wildlife roadkill mitigation system in wetland habitat. Ecological Engineering **53**: 31–38.

Bard, A. M., H. T. Smith, E. D. Egensteiner, R. Mulholland, T. V. Harber, G. W. Heath, W. J. B. Miller and J. S. Weske. 2002. A simple structural method to reduce road-kills of royal terns at bridge sites. Wildlife Society Bulletin **30**: 603–605.

Clevenger, A. P. 2013. Mitigating highways for a ghost: Data collection challenges and implications for managing wolverines and transportation corridors. Northwest Science **87**: 257–264.

Clevenger, A. P., B. Chruszcz and K. Gunson. 2001. Highway mitigation fencing reduces wildlife-vehicle collisions. Wildlife Society Bulletin **29**: 646–653.

Clevenger, A. P. and M. P. Huijser. 2011. Wildlife Crossing Structure Handbook: Design and Evaluation in North America. U.S. Department of Transportation, Federal Highway Administration, Lakewood, CO. http://www.cflhd.gov/programs/techDevelopment/wildlife/documents/01_Wildlife_Crossing_Structures_Handbook.pdf. Accessed 29 January 2015.

Dodd Jr., C. K., W. J. Barichivich and L. L. Smith. 2004. Effectiveness of a barrier wall and culverts in reducing wildlife mortality on a heavily traveled highway in Florida. Biological Conservation **118**: 619–631.

Dodd, N. L., J. W. Gagnon, S. Boe, K. Ogren and R. E. Schweinsburg. 2012. Wildlife-vehicle collision mitigation for safer wildlife movement across highways: State Route 260. Final project report 603, Arizona Department of Transportation Research Center, Phoenix, AZ. Available from http://wwwa.azdot.gov/adotlibrary/publications/project_reports/PDF/AZ603.pdf. Accessed on 16 September 2014.

Dodd, N. L., W. Gagnon, S. Boe and R. E. Schweinsburg. 2007. Role of fencing in promoting wildlife underpass use and highway permeability. In C. L. Irwin, P. Garrett and K. P. McDermott (Eds). 2007 Proceedings of the International Conference on Ecology and Transportation. Center for Transportation and the Environment, North Carolina State University, Raleigh, NC, pp. 475–487.

Falk, N. W., H. B. Graves and E. D. Bellis. 1978. Highway right-of-way fences as deer deterrents. Journal of Wildlife Management **42**: 646–650.

Gagnon, J. W., N. L. Dodd, S. Sprague, K. Ogren and R. E. Schweinsburg. 2010a. Preacher Canyon wildlife fence and crosswalk enhancement project evaluation: State Route 260. Final project report submitted to Arizona Department of Transportation, Phoenix, AZ. Available from http://www.azgfd.gov/w_c/documents/Preacher_Canyon_Elk_Crosswalk_and_Wildlife_Fencing_Enhancement_Project_2010.pdf. Accessed on 16 September 2014.

Gagnon, J. W., N. L. Dodd, S. Boe and R. E. Schweinsburg. 2010b. Using Global Positioning System technology to determine wildlife crossing structure placement and evaluating their success in Arizona, USA. In P. J. Wagner, D. Nelson and E. Murray (Eds). 2009 Proceedings of the International Conference on Ecology and Transportation Center for Transportation and the Environment, North Carolina State University, Raleigh, NC, pp. 452–462.

Gagnon, J. W., N. L. Dodd, S. Sprague, R. Nelson, C. Loberger, S. Boe and R. E. Schweinsburg. 2013. Elk movements associated with a high-traffic highway: Interstate 17. Final project report 647, Arizona Department of Transportation Research Center, Phoenix, AZ. Available from http://wwwa.azdot.gov/adotlibrary/publications/project_reports/PDF/AZ647.pdf. Accessed on 16 September 2014.

Gagnon, J. W., C. D. Loberger, S. C. Sprague, M. Priest, S. Boe, K. Ogren, E. Kombe and R. E. Schweinsburg. 2014. Evaluation of desert bighorn sheep overpasses along US Highway 93 in Arizona, USA. In 2013 Proceedings of the International Conference on Ecology and Transportation. Available from http://www.icoet.net/ICOET_2013/proceedings.asp. Accessed on 5 December 2014, pp. 1–18.

Glista, D. J., T. L. DeVault and J. A. DeWoody. 2009. A review of mitigation measures for reducing wildlife mortality on roadways. Landscape and Urban Planning **91**: 1–7.

Gulsby, W. D., D. W. Stull, G. R. Gallagher, D. A. Osborn, R. J. Warren, K. V. Miller and L. V. Tannenbaum. 2011. Movements and home ranges of white-tailed deer in response to roadside fences. Wildlife Society Bulletin **35**: 282–290.

Grandmaison, D. D. 2011. Wildlife linkage research in Pima County: Crossing structure and fencing to reduce wildlife mortality. Prepared for Pima County Regional Transportation Authority. Arizona Game and Fish Department. Final report prepared for Pima County Regional Transportation Authority by Arizona Game and Fish Department, Phoenix, AZ. Available from http://www.rtamobility.com/documents/RTACulvertFencingStudyFR112111.pdf. Accessed on 29 January 2015.

Huijser, M. P., P. McGowen, J. Fuller, A. Hardy, A. Kociolek, A. P. Clevenger, D. Smith and R. Ament. 2007. Wildlife-vehicle collision reduction study. Report to Congress. U.S. Department of Transportation, Federal Highway Administration, Washington, DC.

Iuell, B., G. J. Bekker, R. Cuperus, J. Dufek, G. Fry, C. Hicks, V. Hlaváč, V. Keller, B. Rosell, T. Sangwine, N. Tørsløv and B. l. M. Wandall (Eds). 2003. COST 341–Wildlife and Traffic: A European Handbook for Identifying Conflicts and Designing Solutions. KNNV Publishers, Brussels.

Jaeger, J. A. G. and L. Fahrig. 2004. Effects of road fencing on population persistence. Conservation Biology **18**: 1651–1657.

Jackson, S. D. and T. F. Tyning. 1989. Effectiveness of drift fences and tunnels for moving spotted salamanders *Ambystoma maculatum* under roads. In T. E. S. Langton (Ed.). Amphibians and Roads, Proceedings of the Toad Tunnel Conference. ACO Polymer Products Ltd., Bedfordshire, pp. 93–99.

Klar, N., M. Herrmann and S. Kramer-Schadt. (2009) Effects and mitigation of road impacts on individual movement behaviour of wildcats. Journal of Wildlife Management **73**: 631–638

Leblond, M., C. Dussault, J. Ouellet, M. Poulin, R. Courtois and J. Fortin. 2007. Electric fencing as a measure to reduce moose-vehicle collisions. Journal of Wildlife Management **71**: 1695–1703.

McCollister, M. F. and F. T. Van Manen. 2010. Effectiveness of wildlife underpasses and fencing to reduce wildlife-vehicle collisions. Journal of Wildlife Management **74**: 1722–1731.

Pepper, H. W., M. Holland and R. Trout. 2006. Wildlife Fencing Design Guide. CIRIA, Classic House, London, p. 60.

Puglisi, M. J., J. S. Lindzey and E. D. Bellis. 1974. Factors associated with highway mortality of white-tailed deer. Journal of Wildlife Management **38**: 799–807.

Reed, D. F., T. M. Pojar and T. N. Woodward. 1974a. Use of one-way gates by mule deer. Journal of Wildlife Management **38**: 9–15.

Reed, D. F., T. M. Pojar and T. N. Woodard. 1974b. Mule deer responses to deer guards. Journal of Range Management **27**: 111–113.

Ruby, D. E., J. R. Spotila, S. K. Martin and S. J. Kemp. 1994. Behavioral responses to barriers by desert tortoises: Implications for wildlife management. Herpetological Monographs **8**: 144–160.

Sawyer, H., C. Lebeau and T. Hart. 2012. Mitigating roadway impacts to migratory mule deer – a case study with underpasses and continuous fencing. Wildlife Society Bulletin **36**: 492–498.

Seamans, T. W. and K. C. VerCauteran. 2006. Evaluation of ElectroBraid™ as a white-tailed deer barrier. Wildlife Society Bulletin **34**: 8–15.

Seamans, T. W. and D. A. Helon. 2008. Evaluation of an electrified mat as a white-tailed deer (*Odocoileus virginianus*) barrier. International Journal of Pest Management **54**: 89–94.

Sielecki, L. E. 2007. The evolution of wildlife exclusion systems on highways in British Columbia. In C. L. Irwin, D. Nelson, and K. P. McDermott (Eds). Proceedings of the 2007 International Conference on Ecology and Transportation. Center for Transportation and the Environment, North Carolina State University, Raleigh, NC, pp. 459–474.

Siemers, J. L., K. R. Wilson and S. Baruch-Mordo. 2014. Wildlife fencing and escape ramp monitoring: Preliminary results for mule deer in southwest Colorado. In 2013 Proceedings of the International Conference on Ecology and Transportation. Available from http://www.icoet.net/ICOET_2013/proceedings.asp. Accessed on 5 December 2014, pp. 1–11.

Smith, D. J. and R. F. Noss. 2011. A reconnaissance study of actual and potential wildlife crossing structures in Central Florida. Final Report. UCF-FDOT Contract No. BDB-10. 154 pp. + appendices.

van der Ree, R. 1999. Barbed wire fencing as a hazard for wildlife. The Victorian Naturalist **116**: 210–217.

Woltz, H. W., J. P. Gibb and P. K. Ducey. 2008. Road crossing structures for amphibians and reptiles: Informing design through behavioral analysis. Biological Conservation **141**: 2745–2750.

Chapter 21

WILDLIFE CROSSING STRUCTURES: AN EFFECTIVE STRATEGY TO RESTORE OR MAINTAIN WILDLIFE CONNECTIVITY ACROSS ROADS

Daniel J. Smith[1], Rodney van der Ree[2] and Carme Rosell[3,4]

[1]Department of Biology, University of Central Florida, Orlando, FL, USA
[2]Australian Research Centre for Urban Ecology, Royal Botanic Gardens Melbourne, and School of BioSciences, The University of Melbourne, Melbourne, Victoria, Australia
[3]MINUARTIA Wildlife Consultancy, Barcelona, Spain
[4]Department of Animal Biology, Universitat de Barcelona, Barcelona, Spain

SUMMARY

Roads, railways and other linear infrastructure are often filters or barriers to the movement of wildlife. Wildlife crossing structures (underpasses and overpasses) improve traffic safety and contribute to the conservation of biodiversity by allowing animals to move safely across roads, thereby reducing the risk of wildlife-vehicle collision. This connectivity between populations on opposite sides of the road allows animals to access resources and mates and facilitates gene flow, thereby improving the viability of wildlife populations. The effectiveness of crossing structures is significantly enhanced when combined with fences, and both measures are usually best implemented together.

21.1 Follow a logical sequence of steps to implement an effective mitigation strategy.

21.2 Wildlife crossing structures are diverse in their design, shape and size; and they must be fully described in plans and reports to avoid confusion.

Handbook of Road Ecology, First Edition. Edited by Rodney van der Ree, Daniel J. Smith and Clara Grilo.
© 2015 John Wiley & Sons, Ltd. Published 2015 by John Wiley & Sons, Ltd.
Companion website: www.wiley.com\go\vanderree\roadecology

21.3 Multi-use structures are a potentially effective approach to increase the permeability of roads for wildlife.
21.4 The selection of structure type depends on the goals of mitigation, target species and engineering constraints.
21.5 The detailed design of crossing structures is critical to success.
21.6 The location and spacing of crossing structures should be guided by the ecological and biological needs of the target species.
21.7 Maintenance, monitoring and adaptive management of crossing structures are needed to assure success.
21.8 Alternatives to wildlife crossing structures are more appropriate in some locations and situations.
 Wildlife crossing structures should be constructed when impacts cannot be avoided or minimised. There is a wide diversity in the type of wildlife crossing structures available. Selecting the appropriate design depends on the impacts to be mitigated, the target species, engineering and other location-related constraints and traffic safety considerations. In addition, effectiveness can be maximised by addressing a number of design, monitoring and maintenance issues.

INTRODUCTION

Wildlife crossing structures increase the permeability of roads and other linear infrastructure for wildlife by allowing animals to safely cross under or over roads and by reducing the risk of wildlife-vehicle collisions (WVC). Wildlife crossing structures are designed primarily and foremost for the movement of *wildlife*, although some allow co-use by people, such as for recreation (Chapter 22). Crossing structures for other purposes, such as for water flow (Chapters 44 and 45) or the movement of stock or vehicles, can be modified to also allow the movement of wildlife (Lesson 21.3). Wildlife crossing structures have been constructed on roads and railways around the world, with the majority in North America and Europe, and more recently in Australia, Asia, Africa and South America. Most of the lessons and concepts about crossing structures across roads in this chapter apply equally to other types of linear infrastructure. Research has clearly demonstrated that many species, ranging from salamanders to elephants, will use them to cross roads (Chapters 28–45; Langton 1989; Bank et al. 2002; Bissonette & Cramer 2008; Clevenger 2012). By reducing the rate and severity of WVC, crossing structures (i) improve road safety for people; (ii) reduce mortality of wildlife; and (iii) improve the permeability of the road for wildlife, thereby reducing the barrier effect of the road, reconnecting animal populations and restoring ecological processes. While their effectiveness at restoring or maintaining connectivity for many species is unequivocal, the effectiveness of wildlife crossing structures at maintaining viable populations is less clear, and further research is required (Chapter 16).
 There are two main categories of wildlife crossing structures – underpasses and overpasses – and the

variation in their size, shape, construction style and materials, and target species is large (Lesson 21.2), with names of structure type often varying regionally. Consequently, it is beyond the scope of this chapter to define each and every type of crossing structure. Rather than attempt a reclassification of structure types, we have largely adopted terms currently used in North America and Europe (after Iuell et al. 2003; Clevenger & Huijser 2011), with a few minor adaptations. Because of the broad geographic and biological scope of this book, this chapter cannot outline species-specific applications of crossing structure types. Instead, generalised guidelines are given along with references to more detailed applications and examples (see Chapter 59). The type of crossing structure to be installed and its location and design and the number of structures and their spacing will be context and location specific. While the overarching goals of wildlife crossing structures are to restore connectivity and reduce WVC, their specific aims will depend on the target species, the impact(s) of the road and whether daily, seasonal or occasional movements of wildlife are to be facilitated. The aims of this chapter are to describe the diversity of structure types and their application for different species groups and highlight the key planning, design and selection processes and principles to achieve successful mitigation.

LESSONS

21.1 Follow a logical sequence of steps to implement an effective mitigation strategy

The first two steps when developing a mitigation strategy is to avoid impacts and minimise them and then consider mitigation measures (Chapter 7). Wildlife crossing structures are the most effective approach to mitigate the barrier effect of roads on wildlife movement, thereby maintaining or restoring landscape connectivity. When installed with fencing to keep animals off the roadway and funnel them towards crossing structures, they can also reduce or eliminate WVC and improve road safety (Chapter 20). The steps to developing an effective mitigation strategy are:

(i) Define the problem: identify and clearly articulate the specific ecological impact(s) of the proposed or existing road.

(ii) Set specific, measurable, achievable, realistic and time-framed (SMART) goals for mitigation: once the problem is known, the goals for mitigation can be developed. Goals should be SMART (Lesson 16.1, Textbox 16.1).

(iii) Plan and design the mitigation strategy: ensure the mitigation strategy addresses the specific problem as not all road impacts are solved by crossing structures (Chapter 9; Lesson 21.8).

(iv) Construct and maintain the mitigation measures: ensure the mitigation is built as designed (Chapters 8 and 9) and maintained appropriately (Chapter 17).

(v) Evaluate and adaptively manage the mitigation measures: monitor and evaluate use and/or effectiveness (Chapters 10, 15 and 16) and adaptively manage the mitigation to ensure the goals are achieved.

Collaboration and cooperation among stakeholders should occur at all stages to facilitate productive input and achieve better outcomes.

21.2 Wildlife crossing structures are diverse in their design, shape and size; and they must be fully described in plans and reports to avoid confusion

Wildlife crossing structures come in a variety of designs, shapes and sizes, and their names vary regionally, complicating the development of a standard set of terms that can be applied globally (Iuell et al. 2003; Clevenger & Huijser 2011). However, despite this complexity, there are two main types, namely, overpasses and underpasses. The greatest difference in naming among regions appears to be whether structures are described by their intended function (e.g. small mammal *crossing* or *passage*) or by its type (e.g. small mammal *culvert* or *tunnel*). Structures have also been variously described according to their target group or species of wildlife and structure type, such as wildlife underpass, small mammal underpass, frog culvert or badger pipe. Unique names, specific to the type of structure, have also been applied, such as ecoduct, ecopassage or amphibian tunnel. The

important consideration when naming and describing structures is to minimise the risk of confusion or misunderstanding. This is particularly relevant (i) during the planning and design phase to ensure the correct structure is installed (Chapter 9), (ii) when reporting crossing rates to allow the effectiveness of different types of structures to be correctly interpreted and (iii) when adopting or adapting designs or recommendations from other regions. All plans and reports should give the following information when describing crossing structures:

• Broad classification: that is, underpass or overpass.
• Structure type: for example, bridge, pipe or culvert.
• Shape: for example, culverts can be round, square, rectangular, elliptical (oval) or arched; and wildlife overpasses can be straight or hourglass shaped.
• Size: length, width and height.
• Construction materials: for example, concrete, corrugated steel, timber or composite material (e.g. polymer concrete).
• Interior design: for example, natural substrate, concrete floor, grated roof or addition of tree stumps, logs, brush or rocks.
• Strategy to funnel or attract wildlife to the structure: for example, fencing, hedgerows or stone walls to guide animals to entrances or ponds to attract wildlife.
• Modified features: for example, wooden or steel shelf on culvert wall or suspended poles/ropes within underpasses.
• Additional uses or features: for example, stream or watercourse, road, railroad or co-use by livestock or people.

Wildlife underpasses

Underpasses allow animals to move under the road and are the most common type of crossing structure. Underpasses built specifically for wildlife (see Lesson 21.3 and Chapter 22 for multi-use structures) include standard engineering structures, for example, bridge or culvert, as well as structures specifically designed and constructed for wildlife movement, for example, amphibian tunnels. It is beyond the scope of this chapter to describe each type in detail; however, we identify three main types, in approximately decreasing order of size: (i) viaducts and long or open-span bridges are typically long or high bridges, usually traversing valleys or riparian areas (Fig. 21.1A); (ii) wildlife underpasses are variable in size and type (e.g. bridge, box, pipe) but typically smaller than open-span bridges/viaducts (Fig. 21.1B); and (iii) amphibian tunnels are small purpose-built structures specifically for amphibians (Fig. 21.1C; Langton 1989).

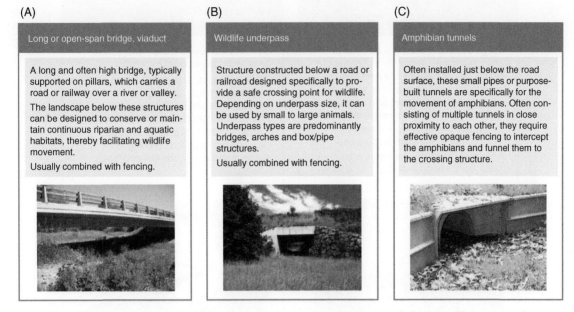

(A)

Long or open-span bridge, viaduct

A long and often high bridge, typically supported on pillars, which carries a road or railway over a river or valley.

The landscape below these structures can be designed to conserve or maintain continuous riparian and aquatic habitats, thereby facilitating wildlife movement.

Usually combined with fencing.

(B)

Wildlife underpass

Structure constructed below a road or railroad designed specifically to provide a safe crossing point for wildlife. Depending on underpass size, it can be used by small to large animals. Underpass types are predominantly bridges, arches and box/pipe structures.

Usually combined with fencing.

(C)

Amphibian tunnels

Often installed just below the road surface, these small pipes or purpose-built tunnels are specifically for the movement of amphibians. Often consisting of multiple tunnels in close proximity to each other, they require effective opaque fencing to intercept the amphibians and funnel them to the crossing structure.

Figure 21.1 The three main types of wildlife underpasses. Source: (A and B) Photographs by C. Rosell/Minuartia and (C) photograph by and reproduced with permission of Miklós Puky.

Depending on their size, shape, construction materials and landscaping, wildlife underpasses permit the movement of small to large species of wildlife, with larger structures serving a greater range of species and smaller structures restricting use by larger-bodied species. Species-specific minimum dimensions are given in many of the regional best-practice guidelines (Chapter 59).

Wildlife overpasses

Overpasses allow animals to cross above the road and include (i) ecoducts, also referred to in some countries as a green bridge, land bridge or landscape bridge (Fig. 21.2A); (ii) wildlife overpasses (Fig. 21.2B); (iii) canopy bridges (Fig. 21.2C); and (iv) glider poles (Fig. 21.2D). Ecoducts and wildlife overpasses are similar, and the terms are often used interchangeably, preventing clear and concise use of terms. Ecoducts are wide, often greater than 50 m wide (minimum recommended width in Europe is 80 m), and tend to 'reconnect habitat and landscapes' without interruption across the road. Wildlife overpasses are typically narrower than ecoducts (sometimes ~20 m). This challenge of defining terms in an international setting demonstrates the importance of comprehensively describing structures as discussed earlier in this lesson. Nevertheless, overpass width affects the extent to which different habitat zones, such as strips of different

vegetation or soil types, can be included on the structure, affecting the number and diversity of species which may use it. Ecoducts and wildlife overpasses can be used by a wide diversity of species, from invertebrates (Chapter 29) to large herbivores (Chapter 42). Canopy bridges (ropes or poles that connect tree canopies) and glider poles (vertical poles that act as artificial trees, providing launch and landing points for jumping and/or gliding species – Chapters 40 and 41) are specifically for arboreal and/or gliding species.

21.3 Multi-use structures are a potentially effective approach to increase the permeability of roads for wildlife

Multi-use (also called multipurpose or multifunction) structures differ from wildlife crossing structures in that wildlife movement is a secondary function or goal. For example, drainage culverts are primarily about the movement of water, while over- or underpasses for stock and forestry access are primarily about access for human-related activities (Fig. 21.3). A limitation of multi-use structures is that they are designed and maintained to achieve their primary function (e.g. drainage, stock or vehicle access), and this will compromise their suitability for some wildlife. However, depending on the timing of each use and the degree of compatibility between the human use and wildlife (van der Grift et al.

(A)

Ecoduct, landscape bridge, land bridge, green bridge

Large overpass, usually > 50 m wide (min. recommended width in Europe 80 m), where habitats are continuous across the road. Due to their width, a diversity of habitat types (e.g. vegetation or soil types) can be included.

The main difference from wildlife overpasses is the width and vegetation cover, however the terms are often used interchangeably.

(B)

Wildlife overpass

Constructed above roads, specifically to provide connectivity for wildlife.

While similar to landscape bridges, they are narrower, limiting the extent to which different habitats and vegetation can be included on the structure. Landscape bridges and wildlife overpasses usually include fencing to funnel animals towards the structures.

(C)

Canopy bridge

Rope, net or pole suspended above the road from vertical poles or trees, for arboreal and scansorial species.

While fencing would improve rates of use, fence designs are yet to be developed due to the climbing ability of the target species.

Similar structures, called hop-overs, have been proposed for bats, with little evidence for success.

(D)

Glider poles

Timber poles, erected on the roadside or within the median, to act as 'artificial trees' for gliding and jumping species to cross the road.

Pole height and spacing is critical, and must be designed according to the gliding or jumping ability of the target species (Fig.40.3)

Effective fence designs have not yet been developed.

Figure 21.2 The four main types of wildlife overpasses. Source: (A) Photograph by and reproduced with permission of GIASA/Junta de Andalucía; (B) Photography by C. Rosell/Minuartia and (C and D) photographs by R. van der Ree.

(A)   (B)   (C)

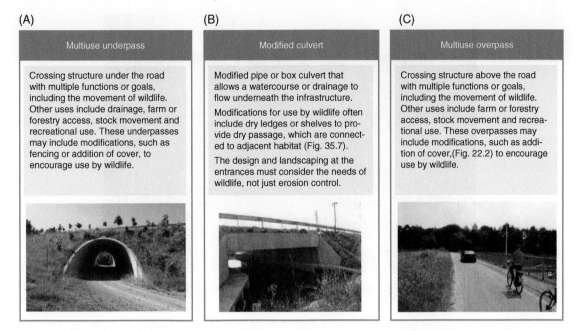

| Multiuse underpass | Modified culvert | Multiuse overpass |

Crossing structure under the road with multiple functions or goals, including the movement of wildlife. Other uses include drainage, farm or forestry access, stock movement and recreational use. These underpasses may include modifications, such as fencing or addition of cover, to encourage use by wildlife.

Modified pipe or box culvert that allows a watercourse or drainage to flow underneath the infrastructure.

Modifications for use by wildlife often include dry ledges or shelves to provide dry passage, which are connected to adjacent habitat (Fig. 35.7).

The design and landscaping at the entrances must consider the needs of wildlife, not just erosion control.

Crossing structure above the road with multiple functions or goals, including the movement of wildlife. Other uses include farm or forestry access, stock movement and recreational use. These overpasses may include modifications, such as addition of cover,(Fig. 22.2) to encourage use by wildlife.

Figure 21.3 The three main types of multi-use crossing structures. Source: (A and B) Photographs by C. Rosell/Minuartia and (C) photograph by and reproduced with permission of Marcel Huijser.

2012), this approach can be a cost-effective strategy to increase the number of opportunities for safe wildlife crossings because of their potential for wide deployment across the landscape. Importantly, multi-use structures must still achieve the specific mitigation goals regardless of their other functions and should not be considered a cheap alternative to wildlife crossing structures.

Multi-use structures may be functional for use by wildlife with or without modification (Mata et al. 2008). For example, drainage culverts may permit the movement of some terrestrial species when dry or mostly dry. Numerous species in the Netherlands have been shown to use overpasses with primarily daytime recreational use by humans (Textbox 22.1). However, the rate of use by wildlife can be increased, often quite dramatically, with relatively minor modifications to their design or maintenance (Chapter 17). For example, drainage culverts can be modified by adding an elevated ledge or shelf on one or both walls of the culvert to provide dry passage for wildlife (Figs. 9.1, 21.3B, 35.7, 39.1 and 39.3; Veenbaas & Brandjes 1999; Villalva et al. 2013). Adding a second culvert with a raised floor adjacent to one with permanent water flow is a simple but effective approach (Figs. 21.4 and 45.5 but see Fig. 9.2). Other strategies include the removal of large rocks (also called riprap) from approaches to drainage structures (Fig. 21.5A, B), the addition of

fencing to funnel wildlife towards the structures (Chapter 20) or the addition of structural elements (also known as 'furniture') such as rocks, tree stumps, brush piles or tubes within the structure or the approaches to provide resources or shelter from predators (Figs. 29.5 and 39.2; Connolly-Newman et al. 2013). Larger modifications may also be required to ensure multi-uses are compatible, such as increasing the size of the structure.

Fish and other aquatic species such as otters, platypus and invertebrates use a range of drainage structures for movement. The most effective structures for aquatic species are designed to maintain natural rates of water flow or provide resting opportunities for fish swimming upstream (Chapters 44 and 45).

21.4 The selection of structure type depends on the goals of mitigation, target species and engineering constraints

The choice of structure depends primarily on (i) the goal of mitigation (Lesson 21.1); (ii) the specific requirements of the target species; and (iii) cost and engineering constraints. Importantly, planners, engineers and wildlife experts must collaborate to design and implement optimal solutions. The goal of mitigation relates in part to

Figure 21.4 The two pipe culverts are elevated above the box culvert used for stream flow, providing dry passage for wildlife. Source: Photograph by D.J. Smith.

Figure 21.5 Example of (A) before and (B) after removal of riprap and creation of dry ledge adjacent to stream to facilitate use by white-tailed deer. Source: Photographs by D.J. Smith.

whether the movement is to occur daily (e.g. to access food), seasonally (e.g. annual migration) or occasionally (e.g. dispersal to maintain gene flow), which will influence the number and spacing of the structures (Lesson 21.5). The ecological or biological requirements of the target species will influence the type, size and design of the structure (e.g. Rosell et al. 1997; Clevenger & Waltho 2000, 2005). For example, the target species must physically fit within the underpass but also be behaviourally comfortable with the size, substrate and surroundings to encourage them to use it. Prey species may be wary of

enclosed spaces that increase their perceived risk of predation (Chapter 42; Brudin 2003; Gordon & Anderson 2003). Certain species of amphibian may avoid structures with concrete or metal floors and prefer underpasses with similar microclimates to natural habitats (Woltz et al. 2008). The details of these species-specific requirements are provided and being continually updated in the regional best-practice guidelines (see Chapter 59 for examples).

The optimal approach to mitigation is to avoid impacts first, minimise second and finally mitigate any remaining impacts (Chapter 7). It may not be feasible to build the optimal structure due to engineering or financial constraints, and alternative options such as multi-use structures or alternatives to crossing structures should be considered. If there are no feasible effective mitigation options available, the road should be realigned to avoid the impact.

21.5 The detailed design of crossing structures is critical to success

The detailed design of crossing structures, including landscaping, structural elements to provide shelter and refuge, screening and substrate type, is just as important in influencing success as selecting the optimal type of structure. Detailed designs must be comprehensive because depending on contractual arrangements for delivery of the project, construction teams may only be obligated to build what has been specified. Therefore, significant alterations to the detailed design after the project has commenced will likely cost more (Chapter 8).

The landscaping of crossing structures and approaches to them should (i) encourage use of the structure for the target species; (ii) guide animals towards the entrances of the structure; and (iii) minimise negative disturbance effects of traffic. As a guide, the habitat on the approaches and the structures should closely resemble the adjacent habitat such that animals do not notice a change. This is not always possible, particularly in underpasses, and other features such as logs and branches, rocks and stone walls can be used to provide cover (Fig. 35.5; Connolly-Newman et al. 2013). Depending on the topography, the ground adjacent to the crossing

structure can be contoured to naturally funnel animals towards the entrances (Lesson 20.3, Figs. 20.8 and 21.6A). Vegetation plantings can be arranged to guide animals towards entrances and also to provide shelter or cover for prey species (Fig. 21.6B). For example, species such as deer prefer a clear sight path to aid with vigilance for potential large predators (Chapter 42). Smaller prey species require cover for protection from predators (Chapters 23, 32 and 39) or adverse environmental conditions (e.g. amphibians, Chapter 31).

Fencing is usually necessary to funnel animals towards the crossing structure and fence design and material is a critical consideration. Chapter 20 provides extensive information about fencing. Screening on and adjacent to crossing structures, such as noise and light walls, dense vegetation plantings or earth berms (Figs 22.1–22.3, 33.3 and 33.4), can be used to reduce the disturbance effects of traffic noise, light and pollution that may lead to avoidance by some species (Chapters 18, 19, 22 and 33; Friedman 1997). The specific design of screening material is dependent on the target species, the impact to be mitigated, local conditions and aesthetic considerations.

The type and quality of substrate used on approaches and crossing structures can influence rates of use by wildlife. Imported soil, crushed rock and other material should be avoided if the target species is sensitive to the surface and subsurface conditions, such as amphibians and fossorial or digging species (Chapters 29, 31, 32 and 39). Construction equipment can also transmit soil-borne pathogens and diseases, as well as compact the soil, all potentially affecting plant growth and the ability of fossorial species to dig. Moisture retention is critical to amphibians, and prolonged exposure can

(A)

(B)

Figure 21.6 (A) Contouring and (B) plantings at entrances to wildlife crossing structures (shown with red circle – leading to underpass) can enhance movement of wildlife that require cover. Source: (A) Photograph by D.J. Smith and (B) © Google Earth.

result in rapid desiccation under sunny conditions. Presence of suitable soils, vegetation, leaf litter and other ground cover at the site can reduce soil evaporation rates and increase moisture levels (Chapter 31). Rocks and woody debris can act as stepping stones for amphibians providing shelter and refugia where moisture retention is more prolonged.

21.6   The location and spacing of crossing structures should be guided by the ecological and biological needs of the target species

The location of crossing structures along a road significantly affects rates of use by wildlife. The primary determinant of the location of crossing structures in the landscape should be biological and ecological (Clevenger et al. 2002; Langen et al. 2009). In other words, crossing structures should be positioned where they will be of maximum benefit for the target species, such as along the preferred movement pathways of the target species (Clevenger et al. 2003; Ramp et al. 2005). Many species move through the landscape along riparian corridors, and crossing structures along waterways will likely be optimal for them (Clevenger et al. 2002). Other locations include where green corridors cross roads, locations between seasonally occupied habitats or where resources are located on opposite sides of the road (Beaudry et al. 2008, 2009). These locations can be identified using models developed with WVC and/or mortality data (Chapter 13), but they can also be at locations where there is no WVC or roadkill due to road avoidance or because the population has already declined. Landscape-scale analyses that identify important habitats and wildlife corridors help planners avoid these important areas when planning new roads and to prioritise locations for mitigation.

The spacing of crossing structures is primarily dependent on the specific goals of the mitigation. For example, crossing structures to provide daily access to foraging resources for all members of a population should be closer together than structures for occasional dispersal of a few individuals. Similarly, crossing structures for species with small home ranges (e.g. salamanders) will need to be closer together than for species with very large home ranges (e.g. bears or wolves). Decisions related to these issues must be based on what is necessary to maintain or restore the health and structure of the populations of the species impacted, rather than simply the cheapest or most convenient location. An objective approach is to base the spacing on the size of the target species' home range (i.e. the area used by an animal during its day to day activities). This area can be converted to a linear distance by calculating its square root (i.e. $\sqrt{HR}$) (Bowman et al. 2002), which Bissonette and Adair (2008) propose is the spacing necessary to create a truly 'permeable' road. They further propose that crossing structures designed to facilitate dispersal should be seven times the linear home range (i.e. $7 \times \sqrt{HR}$).

21.7   Maintenance, monitoring and adaptive management of crossing structures are needed to assure success

Maintenance, monitoring and adaptive management are critical to the long-term success of all mitigation, including wildlife crossing structures. Appropriate maintenance of crossing structures and approaches is an ongoing need and often is not conducted properly (Chapter 17). Maintenance regimes need to evaluate both the structural and functional integrity of the crossing structure and be integrated with routine maintenance programmes (Chapter 17). Fences have particular maintenance requirements which, if neglected, may result in increased rates of WVC and decreased effectiveness of crossing structures (Chapter 20). Vegetation management along roadsides, at approaches to crossing structures and along fencing should also be ongoing and requires special detailed instructions for maintenance crews (Chapters 17, 20 and 46). Monitoring the rate of use of the crossing structure and its effectiveness at meeting the specified goals is essential to (i) evaluate success; (ii) improve the design of future mitigation measures; and (iii) modify the design of existing structures (Fig. 21.7, Chapters 15 and 16). Studies should begin before construction and may need to continue for 5 or even 10 years if the target species is signfiicant (e.g. endangered or high profile) or if acceptance by wildlife is slow (Chapters 10, 15 and 16). If the goals of mitigation have not been reached, corrective actions must be taken.

21.8   Alternatives to wildlife crossing structures are more appropriate in some locations and situations

Wildlife crossing structures are not always the most appropriate mitigation measure (Huijser & McGowen, 2010; Langbein et al. 2011). The critical consideration when developing a mitigation strategy is to clearly identify the specific impact(s) of the road and traffic and determine if crossing structures, which

Figure 21.7 Monitoring and adaptive management were used to improve the function of this structure for management of an important game species, wild turkey. Clockwise from upper left, (A) flooding of approaches and (B) centre of crossing structure that prevented use by turkeys was corrected by (C) installing a drainage culvert which provided a dry pathway and (D) increased use by wild turkey adults and offspring. Source: Photographs by D.J. Smith.

allow the safe movement of wildlife across the road, will address this impact. Impacts other than reduced connectivity or WVC (e.g. reduced habitat quality along a road due to noise, light or chemical pollution) are unlikely to be solved by the installation of crossing structures.

While there are other potential strategies to maintain or restore connectivity that do not require crossing structures, these are usually less effective. For example, at-grade crossings require animals to cross the road between passing cars (Chapters 20 and 24) and are thus inappropriate on roads with high volume or speed of traffic. Traffic calming reduces the speed and/or number of vehicles on the road and is unlikely to succeed on high-speed/high-traffic-volume roads and if alternative roads are not

available. Signage has varying levels of effectiveness at slowing drivers and reducing WVC (Chapter 24). Natural canopy connectivity (Fig. 40.1A) allows animals to climb across the road from branch to branch or jump from tree to tree, but is only possible on relatively narrow roads or those with wide medians and does not entirely eliminate the risk of collision with vehicles (Chapter 40). While still considered somewhat experimental, animal detection systems can be effective for large-bodied animals on relatively straight roads, but are less suitable on roads with high traffic volumes or high-speed vehicles or for relatively small species (Chapter 24). Animal detection systems were originally focused on reducing rates of WVC; however, they are increasingly being used to improve connectivity for wildlife.

CONCLUSIONS

Wildlife crossing structures are a common and frequently applied mitigation measure used to restore or maintain the movement of wildlife across roads. While there are two basic types of crossings structures – underpasses and overpasses – the names given to the same and/or different structure types vary regionally. Consequently, caution must be exercised when adapting or adopting structure recommendations from one region to another. Therefore, we urge that structures be comprehensively described in plans and reports to avoid confusion. The types of structures and their detailed design, location and spacing should primarily be based upon ecological and biological requirements of the target species and the goal of mitigation. Inevitably, the preferred strategy will be difficult or impossible to implement at some locations and compromises must be made. In these situations, planners, engineers and wildlife experts must collaborate to ensure that the effectiveness of the mitigation can still be achieved, thereby ensuring the risk of extinction of the target species does not increase. If mitigation is unsuccessful, it is unlikely that the conditions of approval for the project will have been met, and further mitigation or offsetting work will likely be required. It is important to realise that while crossing structures are effective at restoring connectivity and reducing rates of WVC, other effects of roads are not mitigated with these structures. Importantly, mitigation is the third step in the hierarchy – and impact avoidance and minimisation principles should be adopted first. In situations where crossing structures or other methods are likely to be ineffective at achieving the specific goals of mitigation, the road should be rerouted to avoid such areas.

ACKNOWLEDGEMENTS

We thank Kylie Soanes for comments on an earlier version of the chapter. R. van der Ree is supported by the Baker Foundation.

FURTHER READING

Clevenger and Ford (2010): A clearly written summary of issues around the design, location and construction of wildlife crossing structures and fences, with a North American focus.

Clevenger and Huijser (2011): A detailed and comprehensive account for North America of the planning, design, construction and maintenance of wildlife crossing structures.

Forman et al. (2003): The sixth chapter within the seminal book 'Road Ecology: Science and solution' focuses on mitigation techniques for wildlife, primarily wildlife crossing structures.

Iuell et al. (2003): A European handbook that brought together best-practice in the mitigation of wildlife and traffic conflict. While presenting data and insights from the early 2000s, much of it is still relevant today.

REFERENCES

Bank, F. G., C. L. Irwin, G. L. Evink, M. E. Gray, S. Hagood, J. R. Kinar, A. Levy, D. Paulson, B. Ruediger and R. M. Sauvajot. 2002. Wildlife habitat connectivity across European highways. Report no. FHWA-PL-02-011. U.S. Department of Transportation, Federal Highway Administration, Washington, DC.

Beaudry, F., P. G. deMaynadier and M. L. Hunter Jr. 2008. Identifying road mortality threat at multiple spatial scales for semi-aquatic turtles. Biological Conservation **141**: 2550–2563.

Beaudry, F., P. G. deMaynadier and M. L. Hunter Jr. 2009. Seasonally dynamic habitat use by Spotted (*Clemmys guttata*) and Blanding's Turtles (*Emydoidea blandingii*) in Maine. Journal of Herpetology **43**: 636–645.

Bissonette, J. A. and P. Cramer. 2008. Evaluation of the use and effectiveness of wildlife crossings. NCHRP Report no. 615. National Cooperative Research Program, Transportation Research Board of the National Academies, Washington, DC, , 161 pp.

Bissonette, J. A. and W. A. Adair. 2008. Restoring habitat permeability to roaded landscapes with isometrically-scaled wildlife crossings. Biological Conservation **141**:482–488.

Bowman, J., J. A. G. Jaeger and L. Fahrig. 2002. Dispersal distance of mammals is proportional to home range size. Ecology **83**:2049–2055.

Brudin, C. O. 2003. Wildlife use of existing culverts and bridges in north central Pennsylvania. In C. L. Irwin, P. Garrett and K. P. McDermott, editors. Proceedings of the International Conference on Ecology and Transportation. Center for Transportation and the Environment, North Carolina State University, Raleigh, NC, pp. 344–352.

Clevenger, A. P. 2012. 15 years of Banff research: What we've learned and why it's important to transportation managers beyond the park boundary. In Sustainability in Motion, Proceedings of the 2011 International Conference on Ecology and Transportation, Seattle, WA, pp. 409–423.

Clevenger, A. P. and A. T. Ford. 2010. Wildlife crossing structures, fencing, and other highway design considerations. In J. P. Beckmann et al., editors. Safe Passages: Highways, Wildlife, and Habitat Connectivity. Island Press, New York, pp. 17–49

Clevenger, A. P. and M. P. Huijser. 2011. Wildlife crossings structure handbook: Design and evaluation in North America. U.S. Department of Transportation, Federal Highway Administration. Publication no. FHWA-CFL/TD-11-003. Lakewood CO, 211 pp.

Clevenger, A. P. and N. Waltho. 2000. Factors influencing the effectiveness of wildlife underpasses in Banff National Park, Alberta, Canada. Conservation Biology **14**: 47–56.

Clevenger, A. P. and N. Waltho. 2005. Performance indices to identify attributes of highway crossing structures facilitating movement of large mammals. Biological Conservation **121**:453–464.

Clevenger, A. P., J. Wierzchowski, B. Chruszcz and K. Gunson. 2002. GIS-generated, expert-based models for wildlife habitat linkages and mitigation planning. Conservation Biology **16**:503–514.

Clevenger, A. P., B. Chruszcz and K. Gunson. 2003. Spatial patterns and factors influencing small vertebrate fauna road kill aggregations. Biological Conservation **109**: 15–26.

Connolly-Newman, H. R., M. P. Huijser, L. Broberg, C. R. Nelson and W. Camel-Means. 2013. Effect of cover on small mammal movements through wildlife underpasses along US Highway 93 North, Montana, USA. Proceedings of the 2013 International Conference on Ecology and Transportation. 23–27 June 2013, Scottsdale, AZ, USA, 12 pp. Available from http://www.icoet.net/ICOET_2013/proceedings.asp, Accessed on 22 September 2014.

Forman, R. T. T., et al., editors. 2003. Mitigation for wildlife. In Road Ecology: Science and Solutions, Island Press, Washington, DC, pp. 139–167.

Friedman, D. S. 1997. Nature as infrastructure: The National Ecological Network and wildlife crossing structures in the Netherlands. Report no. 138. DLO Winand Staring Centre, Wageningen.

Gordon, K. M. and S. H. Anderson. 2003. Mule deer use of underpasses in western and southwestern Wyoming. In C. L. Irwin, P. Garrett and K. P. McDermott, editors. Proceedings of the International Conference on Ecology and Transportation. Center for Transportation and the Environment, North Carolina State University, Raleigh, NC. pp. 246–252.

Huijser, M. P. and P. T. McGowen. 2010. Reducing wildlife-vehicle collisions. In: J. P. Beckman et al., editors. Safe Passages. Highways, Wildlife and Habitat Connectivity. Island Press: Washington, DC, pp. 51–74.

Iuell, B., et al. 2003. Wildlife and traffic: A European handbook for identifying conflicts and designing solutions. COST-341 Habitat Fragmentation and Transportation Infrastructure. European Co-operation in the Field of Scientific and Technical Research, KNNV, Brussels.

Langbein, J., R. Putman and B. Pokorny. 2011. Traffic collisions involving deer and other ungulates in Europe and available measures for mitigation. In: R. Putman et al., editors. Ungulate Management in Europe. Cambridge University Press: Leiden, pp. 215–259.

Langen, T. A., K. Ogden and L. Schwarting. 2009. Predicting hotspots of herpetofauna road mortality along highway road networks: model creation and experimental validation. Journal of Wildlife Management **73**:104–114.

Langton, T. E. S., editor. 1989. Amphibians and roads. ACO Polymer Products, Ltd. Bedfordshire, 202 pp.

Mata, C., I. Hervás, J. Herranz, F. Suárez, and J.E. Malo. 2008. Are motorway wildlife passages worth building? Vertebrate use of road-crossing structures on a Spanish motorway. Journal of Environmental Management **88**: 407–415

Ramp, D., J. K.A. Caldwell, D. Edwards, D. Warton and D. B. Croft. 2005. Modeling of wildlife fatality hotspots along the Snowy Mountain Highway in New South Wales, Australia. Biological Conservation **126**:474–490.

Rosell, C., J. Parpal, R. Campeny, S. Jove, A. Pasquina and J. Velasco. 1997. Mitigation of barrier effect of linear infrastructure on wildlife. In K. Canters, A. Piepers and D. Hendriks-Heersma, editors. Habitat Fragmentation and Infrastructure. DWW, Maastricht and The Hague, pp. 367–372.

van der Grift, E. A., F. Ottburg, R. Pouwels, and J. Dirksen. 2012. Multiuse overpasses: does human use impact the use by wildlife? Proceedings of the 2011 International Conference on Ecology and Transportation. 21–25 August 2011, Seattle, Washington, USA, 1080 pp. Available from http://www.icoet.net/ICOET_2011/proceedings.asp. Accessed on 22 September 2014.

Veenbaas, G. and G. J. Brandjes. 1999. The use of fauna passages along waterways under motorways. In J. W. Dover and R. G. H. Bunce, editors. Key Concepts in Landscape Ecology. International Association of Landscape Ecology, Preston, pp. 315–320.

Villalva, P., D. Reto, M. Santos-Reis, E. Revilla and C. Grilo. 2013. Do dry ledges reduce the barrier effect of roads? Ecological Engineering **57**:143–148.

Woltz, H. W., J. P. Gibbs and P. K. Ducey. 2008. Road crossing structures for amphibians and reptiles: Informing design through behavioral analysis. Biological Conservation **141**:2745–2750.

Chapter 22

RECREATIONAL CO-USE OF WILDLIFE CROSSING STRUCTURES

Rodney van der Ree[1] and Edgar A. van der Grift[2]

[1]Australian Research Centre for Urban Ecology, Royal Botanic Gardens Melbourne, and School of BioSciences, The University of Melbourne, Melbourne, Victoria, Australia
[2]Alterra, Wageningen UR, Environmental Science Group, Wageningen, The Netherlands

SUMMARY

There is growing pressure to build crossing structures that facilitate the movement of both people and wildlife across roads. In this chapter, we focus primarily on recreational co-use of wildlife crossing structures, specifically hikers, runners, cyclists and horse riders. This pressure to install co-use structures is most apparent in and around cities and towns and in recreational areas where trails are obstructed by roads. There is little knowledge to determine the appropriateness and design of such co-use, but preliminary findings and information gleaned from other sources are instructive.

22.1 Wildlife crossing structures are expensive, and it is intuitively appealing that they function for wildlife and people.

22.2 Different species respond differently to human co-use of wildlife crossing structures.

22.3 Simple design principles may enable co-use by humans.

22.4 Carefully designed and executed studies are required to confidently determine the efficacy of co-use structures.

It is clear that some species of wildlife, particularly those that readily survive or co-inhabit areas with humans, appear capable of using wildlife crossing structures with high rates of co-use by people. However, it is also evident that some species lower their rate or modify their timing of use of crossing structures which are also used by people. Consistent trends in the use of such crossing structures by wildlife are difficult to discern due to widely varying design, location and methods used in studies. Therefore, the widespread installation of co-use crossing structures is not recommended until rigorous studies quantify the effects of recreational co-use on wildlife behaviour and crossing rates.

Handbook of Road Ecology, First Edition. Edited by Rodney van der Ree, Daniel J. Smith and Clara Grilo.
© 2015 John Wiley & Sons, Ltd. Published 2015 by John Wiley & Sons, Ltd.
Companion website: www.wiley.com\go\vanderree\roadecology

INTRODUCTION

There is unequivocal evidence that well-designed, suitably located and properly maintained wildlife crossing structures (Chapter 21) can reduce the fragmentation effects of roads and other linear infrastructure for wildlife. In the absence of crossing structures, animals are less likely to cross roads and are certainly less likely to do so safely. Roads and railways, especially those with high speed and/or high traffic volumes, are also barriers to the movement of people. The fragmentation effect of roads and railways for people is especially problematic in and around cities and towns and in recreational areas, such as state, provincial and national parks and wilderness areas. What are the effects of having recreational use of wildlife crossing structures? Which species of wildlife will avoid co-use structures? Will species that use them reduce their rate of crossing or change their time of crossing to avoid people? Are these changes large enough to significantly impact population viability? How should co-use structures be designed to minimise the negative impacts on wildlife?

The aims of this chapter are to summarise the evidence about the impacts of recreational co-use of wildlife crossing structures on wildlife and suggest some simple design principles to maximise the ability to effectively achieve both uses. The focus is on the effects of recreational use of wildlife crossing structures (i.e. those designed primarily for wildlife) on wildlife, with minimal insight for other co-use structures (Lesson 21.3), although similar principles probably apply. We conclude that rigorous and well-designed research which investigates the feasibility of co-use structures to achieve connectivity for wildlife and people is a high priority.

LESSONS

22.1   Wildlife crossing structures are expensive, and it is intuitively appealing that they function for wildlife and people

Designing and building crossing structures that are functional for wildlife and people (Fig. 22.1) is an attractive strategy, provided the structure remains effective for the target species. The cost–benefits of building one crossing structure for both wildlife and people, rather than two structures, are also convincing. Importantly, there is a growing recognition that spending time in natural settings, including for recreation, leads to improvement in people's health and well-being (Maller et al. 2005). Unfortunately, easy access to recreational parks around urban areas is often hindered by roads, railways and other developments. There is an increasing demand that crossing structures for wildlife be adapted or designed for co-use by hikers, cyclists and horse riders. Furthermore, the ability to obtain funding and support for recreational co-use of wildlife crossing structures will be greater because of the dual benefits and the larger size of the group lobbying for its construction. However, the extent to which the twin goals of moving wildlife and people can be achieved is largely unknown.

(A)

(B)

Figure 22.1   (A) Recreational use by cyclists of the land bridge at Slabroek, The Netherlands. (B) This wildlife overpass in Brisbane, Australia, has boulders separating the recreational and wildlife zones. Both structures are just 15 m wide, and neither represents best practice in co-use design (Lesson 22.3). Source: Photograph (A) by Rodney van der Ree and (B) by Edgar van der Grift.

22.2 Different species respond differently to human co-use of wildlife crossing structures

Many best-practice guidelines recommend that human use of wildlife crossing structures should be discouraged (e.g. Iuell et al. 2003) on the basis that wildlife are less likely to use structures that are also used by people. But what is the evidence for this recommendation? As noted in Textbox 22.1, there has only been one study that explicitly set out to test the effects of recreationists on the rate, timing and behaviour of crossing by wildlife (van der Grift et al. 2011). The remainder of studies quantified rates of crossing by wildlife and included a measure of human activity (e.g. number of crossings by people, proximity to urban area) as a co-variate when trying to identify factors influencing rate of use. The most convincing of these studies monitored the crossing rates of wildlife over a 35-month period at 11 crossing structures in Banff National Park, Canada (Clevenger & Waltho 2000). Here, the most important factor negatively affecting the rate of crossing by large predators and omnivores (black bear, grizzly bear, cougar and wolf) was human activity, measured as either the number of cyclists, hikers or horse riders or the distance to the nearest town. Human use was also negatively correlated with ungulate crossing rates but less importantly. In the remaining studies, the breadth

of responses by wildlife to human activity, even when restricted to the number of crossings by humans walking, cycling or horse riding, is wide. For example, there were negative correlations found for badger and genet in Portugal (Grilo et al. 2008); wolves during summer months in Banff (Clevenger & Waltho 2005); coyotes in California, United States (Ng et al. 2004); and roe deer in Sweden (Olsson 2007). In contrast, positive relationships were reported for reptiles in Spain (Rodriguez et al. 1996) and raccoons in California, United States (Ng et al. 2004). Many studies failed to detect any relationship between human use of structures and that of wildlife, for example, vertebrates in Spain (Mata et al. 2005), red foxes and wildcat under a high-speed rail in Spain (Rodriguez et al. 1997), cougar in Banff (Gloyne & Clevenger 2001) and moose in Sweden (Olsson 2007), prompting some to suggest that co-use may be possible if human use is restricted to times when wildlife are less likely to use it and if human crossing rates are low.

One limitation in many studies is that proximity to urban areas and rate of crossing by recreationists is likely to be confounded (Clevenger & Waltho 2000), and it is difficult to determine if the reduced rate of crossing by wildlife is due to recreationists or due to changes in habitat quality due to urbanisation. Additionally, drawing a consensus is complicated by the range in (i) habitat types among the studies (e.g. wilderness areas to

---

Textbox 22.1 Investigating co-use of overpasses Zanderij Crailoo and Slabroek in The Netherlands.

Community groups representing the public and recreationists in The Netherlands are strongly pushing for wildlife crossing structures to also be available for hikers, cyclists and horse riders (Fig. 22.1A). In response to this demand, researchers aimed to investigate the impact of human co-use on the (i) rate, (ii) timing and (iii) behaviour of animals crossing two wildlife overpasses (van der Grift et al. 2011). The rate of human use varied between the two structures, with 182,000 and 60,000 hikers, cyclists and horse riders crossing overpasses Zanderij Crailoo and Slabroek per year, respectively. Rate of use by people varied over time, with most people crossing in spring and summer, on Sundays and between 2 and 4pm. No correlation, either negative or positive, was found between crossing rates by roe deer, red fox, rabbit or European hare and the number of people that crossed overpass Zanderij Crailoo. Interestingly, these same species plus red squirrel appeared to avoid the land bridge at Slabroek (width, 15m) but readily used the structure at

Zanderij Crailoo (width, 50m), suggesting that factors other than human co-use (e.g. structure width) are influencing animal crossing rates. Animals appeared to modify their use of overpass Zanderij Crailoo by accessing the structure at approximately 7pm on busy days, about 3 hours later than on days with few human visitors. A greater proportion of roe deer crossed the bridges faster (trot or gallop) (7% at Zanderij Crailoo and 21% at Slabroek) than at Groene Woud wildlife bridge (3%), where human co-use was not allowed. The polecat also appeared to hunt more on bridges without human co-use than those shared with people.

This study concluded that for these species in The Netherlands, human co-use did not prevent animals from crossing. They did, however, stress the importance of increased width of structures to accommodate wildlife and people and careful design to separate wildlife and people. The need for further research across more structures, species and locations was also strongly recommended.

high-intensity agricultural landscapes); (ii) human crossing rates (from <1 person per day to 182,000 per year); and (iii) local differences in animal behaviour (e.g. hunted species are likely to be cautious in areas where legal hunting or poaching occurs). In the meantime, the recreational use of wildlife crossing structures should continue to be limited until reliable studies have been completed (Lesson 22.4).

22.3 Simple design principles may enable co-use by humans

Multi-use crossing structures are likely to be most effective when the target species are relatively tolerant of human presence and when crossing structures explicitly incorporate design features for wildlife and people. Most importantly, people should be restricted to the narrowest possible strip on one side of the crossing structure. This 'recreational' zone should be clearly demarcated from the 'wildlife' zone and have a well-maintained path – enough to contain most people. Fencing, earth berms or plantings which separate the zones (internal screening) are necessary to discourage inquisitive people from exploring beyond the recreational zone and to provide a physical screen between wildlife and people (Figs. 22.2A and B). The effectiveness of the internal screening will vary depending on its design and the amount of simultaneous use by wildlife and people and the sensitivity of the species to human presence. External screens, which reduce the amount of traffic noise and light that penetrates the structure, should always be installed on wildlife crossing structures (Figs. 22.1A and B). External screens

will also likely be required if horse riders use the recreational zone to prevent horses from becoming spooked by passing traffic.

Co-use structures must be wider than the standard recommended for that type of wildlife crossing structure (Fig. 22.3; see also van der Grift et al. 2011). While the amount of extra width required is unknown, it must be at least as wide as the recreational zone plus the area taken up by any internal screening. For example, a 5 m-wide recreational zone (i.e. a 2 m-wide path, plus 1.5 m on each side of mown grass) will increase the width of the structure by at least 5 m, and if the negative effects spill over into the wildlife zone, it may need to be even wider. The additional width should also include the space taken up by any extra screening (i.e. the width taken up by internal screening) if the recreational zone is integrated with the wildlife zone (Fig. 22.3B). This extra width is not necessary where the two zones are not integrated (Fig. 22.3C) because there is no external screen between the recreational trail and the edge of the structure. In all cases, the amount of extra width required will depend upon the nature and extent of the interaction between wildlife and people and its effect on population viability. If the rate of crossing by wildlife is reduced but is still sufficient to maintain viable populations (assuming population viability is a specific goal for the structure), then the extra width can be limited to the width of the recreational zone. If the crossing structure is for a high-profile or endangered species that is intolerant to human presence, it may be appropriate to increase the width of the crossing structure by many times the size of the recreational zone because the risks are so great. However, in these circumstances, it is probably

(A)

(B)

Figure 22.2 Examples of internal screening on land bridges in The Netherlands. (A) Rock gabion walls on Wolfheze land bridge and (B) earth berm with shrubs on the Zanderij Crailoo land bridge. Source: Photographs by Edgar van der Grift.

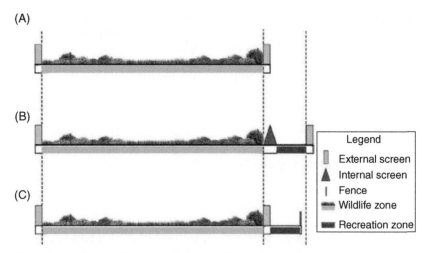

Figure 22.3 Profile view of standard wildlife crossing structure (A) and co-use structures (B and C) with clearly demarcated zones for recreational and wildlife use. The same principles apply for underpasses: (i) ensure functionality for target wildlife (assessed against the specific goals for each structure) is not compromised by reduced width or human activity; (ii) restrict human use to the narrowest width strip on one edge; (iii) clearly demarcate the wildlife and recreational zones; and (iv) provide a buffer or screen between the two zones. Source: Illustration by Zoe Metherell. Reproduced with permission of Zoë Metherell.

inappropriate to install a co-use structure if the consequences of failure are unacceptable.

Consider allowing a number of years for vegetation to establish and wildlife to become accustomed to the crossing structure before allowing use by people. Crossing structures should not contain artificial night lighting for humans, as this can have wide-reaching negative effects on wildlife (Chapter 18). A lack of artificial lighting will also discourage nocturnal use by people, which is a positive outcome because many species of wildlife are more active from dusk till dawn. Human co-use should be carefully considered in situations where hunting or poaching of threatened wildlife is possible or cannot be controlled (Fig. 49.2). Importantly, animals in areas where hunting occurs are likely to be more wary of humans and are thus less likely to use co-use structures.

22.4 Carefully designed and executed studies are required to confidently determine the efficacy of co-use structures

The demands to allow humans to use wildlife crossing structures are growing. Well-designed scientific studies that test the effects of recreational co-use of crossing structures on a range of target species are urgently required before they are widely deployed or dismissed as inappropriate without sufficient supporting evidence.

It is critical to evaluate the extent to which the proposed co-use will affect the specific goals of a structure. A reduction in the rate of crossing or change in timing of use (e.g. delayed from dusk to 2 hours after dusk) may be inconsequential if the goal of the structure (e.g. to maintain gene flow) is not compromised. Future studies must include a wider range of target species and structure types (e.g. underpasses), include areas along a gradient of human disturbance, have adequate replication and experimentally manipulate the number, timing and type of human activities. These experiments can be explicitly incorporated into the goals and design of new structures (i.e. the twin goals are to facilitate connectivity for wildlife and investigate the efficacy of recreational co-use) as well as with existing structures. Including the people who are lobbying for co-use structures in the experiment by engaging them to cross the structure at certain times of day or at certain frequencies will be critical to rigorously evaluating if recreational and wildlife uses are compatible. This also helps to remind users that recreational co-use may be found to be inappropriate and a consequence may be to prevent human access. Chapter 10 provides more insights and direction into study design for research and monitoring projects.

If the impacts of co-use on wildlife are unacceptable, it may be necessary to exclude human use altogether and install a crossing structure for people nearby. Changing the behaviour of people who are accustomed to using specific co-use structures will be difficult because they may consider it 'their right' to use it. Therefore, we recommend a cautious approach where recreational zones are installed on an experimental basis to test the efficacy of co-use while retaining the potential to revert back to 'wildlife-only' use if co-use is incompatible with the specific goals of the structure.

CONCLUSIONS

Limited funds for mitigation and the human health benefits of nature-based recreation are driving the push for more co-use structures. In addition, ongoing clearing of natural vegetation and the expansion of transportation networks in protected areas mean that nature-based recreationists are increasingly coming into contact with linear infrastructure. Until now, most road agencies and ecologists have resisted the urge to allow or encourage human co-use of wildlife crossing structures. However, in general, the evidence for and against co-use is equivocal and likely depends on the species; the rate, timing and type of human use; and the local response of wildlife to people (e.g. due to hunting pressure). An analysis of the risks associated with not achieving the goals for each wildlife crossing structure is essential. A reduced rate of crossing or shift in time of crossing due to human co-use may not matter if the goal of a structure is to maintain gene flow for a common species and sufficient individuals still use the structure. In contrast, the risks associated with recreational co-use may be too great if the target species is rare, threatened or highly sensitive to human presence, and recreational use should be avoided. This issue will not be resolved until high-quality research and monitoring that include experimental manipulation of human activity are undertaken.

ACKNOWLEDGEMENTS

Rodney van der Ree is supported by the Baker Foundation.

FURTHER READING

Clevenger and Waltho (2000): A high-quality study that quantified the rate of use of 11 underpasses in Banff National Park and found that human activity strongly negatively affected the rate of use by large carnivores and, to a lesser extent, ungulates.

van der Grift et al. (2011): The first study to evaluate the effect of human use of two land bridges in The Netherlands on the rate, timing and behaviour of wildlife.

REFERENCES

Clevenger, A. P. and N. Waltho. 2000. Factors influencing the effectiveness of wildlife underpasses in Banff National Park, Alberta, Canada. Conservation Biology **14**:47–56.

Clevenger, A. P. and N. Waltho. 2005. Performance indices to identify attributes of highway crossing structures facilitating movement of large mammals. Biological Conservation **121**:453–464.

Gloyne, C. C. and A. P. Clevenger. 2001. Cougar *Puma concolor* use of wildlife crossing structures on the Trans-Canada highway in Banff National Park, Alberta. Wildlife Biology **7**:117–124.

Grilo, C., J. A. Bissonette and M. Santos-Reis. 2008. Response of carnivores to existing highway culverts and underpasses: implications for road planning and mitigation. Biodiversity & Conservation **17**:1685–1699.

Iuell, B., G. J. Bekker, R. Cuperus, J. Dufek, G. Fry, C. Hicks, V. Hlaváč, V. Keller, B. Rosell, T. Sangwine, N. Tørsløv and B. l. M. Wandall, editors. 2003. COST 341 – wildlife and traffic: a European handbook for identifying conflicts and designing solutions. KNNV Publishers, Utrecht, The Netherlands.

Maller, C., M. Townsend, A. Pryor, P. Brown and L. St Leger. 2005. Healthy nature healthy people: 'contact with nature' as an upstream health promotion intervention for populations. Health Promotion International **21**:45–54. doi:10.1093/heapro/dai1032.

Mata, C., I. Hervas, J. Herranz, F. Suarez and J. E. Malo. 2005. Complementary use by vertebrates of crossing structures along a fenced Spanish motorway. Biological Conservation **124**:397–405.

Ng, S. J., J. W. Dole, R. M. Sauvajot, S. P. D. Riley and T. J. Valone. 2004. Use of highway undercrossings by wildlife in southern California. Biological Conservation **115**:499–507.

Olsson, M. 2007. The use of highway crossings to maintain landscape connectivity for moose and roe deer. Faculty of Social and Life Sciences, Karlstad University, Sweden.

Rodriguez, A., G. Crema and M. Delibes. 1996. Use of non-wildlife passages across a high speed railway by terrestrial vertebrates. Journal of Applied Ecology **33**:1527–1540.

Rodriguez, A., G. Crema and M. Delibes. 1997. Factors affecting crossing of red foxes and wildcats through non-wildlife passages across a high-speed railway. Ecography **20**:287–294.

van der Grift, E. A., F. G. W. A. Ottburg, R. Pouwels and J. Dirksen. 2011. Multi-use overpasses: does human use impact the use by wildlife? Pages 125–133 in P. J. Wagner, D. Nelson and E. Murray, editors. 2011 Conference on ecology and transportation. Centre for Transportation and the Environment, North Carolina State University, USA.

Chapter 23

PREDATOR-PREY INTERACTIONS AT WILDLIFE CROSSING STRUCTURES: BETWEEN MYTH AND REALITY

Cristina Mata[1], Roberta Bencini[2], Brian K. Chambers[2] and Juan E. Malo[1]

[1]Departamento de Ecología, Universidad Autónoma de Madrid, Madrid, Spain
[2]Faculty of Science, School of Animal Biology, The University of Western Australia, Crawley, Western Australia, Australia

SUMMARY

The potential for predators to use wildlife crossing structures for hunting could result in some species being preyed upon more frequently than elsewhere in the landscape. What would the consequences be if predators learn that hunting success is greater at wildlife crossing structures and develop preferences for these locations? Researchers and ecologists are frequently asked this question during the planning and design stages of road projects. Because there is no definitive answer to this question, the usual response is 'it depends', and the potential relevance of this situation should be assessed on a case-by-case basis.

23.1 Wildlife crossing structures will be less effective for prey if their use is adversely influenced by predator-prey interactions.

23.2 The outcome of predator-prey interactions at wildlife crossing structures may depend on the degree of co-evolution between them.

23.3 Our knowledge of specific interactions between predators and prey at crossing structures is scarce, and research is urgently needed.

23.4 Roads and crossing structures should be designed to minimise potentially deleterious predator-prey interactions.

Handbook of Road Ecology, First Edition. Edited by Rodney van der Ree, Daniel J. Smith and Clara Grilo.
© 2015 John Wiley & Sons, Ltd. Published 2015 by John Wiley & Sons, Ltd.
Companion website: www.wiley.com\go\vanderree\roadecology

Altogether, the increased connectivity benefits provided by wildlife crossing structures for a wide range of species are positive, and the potential for predation should not be seen as a reason not to install such mitigation measures. The current evidence for predators using crossing structures in a systematic way to increase hunting effectiveness is scarce and controversial. However, the potential for such predation at or near crossing structures exists, and road planners and wildlife managers should aim to reduce this risk, particularly for rare species of prey.

INTRODUCTION

Habitat fragmentation from roads and other linear infrastructure spatially transforms the landscape and affects interactions among species, including those between predator and prey (Schneider 2001). Wildlife crossing structures reduce the barrier effect of transport infrastructure, and numerous studies have demonstrated use by a wide variety of species on almost every continent and often in large numbers (e.g. Mata et al. 2005; Bond & Jones 2008; Clevenger 2012; Chapter 21). However, since crossing structures, together with fences (Chapter 20), funnel animals from surrounding areas towards precise locations in the landscape, it is conceivable that predators may use them to ambush prey. This unintended ramification challenges their effectiveness as mitigation measures; but is it a myth or is it a reality? Research on the behaviour of predators and prey at crossing structures is required to answer this question.

The aims of this chapter are to review the current evidence of predator and prey interactions at crossing structures, provide practical recommendations for road planners and designers and make recommendations for future research.

LESSONS

23.1   Wildlife crossing structures will be less effective for prey if their use is adversely influenced by predator-prey interactions

It has been suggested that wildlife crossing structures may be used by predators to ambush prey (Little et al. 2002). The consequences of increased predation at crossing structures potentially include (i) avoidance of wildlife crossing structures and increased habitat fragmentation

for prey species; (ii) a reduction in the effectiveness of these mitigation measures for prey species; (iii) altered behaviour and population dynamics of prey and predator species living adjacent to roads; and (iv) increased risk to the long-term viability of the populations of prey species.

Three alternative scenarios are possible:

(i) Prey exclusion/avoidance scenario: Prey species avoid crossing structures that are also used by predators (Doncaster 1999), leading to a diminished rate of use by prey. This avoidance phenomenon is possible for some species as demonstrated by field and laboratory studies that showed certain prey species change their activity patterns and even avoid places where they detect the presence of predators (Apfelbach et al. 2005). Prey may completely avoid or be excluded from crossing structures, or they may use the structures at times when predators are not nearby.

(ii) Prey-trap scenario: Crossing structures are optimal places for hunting by ambush predators, and we find predators more frequently in structures used by prey. Researchers have speculated that certain predators may use crossing structures as traps for prey (Hunt et al. 1987; Foster & Humphrey 1995; Little et al. 2002), and apparent predation in crossing structures has been observed (e.g. Fig. 23.1).

(iii) Null effect scenario: Predators do not use crossing structures specifically to ambush prey, and prey do not avoid structures used by predators. The coincidence of predators and prey within or on a structure would be random, and the outcome of such encounters would not change their natural dynamics in the area.

These different scenarios may eventuate depending on how prey species detect the presence of potential predators and on their capacity to react to them. In Canada, prey such as deer and elk avoided underpasses that were used by potential predators such as wolves or cougars (Clevenger & Waltho 2000). However, a more recent analysis of data from Banff National Park found that the proximity of ungulate kill sites to the highway was

Figure 23.1 Apparent predation event in a circular culvert under the A-52 motorway (NW Spain): a stoat with a water vole in its jaws. Water voles in this forest area only live in small creeks such as this one passing through the culvert. Source: Photograph by C. Mata.

similar before and after the construction of fencing and crossing structures and there was no evidence that prey movements affected predator behaviour at crossing structures (Ford & Clevenger 2010). Therefore, interactions between large carnivores and their prey at wildlife crossing structures in Banff National Park do not support the prey-trap scenario.

Conversely, recent work on small- and medium-sized mammalian predator and prey species in NW Spain showed different results (Mata et al. in review). Both predator and prey species used the same structures to cross fenced roads, but their use of crossing structures did not occur independently. In fact, some predators used crossing structures on the same day as prey more frequently than expected by chance, almost as if they visited them in search of prey. Thus, patterns of temporal co-occurrence were detected for the smallest prey with the smallest predators and for the largest prey with larger carnivores. Additionally, some prey species avoided using crossing structures on the same date that predators did (Table 23.1).

The non-random occurrence of certain predators and prey at wildlife crossing structures in NW Spain (Table 23.1) suggests that some predator-prey effects are present at these crossing structures. The results suggest that some predators 'win the race' against their preferred prey (those species closest in size to them), as they seem able to track them in crossing structures and potentially use these structures as prey traps. In contrast, some small-sized Mediterranean prey species

Table 23.1 Patterns of interspecific coincidence between prey and predator species in NW Spain as denoted by two-species occupancy modelling applied to the daily use of crossing structures.

Potential prey	Potential predator				
	Weasel and stoat	**Cat**	**Eurasian Badger**	**Red fox**	**Large canid**
Micromammals	(+)	(+)	(+)	(−)	(−)
Rat-sized rodents	(−)	n.s.	n.s.	(−)	n.s.
Rabbits and hares	n.s.	(−)	(+)	(+)	n.s.

Source: Adapted from Mata et al. (in review).
Type of interactions: (+), attraction or tendency for co-occurrence; (−), avoidance; n.s., no significant deviation from random coincidence.
Species are in increasing body size order from left to right and up to down.

may reduce their use of crossing structures in the presence of predators. The extent of population effects arising from this situation remains to be analysed, but it could be of concern in the case of threatened prey species. Further research to understand if these interactions occur more frequently near crossing structures than elsewhere in the landscape is also required.

In conclusion, if predator-prey interactions at crossing structures lead to the prey exclusion or the prey-trap scenarios, their effectiveness against fragmentation may be hampered, and, at least under some situations, this does happen.

23.2 The outcome of predator-prey interactions at wildlife crossing structures may depend on the degree of co-evolution between them

The outcome of predator-prey interactions in wildlife crossing structures depends on the capacity of species to detect, track and respond to the presence of their predators or prey. Such capacity is deeply rooted in the evolutionary history of the species, and if predator and prey have shared habitats along evolutionary times, the responses between them may be tightly matched and may result in the avoidance scenario. However, if predator and prey species did not co-evolve together, the prey species might be naïve, and this will most probably result in the prey-trap scenario.

The lack of co-evolution between predators and prey is particularly serious in Australia, where introduced predators, particularly the red fox and cat, have been implicated in the extinction of numerous species of native wildlife (Burbidge & McKenzie 1989). Predation by foxes and cats is listed as a major threatening process because Australian marsupials in the critical weight range of 50–5000 g are generally naïve to these predators and might not recognise their scent (Dickman & Doncaster 1984; Hayes et al. 2006). However, our knowledge of predator recognition by native species is scarce, and in some studies, native species were able to recognise the smell of introduced predators (e.g. Mella et al. 2010). These different responses might be due to the fact that some Australian marsupials respond to predators even if they have not shared a long evolutionary history (McLean et al. 1996), and in some cases, they may learn to recognise introduced predators if they are exposed to them (McLean et al. 2000).

In Australia, foxes and cats use underpasses regularly, but it is unclear if they target them to ambush prey. Foxes use roads and the disturbed roadsides to move around, and roads also provide them with food in the form of scavenged roadkills (Ramp et al. 2006). Therefore, foxes are likely to encounter crossing structures and as a result will learn quickly if they are a reliable source of prey (Fig. 23.2). This is likely to result in increased encounters between predators and prey around roads similar to relationships found between predators and prey in North America (Whittington et al. 2011). As a result, native

Figure 23.2 A red fox with an European rabbit in its jaws uses a wildlife underpass along the Calder Freeway, south-east Australia. There was no evidence that this rabbit was killed while using the underpass. Source: Photograph by R. van der Ree.

Textbox 23.1 **Fox predation at a wildlife underpass causes the local extinction of a population of southern brown bandicoots (Harris et al. 2010)**

In 2010, 56 southern brown bandicoots were trapped near a section of the Roe Highway extension in Perth, Western Australia. All captured bandicoots were microchipped, and a microchip decoder was installed in an underpass to establish if multiple individuals used it. Eight bandicoots were recorded using the underpass during the study (Fig. 23.3). After foxes were detected using the underpass and built a den near its entrance, the bandicoots were no longer observed using the underpass. Despite extensive trapping in the area, the unsuccessful recapture of any bandicoots, microchipped or not, suggested that foxes had caused the local extinction of the bandicoot population. This work demonstrates that although underpasses have the potential to reconnect populations, they could also be detrimental to various target species if certain predators are not monitored and controlled.

Figure 23.3 A southern brown bandicoot uses an underpass under the Roe Highway, Western Australia. An entire population of southern brown bandicoots using this underpass became locally extinct, most likely due to fox predation. Source: Photograph by B. Chambers.

species may become locally extinct (Harris et al. 2010; Textbox 23.1). This might be of particular concern in the case of rare or endangered fauna.

23.3   Our knowledge of specific interactions between predators and prey at crossing structures is scarce, and research is urgently needed

Evidence presented in Lessons 23.1 and 23.2 applies to individual cases and cannot be generalised to all situations. More research is urgently needed to identify under what circumstances, if any, predation or predator avoidance around crossing structures changes the natural dynamics of populations. Until now, few experimental studies have addressed potential impacts on prey species (see review in Little et al. 2002), possibly due to the inherent difficulties in designing such complex experiments (Roedenbeck et al. 2007; Chapter 10). GPS collars and proximity radio-transmitters coupled with behavioural studies could be used to determine if predation events occur at a higher rate at underpasses than elsewhere in the landscape and to what extent prey species avoid crossing structures.

23.4 Roads and crossing structures should be designed to minimise potentially deleterious predator-prey interactions

If predators use wildlife crossing structures to their advantage to ambush prey, can these structures be modified to allow prey species to escape the increased risk of predation?

While more research on predator and prey interactions adjacent to roads and crossing structures is needed, some general recommendations for the planning and design of wildlife crossing structures can be adopted to reduce potential impacts on species vulnerable to ambush predators. These include:

(i) Install wider crossing structures at relatively high densities (e.g. the Spanish Ministerio de Medio Ambiente (2006) recommends one underpass at least 7 m-wide spaced <1 km apart in fauna rich areas) to decrease the risk of spatial and temporal coincidence of predator and prey (Clevenger & Waltho 2005);

(ii) Include 'furniture' within structures and at their entrances (e.g. logs, vegetation, rocks, brush piles, tubes) to provide cover for prey species and allow them to escape from predators (Figs 9.1, 39.1 and 40.5);

(iii) Where possible, preserve or re-establish the original habitats in the surroundings of crossing structures to avoid the increase in opportunistic predators;

(iv) Provide – if possible – short, wide and high underpasses to protect large herbivores (e.g. elk, deer) from large predators (e.g. wolves; Little et al. 2002; Chapter 42). Avoid underpasses with ledges (such as bridge underpasses with flat sections above abutment walls) where ambush predators could sit (Fig. 23.4).

The precise design of these recommendations should be based on the needs of the target species since cover for some prey species may also assist ambush predators.

Ongoing and large-scale control of introduced predators using poison baits and trapping is a common approach in many Australian conservation reserves to protect native wildlife. A similar approach should be applied to limit the size of fox and cat populations around new and existing roads and crossing structures in Australia. Similar actions could be implemented in other areas where (i) prey species are naïve to introduced predators; (ii) it has been demonstrated that predators are using crossing structures to improve their hunting efficiency; or (iii) the predation is of a rare or endangered prey species. In all cases, a careful case-by-case analysis is required.

Thus, the analysis of risks linked to changes in predator-prey interactions, as well as the potential need for the control of introduced predators, must be considered when planning the installation of wildlife crossing structures. Additionally, post-construction monitoring should assess the impacts of construction activities

Figure 23.4 Avoid abutments or ledges in underpasses such as those shown here on State Route 260, Arizona, United States, where ambush predators may hunt from. Source: Photograph by R. van der Ree.

and the new road on the presence and densities of predators and the potential cascading effects of increased predation on prey.

CONCLUSIONS

The most important message from this chapter is the need for further research, particularly to clarify if predator-prey interactions detected at wildlife crossing structures parallel the same level of interactions at the landscape scale or if they show a shift in the balance between species (Little et al. 2002). To explore this issue, predation rates and the population-level effects on prey populations should be evaluated before and after the construction of roads with and without crossing structures. Experiments that manipulate predator populations or the design of crossing structures (e.g. size of structure, vegetation cover) could also help to quantify the predator-prey interaction and identify management solutions (see Chapter 10 for further discussion of experimental designs, including BACI). It is particularly important to separate the potential for increased predation risk due to the presence of linear infrastructure (Whittington et al. 2011) versus that caused by the presence of crossing structures. If predation risk is increased by the presence of crossing structures, researchers should assess whether the intensity of such effects has substantially reduced the probability of survival of local populations of prey species. Current evidence demonstrates that wildlife crossing structures and fencing can mitigate population fragmentation and roadkill, although some attention should also be given to their potential side effects.

ACKNOWLEDGEMENTS

The research of the TEG-UAM benefitted from the financial support of the REMEDINAL-2 research network (Fondo Social Europeo-Comunidad de Madrid, S-2009/AMB/1783) and two projects funded by the Spanish Ministry of the Environment, the Centro de Estudios y Experimentación (CEDEX), the Spanish Ministry of Science and Innovation (CDTI) and a consortium of companies (OASIS CENIT-2008 1016)

The University of Western Australia's Wildlife Research Group is supported by funding from Main Roads Western Australia. We thank Gerry Zoetlief and Alan Grist for their support. Our work would be difficult to conduct without our research students: we are thankful to Kaori Yokochi, Alexia Jankowski, Veronica Phillips, Rachael Glasgow, Ian Harris, Faradilla Roselan and Jessica Hunter.

FURTHER READING

Barbosa and Castellanos (2005): A keystone review necessary to achieve a thorough understanding of predator and prey interactions.

Dickman (1996): This article explains why generalist predators like the red fox have had such a devastating impact on the naïve native fauna of Australia.

Little et al. (2002): In this article, the authors first proposed the prey-trap hypothesis after reviewing the evidence suggesting that predators might use wildlife crossing structures to ambush prey.

Roedenbeck et al. (2007): This article emphasises the importance of applying rigorous science to the applied aspects of monitoring the effectiveness of fauna crossing structures.

REFERENCES

Apfelbach, R., C. D. Blanchard, R. J. Blanchard, R. A. Hayes and I. S. McGregor. 2005. The effects of predator odors in mammalian prey species: a review of field and laboratory studies. Neuroscience and Biobehavioral Reviews **29**:1123–1144.

Barbosa, P. and I. Castellanos (eds). 2005. Ecology of Predator-prey Interactions. New York: Oxford University Press.

Bond, A. R. and D. N. Jones. 2008. Temporal trends in use of fauna-friendly underpasses and overpasses. Wildlife Research **35**:103–112.

Burbidge, A. A. and N. L. McKenzie. 1989. Patterns in the modern decline of Western Australia's vertebrate fauna: causes and conservation implications. Biological Conservation **50**:143–198.

Clevenger, A. P. 2012. 15 Years of Banff research: what we've learned and why it's important to transportation managers beyond the park boundary. Pages 409–423. In Proceedings of the International Conference on Ecology and Transportation. Center for Transportation and the Environment, North Carolina State University, Raleigh, NC.

Clevenger, A. P. and N. Waltho. 2000. Factors influencing the effectiveness of wildlife underpasses in Banff National Park, Alberta, Canada. Conservation Biology **14**:47–56.

Clevenger, A. P. and N. Waltho. 2005. Performance indices to identify attributes of highway crossing structures facilitating movement of large mammals. Biological Conservation **121**:453–464.

Dickman, C. R. 1996. Impact of exotic generalist predators on the native fauna of Australia. Wildlife Biology **2**:185–195.

Dickman, C. R. and C. P. Doncaster. 1984. Responses of small mammals to red fox (*Vulpes vulpes*) odour. Journal of Zoology **204**:521–531.

Doncaster, C. P. 1999. Can badgers affect the use of tunnels by hedgehog? A review of the literature. Lutra **42**:59–64.

Ford, A. T. and A. P. Clevenger. 2010. Research note: validity of the prey-trap hypothesis for carnivore-ungulate interactions at wildlife-crossing structures. Conservation Biology **24**:1679–1685.

Foster, M. L. and S. R. Humphrey. 1995. Use of highway underpasses by Florida panthers and other wildlife. Wildlife Society Bulletin **23**:95–100.

Harris, I. M., H. R. Mills and R. Bencini. 2010. Multiple individual southern brown bandicoots (*Isoodon obesulus fusciventer*) and foxes (*Vulpes vulpes*) use underpasses installed at a new highway in Perth, Western Australia. Wildlife Research **37**:127–133.

Hayes, R. A., H. F. Nahrung and J. C. Wilson. 2006. The response of native Australian rodents to predator odours varies seasonally: a by-product of life history variation? Animal Behaviour **71**:1307–1314.

Hunt, A., H. J. Dickens and R. J. Whelan. 1987. Movements of mammal through tunnels under railway lines. Australian Zoologist **24**:89–93.

Little, S. J., R. G. Harcourt and A. P. Clevenger. 2002. Do wildlife passages act as prey-traps? Biological Conservation **107**:135–145.

Mata, C., I. Hervás, J. Herranz, F. Suárez and J. E. Malo. 2005. Complementary use by vertebrates of crossing structures along a fenced Spanish motorway. Biological Conservation **124**:397–405.

Mata, C., F. Suárez and J. E. Malo. (in review) Attraction and avoidance between predators and prey at wildlife crossings on roads.

McLean, I. G., G. Lundie-Jenkins and P. J. Jarman. 1996. Teaching an endangered mammal to recognise predators. Biological Conservation **75**:51–62.

McLean, I. G., N. T. Schmitt, P. J. Jarman, C. Duncan and C. D. L. Wynne. 2000. Learning for life: training marsupials to recognize introduced predators. Behaviour **137**:1361–1376.

Mella, V. S., C. E. Cooper and S. J. Davies. 2010. Ventilatory frequency as a measure of the response of tammar wallabies (*Macropus eugenii*) to the odour of potential predators. Australian Journal of Zoology **58**:16–23.

Ministerio de Medio Ambiente. 2006. Prescripciones Técnicas para el diseño de pasos de fauna y vallados perimetrales. Documentos para la reducción de la fragmentación de hábitats causada por infraestructuras de transporte, número 1. O.A. Parques Nacionales. Ministerio de Medio Ambiente, Madrid, Spain. 108 pp.

Ramp, D., V. K. Wilson and D. B. Croft. 2006. Assessing the impacts of roads in peri-urban reserves: road-based fatalities and road usage by wildlife in the Royal National Park, New South Wales, Australia. Biological Conservation **129**:348–359.

Roedenbeck, I., L. Fahrig, C. S. Findlay, J. E. Houlahan, J. A. G. Jaeger, N. Klar, S. Kramer-Schadt and E. A. van der Grift. 2007. The Rauischholzhausen agenda for road ecology. Ecology and Society **12**:11.

Schneider, M. F. 2001. Habitat loss, fragmentation and predator impact: spatial implications for prey conservation. Journal of Applied Ecology **38**:720–735.

Whittington, J., M. Hebblewhite, N. J. DeCasare, L. Neufeld, M. Bradley, J. Wilmshurst and M. Musiani. 2011. Caribou encounters with wolves increase near roads and trails: a time-to-event approach. Journal of Applied Ecology **48**:1535–1542.

Chapter 24

WILDLIFE WARNING SIGNS AND ANIMAL DETECTION SYSTEMS AIMED AT REDUCING WILDLIFE-VEHICLE COLLISIONS

Marcel P. Huijser[1], Christa Mosler-Berger[2], Mattias Olsson[3] and Martin Strein[4]

[1]Western Transportation Institute (WTI), Montana State University, Bozeman, MT, USA
[2]WILDTIER SCHWEIZ, Swiss Wildlife Information Service, University of Zurich, Zurich, Switzerland
[3]EnviroPlanning AB, Göteborg, Sweden
[4]Forest Research Institute Baden-Wurttemberg (FVA), Freiburg, Germany

SUMMARY

Wildlife warning signs are among the most frequently used mitigation measures aimed at reducing wildlife-vehicle collisions (WVC). Road agencies have been using these signs for many decades, and their use has become standard practice in most parts of the world.

24.1 Warning signs are intended to reduce the rate and severity of WVC, not the barrier effect of roads and traffic.

24.2 Warning signs must be reliable if they are to be effective.

24.3 Standard and enhanced warning signs are unlikely to be effective in reducing collisions.

24.4 Warning signs that are place and time specific can be effective in reducing collisions.

24.5 Adopt a stepwise approach when implementing an animal detection system.

24.6 Warning signs can be used with other mitigation measures.

Handbook of Road Ecology, First Edition. Edited by Rodney van der Ree, Daniel J. Smith and Clara Grilo.
© 2015 John Wiley & Sons, Ltd. Published 2015 by John Wiley & Sons, Ltd.
Companion website: www.wiley.com\go\vanderree\roadecology

While standard and enhanced wildlife warning signs are frequently used, they appear to be ineffective in reducing WVC. Their widespread use may be primarily because of engrained practices, their relatively low cost, a desire to inform the public about the impact of WVC on human safety and nature conservation and possible litigation concerns, rather than a proven substantial reduction in these types of collisions. However, signs that are more place and time specific can be effective in reducing WVC. There is a wide range in the effectiveness of temporal warning signs and animal detection systems; however, neither of these reduces the barrier effect of highways and traffic.

INTRODUCTION

Wildlife warning signs are aimed at reducing wildlife-vehicle collisions (WVC) by warning drivers about the potential or actual presence of wild animals on the road. They vary significantly and can be categorised as (i) standard wildlife warning signs; (ii) enhanced wildlife warning signs; (iii) temporal wildlife warning signs; or (iv) animal detection systems.

Standard wildlife warning signs are typically manufactured in the same style as other traffic warning signs (Fig. 24.1), which is often country dependent. Most signs depict a stylised large mammal and usually species which are common, widespread and large enough to be a safety concern for motorists (e.g. red deer in Europe, white-tailed deer or mule deer in the United States). Other species may be depicted regardless of their size and potential threat to human safety if:

• The road dissects habitat of a species that is of conservation concern (e.g. Fig. 24.2).

• Agencies or the public would like to see a reduction in road mortality for a particular species (e.g. Fig. 24.2).

Standard signs are normally installed at road sections that had (or have) a relatively high number of WVC. Most standard wildlife warning signs are not very specific in time or place. The distance to which the warning applies may be many kilometres; however, only 5–10% of the drivers that were stopped 200 m after passing a warning sign were able to recall it (Drory & Shinar 1982), and a dummy of a moose along a roadway either was not or barely detected by drivers (Åberg 1981). Another factor that contributes to the abundance of wildlife warning signs along roads is that once a sign has been installed, it is rarely removed, even if the problem no longer exists.

Enhanced wildlife warning signs (Fig. 24.3) tend to be larger than standard signs, they may have flashing lights or bright flags attached to them, and they may also include eye-catching or perhaps even disturbing illustrations, images of certain species that the warning relates to, WVC statistics or other customised text (e.g. Fig. 24.3).

(A)

(B)

Figure 24.1 Standard deer warning sign in (A) Flevoland, the Netherlands, and in (B) Colorado, United States. Source: Photographs by Marcel Huijser.

(A)

(B)

Figure 24.2 Standard warning signs for species of conservation concern (A, nene in Hawaii) or public interest (B, yellow-bellied marmots near Crested Butte, Colorado, United States). Source: Photographs by Marcel Huijser.

(A)

(B)

Figure 24.3 Enhanced wildlife-vehicle collision warning sign in (A) British Columbia, Canada, and the endangered Florida key deer in (B) Florida, United States. Source: Photographs by Marcel Huijser.

These characteristics aim to capture the attention of motorists and educate them about the safety and nature conservation impact of WVC. Enhanced warning signs are generally more frequently observed and recalled by drivers than standard warning signs (Summala & Hietamaki 1984). They are normally installed at road sections that have a relatively high number of WVC or in areas where species of conservation concern occurs.

Temporal wildlife warning signs warn drivers of wildlife presence during specific times of the year or day (Fig. 24.4). These signs tend to be species specific and may only be visible to drivers (e.g. signs that fold in half and are removed in the off season or variable message signs (i.e. electronic signs with programmable text or symbols)) during the most potentially hazardous time of the year or day. Seasonal warning signs may be placed where roads intersect migration corridors (e.g. mule deer migration routes in the western United States; Fig. 24.4A) or where

species are attracted to the highway during specific times of the year (e.g. bighorn sheep licking road salt in specific areas in the Rocky Mountains in North America). Seasonal warning signs can also apply to smaller animals such as amphibians (Fig. 24.4B) that leave their winter hibernacula to move to breeding habitat in large numbers during a short period in spring (Chapter 31). If the warning relates to certain hours of the day, the signs may be permanent, but their message may be enhanced during the time of the day with peak wildlife activity (e.g. flashing lights around dusk and dawn).

Animal detection systems use electronic sensors to detect large animals (i.e. deer size and larger) that approach the road; signs are then activated to warn drivers (Fig. 24.5). These signs are very specific in time and place. However, current animal detection systems are unlikely to work for small- to medium-sized animals because they are more difficult to detect reliably.

Figure 24.4 Seasonal wildlife warning sign for (A) deer migration on variable message sign, California, United States, and (B) pay attention (common), toad migration, Limburg, the Netherlands. Source: Photographs by Marcel Huijser.

Figure 24.5 Activated warning signs on animal detection systems: (A) 'wildlife detected' Colorado, United States, and (B) activated warning sign and advisory speed limit reduction, the Netherlands. Source: Photographs by Marcel Huijser.

LESSONS

24.1 Warning signs are intended to reduce the rate and severity of WVC, not the barrier effect of roads and traffic

Wildlife warning signs aim to warn drivers and urge them to be more attentive to wildlife that may be on or near the road and/or reduce their speed (Fig. 24.6). The primary goal of wildlife warning signs is to improve human safety by reducing the rate and severity of WVC. The concern is not explicitly nor primarily with providing safe and effective crossing opportunities for wildlife because:

• Wildlife warning signs do not make it any more attractive for wildlife to approach and cross the road. Warning signs do not change the fact that roads are linear open areas without cover with an unnatural substrate (usually asphalt or concrete) and traffic.
• Wildlife warning signs do not reduce the traffic volume and animals still have to avoid vehicles while crossing the road. However, depending on the type of sign, drivers may be more attentive and may (slightly) reduce their speed. This may increase the rate of successful road crossings for some wildlife. However, in some situations, drivers feel that an evasive manoeuvre would be too dangerous to them or other humans, and they may choose to hit the animal. Other drivers will

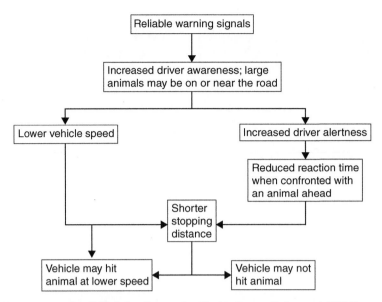

Figure 24.6 Warning signs must be reliable before they can be effective. Source: Huijser et al. (2006).

actually aim to hit and kill certain species (e.g. reptiles; Lesson 32.2), and they may use the information provided by the warning signs to be more alert and try and hit the animals (Ashley et al. 2007).

• Wildlife warning signs should be located at WVC hotspots. These are not necessarily the same locations where wildlife crosses the road successfully or locations that need improved connectivity to enhance population viability or where no or few animals cross the road currently (Chapter 13).

24.2 Warning signs must be reliable if they are to be effective

For wildlife warning signs to be effective (i.e. result in fewer or less severe WVC), drivers need to respond to the warning, which occurs when drivers observe, understand and take the signs seriously. Reliability of the sign influences driver response (Fig. 24.6), which is primarily affected by two factors:

(i) *Location* – Are wildlife warning signs installed at the road sections that have had the highest numbers of WVC or at road sections where animals cross the road most frequently? Do drivers perceive these sections as high-risk areas?

(ii) *Time of year or day* – Is this the season or time of the day when drivers are most at risk of WVC? Do drivers perceive an increase in movements across the road by the target species at the time of year or day indicated?

Reliability is primarily achieved by installing signs at the correct location, which is usually based on reported crashes from law enforcement agencies or carcass removal data from highway maintenance personnel (Chapters 11, 12 and 13). However, for the location or road section to be perceived as correct, drivers need to have confidence in the organisations that installed the signs (with or without providing supporting data to the public) or to regularly see the target species dead or alive on or near the road.

If the warning signs are effective, the number of crashes and carcasses should decrease, which ironically may give the impression that the signs are installed at the wrong location. This can be somewhat negated if the drivers still see relatively high numbers of animals alive on or near the road or if they have great confidence in the organisations that installed the signs. However, many large mammal species are most active from dusk to dawn when visibility for drivers is reduced, reducing the probability of observing wildlife (Mastro et al. 2010).

In some areas, WVC data are assessed regularly and the locations of signs are adjusted as necessary, including the removal of signs where they have been successful in reducing the rate of WVC. After sign removal, the collision rate can be expected to increase again so that the signs may need to be reinstalled after the next evaluation. This illustrates that selecting the 'correct' locations for signs should be based on more than just the number of crashes or carcasses in recent years (see also Krisp & Durot 2007). While difficult to measure, it should ideally be based on the number of animals that cross the road and that would be hit by traffic if signs were absent. The specific challenges vary for the different types of signs:

• Once a standard sign is installed, it is typically never removed, regardless of whether the location or road section has continued to be a WVC hotspot or whether wildlife still cross the road at relatively high numbers (Krisp & Durot 2007). For regionally abundant target species (e.g. white-tailed deer in most of North America, kangaroos in some parts of Australia), this means that standard signs are virtually everywhere and inevitably become more abundant and increasingly ignored by drivers over time (Krisp & Durot 2007).

• Standard, enhanced and temporal wildlife warning signs tend to relate to road sections rather than discrete locations. The longer the road section, the more likely it is that drivers will forget the sign and the less likely they will continue limiting their speed for the full length of the road section (Fig. 24.7). Repeating the signs at relatively short intervals is not a solution as this leads to oversaturation of signs which, as a consequence, are then ignored by drivers. Enhanced signs attempt to attract more attention from drivers, which is in direct conflict with the need to standardise traffic signs so that people recognise them quickly, interpret them correctly and keep their attention focussed on the road and traffic. Temporal signs have the advantage of relating to specific species that move through a defined area at certain times and are thus more likely to be perceived as reliable compared to standard or enhanced signs.

• Animal detection systems are time and place specific and the associated warning signs should only be activated when the target species has been detected. Therefore, correctly functioning animal detection systems have the potential to be reliable, which can be measured (Huijser et al. 2009a). However, actual and perceived reliability can differ as drivers may rarely see animals on or along the road when the warning signs are activated (Sharafsaleh et al. 2012), or they may see animals in the proximity of the road with the warning signs turned off as the animals are beyond the range of the sensors. Regardless, in order to inform the driver adequately, it is important that the warning signs are relatively close together. A driver should not pass a warning sign without being able to see and interpret the next warning sign should it be activated. This may require a modification of the guidelines for sign placement which tend to be based on static signs rather than signs that display no message at all unless a danger has been detected. Many animal detection systems have a portion of the warning signs visible all the time (e.g. an additional flashing light is activated after a detection has occurred). It is best if no message is displayed unless an animal has been detected to minimise the likelihood that drivers ignore activated signs and to avoid oversaturating the roadside with signs. Additional standard signs spaced at relatively great distances can then still address potential liability issues.

Figure 24.7 Enhanced warning sign for kangaroos and cattle that applies to a very long distance (next 420 km), Western Australia, Australia. Source: Photograph by Rodney van der Ree.

24.3 Standard and enhanced warning signs are unlikely to be effective in reducing collisions

Wildlife warning signs are typically considered effective if they result in a reduction in the number of WVC. Other parameters may also be used to measure effectiveness, such as a reduction in vehicle speed or other driver responses such as touching the brakes or being more alert. While drivers may reduce vehicle speed in response to standard and enhanced signs (Pojar et al. 1975; Al-Ghamdi & AlGadhi 2004; Rogers 2004; Sullivan et al. 2004), the majority of studies of the effectiveness of these sign types in reducing WVC concluded that they were not effective (e.g. Pojar et al. 1975; Coulson 1982; Rogers 2004; Meyer 2006; Bullock et al. 2011). However, some have found standard warning signs to be effective (34% reduction in collisions) immediately after installation (Found & Boyce 2011) or at a gap in a fence with a crosswalk painted on the road surface (37–43%) (Lehnert & Bissonette 1997). Regardless, implementing standard wildlife warning signs may still be required or desirable to limit liability concerns. While standard and enhanced signs have some educational value, one could also argue that drivers may wrongfully think that these sign types reduce collisions and consequently do not support more effective mitigation measures that may be more expensive.

24.4 Warning signs that are place and time specific can be effective in reducing collisions

Temporal warning signs tend to be more place specific than standard or enhanced warning signs, and, by definition, they are also more time specific. Animal detection systems are extremely specific in place and time. Data on the effectiveness of temporal signs and animal detection systems suggest that they can be effective in reducing WVC.

Temporal warning signs can reduce collisions, although effectiveness varies substantially (9–50%) (Sullivan et al. 2004; CDOT 2012). The effectiveness of animal detection systems is also variable, but they appear to reduce WVC more than temporal signs: 33–97% reduction in collisions with large mammals (Mosler-Berger & Romer 2003; Huijser et al. 2006; Dai et al. 2009; Gagnon et al. 2010; Strein 2010; MnDOT 2011; Sharafsaleh et al. 2012). Since the risk of severe crashes (WVC) increases exponentially with increasing vehicle speed (Kloeden et al. 1997), it is useful to also evaluate the potential effect of

activated warning signs associated with animal detection systems on vehicle speed. Drivers tend to reduce their speed somewhat (<5 km/h) (Kistler 1998; Muurinen & Ristola 1999; Hammond & Wade 2004; Huijser et al. 2006) or more substantially (≥5–22 km/h) in response to activated signs of animal detection systems (Kistler 1998; Kinley et al. 2003; Gordon et al. 2004; Gagnon et al. 2010; Sharafsaleh et al. 2012). The greatest reductions in vehicle speed seem to occur when the signs are associated with advisory or mandatory speed limit reductions or if road conditions and visibility for drivers are poor (Kistler 1998; Muurinen & Ristola 1999).

24.5 Adopt a stepwise approach when implementing an animal detection system

The implementation of animal detection systems requires greater planning and investment than standard, enhanced or temporal wildlife warning signs. Therefore, we recommend that organisations follow these steps when installing an animal detection system (adapted from Huijser et al. 2009a, 2009b).

Step 1. Define the problem and the goal.

Define the problem to be solved (e.g. too many WVC with certain species) and the goal including parameters of effectiveness (e.g. obtain an 80% reduction in WVC). Animal detection systems will not be an effective solution if the problem cannot be addressed by this approach (e.g. the barrier effect of roads and traffic) or if the target species is too small to be reliably detected (e.g. smaller than deer).

Step 2. Decide on the strategy.

A typical approach to dealing with ecological issues along roads is to adopt the mitigation hierarchy: avoid, minimise, mitigate, and then finally offset or compensate (Cuperus et al. 1999; Chapter 7). Animal detection systems can mitigate or reduce WVC but are probably inappropriate if the strategy is to avoid or compensate WVC.

Step 3. Decide on the method or measure.

If the problem (see Step 1) is 'too many WVC with large mammals' (e.g. deer size and larger) and if the strategy is to reduce the number of WVC, then animal detection systems, either as a stand-alone measure or in combination with wildlife fencing, should be considered. However, it is advisable to compare these systems to other measures, especially wildlife fencing in combination with crossing structures for wildlife (Huijser et al. 2009c; Chapters 20 and 21). The following factors should be considered when making a decision on what measures to implement:

• The level and range of effectiveness of the proposed measures in reducing WVC with large species (33–97% for detection systems; 80% to nearly 100% for wildlife fencing and crossing structures).

• Animal detection systems should still be considered experimental, while wildlife fencing and crossing structures can substantially reduce WVC if implemented correctly. Detection systems are experimental with regard to the level of certainty that a system will be operating as desired by a particular date – especially in detecting the target species with sufficient reliability – and a relatively wide and variable range of effectiveness in reducing WVC. The latter is probably associated with the different types of detection technologies and the great variability in the signs presented to drivers.

• Is reducing WVC the only goal, or is it also to reduce the barrier effect of roads and traffic to wildlife? Detection systems do not address the barrier effect in the same way as fencing and crossing structures.

• The costs of the measures. If WVC occur in high enough numbers, it may be more cost-effective to implement mitigation than to have WVC continue to occur (Huijser et al. 2009c). Despite relatively low initial costs, detection systems may be more expensive in the long term than fencing with crossing structures. This is primarily because the latter have a much greater projected lifespan.

If the decision is to explore the potential installation of an animal detection system, then proceed with the following steps:

Step 4. Identify system and project requirements.

Identify the desired level of effectiveness (e.g. ≥80% reduction in WVC, minimum norm for system reliability in detecting the target species, a certain maximum operation and maintenance effort). This is important to help select a system and associated signs, and it also provides a reference for the functioning of the animal detection system and the success of the project. Animal detection systems are still experimental and success should also relate to increasing the knowledge about the implementation, reliability and effectiveness of these systems, rather than just a substantial reduction in WVC. Animal detection systems can be applied as

stand-alone mitigation measures or used in combination with other measures, such as fencing and crossing structures (Chapters 20 and 21).

Step 5. Identify site characteristics and requirements.

Identify the site-specific conditions and potential associated requirements:

• *WVC*: The site should have a history of a relatively high number of WVC with large animals because: (i) animal detection systems can reduce WVC but not the barrier effect of roads and traffic; (ii) animal detection systems are typically designed to detect large mammals only; and (iii) the costs of the animal detection system may be compensated by the savings associated with reduced WVC and the savings are greatest if the WVC relate to large animals that pose the highest risk to vehicles and human safety. Ideally, site selection is the outcome of a regional prioritisation process based on current WVC hotspots.

• *Historic WVC data and control road section*: A before–after–control–impact (BACI) study (see Chapter 10) allows for the best type of evaluation of the effectiveness of animal detection systems and other mitigation measures (van der Grift et al. 2013). Therefore, historic WVC data is important ('before'), as is a road section without an animal detection system ('control').

• *Animal movements*: The site should be located in an area where many large animals are known to cross the road (daily or seasonally). Only a small proportion of the animal movements across a road may currently result in WVC, and a system at such sites may protect motorists and wildlife against potential future WVC. Note that WVC hotspots may not be the locations where wildlife populations may benefit most from the implementation of an animal detection system; a small population size that suffers just a few WVC may be more of a conservation concern than a large population with many WVC.

• *Traffic volume*: As traffic volume increases, it becomes less desirable to have large animals crossing the road at grade as sudden braking may result in rear-end vehicle collisions. In addition, a threshold in traffic volume may be reached where the barrier effect may be close to absolute with few animals attempting to cross the road. While there is no established maximum traffic volume for animal detection systems, it appears that when traffic volume is ≥20,000 vehicles per day, an animal detection system is almost always undesirable. The lower the traffic volume (e.g. up to a few thousand vehicles per day), the more appropriate an animal detection system may be.

• *Type of traffic*: Trucks and buses are less able or less willing to respond to the activated signs and slow down. Similarly, commuters may drive relatively fast and also be less willing to reduce vehicle speed in response to activated signs. Thus, animal detection systems may be most suitable along highways with a high percentage of passenger cars and a high percentage of drivers that are not in a particular rush (e.g. tourists).

• *Traffic delays during construction*: Compared to wildlife underpasses or overpasses, traffic delays and traffic control may be minimal when installing an animal detection system.

• *Terrain*: The terrain along the road edge must allow for the installation of an animal detection system. For example, numerous ridges, gullies and rocky outcrops may make a location less suitable for an animal detection system, especially break-the-beam systems that require a line of sight between a transmitter and a receiver. Difficult terrain will also require more sensors and other equipment than relatively flat areas.

• *Snow depth*: Some systems, especially break-the-beam systems, can be buried under deep snow and must be shut down in winter. Systems with variable sensor height could solve this problem.

• *Curves and access roads*: The number of curves and access roads should be low to minimise the number of sensors and to avoid gaps (blind spots) or excessive false positives caused by traffic turning on or off the road, depending on what sets off the sensors. More sensors increase costs, require more sophisticated system integration and increase the probability of system failure.

• *Traffic on access roads*: Some detection systems are triggered by vehicles turning off or on the road, for example, at access roads. These false detections may be reduced or eliminated by installing vehicle detection loops at access roads. A system may be programmed to not turn on the warning signs when the wildlife sensors and the vehicle detection loop report detections simultaneously.

• *Vegetation management*: Trees and bushes on the edge of the pavement increase the probability of false triggers. Depending on the type of system and signal types, the vegetation may need to be mowed or cut in the detection area ('area cover sensors') or in between a transmitter and a receiver ('break-the-beam' sensors).

• *Minimum distance of the sensors from the road*: While there are no standards currently, sensors are typically placed within 2–3 m from the edge of the pavement. The equipment must have breakaway provisions if the equipment is installed in the clear zone. Placing sensors further away from the road results in an earlier detection and increased warning time for drivers. However, animals feeding in the right of way may remain after the warning signs have turned off.

• *Landscape aesthetics*: Some detection systems require substantial equipment in the right of way (e.g. sensors, solar panels and associated posts and cabinets), which may affect landscape aesthetics. However, system hardware is decreasing in size with ongoing research and development.

• *Sign activation period*: While there are no standards currently, warning signs are typically activated for several minutes (e.g. 3) after a detection has occurred. If a subsequent detection occurs while the warning signs are still activated, the clock for the warning signs is set back to zero and the signs are activated for an additional few minutes. Ideally, the time the warning signs are turned on is based on the behaviour of the target species on or near the road and how much time they take to cross the road under different traffic volumes.

• *Length of road section and length of potential evaluation*: The road length over which detection systems are installed varies from several tens of metres (e.g. 30 m) up to several kilometres. Shorter road sections (e.g. up to 200 m) may only be useful if the system is located at a gap in a fence or at a fence end as it takes time for a driver to observe and respond to a warning sign. Another consideration is whether the effectiveness in reducing WVC is to be evaluated. While the road section may qualify as a WVC hotspot, the absolute number of pre-mitigation WVC per year may not be very high. Since the number of pre-mitigation WVC increases with road length, it becomes more likely that the effectiveness of a detection system can be evaluated if the mitigated road section is relatively long (e.g. at least one or several kilometres) and spans at least 3–5 years (both before and after implementation) as the data are often aggregated per year to accommodate for seasonal patterns in WVC. The mitigated road section can consist of a stand-alone detection system or a detection system at a fence gap or fence end. Another consideration with regard to evaluating the effectiveness of a detection system is the spatial resolution associated with WVC data which typically vary between 0.1 and 1.0 km or mile. If the mitigated road section is shorter than the spatial resolution of the WVC data, no evaluation of the effectiveness can be conducted. Longer road sections with a stand-alone detection system are more challenging with regard to system integration and operation and maintenance than shorter road sections at a fence gap or fence end. On the other hand, a stand-alone detection system does not restrict where animals can cross the road.

• *Changes in road or landscape*: The road and surrounding landscape should not be expected to undergo major changes within the lifespan of the mitigation measure – for animal detection systems perhaps about 10 years. However, the detection system may need to be moved or extended if the location of successful road crossings or location of WVC changes.

• *Project partners*: All stakeholders at the site should support the project, including permitting, installation, operation and maintenance.

• *Travel costs*: The site should preferably be close to where operation and maintenance personnel have their offices. This reduces costs for travel and stay and reduces the probability that the managing agency will abandon the project.

• *Utilities and communication*: The site should allow for either solar power or a connection to the power grid. Communication facilities (e.g. land phone line or cell phone reception) is very useful.

• *Safety for personnel*: The site should preferably have a safe pull-out location for vendors and maintenance and research personnel.

• *Vandalism and theft*: The site should preferably have a low risk of theft and vandalism, for example, a controlled access road only intended for motorised traffic where stopping vehicles is not permitted.

Step 6. Re-evaluate animal detection system option versus other measures.

Given the site characteristics and requirements, re-evaluate whether an animal detection system is the most appropriate mitigation measure. While there are dozens of mitigation measures aimed at reducing WVC, animal detection systems and wildlife fencing in combination with wildlife crossing structures are the most effective.

Step 7. Obtain an overview of available animal detection systems.

Obtain a current overview of available animal detection systems – their vendors and system reliability, effectiveness and operation and maintenance effort,

Step 8. Select a detection system.

Select a system that meets the goals and the site conditions. Ideally, systems considered should meet minimum standards for system reliability (see Huijser et al. 2009a). If no reliability data are available,

consider a two-phased contract with the vendor. The first phase would entail a beta test of the system in a smaller temporary installation to measure system reliability prior to a more permanent roadside installation in the second phase.

Step 9. Select the warning signs.

Identify the most effective warning signs and be aware of potential future standards for warning signs for animal detection systems. We recommend warning signs that only display a message when a detection has occurred. Consider pairing a warning with an advisory or mandatory speed limit as drivers reduce their vehicle speed more substantially (Fig. 24.8).

Step 10. Take lessons from other projects into account.

Take the lessons learned from other projects into account (see Huijser et al. 2006, 2009a; Sharafsaleh et al. 2012) when preparing project descriptions, contracts and other agreements with vendors, installation contractors, researchers and other project partners.

Step 11. Prepare for technical difficulties, delays and maintenance.

Prepare for technological difficulties and substantial delays following the installation of an animal detection system. It may take many months or years before an animal detection system becomes fully operational. Even systems that are initially successful will fail without proper monitoring and maintenance. Also prepare for potential abandonment of the project and system removal.

Step 12. Make a realistic risk assessment.

Make a realistic risk assessment for potential delays, technological challenges, the financial situation of a vendor and political support for the project. If the outcome of the assessment is not acceptable, consider alternative mitigation measures. Reliability and effectiveness data from some systems seem to meet general expectations. Nonetheless, animal detection systems should still be considered experimental rather than a tried and proven 'plug and play' mitigation measure. This means that if the goal is to reduce WVC by a certain percentage by a set date, then a

Figure 24.8 Animal detection system with inactivated and 'invisible' warning signs (A) and activated warning sign and mandatory speed limit reduction to 40 km/h (B). Source: Photograph by Christa Mosler.

more reliable and proven technique may be to install wildlife fencing in combination with crossing structures. However, if the goal is also to further develop animal detection systems, then consider initiating the project.

Step 13. Conduct system acceptance tests.

The basic system functioning of the system should be evaluated before the warning signs or lights are attached and drivers are exposed to the warning signs. These 'system acceptance tests' should ensure that each detection zone is operational and identify any blind spots that may be present. In addition, the detection data should be regularly analysed for unexpected patterns (e.g. detections that do not match expectations of when and how frequently animals move in the road section). To facilitate system acceptance tests, all individual detections should be saved with a date and time stamp and detection zone in a log. Efficient monitoring of the system can be achieved by having remote access to the detection log and automatic screening and summaries of the reliability and status parameters of the system.

Step 14. Document and publish experiences and study results.

Document and publish the experiences and lessons learned, including data on the reliability, effectiveness and robustness of the system, regardless of the success of

the project. This will allow transportation agencies to make better informed decisions in the future.

24.6 Warning signs can be used with other mitigation measures

The actual and perceived reliability or effectiveness of all wildlife warning signs can be improved when used in combination with other mitigation measures, especially fencing and crossing structures (Fig. 24.9). For example, all types of signs can be installed at a gap in a fence where animals are allowed to cross the road at grade (Lehnert & Bissonette 1997; Gagnon et al. 2010; Fig. 24.10; Chapter 20), encouraging drivers to pay increased attention to wildlife at that location. Wildlife should be prevented from straying onto the roadway at fence gaps or fence ends using wildlife guards or electric mats (Fig. 20.11; Lesson 20.5). However, at-grade crossings expose drivers and wildlife to potential collisions, while crossing structures physically separate traffic and wildlife, thereby substantially reducing WVC.

All types of wildlife warning signs may be combined with advisory or mandatory speed limit reduction. However, requesting or demanding lower vehicle speeds over long distances on highways that have a design speed that is higher than the advisory or mandatory speed limit is not advised. This may result in speed dispersion and an increase in high-risk passing behaviour and head-on collisions as a result of mixing

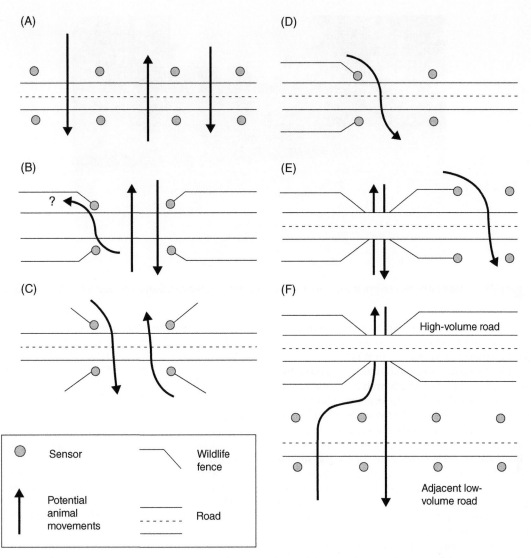

· **Figure 24.9** Schematic representation of potential applications of animal detection systems along a road. (A) System installed over a relatively long road section without wildlife fencing. (B) System installed in a gap with extensive wildlife fences on either side, including illustration of potential wildlife movements into the fenced right of way. (C) System installed in a gap with limited wildlife fences on either side aimed at funnelling the animals towards the road section with the system. (D) System installed at the end of extensive wildlife fencing. (E) System installed at the end of extensive wildlife fencing aimed at funnelling the animals through an underpass. (F) System installed at a low-volume frontage road adjacent to a high-volume highway with a wildlife underpass and wildlife fencing. Source: Huijser et al. (2006).

slow-moving vehicles (drivers who adhere to the lower posted speed limit) with fast-moving vehicles (drivers who drive a speed consistent with the design speed). Advisory or mandatory speed limits should only be considered at specific locations and times when and where the risk for WVC is greatest.

CONCLUSIONS

Standard and enhanced wildlife warning signs are unlikely to be effective in reducing WVC. Their abundance appears to be based on engrained practices, their relatively low cost, a desire to inform the public about

Figure 24.10 Moose detection system at a gap in a wildlife fence, southern Sweden. Source: Photograph by Mattias Olsson.

the impact of WVC on human safety and nature conservation, and possible litigation concerns, rather than a proven substantial reduction in WVC.

Temporal warning signs and animal detection systems are more time and place specific with their messages and can effectively reduce WVC. However, in contrast to the other sign types, the installation of detection systems takes considerable effort and many detection system projects experience technological or management problems. While some detection systems are reliable and substantially reduce collisions with large mammals, the approach should still be considered experimental and a stepwise approach when implementing these systems – as outlined in this chapter – is highly recommended. This includes defining measurable goals for the project and making a realistic assessment of the likelihood of success.

Warning signs, especially animal detection systems, can be combined with other mitigation measures such as wildlife fencing and wildlife crossing structures. For example, detection systems can be installed at a gap in a fence or at a fence end. In such situations, it is suggested to add wildlife guards or electric mats embedded in the road to discourage animals from straying into the fenced right of way.

FURTHER READING

Gagnon et al. (2010): Successful implementation of an animal detection system and electric mat embedded in the road surface at a fence end. Collisions with elk were all but eliminated from the mitigated road section (combination of wildlife fence, detection system and electric mat).

Huijser et al. (2009a): Reports on the reliability of a range of different detection systems from different manufacturers in a controlled environment with livestock as models for wild ungulates. Also contains a basis for minimum norms for system reliability.

Huijser et al. (2009b): Assessment of a detection system project and lessons learned. Includes data on the reliability and effectiveness of the system in reducing collisions with large mammals.

Mosler-Berger and Romer (2003): Successful implementation of several detection systems in Switzerland that reduced WVC with large ungulates substantially. Suggests that signs combined with advisory or mandatory speed limit reduction are most effective in reducing vehicle speed.

REFERENCES

Åberg, L. 1981. The human factor in game-vehicle accidents: a study of drivers' information acquisition. PhD Thesis, Dalarna University, School of Health and Social Studies, Psychology, Sweden.

Al-Ghamdi, A.S. & S.A. AlGadhi. 2004. Warning signs as countermeasures to camel-vehicle collisions in Saudi Arabia. Accident Analysis & Prevention **36**: 749–760.

Ashley, E.P., A. Kosloski & S.A. Petrie. 2007. Incidence of intentional vehicle–reptile collisions. Human Dimensions of Wildlife **12**: 137–143.

Bullock, K.L., G. Malan & M.D. Pretorius. 2011. Mammal and bird road mortalities on the Upington to Twee Rivieren main road in the southern Kalahari, South Africa. African Zoology **46**: 60–71.

Colorado Department of Transportation (CDOT). 2012. Wildlife crossing zones report. Colorado Department of Transportation, to the House and Senate Committees on Transportation, Denver, CO.

Coulson, G.M. 1982. Road kills of macropods on a section of highway in central Victoria Australia. Australian Wildlife Research **9**: 21–26.

Cuperus, R., K.J. Canters, H.A. Udo de Haes & D.S. Friedman. 1999. Guidelines for ecological compensation associated with highways. Biological Conservation **90**: 41–51.

Dai, Q., R. Young & S. Vander Giessen. 2009. Evaluation of an active wildlife-sensing and driver warning system at Trapper's Point. FHWA-WY-09/03F. Department of Civil and Architectural Engineering, University of Wyoming, Laramie, WY.

Drory, A. & D. Shinar. 1982. The effects of roadway environment and fatigue on sign perception. Journal of Safety Research **13**: 25–32.

Found, R. & M.S. Boyce. 2011. Warning signs mitigate deer–vehicle collisions in an urban area. Wildlife Society Bulletin **35**: 291–295.

Gagnon, J.W., N.L. Dodd, S.C. Sprague, K. Ogren & R.E. Schweinsburg. 2010. Preacher canyon wildlife fence and crosswalk enhancement project evaluation. State Route 260. Final report – Project JPA 04-088. Arizona Game and Fish Department, Phoenix, AZ.

Gordon, K.M., M.C. McKinstry & S.H. Anderson. 2004. Motorist response to a deer-sensing warning system. Wildlife Society Bulletin **32**: 565–573.

Hammond, C. & M.G. Wade. 2004. Deer avoidance: the assessment of real world enhanced deer signage in a virtual environment. Final report. Minnesota Department of Transportation, St. Paul, MN. Available from: http://www.lrrb.gen.mn.us/pdf/200413.pdf (accessed 24 September 2014).

Huijser, M.P., P.T. McGowen, W. Camel, A. Hardy, P. Wright, A.P. Clevenger, L. Salsman & T. Wilson. 2006. Animal vehicle crash mitigation using advanced technology. Phase I: review, design and implementation. SPR 3(076). FHWA-OR-TPF-07-01. Western Transportation Institute – Montana State University, Bozeman, MT. Available from: http://www.oregon.gov/ODOT/TD/TP_RES/ResearchReports.shtml (accessed 24 September 2014).

Huijser, M.P., T.D. Holland, M. Blank, M.C. Greenwood, P.T. McGowen, B. Hubbard & S. Wang. 2009a. The comparison of animal detection systems in a test-bed: a quantitative comparison of system reliability and experiences with operation and maintenance. Final report. FHWA/MT-09-002/5048. Western Transportation Institute – Montana State University, Bozeman, MT. Available from: http://www.westerntransportationinstitute.org/documents/reports/4W0049_Final_Report.pdf (accessed 24 September 2014).

Huijser, M.P., T.D. Holland, A.V. Kociolek, A.M. Barkdoll & J.D. Schwalm. 2009b. Animal-vehicle crash mitigation using advanced technology. Phase II: system effectiveness and system acceptance. SPR3(076) & Misc. contract & agreement no. 17,363. Western Transportation Institute – Montana State University, Bozeman, MT. Available from:

http://www.oregon.gov/ODOT/TD/TP_RES/docs/Reports/2009/Animal_Vehicle_Ph2.pdf (accessed 24 September 2014).

Huijser, M.P., J.W. Duffield, A.P. Clevenger, R.J. Ament & P.T. McGowen. 2009c. Cost–benefit analyses of mitigation measures aimed at reducing collisions with large ungulates in the United States and Canada; a decision support tool. Ecology and Society **14**(2): 15. Available from: http://www.ecologyandsociety.org/viewissue.php?sf=41 (accessed 24 September 2014).

Kinley, T.A., N.J. Newhouse & H.N. Page. 2003. Evaluation of the wildlife protection system deployed on Highway 93 in Kootenay National Park during autumn, 2003. Sylvan Consulting Ltd., Invermere, British Columbia, Canada.

Kistler, R. 1998. Wissenschaftliche Begleitung der Wildwarnanlagen Calstrom WWA-12-S. July 1995–November 1997. Schlussbericht. Infodienst Wildbiologie & Oekologie, Zürich, Switzerland.

Kloeden, C.N., A.J. McLean, V.M. Moore & G. Ponte. 1997. Traveling speed and the risk of crash involvement. Volume 1 – findings. NHMRC Road Accident Research Unit, University of Adelaide, Adelaide, Australia.

Krisp, J.M. & S. Durot. 2007. Segmentation of lines based on point densities – an optimisation of wildlife warning sign placement in southern Finland. Accident Analysis & Prevention **39**: 38–46.

Lehnert, M.E. & J.A. Bissonette. 1997. Effectiveness of highway crosswalk structures at reducing deer-vehicle collisions. Wildlife Society Bulletin **25**: 809–818.

Mastro, L.L., M.R. Conover & S.N. Frey. 2010. Factors influencing a motorist's ability to detect deer at night. Landscape and Urban Planning **94**(3–4): 250–254.

Meyer, E. 2006. Assessing the effectiveness of deer warning signs. Final report. Report no. K-TRAN: KU-03-6. The University of Kansas, Lawrence, KS.

Minnesota Department of Transportation (MnDOT). 2011. Deer detection and warning system. MnDOT, St. Paul, MN. Available from: http://www.dot.state.mn.us/signs/deer.html (accessed October 2013).

Mosler-Berger, Chr. & J. Romer. 2003. Wildwarnsystem CALSTROM. Wildbiologie **3**: 1–12.

Muurinen, I. & T. Ristola. 1999. Elk accidents can be reduced by using transport telematics. Finncontact **7**(1):7–8. Available from: http://www.tiehallinto.fi/fc/fc199.pdf (accessed 8 August 2003).

Pojar, T.M., R.A. Prosence, D.F. Reed & T.N. Woodard. 1975. Effectiveness of a lighted, animated deer crossing sign. The Journal of Wildlife Management **39**: 87–91.

Rogers, E. 2004. An ecological landscape study of deer–vehicle collisions in Kent County, Michigan. Report prepared for Kent County Road Commission, Grand Rapids, Michigan. White Water Associates, Amasa, MI.

Sharafsaleh, M.(A.), M. Huijser, C. Nowakowski, M.C. Greenwood, L. Hayden, J. Felder & M. Wang. 2012. Evaluation of an animal warning system effectiveness. Phase Two – final report, California PATH research report,

UCB-ITS-PRR-2012-12. California PATH Program, University of California at Berkeley, Berkeley, CA.

Strein, M. 2010. Restoring permeability of roads for wildlife: wildlife warning systems in practice. p. 77. Programme and book of abstracts. 2010 IENE international conference on ecology and transportation: improving connections in a changing environment. 27 September–1 October 2010, Velence, Hungary.

Sullivan, T.L., A.F. Williams, T.A. Messmer, L.A. Hellinga & S.Y. Kyrychenko. 2004. Effectiveness of temporary warning signs in reducing deer-vehicle collisions during mule deer migrations. Wildlife Society Bulletin **32**: 907–915.

Summala, H. & J. Hietamaki. 1984. Drivers' immediate responses to traffic signs. Ergonomics **27**: 205–216.

van der Grift, E.A., R. van der Ree, L. Fahrig, S. Findlay, J. Houlahan, J.A.G. Jaeger, N. Klar, L.F. Madriñan & L. Olson. 2013. Evaluating the effectiveness of road mitigation measures. Biodiversity and Conservation **22**: 425–448.

Chapter 25

USE OF REFLECTORS AND AUDITORY DETERRENTS TO PREVENT WILDLIFE– VEHICLE COLLISIONS

Gino D'Angelo[1] and Rodney van der Ree[2]

[1]Minnesota Department of Natural Resources, Farmland Wildlife Populations & Research Group, Madelia, MN, USA
[2]Australian Research Centre for Urban Ecology, Royal Botanic Gardens Melbourne, and School of BioSciences, The University of Melbourne, Melbourne, Victoria, Australia

SUMMARY

Roads present an unnatural and confusing environment for wildlife and humans alike. Wildlife reflectors and auditory deterrents aim to modify the behaviour of wildlife on or adjacent to the road. Reflectors are designed to redirect the light from oncoming vehicles into the adjacent verge, while auditory deterrents are designed to cause pain, irritation or masking of other biologically relevant noises. Both techniques attempt to warn animals and discourage them from attempting to cross roads in front of approaching vehicles, with the ultimate goal of reducing the rate of wildlife-vehicle collisions (WVC).

25.1 Most studies of the effectiveness of wildlife warning reflectors have been poorly designed and are inconclusive.

25.2 The colour and intensity of light produced by reflectors may not elicit a response in the target species.

25.3 Wildlife warning reflectors are unlikely to alter animal behaviour and prevent WVC.

25.4 Auditory deterrents, typically mounted to the front of vehicles, appear ineffective at modifying animal behaviour and are unlikely to significantly reduce the rate of WVC.

Wildlife reflectors and auditory deterrents may add to the complexity of the roadway environment without achieving the intended effect of preventing WVC. Given the unproven effectiveness of these techniques and the potential negative consequences of using the devices, the implementation of wildlife warning reflectors and auditory deterrents is not recommended.

Handbook of Road Ecology, First Edition. Edited by Rodney van der Ree, Daniel J. Smith and Clara Grilo.
© 2015 John Wiley & Sons, Ltd. Published 2015 by John Wiley & Sons, Ltd.
Companion website: www.wiley.com\go\vanderree\roadecology

INTRODUCTION

Wildlife warning reflectors and vehicle-mounted auditory deterrents (e.g. wildlife warning whistles) are intended to reduce WVC with primarily deer species and other large herbivores by modifying their behaviour. Manufacturers claim that the reflectors deter animals from attempting road crossings by altering and deflecting light from oncoming vehicle headlights across the road and into the roadside verge to provide a visual warning (Strieter Corp., unpublished instruction manual: 3). Reflectors are mounted on posts along roadsides and consist of a housing with reflective mirrors, which redirect light through coloured lenses, usually red (Fig. 25.1). The reflectors are staggered on both sides of the road, and the headlights of approaching vehicles shine on the reflectors, and light travels diagonally across the road to the next reflector in the installation (Fig. 25.2). Reflectors may also be added on the backs of the posts to direct additional light away from the roadway.

Figure 25.1 Three different types of roadside reflectors. Source: Photographs by Marcel Huijser.

Headlights from approaching vehicle strike the reflector mounted on a post, directing light across the road.
The suggested reason this might be effective is that animals may be disturbed by the light and are less likely to proceed toward the road. Studies have not yet provided evidence for this. Reflectors are spaced so that light reflects through a brightening arc and will, to some extent, cover the whole of the roadside. Light would be perceived as a series of one or more flashes.

Figure 25.2 There is currently no evidence that roadside reflectors effectively reduce wildlife–vehicle collisions. This schematic shows the theoretical application of roadside reflectors, where light from oncoming vehicles are reflected across the road (and possibly onto the opposite reflectors) and into both verges (although light only shown on one verge to improve clarity), thereby discouraging wildlife from attempting to cross the road. Source: Reproduced with permission of Scott Watson.

Wildlife warning whistles are mounted to the front of the vehicle and emit noise as the vehicle is driven. Invented in 1979 in Austria (Romin & Dalton 1992), they are inexpensive, readily available and widely used in North America and Europe for deer and in Australia for kangaroos. More recently, battery-operated devices have been developed that produce and propagate noises, even when the vehicle is stationary.

Wildlife warning reflectors and auditory deterrents are *marketed* as a proven and humane technique for reducing WVC (e.g. www.strieter-lite.com, www.shuroo.com.au). Planners find it difficult to decide whether reflectors and/or whistles are a good investment for reducing the risk of WVC. The aim of this chapter is to raise awareness of past experiences with these devices, possible explanations for why reflectors and whistles remain unproven and insights into the limitations of altering the behaviour of wildlife in the immediate vicinity of roads.

LESSONS

25.1 Most studies of the effectiveness of wildlife warning reflectors have been poorly designed and are inconclusive

Studies of the effectiveness of a range of wildlife warning reflector models have produced variable results (Gilbert 1982; Armstrong 1992; Reeve & Anderson 1993; Pafko & Kovach 1996; Gulen et al. 2006). Researchers have used a diversity of methods with various levels of scientific validity to study the effectiveness of reflectors (D'Angelo et al. 2005). However, there remains a limited understanding of reflector efficacy. Most reflector evaluations were based on counts of WVC within test sections that were either (i) pre- and post-installation of reflectors (Ingebrigtsen & Ludwig 1986; Waring et al. 1991; Pafko & Kovach 1996); (ii) when reflectors were covered versus uncovered (Schafer & Penland 1985; Woodham 1991; Armstrong 1992); or (iii) within sections of roads with reflectors as compared to adjacent control sections without reflectors (Reeve & Anderson 1993; Sielecki 2001). Such methods failed to consider changes in animal densities, seasonal movements or traffic patterns. Little is known about how animals react to reflector activation or if individual animals become habituated to the devices over time. Beyond differences in experimental design, comparison of results among different reflector studies is further confounded by the variety

of reflector models tested and the distinct spectral properties of the different devices (D'Angelo et al. 2005).

Studies were also often limited by sample size and poor experimental design. In most cases, animal carcasses along roads were counted, but rarely were quality controls such as video surveillance of test sections, driver surveys or accident reports used to account for collisions where animals are left on the roadside. Most reflector studies also provided little data on the behaviour of free-ranging animals to reflector activation, a significant omission, given that these behavioural reactions constitute the basis for the purported effectiveness of reflectors.

25.2 The colour and intensity of light produced by reflectors may not elicit a response in the target species

Examinations of the visual abilities in white-tailed deer and fallow deer have shown that peak sensitivity of colour vision is well below the long wavelength of red (Jacobs et al. 1994), which is the most commonly marketed colour of wildlife warning reflectors. Most marsupials have dichromatic vision, lacking sensitivity to long wavelengths (Fig. 18.2) (Hemmi et al. 2000; Deeb et al. 2003), which would likely render red reflectors ineffective in deterring collisions with these species as well.

The effectiveness of four wildlife warning reflector lens colours (blue-green, amber, red and white) was evaluated for altering the behaviour of white-tailed deer along roads (D'Angelo et al. 2006). Based on characteristics of deer colour vision and the assumption that reflectors are effective, they hypothesised that short-wavelength (i.e. blue-green) reflector lens colours would be the most effective and long-wavelength (i.e. red) lens colours would be the least effective for preventing deer–vehicle collisions. The experiment demonstrated nearly opposite results. The highest level of deer–vehicle collision risk, based on deer behaviour along roadways, was observed during the blue-green reflector treatments with slightly lower levels of risk during the amber, red and white reflector treatments, in respective order of decreased risk. These results suggest that negative responses by animals may directly increase with greater sensitivity to different colours of light from reflectors.

Evidence for animals with nocturnal visual systems suggests that the rapidity of their visual adaptation from darkness to abrupt increases in light (e.g. vehicle headlights) may be considerably slower than that of daylight-active species like humans (Ali & Klyne 1985).

A possible explanation for the increase in WVC in areas where reflectors were installed in some studies may be that light from reflectors, in combination with vehicle headlights, overwhelmed the animals' visual system.

25.3 Wildlife warning reflectors are unlikely to alter animal behaviour and prevent WVC

Few descriptions from the scientific literature exist that describe animal behaviour in direct interaction with vehicles when reflectors are activated. Observations of the response of white-tailed deer to vehicles suggest that deer tend to avoid crossing roads in the presence of vehicles, regardless of whether reflectors are in place (D'Angelo et al. 2006). Likewise, Waring et al. (1991) observed that greater than 70% of crossings by white-tailed deer were completed without a deer–vehicle interaction on a two-lane highway with regular traffic.

Fallow deer quickly became habituated to repeatedly occurring light reflections from a red WEGU reflector (Walter Dräbing KG, Kassel, Germany) placed directly in front of a bait site (Ujvári et al. 1998). During the first experimental night, fallow deer fled from the stimulus in 99% of cases but exhibited increasing indifference to reflections over the remaining 16 nights. This was interpreted as habituation of the deer to the stimulus. Similarly, captive red kangaroos and red-necked wallabies showed a negligible behavioural response to a simulated roadway environment with wildlife warning reflectors activated by a series of lights (Ramp & Croft 2006).

The primary intent of using wildlife warning reflectors is to elicit a response from an animal in the immediate vicinity of moving vehicles. Behavioural responses to stimuli may differ among species and individuals depending on several factors (e.g. number of animals in a group, season, roadside characteristics, number of traffic lanes, traffic volume, traffic speed). Two possible behavioural responses to reflectors could be directional flight or vigilance (i.e. stop and observe). Depending on the location of the animal relative to the roadway at the time of the behavioural response and their direction of travel and also based on the reaction of the animal, a WVC could either be averted or occur. Given the unpredictable behaviour of wildlife under various roadway conditions, the theoretical basis for using wildlife warning reflectors and similar devices to prevent WVC is questionable. Therefore, in the interest of motorist safety, we recommend that planners do not consider the use of wildlife warning reflectors.

25.4 Auditory deterrents are ineffective at modifying the behaviour of wildlife and unlikely to reduce the rate of WVC

Auditory deterrents are used in a wide range of situations as a non-lethal method to modify animal behaviour and are promoted as humane, inexpensive, scientifically proven and easy to use (Bomford & O'Brien 1990; Bender 2001). However, there is no published scientific evidence to indicate they are effective at modifying animal behaviour or reducing the rate of WVC (Romin & Dalton 1992; Ujvári et al. 1998; Bender 2001; Scheifele et al. 2003; Valitski et al. 2009).

There are a number of compelling reasons why vehicle-mounted auditory deterrent systems are unlikely to ever be effective. First, most whistles are designed (and purported) to emit noise in the 16–25 kHz range (i.e. ultrasonic), but most tests have demonstrated that their actual performance differs from that stated by manufacturers (e.g. Bender 2001; Scheifele et al. 2003). Second, sound attenuates (i.e. becomes quieter) with increasing distance from the source, and higher-frequency sounds attenuate more quickly than lower frequencies. Therefore, the effective distance of high-frequency ultrasonic noises generated by whistles or other vehicle-mounted devices is unlikely to extend far enough in front of fast-moving vehicles to give the animal time to respond and move away from the road (Bender 2001; Scheifele et al. 2003). In addition, environmental conditions (e.g. weather, topography, vegetation) and road design (e.g. bends or cuttings) will further reduce the effective distance of ultrasonic deterrents. This is potentially problematic, because they are designed to work on the premise that animals are alerted with sufficient time to respond and move away, rather than be startled and move onto the road in front of oncoming vehicles. Third, the designed frequency spectrum of the generated noise will be compromised by other engine and road noises, potentially masking the high-frequency sounds that are purported to affect animal behaviour. Fourth, animals typically habituate quickly to noises (Bomford & O'Brien 1990), and the response to vehicle-mounted auditory deterrents should be no different, and unless there is an associated negative stimulus, animals will habituate to the whistles. Finally, whistles mounted to the front of vehicles are easily blocked by insects, thus not producing any sound at all and potentially giving drivers a false sense of protection.

CONCLUSIONS

The road-crossing success of animals in localised areas may be impacted by factors such as vehicle speed, traffic volume and patterns, vehicle types, motorist awareness of wildlife, weather conditions, ambient and vehicle-produced light and noise levels, characteristics of the habitat–roadway interface and mitigation measures (D'Angelo et al. 2005). Irrespective of the unproven effectiveness of reflectors or questions of transmission of light (intensity and wavelength) from reflectors and noise from whistles, attempting to affect the behaviour of animals in the immediate vicinity of the roadway is a poor strategy. The myriad of stimuli already present on vehicles and along roads do not prevent animals from crossing roads in the presence of moving vehicles. For these same reasons, as well as logistical challenges of delivery in the field, approaches using predator odours and diversionary feeding are similarly unlikely to be effective across large spatial scales. Until effective science-based strategies become available, management efforts should focus on: (i) avoiding the construction of new roads in areas with large-bodied animals; (ii) using fencing and wildlife crossing structures (Chapters 20 and 21) to prevent access by wildlife to the roadway and facilitate connectivity; (iii) proper population management programmes; (iv) controlling roadside vegetation to minimise its attractiveness to wildlife and to maximise visibility for motorists (Chapter 46); (v) increasing motorist awareness of danger associated with WVC; and (vi) minimisation of unnecessary stimuli in the roadway environment that may increase confusion of wildlife. Our understanding of animal senses continues to expand and be refined. Future development of animal deterrent strategies should be guided by our knowledge of animal senses and their behaviours in roadway environments and be subject to thorough independent testing prior to deployment.

FURTHER READING

Blackwell and Seamans (2008): Vehicles present a complex array of stimuli that can confuse wildlife along the road verge. This research demonstrated that specific vehicle headlight designs with light transmissions that better complemented the peak visual capabilities of white-tailed deer at night yielded an earlier flight by deer from an approaching vehicle.

D'Angelo et al. (2006): An in-depth examination of white-tailed deer behaviour in close proximity to roads during activation of wildlife reflectors by vehicles which showed that wildlife warning reflectors did not alter deer behaviour such that deer–vehicle collisions might be prevented.

Grandin and Johnson (2005): Presents unique perspectives about animal behaviours that result from the animal's biology, evolution, and situational awareness. The lead author, Temple Grandin, is an animal scientist who uses her own experiences of living with autism to explain how animals perceive the world.

Bomford and O'Brien (1990): A detailed overview of the use of auditory deterrents for wildlife, including a summary of use and effectiveness in different situations, the theoretical basis for their use and the many problems associated with their deployment.

REFERENCES

Ali, M.A. and M.A. Klyne. 1985. Vision in vertebrates. Plenum, New York.

Armstrong, J.J. 1992. An evaluation of the effectiveness of Swareflex deer reflectors. Ontario Ministry of Transportation, Research and Development Branch, Report No. MAT-91-12, Downsview, Canada.

Bender, H. 2001. Deterrence of kangaroos from roadways using ultrasonic frequencies – efficacy of the Shu Roo. University of Melbourne, Report to NRMA.

Blackwell, B.F. and T.W. Seamans. 2008. Enhancing the perceived threat of vehicle approach to deer. Journal of Wildlife Management **73**:128–135.

Bomford, M. and P.H. O'Brien. 1990. Sonic deterrents in animal damage control: a review of device tests and effectiveness. Wildlife Society Bulletin **18**:411–422.

D'Angelo, G.J., S.A. Valitzski, K.V. Miller, G.R. Gallagher, A.R. DeChicchis and D.M. Jared. 2005. Thinking outside the marketplace: a biologically based approach to reducing deer-vehicle collisions. Pages 662–665 in Proceedings of the International Conference on Ecology and Transportation. C.L. Irwin, P. Garrett and K.P. McDermott, editors. Center for Transportation and the Environment, North Carolina State University, Raleigh, NC.

D'Angelo, G.J., J.G. D'Angelo, G.R. Gallagher, D.O. Osborn, K.V. Miller and R.J. Warren. 2006. Evaluation of wildlife warning reflectors for altering white-tailed deer behavior along roadways. Wildlife Society Bulletin **34**:1175–1183.

Deeb, S.S., M.J. Wakefield, T. Tada, L. Marotte, S. Yokoyama and J.A.M. Graves. 2003. The cone visual pigments of an Australian marsupial, the tammar wallaby (*Macropus eugenii*): sequence, spectral tuning, and evolution. Molecular Biology and Evolution **20**:1642–1649.

Gilbert, J.R. 1982. Evaluation of deer mirrors for reducing deer-vehicle collisions. U.S. Federal Highway Administration Report No. FHWA-RD-82-061, Washington, DC.

Grandin, T. and C. Johnson. 2005. Animals in translation: using the mysteries of autism to decode animal behavior. Scribner, New York.

Gulen, S., G. McCabe, I. Rosenthal, S. Wolfe and V. Anderson. 2006. Evaluation of wildlife reflectors in reducing vehicle deer collisions on Indiana Interstate 80/90. U.S. Federal Highway Administration Report No. FHWA/IN/JTRP-2006/18, Washington, DC.

Hemmi, J.M., T. Maddess and R.F. Mark. 2000. Spectral sensitivity of photoreceptors in an Australian marsupial, the tammar wallaby (*Macropus eugenii*). Vision Research **40**:591–599.

Ingebrigtsen, D.K. and J.R. Ludwig. 1986. Effectiveness of Swareflex wildlife warning reflectors in reducing deer-vehicle collisions in Minnesota. Minnesota Department of Natural Resources Wildlife Report No. 3.

Jacobs, G.H., J.F. Deegan, J. Neitz, B.P. Murphy, K.V. Miller and R.L. Marchinton. 1994. Electrophysical measurements of spectral mechanisms in the retinas of two cervids: white-tailed deer (*Odocoileus virginianus*) and fallow deer (*Dama dama*). Journal of Comparative Physiology **174**:551–557.

Pafko, F. and B. Kovach. 1996. Minnesota experience with deer reflectors. Pages 130–139 in Proceedings of the International Conference on Ecology and Transportation. G.L. Evink, P. Garrett, D. Zeigler and J. Berry, editors. Center for Transportation and the Environment, North Carolina State University, Raleigh, NC.

Ramp, D. and D.B. Croft. 2006. Do wildlife warning reflectors elicit aversion in captive macropods? Wildlife Research **33**:583–590.

Reeve, A.F. and S.H. Anderson. 1993. Ineffectiveness of Swareflex reflectors at reducing deer-vehicle collisions. Wildlife Society Bulletin **21**:127–132.

Romin, L.A. and L.B. Dalton. 1992. Lack of response by mule deer to wildlife warning whistles. Wildlife Society Bulletin **20**:382–384.

Schafer, J.A. and S.T. Penland. 1985. Effectiveness of Swareflex reflectors in reducing deer-vehicle accidents. Journal of Wildlife Management **49**:774–776.

Scheifele, P.M., D.G. Browning and L.M. Collins-Scheifele. 2003. Analysis and effectiveness of deer whistles for motor vehicles: frequencies, levels and animal threshold responses. Acoustics Research Letters Online **4**:71–76.

Sielecki, L.E. 2001. Evaluating the effectiveness of wildlife accident mitigation installations with the Wildlife Accident Reporting System (WARS) in British Columbia. Pages 473–489 in Proceedings of the International Conference on Ecology and Transportation. C.L. Irwin, P. Garrett and K.P. McDermott, editors. Center for Transportation and the Environment, North Carolina State University, Raleigh, NC.

Ujvári, M., H.J. Baagøe and A.B. Madsen. 1998. Effectiveness of wildlife warning reflectors in reducing deer-vehicle collisions: a behavioral study. Journal of Wildlife Management **62**:1094–1099.

Valitski, S.A., G.J. D'Angelo, G.R. Gallagher, D.A. Osborn, K.V. Miller and R.J. Warren. 2009. Deer responses to sounds from a vehicle-mounted sound-production system. Journal of Wildlife Management **73**:1072–1076.

Waring, G.H., J.L. Griffis and M.E. Vaughn. 1991. White-tailed deer roadside behavior, wildlife warning reflectors, and highway mortality. Applied Animal Behaviour Science **29**:215–223.

Woodham, D.B. 1991. Evaluation of Swareflex wildlife warning reflectors. Colorado Department of Transportation Final Report No. CDOT-DTD-R-91-11.

Chapter 26

ECOLOGICAL EFFECTS OF RAILWAYS ON WILDLIFE

Benjamin Dorsey[1], Mattias Olsson[2] and Lisa J. Rew[1]

[1]Department of Land Resources and Environmental Sciences,
Montana State University, Bozeman, MT, USA
[2]EnviroPlanning AB, Göteborg, Sweden

SUMMARY

Compared to roads, relatively little is known about the ecological effects of railways on wildlife. Railways and roads are frequently co-aligned in the same corridor, but most road ecology projects ignore parallel railways due to landownership issues, road-specific funding or perceptions that railway impacts are negligible. Railways and trains can negatively affect wildlife and the environment in ways similar to roads and vehicles (including wildlife mortality, habitat loss and fragmentation), but the degree of these impacts may differ. Most research has focused on evaluating the impacts of trains and railways, with little attention towards mitigation, and most research projects have focused solely on moose and bears.

26.1 Railways and trains affect wildlife in similar ways to roads and vehicles, but the degree of impact differs.

26.2 Railways impact wildlife in unique ways that are not well understood.

26.3 Wildlife–train collisions are a multifactor problem.

26.4 Mitigation options exist but some are more proven than others.

Many of the solutions developed for roads can probably be applied successfully to railways, but in some cases, new approaches and technologies are needed.

INTRODUCTION

Railways alter landscapes in simple yet profound ways. The construction of a railway line converts a relatively narrow strip of land into a rocky or concrete surface with two or more steel rails. This change could be considered minor in environments where the ground is naturally rocky or barren, but in forests and other well-vegetated landscapes, the gap created by the railway can be a conduit for

Handbook of Road Ecology, First Edition. Edited by Rodney van der Ree, Daniel J. Smith and Clara Grilo.
© 2015 John Wiley & Sons, Ltd. Published 2015 by John Wiley & Sons, Ltd.
Companion website: www.wiley.com\go\vanderree\roadecology

poachers, weeds and invasive species as well as a barrier to the movement of wildlife. Railways can also fragment natural areas into ever smaller pieces, modifying the behaviour and reducing the survival of some wildlife species. Changes can also occur when the disruption or increased mortality of one species affects another, with potentially cascading effects. Thus, it is important to consider ecosystems and not just individual species when quantifying impacts of railways and trains and designing mitigation strategies.

A better understanding of the ecological effects of railways is needed because the global rail network is extensive (about 1.4 million km globally) and many nations are extending and/or increasing the capacity of their railway networks (UIC 2013). Most efforts will focus on building new high-speed railways (HSR), which are projected to double in total track length from the current 21,472 km in operation today over the next 15 years (UIC 2013). Both HSR and traditional systems (speeds <200 km/h) are being increasingly relied upon because they are more efficient than automobiles at moving goods and people. Even traditional railways can transport up to 500 ton-miles/gallon (0.54 l/100 km for 0.89 tonnes), resulting in two to five times higher fuel efficiency than truck transport, as well as using 21% less energy per passenger mile (Welbes 2011). Thus, moving freight and people by rail can reduce carbon dioxide emissions (AASHTO 2007), reduce traffic congestion and yield other environmental benefits, compared to road transportation.

In this chapter, we describe the negative effects of railways and trains on wildlife and suggest ways to minimise and mitigate those impacts. Wildlife often provide the best indicator of railway impacts on an ecosystem because they are easy to measure and are often of public or economic interest due to train strikes and the loss of harvestable resources (see also Chapter 42). However, some species are incredibly adaptable to changes in their environment, and wildlife populations and behaviour cannot be used as the only indicator of railway impacts on ecosystems. Likewise, individual wildlife species may contradict generalities regarding specific impacts. For example, moose and fox have been documented to utilise railways for movements (Child 1983; Kolb 1984) in contrast to many species that avoid railways altogether. As a result, this chapter describes specific scenarios where impacts have been discovered and their possible solutions.

LESSONS

26.1 Railways and trains affect wildlife in similar ways to roads and vehicles, but the degree of impact differs

Railways and trains, like roads and vehicles, have the potential to negatively affect wildlife. Impacts are generally similar along roads and railways because they both convert a strip of land into an area that has moving trains and cars that can strike animals and emit noise, light and chemical pollution. Four broad types of impacts are evident (Dorsey 2011), with the most obvious being direct mortality due to wildlife-train collision (WTC). The remaining impacts are habitat alteration, habitat fragmentation and barrier effects (van der Grift 1999). The earliest road ecology research speculated that the same impacts were occurring along railways, although likely at lower levels (Forman et al. 2003). However, closer inspection of impacts is needed because it appears this rule of thumb may not always hold true. In some cases, the impacts may be greater than those occurring along roads, and in almost all cases, railway impacts are more difficult to detect.

Wildlife mortality occurs along railways due to WTC, electrocution, rail entrapment and wire strikes. In some scenarios, wildlife can become trapped between the two rails and overheat due to excessive exposure to the sun and the heating up of the rails and rock ballast (Kornilev et al. 2006; Fig. 26.1). WTC are the most common cause of mortality and have been documented for at least 84 different species ranging from Asian elephants to box turtles, lizards and birds (Fig. 26.2) (Dorsey 2011). In some cases, the rate of WTC has clearly impacted wildlife populations. For example, a 70% decline in a moose population in Alaska was attributed to WTC (Schwartz & Bartley 1991). In another example, the rate of WTC of brown bears along co-aligned transportation features was equal to or higher than roadkill rates (Huber et al. 1998). In one case, bear–train collisions were twice as frequent on a railway compared to a parallel road and were the second highest cause of mortality after poaching (Waller & Servheen 2005). In these circumstances, railways likely affected the population to a greater extent than roads. WTC are common because many wildlife species are able to traverse or utilise the land altered by railways (Fig. 26.3) and are therefore at risk of collision.

Habitat alteration occurs through the construction and operation of railways. One approach to determine the amount of habitat altered is based on the width of

Figure 26.1 Turtles enter railway tracks, usually at locations where the tracks are level with the ground, such as level-train crossings, and, after moving along the tracks, are unable to climb out. Thirty of the 128 spotted pond turtles found along a 33 km section of railway through Dudhwa Tiger Reserve, India, in 2012 were dead. Source: Photographs by and reproduced with permission of Subrat Behera, Wildlife Trust of India.

(A) (B)

Figure 26.2 (A) Elephant–train collisions are common in parts of Asia where railways dissect areas populated by elephants. (B) Mortality due to wildlife–train collisions often attracts scavengers who are also at risk. In this example, a northern goshawk was killed after being struck by a train while scavenging on a dead moose. Source: (A) Photograph by and reproduced with permission of A. S. Negi; and (B) photograph by M. Olsson.

a standard gauge rail line (1435 mm), plus the right of way (ROW), typically totalling 15 m per track (Carpenter & Lewis 1994). This approach is overly simplistic because some species, such as Mongolian gazelles, may avoid railways at much greater distances, even as much as 300 m (Ito et al. 2005; Chapter 58).

Second, it ignores the effects of noise, human presence and other factors that may operate outside the boundaries of the physical footprint.

Habitat fragmentation occurs when a patch of habitat is broken into smaller pieces, subdividing populations (Forman & Deblinger 2000) and often resulting

(A)

(B)

Figure 26.3 (A) Moose tracks show evidence of foraging on vegetation along a railway right of way. In this case, the fence, designed for cattle, is not a barrier to moose. (B) Wolves use the Canadian Pacific Railway in Banff National Park to move efficiently across this forested landscape. Source: (A) Photograph by and reproduced with permission of Evelina Augustsson; and (B) photograph by and reproduced with permission of Alli Banting.

in decreased biodiversity. Fragmentation can affect populations because the environmental conditions within the gaps between the original habitat may change, and subsequently be utilised by different species; for example open-habitat species versus closed habitat specialists. Depending on the type and location of fragmentation increased rates of WTC may occur.

Barrier effects may result from (i) species-specific interactions with railway design (e.g. fencing, railway tracks, rock ballast); (ii) lack of natural cover (e.g. gaps in forest canopy or other vegetation); and/or (iii) avoiding other conditions (e.g. sounds, sights or human presence). Reduced movements have been documented for wary and limited-mobility species such as the eastern box turtle which are physically unable to climb over standard gauge rails (Kornilev et al. 2006). Other species may be physically able to cross but do not because of exposure to risky conditions (predation or thermal maximums) or perceived risk (e.g. wariness).

The strength of a barrier is location and species specific and likely varies with train traffic volume. While some species such as moose preferentially move along railways (Fig. 26.3), others like the Mongolian gazelle rarely cross despite being physically able to leap great heights and travel large distances (Ito et al. 2005). In fact, many wildlife species appear able to move across and along the land altered by railways (Figs 26.3 and 26.4), and the exceptions appear to be species with extreme wariness and/or limited mobility. Wildlife have been documented using railways for foraging, accessing critical resources, migrating and dispersing. However, habitat use often coincides with other impacts such as increased WTC; thus, the perceived

benefits of habitat use need to be considered relative to other impacts.

The loss and fragmentation of habitat and barrier effects due to railways do not appear to affect wildlife to the same degree as roads. In many cases, this is likely because the footprint of railways is narrower and volume of train traffic lower. Also, railway networks are much less extensive than for roads and are often co-aligned with road corridors. However, each existing and proposed railway has the potential to effect wildlife in a range of deleterious ways and should therefore be thoroughly assessed in its own right and in concert with adjacent roads.

26.2   Railways impact wildlife in unique ways that are not well understood

Most of our understanding about railway impacts on wildlife comes from a small number of studies from North America and Europe on a few species, primarily moose and bears (see Dorsey 2011). Large mammals such as moose, bears and elephants have likely received the most attention because they hold special economic or conservation status and in some cases can cause significant damage to trains (Fig. 26.2A). The countries with the largest rail networks, including China, are just beginning to examine the impacts of railways on wildlife (Ito et al. 2005). Consequently, our understanding of the ecological impacts of railways and trains is incomplete. Important areas for future research include the effects of train noise (see also Chapter 19) and the cumulative effects from the interactions

Figure 26.4 A grizzly bear in Banff National Park, Canada, eating wheat leaked from moving trains. Source: Photograph by B. Dorsey.

between railways and other landscape features, especially co-aligned roads.

Railways and roads may interact or have cumulative impacts that are not typically addressed in most case studies. The cumulative impact of multiple transportation features may exacerbate impacts or result in unique impacts. In one case, the number of linear features including roads and railways that separated breeding habitat for moor frogs better explained the genetic differences than geographic distance (Reh & Seitz 1990). The cumulative effect of railways and roads has resulted in increased barrier effects and WTC for a range of species (Skogland 1986; Vos et al. 2001). For example, grizzly bears in Montana, United States, crossed a transportation corridor when highway traffic volume was lowest but train traffic was highest (Waller & Servheen 2005), resulting in higher bear mortalities on the railroad compared to the highway. These potential interactions need to be considered when studying impacts and designing mitigation. However, clustering transportation features is likely preferable to spreading them out because habitat fragmentation is reduced while transportation efficiency is improved. Co-aligning multiple features also allows for cost-effective mitigation by concentrating fencing and wildlife crossing structures to the same location. However, there is likely a limit to the effectiveness of crossing structures associated with increasing length as they pass under multiple linear features.

Trains produce noise, often at high frequency and high intensity, but typically of short duration at any one location. In the Netherlands, noise level and duration have been correlated with decreased avian density and may affect wildlife behaviour (Waterman et al. 2002). The disturbance effects due to train noise are likely species specific and vary with sound characteristics, speed, train traffic and design (see also Chapter 19). High-speed trains typically generate high-intensity noise across a range of frequencies, and the impacts are of particular importance because of the widespread and rapid expansion of HSR around the world.

Railways also create indirect or secondary impacts such as increased human presence and related disturbance, similar to roads (e.g. poaching and land use change; Chapters 2, 3 and 56). Indirect impacts occur while the railway is operational and after it has been decommissioned and thus can affect wildlife over long time periods. Indirect long-term impacts need research attention and opportunity for research on these impacts exists in countries like Brazil and China, where new railways are being built through relatively undeveloped landscapes (Chapters 51 and 57).

26.3 Wildlife–train collisions are a multifactor problem

It may be surprising to learn wildlife are killed by trains, sometimes in large groups (e.g. 270 individuals simultaneously), because trains are usually thought of as loud, large and slow moving (Associated Press 2011). Even though most of the world's railways operate at speeds less than 200 km/h, they can be relatively silent, particularly when descending a grade, often resulting in WTC. Additionally, trains cannot evade wildlife nor stop quickly, resulting in many collisions that may have been avoided by vehicles on a road.

There has been much research focused on quantifying the rates of WTC and the factors influencing it (Seiler & Helldin 2005). The abundance of wildlife in habitat adjacent to a railroad is a primary factor because higher wildlife abundance exposes more individuals to approaching trains. Strikes for ungulates and carnivores are common where high-quality wildlife habitat and railways intersect (Modafferi 1991). More herbivores than carnivores are usually struck, which probably reflects their relative abundance as well as a tendency for herbivores to browse or graze within railway easements (Fig. 26.3A) (Andersen et al. 1991). However, some carnivores frequently encounter railways due to their large home ranges and/or move along them (Fig. 26.3B), increasing their risk of collision. Some wildlife also behave in ways that increase their likelihood of being struck. For example, moose flee down the track when a train approaches, typically resulting in WTC and death for the individual (Child 1983; Rea et al. 2010). Forage including natural vegetation, carrion and

agricultural products spilled from train cars during loading, transport and/or derailments (Fig. 26.4) and other wildlife attractants on and along railroads can increase WTC rates (Huber et al. 1998; Wells et al. 1999; Waller & Servheen 2005).

The alignment and design of a railway influence the rates of WTC by affecting an animal's ability to detect and evade oncoming trains. Cuttings, bridges, bends and dense vegetation all reduce an animal's visibility, thereby increasing the risk of collision. The most obvious example of poor railway design was revealed by research on turtles, many of which are unable to climb over the railway tracks (e.g. Kornilev et al. 2006) (Fig. 26.1). For most species, it is likely that a combination of these factors influences the rates of WTC and mortality. The volume of train traffic is also likely an important factor influencing the rate of WTC. In Sweden, the highest numbers of moose and roe deer accidents were reported on railways with moderate train traffic (50–130 average daily traffic), while lower numbers were reported on stretches with higher train traffic (Seiler 2011) (Textbox 26.1). The underlying reason for this pattern may be that higher traffic volumes deter wildlife from attempting to cross.

Textbox 26.1 Ungulate–train collisions in Sweden.

Wildlife–train collisions have become more frequent in Sweden during the past decade. Collisions with moose appear to be twice as frequent on railways compared to roads, and the costs are significant. Nevertheless, empirical knowledge about train collisions in Sweden is poor. Long-term trends and large-scale patterns in collisions appear to primarily reflect the distribution and abundance of wildlife, with year-to-year variation affected by snow accumulation. Collision hotspots are a result of other, smaller-scaled factors such as train traffic volume, habitat management and landscape structure.

Based on the results of a literature review, spatial analyses and train-driver questionnaires, it was concluded that traffic volume, track design and vegetation management alongside tracks affect collision rates (Seiler 2011). Spatial analyses revealed that moose–train collisions were more frequent in areas with more clear-cuts, more forests, more bridges and tunnels but fewer roads in the vicinity of the railway. Collisions with roe deer were more common on tracks in areas with relatively open (agricultural) habitat, more water

courses and fewer highways. Fewer collisions occurred on both the busiest railways (>200 trains/day) and the least active railways (<10 trains/day).

Some of these findings were also indicated in a train-driver questionnaire. Train drivers suggested the risk of WTC was lower in deforested or cleared railway corridors compared to densely vegetated areas. Drivers perceived an increased collision risk near dense vegetation and where visibility was reduced. Our work and others indicate that that removal of vegetation can reduce the risk of collision; however, the risk may increase if applied at intervals greater than 3–4 years.

We recommend intensified studies of the interactions between temporal and spatial factors, comparative field studies of collision hotspots, assessments of railway bridges as potential wildlife passages, behavioural studies of radio-tracked wildlife along railways with vegetation management regimes and scientific experiments with warning signals. There is further need for a comprehensive evaluation of the societal costs of wildlife–train collisions, as well as an estimation of the accuracy of collision records.

26.4 Mitigation options exist but some are more proven than others

Strategies to reduce the impacts of railways and trains span the mitigation hierarchy – avoid, minimise, mitigate and offset – and are similar to those used along roads (Chapter 7). However, most effort has focused on minimising or mitigating impacts on existing railways and primarily WTC. Greater attention needs to focus on avoiding and minimising impacts during planning and design, and the expansion of the HSR network should urgently consider this. Current mitigation options include wildlife crossing structures, habitat alteration, aversion/exclusion systems, reduced train speeds and supplemental feeding. Vegetation management (i.e. mowing or pruning vegetation that provides forage or cover) is one of the more proven techniques, and it has reduced moose WTC by as much as 56% (Jaren et al. 1991; Andreassen et al. 2005). Vegetation management is an ongoing technique because strike rates will increase if not maintained (Seiler 2011).

Wildlife crossing structures (Chapter 21) allow wildlife safe passage across railways and are one of the best strategies to reduce WTC and barrier effects for a large number of species. Wildlife use of structures is affected by many factors including the size of the structure, the vegetation at structure entrances and the presence and design of fences. Some species will use drainage culverts, indicating even basic structures may provide passage. Simple strategies for turtles and other small species include the excavation of rock ballast between pairs of railway sleepers, providing a shallow depression below the tracks for animal use (Fig. 26.5)

(Pelletier et al. 2006). While it is cheaper to install crossing structures or fencing singly, it is less effective and not recommended (Chapters 20 and 21). It is important to understand that the use of fences and crossing structures requires a large initial investment but may prove more effective and easier to maintain over the long term.

Electronic systems to detect and deter animals can be mounted to the front of trains or installed as stationary systems along railways. Stationary systems are used to exclude wildlife from fenced areas and on bridges using a motion-activated sensor with an audible or visual signal (lights and horns) to frighten animals. As with other auditory deterrents, habituation can occur (Chapter 25). Other systems use an electrified pad across the train track at breaks in fences, similar to wildlife guards along roads (Lesson 20.5; Fig. 20.11). Train-based solutions include modified train warning signals for humans (i.e. lights and horns), and train drivers report that wildlife usually flee from approaching trains when multiple blasts are used rather than a prolonged single warning signal (Seiler 2011). Improved signals are also needed to combat the tendency of some wildlife to flee down track.

Many mitigation projects target a single species (e.g. moose) and multi-species strategies are not yet well established, despite being an efficient use of resources. Further, efficient use of mitigation funding should target corridors where surface transportation and industrial linear features are co-aligned. Unfortunately, most railways, roads, power lines and pipelines have their own easement and are managed by different agencies or companies, complicating planning and mitigation

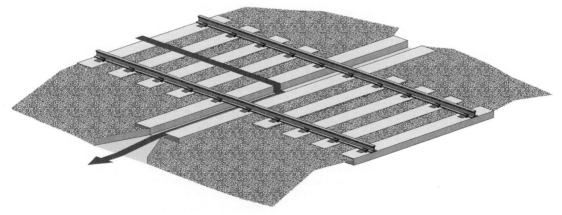

Figure 26.5 Diagram of simple crossing structure under railway tracks for turtles and other small animals. The space between the railway sleepers allows animals to cross fully from one side to the other, as well as allow animals, such as turtles, that may be trapped between the tracks to drop into the structure and escape. Source: Reproduced with permission of Scott Watson.

(Chapter 27). For this reason, detailed planning and cooperation are urgently required to integrate the existing and future infrastructure and mitigation within these corridors. Consolidated regional networks with representation from relevant government agencies and infrastructure companies are essential for long-term planning and cost-effective regional mitigation.

CONCLUSIONS

The emerging field of rail ecology is of global importance because railways are an effective strategy to reduce CO_2 emissions from transportation and reduce traffic congestion by transporting goods and people more efficiently than trucks and cars. Much less is known about railway impacts on wildlife compared to roads and more research is needed. Impacts of HSR are of particular importance because this type of railway is expanding rapidly. Impacts will likely differ along HSR compared to freight rail due to differences in noise levels, speeds and the common practice of fencing high-speed lines. Future work needs to move beyond the primary focus of impacts on moose and bear to assessing the overall effects on biodiversity and ecosystems. In addition, the cumulative impacts and interactions with other linear infrastructure require attention. In the short term, there is an urgent need for increased dialogue between wildlife and railway managers and integration of transportation planning to protect wildlife.

FURTHER READING

Child (1983): One of the earliest papers describing conflict between railways and wildlife and still relevant today.
van der Grift (1999): A detailed review of the impacts of railways and trains on mammals.
Kaczensky et al. (2003): A straightforward analytical approach to the data commonly available and revels the impacts from railways can be surprisingly high.

REFERENCES

American Association of State Highway and Transportation Officials (AASHTO). 2007. Transportation invest in our future. A new vision for the 21st century. AASHTO, Washington, DC, 93pp.

Andersen, R., B. Wiseth, P.H. Pedersen and V. Jaren. 1991. Moose-train collisions: effects of environmental conditions. Alces **27**:79–84.

Andreassen, H.P., H. Gundersen and T. Storaas. 2005. The effect of scent-marking, forest clearing, and supplemental feeding on moose-train collisions. Journal of Wildlife Management **69**:1125–1132.

Associated Press. 2011. Trains kill more than 800 antelope and deer on Montana tracks this winter. The Missoulian. Available from http://missoulian.com/news/state-and-regional/article_3d955d6a-4831-11e0-84f6-001cc4c03286.html (accessed 11 December 2013).

Carpenter, T.G. and M. Lewis. 1994. The environmental impact of railways. John Wiley & Sons, Ltd, Chichester, UK.

Child, K. 1983. Railways and moose in the central interior of BC: a recurrent management problem. Alces **19**:118–135.

Dorsey, B. 2011. Factors affecting bear and ungulate mortalities along the Canadian Pacific Railroad through Banff and Yoho National Parks. Master's thesis, Montana State University. Available from http://scholarworks.montana.edu/xmlui/handle/1/1190 (accessed 11 December 2013).

Forman, R.T.T. and R.D. Deblinger. 2000. The ecological road effect zone of a Massachusetts (USA) suburban highway. Conservation Biology **14**:36–46

Forman, R.T.T., D. Sperling, J.A. Bissonette, et al. 2003. Road ecology: science and solutions. Island Press, Washington, DC.

Huber, D., J. Kusak and A. Frkovic. 1998. Traffic kills of brown bears in Gorski Kotar, Croatia. Ursus **10**:167–171.

Ito, T.Y., N. Miura, B. Lhagvasuren, D. Enkhbileg, S. Takatsuki, A. Tsunekawa and Z. Jiang. 2005. Preliminary evidence of a barrier effect of a railroad on the migration of Mongolian gazelles. Conservation Biology **19**:945–948.

Jaren, V., R. Andersen, M. Ulleberg, P. Pedersen and B. Wiseth. 1991. Moose-train collisions: the effects of vegetation removal with a cost-benefit analysis. Alces **27**:93–99.

Kaczensky, P., F. Knauer, B. Krze, M. Jonozovic, M. Adamic and H. Gossow. 2003. The impact of high speed, high volume traffic axes on brown bears in Slovenia. Biological Conservation **111**:191–204. doi:10.1016/S0006-3207(02)00273-2.

Kolb, H.H. 1984. Factors affecting the movements of dog foxes in Edinburgh. The Journal of Applied Ecology **21**:161–173.

Kornilev, Y., S. Price and M. Dorcas. 2006. Between a rock and a hard place: responses of eastern box turtles (Terrapene carolina) when trapped between railroad tracks. Herpetological Review **37**:145–148.

Modafferi, R.D. 1991. Train moose-kill in Alaska: characteristics and relationship with snowpack depth and moose distribution in lower Susitna Valley. Alces **27**:193–207.

Pelletier, S.K., L. Carlson, D. Nein and R.D. Roy. 2006. Railroad crossing structures for spotted turtles: Massachusetts Bay Transportation Authority – Greenbush rail line wildlife crossing demonstration project. In C.L. Irwin, P. Garrett and K.P. McDermott (eds). Proceedings of the 2005 International Conference on Ecology and Transportation, pp. 414–425. Center for Transportation and the Environment, North Carolina State University, Raleigh, NC.

Rea, R., K. Child and D. Aitken. 2010. YOUTUBE™ insights into moose-train interactions. Alces **46**:183–187.

Reh, W. and A. Seitz. 1990. The influence of land use on the genetic structure of populations of the common frog *Rana temporaria*. Biological Conservation **54**:239–249.

Schwartz, C. and B. Bartley. 1991. Moose conference workshop, Anchorage, May 17 reducing incidental moose mortality: considerations for management. Alces **27**:93–99.

Seiler, A. (ed). 2011. Klövviltolyckor på järnväg: kunskapsläge, problemanalys och åtgärdsförslag (in Swedish). Trafikverket rapport 2011:048. In English – Ungulate-train collisions in Sweden – literature review, GIS-analyses of mortality data and train drivers' experiences.

Seiler, A. and J.O. Helldin. 2005. Mortality in wildlife due to transportation. In J. Davenport and J.L. Davenport (eds). The ecology of transportation: managing mobility for the environment. Springer, Dordrecht, the Netherlands.

Skogland, T. 1986. Movements of tagged and radio-instrumented wild reindeer in relation to habitat alteration in the Snøhetta region, Norway. Rangifer **1**:267–272.

International Union of Railways (UIC). 2013. High speed lines in the world. UIC, Paris, France. Available from http://www.uic.org/spip.php?article573 (accessed 11 December 2013).

van der Grift, E. 1999. Mammals and railroads: impacts and management implications. Lutra **42**:77–98.

Vos, C.C., A.G. Antonisse-De Jong, P.W. Goedhart and M.J. Smulders. 2001. Genetic similarity as a measure for connectivity between fragmented populations of the moor frog (*Rana arvalis*). Heredity **86**:598–608.

Waller, J.S. and C. Servheen. 2005. Effects of transportation infrastructure on grizzly bears in northwestern Montana. Journal of Wildlife Management **69**:985–1000.

Waterman, E., I. Tulp, R. Reijnen, K. Krijgsveld and C. Braak. 2002. Disturbance of meadow birds by railway noise in The Netherlands. Geluid **1**:2–3.

Welbes, M.J. 2011. Rail modernization study: report to Congress. Diane Publishing, Darby, PA.

Wells, P., J.G. Woods, G. Bridgewater and H. Morrison. 1999. Wildlife mortalities on railways; monitoring methods and mitigation strategies. In G. Evink, P. Garrett and D. Zeigler (eds). Proceedings of the Third International Conference on Wildlife Ecology and Transportation, pp. 237–246. Florida Department of Transportation, Tallahassee, FL.

Chapter 27

IMPACTS OF UTILITY AND OTHER INDUSTRIAL LINEAR CORRIDORS ON WILDLIFE

A. David M. Latham[1] and Stan Boutin[2]

[1]Wildlife Ecology and Management Team, Manaaki Whenua Landcare Research, Lincoln, New Zealand
[2]Department of Biological Sciences, University of Alberta, Edmonton, Alberta, Canada

SUMMARY

Roads are only one type of man-made linear corridors in the landscape. Utility easements and other industrial linear corridors (hereafter ILC) provide access and energy and support economic growth. They are pervasive in many regions and can have wide-ranging environmental and biological effects. Power lines and pipelines (i.e. ILC associated with utility easements) usually occur at low densities compared to roads. However, they can be as wide as roads, are usually kept clear of trees and woody shrubs and can cover long distances across a wide range of habitats. Other ILC, such as seismic exploration lines, can occur at densities similar to or greater than roads.

27.1 Roads are only one component of the man-made network of linear corridors in a landscape.

27.2 Industrial linear corridors can have environmental and biological effects that are often complex and difficult to predict.

27.3 Some landscapes and species are more vulnerable than others to the effects of ILC.

27.4 Effective mitigation involves strategies that reduce the number and duration of ILC in the landscape and the impacts caused by associated structural elements.

27.5 Much remains unknown about the effect of ILC on species and ecosystems, particularly in remote wilderness areas and developing countries.

The demand for energy is increasing at an alarming pace throughout the world. Industrial linear corridors are being built to meet this demand but generally with scant knowledge of possible impacts to the environment and inadequate mitigation measures. Three types of effects need to be considered prior to the construction of ILC: impacts to the physical environment, direct biological effects and indirect biological effects. We suggest a large-scale, integrated approach to land management and mitigation for companies involved in natural resource extraction, the provision of energy and ILC construction.

Handbook of Road Ecology, First Edition. Edited by Rodney van der Ree, Daniel J. Smith and Clara Grilo.
© 2015 John Wiley & Sons, Ltd. Published 2015 by John Wiley & Sons, Ltd.
Companion website: www.wiley.com\go\vanderree\roadecology

INTRODUCTION

In some regions of the world, the network of non-road ILC that criss-cross the landscape can occur at higher densities than roads alone (Lee & Boutin 2006). This network includes utility easements (e.g. power lines, telephone lines and pipelines) and other ILC (e.g. seismic exploration lines) that are usually associated with energy exploration, extraction and provision. Paved and unpaved roads, railways and recreational trails for walking, cycling and horse riding are not ILC. There are often complex legal rights associated with utility and other ILC; however, this chapter is primarily concerned with the impacts of the linear footprint and infrastructure that remain following their construction.

Similar to roads, the footprint of ILC usually intensifies as industrial activity and urbanisation increase, with many ILC becoming 'permanent' corridors/barriers across the landscape. Although the effects of ILC can be similar to roads, railways and recreational trails, there are also unique differences. Often, the biological impacts, particularly indirect effects, associated with ILC are not considered during environmental impact assessments (EIA) or the planning process. Because of the economic urgency associated with industrial expansion and energy provision, conservation managers are often forced to play catch-up once developments have been constructed in order to mitigate their negative impacts.

The aims of this chapter are to (i) briefly review some of the literature on the impacts of ILC; (ii) highlight the complex biological issues that need to be considered by planning and design teams; and (iii) discuss ILC mitigation from an environmental and biological perspective. We use examples from systems impacted by various ILC from around the world but especially focus on impacts in the boreal forest in northern North America as a case study (Textbox 27.1).

LESSONS

27.1   Roads are only one component of the man-made network of linear corridors in a landscape

Although roads or trails often represent the first incursion into a new region, industry often develops a far more extensive network of ILC. Roads differ from ILC in that they usually have a hard compacted and often paved surface that is buffered by a grassy verge. The direct and indirect impacts of roads and traffic are well studied and documented. However, the negative impacts of ILC, although arguably less obvious or well studied, can also be significant, particularly where they are abundant in the landscape.

Power line easements can be 100 m or more in width and are constructed to provide industry and the public with electricity. They are particularly common in highly urbanised areas (where they often parallel roads), but they can also span long distances across relatively pristine habitat (Fig. 27.1) to provide electricity from power stations to large urban centres and remote communities. Both the linear clearing that results from the construction of the utility and the structural elements (e.g. pylons, cables and insulators) of the power line can negatively impact species of plants and/or animals (Bevanger 1994; Goosem & Marsh 1997; Guil et al. 2011).

The most extensive networks of ILC in many parts of the world are associated with the energy sector. Physically, oil and gas development starts with seismic exploration to identify and map oil and gas deposits suitable for extraction. Conventional seismic lines are 5–8 m in width and are cleared of vegetation using a bulldozer (Gibson & Rice 2003). The so-called 'low-impact' seismic lines of 2–3 m in width are being adopted in some areas because they are cut with a chainsaw or mulcher and can thus meander more easily around valuable stands of vegetation and reduce the loss of merchantable timber (Gibson & Rice 2003). Once exploration has been completed, vegetation on seismic lines is usually allowed to regenerate, although the ILC can remain obvious for many decades after construction (Lee & Boutin 2006; Jorgenson et al. 2010). Seismic lines are often the most abundant ILC in a landscape, for example, they occur at an average density of 1.5 km per km^2 in northern Alberta, Canada (Fig. 27.2) (cf. a road network of 1 km per km^2 in the developed agricultural zone of the province; Lee & Boutin 2006). Following exploration, unpaved roads are built to enable drilling of wells (on a pad of ~1 ha in size) for the extraction of oil or gas. Above- or below-ground pipelines are then constructed to transport oil or gas to processing plants. Pipeline clearings tend to be approximately 30 m in width, and their maintenance requires prevention of tree and woody shrub regrowth within the right of way.

Roads and ILC are often inextricably linked. Power lines, pipelines and roads often occur immediately adjacent to each other in the same linear clearing. This is positive in that it reduces overall fragmentation of the landscape, but it can exacerbate barrier effects if the cleared corridor becomes increasingly wide to accommodate these features. Similarly, pipelines and seismic lines are often graded and used as tracks to provide

Figure 27.1 A power line spanning savannah in Kruger National Park, South Africa. Both the linear clearing and the power line have been shown to negatively impact some species of plants and animals. Note the game trail running along the left side of the linear clearing and the prevention of tree and woody shrub regrowth on it. Source: Photograph by and reproduced with permission of David M. Forsyth.

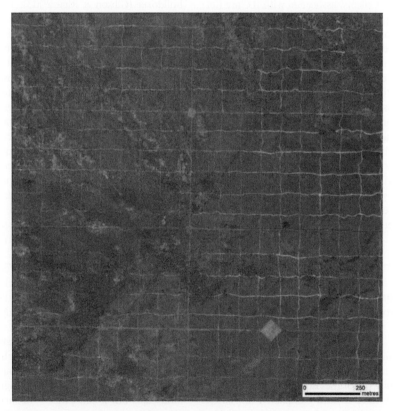

Figure 27.2 A satellite view showing an intensive network of seismic exploration lines in the boreal forest, Alberta, Canada. The diamond-shaped clearing in the bottom right of the map is a well pad (~1 ha in size) that is connected to a pipeline right of way. Source: Map by David Latham.

vehicles with access to remote areas. Subsequently, they are often further developed and become unpaved or paved roads.

27.2   Industrial linear corridors can have environmental and biological effects that are often complex and difficult to predict

Industrial linear corridors can change the abundance or distribution of a species because of changes in available habitat (e.g. available light, temperature, moisture or nesting sites) caused during ILC construction. For example, the composition of small mammal communities can change due to changes in vegetation structure on power lines, and the ILC can act as a barrier to animal movement (e.g. Goosem & Marsh 1997). These are 'direct biological effects' caused by ILC on a species or suite of species (Murcia 1995). Changes driven by the construction of the ILC might also affect the dynamics of interactions between different species on or adjacent to an ILC. For example, the habitat on an ILC may be preferred for nesting by one bird species, causing it to become more common and to attract more nest predators to the area. A second bird species, which is not directly affected by the habitat on the ILC, may decline in numbers because it is more vulnerable to being killed by the increased number of nest predators hunting in the area. These are termed 'indirect biological effects' (Murcia 1995). Indirect biological effects can be particularly complex and difficult to predict, yet they can have major ramifications for conservation and management. EIA must consider and investigate both direct and indirect biological effects (Chapter 5), as well as the effects of ILC on the physical environment (e.g. soil moisture, water flow, light intensity and air temperature).

27.3   Some landscapes and species are more vulnerable than others to the effects of ILC

Natural and man-made edge habitat can be beneficial for some edge-dwelling species of plants and animals and can result in high species diversity near the edge (Murcia 1995). White-tailed deer, for example, have become hyper-abundant across much of North America following an increase in man-made edge habitat because of improved forage and browse on these features (Alverson et al. 1988). However, it is also recognised that edges, particularly man-made edges along ILC, can have detrimental effects for many species and

result in complex changes in plant and animal communities (Murcia 1995). This is particularly the case for species that require large undisturbed areas, because these areas are effectively reduced in size by the presence of ILC. For example, interior forest birds like ovenbirds tend not to incorporate man-made edges into their territories, but rather use them as territorial boundaries (Bayne et al. 2005). Edges can also disrupt movement patterns and decrease gene flow and genetic diversity for interior species (Yahner 1988).

Some species of animals are vulnerable to noise disturbances associated with the construction or use of ILC (Bayne et al. 2008; Chapter 19). For example, elephants and great apes were more likely to avoid human activity and noise associated with oil exploration in Gabon than were small antelopes or monkeys (Rabanal et al. 2010). Further, some animal species are more likely to be targeted by hunters or poachers who use ILC to enter previously difficult-to-access areas (Laurance et al. 2009; Chapters 2, 51 and 56).

Tropical rainforests are highly susceptible to the impacts of ILC (Laurance et al. 2009; Chapter 49). Biologically, rainforests are characterised by a stable microclimate that is humid, dark and architecturally complex. They contain high biodiversity, and many species within rainforests are interior specialists that avoid forest edges, including ILC. Much of the world's rainforests occur in developing nations which are eagerly pursuing natural resource exploitation for economic growth, arguably with minimal forethought or consideration regarding environmental impacts or appropriate planning (Laurance et al. 2006; Chapter 51).

Slow-growing tundra vegetation and underlying permafrost soils in the Arctic are very sensitive to industrial disturbance (Jorgenson et al. 2010). Scars from oil exploration in Alaska in the early 1940s are still visible today due to the slow regeneration of vegetation. Similarly, some arctic and boreal forest animal species are believed to be detrimentally impacted by ILC, either directly, indirectly or both (e.g. Textbox 27.1).

27.4   Effective mitigation involves strategies that reduce the number and duration of ILC in the landscape and the impacts caused by associated structural elements

ILC rarely exist in a landscape as single features. Usually, there are multiple man-made corridors fulfilling a number of purposes or, in the case of seismic exploration lines, a dense network of ILC (i.e. Fig. 27.2). Mitigation measures may involve reducing local-scale impacts of existing ILC, better planning of new ILC, and co-planning and coordinating the activities among companies to reduce impacts at a landscape scale (Schneider 2002; Laurance et al. 2009).

The longevity of the impacts caused by an ILC is a function of the intensity of disturbance during its creation and use and the degree to which it is actively reclaimed after it becomes inoperative. It may prove difficult to fully mitigate the negative effects caused by active ILC. For example, the structural components of active power lines and pipelines, such as pylons, overhead wires and aboveground pipelines, remain necessary despite their potential to negatively impact wildlife. Similarly, active power lines and pipelines will typically need to be kept clear of tree and woody shrub species to provide maintenance access and to prevent damage and disturbance to the system facilities. However, a compromise between the management of vegetation on power lines or pipelines and outcomes for biodiversity may be possible. For example, shrubby vegetation that had regenerated to an intermediate amount on power line easements in Victoria, Australia, provided good habitat for some native species but could also be managed as an acceptable fire risk (Clarke et al.

2006). Similarly, mitigation measures that improve the structural design of power lines have been developed to minimise the number of large birds, such as birds of prey, that die from collisions and electrocution (Bevanger 1994). If an ILC has yet to be constructed, many of the negative impacts of ILC may be reduced by avoiding routes that intersect critical foraging and breeding habitat for wildlife (Guil et al. 2011).

The aim of reclamation of existing ILC that are no longer used should be to return the linear clearing to its 'original' native vegetated state (Chapter 3). This should eliminate or reduce the impact of invasive species and/or prevent it from being used as a corridor by humans (for hunting or recreation) or opportunistic predators. Where the humus and surface soil have not been badly damaged during construction, the vegetation will likely regenerate naturally if left undisturbed by humans. If the surface layer has been badly damaged, it is probable that early colonising plant species (often non-natives) will dominate the clearing. This could potentially retard regeneration of native species for an extended and unknown period of time (Fig. 27.3). In such cases, extensive seeding or planting with native species might be needed. Human access along ILC has been successfully reduced or eliminated, for example, by using a bulldozer to push logs, branches or other natural materials across the corridor to prevent

Figure 27.3 A 6 m wide conventional seismic exploration line transitioning from peat bog to upland forest in Alberta, Canada. Although this line is about 7 years old, regeneration of vegetation on the line has been limited. Seismic lines and other man-made corridors allow hunters, on foot or all-terrain vehicle, to access previously remote areas. Source: Photograph by and reproduced by permission of Pierre-Olivier Côte.

Textbox 27.1 The effects of industrial linear corridors on the boreal forest of North America.

Forestry and oil and gas activity has increased substantially in the boreal forest in recent decades (Schneider 2002). Logged forestry blocks, roads, industrial linear corridors (ILC) and infrastructure associated with industry are now common throughout large areas of the boreal forest. For example, in 2002, there was an average of 1.8 km per km^2 of combined roads and ILC or over 1.5 million km of seismic exploration lines in the forested area of Alberta, Canada (Schneider 2002). Forest loss associated with seismic exploration rivalled that caused by the logging industry in the same region and time frame (Schneider 2002). Although the total footprint associated with these features is only about 5% or less of the total forested area of Alberta (e.g. Latham et al. 2011), the amount of habitat that is functionally lost to animals can be much higher (e.g. 48% for woodland caribou; Dyer et al. 2001).

ILC often become persistent features in the landscape, for example, more than 60% of conventional seismic lines in Alberta remain obvious 35 years after their construction (Lee & Boutin 2006). Increasing numbers of off-road vehicle enthusiasts access ILC and further prevent them from regenerating. Subsequently, harassment to some species of animals has increased from vehicle noise and an increase in hunting pressure in previously remote areas. ILC in the boreal forest have also been linked to exotic plant and animal invasions. For example, aggressive weed species were often used to revegetate pipelines (Schneider 2002), and because

these were often preferred seasonal foods for some herbivores (e.g. white-tailed deer), they contributed to changes in the distribution of some animal species and the spread of others into the boreal forest (e.g. coyotes; Latham et al. 2013).

ILC can disrupt animal migration and movement patterns (Galpern et al. 2012). An important example in North America involves threatened woodland caribou, which avoid these features and perceive roads as barriers to movement (Dyer et al. 2001, 2002). Evidence is also mounting to demonstrate an indirect biological effect of ILC on caribou through increased rates of predation by wolves (Fig. 27.4; Festa-Bianchet et al. 2011). Wolves are believed to benefit from using ILC because they are able to travel more efficiently on lines than off them and thus encounter prey (including caribou) more often (James & Stuart-Smith 2000). Latham et al. (2011) found that wolves most often used ILC to move into and around caribou habitat in summer, which was also the time of year when most caribou were killed by wolves. A greater risk of predation by wolves as a result of increased hunting efficiency on or near ILC explains why caribou avoid otherwise suitable habitat in close proximity to ILC. Thus, ILC are believed to be partially responsible for current declines in woodland caribou numbers (Latham et al. 2011). This example also highlights the inherent complexities facing biologists tasked with conducting EIA for proposed ILC.

Figure 27.4 Woodland caribou on a power line clearing in the boreal forest, Alberta, Canada. Power line clearings and other ILC are dangerous places for caribou because wolves, their main predator, use these features to move around and often kill prey (including caribou) on or near them. Source: Photograph by David Latham.

motorized vehicle access. This method appears to be less successful at excluding species such as wolves, which can benefit from increased hunting efficiency until vegetation has regenerated on the ILC (James & Stuart-Smith 2000). A more insurmountable obstacle to mitigation efforts is the overabundance of these features in some areas, including many ILC that need to be maintained because their use is still required by industry. Thus, even if 50% of abandoned features were successfully reclaimed, which may be adequate to mitigate some negative impacts, hunters, wolves or invasive species could simply use (or reinvade from) the next, relatively close feature in the landscape. Compared to conventional seismic lines, the use of narrow, meandering, low-impact seismic lines has shown promise in reducing the total amount of forest lost during oil and gas exploration (Gibson & Rice 2003). Further, vegetation regenerates relatively quickly on them, and they are used less by humans (and possibly opportunistic predators like wolves).

ILC are clearly expensive to construct. The activities of multiple resource extraction companies operating concurrently in the same area have been additive, and a lack of communication between them has led to scenarios whereby power lines, pipelines, unpaved roads and seismic lines have been placed within metres of each other. Co-planning among companies operating in a region can reduce the total footprint required by substantial amounts (15–30%), thereby reducing costs and biological impacts (Schneider 2002). This requires that the traditional lack of communication within and among companies, resource sectors and government agencies be overcome. In general, there have been limited but laudable examples (such as seismic data sharing agreements) of cooperation between companies to coordinate their efforts and minimise the number of ILC that are constructed in an area (Tillman 1976; Gibson & Rice 2003).

27.5 Much remains unknown about the effect of ILC on species and ecosystems, particularly in remote wilderness areas and developing countries

The ever increasing demand by human society for natural resources has led to an unprecedented rate of exploration, development and extraction. The frenetic pace at which this is occurring is a global problem, and one that is unlikely to decline any time soon. Vulnerable ecosystems, such as the rainforest in the Amazon basin, are undergoing rapid oil and

gas exploration and extraction which threaten biodiversity and indigenous peoples on a massive scale (Finer et al. 2008; Chapter 51). Demand for natural resources and the desire for economic growth are likely to outpace the development of effective mitigation strategies. If we are to prevent this, resource extraction companies, environmental agencies and permitting agencies need to further develop operating and management practices that avoid and minimise the footprint and impacts associated with their activity. Similarly, biologists need to consider biological uncertainty relating to the effects of ILC and how best to deal with it, for example, recommend precautionary risk avoidance approaches where uncertain scientific knowledge exists but where impacts to wildlife could be high (Chapter 5).

Although there is much still to learn about the effects of ILC on species, communities and ecosystems, there has been considerable research and advances in our understanding of these factors, dating back to at least 1976 when the First National Symposium on Environmental Concerns in Rights-of-Way Management was held (Tillman 1976). Since then, there have been a number of symposia on the environmental impacts of ILC (e.g. http://www.rights-of-way.org/), and this and similar professional groups provide an ideal forum to continue to raise awareness of research areas, management concerns and new methods of mitigating the impacts of ILC.

CONCLUSIONS

Industrial developments are increasingly becoming a global environmental and biological threat. ILC should be constructed so that they run through less biologically sensitive areas, and the total number of lines constructed should be minimised through integrated planning. Although there are difficulties associated with this approach, it is easier than trying to mitigate the myriad impacts once ILC have been constructed. Small advances have been made in this direction, but greater impetus is needed. A complementary and attainable method of preventing/minimising the impacts of ILC is to protect ecologically sensitive areas of high conservation value from industrial activity and to construct ILC to avoid these areas. In the meantime, biologists must endeavour to persevere with the unenviable task of evaluating and understanding the complex array of impacts associated with new developments and attempting to mitigate these accordingly.

ACKNOWLEDGEMENTS

Thanks to Cecilia Latham and Bruce Warburton for insightful comments on this chapter.

FURTHER READING

Bevanger (1998): A review identifying and describing the characteristics that make some species of birds more susceptible to collisions with power lines or electrocution than others. This information is critical for designing species- and/or site-specific mitigation measures (also see Bevanger (1994)).

Gibson and Rice (2003): Provides a perspective from industry on how best to promote environmental responsibility during exploration for oil and gas reserves.

Schneider et al. (2003): This case study from Alberta, Canada, uses a modelling approach to weigh current management options from the point of view of their long-term effects of industry on the boreal forest. The aim is to show that a suite of 'best practices' can provide some balance between conservation and economic objectives.

REFERENCES

Alverson, W. S., D. M. Waller and S. L. Solheim. 1988. Forests too deer: edge effects in northern Wisconsin. Conservation Biology **2**:348–358.

Bayne, E. M., S. L. Van Wilgenburg, S. Boutin and K. A. Hobson. 2005. Modeling and field-testing of ovenbird (*Seiurus aurocapillus*) responses to boreal forest dissection by energy sector development at multiple spatial scales. Landscape Ecology **20**:203–216.

Bayne, E. M., L. Habib and S. Boutin. 2008. Impacts of chronic anthropogenic noise from energy-sector activity on abundance of songbirds in the boreal forest. Conservation Biology **22**:1186–1193.

Bevanger, K. 1994. Bird interactions with utility structures: collision and electrocution, causes and mitigating measures. Ibis **136**:412–425.

Bevanger, K. 1998. Biological and conservation aspects of bird mortality caused by electricity power lines: a review. Biological Conservation **86**:67–76.

Clarke, D. J., K. A. Pearce and J. G. White. 2006. Powerline corridors: degraded ecosystems or wildlife havens? Wildlife Research **33**:615–626.

Dyer, S. J., J. P. O'Neill, S. M. Wasel and S. Boutin. 2001. Avoidance of industrial development by woodland caribou. Journal of Wildlife Management **65**:531–542.

Dyer, S. J., J. P. O'Neill, S. M. Wasel and S. Boutin. 2002. Quantifying barrier effects of roads and seismic lines on movements of female woodland caribou in northeastern Alberta. Canadian Journal of Zoology **80**:839–845.

Festa-Bianchet, M., J. C. Ray, S. Boutin, S. D. Côte and A. Gunn. 2011. Conservation of caribou (*Rangifer tarandus*) in Canada: an uncertain future. Canadian Journal of Zoology **89**:419–434.

Finer, M., C. N. Jenkins, S. L. Pimm, B. Keane and C. Ross. 2008. Oil and gas projects in the western Amazon: threats to wilderness, biodiversity, and indigenous peoples. PLoS One **3**:e2932.

Galpern, P., M. Manseau and P. Wilson. 2012. Grains of connectivity: analysis at multiple spatial scales in landscape genetics. Molecular Ecology **21**:3996–4009.

Gibson, D. and S. Rice. 2003. Promoting environmental responsibility in seismic operations. Oilfield Review **15**:10–21.

Goosem, M. and H. Marsh. 1997. Fragmentation of a small-mammal community by a powerline corridor through tropical rainforest. Wildlife Research **24**:613–629.

Guil, F., M. Fernández-Olalla, R. Moreno-Opo, I. Mosqueda, M. E. Gómez, A. Aranda, Á. Arredondo, J. Guzmán, J. Oria, L. M. González and A. Margalida. 2011. Minimising mortality in endangered raptors due to power lines: the importance of spatial aggregation to optimize the application of mitigation measures. PLoS One **6**:e28212.

James, A. R. C. and A. K. Stuart-Smith. 2000. Distribution of caribou and wolves in relation to linear corridors. Journal of Wildlife Management **64**:154–159.

Jorgenson, J. C., J. M. Ver Hoef and M. T. Jorgenson. 2010. Long-term recovery patterns of arctic tundra after winter seismic exploration. Ecological Applications **20**:205–221.

Latham, A. D. M., M. C. Latham, M. S. Boyce and S. Boutin. 2011. Movement responses by wolves to industrial linear features and their effect on woodland caribou in northeastern Alberta. Ecological Applications **21**:2854–2865.

Latham, A. D. M., M. C. Latham, M. S. Boyce and S. Boutin. 2013. Spatial relationships of sympatric wolves (*Canis lupus*) and coyotes (*C. latrans*) with woodland caribou (*Rangifer tarandus caribou*) during the calving season in a human-modified boreal landscape. Wildlife Research **40**:250–260.

Laurance, W. F., B. M. Croes, L. Tchignoumba, S. A. Lahm, A. Alonso, M. E. Lee, P. Campbell and C. Ondzeano. 2006. Impacts of roads and hunting on central African rainforest mammals. Conservation Biology **20**:1251–1261.

Laurance, W. F., M. Goosem and S. G. W. Laurance. 2009. Impacts of roads and linear clearings on tropical forests. Trends in Ecology and Evolution **24**:659–669.

Lee, P. and S. Boutin. 2006. Persistence and developmental transition of wide seismic lines in the western Boreal Plains of Canada. Journal of Environmental Management **78**:240–250.

Murcia, C. 1995. Edge effects in fragmented forests: implications for conservation. Trends in Ecology and Evolution **10**:58–62.

Rabanal, L. I., H. S. Kuehl, R. Mundry, M. M. Robbins and C. Boesch. 2010. Oil prospecting and its impact on large rainforest mammals in Loango National Park, Gabon. Biological Conservation **143**:1017–1024.

Schneider, R. R. 2002. Alternative futures: Alberta's boreal forest at the crossroads. Federation of Alberta Naturalists, Edmonton, Alberta.

Schneider, R. R., J. B. Stelfox, S. Boutin and S. Wasel. 2003. Managing the cumulative impacts of land uses in the Western Canadian Sedimentary Basin: a modeling approach. Conservation Ecology **7**:8. http://www.consecol.org/vol7/iss1/art8/ (accessed 26 September 2014).

Tillman, R., editor. 1976. Proceedings of the First National Symposium on Environmental Concerns in Rights-of-Way Management. Mississippi State University, Mississippi State, MS.

Yahner, R. H. 1988. Changes in wildlife communities near edges. Conservation Biology **2**:333–339.

Chapter 28

THE IMPACTS OF ROADS AND TRAFFIC ON TERRESTRIAL ANIMAL POPULATIONS

Trina Rytwinski and Lenore Fahrig

Geomatics and Landscape Ecology Research Laboratory, Department of Biology, Carleton University, Ottawa, Ontario, Canada

SUMMARY

There is growing evidence that roads and traffic reduce populations of many species and efforts to mitigate road effects are now common. To maximise understanding of road impacts and for conservation of particular species, we need to know how roads affect the viability of a group of individuals of the species rather than a single individual. Roads and traffic affect wildlife populations in three major ways, by (i) increasing mortality, (ii) decreasing habitat amount and quality and (iii) fragmenting populations into smaller sub-populations which are more vulnerable to local extinction. To ensure mitigation is effective, we need to identify the species most affected, and the cause(s) of the effects, so that appropriate mitigation can be tailored to those species.

28.1 Mammals: Larger, more mobile species with lower reproductive rates are more susceptible to road mortality, and species that avoid roads from a distance due to traffic-related disturbance are susceptible to habitat fragmentation, loss and degradation.

28.2 Birds: Species that have large territories and possibly species that are low flying, ground dwelling and/or heavy relative to their wing size are more susceptible to road mortality.

28.3 Amphibians and reptiles: All species, regardless of life history traits, are prone to negative road effects as they are particularly susceptible to road mortality and habitat fragmentation by roads.

28.4 A species response to roads and traffic will vary depending on its conservation status, geographical location, habitat preferences, road type and/or traffic volume.

28.5 There are still many species for which we do not know the population-level effects of roads. To ensure mitigation will be effective for as many species as possible, research is needed on the effects of roads on a broader range of species.

This chapter provides a high-level overview of the population-level effects of roads on animals using the available data from 75 studies. For more detailed information on specific species groups, please refer to Chapters 29–45.

Handbook of Road Ecology, First Edition. Edited by Rodney van der Ree, Daniel J. Smith and Clara Grilo.
© 2015 John Wiley & Sons, Ltd. Published 2015 by John Wiley & Sons, Ltd.
Companion website: www.wiley.com\go\vanderree\roadecology

INTRODUCTION

There are many studies on the effects of roads on animal movement and mortality, neither of which allows for strong inference about the impacts of roads on population persistence; for example, it is possible that increased reproduction rates counterbalance losses caused by road mortality (Roedenbeck et al. 2007). For conservation of a particular species, we need to know how roads affect the viability of a group of individuals (i.e. the population) rather than a single individual. The main question is therefore: can roads and/or traffic reduce or even eliminate a population, and how? Roads and traffic affect wildlife populations in three major ways, by (i) increasing mortality, (ii) decreasing habitat amount and quality and (iii) fragmenting populations into smaller sub-populations that each are more vulnerable to local extinction than a large population.

The vulnerability of a species to roads and/or traffic is influenced by its ecological traits and behavioural responses (Table 28.1). Important ecological traits are its reproductive rate (a higher reproductive rate allows populations to recover from road mortality) and its mobility (a more mobile species will encounter roads more often than species that are more sedentary). Four types of behaviour influence whether roads or traffic affects animal populations: (i) avoidance of the road surface; (ii) avoidance of traffic disturbance (noise, lights, chemical emissions); (iii) vehicle avoidance (the ability to move out of the path of an oncoming vehicle);

and (iv) attraction to roads (Fig. 28.1). Species that avoid the road surface are less likely to be killed on roads because they rarely attempt to cross it, but they may have trouble accessing important habitats or resources on the other side of the road. Similarly, animals that avoid traffic disturbance are less susceptible to road mortality, but their populations may be fragmented into smaller, partially isolated populations that may be more vulnerable to extinction. Avoidance of traffic disturbance also reduces the amount of habitat since the area near roads becomes unsuitable (i.e. road effect zone; Lesson 1.2). Species that can move out of the path of an oncoming vehicle should be less susceptible to road mortality and may be able to cross the road when traffic volumes are not too high. Lastly, some species can be attracted to roads for a resource such as carrion (e.g. some birds) and nesting sites (e.g. some turtles) or to bask (e.g. some snakes) which can make them vulnerable to road mortality (Chapters 32 and 33).

The insights in this chapter are based on a formal review of 75 studies published during 1979–early 2011 that measured the relationship between roads and/or traffic and population size of a species. Studies were predominantly in North America (49 studies) or Europe (19), but a few were from Oceania (3), Africa (2), and Asia (2). For each study, the raw data were either provided in the paper (e.g. from graphs or figures) or they were provided directly by the authors. To determine whether a species was negatively or positively affected or unaffected (neutral effect) by roads,

Table 28.1 Characteristics that can affect a species vulnerability to the major impacts of roads.

Characteristics that affect a species vulnerability to road effects	Effects of roads and/or traffic		
	Road mortality	Habitat loss/reduced habitat quality	Habitat fragmentation/reduced connectivity
Low reproductive rate	x	x	x
Young age at sexual maturity	x	x	x
Long generation time (lifespan)	x	x	x
High intrinsic mobility	x		
Large area requirements/low natural density	x	x	x
Large body size	x	x	x
Multiple resource needs	x		x
Attraction to roads	x		
Road surface avoidance			x
Traffic disturbance avoidance		x	x
No road or traffic disturbance avoidance	x		

Source: Adapted from Forman et al. (2003).

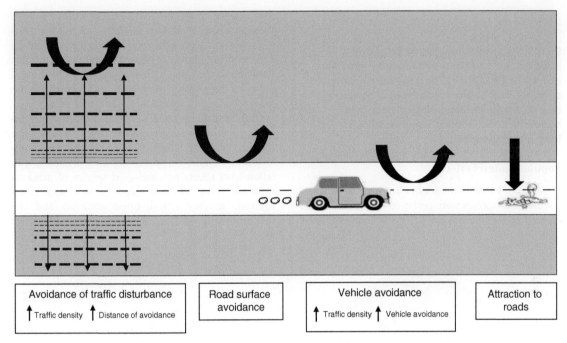

Avoidance of traffic disturbance	Road surface avoidance	Vehicle avoidance	Attraction to roads
↑ Traffic density ↑ Distance of avoidance		↑ Traffic density ↑ Vehicle avoidance	

Figure 28.1 Illustration of species behavioural responses to roads and traffic. 'Avoidance of traffic disturbance' is avoidance of roads from a distance due to traffic disturbance (e.g. lights, noise, chemical emissions). As traffic density increases, the distance at which a species avoids the road (represented as black dashed lines) increases resulting in more habitat effectively lost to a species (strength of effect represented by thickness of dashed lines). 'Road surface avoidance' is a short distance avoidance of the road surface itself due to a lack of cover and/or to the character of embankment and pavement which is different from natural habitat. 'Vehicle avoidance' is the avoidance of oncoming vehicles. 'Attraction to roads' is when animals are attracted to a road for a resource (e.g. for food, a nesting site, a mate or thermoregulation). Source: Adapted from Jaeger et al. (2005).

the data from each study were converted into a common measure, the Pearson correlation coefficient r, a measure of the strength of the relationship between roads and an animal's population abundance. The coefficient, r, ranges from -1.00 (largest negative effect) through 0 (no effect) to $+1.00$ (largest positive effect). To determine species traits that make them prone to negative road and/or traffic effects, we considered traits that are related to population abundance: reproductive rate and/or age at sexual maturity, species mobility and body size. Full details of the methods are in Rytwinski and Fahrig (2012). We limit the discussion here to four groups of vertebrates – mammals, birds, amphibians and reptiles that spend at least part of their life cycle on land. Invertebrates were not included in this discussion because there were too few population-level studies, but see Chapters 29 and 30 for more details on this group.

When reading this chapter, two points are important. First, when more than one study was conducted on a particular species, we determined the average direction and size of the road effect. While this provides an indication of the overall effect of roads on the species, studies conducted in different locations or habitats may actually measure different road effects on a species (Lesson 28.4). Second, we present information based only on studies that have measured the effect of roads on at least one population. Many other species may be affected by roads but have not yet been studied (Lesson 28.5).

The aims of this chapter are to identify (i) the animals whose populations are most vulnerable to road impacts, (ii) species traits and behavioural responses to roads that make animals vulnerable to road impacts and (iii) the likely causes of those impacts, so that appropriate road mitigation measures can be identified. For mitigation to be effective, the cause of the impact must be specifically addressed. For example, if a species is mainly affected by road mortality, mitigation should be directed towards preventing animals from moving onto roads. In this case, installing wildlife crossing structures would not adequately address the main issue of road mortality, unless fencing was also installed.

LESSONS

28.1 Mammals: Larger, more mobile species with lower reproductive rates are more susceptible to road mortality, and species that avoid roads from a distance due to traffic-related disturbance are susceptible to habitat fragmentation, loss and degradation

Population-level effects

A total of 34 studies from 12 countries that included 84 mammal species were reviewed. From these, 127 records of road and/or traffic effects were extracted.

Most studies of mammals at the population level to date have been conducted on three orders: (i) rodents (27 species), (ii) hoofed mammals (more specifically even-toed ungulates) (16 species) and (iii) carnivores (24 species) (Fig. 28.2). On average, rodent and hoofed mammal populations increase, and carnivore populations decrease in response to roads (Fig. 28.2).

Of the rodents studied, only a few species are negatively affected by roads compared to a much higher number that are either positively affected or unaffected (Fig. 28.3A(i)). Species showing negative population-level effects are mid-sized species of arboreal squirrels (grey squirrel, Lord Derby's scaly-tailed squirrel, Beecroft's scaly-tailed squirrel) and the California vole. Rodents showing positive or neutral

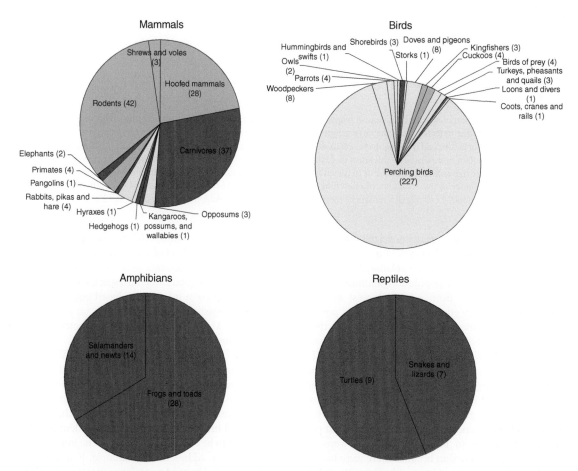

Figure 28.2 Proportion of the total number of road effects extracted for review for the various animal orders for each class. Numbers in brackets correspond to the number of road effects extracted for each animal order out of a total of 127 for mammals, 270 for birds, 42 for amphibians and 16 for reptiles. Colours correspond to the direction in which populations are responding on average to road impacts: red, negatively affected (population abundance decreasing); yellow, unaffected (no change in population abundance); and green, positively affected (population abundance increasing).

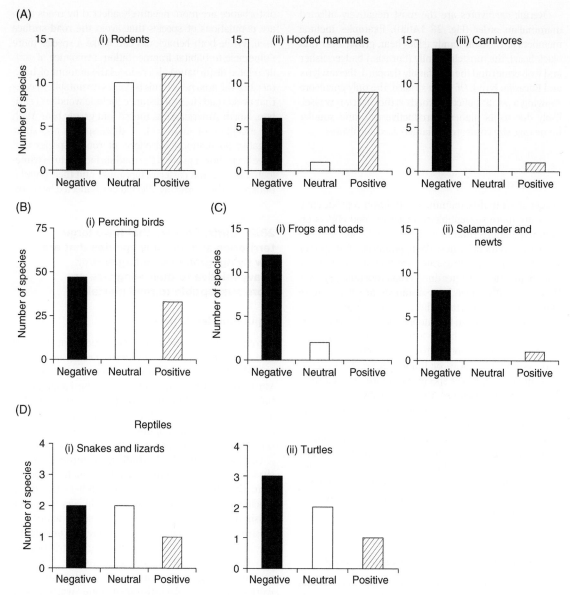

Figure 28.3 The number of species within each animal order showing on average negative, neutral, or positive effects of roads on their population abundance for (A) mammals, (B) birds, (C) amphibians and (D) reptiles.

effects of roads are typically smaller: for example, white-footed mouse and least chipmunk.

For the hoofed mammals, the number of studied species showing positive population-level effects of roads is nearly double the number showing negative effects (Fig. 28.3A(ii)). Species showing negative effects include North American elk, wild boar, European roe deer, woodland caribou and mule deer.

Species showing positive effects include white-tailed deer, moose, Peter's duiker and yellow-backed duiker. It has been suggested that the positive effects of roads on large herbivores such as white-tailed deer (and indeed small mammals as well) may be due to predation release, as populations of many of their main predators are reduced in areas of high road density (Munro et al. 2012).

Overall, carnivores are the most negatively affected mammalian order (Fig. 28.3A(iii)). Examples include members of the bear family (sloth bear, grizzly bear and black bear), the mustelid family (Eurasian badger, fisher and wolverine) and the felid family (leopard, Iberian lynx and Eurasian lynx). Of those studied, the only carnivore showing a positive effect of roads is the Siberian weasel, likely due to its higher reproductive rate and smaller home rage size compared to larger-sized carnivores.

Species traits

Larger, more mobile mammals with lower reproductive rates are more susceptible to negative road effects on their populations than smaller, less mobile species with higher reproductive rates. We hypothesise that species with lower reproductive rates are less able to recover from population declines due to road mortality. Species that frequently move long distances are likely more affected by road mortality, because they interact with roads more often than less mobile species. For the same reasons, species with larger territories or home ranges are also more susceptible to road effects than those with smaller territories or home ranges. This means that, in general, larger species are more affected than smaller species because they generally have lower reproductive rates and are more mobile than smaller species (Chapter 39).

There can however be exceptions to the aforementioned generalities. Hypothetically, if a species is locally abundant but of limited geographic range and/or dispersal capability, the population may be vulnerable to road impacts despite it having a high reproductive rate and/or being less mobile.

Behavioural responses to roads and/or traffic

Anecdotal observations of an animal reacting to a road or vehicle are common. However, there are actually very few quantitative studies documenting such behaviours. A scarcity of animals in areas of high road density is sometimes assumed to indicate road avoidance, but this assumption may not be valid because such a scarcity could also be caused by mortality. Even documenting movement paths of animals near roads cannot tell us whether the animal is avoiding the road itself or the traffic on it, unless animals only cross the road when traffic volume is very low.

From the available studies, populations of mammals that avoid roads from a distance due to traffic disturbance are more negatively affected by roads than are populations of species that avoid the road surface itself. While both behaviours can make a species more vulnerable to habitat fragmentation, avoidance of traffic-related disturbance also reduces the amount of habitat since the area near roads becomes unsuitable. Species that avoid roads from a distance include woodland caribou, North American elk, moose and grizzly bear. With the exception of moose, all of these species also show negative population-level effects of roads. Species that have been shown to avoid the road surface include white-footed mouse and eastern chipmunk, and their populations are either positively affected or unaffected by roads.

28.2 Birds: Species that have large territories and possibly species that are low flying, ground dwelling and/or heavy relative to their wing size are more susceptible to road mortality

Population-level effects

A total of 16 studies from 8 countries that included 194 bird species were reviewed. From these, 270 records of road and/or traffic effects were extracted. Most studies of the effects of roads on bird populations have been on perching birds, that is, passerines (153 of 194 species) (Fig. 28.2). While some species within this group show negative population effects of roads, there is no strong overall effect (Fig. 28.2).

Of the perching birds studied, examples of those showing negative population effects include species from the chats and old world flycatchers (northern wheatear and European robin), sandpipers (common redshank), wrens (winter wren and sedge wren) and Australian treecreepers (brown treecreeper and white-throated treecreeper). Species showing neutral or positive effects are primarily from the buntings, American sparrows (song sparrow and rock bunting) and new world warblers (black-throated blue warbler and Nashville warbler).

Species traits

In general, more mobile birds (i.e. species with larger territories) are more susceptible to road effects than are less mobile species. While this is the only species trait found to explain variation in bird population-level effects of roads in our literature review, researchers have suggested other potentially important traits that were not included in our analyses. For example, ground-dwelling birds have been suggested to be at greater risk of wildlife-vehicle collisions (WVC) because they spend longer time on the road surface and in low flight (Jacobson 2005). Furthermore, birds that are heavy relative to their wing size (e.g. female owls) or have a low take-off trajectory may also be more vulnerable to WVC (Kociolek & Clevenger 2011). Species that need to move between different habitat types (e.g. some woodland birds and wintering water birds) are likely more sensitive to road impacts (Chapter 33).

Behavioural responses to roads and/or traffic

There are very few studies documenting behavioural responses to roads and/or traffic in birds. Road attraction behaviour has been shown in two species (common raven and black kite). For both species, populations have been found to be unaffected by roads, even though mortality does occur (Palomino & Carrascale 2007). Although there are no quantitative studies of vehicle avoidance in these species, if they do show vehicle avoidance and if they benefit from the carrion resource on roads, a positive effect of this food resource on reproduction could balance or even outweigh negative effects of road mortality, the net effect being the observed neutral road effects on their populations.

Many authors have either argued or assumed that traffic noise is the main cause of negative road effects on bird populations. Traffic noise could interfere with the ability to communicate by song which could make it hard for some species of birds to attract mates and/or defend territories (Rheindt 2003). Traffic noise could also distract individuals making them more vulnerable to predation. These conclusions are based mainly on observations of both lower bird occurrence and higher traffic noise in locations closer to roads (Chapter 19). However, in addition to traffic noise, road mortality should be higher closer to roads, so it is not clear whether noise or mortality (or both) is the real cause of the negative effects on bird populations. Distinguishing these is important for designing appropriate mitigation (Summers et al. 2011). The finding that more mobile birds are more prone to road effects than less mobile species indirectly supports the mortality hypothesis over the noise disturbance hypothesis. In addition, some of the studies of road effects on birds were designed such that the effects of distance from the road and distance from habitat edge are confounded, which means that apparent road effects could be partly or even mainly due to negative edge effects (Delgado García et al. 2007; but see Summers et al. 2011).

28.3   Amphibians and reptiles: All species, regardless of life history traits, are prone to negative road effects as they are particularly susceptible to road mortality and habitat fragmentation by roads

Population-level effects

For amphibians, 16 studies from 6 countries that included a total of 23 species were reviewed. From these, 42 records of road and/or traffic effects were extracted. On average, roads and traffic reduce populations of frogs and toads and salamanders (Fig. 28.2). For reptiles, 9 studies from 3 countries that included a total of 11 species were reviewed, from which 16 records of road effects were extracted. On average, populations of turtles and snakes and lizards are negatively affected by roads (Fig. 28.2).

Although amphibians and reptiles have significantly more species at risk than either mammals or birds (IUCN 2010), there are relatively few studies of the effects of roads on their populations. Those that do exist suggest that amphibians are in general negatively affected by roads, with only one species showing a positive effect (northern two-lined salamander) (Fig. 28.3C(i) and (ii)). Frogs showing negative population-level effects include the spring peeper, European tree frog, northern leopard frog, wood frog and common spadefoot toad. Salamanders showing negative effects include the tiger salamander, blue-spotted salamander, red-backed salamander, seal salamander and eastern newt.

There are only 11 reptile species for which the population-level effects of roads have been evaluated. About equal numbers of snake species show negative and neutral effects of roads, and one species showed a weak positive effect (eastern diamondback rattlesnake) (Fig. 28.3D(i)). Snakes showing negative population-level effects include the lava lizard and timber rattlesnake, and those showing neutral effects include eastern massasauga rattlesnake and eastern hognosed snake.

Population-level effects of roads on turtles are mixed with three species showing negative effects (desert tortoise, wood turtle and spotted turtle), two showing neutral effects (common snapping turtle and common musk turtle) and one showing a positive effect (painted turtle) (Fig. 28.3D(ii)).

Species traits

In general, populations of amphibian species with lower reproductive rates are more susceptible to negative road effects than species with higher reproductive rates.

Many reptiles are long-lived with high natural year-to-year survival of the adults, and many make long movements over land searching for nests or mates. These characteristics along with their slow movements across roads make reptile populations particularly vulnerable to road mortality. There may be more negative effects of roads on reptiles than suggested by studies to date because it is difficult to estimate reptile population sizes, which would make it hard to detect effects (Chapter 32). Also, for species that nest along roads (e.g. painted turtles), the negative effect of road mortality may be compensated by lower rates of nest predation (Langen 2009).

Species that need to move among different habitats are also particularly susceptible to road mortality and landscape fragmentation by roads. For example, many frogs and salamanders need to move among aquatic breeding habitats, upland feeding habitats and specialised overwintering habitats to complete a life cycle. When these habitats are not adjacent, amphibians must move long distances, sometimes several kilometres, to find them. At high road density, the chance of all these habitats occurring within an area absent of roads is unlikely. In some cases, such as when roads run adjacent to a river or stream, all animals in the population must cross roads to reach other habitats, resulting in a very high mortality rate (Chapter 31).

Road mortality also affects amphibian and reptile populations indirectly by reducing reproductive rate. Reproductive rate of amphibians and reptiles increases with age because larger animals have more eggs and they keep growing as they age. Roadkill results in a shift in age within the population towards younger individuals, which are smaller, and this reduces the overall reproductive rate of the population (Karraker & Gibbs 2011).

Behavioural responses to roads and/or traffic

There are not many studies of amphibian and reptile behavioural responses to roads. Three snakes, the timber rattlesnake, the eastern hog-nosed snake and the eastern massasauga rattlesnake, avoid the road surface (Andrews & Gibbons 2005), and of these three species, only the timber rattlesnake showed a negative population-level effect of roads. There is one study of the behavioural response of frogs to roads; the northern leopard frog showed no behavioural avoidance of roads or traffic (Bouchard et al. 2009), which probably explains its negative population-level response to roads, likely due to abundant road mortality.

28.4 A species response to roads and traffic will vary depending on its conservation status, geographical location, habitat preferences, road type and/or traffic volume

In our review, we determined the average direction and size of the road effect when more than one study was conducted on a particular species. While this provides an indication of the overall effect of roads on a particular species, the individual studies may have been conducted in different locations or habitat types, for example, Florida versus California, United States, using different road measures, for example, road density versus traffic density, or road types, for example, highways versus gravel roads, which may result in different road effects on population abundance. For example, grey wolves respond negatively to increasing road density in northern Wisconsin and upper Michigan, United States (Mladenoff et al. 1995), but positively in the boreal forest of northern Ontario, Canada (Bowman et al. 2010). On average, the wolf response is neutral, but this hides these different positive and negative effects. The regional difference could be because most roads in northern Ontario are lightly used gravel logging roads, whereas in northern Wisconsin and Michigan, they are paved roads with higher traffic volumes. Therefore, effects of roads may be context or location dependent so extrapolation of road effects for a species from one region to another should be carefully scrutinised.

Road effects may also be dependent on the conservation status of the species or its local population. For example, it is possible that a species with traits that would normally make it resilient to road effects may already be so depleted in an area from other causes that a new road, even with relatively low rates of mortality or reduced habitat connectivity, may be sufficient to cause it to decline further, possibly to local extinction.

28.5   There are still many species for which we do not know the population-level effects of roads. To ensure mitigation will be effective for as many species as possible, research is needed on the effects of roads on a broader range of species

There are large biases towards studies on certain groups of mammals and birds, leaving gaps in knowledge on population-level effects of roads for many species and species groups. Most studied mammals belong to either the rodent, hoofed mammal or carnivore orders (i.e. 67 of the 84 studied species), highlighting the need for more population-level studies for other orders. Furthermore, the majority of population-level bird studies have been conducted on perching birds (passerines) (153 of 194 species studied). On average, there was no strong overall effect of roads found for this group. If all perching birds were found to have a trait that makes them tolerant to road effects, this could explain this lack of effect, suggesting that more studies on a wider range of bird orders are needed. Some of the empirical studies reporting road effects on birds were designed such that the effects of distance from the road and distance from habitat edge are confounded, which means that apparent road effects could be partly or even mainly because of habitat edge effects on birds (but see Summers et al. 2011). With a combined species total of 34, amphibians and reptiles were the least represented animal groups in this review, suggesting more population-level studies are needed.

To better facilitate future reviews such as this one or to estimate potential effects for new road projects, we have the following recommendations. First, when reporting an effect of roads and/or traffic on a species population abundance, authors should include (i) the test statistic for the effect (e.g. F or r^2) and/or summary statistics (e.g. mean and variance) from which an effect size can be calculated and (ii) the sample size (or the P value of the test if a test statistic was reported). The number of studies that could be included in our review was often limited by the lack of statistical information provided. Second, authors should provide a brief description of the ecology of the species of interest, including information on species traits for the geographical location of the study, along with references, as this information is often lacking or difficult to obtain for researchers living in different regions. Third, authors should include maps with a scale or provide GPS coordinates of study locations/sites to allow the potential of further analyses of landscape variables or evaluation of spatial independence of study sites.

CONCLUSIONS

From the available literature, there is evidence that road mitigation should be considered for wide-ranging large mammals with low reproductive rates; birds with larger territories; possibly birds that are low flying, ground dwelling and/or heavy relative to their wing size; all amphibians and reptiles (due to road mortality); and species that do not avoid roads or are known to be disturbed by traffic. For species that are mainly affected by roads through road mortality, mitigation should focus on preventing animals from moving onto roads (e.g. fences; Chapter 20). For species that are disturbed by traffic, road effects can be mitigated by measures aimed at reducing road and traffic density in the landscape (e.g. by closing some roads (Chapter 3) or increasing the capacity of roads outside important wildlife areas). In addition, engineering solutions to reduce traffic noise (e.g. changes to pavement or tyres) could partially mitigate the disturbance effects. For species that are mainly affected by roads through habitat fragmentation, mitigation should focus on improving habitat connectivity by installing wildlife crossing structures (Chapter 21).

When there is an endangered species present or when a population is declining or at risk of local extinction due to other disturbances or modifications to the environment, roads should be mitigated even if they are not the main reason for the species' endangerment or decline. Even if the rate of road mortality on such a species is low, any additional mortality or reduced connectivity can drive it to extinction. Furthermore, species responses to roads are sometimes context (e.g. habitat type studied, road/traffic measure used and/or road type studied) or location dependent, so road impacts on species for a given location of interest should be considered carefully before new roads are constructed or modified.

While our review included 312 species and 455 data sets on population-level effects of roads, large biases towards studies on certain groups of mammals (i.e. rodents, hoofed mammals and carnivores) and birds (i.e. perching birds) were uncovered, highlighting the need for more population-level studies for other species groups.

ACKNOWLEDGEMENTS

We thank the Natural Sciences and Engineering Research Council of Canada for support.

FURTHER READING

Fahrig and Rytwinski (2009): Provides our preliminary review findings of the effects of roads on animal abundance and examples of some of the common issues associated with road ecology study designs.

Roedenbeck et al. (2007): Based on discussions during the 'Landscape-scale effects of roads and biodiversity' workshop in Germany in 2005, this paper identifies the questions in road ecology of most direct relevance to the decision-making process and then provides suggestions for designing studies that have high inferential strength to address those questions.

Rytwinski and Fahrig (2012): This paper formed the basis of the information provided within this chapter. Further information on the methodology used to carry out the review as well as further discussion on its findings and the actual data itself can be retrieved within this paper and its supporting information.

REFERENCES

Andrews, K. M. and J. W. Gibbons. 2005. How do highways influence snake movement? Behavioral responses to roads and vehicles. Copeia **2005**:772–782.

Bouchard, J., A. T. Ford, F. E. Eigenbrod and L. Fahrig. 2009. Behavioral responses of northern leopard frogs (*Rana pipiens*) to roads and traffic: implications for population persistence. Ecology and Society **14**:23. Available from http://www.ecologyandsociety.org/vol14/iss2/art23/ (accessed 28 September 2014).

Bowman, J., J. C. Ray, A. J. Magoun, D. S. Johnson and F. N. Dawson. 2010. Roads, logging, and the large-mammal community of an eastern Canadian boreal forest. Canadian Journal of Zoology **88**:454–467.

Delgado García, J. D., J. R. Arévalo and J. M. Fernández-Palacios. 2007. Road edge effect on the abundance of the lizard *Gallotia galloti* (*Sauria: Lacertidae*) in two Canary Islands forests. Biodiversity and Conservation **16**:2949–2963.

Fahrig, L. and T. Rytwinski. 2009. Effects of roads on animal abundance: an empirical review and synthesis. Ecology and Society **14**:21. Available from http://www.ecologyandsociety.org/vol14/iss1/art21/ (accessed 28 September 2014).

Forman, R. T. T., D. Sperling, J. A. Bissonette, A. P. Clevenger, C. D. Cutshall, V. H. Dale, L. Fahrig, R. France, C. R. Goldman, K. Heanuem, J. A. Jones, F. J. Swanson, T. Turrentine and T. C. Winter. 2003. Road ecology: science and solutions. Island Press, Washington, DC.

International Union for the Conservation of Nature (IUCN). 2010. IUCN red list of threatened species. IUCN, Gland, Switzerland. Available from http://www.iucn.org (accessed 28 September 2014).

Jacobson, S. J. 2005. Mitigation measures for highway-caused impacts to birds. General technical report PSW-GTR-191. USDA Forest Service, Washington, DC.

Jaeger, J. A. G., J. Bowman, J. Brennan, L. Fahrig, D. Bert, J. Bouchard, N. Charbonneau, K. Frank, B. Gruber and K. Tluk von Toschanowitz. 2005. Predicting when animal populations are at risk from roads: an interactive model of road avoidance behavior. Ecological Modeling **185**:329–348.

Karraker, N. E. and J. P. Gibbs. 2011. Contrasting road effect signals in reproduction of long- versus short-lived amphibians. Hydrobiologia **66**:213–218.

Kociolek, A. V. and A. P. Clevenger. 2011. Effects of paved roads on birds: a literature review and recommendations for the Yellowstone to Yukon ecoregion. Technical report #8. Yellowstone to Yukon Conservation Initiative, Canmore, Alberta. Available from http://y2y.net/files/979-y2y-technical-report-8-effects-of-paved-roads-on-birds.pdf (accessed 28 September 2014).

Langen, T. A. 2009. Design and testing of prototype barriers and tunnels to reduce the impact of roads on turtle survival and reproductive success. Final report MOU AM05405. New York State Department of Environmental Conservation US Fish & Wildlife Service, Albany, NY.

Mladenoff, D. J., T. A. Sickley, R. G. Haight and A. P. Wydeven. 1995. A regional landscape analysis and prediction of favourable gray wolf habitat in the northern Great-Lakes region. Conservation Biology **9**:279–294.

Munro, K. G., J. Bowman and L. Fahrig. 2012. Effect of paved road density on abundance of white-tailed deer. Wildlife Research **39**:478–487.

Palomino, D. and L. M. Carrascale. 2007. Threshold distances to nearby cities and roads influence the bird community of a mosaic landscape. Biological Conservation **140**:100–109.

Rheindt, F. E. 2003. The impact of roads on birds: does song frequency play a role in determining susceptibility to noise pollution? Journal für Ornithologie **144**:295–306.

Roedenbeck, I. A., L. Fahrig, C. S. Findlay, J. E. Houlahan, J. A. G. Jaeger, N. Klar, S. Kramer-Schadt and E. A. van der Grift. 2007. The Rauischholzhausen agenda for road ecology. Ecology and Society **12**:11–32.

Rytwinski, T. and L. Fahrig. 2012. Do species life history traits explain population responses to roads? A meta-analysis. Biological Conservation **147**:87–98.

Summers, P. D., G. M. Cunnington and L. Fahrig. 2011. Are negative effects of roads on breeding birds caused by traffic noise? Journal of Applied Ecology **48**:1527–1534.

Chapter 29

INSECTS, SNAILS AND SPIDERS: THE ROLE OF INVERTEBRATES IN ROAD ECOLOGY

Heinrich Reck[1] and Rodney van der Ree[2]

[1]Department of Landscape Ecology, Institute for Natural Resource Conservation, Kiel University, Kiel, Schleswig-Holstein, Germany
[2]Australian Research Centre for Urban Ecology, Royal Botanic Gardens Melbourne, and School of BioSciences, The University of Melbourne, Melbourne, Victoria, Australia

SUMMARY

Animals that lack a backbone are called invertebrates, and they account for 95–99% of the animal species on earth. Invertebrates are important components of ecosystems and play key roles in the functioning of all ecosystems, such as pollination, decomposition and nutrient cycling. Many species are subject to mortality due to wildlife-vehicle collisions (WVC) and are negatively affected by traffic pollution and habitat fragmentation. However, roadside vegetation in otherwise cleared or modified landscapes provides important habitats and corridors for some species.

29.1 Invertebrates are critical to the healthy functioning of ecosystems and the conservation of biodiversity.

29.2 Mortality rates due to traffic or artificial lighting can be very high.

29.3 Traffic-related pollution leads to population declines and habitat degradation.

29.4 Barrier effects are a significant threat to the survival of flightless species.

29.5 Invertebrates should be included in environmental impact assessment, and using indicator groups is a practical yet comprehensive approach.

29.6 Mitigation measures for invertebrates are urgently needed, and although similar to those used for mammals, reptiles or amphibians they differ in important details.

29.7 Invertebrates benefit indirectly from effective mitigation for vertebrates because of the ecosystem function these larger species perform.

29.8 Sympathetic management of roadside habitats can help the survival of some threatened species of invertebrates.

Handbook of Road Ecology, First Edition. Edited by Rodney van der Ree, Daniel J. Smith and Clara Grilo.
© 2015 John Wiley & Sons, Ltd. Published 2015 by John Wiley & Sons, Ltd.
Companion website: www.wiley.com\go\vanderree\roadecology

> Invertebrates are critical to the survival of life on earth. Greater consideration of invertebrates in the planning, design, construction and operation of roads is urgently required, including their specific inclusion in environmental impact assessments and mitigation.

INTRODUCTION

Worldwide, approximately 1.3 million invertebrate species (excluding protozoa) have been identified, compared to approximately 63,000 vertebrate species (Fig. 29.1). However, the proportion of animal species on earth that are invertebrates is likely closer to 99%, because many such species remain undiscovered or uncatalogued. Of these invertebrates, about 75% are insects (e.g. beetles, bees, butterflies or crickets), and the remainder are molluscs (7%), crustaceans (4%) (see Chapter 30), spiders (8%) and other invertebrates like worms (5%).

It is almost impossible to give a short overview of what invertebrates are and how they differ from each other because of their incredible diversity. For example, the difference between a humming bird and a whale is small in comparison to the difference between a lightning bug and a ground beetle, an earthworm and a bee, or a spider and clam. The differences among invertebrates exceed the differences among vertebrates in relative size, social behaviour, foraging, mobility and reproduction.

The aims of this chapter are to highlight: (i) the great diversity in form and function of invertebrates; (ii) their important role in maintaining healthy ecosystems; (iii) the impacts of roads and traffic on invertebrates; and (iv) important considerations for impact assessment and mitigation.

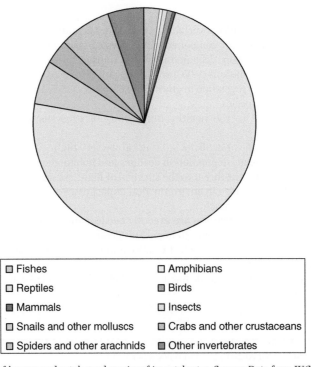

☐ Fishes	☐ Amphibians
☐ Reptiles	◼ Birds
◼ Mammals	☐ Insects
☐ Snails and other molluscs	☐ Crabs and other crustaceans
☐ Spiders and other arachnids	◼ Other invertebrates

Figure 29.1 The number of known and catalogued species of invertebrates. Source: Data from IUCN (2010).

LESSONS

29.1 Invertebrates are critical to the healthy functioning of ecosystems and the conservation of biodiversity

Invertebrates are critical to maintaining healthy and functional ecosystems and play key roles as pollinators, decomposers, seed transporters and nutrient recyclers (e.g. Christmas Island crabs, Chapter 30) but also as pests and in pest control. In short, they have evolved to be an irreplaceable and integral part of earth's recent life. For example, decomposition is essential in reducing the build-up of leaf litter and recycling nutrients back into the soil. In the 1980s, the rate of leaf-litter decomposition along motorways in Central Europe slowed by 50–80% due to the use of leaded fuels. The lead concentration in soil was about eight times higher than average and could be found up to 200 m on either side of roads with 100,000 vehicles per day (Klemens et al. 1997). The diet of many species of mammals, birds, amphibians and reptiles consists partially or entirely of invertebrates, and many predators rely entirely on specific species. Invertebrates often accumulate traffic-related pollutants (e.g. worms near roads have exhibited cadmium levels 500 times greater than expected (LFU 1994)). These pollutants can be passed from invertebrate prey to vertebrate predator, sometimes accumulating until toxic levels are reached, causing disease or death of individuals higher in the food chain.

29.2 Mortality rates due to traffic or artificial lighting can be very high

The available statistics indicate that mortality rates of invertebrates far exceed those recorded for vertebrates. Threateningly high mortality rates of bumblebees and dragon flies were recognised as early as the 1980s (e.g. Hagen 1984; Donath 1986), and rates numbering in the thousands have been reported (Gepp 1973; Hayward et al. 2010). For example, up to 2000 moths per km have been found dead along Swiss roads (Ruckstuhl cited in SBN 1987: 86), and 2–35 dragonflies were killed per kilometre per day in Illinois, United States (Soluk et al. 2011), and approximately 20 million individual butterflies and moths were estimated to be killed per week when in flight (McKenna et al. 2001). These statistics and estimates confirm that Gepp's (1973) estimate of insect mortality for Germany due to collision with vehicles of 100–3000 individuals per vehicle

per kilometre was probably realistic, eventually leading to the loss of trillions of individuals and hundreds of tons of invertebrate biomass annually. For example, a local population of the threatened rattle grasshopper adjacent to a road in the German Jurassic Mountains suffered losses of 30% of the local population due to road mortality (Weidemann & Reich 1995). Although many butterfly species avoid crossing roads, those individuals that do often suffer high rates of mortality, in some cases one out of three individuals will be killed per crossing attempt (Pfister et al. 1997). Rates of mortality of red crabs on Christmas Island can number in the hundreds of thousands annually (Chapter 30). While population-level studies on invertebrates next to roads are rare, one can observe conspicuously low population densities of many species, and especially moths, in areas with high traffic volumes. The challenge with studies of invertebrate road mortality is that it is extremely difficult to detect dead animals (e.g. Fig. 29.2), and in most cases, the impact on vehicles is negligible.

Insect mortalities also occur from traffic lighting, and in some places, the accumulated depth of dead insect bodies under street lights may be several centimetres thick. Insects are particularly attracted to lights with high wavelengths (see Chapter 18). Insect mortality from lighting can be avoided by using lighting only when and where necessary, by using lights with less attractive wavelengths and by shielding lights to prevent spillover (Chapter 18). Invertebrates can also be killed due to incorrect verge maintenance (e.g. using suction mowers at certain times of the year) or roadside kerbing that can trap invertebrates.

29.3 Traffic-related pollution leads to population declines and habitat degradation

Invertebrates are extremely susceptible to environmental pollution. Pollution from vehicles (e.g. exhaust fumes and tyre wear) and roads (e.g. dust from unsealed roads or de-icing chemicals) can accumulate in soil and vegetation adjacent to roads and impact invertebrate communities. For example, the population density of earthworms within 30 m of a road with 3000 cars per day near Moscow, Russia, was 50% lower than in a similar habitat 200 m away (Bykov & Lysikov 1991).

Other chemicals, such as ozone from traffic, appear to negatively affect parasitic wasps that can play an important role in regulating pest invertebrate species (e.g. aphids) (Gate et al. 1995). Indirectly, pollutants can change the composition of plant species along roadsides, thereby changing the suitability of the habitat for

Figure 29.2 Dead insects, such as this great green bush cricket, are detectable for only very short periods of time before they are completely smashed or eaten by scavengers. Source: Photograph by H. Reck.

different species. Invertebrates have benefitted from efforts to reduce chemical pollution for human health (e.g. removal of lead from fuel). However, the effects of traffic pollution on invertebrates is still poorly understood and rarely included in environmental impact assessments (EIA). Aquatic invertebrates are often at risk during road construction or after heavy rainfall events due to erosion of sediment that can cause siltation of waterways. In the Rhön Mountains in central Germany, one of the few remaining populations of the critically endangered freshwater pearl mussel became extinct due to sedimentation caused by poorly managed road construction (Groh & Jungbluth 1993).

29.4   Barrier effects are a significant threat to the survival of flightless species

Roads and other linear infrastructure can be a barrier or filter to the movement of invertebrates (Mader 1979). The species most affected are those that cannot fly, are slow moving, or avoid roads. For example, areas isolated by one road had up to 50% fewer ground beetle species than expected based on known habitat preferences, increasing to an 80% reduction in species in areas with many roads; and flightless species experienced disproportionately higher losses (Pfister et al. 1997: 344). Some species that persist in roaded areas

are additionally affected by a significant reduction in genetic variability, further threatening their viability (Keller & Largiadèr 2003; Chapter 14). Numerous factors increase the barrier effect, including traffic volume, traffic noise and road design (e.g. kerb height, noise walls, vegetation density). Figure 29.3 shows the movement of one species of ground beetle before and after the construction of a new road. After construction, the rate of capture was much reduced, and there were no movements of individuals across the road. It is unclear if this is due to the width of the road or traffic volume, but the reduction in the rate of crossing is striking. Even narrow roads can act as barriers, inhibiting the movements of flightless grasshoppers (Reck & Kaule 1993: 89–114) or snails (Martin & Roweck 1988). However, the barrier effect may be reduced for some species of invertebrate if they are transported while on or attached to other larger-bodied species (e.g. Fig. 29.6).

Traffic noise may also affect rates of crossing because certain species will move away from roads because of acoustic interference. For example, adult steppe grasshoppers, which rely on acoustic communication, avoid roads (Pfister et al. 1997: 426) despite being able to alter their calls in an attempt to be heard over traffic noise (Lampe et al. 2012; see also Chapter 19).

Many species of invertebrate have specific habitat requirements (e.g. vegetation structure, food species of

(A)

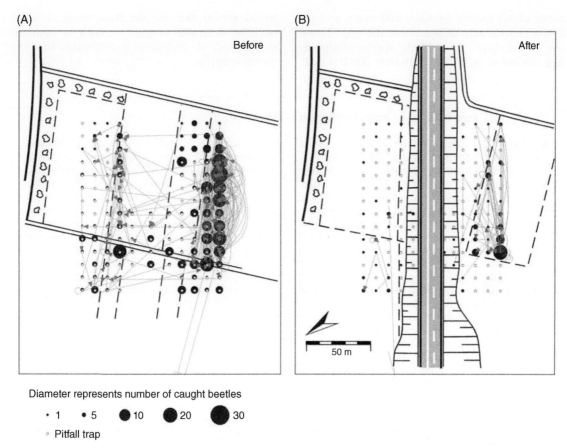

Diameter represents number of caught beetles

· 1   • 5   ● 10   ● 20   ● 30

○ Pitfall trap

Figure 29.3 The effect of a three-lane road on the movement and abundance of the grain wart ground beetle. Before construction (A), many pitfall traps captured more than 20 males, with frequent movements across the road zone. After construction (B), the capture rate and movements were significantly lower. Source: Adapted from Pfister et al. (1997): 395.

plant, soil type, soil moisture), and they need to move around the landscape as habitats change seasonally or due to succession, often on relatively short timescales. Local extinctions can occur if the original habitat becomes unsuitable and roads act as a barrier to movements, which occurred to a population of the rotund disc land snail in Rhineland-Palatinate, Germany (Martin & Roweck 1988).

29.5 Invertebrates should be included in environmental impact assessment, and using indicator groups is a practical yet comprehensive approach

The effects of proposed road projects on invertebrates should be assessed as part of the EIA because significant impacts are likely (Chapter 5). It is not appropriate to assume that the response and needs of invertebrates will be addressed if vertebrates are considered because both groups may respond differently. Many countries have legislation that protects rare and endangered species, including invertebrates, and these species, under law, must be included in EIAs. However, even the common species (which are rarely assessed) should be represented in the EIA because invertebrates are critically important to the healthy functioning of ecosystems and the loss of them may affect higher-order vertebrates, some of which may be threatened. Furthermore, the Convention on Biological Diversity, which has been ratified by most countries around the world, recognises the importance of conserving all biodiversity.

The massive species diversity and range of life forms of invertebrates (Fig. 29.1) suggest that a one-size-fits-all approach to invertebrate EIAs will not work, nor will an approach that considers the needs of all species for each and every project. While considerable debate about the appropriate use of 'indicator' or 'umbrella' species continues (e.g. Baumann et al. 1999; Haslett 2007: 41), it would appear that for the time being at least, no other tractable solution remains to optimise the process of including invertebrates in EIAs (Textbox 29.1).

Textbox 29.1 Invertebrates as indicator species in EIA

It is always best to directly measure the species or environmental condition that we are interested in. However, this is not always possible (Chapter 16) and indicators are an alternative approach. An indicator species is 'an organism whose characteristics (e.g. presence or absence, population density, dispersion, reproductive success) are used as an index of attributes too difficult, inconvenient, or expensive to measure for other species or environmental conditions of interest' (Landres et al. 1988: 317). There are many kinds of indicator species (e.g. umbrella, bioindicator, keystone, pollution indicator), each with a different meaning and intent (Lindenmayer & Burgman 2005). Invertebrates are frequently used as indicator species, including ants to measure post-mining rehabilitation, bees as a measure of pollination, butterflies and moths for ecosystem health and aquatic invertebrates for waterbody health (see references in Lindenmayer and Burgman (2005)). In practice, indicator species are usually a suite or group of species, rather than a single species, and it is important to distinguish if the indicator species is to reflect an environmental condition (e.g. pollution levels), a specific species (or group of species) or a combination of the two.

This approach assumes that by measuring the relevant parameter (e.g. presence, abundance, survival) for the indicator species, the other species are also accommodated for or the environmental condition is accurately assessed. Therefore, in the context of EIA, the species in the indicator groups must be highly responsive to the proposed impact (e.g. road mortality, barrier effect, traffic noise) and occur at the location of the proposed development, and its response be representative of the actual target species/environmental condition. The appropriate indicator species or group to be used in an EIA also depends on the habitat type to be affected (Table 29.1). There are a number of approaches to selecting indicator species to ensure they accurately represent other species or environmental conditions, the detail of which is beyond the scope here. In its simplest form, detailed EIA and mitigation planning should identify indicators that are highly sensitive/demanding and, if possible, be locally important (e.g. threatened or iconic) to raise its profile and provide public support. It appears possible to reduce the enormous variety and diversity of invertebrates into a practical number of groups and subsequent target species to save on costs and simplify an otherwise impossibly complex process. However, the details of the approach and its success still require significant region-specific testing and evaluation to identify the appropriate indicators.

Table 29.1 List of indicator groups of species used in Central European environmental impact assessments.

Broad habitat type	Potential combinations of indicator species; partially redundant (non-invertebrates given in parentheses)
Waterbodies and banks	Dragon flies, macrozoobenthos, crayfish, molluscs, ground beetles (chandelier algae, vascular plants, fish, birds, reptiles, amphibians, mammals)
Agricultural land (arable fields)	Ground beetles, grasshoppers/crickets, spiders (vascular plants, birds, mammals)
Open habitats (incl. meadows and pastures) and forest edges	Ground beetles, butterflies, grasshoppers/crickets, wood-inhabiting beetles, bees, ants, moths, spiders (vascular plants, birds, reptiles, amphibians, mammals)
Forest and woodland	Ground beetles, butterflies, grasshoppers/crickets, wood-inhabiting beetles, bees, ants, moths, spiders (vascular plants, birds, amphibians, mammals)
Caves	Ground beetles (vascular plants, bats)
Springs	Dragonflies, snails (vascular plants)
Subterranean waterbodies	Snails, crustaceans

29.6 Mitigation measures for invertebrates are urgently needed, and although similar to those used for mammals, reptiles or amphibians, they differ in important details

Invertebrates, just like mammals, amphibians, birds and reptiles, need effective measures to reduce road mortality and restore habitat connectivity (Fig. 29.4). Unfortunately, most monitoring programmes fail to assess the use of crossing structures by invertebrates and those that have suggest that land bridges have the highest rates of use compared to small or dark underpasses (Sporbeck et al. 2013). However, many land bridges are not suitable for some invertebrates because they were designed only for vertebrates. Invertebrates have different requirements, including specific soil type, soil profile or vegetation structure. In addition, invertebrates with low mobility will need crossing structures in close proximity to their habitats and movement corridors. In summary, a barrier for invertebrates may not be a barrier for large vertebrates, and vice versa, and this means achieving multifunctional passages which will require careful planning (Textbox 29.2).

Figure 29.4 The importance of dung beetles has been recognised in South African National Parks.
Source: Photograph by R. van der Ree.

29.7 Invertebrates benefit indirectly from effective mitigation for vertebrates because of the ecosystem function these larger species perform

Many vertebrates (and indeed some invertebrates) are important for the survival and movement of invertebrates by creating suitable habitat conditions. Large herbivores (Chapter 42) are particularly important through grazing, trampling, burrowing and defecating. For example, the activity of ground beetles on overpasses with tall grass in Germany was on average threefold higher on the trampled vegetation of deer paths than at control sites (Reck 2013). In addition, many species are directly involved in the transport of individual animals and plant propagules (Fig. 29.6), potentially over great distances (Chapter 56).

29.8 Sympathetic management of roadside habitats can help the survival of some threatened species of invertebrates

The vegetation of roadside verges is often distinctly different to that of the surrounding landscape (Chapter 46). Within forests, the roadside is a clearing, and within agricultural or urban landscapes, the roadside may support trees, shrubs or grasses that no longer exist in the

area. The role of verges as 'green infrastructure' for biological diversity within intensively managed landscapes is increasingly appreciated (Verstrael et al. 2000). Old trees along European roadsides are important refuges for wood-inhabiting beetles, including the impressive stag beetles (Gürlich 2009) and the endangered hermit beetle. Roadside trees provide habitat and movement corridors for moths and many species that are poor flyers (Oleksa & Tyszko-Chmielowiec 2012; Roloff 2012). Similarly, verges with native plants can provide habitat and act as a corridor for ground and grassland species (Vermeulen 1994; Kiss et al. 2012; Schaffers et al. 2012). If mowing is conducted at the right frequency and time of year, flowering can be optimised for flower-visiting insects, thereby positively affecting the surrounding ecosystem due to the presence of pollinators (Noordijk 2009).

A number of simple principles should direct management of roadside verges to reduce costs and improve aesthetics and suitability for invertebrates (Verstrael et al. 2000). Management should:
• Aim to maximise benefits for as many species as possible;
• Avoid frequent mowing and include areas where grasses can grow fully;
• Maintain natural soil types, profiles and nutrient levels according to the needs of the target species;

Textbox 29.2 Designing wildlife crossing structures for invertebrates

The most effective crossing structures for invertebrates are likely to be those where the individual animal does not notice a significant change in its preferred habitat as it crosses over or under the road (e.g. Fig. 29.5). Crossing structures can also be designed specifically for invertebrates (e.g. red crabs on Christmas Island, Figs 30.2 and 30.3). Crossing structures for large vertebrates can probably be slightly modified to also be suitable for a wide range of invertebrate species. There are a number of principles to inform the design and management of wildlife crossing structures for invertebrates:

• They should be located sufficiently close to the target populations; otherwise, corridors or stepping stones are required.

• They should be large enough to support the preferred habitat of the target species for movement and/or permanent occupation. Different species will likely require different habitats, and these may need to be distributed across multiple crossing structures.

• Crossing structures should contain a suitable substrate (e.g. soil type and depth, level of compaction), which is particularly important for burrowing species but more so for its affect on plant growth (Chapter 21). Depending on the needs of the target species, crossing structures could include longitudinal strips or patches of bare earth to facilitate movement and species-specific habitat components (e.g. rocks, logs, grasses or a dense leaf canopy, etc.) will also assist. The greater the diversity of microhabitats, the more effective the structure will be for a greater number of species.

• Crossing structures with concrete floors (e.g. culverts and pipes) will only be suitable for a small number of species, most likely generalists.

• Wildlife overpasses must be designed to retain sufficient moisture to support plant growth without compromising the structural integrity of the structure if soils become waterlogged. On some structures, a wet and dry habitat type may be required for different species.

• Underpasses must be relatively open and have visible light because some species avoid dark spaces. Similarly, invertebrates that require continuous vegetation cover are unlikely to use underpasses without such cover.

• Water passing through underpasses should mimic the natural flow of waterways (i.e. varying velocities, natural pools and riffles, substrate with logs and rocks for shelter and no barriers) (Chapters 44 and 45). Some species (e.g. clams) are dependent on fish migration because their larvae are attached to them.

• Artificial lighting should be avoided wherever possible. Where required, use globes with a narrow spectrum of low wavelength light (monochromatic yellow light about 590 nm) and install them at the lowest possible height. Lamp boxes should be enclosed to prevent invertebrates from accessing the globe and focus the light only where it is needed (i.e. prevent unnecessary light spillage, Chapter 18).

• Avoid barriers and traps for flightless species, such as kerbs and drains.

Figure 29.5 Wildlife overpass for mammals, reptiles, amphibians and invertebrates in Germany. Designed to maximise habitat heterogeneity, this structure has three different soil types and ponds on both approach ramps and includes grassy and shrubby vegetation. A range of invertebrate species can use the bridge, including those preferring drier or moister environments and those that prefer grasses or shrubs. The two wire exclosures are part of an experiment to test the effect of ungulates on plant growth. Source: Photograph by H. Reck.

(A)   (B)

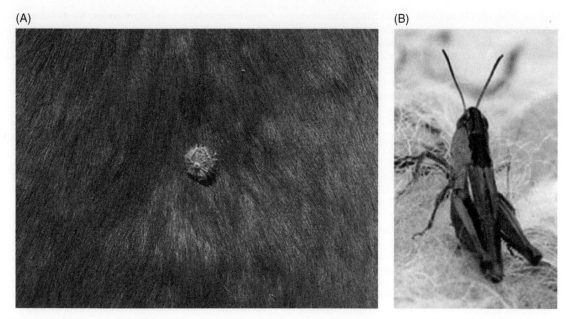

Figure 29.6 Vertebrates are important vectors for the movement of invertebrates: (A) limpet attached to red deer and (B) grasshoppers being transported by sheep. Source: Photograph by and reproduced with permission from B. Stöcker (A) and B. Schulz (B).

• Maintain a naturally diverse mix of plant species and allow for natural succession; and

• Design roadsides to be as natural as possible, with open ditches and slopes rather than impermeable surfaces, and of course avoid installing barriers or traps.

CONCLUSIONS

Invertebrates are the most diverse and abundant group of wildlife on earth and play critically important roles to maintain ecosystem health. Roads and the traffic they contain and other linear infrastructure impact invertebrates in a range of ways, including the mortality of vast numbers of individuals, restriction of movement and reduction in the amount and quality of habitat. Compared to vertebrates, there has been significantly less research on the impacts of roads and traffic on invertebrates and even less on mitigation solutions. Future road projects should explicitly consider invertebrates during the EIA and design stages to ensure future roads do not further endanger invertebrates. Critical aspects to consider include (i) further developing indicator species approaches to EIA and mitigation planning; (ii) maintaining connectivity among populations for movement and gene flow; and (iii) identifying and protecting species with key functional roles in ecosystem function. Efforts should take a landscape-scale focus on recovering populations of threatened species by reducing road mortality and restoring connectivity through the construction of appropriate crossing structures (Chapter 21) and movement corridors across fragmented landscapes.

ACKNOWLEDGEMENTS

Thanks to the many project partners and funders, especially Jörg Rietze, Henning Nissen and Jörn Krütgen and the Bundesamt für Naturschutz, the Bundesanstalt für Straßenwesen and the Landesbetrieb Straßenbau Schleswig-Holstein. R. van der Ree is supported by the Baker Foundation.

FURTHER READING

European Commission (2013): This report promotes integrated approaches (including all species and ecosystem services) to achieve sustainable solutions. This approach is relevant for green infrastructure, where functioning ecological networks across and along roads and other linear infrastructure are critical.

Haslett (2007): A European strategy for decision-makers, governments and conservationists that provides basic ecological facts about invertebrates and presents a clear vision, goals and objectives to conserve invertebrates and maintain the crucial ecosystem services they provide.

Jongman et al. (2011) and Reck et al. (2010): These two strategies promote ecosystem-based planning at international (European) and national (Germany) scales, respectively.

Samways (1994): This readable book is still relevant today and gives a profound and convincing introduction on the importance and practice of the conservation of invertebrates ranging from single species to landscape conservation approaches.

REFERENCES

Baumann, T., Biedermann, R. and E. Hoffmann 1999. Mitnahmeeffekte wirbelloser Zielarten am Beispiel von Trockenstandorten [Umbrella effect of invertebrate target species of dry grassland ecosystems]. In Amler, K. et al. (eds). Populationsbiologie in der Naturschutzpraxis [Population biology and practical nature conservation]: 37–45; Verlag Eugen Ulmer, Stuttgart.

Bykov, A.V. and A.B. Lysikov 1991. Mole burrows and pollution of forest soils adjacent to highways. Pochvovedenie **8**: 31–39.

Donath, H. 1986. Der Straßentod als bestandsgefährdender Faktor für Hummeln (Insecta, Hymenoptera, Bombidae) [Road mortality as threat to bumble bee populations]. Potsdam **22**: 39–43.

European Commission. 2013. Green Infrastructure – enhancing Europe's natural capital. European Commission, Brussels. Available at http://ec.europa.eu/environment/nature/ecosystems/ (accessed 17 February 2014).

Gate, I.M., McNeill, S. and M.R. Ashmore 1995. Effects of air pollution on the searching behaviour of an insect parasitoid. Water Air Soil Pollution **85**: 1425–1430.

Gepp, J. 1973. Kraftfahrzeugverkehr und fliegende Insekten [Car traffic and winged insects]. Natur und Landschaft **59**: 127–129.

Groh, K. and J.H. Jungbluth 1993. Aktionsräume und Neubesiedlung von Lebensräumen am Beispiel von Weichtieren – Konsequenzen für die Ausgleichbarkeit von Eingriffen [Migration range and patch colonization by the example of molluscs – implications for impact regulation]. Forschung Straßenbau und Straßenverkehrstechnik **636**: 183–189.

Gürlich, S. 2009. Die Bedeutung alter Bäume für den Naturschutz – Alt- und Totholz als Lebensraum für bedrohte Artengemeinschaften [The importance of old trees for nature conservation – matured timber and deadwood as habitat for endangered species communities]. In: LLUR (ed.). Historische Allen in Schleswig-Holstein – geschützte Biotope und grüne Kulturdenkmale. Abschlusspublikation des DBU-geförderten Modellprojekts 2005-2009 [Historic tree rows in the state Schleswig-Holstein – protected habitats and green cultural monuments. Results of the DBU (German Federal Environmental Foundation) – pilot project 2005–2009]: 49–82. Pirwitz Druck & Design, Kronshagen.

Hagen, H. 1984. Unfalltod bei *Calopteryx splendens* Harris [Road kill of *Calopteryx splendens*]. Libellula **3**: 100–102.

Haslett, J.R. 2007. European Strategy for the conservation of invertebrates. Nature and Environment, No. 145. Council of Europe Publishing, Strasbourg.

Hayward, M.W., Hayward, G.J. and G.I.H. Kerley 2010. The Impact of upgrading roads on the conservation of the threatened flightless dung beetle, *Circellum bacchus* (F.) (Coleoptera: Scarabaeidae). The Coleopterists Society **64**: 75–80.

International Union for Conservation of Nature (IUCN). 2010. IUCN Red List version 2010.4: Table 1. IUCN, Switzerland Available at (Numbers of threatened species by major groups of organisms (1996–2010)). http://www.iucnredlist.org/documents/summarystatistics/2010_4RL_Stats_Table_1.pdf and Table 3a (Status category summary by major taxonomic group (animals)). http://www.iucnredlist.org/documents/summarystatistics/2010_4RL_Stats_Table_3a.pdf (accessed 27 September 2014).

Jongman, R.H.G., Bouwma, I.M., Griffioen, A., Jones-Walters, L. and A.M.M. Van Doorn 2011. The Pan European Ecological Network – PEEN. Landscape Ecology **26**: 311–326.

Keller, I. and C.R. Largiadèr 2003. Recent habitat fragmentation caused by major roads leads to reduction of gene flow and loss of genetic variability in ground beetles. Proceedings of the Royal Society B: Biological Sciences **270**: 417–423.

Kiss, B., Illyes, E., Kozar, F. and E. Szita 2012. Biodiversity survey in Hungarian highway margins [a74]. In: IENE 2012 International Conference, October 21–24, 2012; Berlin-Potsdam, Germany. Publisher: Swedish Biodiversity Centre 2012. ISBN 978-91-89232-80-8. http://iene2012.iene.info/wp-content/uploads/2013/07/IENE2012_Proceedings.pdf. Accessed 18 January 2015.

Klemens, K., Straub, H.-P. and R. Umlauff-Zimmermann 1997. Wie beeinflusst der Straßenverkehr Oekosysteme? [How does road traffic effect ecosystems?] LFU Jahresbericht 1996/97, 50–53, Mannheim.

Lampe, U., Schmoll, T., Franzke, A. and K. Reinhold 2012. Staying tuned: grasshoppers from noisy roadside habitats produce courtship signals with elevated frequency components. Functional Ecology **26**: 1348–1354.

Landesanstalt für Umweltschutz Baden-Württemberg (LFU) (eds). 1994. 10 Jahre ökologisches Wirkungskataster Baden-Württemberg [10 years ecological impact register]. State Institute for Environment, Measurements and Nature Conservation Baden-Württemberg, Ettlingen.

Landres, P. B., J. Verner, and J. W. Thomas. 1988. Ecological uses of vertebrate indicator species. Conservation Biology **2**: 316–328.

Lindenmayer, D.B. and M.A. Burgman. 2005. Practical conservation biology. CSIRO Publishing, Collingwood, Australia.

Mader, H.J. 1979. Die Isolationswirkung von Verkehrsstraßen auf Tierpopulationenuntersucht am Beispiel von Arthropoden und Kleinsäugern der Waldbiozönose [The isolation effects of roads on arthropod populations and small mammals of forests]. Schriftenreihe für Landschaftspflege und Naturschutz **19**: 131.

Martin, K. and H. Roweck 1988. Zur anthropogenen Isolierung von Landschnecken-Populationen [Athropogenic isolation of snail populations]. Landschaft und Stadt **20**: 151–155.

McKenna, D.D., McKenna, K.M., Malcolm, S.B. and M.R.

Berenbaum 2001. Mortality of Lepidoptera along roadways in central Illinois. Journal of the Lepidopterists' Society **55**: 63–68.

Noordijk, J. 2009. Arthropods in linear elements – occurrence, behaviour and conservation management. Dissertation. Wageningen University, Wageningen.

Oleksa, A. and P. Tyszko-Chmielowiec. 2012. Transport infrastructure as green infrastructure: tree-lined roads as habitats and ecological corridors for Hermit Beetle and other organisms [a75]. In: IENE 2012 International Conference, October 21–24, 2012; Berlin-Potsdam, Germany. Publisher: Swedish Biodiversity Centre 2012. ISBN 978-91-89232-80-8. http://iene2012.iene.info/wp-content/uploads/2013/07/IENE2012_Proceedings.pdf. Accessed 18 January 2015.

Pfister, H.P., Keller, V., Reck, H. and B. Georgii 1997. Bio-ökologische Wirksamkeit von Grünbrücken über Verkehrswege [Bio-ecological effectiveness of green bridges across transport infrastructure]. Forschung Strassenbau und Strassenverkehrstechnik, **756**. German Federal Ministry of Transport.

Reck, H. 2013. Die ökologische Notwendigkeit zur Wiedervernetzung und Anforderungen an deren Umsetzung/ The ecological necessity and practical demands for defragmentation in Germany. Natur und Landschaft **88**: 486–496.

Reck, H. and G. Kaule 1993. Straßen und Lebensräume [Roads and habitats]. Forschung, Straßenbau und Straßenverkehrstechnik **654**: 170–219.

Reck, H., Hänel, K. and A. Huckauf 2010. Nationwide priorities for re-linking ecosystems: overcoming road-related barriers. Bundesamt für Naturschutz, Bonn. Available at http://www.bfn.de/fileadmin/MDB/documents/themen/landschaftsplanung/wiedervernetzung_oekosysteme-en.pdf; http://www.bfn.de/fileadmin/MDB/documents/themen/landschaftsplanung/wiedervernetzung_inter-nationalen.pdf (accessed 3 October 2014).

Roloff, A. (ed.) 2012. Aktuelle Fragen der Baumpflege, Planung, Wertschätzung und Wirkung von Stadtbäumen. Tagungsband, Dresdner StadtBaumtage in Dresden, Tharandt 15./16.03.2012 [Current questions on maintenance, planning, valuation and effects of urban trees.

Proceedings of the Dresdner town tree conference]. Forstwissenschaftliche Beiträge Tharandt Beiheft 13. Institut für Dendrochronologie Baumpflege und Gehölzmanagement, Tharandt.

Samways, M.J. 1994. Insect conservation biology. Chapman and Hall, London.

Schaffers, A.P., Raemakers, I.P. and V. Sýkora 2012. Successful overwintering of arthropods in roadside verges. Journal of Insect Conservation **16**: 511–522.

Schweizerischer Bund für Naturschutz (SBN). 1987. Tagfalter und ihre Lebensräume. Arten, Gefährdung, Schutz [Butterflies and their habitats. Species, threats, conservation]. First edition. Egg, Druck und Verlag Fotorotar AG, Basel.

Soluk, D.A., Zercher, D.S. and A.M. Worthington 2011. Influence of roadways on patterns of mortality and flight behavior of adult dragonflies near wetland areas. Biological Conservation **144**: 1638–1643.

Sporbeck, O., Meinig, H., Herrmann, M., Ludwig, D. and J. Lüchtemeier 2013. Vernetzung von Lebensräumen unter Brücken [Development of technical methods of linking biotopes beneath bridges]. Forschung Strassenbau und Strassenverkehrstechnik, **1101**. German Federal Ministry of Transport.

Vermeulen, H.J.W. 1994. Corridor function of a road verge for dispersal of stenotopic heathland ground beetles carabidae. Biological Conservation **69**: 339–349.

Verstrael, T., Van Den Hengel, B., Keizer, P.-J., Van Schaik, T., De Vries, H. and S. Van Den Berg. 2000. National highway verges...national treasures! Drukkerij Ronaveld, Den Haag.

Weidemann, G. and M. Reich 1995. Auswirkungen von Straßen auf Tiergemeinschaften der Kalkmagerrasen unter besonderer Berücksichtigung der Rotflügeligen Schnarrschrecke (*Psophus stridulus*) und des Schachbretts (*Melanargia galathea*) (Saltatoria, Acrididae und Lepidoptera, Satyridae) [Effects of roads on of dry grassland animal communities with specific respect to the rattle grasshopper and the marbled white]. Beiheft Veröffentlichungen für Naturschutz und Landschaftspflege Baden-Württemberg **83**: 407–424.

Chapter 30

CASE STUDY: PROTECTING CHRISTMAS ISLAND'S ICONIC RED CRABS FROM VEHICLES

Rob Muller and Mike Misso

Christmas Island and Pulu Keeling National Parks, Christmas Island, Indian Ocean, Australia

One of Christmas Island's most ecologically distinct and significant characteristics is its land crabs (Fig. 30.1). The island, positioned some 1500 km NW of Australia in the Indian Ocean, has the largest and most diverse land crab community in the world. The most conspicuous crab species is the endemic red crab, with an estimated population of 45 million in 2011. Red crabs play a vital role in recycling nutrients and shaping and maintaining the structure and plant species composition of the rainforests by consistently controlling the dynamics of seedling recruitment.

Each year, at the beginning of the wet season (usually October to November), most adult red crabs begin a remarkable migration from the forests to the coast to breed. It is one of the world's natural spectacles, attracting national and international visitors alike. During the peak migration periods, it's possible to walk among a moving sea of tens of thousands of crabs. Unfortunately, hundreds of thousands of red crabs can be killed annually during their breeding migration by vehicles. While two thirds of the island's 135 km² is protected as a national park (Christmas Island National Park), high levels of crab mortality still occur, especially outside the park. In cooperation with key stakeholders and the community, Parks Australia implements a range of mitigation measures to significantly reduce red crab mortality rates.

A key mitigation measure is the installation of small underpasses topped with cattle guards (Fig. 30.2) on selected roads with high traffic volume that cross major crab migration routes. Low plastic and (more recently) steel fencing along the road edges funnels crabs to the culverts, enabling them to safely cross (Fig. 30.2). An overpass, known locally as 'the crab bridge', was installed on one road for the 2012–2013 red crab migration (Fig. 30.3). While crossing structures and fencing for crabs are effective at reducing mortality rates, they do not eliminate all mortality as some crabs manage to breach the fence and the structures require ongoing maintenance.

The most effective and efficient approach is to close certain roads during peak migration periods, particularly those within the national park. However, the island is home to around 2000 human residents, and it is problematic to close roads elsewhere on the island due to high traffic volumes, proximity to residential areas and a lack of alternative crab-free routes. To tackle this problem, park and local shire staff temporarily close selected roads and redirect traffic onto roads with fewer crabs. Community and organisational education and support are also critical, and by working together, the community on Christmas Island is saving the lives of thousands of crabs annually and helping to maintain the ecosystem services they perform.

Handbook of Road Ecology, First Edition. Edited by Rodney van der Ree, Daniel J. Smith and Clara Grilo.
© 2015 John Wiley & Sons, Ltd. Published 2015 by John Wiley & Sons, Ltd.
Companion website: www.wiley.com\go\vanderree\roadecology

Figure 30.1 The red crab is endemic to Christmas Island, with an estimated population size of 45 million. Source: Photograph by and reproduced with permission of Di Masters.

(A)   (B)

Figure 30.2 Small, open-topped underpasses (A) with funnel fencing (B) are the key mitigation measures to reducing mortality and maintaining crab movements. Source: Photographs reproduced with permission of Parks Australia.

Figure 30.3 Crab overpass, Christmas Island. Source: Photograph reproduced with permission of Parks Australia.

ACKNOWLEDGEMENTS

We acknowledge the Christmas Island National Park staff, particularly Rangers (Azmi Yon, Eddly Johari and Max Orchard) for their efforts to conserve Christmas Island's red crabs over many years. We also thank those who support the efforts to reduce crab mortality rates, in particular the shire of Christmas Island, as well as the Christmas Island community.

FURTHER READING

Orchard, M. (2012). Crabs of Christmas Island. Christmas Island Natural History Association, Christmas Island.

Chapter 31

MAKING A SAFE LEAP FORWARD: MITIGATING ROAD IMPACTS ON AMPHIBIANS

Andrew J. Hamer[1], Thomas E. S. Langton[2] and David Lesbarrères[3]

[1]Australian Research Centre for Urban Ecology, Royal Botanic Gardens Melbourne, and School of BioSciences, The University of Melbourne, Melbourne, Victoria, Australia
[2]Transport Ecology Services (HCI Ltd.), Suffolk, UK
[3]Genetic & Ecology of Amphibians Research Group (GEARG), Centre for Evolutionary Ecology and Ethical Conservation, Faculty of Graduate Studies, Laurentian University, Sudbury, Ontario, Canada

SUMMARY

Amphibian populations are at risk of adverse impacts from roads and traffic. Roads constructed in the vicinity of wetlands and streams often interrupt amphibian movement pathways and can prevent individuals from accessing critical habitats. High numbers of amphibians are either deterred from crossing or killed by traffic, contributing to population declines. Although information on how best to mitigate the impacts of roads on amphibians is lacking in many areas of the world, several important lessons can be identified.

31.1 Roads and traffic contribute to amphibian mortality and population declines.

31.2 Planning the location of new roads to avoid amphibian habitat is critical.

31.3 Road construction should be timed to avoid periods of high amphibian activity.

31.4 Good design and placement of wildlife crossing structures is paramount.

31.5 Fencing must be designed to keep amphibians off the road.

31.6 Construction of replacement ponds as mitigation and compensation measures may provide solutions.

31.7 Wildlife crossing structures and fences must be maintained at least annually and to a high standard.

Handbook of Road Ecology, First Edition. Edited by Rodney van der Ree, Daniel J. Smith and Clara Grilo.
© 2015 John Wiley & Sons, Ltd. Published 2015 by John Wiley & Sons, Ltd.
Companion website: www.wiley.com\go\vanderree\roadecology

31.8 Some circumstances involve management of traffic flow and assistance with amphibian migrations.
31.9 The effectiveness of mitigation for amphibians should be studied pre- and post-construction.

Owing to the uncertainty surrounding the ability of crossing structures to mitigate road impacts for many amphibians, design and installation must be done under supervision by experienced and qualified biologists, and effectiveness demonstrated through monitoring. This approach is especially warranted in areas outside of Europe and North America, where there is little information on amphibian usage rates of crossing structures.

INTRODUCTION

Amphibians (frogs, toads, newts and salamanders) require interconnected areas of land and freshwater habitat to fulfil their complex life cycles, including areas in which to forage, shelter, breed and disperse. Roads and traffic can disrupt their life cycles and lead to population declines at local and regional scales. These declines are part of the continuing global decline in amphibian populations, with some declines attributed to habitat loss and fragmentation arising from road impacts (Beebee & Griffiths 2005; Beebee 2013). It is therefore critical to incorporate effective measures to avoid or mitigate the impact of new and existing roads on amphibians if the loss of amphibian diversity is to be reversed. The aims of this chapter are to identify the main impacts of roads and traffic on amphibians and highlight practical and effective solutions.

LESSONS

31.1 Roads and traffic contribute to amphibian mortality and population declines

Amphibians depend on closely distributed patches of aquatic and terrestrial habitat to complete their life cycles (Semlitsch 2002). Different species of frogs, toads, newts and salamanders require either still water (ponds and wetlands) or running water (creeks and streams) for breeding and larval development. At metamorphosis, the young emerge from the water and often disperse several kilometres. In temperate regions of the world, there are often seasonal cycles to amphibian movements. Species overwinter (hibernate) in terrestrial habitats such as forests, then migrate along specific pathways to breeding sites in spring. In warm climates (e.g. tropical parts of Australia, Africa, South America

and Southeast Asia), amphibian movement to breeding sites is often initiated by rainfall (see also Textbox 6.1). Fluctuating water levels can also trigger amphibian movements; in these instances, movement is not restricted to specific pathways and appears random. Amphibians therefore need to be able to move among aquatic and terrestrial habitats for long-term persistence of metapopulations (Smith & Green 2005). Reliance on both the aquatic and terrestrial environments is the main factor that makes many amphibian species sensitive to the fragmentation effects of roads (Marsh et al. 2008).

The construction of a road can destroy and modify amphibian habitat and can fragment movement pathways. Amphibians are particularly vulnerable to the negative effects of roads and traffic due to their relatively slow rate of movement, their moist delicate skin which is prone to desiccation and, for most species, an aversion to light and noise. The frequent positioning of roads close to wetlands can result in the mass mortality of individuals when they are moving. For example, Ashley and Robinson (1996) recorded mortality of 30,034 frogs over a 4-year period on a 3.6 km section of a two-lane road adjacent to a wetland on Lake Erie, Canada. Roads and traffic can impede amphibian movement and inflict high mortality, ultimately resulting in reduced population sizes (Fahrig et al. 1995; Hels & Buchwald 2001; Chapter 28). Indeed, amphibians are often the vertebrate group with the highest rates of road mortality (Glista et al. 2008; Textboxes 13.1 and 54.2). Amphibian road mortality rates are dependent on the volume and timing of traffic, as well as the season and weather conditions (particularly rainfall) which often influence migratory behaviour.

There are a range of direct and indirect effects during road construction and the operational phase that usually impact amphibians. For example, roads can degrade habitat through impacts on water flow levels, flow patterns, the quality of groundwater and altered water table fluctuation and by pollutants on roads and in roadside habitats (Chapters 44 and 45). To ensure the ability of amphibian populations to function, a critical challenge is thus to prevent roadkills and maintain habitat connectivity. Practical methods to reduce and to prevent amphibian road mortality, including road signs, speed reduction, temporary fencing, road closures, replacement habitats, road removal, wildlife crossing structures, are reviewed by Schmidt and Zumbach (2008).

31.2   Planning the location of new roads to avoid amphibian habitat is critical

The magnitude of the impact of roads on amphibian populations will largely depend on the location of the road. The most important aspect in planning new roads is to avoid placing them in important habitats and to accommodate the needs of amphibians. Ecological planning is vital at the earliest stage possible to inform the design and to determine how cost and safety considerations can be managed around essential ecological needs (Chapters 4 and 9).

There is a strong need to engage amphibian experts at the very earliest planning stages to ensure the best outcomes (Chapter 9). Good planning will require the identification of amphibian habitat and dispersal pathways. Care should be taken to consider all possible breeding habitats for amphibians during the initial planning stage. Field surveys for amphibian species should also consider that some species may be hard to find, and so it is always best to conduct repeat surveys using multiple survey techniques (Dodd 2010; Textbox 6.1).

After a road project has identified all available amphibian habitat and populations, the subsequent environmental impact assessment (Chapter 5) must consider the effect of the road and traffic within a broader landscape perspective and that the species may be part of a larger regional metapopulation. Particular attention needs to be paid to rare and declining species and species that are sensitive to habitat fragmentation (e.g. species that disperse widely) or habitat disturbance (e.g. species that are habitat specialists), as well as common species.

31.3   Road construction should be timed to avoid periods of high amphibian activity

Construction machinery and earthworks can kill or injure amphibians, and adequate controls must be in place during construction to reduce mortality.

Amphibians occurring in close proximity to the site may need to be removed before construction and transferred (translocated) to replacement habitat, preferably created within dispersal distance of the original site. Translocations must adhere to strict protocols to minimise the risk of spreading the amphibian chytrid fungus or other pathogens such as ranaviruses and must ensure that local gene pools are maintained. Guidelines on translocation of amphibians are available (IUCN/SSC 2013).

Temporary exclusion fencing should be used to 'trap out' an area prior to clearing and to direct movements of amphibians away from construction zones. Silt fencing can be used to minimise sedimentation of breeding ponds and streams. The effectiveness of these actions, however, will depend on the time of year that they are undertaken relative to the amphibian activity cycle. For example, when a breeding site is to be destroyed, trapping of a large proportion of adults may be more successful during periods of high activity, for example, when breeding congregations have formed and individuals can easily be collected. Alternatively, a 'doomed' pond may best be fenced off and destroyed when amphibians are not present, for example, during winter if all species are spring breeding and known to hibernate in upland habitats away from the pond.

It is important to start pre-construction translocation early as this process is often too short because construction often commences soon after the release of funding. Awarding separate contracts for the ecological and construction work often produces a better result than combining the two but only if the ecology and construction teams collaborate closely. Amphibian translocation must consider animal welfare and often requires scientific permits, which must be taken into consideration as it may delay pre-construction surveys. Translocation is also labour intensive. Most importantly, translocation may be fraught with uncertainty and should only be pursued when there are no other options for avoiding road impacts.

31.4   Good design and placement of wildlife crossing structures is paramount

Wildlife crossing structures and fencing have been shown to reduce road mortality and facilitate the seasonal migration of a range of amphibian species in several continents (Chapter 20). However, the location of crossing structures is critical to their ability to mitigate road impacts on amphibians. A typical approach is to use roadkill data to identify specific locations where animals attempt to cross roads and are killed (Chapter 13). While such hotspots are generally associated with good-quality habitat next to the road, recent population counts or roadkill data can be misleading if the mortality has already reduced local populations (Fahrig et al. 1995). Therefore, complementary studies, such as simulation modelling including animal movement behaviour and habitat distribution, could be used to identify the most likely crossing sites and hence location for crossing structures.

The most common means of maintaining connectivity for populations of amphibians in Europe and North America has been small road underpasses (Textbox 31.1; Fig. 21.1C). Evidence from trials of early designs of amphibian tunnel and fence systems suggested that it was extremely difficult to maintain amphibian dispersal patterns that were present prior to road building (Langton 1989). However, greater retention of original dispersal patterns has been achieved using modern designs of small tunnel and fence systems and using

---

Textbox 31.1 Designing the Besthorpe–Wymondham bypass, Norfolk, United Kingdom, for amphibians

This project comprehensively addressed the impact of a major new road (four lanes) (Fig. 31.1) for the great crested newt (endangered species under the UK Wildlife and Countryside Act 1981) and the common toad (Highways Agency 2001). The road would affect an old gravel pit containing a large pond and surrounded by intensive agricultural land. The pond was reduced in size but was largely retained next to the road, where amphibian fencing was installed on either side and maintained. Mitigation measures were initiated 2 years before construction began and included:
• Two replacement ponds constructed within 200 m of former habitat and one pond restored. Road edge fenced with permanent amphibian barrier.

• Arable fields converted to 6 ha of grassland; hedges, scrub and woodland planted to replace lost terrestrial habitat and along the road to screen light, noise, spray and dust from the road.
• Underpass ledge built into a river bridge and a road bridge constructed to maintain movement pathways for amphibians to facilitate genetic exchange.
Selected areas were also placed under nature conservation management. Newt and toad species are still breeding in habitats on both sides of the road after 20 years. The replacement habitat now supports a large population of great crested newts.

Figure 31.1 The Wymondham location in arable farmland showing retained, restored and constructed habitats for amphibians. Source: Photograph by and reproduced with permission from Mike Page.

Table 31.1 Recommended minimum tunnel dimensions for underpasses (width and area of the entrance) of different lengths and shapes, derived from the European (Iuell et al. 2003) and North American (Clevenger & Huijser 2009) guidelines.

Shape	Tunnel length				
	20 m	**20–30 m**	**30–40 m**	**40–50 m**	**50–60 m**
Rectangular	1.0 × 0.75 m **0.75 m²**	1.5 × 1.0 m **1.5 m²**	1.75 × 1.2 m **2.1 m²**	2.00 × 1.5 m **3 m²**	2.3 × 1.75 m **4 m²**
Circular	**1.0 m²**	**1.2 m²**	**1.6 m²**	**2 m²**	**2.5 m²**
Dome	1.0 × 0.7 m **c.0.50 m²**	1.4 × 0.7 m **c.0.70 m²**	1.6 × 1.1 m **c.1.3 m²**	—	—

Approximate surface area of underpass entrance is shown in bold.

wildlife overpasses and wider 'cut-and-cover' road tunnels (some around 500 m wide). In some instances, however, road tunnelling through hills and mountains may be the only way that original patterns of amphibian dispersal are fully retained. This is because even if large numbers of animals currently use the tunnels, dispersal patterns will change naturally over time, resulting in tunnels that are temporarily or no longer used. Ultimately, the aim is to avoid or minimise any change to dispersal patterns and to sustain similar species diversity, distribution and abundance to that prior to road construction.

While a 'one-size-fits-all' approach may not be applicable for amphibians, some common design specifications can be outlined. Tunnel shape, length and width, floor substrate type, moisture, temperature, humidity, air and water flow and day and night light levels are all important. Road underpasses designed specifically to facilitate amphibian movement include:

• Purpose-built underpasses of around 0.5 m width, mostly with surface slots to allow water and light in, keeping the tunnel moist and aired, with an internal temperature similar to the atmospheric temperature;
• Small underpasses under 3 m width including concrete rectangular box culverts.

Underpasses have been the most commonly used crossing structures for amphibians. Experimental results show that, because of a species-specific preference for tunnel use, it may be difficult to propose a single protective measure that works equally well for all species (Lesbarrères et al. 2004; Hamer et al. 2014). From early studies in Europe (Stolz & Podloucky 1983), it was recognised that bare concrete tunnels and culverts were not acceptable for many amphibians, particularly salamanders. One proposed improvement was placement of soil along the tunnel floor where soil thickness and water ingress keeps the inside damp along its length, but

examples of this being achieved are still few. Amphibian tunnel size suggestions for the European Union have been developed (Iuell et al. 2003) and recommended for use in North America (Clevenger & Huijser 2009). Although quantitative research on amphibian response to tunnel characteristics has been lacking, Table 31.1 shows tunnel sizes and surface areas according to these guidelines. It is clear that small (0.5 m width) surface tunnels can facilitate amphibian movement (Pagnucco et al. 2011). The precautionary principle should always be applied where there is a lack of information on acceptance rates of these structures: some tunnels larger in volume than that predicted is advised until the appropriate type has been accurately determined. The distance between tunnels also requires consideration, but it is generally recommended that amphibian tunnels should be spaced less than approximately 50 m apart (Ryser & Grossenbacher 1989). Current German guidelines state that inter-tunnel distances should be no more than 30 m, and the distance from the last tunnel to the end of the barrier wall or fence should be no less than 50 m, although this will vary (FMTBH 2000).

31.5   Fencing must be designed to keep amphibians off the road

The choice of fencing is equally as important as the design of the crossing structures (Figs. 20.5 and 20.6). Fencing should keep amphibians off the road and direct individuals towards the crossing structures. Low fencing without overhangs may not work well for tree frogs or other climbing or jumping amphibians. In Europe, effective fences are usually about 500 mm high and made of a wide range of materials with widely varying permanence: sheet polythene plastic for up to 3 years and shade cloth (woven polypropylene monofilament strands) and heavier plastics or metal mesh (galvanised and stainless steel) for longer periods. In Australia, effective (albeit with a short lifespan) frog fencing is usually comprised of shade cloth up to 1.2 m high, with an overhang (about 300 mm) to discourage climbing frogs from scaling the top of the fence. The fence is buried about 200 mm into the ground. The lifespan of fences varies significantly depending on the type of material and construction method and the harshness of the landscape (e.g. snow or sand drift, wind strength and UV levels, etc.). The size of mesh used must be small enough (<3 mm²) to prevent trespass by the smallest of amphibians – typically newly metamorphosed individuals. Current practice (and one that works well) is to secure shade cloth to metal star pickets or wood fence posts using high tensile fencing wire.

Fence type will ultimately be determined by the target species, and combination fencing for multiple species is a cost-effective option (Fig. 20.5). Longevity of fences is a vital cost issue, and calculation of construction and maintenance costs over long periods (>25 years) may reveal the most cost-effective solution. Robust fence designs are best; fences should be capable of withstanding some contact with maintenance machinery, including mowers and snow ploughs, as well as impact from windblown sand and snow. A flat platform in front of the fence may prevent vegetation from growing in immediate contact with the fence, which would otherwise allow individuals to get over the fence.

31.6 Construction of replacement ponds as mitigation and compensation measures may provide solutions

In the last two decades, amphibian habitat loss has been mitigated by the construction of replacement ponds in the vicinity of where the original ponds were located. While the dynamics of the colonisation process is often species specific (Lesbarrères et al. 2010), successful replacement ponds can be designed around simple habitat features, such as providing aquatic and terrestrial vegetation and keeping ponds free of predatory fish (Semlitsch 2002). For example, logs and mulch from felled trees and rocks removed during construction can be used as refuge sites around replacement ponds. However, even if replacement ponds are well connected to existing ponds with amphibian populations, it may take a few years for the ponds to become fully established (Textbox 31.2).

31.7 Wildlife crossing structures and fences must be maintained at least annually and to a high standard

Wildlife crossing structures and fences require regular inspections and maintenance to ensure they are functioning effectively, especially before the start of the breeding season of the target species (Chapter 17). Fences can be torn or broken by vehicles, fallen vegetation, vandalism, machinery and animals and should be routinely checked and mended if necessary at least twice a year (to coincide with seasonal migration to and from the wetland). Fences fall over if not supported properly (Fig. 31.4). Cracks and tears in fences need to be repaired rapidly, especially during periods of high amphibian movement (e.g. spring). Tall grasses and shrubs along the fence must be kept mown or pruned to prevent individuals from climbing over.

Underpasses and their entrances require checking for overgrown vegetation, accumulated silt and debris, erosion and animal excavation of soil floors and other problems. Blocked entrances may discourage amphibians from entering (Fig. 31.5). In cold temperate regions, road authorities should use non-caustic de-icing materials such as sand near slotted/open-topped tunnels. Surface-slotted tunnels can be washed out using a fire hose. If replacement ponds or ponds at entrance points to tunnels have been constructed, the water levels need to be maintained according to the specific requirements of the amphibian community present. In essence, annual fence and underpass maintenance is essential.

31.8 Some circumstances involve management of traffic flow and assistance with amphibian migrations

Alternative options to crossing structures and fences may be more feasible when funding for road mitigation is insufficient or landscape type (soils and gradient) makes under-road tunnelling impractical. Such measures include temporary road closures during the migration or movement season, reduced speed zones at crossing hotspots and the so-called bucket brigades of volunteers who carry migrating amphibians across roads. Installing temporary barrier fences on one side of the road can also act to intercept dispersing amphibians before they cross the road, and these individuals can then be released on the other side of the road by volunteers. These methods have public education and

Textbox 31.2 **Replacement ponds: Do they work?**

Following the construction of Highway A87 in western France, a restoration project was initiated in 1999, and the success of restoration has been monitored until 2011 (Fig. 31.2; Lesbarrères et al. 2010). The presence of amphibian species was recorded in eight replacement ponds, and this was compared to the original amphibian community in the area. Species richness initially declined following construction of the replacement ponds but generally returned to pre-construction levels (Fig. 31.3). The most significant habitat characteristics explaining amphibian species richness were pond surface area, pond depth and sun exposure. Moreover, high frog and toad species richness was associated with an increase in the amount of vegetation in and around the pond.

Figure 31.2 Replacement ponds constructed as part of the restoration project, western France 1999. Source: Photographs by D. Lesbarrères.

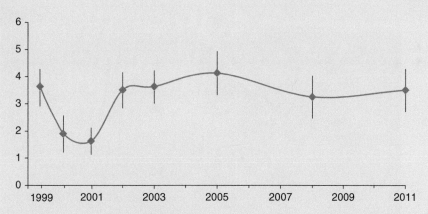

Figure 31.3 Average species richness (±1 SE) in eight replacement ponds for 12 years post-construction. Source: Adapted from Lesbarrères et al. (2010).

awareness benefits and do not require permanent mitigation structures to be installed, but they do require the approval of road agencies and high standards of road safety awareness. Experience also shows that volunteers can lose interest in the work over time.

Decisions regarding the location and timing of road closures and low-speed zones are based upon traffic volumes and road mortality thresholds, which indicate when critical levels are being reached (see review by Schmidt and Zumbach (2008)).

Figure 31.4 Amphibian fencing that has fallen over and is no longer functional. Frogs are able to cross onto the busy freeway, risking death or injury. The crossing structure accompanying the fence is shown in Figure 31.5. Source: Photograph by R. van der Ree.

Figure 31.5 Poorly designed and maintained underpasses for the growling grass frog, an endangered species in south-east Australia. The fence is not flush with the tunnel entrance and does not direct frogs into the tunnel. The culvert also has no ventilation slits and is probably dry. The entrance is overgrown with vegetation. There is no evidence that frogs move through this structure. Photograph by R. van der Ree.

31.9   The effectiveness of mitigation for amphibians should be studied pre- and post-construction

The success of mitigation measures can be assessed at different levels, including as a reduction in road mortality to the maintenance of genetic connectivity between populations. The best means of assessing the effectiveness of mitigation is to implement and evaluate them scientifically (Lesbarrères & Fahrig 2012; Chapter 10). Movement and mortality data should be collected before roads and mitigation structures are built and then after construction. Long lead-up times for research programmes (3–5 years) may be required because amphibian movements are often dependent on rainfall which can vary year to year and information collected over only one season may not adequately describe the annual movement patterns of the target species.

Because metapopulation processes are critical to many amphibian populations (Semlitsch 2002), effective monitoring requires surveys of populations near the road and in replacement habitats and also in the wider landscape (i.e. 100s of metres from the road). Surveying multiple sites over a large area and over multiple field seasons will cost more but is more likely to produce a robust assessment of the population. Unfortunately, it is unlikely to be achieved when road agencies are not mandated to fund these studies. Amphibian studies have used the density of roads within a certain distance of an amphibian monitoring site (e.g. 100–5000 m) to analyse the effect of roads on populations (e.g. Carr & Fahrig 2001; Eigenbrod et al. 2008). This approach can be used in modelling the effect of a new road on amphibian metapopulations.

Long-term (multiple generations) monitoring is needed as the negative impacts of roads may not become evident for several generations. The quality of habitat next to roads may decline rapidly or gradually over several years or more. If monitoring is conducted as part of a long-term research experiment, the potential to modify and adapt the mitigation over time may be better informed; that is, the mitigation is guided by adaptive management (Chapter 10). This approach will also guide future designs of mitigation structures.

Monitoring programmes that include counting amphibian roadkill may be misleading because the delicate carcasses of amphibians deteriorate rapidly or are quickly eaten by scavengers (Chapter 12). Roadkill surveys are best performed on nights in conditions where target species are likely to be moving and therefore prone to being killed by passing traffic (Mazerolle 2004).

CONCLUSIONS

Roads and traffic negatively affect amphibian populations worldwide, yet we have little experience of how best to mitigate these impacts outside of Europe and North America. Lessons learned in temperate regions of the northern hemisphere should be experimentally tested in other parts of the world to evaluate their local effectiveness. Wildlife crossing structures are most successful when they are installed at locations where amphibians cross roads in large numbers, for example, during spring migration events to breeding ponds. However, many species (e.g. in the tropics) move along unpredictable pathways, and so deciding where to install crossing structures in these cases should be guided by research into their movement patterns. The population-level effectiveness of mitigation for amphibians can only be assessed using long-term monitoring performed within a framework of adaptive management. Regular maintenance is also essential – torn fences and blocked underpasses do not prevent roadkill nor facilitate connectivity.

Considerable resources are required to effectively mitigate the impacts of roads on amphibians: planning several years before construction is essential; appropriately qualified personnel need to provide input into the project; rigorous monitoring takes time; and maintenance requires commitment. While these costs are minimal in relation to the budget of most large road projects, they still require justification. Road agencies must collaborate with appropriate experts and implement mitigation measures for amphibians in new road projects and by retrofitting existing roads, because failure to do so will only further add to the declines in amphibian populations.

ACKNOWLEDGEMENTS

The Baker Foundation and the Natural Sciences and Engineering Research Council of Canada provided generous support to A. J. Hamer and D. Lesbarrères, respectively.

FURTHER READING

Langton (1989): The first compilation of applied research into the effects of roads on amphibians and the effectiveness of mitigation measures.

Lesbarrères and Fahrig (2012): A review outlining previous attempts at mitigating road effects on animals and suggesting ways of improving studies into the effectiveness of crossing structures.

Marsh and Trenham (2001): Reviews amphibian spatial dynamics and highlights the importance of maintaining connections between aquatic and terrestrial habitat to maintain amphibian populations.

Schmidt and Zumbach (2008): Reviews methods to reduce road mortality in Europe since the early 1970s, suggests measures to evaluate the efficiency of tunnels and barrier walls and identifies priority areas of future research.

REFERENCES

Ashley, E. P. and J. T. Robinson. 1996. Road mortality of amphibians, reptiles and other wildlife on the Long Point Causeway, Lake Erie, Ontario. The Canadian Field-Naturalist **110**:403–412.

Beebee, T. J. C. 2013. Effects of road mortality and mitigation measures on amphibian populations. Conservation Biology **27**:657–668.

Beebee, T. J. C. and R. A. Griffiths. 2005. The amphibian decline crisis: a watershed for conservation biology? Biological Conservation **125**:271–285.

Carr, L. W. and L. Fahrig. 2001. Effect of road traffic on two amphibian species of differing vagility. Conservation Biology **15**:1071–1078.

Clevenger, A. P. and M. P. Huijser. 2009. Handbook for design and evaluation of wildlife crossing structures in North America. Department of Transportation, Federal Highway Administration, Washington, DC.

Dodd, C. K. Jr., editor. 2010. Amphibian ecology and conservation. A handbook of techniques. Oxford University Press, Oxford.

Eigenbrod, F., S. Hecnar and L. Fahrig. 2008. Accessible habitat: an improved measure of the effects of habitat loss and roads on wildlife populations. Landscape Ecology **23**:159.

Fahrig, L., J. H. Pedlar, S. E. Pope, P. D. Taylor and J. F. Wegner. 1995. Effect of road traffic on amphibian density. Biological Conservation **73**:177–182.

Federal Ministry of Transport, Building and Housing, Road Engineering and Road Traffic (FMTBH). 2000. Merkblatt zum Amphibienschutz an Straßen [Guidelines for Amphibian protection on Roads], 28 p. Germany.

Glista, D. J., T. L. DeVault and J. A. DeWoody. 2008. Vertebrate road mortality predominantly impacts amphibians. Herpetological Conservation and Biology **3**:77–87.

Hamer, A. J., R. van der Ree, M. J. Mahony and T. Langton. 2014. Usage rates of an under-road tunnel by three Australian frog species: implications for road mitigation. Animal Conservation **17**:379–387.

Hels, T. and E. Buchwald. 2001. The effect of road kills on amphibian populations. Biological Conservation **99**:331–340.

Highways Agency. 2001. Nature conservation management advice in relation to amphibians. Design Manual for Roads and Bridges, Volume 10, Section 4, Part 6 – HA 98/01. Highways Agency, London.

International Union for the Conservation of Nature/Species Survival Commission (IUCN/SSC). 2013. Guidelines for reintroductions and other conservation translocations. Version 1.0.

IUCN Species Survival Commission, Gland, Switzerland. www.iucnsscrsg.org (accessed 27 September 2014).

Iuell, B., G. J. Bekker, R. Cuperus, J. Dufek, G. Fry, C. Hicks, V. Hlavac, V. Keller, C. Rosell, T. Sangwine, N. Torslov and B. Wandall, editors. 2003. COST 341 wildlife and traffic: a European handbook for identifying conflicts and designing solutions. KNNV Publishers, Brussels.

Langton, T. E. S., editor. 1989. Amphibians and roads. Proceedings of the Toad Tunnel Conference Rendsburg, Federal Republic of Germany. ACO Polymer Products Ltd., Shefford.

Lesbarrères, D. and L. Fahrig. 2012. Measures to reduce population fragmentation by roads: what has worked and how do we know? Trends in Ecology and Evolution **27**:374–380.

Lesbarrères, D., T. Lodé and J. Merilä. 2004. What type of amphibian tunnel could reduce road kills? Oryx **38**:220–223.

Lesbarrères, D., M. S. Fowler, A. Pagano and T. Lodé. 2010. Recovery of anuran community diversity following habitat replacement. Journal of Applied Ecology **47**:148–156.

Marsh, D. M. and P. C. Trenham. 2001. Metapopulation dynamics and amphibian conservation. Conservation Biology **15**:40–49.

Marsh, D. M., R. B. Page, T. J. Hanlon, R. Corritone, E. C. Little, D. E. Seifert and P. R. Cabe. 2008. Effects of roads on patterns of genetic differentiation in red-backed salamanders, *Plethodon cinereus*. Conservation Genetics **9**:603–613.

Mazerolle, M. J. 2004. Amphibian road mortality in response to nightly variations in traffic intensity. Herpetologica **60**:45–53.

Pagnucco, K. S., C. A. Paszkowski and G. J. Scrimgeour. 2011. Using cameras to monitor tunnel use by long-toed salamanders (*Ambystoma macrodactylum*): an informative, cost-efficient technique. Herpetological Conservation and Biology **6**:277–286.

Ryser, J. and K. Grossenbacher. 1989. A survey of amphibian preservation at roads in Switzerland. Pages 7–13 in T. E. S. Langton, editor. Amphibians and roads. Proceedings of the Toad Tunnel Conference Rendsburg, Federal Republic of Germany. ACO Polymer Products Ltd., Shefford.

Schmidt, B. R. and S. Zumbach. 2008. Amphibian road mortality and how to prevent it: a review. Pages 157–167 in J. C. Mitchell, R. E. Jung Brown and B. Bartholomew, editors. Urban herpetology. Herpetological Conservation, Number 3. Society for the Study of Amphibians and Reptiles, Salt Lake City.

Semlitsch, R. D. 2002. Critical elements for biologically based recovery plans of aquatic-breeding amphibians. Conservation Biology **16**:619–629.

Smith, M. A. and D. M. Green. 2005. Dispersal and the metapopulation paradigm in amphibian ecology and conservation: are all amphibian populations metapopulations? Ecography **28**:110–128.

Stolz, F. M. and R. Podloucky. 1983. Krötentunnel als Schutzmaßnahme für wandernde Amphibien, dargestellt am Beispiel von Niedersachsen. [Toad tunnels as a protective measure for migrating amphibians in Lower Saxony]. Lower Saxony State Administration Office, Fachbeh. Nature Conservation: Conservation Information Service 3, No. 1, 20 pp.

Chapter 32

REPTILES: OVERLOOKED BUT OFTEN AT RISK FROM ROADS

Kimberly M. Andrews[1], Tom A. Langen[2] and Richard P. J. H. Struijk[3]

[1]Savannah River Ecology Lab, University of Georgia, Aiken, SC, USA
[2]Dept. of Biology, Clarkson University, Potsdam, NY, USA
[3]RAVON Foundation, Nijmegen, Netherlands

SUMMARY

Reptiles include many important prey and predator species and are integral to healthy ecosystem function. Reptiles encounter roads during seasonal migrations as well as during daily activities. Some reptiles avoid crossing roads, and others are attracted to them. Many species are prone to population declines due to wildlife-vehicle collision or the barrier effects of roads. Variability in how roads affect different reptiles necessitates a diverse set of management and mitigation tools. Cases of management successes at seasonal hotspots for reptile road crossings exist, but development of best management practices and mitigation techniques has lagged behind those for larger terrestrial vertebrates.

32.1 Many reptiles encounter roads when making seasonal movements, while other species are attracted to roads for foraging, nesting or temperature regulation.

32.2 Most reptiles are poor at evading oncoming vehicles, and many drivers are poor at detecting or avoiding reptiles.

32.3 Turtles, many snakes and some lizards have biological traits that cause their populations to be highly vulnerable to road mortality and barrier effects of roads.

32.4 Because reptiles are cryptic and attract little attention, reptile population declines caused by roads may be frequent but infrequently detected.

32.5 While fencing and crossing structures or periodic road closures can reduce reptile road mortality and maintain habitat connectivity, they rarely mitigate all barrier effects caused by road avoidance.

32.6 Novel approaches for maintaining reptile populations include predictive models of crossing locations for mitigation planning and management of roadside habitat to modify reptile behaviour.

Road planners, ecological consultants and wildlife researchers should presume that roads and traffic present a risk to all reptile populations because the behaviour and demography of most species make them susceptible to road mortality and fragmentation. Some risks can be eliminated with proactive management,

Handbook of Road Ecology, First Edition. Edited by Rodney van der Ree, Daniel J. Smith and Clara Grilo.
© 2015 John Wiley & Sons, Ltd. Published 2015 by John Wiley & Sons, Ltd.
Companion website: www.wiley.com\go\vanderree\roadecology

while others may only be reduced because the road and traffic still represent a significant modification to habitat. Furthermore, many effects on reptiles are difficult or impossible to mitigate retroactively because some species will always avoid open habitats, and in many instances, proposed mitigation options will not currently be supported due to additional costs, logistical challenges and lack of public support.

INTRODUCTION

Turtles, lizards, snakes and crocodilians are major components of terrestrial and aquatic food webs, especially in warmer climates. The ecology and behaviour of reptile species is highly diverse and poorly known for many species, but certain generalisations are possible. Many reptiles undergo periodic movements to locate mates or nesting sites, discover new foraging grounds or return to a hibernation or aestivation shelter. Reptiles often move slowly and tend to freeze or take a defensive posture when they detect danger – ineffective responses for oncoming vehicles. Turtles, crocodilians and many snakes take years to reach sexual maturity but compensate by naturally high adult survivorship; therefore, populations are especially vulnerable to adult mortality from new roads or increases in traffic volume.

Many of the world's reptile species are in decline (Gibbons et al. 2000). The loss, fragmentation and degradation of habitat are the biggest drivers, but pollution, over-exploitation, invasive species, new pathogens and climate change also contribute. There is accumulating evidence that roads and traffic play a role in the decline of many populations of reptiles, often in synergy with other stressors. While roads are a pervasive type of habitat fragmentation, degradation and loss, they are also significant drivers of landscape-level habitat conversion and development (Chapters 2, 3, and 51). The aims of this chapter are to provide the biological and ecological basis explaining when, where and why reptiles encounter roads and present some mitigation options to reduce the impacts of roads and traffic.

LESSONS

32.1 Many reptiles encounter roads when making seasonal movements, while other species are attracted to roads for foraging, nesting or temperature regulation

Reptiles can be surprisingly wide-ranging, and the likelihood of them encountering a road can be high. For example, wetland-associated reptiles (turtles,

crocodilians, some snakes and a few lizards) will move large distances between permanent and seasonal wetlands. Mass movements can occur during droughts as wetlands dry, resulting in large numbers of reptiles crossing roads (Aresco 2005a; Rees et al. 2009). Reptiles in highly seasonal climates have a period of hibernation or aestivation, and many species hibernate communally in dens of tens to thousands of individuals. For example, spring migration from hibernation dens to summer habitat in Manitoba, Canada, resulted in the death of roughly 10,000 red-sided garter snakes annually on a 3 km segment of road (Seburn & Seburn 2000).

Some reptiles are characterised by sex-specific long-distance movements associated with breeding. For many snakes, lizards, tortoises and other terrestrial reptiles, it is the males that travel long distances in search of potential mates. Females may also migrate in search of appropriate sites to deposit eggs; in the Netherlands, a female grass snake migrated 6.1 km to do so (Janssen 2003).

Road verges usually have good sun exposure, low vegetation, areas of bare soil and radiant heat from the road, making them highly attractive to reptiles for nesting (e.g. aquatic turtles, Steen et al. 2006; Fig. 32.1) and foraging sites (Daniel J. Smith, personal observations). Some scavenging species are attracted to roads to feed on roadkill (e.g. Smith & Dodd 2003). Rock-associated reptiles may colonise roadsides that are 'armoured' with rocks, and wetland-associated reptiles may be attracted to roadside drainage swales. It is usually unclear whether this created habitat along roadsides results in a larger population of reptiles than might otherwise exist in the landscape or else creates an ecological trap, luring animals into areas where survival or breeding success is low.

Reptiles actively regulate their body temperature by moving to thermally favourable microclimates. Road pavement and shoulders warm up during the day and radiate heat at night, and some species may be attracted to absorb the heat. The risk of wildlife-vehicle collision (WVC) and mortality WVC can be high because animals may remain on the road for prolonged periods.

(A)

(B)

Figure 32.1 (A) Diamondback terrapin nesting on the road shoulder and (B) in a landscaped median along the causeway to the barrier island, Jekyll Island, Georgia, United States. Source: Photographs by and reproduced with permission from B. A. Crawford.

32.2 Most reptiles are poor at evading oncoming vehicles, and many drivers are poor at detecting or avoiding reptiles

Many reptiles are killed on roads because they are too slow to avoid vehicles or are too small for drivers to see and avoid (Fig. 32.2). Additionally, species typically respond to oncoming vehicles with their primary defensive behaviour; species that rely on camouflage to avoid danger typically immobilise, while many turtle species tuck inside their shell. Immobilisation behaviour increases the time spent on the road, thereby increasing an animal's risk of mortality. This risk is not only from vehicles, but also from bicycles and pedestrians (Fig. 32.3).

Negative cultural perceptions of reptiles, and resulting driver behaviour, strongly affect the safety of reptiles on roads. Deliberate killing of snakes on roads is common around the world, and some people even swerve to hit turtles (e.g. Langley et al. 1989; Bush et al. 1991).

Figure 32.2 Reptiles, in this case a slow worm, often cross roads, making them vulnerable to road mortality. Source: Photograph by R. P. J. H. Struijk.

32.3 Turtles, many snakes and some lizards have biological traits that cause their populations to be highly vulnerable to road mortality and barrier effects of roads

Animals that reproduce from a young age and have high fecundity are more resilient against losses due to roadkill (Chapter 28). In contrast, turtles, many snakes and some lizards are at greater risk of population decline because of their low fecundity and late age of reproduction. Some lizards and snakes do not reach

sexual maturity until after age 5, and some turtles may not begin breeding until in their mid-teens, don't reproduce every year and may only lay a few eggs in each clutch.

When one sex more frequently encounters roads, as is typically the case with movements for breeding, road mortality is correspondingly sex biased, and the adult sex ratio may become skewed (Aresco 2005b; Andrews & Gibbons 2008). Using a simulation approach, it was

Figure 32.3 Mortality of sand lizard along bike path in the Netherlands. Source: Photograph by R. P. J. H. Struijk.

concluded that some turtle populations in eastern North America are likely to be in decline due to adult female road mortality (Gibbs & Shriver 2002).

Many reptiles are distributed in small, localised populations and are less mobile than large mammals. Therefore, they experience a greater risk of local extinction and reduced chances of recolonisation because roads are often a barrier to movement (e.g. Andrews et al. 2008). Typifying this, the extinction of a small population of black rat snakes in Ontario, Canada, seems likely under current levels of road mortality, even though the number of roadkill is relatively small (<10 per year out of a population of 340 adults; Row et al. 2007). While roads may reduce the movement and mating success of reptiles, resulting in decreased genetic diversity, we don't know how frequently this leads to population decline and local extinction.

32.4 Because reptiles are cryptic and attract little attention, reptile population declines caused by roads may be frequent but infrequently detected

The density of reptile populations is challenging to estimate, and changes in population size are difficult to detect. Most reptiles are small, cryptic and nocturnal or inhabit places that are difficult for humans to access. Furthermore, reproduction is not as detectable as it is

for other species (e.g. amphibians), and breeding behaviour is unknown for even some common reptile species. These challenges complicate population-level assessments for many reptile species. However, despite these challenges, numerous examples of population declines due to roads and traffic exist. For example, turtle abundance was lower and adult sex ratios were male-biased in ponds located in areas of high road density in New York, United States; this sex bias was attributed to female road mortality during nesting (Steen & Gibbs 2004). In Ontario, Canada, reptile species richness at wetlands was negatively correlated with local road density, and the decrease was more dramatic where roads have existed for decades than where roads were more recently constructed (Findlay & Houlahan 1997). The density of lava lizards was reduced near a highway across Santa Cruz Island in the Galapagos Islands, Ecuador (Tanner & Perry 2007). Similarly, desert tortoises in the Mojave Desert of California, United States, also had reduced populations up to 400 m from a highway (Boarman & Sazaki 2006). A recent review of the response of reptile populations to roads and traffic found only eight published studies on the topic, many fewer than for amphibians, mammals or birds, and they concluded reptile populations typically decline near roads (Rytwinski & Fahrig 2012; Chapter 28).

The decline of populations near roads may be due to road mortality, barrier effects, increases in human disturbance (including collecting or killing reptiles), pollution of soils and water bodies or spread of invasive species (Andrews et al. 2015). Because these stressors may act synergistically, managers should consider multiple risks when designing mitigation for reptiles.

32.5 While fencing and crossing structures or periodic road closures can reduce reptile road mortality and maintain habitat connectivity, they rarely mitigate all barrier effects caused by road avoidance

Strategies to reduce mortality of reptiles and maintain connectivity include fences, wildlife crossing structures and periodic road closures. Fences and various types of crossing structures are effective at reducing reptile mortality (e.g. reviewed in Struijk (2011); Langen (2012); Andrews et al. 2015). Recent guides provide specifications on fencing and crossing structures for reptiles (Iuell et al. 2003; Andrews et al. 2008; Clevenger & Huijser 2011; Andrews et al. 2015; references in Chapter 59), although more field testing and monitoring are needed to thoroughly evaluate effectiveness.

Barrier fencing installed adjacent to the road can effectively reduce reptile mortality by excluding animals from the road. Unfortunately, this technique can be so effective that the road becomes a complete barrier,

preventing all exchange and gene flow across the road. Funnel fencing used in conjunction with underpasses (e.g. culverts and tunnels) is the most frequent application. Fencing materials must be durable and withstand wear from severe environmental conditions and roadside maintenance. Fencing for reptiles has been constructed of prefabricated galvanised steel, (polymer) concrete and silt-screen sheeting affixed to the bottom of existing or stand-alone fences (Fig. 20.5). The base of the fence should be buried for stability and to prevent trespass underneath the fence by burrowing animals and to reduce exposed gaps due to erosion (Chapter 20).

Effectiveness is mainly influenced by the construction materials (durability, smoothness) and design (height, length, shape) along with maintenance of the fence and adjacent vegetation (Struijk 2010). While the use of silt-screening and other similar materials has demonstrated temporary effectiveness with reptiles, these materials are easily damaged and require extra maintenance, and should therefore not be considered for permanent use (Fig. 31.4, Chapters 17, 20 and 31). Additionally, these fences are less effective for climbing species and can result in entrapment of smaller species and juveniles who can try to crawl through finer mesh. In the Netherlands, fencing (and crossing structures) reduced reptile roadkill, until regular mowing ceased, allowing animals to climb over the fence via vegetation (Fig. 32.4A) (Struijk 2011). The installation of a 'lip' can prevent many smaller reptiles from crawling over

(A)

(B)

Figure 32.4 Barrier fence for reptiles with (A) grass that needs mowing to prevent animals from climbing over and (B) a well-maintained and durable fence with wide base, smooth wall and overhanging lip. Source: Photographs by R. P. J. H. Struijk.

the fence (Figs 20.4 and 32.4B). Since fences cannot always be continuous because of access roads or the expense and aesthetics of installing extended lengths of fencing, such fence breaks provide a way for animals to reach the road and be killed (Dodd et al. 2004). It is vital that fences are properly designed and adequately tested to ensure their effectiveness under different conditions.

Various types of crossing structures are used by reptiles, and their effectiveness is influenced by structure size, whether it has an open (Figs. 35.2A, B) or closed (Fig. 31.5) top, other species-specific needs and traffic volume. The most effective crossing structures are likely to be those where the habitat for the target species can extend across the structure, providing a continuous strip of cover. While underpasses are more typically used for small vertebrate species that are hesitant to traverse open areas with higher predation risk (e.g. Andrews et al. 2015), lizard and snake species have also been observed on various wildlife overpasses in Europe (e.g. Teufert et al. 2005; Puky et al. 2007; Struijk 2011). Structures that were not originally designed for wildlife, such as viaducts and storm-water culverts, can be adapted to enhance road crossing by reptiles by adding tree stumps (e.g. Fig. 32.6) and soil substrate (Struijk 2011).

(A) (B)

Figure 32.5 Underpasses for reptiles should aim to maximise openness. (A) This underpass with grates at Drents-Friese Wold, Netherlands, can be installed on roads with lower traffic densities, and is used by common lizard and common European adder. (B) This large underpass is open in the median to allow sunlight to enter. Source: (A) Photograph by R. P. J. H. Struijk and (B) Photograph by and reproduced with permission from S. Jansen.

(A) (B)

Figure 32.6 (A) View of the Terlet wildlife overpass above the A50 Highway, Netherlands, with a row of tree stumps that have been added to improve crossing rates by wildlife. (B) At least three species of lizard, including this sand lizard, and one species of snake have used this crossing structure. Source: Photographs by R. P. J. H. Struijk.

Road closures and enhanced signs have been used to reduce road mortality of reptiles where the rate of road crossing is high, localised, episodic and predictable (Fig. 32.7; Chapter 24). For example, a road in Illinois, United States, is closed for 2 months in spring and autumn to provide safe passage for massasauga rattlesnakes where it passes between overwintering and summer habitat (Shepard et al. 2008). Road closures are unlikely to be practical in most instances unless there is a convenient, alternate route for motorists (Chapter 3).

Habitat fragmentation and barrier effects can result in severe population-level impacts for reptiles that avoid roads and are attracted to roads but don't cross or those that experience such high rates of mortality that gene flow is reduced. Some species may even avoid roads without traffic because of their aversion to the open road. Data on these effects on reptiles are among the most challenging to acquire as they must be collected at a landscape scale, and avoidance behaviour is difficult to detect. However, these effects likely cause large-scale conservation problems and are a priority for research and mitigation because many behavioural effects on reptiles cannot be mitigated after a road is constructed. In addition, mitigation options for reptiles

Figure 32.7 A diamondback terrapin crossing the road at a hotspot along the causeway to Jekyll Island, Georgia, United States. Signage equipped with a flasher that activates at peak movement times alerts motorists to look out for turtles. Source: Photograph by and reproduced with permission from B. A. Crawford.

that encounter roads at low densities over broad expanses of roadway are much more limited than for species with localised hotspots (see Chapter 13).

32.6 Novel approaches for maintaining reptile populations include predictive models of crossing locations for mitigation planning and management of roadside habitat to modify reptile behaviour

To reduce the negative impact of roads on reptiles, road planners need to prioritise efforts for mitigation at locations critical to population viability, such as where individuals need to cross the road or locations with high rates of mortality. Local knowledge may help pinpoint locations with high rates of WVC; however, most reptiles are inconspicuous, so roadkill usually goes unnoticed. Furthermore, for reptile species that are deterred by roads or road traffic, barriers to movement may not be detected because animals don't enter the roadway. The combination of small carcass size and inconspicuous ecological impacts mean that the true effects of roads on reptiles is often underestimated and therefore overlooked in transportation planning and design. Road planners should contact herpetologists about the behaviour and habitat preferences of reptile species of concern so that locally appropriate habitat management can be designed.

A reliable way to detect critical road segments for reptiles is by using predictive hotspot models (Chapter 13; Langen 2010; Andrews et al. 2015). Some models use data on where reptiles occur, their habitat preferences or behaviour to predict where most animals will encounter roads (e.g. Beaudry et al. 2008). Others use important habitat features (e.g. proximity to wetlands, habitat type, traffic volume, proximity to nesting substrate or hibernacula) associated with critical road locations for reptiles (e.g. roadkill hotspots, blocked movement corridors) as predictors (e.g. Langen et al. 2012). Hotspot models are mostly useful for reptile species that encounter roads in high numbers and at specific locations.

Reptiles often have strong preferences for habitat features, avoiding some and preferentially moving through or inhabiting others. Although there is little published research on this topic, there are good reasons to consider vegetation and substrate management as options for encouraging or discouraging road crossing or guiding animals to passage structures. For example, many snakes and lizards prefer to remain close to cover such as logs or rocks. A strip of natural cover or vegetation parallel to the road in the roadside verge could guide reptiles from the surrounding landscape to a crossing structure (Fig. 39.2). Permitting growth of the tree canopy over a road (Fig. 40.1A) can make the local microclimate cooler and more humid and thus more suitable for forest-living reptiles while additionally providing canopy connectivity for arboreal reptiles (Chapter 40).

Much research is needed to field-test solutions to behavioural road avoidance and roadside nesting by turtles where fencing to keep them off the road is not possible. These responses to roads cannot be resolved through the placement of crossing structures because connectivity is not the issue. In addition, linear features offer a convenient and easy way for egg-eating predators to depredate roadside nests, potentially resulting in greater losses of recruitment than that attributed solely to road mortality.

CONCLUSIONS

Many reptile species are at risk of population declines due to excessive road mortality or reduced habitat connectivity. However, reptiles receive less attention from the public and managers than other vertebrates, and population declines caused by roads and traffic are poorly documented. The magnitude of the impacts of roads on reptile populations is uncertain, but likely to be serious, especially where road densities and traffic volumes are high.

There has been some recent progress towards developing predictive models of roadkill hotspots or critical locations for connectivity to use as tools in locating mitigation measures. Road planners have had some successes at reducing reptile road mortality and maintaining habitat connectivity by using fences to funnel reptiles towards crossing structures. Vegetation and substrate management along verges may also serve to guide animals to crossing structures. Nevertheless, there remains much to be done to identify populations at risk of decline caused by roads, and towards implementing management and mitigation practices that prevent or reverse such declines.

Some road effects, such as behavioural avoidance of open spaces, cannot be mitigated as long as the road is in place and habitat is disrupted. While many road effects can be mitigated and reduced, others are permanent consequences and should be proactively considered prior to construction. For some species, effective mitigation is unlikely, and we should avoid building roads near populations of these sensitive species.

ACKNOWLEDGEMENTS

We are grateful to the researchers, habitat managers and engineers who have contributed to knowledge of the effects of roads on reptiles and in testing potential solutions. We thank the transportation and environmental agencies that have funded initiatives to understand ecological effects of roads on reptiles and to test mitigation design options.

FURTHER READING

Andrews et al. (2015): A recently published book that focuses on the effects of roads and traffic on small animals and the latest information on effective mitigation options, with a primarily North American focus.

Clevenger and Huijser (2011): A manual identifying road and mitigation design features and specifications for wildlife, including for reptiles and other small-animal groups.

Graeter et al. (2013): A manual with survey techniques for reptiles and habitats that could direct priority species and techniques for road assessments.

McDiarmid et al. (2012): A book that presents techniques for capturing and marking reptiles and analysis for population assessments.

SETRA (2005): A manual outlining crossing structure types and design specifications for small-animal groups, including reptiles. Available for free download in French and English.

REFERENCES

Andrews, K. M. and J. W. Gibbons. 2008. Roads as catalysts of urbanization: snakes on roads face differential impacts due to inter- and intraspecific ecological attributes. Pages 145–153 in J. C. Mitchell, R. E. Jung and B. Bartholomew, editors. Urban herpetology. Herpetological Conservation, Vol. 3, Society for the Study of Amphibians and Reptiles, Salt Lake City, UT.

Andrews, K. M., J. W. Gibbons and D. M. Jochimsen. 2008. Ecological effects of roads on amphibians and reptiles: a literature review. Pages 121–143 in J. C. Mitchell, R. E. Jung and B. Bartholomew, editors. Urban herpetology. Herpetological Conservation, Vol. 3, Society for the Study of Amphibians and Reptiles, Salt Lake City, UT.

Andrews, K. M., P. Nanjappa and S. P. D. Riley, editors. 2015. Roads and ecological infrastructure: concepts and applications for small animals. Johns Hopkins University Press, Baltimore, MD.

Aresco, M. J. 2005a. Mitigation measures to reduce highway mortality of turtles and other herpetofauna at a north Florida lake. Journal of Wildlife Management **69**:549–560.

Aresco, M. J. 2005b. The effect of sex-specific terrestrial movements and roads on the sex ratio of freshwater turtles. Biological Conservation **123**:37–44.

Beaudry, F., P. G. deMaynadier and M. L. Hunter, Jr. 2008. Identifying road mortality threat at multiple spatial scales for semi-aquatic turtles. Biological Conservation **141**:2550–2563.

Boarman, W. I. and M. Sazaki. 2006. A highway's road-effect zone for desert tortoises (*Gopherus agassizii*). Journal of Arid Environments **65**:94–101.

Bush, B., R. Browne-Cooper and B. Maryan. 1991. Some suggestions to decrease reptile roadkills in reserves with emphasis on the western Australian wheatbelt. Herpetofauna **21**:23–24.

Clevenger, A. P. and M. P. Huijser. 2011. Wildlife crossing structure handbook – design and evaluation in North America. Federal Highway Administration FHWA-CFL/TD-11-003. Available from http://www.cflhd.gov/programs/techdevelopment/wildlife/documents/01_Wildlife_Crossing_Structures_Handbook.pdf (accessed 9 January 2015).

Dodd, C. K., Jr., W. J. Barichivich and L. L. Smith. 2004. Effectiveness of barrier wall and culverts in reducing wildlife mortality on a heavily traveled highway in Florida. Biological Conservation **118**:619–631.

Findlay, C. S. and J. Houlahan. 1997. Anthropogenic correlates of species richness in southeastern Ontario wetlands. Conservation Biology **11**:1000–1009.

Gibbons, J. W., D. E. Scott, T. J. Ryan, K. A. Buhlmann, T. D. Tuberville, B. S. Metts, J. L. Greene, T. Mills, Y. Leiden, S. Poppy and C. T. Winne. 2000. The global decline of reptiles, deja vu amphibians. Bioscience **50**:653–666.

Gibbs, J. P. and W. G. Shriver. 2002. Estimating the effects of road mortality on turtle populations. Conservation Biology **16**:1647–1652.

Graeter, G. J., K. A. Buhlmann, L. R. Wilkinson and J. W. Gibbons, editors. 2013. Inventory and monitoring: recommended techniques for reptiles and amphibians. Partners in Amphibian and Reptile Conservation Technical Publication IM-1, Birmingham, AL.

Iuell, B., G. J. Bekker, R. Cuperus, J. Dufek, G. Fry, C. Hicks, V. Hlavác, V. Keller, C. Rosell, T. Sangwine, N. Tørsløv and B. L. M. Wandall. 2003. Cost 341: habitat fragmentation due to transportation infrastructure, wildlife and traffic: a European handbook for identifying conflicts and designing solutions. European Co-operation in the Field of Scientific and Technical Research, Brussels.

Janssen, I. 2003. De ringslang als zwerver. RAVON **16**:1–3.

Langen, T. A. 2010. Predictive models of herpetofauna road mortality hotspots in extensive road networks: three approaches and a general procedure for creating hotspot models that are useful for environmental managers. Pages 475–486 in P. J. Wagner, D. Nelson and E. Murray, editors. Proceedings of the 2009 International Conference on Ecology and Transportation. Available from http://www.icoet.net/ICOET_2009/09proceedings.asp (accessed 9 January 2015).

Langen, T. A. 2012. Design considerations and effectiveness of fencing for turtles: three case studies along northeastern New York state highways. Pages 545–556 in P. J. Wagner, D. Nelson and E. Murray, editors. Proceedings of the 2011 International Conference on Ecology and Transportation.

Center for Transportation and the Environment, North Carolina State University, Raleigh, NC. Available from http://www.icoet.net/ICOET_2011/proceedings.asp (accessed 19 November 2014).

Langen, T. A., K. Gunson, C. Scheiner and J. Boulerice. 2012. Road mortality in freshwater turtles: identifying causes of spatial patterns to optimize road planning and mitigation. Biodiversity and Conservation **21**:3017–3034.

Langley, W. M., H. W. Lipps and J. F. Theis. 1989. Responses of Kansas motorists to snake models on a rural highway. Transactions of the Kansas Academy of Science **92**:43–48.

McDiarmid, R. W., M. S. Foster, C. Guyer, J. W. Gibbons and N. Chernoff. 2012. Reptile biodiversity: standard methods for inventory and monitoring. University of California Press, Berkeley, CA.

Puky, M., J. Farkas and M. T. Ronkay. 2007. Use of existing mitigation measures by amphibians, reptiles, and small to medium-size mammals in Hungary: crossing structures can function as multiple species-oriented measures. Pages 521–530 in C. Leroy Irwin, D. Nelson and K. P. McDermott, editors. Proceedings of the 2007 International Conference on Ecology and Transportation. Center for Transportation and the Environment, North Carolina State University, Raleigh, NC.

Rees, M., J. H. Roe and A. Georges. 2009. Life in the suburbs: behavior and survival of a freshwater turtle in response to drought and urbanization. Biological Conservation **142**:3172–3181.

Row, J. R., G. Blouin-Demers and P. J. Weatherhead. 2007. Demographic effects of road mortality in black ratsnakes (*Elaphe obsoleta*). Biological Conservation **137**:117–124.

Rytwinski, T. and L. Fahrig. 2012. Do species life history traits explain population responses to roads? A meta-analysis. Biological Conservation **147**:87–98.

Seburn, D. and S. Seburn. 2000. Conservation priorities for the amphibians and reptiles of Canada. World Wildlife Fund Canada, Toronto, ON, and the Canadian Amphibian and Reptile Conservation Network, Delta, British Columbia.

Service d'Etudes techniques des routes et autoroutes (SETRA). 2005. Facilities and measures for small fauna, technical guide. SETRA, Ministere de I'Ecologie du Developpment et de I'Amenagement durables, France.

Shepard, D. B., M. J. Dreslik, B. C. Jellen and C. A. Phillips. 2008. Reptile road mortality around an oasis in the Illinois corn desert with emphasis on the endangered eastern massasauga. Copeia **2008**:350–359.

Smith, L. L. and C. K. Dodd, Jr. 2003. Wildlife mortality on U.S. highway 441 across Paynes Prairie, Alachua County, Florida. Florida Scientist **66**:128–140.

Steen, D. A. and J. P. Gibbs. 2004. Effects of roads on the structure of freshwater turtle populations. Conservation Biology **18**:1143–1148.

Steen, D. A., M. J. Aresco, S. G. Beilke, B. W. Compton, C. K. Dodd, Jr., H. Forrester, J. W. Gibbons, J. L. Greene-McLeod, G. Johnson, T. A. Langen, M. J. Oldham, D. N. Oxier, R. A. Saumure, F. W. Schueler, J. Sleeman, L. L. Smith, J. K. Tucker and J. P. Gibbs. 2006. Relative vulnerability of female turtles to road mortality. Animal Conservation **9**:269–273.

Struijk, R. P. J. H. 2010. Rasters voor reptielen. Een verkennende studie. Stichting RAVON, Nijmegen: 37 pp.

Struijk, R. P. J. H. 2011. Het gebruik van faunapassages door reptielen. De Levende Natuur **112**:108–113.

Tanner, D. and J. Perry. 2007. Road effects on abundance and fitness of Galápagos lava lizards (*Microlophus albemarlensis*). Journal of Environmental Management **85**:270–278.

Teufert, S., M. Cipriotti and J. Felix. 2005. Die Bedeutung von Grünbrücken für Amphibien und Reptilien – Untersuchungen an der Autobahn 4 bei Bischofswerda/Oberlausitz (Sachsen). Zeitschrift für Feldherpetologie **12**:101–109.

Chapter 33

FLIGHT DOESN'T SOLVE EVERYTHING: MITIGATION OF ROAD IMPACTS ON BIRDS

Angela Kociolek[1], Clara Grilo[2] and Sandra Jacobson[3]

[1]Western Transportation Institute, Montana State University, Bozeman, MT, USA
[2]Departamento de Biologia & CESAM, Universidade de Aveiro, Aveiro, Portugal
[3]Pacific Southwest Research Station, USDA Forest Service, Davis, CA, USA

SUMMARY

Roads and traffic are typically more of a threat to the conservation of birds rather than a safety issue for motorists. Some bird species have biological features and life history traits that make them particularly vulnerable to habitat loss from roads and mortality due to wildlife-vehicle collisions (WVC). Road planning that proactively considers the biological needs of birds will help avoid project delays and extra costs for mitigation, as well as achieve positive outcomes for birds. Several strategies effectively avoid or mitigate the negative effects of roads on birds.

33.1 Roads can adversely affect birds despite the common assumption that birds avoid mortality and barrier effects because they can fly.

33.2 Wildlife-vehicle collisions kill millions of birds annually.

33.3 Planning the timing and location of road construction and maintenance is crucial for the survival and conservation of birds.

33.4 Flight diverters may reduce the likelihood of vehicle collisions with birds.

33.5 Wildlife crossing structures can decrease the barrier effect.

33.6 Structural changes along roads can reduce noise impacts.

33.7 Roadsides should be managed to make them less attractive to birds.

Implementing design features that separate birds from traffic, reducing resources that attract birds to the roadway and minimising disruptive light and noise emanating from the roadway are the main mitigation measures for birds. However, more research is needed to quantify the various effects of roads and the cumulative effect of road networks on birds and, perhaps more critically, to explore ways to prioritise and effectively mitigate the most negative impacts.

Handbook of Road Ecology, First Edition. Edited by Rodney van der Ree, Daniel J. Smith and Clara Grilo.
© 2015 John Wiley & Sons, Ltd. Published 2015 by John Wiley & Sons, Ltd.
Companion website: www.wiley.com\go\vanderree\roadecology

INTRODUCTION

Human activities have caused hundreds of bird species to go extinct over the past five millennia (Pimm et al. 2006). Extinction risk is related to a suite of factors, the dominant ones being susceptibility to persecution, introduced predators and habitat loss. In the last three to four decades, the massive and expanding surface transportation network has become a new threat to many avian populations globally through habitat loss and direct mortality (Chapter 28). Fortunately, not all taxa are vulnerable to the effects of roads, and many of the responses are related to species-specific traits. Some basic themes apply for avoiding, minimising and mitigating impacts to birds when constructing a new road or when responding to a problem on an existing road. The aims of this chapter are to summarise the adverse effects caused by roads and traffic on bird populations and to suggest potential solutions.

LESSONS

33.1   Roads can adversely affect birds despite the common assumption that birds avoid mortality and barrier effects because they can fly

Birds are typically perceived as being able to avoid road impacts by flying away or flying higher than traffic. Innovative measures for reducing road impacts on birds have lagged behind those of other animal groups. This is probably because the problem is not fully understood nor considered a priority and perhaps also because, in the case of birds, mitigation is not always as conventional as providing standard measures, such as an underpass or a fence. Compared to the size of vehicles, birds are small and bird–vehicle collisions are not typically a safety concern for humans. There are exceptions, such as large birds of prey feeding on roadkilled carcasses which are unable to take off quickly enough to avoid collision. Road impacts to birds are species specific, so it is important to know which species occur near a road or proposed road to determine the impacts and appropriate mitigation measures.

Birds can avoid some impacts by flying above or away from vehicles. However, not all impacts can be readily avoided. Clearing of vegetation for road construction results in habitat loss and creates open areas that may fragment populations of forest-dwelling species (Reijnen et al. 1995). Roads may attract open-country or light-demanding species while causing other birds to avoid the area altogether (Benítez-López et al. 2010). Since birds

rely heavily on acoustic communication, traffic noise is suspected to have a widespread indirect impact on many birds by reducing habitat use and, therefore, population size (Reijnen et al. 1995; Palomino & Carrascal 2007; McClure et al. 2013; Chapter 19; but see Summers et al. 2011). Kociolek et al. (2011) lists the negative effects of human-caused noise on avian community structure, breeding cycles, foraging, communication and brain response. Noise impacts to birds depend on the frequency and amplitude of the noise and of their species-specific calls and songs. While some bird species move away from noisy roads, several species are attracted to roadside verges with consequent high mortality rates due to WVC. Studies have shown that bird species with relatively high reproductive rates, high flight, small home range sizes and small body size are typically less vulnerable to road impacts (Rytwinski & Fahrig 2012; Chapter 28).

33.2   Wildlife-vehicle collisions kill millions of birds annually

WVC are estimated to kill 80 million birds annually in the United States, but true numbers may be an order of magnitude higher (Erickson et al. 2005). Species at high risk of WVC include those that hunt prey adjacent to roads (e.g. owls; Boves & Belthoff 2012, Fig. 33.1), scavenge road-killed carcasses (e.g. corvids, raptors), roost near roads (e.g. passerines) (Fig. 33.2), forage in roadside ditches or drainage retention ponds (e.g. wading birds) or nest near roads (e.g. some species of ground-dwelling birds). Grassland birds and waterfowl nesting in verges are vulnerable to WVC and to mowing practices that can directly kill eggs, fledglings and adults attending nests. Other factors may make birds particularly vulnerable to traffic. The typically low-flight behaviour of owls increases the risk of WVC (Fig. 33.1; Grilo et al. 2012), and the turbulence from passing vehicles can break bones and kill fragile birds (Orlowski & Seimbieda 2005). Vehicle headlights can also stun nocturnal species, leaving them vulnerable to WVC (Rich & Longcore 2006; Chapter 18). Moreover, lights can affect avian patterns of breeding, nestling maturation, singing and moulting (Molenaar et al. 2006). Artificial lighting can also mimic features of the night sky and attract migrating birds, increasing the likelihood of collision with structures (e.g. bridges, utility poles) (Chapter 18). Confusion from artificial lighting can increase flight time and deplete energy stores, thereby reducing body condition and making it more difficult to evade predators.

Birds are often attracted to the road and roadsides by fruit- and seed-bearing plants (Chapter 46), granular de-icing agents and roadway lighting. Vegetated medians enhance aesthetics and driver safety but

Figure 33.1 A little owl carcass on a national road in Portugal. Source: Photograph by and reproduced with permission from Joaquim Pedro Ferreira.

Figure 33.2 A Bullock's oriole carcass on a state highway in Idaho, United States. Source: A. Kociolek, Western Transportation Institute-Montana State University.

may increase the collision risk to birds because nutrient-rich food resources attract birds and their predators. Equally, mineral-rich de-icing salts increase collision risk because birds congregate on the road to ingest the salt to satisfy mineral deficiencies or to aid in grinding food. Ingesting road salt may also result in toxicity or death (Mineau & Brownlee 2005).

Road verges and medians of divided highways often have clear zones with mown grass or low vegetation to improve visibility and to provide space for drivers to recover if they lose control of their vehicles. These cleared areas may provide habitat and are often attractive to certain birds and other animals, typically including species that prefer edges. For example, hawks and owls may hunt small mammals living in verges but in so doing increase their risk of WVC (Chapters 39 and 46).

33.3   Planning the timing and location of road construction and maintenance is crucial for the survival and conservation of birds

Birds are more vulnerable to disturbance at certain times of the year, such as during the breeding season when birds stake out their territories and young fledge and disperse. Migration periods are similarly important, when stopover habitats, which may be used only briefly, provide critical food resources for migrating species. It is important to avoid scheduling construction or maintenance activities during these times that may create visual threats, noise and dust and can harm or kill. Many species of ground nesting birds are affected by repeated disturbance, causing them to abandon their nests with eggs or young, and they are also at risk of being run over. Mowing costs can be reduced by modifying schedules to be less frequent or occur after the breeding season to protect nesting birds.

Identifying and protecting important habitat features can minimise the cumulative impact of WVC on birds. In general, the frequency of WVC with birds tends to be higher near waterbodies and watercourses (Erritzoe et al. 2003) because many bird species depend on these resources and are found in higher densities near water. Therefore, it is good practice to avoid planning new roads or widening existing roads near streams, rivers, lakes and bays, especially because there are few options to mitigate for WVC in these important habitats (Chapter 44). It is also important to avoid locating new roads or upgrading existing roads near other important habitat features (e.g. preferred roosting and nesting sites such as cliff walls or large old trees with hollows).

33.4   Flight diverters may reduce the likelihood of vehicle collisions with birds

Mitigation measures to reduce WVC with birds are not as widely developed or deployed as for larger animals, and research is needed to further develop effective approaches. Generally, mitigation measures to reduce WVC with ground birds will be similar to mitigation measures for larger, ground-based mammals, whereas measures to reduce mortality for flying birds will be similar to measures for bats (Chapter 34) and butterflies. Nevertheless, it is essential to consider the movement patterns of birds and their sensitivity to noise and light when designing roads.

Structural elements can encourage birds to fly above traffic or below the road through bridges or culverts. Flight diversion works best for species with direct, rapid flight rather than for those species with slower or meandering flight. Poles that produce an illusion of a solid barrier were effective in reducing bird roadkill in open coastal areas for royal terns and brown pelicans (Bard et al. 2002), and the concept would probably work in similar locations such as marshlands (Lesson 20.3). Flags or wider posts may also be effective. Fencing aimed at keeping large mammals off the road can serve as flight diverters for birds but can cause large, less manoeuvrable birds such as sage-grouse and Gambel's quail to die in fence collisions (Stevens 2011); flagging of the fence to increase visibility may help some species, but this needs to be tested (Fig. 20.2). No mitigation measures have been devised for species with low or erratic flight patterns, such as barn swallows, which are common roadkill casualties (Erritzoe et al. 2003).

Roads with soil berms that are higher than the road grade may encourage birds to fly up and over the road and traffic (Pons 2000; Grilo et al. 2012). Solid walls (e.g. Fig. 33.3) may also encourage birds to fly up and over the road and traffic, but further testing is required before it can be recommended as an effective approach. Importantly, the installation of these walls may result in bird-wall collisions, can increase the barrier effect for many other species of wildlife and can be aesthetically inappropriate in natural areas. (Pons 2000; Grilo et al. 2012). Similarly, tall trees next to the road may encourage higher flight for canopy-dwelling birds (e.g. Rosell & Velasco 2001) but may also increase the risk to species that live and fly closer to the ground in woody vegetation. Thus, consider adding fencing or walls on bridges adjacent to tall vegetation to encourage birds and bats to fly above traffic or under the bridge (Fig. 33.4). These walls may also double as sound and light walls, reducing penetration into adjacent areas (Figs 33.3 and 33.4).

Reducing the volume or speed of traffic will lessen impacts but are often logistically difficult to implement. Focusing traffic onto fewer high-volume roads rather than distributing vehicles over many roads is a strategic approach to conserve roadless or low-traffic areas (Jaarsma & Willems 2002; Chapter 3). This approach may require the closure of some roads and the upgrading or improvement of others. Lowering traffic speed may be warranted in places where population viability is a concern for at-risk species, such as ground-dwelling or nocturnal birds. Speed control is sometimes difficult to implement even for human safety reasons, but is

Figure 33.3 The elevated road plus 3–4 m tall solid walls along Peninsula Link in Victoria, Australia, may force some birds to fly up and over the traffic, avoiding WVC. However, the efficacy of this approach has not been evaluated, and this road and walls may be a barrier to movement for some species. Source: Photograph by Rodney van der Ree.

Figure 33.4 Fencing and walls on bridges can force birds to fly below the bridge or above the walls, hopefully avoiding traffic. In this example, in Australia, the coloured glass panels also act as sound and light walls. Walls of clear glass should be avoided to reduce the risk of bird-wall collision. Source: Photograph by Rodney van der Ree.

increasingly being adopted within protected areas (e.g. Jones 2000). Although some birds can reduce collision risk by adjusting their flight distances in response to vehicle speed (Legagneux & Ducatez 2013), there has been no systematic investigation of appropriate speeds to reduce WVC with birds of differing levels of mobility.

33.5 Wildlife crossing structures can decrease the barrier effect

Although most birds are physically capable of flying over roads, species vary in their willingness to cross roads. For example, some forest-dwelling species are unlikely to cross gaps in forest cover greater than 50 m (Desrochers & Hannon 1997). Although typically designed for mammals, reptiles and amphibians, birds also benefit from wildlife overpasses and underpasses. Some bird species prefer to cross at these locations rather than over traffic (e.g. Jones & Bond 2010). Waterfowl have been recorded using drainage culverts as small as one metre in diameter across a four-lane highway (S. Jacobson, personal observations), although larger structures would likely service more species. In general, larger drainage structures that allow for natural streamflow are most desirable for more wildlife species (e.g. Fig. 45.4). Many species of birds have been recorded using underpasses, especially larger-diameter structures with streams (Foster & Humphrey 1995). Some ground-dwelling species such as quail and wild turkey may incorporate these structures into their traditional pathways if located appropriately (Smith & Noss 2011). Open-span bridges (Figs 21.1A, 33.4, 44.6 and 45.4) are likely to have higher rates of use by birds than enclosed culverts because they are more open and may be perceived as safer because birds tend to fly upwards when in danger. As with other species, crossing structures with more natural features are likely to be more acceptable to birds. There are currently no well-tested guidelines on designs or recommended dimensions of crossing structures for birds.

Some bridge and underpass structures provide nesting and roosting habitat for birds as diverse as peregrine falcons, guillemots, pigeons and swallows. Some artificial structures may have a net positive population effect for certain species that appear to avoid traffic by nesting on ledges above or below the bridge deck, although further research is required to confirm this for the wide range of species that use bridges (Fig. 33.5). Bridges can be difficult to maintain without disturbing nesting or roosting birds if maintenance is conducted at the same time that birds are nesting (Lesson 33.3). Some states in the United States have detailed maintenance plans to minimise this type of disturbance (Carey 2007).

33.6 Structural changes along roads can reduce noise impacts

Traffic noise can explain some declines in bird abundance (McClure et al. 2013). Several strategies can reduce road and vehicle noise (Chapter 19) so that birds can better utilise habitat adjacent to roads. That said, reducing the mortality of birds by minimising the attractiveness of roads and roadsides and forcing them to fly above traffic is also important to maintain bird populations (Summers et al. 2011). Reductions in noise levels can improve habitat use by birds as well as being appreciated by humans; and noise-absorbing road surfaces and modifying tyre designs may be the least expensive options. Solid wall sound barriers (Figs 33.3 and 33.4) can moderate noise, but it is important to ensure the design does not create a more significant barrier to wildlife movement or increase mortality through direct collision. Any structure that is not easily detectable, such as clear glass or certain

(A)   (B)   (C)

Figure 33.5 Ospreys frequently nest on bridge structures across rivers in northern coastal New South Wales, Australia (A), posing a traffic hazard and mortality risk to the birds. To alleviate this, artificial nesting platforms were erected at the highest point of bridges but suspended above the river (B). It soon became apparent that nesting by a threatened bird species could prevent maintenance during the breeding season, and nesting platforms are now erected on poles away from the bridge (C). Ospreys have successfully nested on many of these platforms and fledged young, allowing routine bridge maintenance to occur year-round. Source: Photographs by Kate Dallimore, Roads and Maritime Services, New South Wales.

types of netting that birds or bats collide with, should be avoided. Vegetation can absorb sound in direct proportion to its density, and dense plantings can reduce noise penetration. However, planting vegetation that is attractive to birds should be avoided as it may increase mortality rates. In contrast to Figs. 33.3 and 33.4 which show elevated roads, constructing the highway below grade or adding berms above grade can reduce sound travelling to adjacent habitats and provide some protection from WVC since birds may tend to fly higher over traffic. Long-term policy solutions include regulations for quieter vehicles, such as electric vehicles, and more effective mufflers. However, further research will be required to determine if the rate of WVC increases as vehicles become quieter potentially making it more difficult for birds to detect oncoming vehicles.

33.7 Roadsides should be managed to make them less attractive to birds

The best practice of bird conservation along roads is to avoid attracting birds to the road or roadside from the earliest planning stages. Roadsides and medians can be less of an attractant to birds if plant species that provide resources (e.g. food and nesting opportunities) are avoided. The attractiveness of the road and verge can be reduced by modifying the maintenance programme. Regularly mowing verges can reduce the attraction for some species and can be seasonally timed to avoid destroying nests. Under certain conditions (such as localised mortality hotspots of at-risk or high-profile species), it may be appropriate to reduce the attractiveness of the verge by converting it to gravel or other non-vegetative surface. Road sands and salts used as de-icing agents can be reduced through the use of ice-detecting technology or by using alternatives that are less attractive to birds (Lesson 33.2). Roadkilled carcasses, especially of larger-bodied animals that provide large quantities of food, should be promptly removed to avoid attracting scavenging birds. Artificial lighting should be avoided and reflective posts or reflectors embedded in the road surface are a low-cost alternative to identify road edges. Where lighting is required, use colours and designs that are less attractive for wildlife (e.g. blue/green lighting may be less attractive to nocturnally migrating birds; Poot et al. 2008; Chapter 18). Roadsides have multiple values and uses, and proposed management actions should be assessed against their potential impacts for other species and uses.

CONCLUSION

Every participant in road planning, construction and management can contribute to building a more sustainable road network and accounting for the biological needs of birds. Annually, hundreds of millions of birds die on roads globally; one way to reduce this loss is to separate birds from roads and traffic. Avoiding areas with high bird densities or rare and threatened species is the optimal approach to solving this problem. Strategically placed infrastructure, such as overpasses or flight diverters that encourage birds to fly higher than traffic or under bridges, provides opportunities for birds to safely traverse roads. Reducing resources that attract birds to the roadway is important to reduce collisions with foraging or nesting birds. Since artificial lighting and road noise can disturb birds well beyond the roadway, limiting their penetration can have positive benefits for birds as well as humans. More research is needed to quantify the various effects of road networks on birds (Guinard et al. 2012) and, perhaps more critically, to explore ways to prioritise and effectively mitigate the most negative impacts.

ACKNOWLEDGEMENTS

Thanks for the encouragement of our respective institutions to be a part of the team writing this chapter. C. Grilo was supported by the Fundação para a Ciência e Tecnologia through postdoctoral grant (SFRH/BPD/64205/2009).

FURTHER READING

Bujoczek et al. (2011): A study from Poland that compared body condition of birds killed by predators with birds killed by vehicle collision; roadkilled birds were in significantly better condition than those killed by raptors.

Fahrig and Rytwinski (2009): A review of the literature on the effects of roads and traffic on animal abundance and distribution revealing that birds showed mainly negative or no effects, with a few positive effects for some small birds and for vultures.

Jacobson (2005): A report outlining solutions to mitigate the negative effects of roads on birds, including crossing structures, flight diverters, modified mowing regimes, roadkill removal, appropriate median vegetation and modified de-icing agents.

Orlowski (2008): This paper shows that a disproportionately high mortality of birds was recorded near tree belts, hedgerows and built-up areas, while it was much lower in open farmland.

Parris and Schneider (2008): A study showing traffic noise hampered detection of song by conspecifics, making it more difficult for birds to establish and maintain territories and attract mates and possibly leading to reduced breeding success.

REFERENCES

Bard, A. M., H. T. Smith, E. D. Egensteiner, R. Mulholland, T. V. Harber, G. W. Heath, W. J. B. Miller and J. S. Weske. 2002. A simple structural method to reduce road-kills of royal terns at bridge sites. Wildlife Society Bulletin **30**: 603–605.

Benítez-López, A., R. Alkemade and P. A. Verweij. 2010. The impacts of roads and other infrastructure on mammal and bird populations: a meta-analysis. Biological Conservation **143**: 1307–1316.

Boves, T. J. and J. R. Belthoff. 2012. Roadway mortality of barn owls in Idaho, USA. The Journal of Wildlife Management **76**: 1381–1392.

Bujoczek, M., M. Ciach and R. Yosef. 2011. Road-kills affect avian population quality. Biological Conservation **144**: 1033–1039.

Carey, M. 2007. Washington State Department of Transportation bridge maintenance and inspection guidance for protected terrestrial species. Pp. 246–248 in Proceedings of the 2007 International Conference on Ecology and Transportation. C. Leroy Irwin, D. Nelson and K. P. McDermott, editors. Center for Transportation and the Environment, North Carolina State University, Raleigh, NC.

Desrochers, A. and S. J. Hannon. 1997. Gap crossing decisions by forest songbirds during the post-fledging period. Conservation Biology **11**: 1204–1210.

Erickson, W. P., G. D. Johnson and D. P. Young Jr. 2005. A summary and comparison of bird mortality from anthropogenic causes with an emphasis on collisions. USDA Forest Service Gen. Tech. Rep. PSW-GTR-191.

Erritzoe, J., T. D. Mazgajski and T. Rejt, 2003. Bird casualties on European roads – a review. Acta Ornithologica **38**: 77–93.

Fahrig, L. and T. Rytwinski. 2009. Effects of roads on animal abundance: an empirical review and synthesis. Ecology and Society **14**(1): 21. http://www.ecologyandsociety.org/vol14/iss1/art21/ (accessed 29 September 2014).

Foster, M. L. and S. R. Humphrey. 1995. Use of highway underpasses by Florida panthers and other wildlife. Wildlife Society Bulletin **23**: 95–100.

Grilo, C., J. Sousa, F. Ascensão, H. Matos, I. Leitão, P. Pinheiro, M. Costa, J. Bernardo, D. Reto, R. Lourenço, M. Santos-Reis and E. Revilla. 2012. Individual spatial responses towards roads: implications for mortality risk. PLoS One. **7**: e43811. http://dx.plos.org/10.1371/journal.pone.0043811 (accessed 29 September 2014).

Guinard, E., J. Romain and C. Barbraud. 2012. Motorways and bird traffic casualties: carcasses surveys and scavenging bias. Biological Conservation **147**: 40–52.

Jaarsma, C. F. and G. P. A. Willems. 2002. Reducing habitat fragmentation by minor rural roads through traffic calming. Landscape and Urban Planning **58**: 125–135.

Jacobson, S. 2005. Mitigation measures for highway-caused impacts to birds. USDA Forest Service Gen. Tech. Rep. PSW-GTR-191.

Jones, M. E. 2000. Road upgrade, road mortality, and remedial measures: impacts on a population of eastern quolls and Tasmanian devils. Wildlife Research **27**: 289–296.

Jones, D. N. and A. R. F. Bond. 2010. Road barrier effect on small birds removed by overpasses in South East Queensland. Ecological Management & Restoration **11**: 65–67.

Kociolek, A., A. P. Clevenger, C. C. St Clair and S. Proppe. 2011. Effects of road networks on bird populations. Conservation Biology **25**: 241–249.

Legagneux, P. and S. Ducatez. 2013. European birds adjust their flight initiation distance to road speed limits. Biology Letters **9**(5): 20130417.

McClure, C. J. W., H. E. Ware, J. Carlisle, G. Kaltenecker and J. R. Barber. 2013. An experimental investigation into the effects of traffic noise on distributions of birds: avoiding the phantom road. Proceedings of the Royal Society B **280**: 20132290. http://dx.doi.org/10.1098/rspb.2013.2290 (accessed 29 September 2014).

Mineau, P. and L. Brownlee. 2005. Road salts and wildlife – an assessment of the risk with particular emphasis on winter finch mortality. Wildlife Society Bulletin **33**: 835–841.

Molenaar, J. G. de, M. E. Sanders and D. A. Jonkers. 2006. Roadway lighting and grassland birds: local influence of road lighting on a black-tailed godwit population. Pp. 114–136 in Ecological consequences of artificial night lighting. C. Rich and T. Longcore, editors. Island Press, Washington, DC.

Orłowski, G. 2008. Roadside hedgerows and trees as factors increasing road mortality of birds: implications for management of roadside vegetation in rural landscapes. Landscape and Urban Planning **86**: 153–161.

Orlowski, G. and J. Siembieda. 2005. Skeletal injuries of passerines caused by road traffic. Acta Ornithologica **40**: 15–19.

Palomino, D. and L. M. Carrascal. 2007. Threshold distances to nearby cities and roads influence the bird community of a mosaic landscape. Biological Conservation **140**: 100–109.

Parris, K. M. and A. Schneider. 2008. Impacts of traffic noise and traffic volume on birds of roadside habitats. Ecology and Society **14**(1): 29. http://www.ecologyandsociety.org/vol14/iss1/art29/ (accessed 29 September 2014).

Pimm, S. L., P. Raven, A. Peterson, C. H. Sekercioglu and P. R. Ehrlich. 2006. Human impacts on the rates of recent, present, and future bird extinctions. Proceedings of the National Academy of Sciences of the United States of America **103**: 10941–10946.

Pons, P., 2000. Height of road embankment affects probability of traffic collision by birds. Bird Study **47**: 122–125.

Poot, H., B. J. Ens, H. de Vries, M. A. H. Donners, M. R. Wernand and J. M. Marquenie. 2008. Green light for nocturnally migrating birds. Ecology and Society **13**: 47. http://www.

ecologyandsociety.org/vol13/iss2/art47/ (accessed 29 September 2014).

Reijnen, R., R. Foppen, C. ter Braak and J. Thissen. 1995. The effects of car traffic on breeding bird populations in Woodland. III. Reduction of density in relation to the proximity of main roads. Journal of Applied Ecology **32**: 187–202.

Rich, C. and T. Longcore. 2006. Ecological consequences of artificial night lighting. Island Press, Washington, DC.

Rosell, C. and J. Velasco. 2001. Manual de prevenció i correcció dels impactes de les infraestrutures viàries sobre la fauna – Documents dels Quaderns de medi ambient, no 4; 2a edição. Generalitat de Catalunya Departament de Medi Ambient. Catalunya.

Rytwinski, T. and L. Fahrig. 2012. Do species life history traits explain population responses to roads? A meta-analysis. Biological Conservation **147**: 87–98.

Smith, D. J. and R. F. Noss. 2011. A reconnaissance study of actual and potential wildlife crossing structures in Central Florida, Final report. UCF-FDOT Contract No. BDB-10. 154 pp. + appendices.

Stevens, B. S. 2011. Impacts of fences on greater sage-grouse in Idaho: collision, mitigation, and spatial ecology. Thesis, University of Idaho, Moscow.

Summers, P., G. M. Cunnington and L. Fahrig. 2011. Are the negative effects of roads on breeding birds caused by traffic noise? Journal of Applied Ecology **48**: 1527–1534.

Chapter 34

BATS AND ROADS

Isobel M. Abbott[1], Anna Berthinussen[2], Emma Stone[3], Martijn Boonman[4], Markus Melber[5] and John Altringham[2]

[1]Abbott Ecology, 15 Melbourne Court, Model Farm Road, Cork, Ireland
[2]School of Biology, University of Leeds, Leeds, UK
[3]African Bat Conservation/Bat Ecology and Bioacoustics Lab, School of Biological Sciences, University of Bristol, Bristol, UK
[4]Bureau Waardenburg, Culemborg, Netherlands
[5]Applied Zoology and Nature Conservation Research Group, University of Greifswald, Greifswald, Germany

SUMMARY

Bats are long-lived mammals with low reproductive rates, making them susceptible to developments that reduce reproductive output or increase mortality. Roads destroy, degrade and fragment habitats, reducing the ability of bats to roost, feed and reproduce. Current mitigation techniques have not been proven to be effective at conserving bats at the population level.

34.1 Road effects vary among species depending on flight style and habitat use, but many species are affected.

34.2 Roads can act as barriers to the movement of bats, and many species suffer traffic mortality attempting to cross roads.

34.3 Artificial light deters some bat species while attracting others, and a policy of 'no lighting' is recommended for bats.

34.4 Traffic noise may reduce the flight activity and foraging efficiency of bats.

34.5 Protection of roosts and foraging sites during and after roadworks is critical to bat survival.

34.6 Underpasses can effectively reduce the barrier effect and reduce the number of roadkills for some bat species.

34.7 Other attempts to reduce the barrier and mortality effects for bats are unproven, and further research is required before widespread implementation.

It is important to conduct thorough pre-construction bat surveys and bring objectivity and rigour into the design and testing of mitigation features.

Handbook of Road Ecology, First Edition. Edited by Rodney van der Ree, Daniel J. Smith and Clara Grilo.
© 2015 John Wiley & Sons, Ltd. Published 2015 by John Wiley & Sons, Ltd.
Companion website: www.wiley.com\go\vanderree\roadecology

INTRODUCTION

The world's 1250+ species of bats represent more than a fifth of all known mammal species. They are the only mammals capable of powered flight, and this has allowed them to evolve into arguably the most ecologically diverse group of mammals (Fig. 34.1).

Bats perform vital ecological roles such as control of insect populations, pollination, seed dispersal and nutrient redistribution. Most species are small insect eaters that find their food using echolocation (insectivorous bats), and nearly all other species are larger fruit bats (e.g. flying foxes) which rely on good night vision to navigate and find food. Lastly, a small

Figure 34.1 Bat diversity. Clockwise from top left: Brazilian free-tailed bats emerging from a cave roost. Roosting straw-coloured fruit bats. Hoary bat, a fast, open-air forager. Nectar-feeding lesser long-nosed bat at the flowers of a saguaro cactus. Tent-making white bats, small fruit eaters. Gleaning pallid bat with large centipede. Source: Photographs © Merlin D. Tuttle, Bat Conservation International, www.batcon.org. Reproduced with permission.

number of species eat frogs, lizards, fish, blood, birds or other mammals.

Bats are remarkably long-lived for their small size (10–20 years is not unusual) and have low reproductive rates, with females usually producing only one pup per year. Thus, bats cannot easily rebound from population crashes. Unfortunately, bat populations have declined dramatically due to the unrelenting pressures placed on their environment by human activities. In particular, forest destruction, intensification of agriculture, urban development and infrastructure construction and operation all destroy and degrade roosting and feeding resources and fragment habitat. Furthermore, a recently introduced fungal disease, white-nose syndrome, is currently devastating bat populations in North America.

Roads impact bats through roadkill, roost and habitat destruction, fragmentation of habitats when roads act as barriers to movement, and light, noise and perhaps other pollutants. The legal protection of bats in Europe and North America has led to enforced efforts to mitigate the impacts of roads. However, bats remain unprotected in many other parts of the world, including those with the greatest bat diversity, such as the tropics and developing countries (Chapter 49).

In the absence of published studies on the effects of roads on fruit bats and other tropical species, this chapter relies on studies of small, insectivorous species from temperate regions. Nevertheless, the aims of this chapter are to explain how roads threaten the survival of bats, describe and evaluate current mitigation practice, outline knowledge gaps and make recommendations for management and research.

LESSONS

34.1 Road effects vary among species depending on flight style and habitat use, but many species are affected

Colonies of most bat species make use of several roosts and exploit numerous dispersed feeding sites on a nightly basis. Roads potentially restrict bats' movements among these sites due to light and noise disturbance, traffic mortality and the reluctance of some species to fly over open ground.

Body size and wing shape, feeding ecology and the structure of echolocation calls determine how bats fly and use the landscape. Therefore, susceptibility to road effects varies among species. Larger, fast-flying species, adapted to foraging in the open, are probably less affected by roads, as they expend less energy over long distance flights and often fly high above the ground. Smaller, forest-adapted species are more manoeuvrable but less efficient flyers. They are more reluctant to fly out in the open and tend to commute along landscape features such as treelines, waterways and forest edges. These features provide protection from weather and predators, are sources of insect prey and provide conspicuous acoustic and visual landmarks for orientation.

The main patterns of flight and habitat use of insectivorous bats are shown in Figure 34.2. Larger fruit bats usually fly high above roads and are unlikely to be significantly affected, unless their daytime roosts or food resources are adjacent to or in the path of a new road. Smaller fruit bats frequently fly lower to the ground and closer to the cover of vegetation. It is unfortunate that the species most likely to be affected by roads, the small, slow-flying, forest-adapted bats, are also generally those that have already suffered most from human activity and many are locally or globally threatened with extinction.

34.2 Roads can act as barriers to the movement of bats, and many species suffer traffic mortality attempting to cross roads

Major roads can act as barriers to bat movement. For example, a busy motorway in Germany severed the habitat of two threatened woodland species but had much stronger barrier effects on the less mobile, more forest-adapted species (Bechstein's bat) by reducing home range and reproductive success (Kerth & Melber 2009). Likewise, bat activity and species diversity (determined from echolocation call recordings) declined with proximity to a 40-year-old, 6-lane motorway (30–40,000 vehicles per day) in England (Berthinussen & Altringham 2012a). The most likely cause was long-term barrier or mortality effects driving colonies away from the road.

Even small gaps (<5 m) in tree or shrub cover along flight routes may interrupt bat commuting movements (Bennett & Zurcher 2013). Bats, including low-flying species, attempt road crossings where flight routes along hedgerows have been bisected by wide motorway gaps of >50 m (Abbott et al. 2013a). However, the rate of bat crossings decreases with increasing distance between hedgerows, indicating that the wider the gap, the greater the barrier effect (Abbott 2012).

Figure 34.2 Contrasting flight styles and habitat use by open-adapted, edge-adapted and clutter-adapted insectivorous bats influence their interaction with road infrastructure and traffic. Source: Drawn by and reproduced with permission of Tom McOwat.

Bats that do not abandon road-severed flight routes risk collision with vehicles. Indeed, hotspots for mortality are found where flyways intersect with roads, as evidenced by studies in Europe (Lesiński et al. 2010; Medinas et al. 2013) and North America (Russell et al. 2009).

Although road mortality rates and their effects on bat populations are not yet known, even conservative estimates of mortality by Altringham (2008) and Russell et al. (2009) suggest they are high enough to lead to population decline over the long term. Bat carcasses are difficult to find because they are small, get thrown into roadside vegetation upon collision and persist on roads for a very short time (<1 day) due to scavenging or disintegration (Santos et al. 2011; Chapter 12). Thus, road mortality studies under-record true rates of mortality. Nonetheless, most bat species in Europe have been documented as roadkill (Lesiński et al. 2010; Medinas et al. 2013). Mortality rates due to wildlife-vehicle collision (WVC) are highest near roosts and active foraging habitats (Medinas et al. 2013). Forest-adapted species are likely to have the highest risk of mortality due to their characteristic low and slow flight. The effects of traffic speed, traffic volume, road width and height, adjacent habitat and vegetation height on rates of bat roadkill remain major knowledge gaps.

34.3 Artificial light deters some bat species while attracting others, and a policy of 'no lighting' is recommended for bats

Artificial light affects the foraging, breeding, social and spatial behaviour of bats, their insect prey and other wildlife (Chapter 18). Slow- and low-flying bat species are generally deterred by light, while faster-flying bats often exploit insects attracted to lights (Rydell 1992; Chapter 29). Bats attracted to forage around street lights may be vulnerable to WVC (Lesiński et al. 2010). Lighting may also exacerbate the barrier effect of roads for some species.

Street lighting, both high-pressure sodium and white LED, deters forest-adapted species from their traditional flight paths, even at low light intensity (Stone et al. 2009, 2012; Lesson 18.3). Any attempts to use light to purposely deflect bats away from a dangerous flight route towards an intended crossing point need to be tested for effectiveness well in advance of road construction. Such attempts have not yet been proven to work and may have negative impacts on other wildlife. However, directing lighting towards the road surface and minimising light spill into the surroundings can reduce the potential for disturbance of roosts, flight routes and feeding sites. Restricting lighting in crossing structures designed for co-use by humans will maximise their use by bats (Chapter 22). Light spill at river crossings should always be avoided, as these are particularly important foraging areas and commuting routes for bats (Abbott et al. 2012a).

34.4   Traffic noise may reduce the flight activity and foraging efficiency of bats

Traffic noise can interfere with bat echolocation and hearing. Many insectivorous bats rely on hearing echoes returning from their ultrasonic calls to orientate, communicate and detect prey. Others may hunt silently, pinpointing their target by listening for the rustling or mating sounds of prey (passive listening). During indoor flight room experiments, simulated traffic noise reduced the feeding efficiency of the greater mouse-eared bat, which typically hunts for ground-running insects by listening for prey-generated sounds (Siemers & Schaub 2011). It is likely that habitats adjacent to noisy roads would therefore be unattractive as feeding areas for species that use passive listening.

Vehicle noise also exacerbates the barrier effect, and bats are less likely to fly across a road as traffic noise increases (Bennett & Zurcher 2013). Many bat species advertise for mates acoustically and their breeding success is probably also reduced around noisy roads. Currently, there are no published field studies that have assessed the effect of traffic noise on bat diversity, abundance or reproductive success (cf. birds and frogs in Chapter 19).

34.5   Protection of roosts and foraging sites during and after roadworks is critical to bat survival

By day, bats shelter primarily in roosts in trees, buildings and caves. They may roost singly or form congregations of tens to many thousands. A variety of roost types may be critical to survival and reproduction, with different conditions suited to different life cycle phases, and great distances may separate the roosts used by individuals of the same colony. Many bat species are limited by availability of suitable roost sites and return to the same roosts again and again over many years and generations. This means that roost destruction and habitat fragmentation can have a profound effect on local bat populations if alternative sites are not available or not found by the bats (Kerth & Melber 2009).

The priority is therefore to avoid roost destruction or disturbance, placing emphasis on pre-construction bat surveys to locate roosts and the associated flyways and feeding sites. Roosting places, especially in forests, are not always obvious, so bat surveys must be thorough (see Hundt 2012 for survey guidelines). Roosts also require protection from artificial light spill, which can cause delayed emergence or even roost abandonment. If roosts are destroyed, replacement roosts can be built, but there is a lack of evidence of the effectiveness of such roosts in mitigating impacts on long-term survival and breeding success (Stone et al. 2013; Berthinussen et al. 2014). Modern bridge and culvert structures often lack crevices suitable for roosting, and we encourage road engineers to actively incorporate bat-friendly roosting spaces into new structures where there is unlikely to be a risk of collision mortality. Brazilian free-tailed bats make extensive use of the expansion joints of modern bridges in the United States (Fig. 34.3). These roosts appear to be as suitable as natural roosts in caves (Allen et al. 2011). See Section 'Further Reading' for resources relating to protection and creation of bat roosts.

34.6   Underpasses can effectively reduce the barrier effect and reduce the number of roadkills for some bat species

Many studies show that a wide range of bat species use underpasses (Fig. 34.4) to cross beneath roads (e.g. Bach et al. 2004; Kerth & Melber 2009; Boonman 2011; Abbott et al. 2012a). However, occasional use by an unknown proportion of bats does not guarantee either habitat accessibility or safe crossing routes for the bat population or community as a whole. For example, despite the existence of three traffic underpasses along 5 km of motorway that bisected a forest, Bechstein's bat, a woodland species with small home ranges, rarely used them and lost access to important roosting and feeding habitat (Kerth & Melber 2009).

Figure 34.3 Hundreds of people gather to watch the spectacular emergence of Brazilian free-tailed bats from a bridge roost in Austin, Texas. Source: Photograph © Merlin D. Tuttle, Bat Conservation International, www.batcon.org. Reproduced with permission.

Figure 34.4 Bat-friendly underpasses (top row of images) are spacious and connected with flight paths along streams, woodland lanes or hedgerows. Underpasses that are too small, blocked with vegetation or grilles, situated in open ground, wide but too low, flood prone or disturbed by lighting (e.g. bottom row of images) are much less beneficial to bats. Source: (top left, top right, bottom left, bottom middle, bottom right) Photographs by Isobel Abbott; and (top middle) Photograph by Markus Melber.

Approximately, a third of road crossings by another threatened woodland species, the lesser horseshoe bat, were directly over a motorway (Isobel Abbott, unpublished data), despite its regular use of three underpasses along a 1 km stretch (Abbott et al. 2012b). It cannot be taken for granted that the reluctance of forest-adapted bats to fly in the open means that they will take detours to use underpasses. Some individuals will make detours (e.g. Kerth & Melber 2009; and references cited in Bach et al. (2004)), but there is little evidence that this is the norm, and many bat species appear reluctant to deviate from their original flight paths after road severance (Kerth & Melber 2009; Abbott 2012; Berthinussen & Altringham 2012b). When a road cuts through a dense network of flight routes, many closely spaced underpasses may be required to provide the population as a whole with safe crossing points. Efforts to reroute bat flight paths, for example, by planting new hedgerows towards underpasses, should be undertaken well in advance of habitat clearance and tested for effectiveness before road opening. Bats have not yet been effectively diverted to underpasses using revegetation in some UK trials (Berthinussen & Altringham 2012b).

Connectivity to the surrounding landscape, the presence of water courses and increased size of underpasses are all factors that encourage underpass use (Boonman 2011; Abbott 2012). In general, underpasses should be located on pre-construction flight routes and sized so

that the target bat species can pass without changing flight height or direction. Underpass height, more than width, is the critical dimension in persuading bats to fly through. Even at the site of an underpass, a high proportion of bats may ignore the underpass and fly over the road above it, if the underpass is too small or low (Abbott 2012). If in doubt, bigger is better. Required minimum heights of underpasses will generally be lower for forest-adapted species (~3 m as a rough guide) compared to generalist edge-adapted species (~6 m as a rough guide), while open-air species are more likely to fly high above roads (Fig. 34.5). It is necessary to observe and measure the flight routes and typical flight altitudes of the local bat species when deciding appropriate underpass sizes and locations. Mitigation practice would benefit by objectively testing and reporting whether designed underpasses are actually providing safe passage for bat populations.

The nocturnal nature of bats means that they can probably make use of underpasses that are mostly used by people during daytime, such as for pedestrian access, minor road traffic, train, forestry or agricultural activities (Chapter 22). Bats will use underpasses that are wet or dry, and long or short, provided connectivity and minimum size requirements are met. Several measures are likely to benefit bats, such as restricting lighting in and around underpasses, placing underpasses at mature tree and hedge lines (which bats naturally follow) and increasing the size of routinely

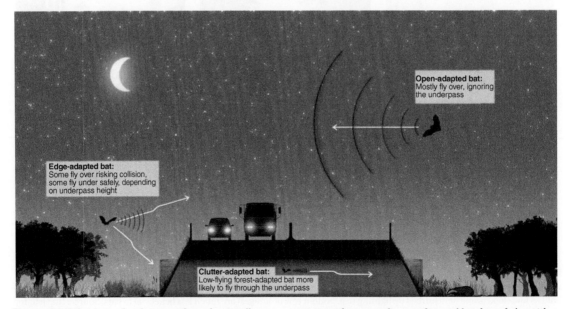

Figure 34.5 Bats' use of underpasses depends critically on connectivity with surroundings and tunnel height and also with the typical flight height of the species. Source: Drawn by and reproduced with permission of Ruadhrí Brennan.

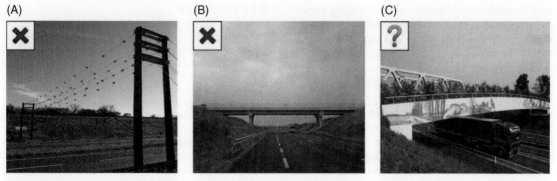

Figure 34.6 Over-the-road-crossing structures: gantries (A) and conventional overpasses (B) are not effective. Wildlife overpasses (C) have potential but need more research. Source: (A) Photograph by Anna Berthinussen, (B) Photograph by Isobel Abbott; and (C) Photograph © Herman Limpens/Zoogdiervereniging. Reproduced with permission.

incorporated structures, such as drainage or badger tunnels, to also accommodate bats. Provision of well-placed, spacious underpasses should be integral to the overall design of road mitigation near major roosts. Roads that are above grade may be particularly dangerous when they sever treelines, since bats appear to maintain flight height on leaving the treeline, increasing the risk of WVC (Fig. 34.5). These sites are ideal candidates for underpasses.

34.7 Other attempts to reduce the barrier and mortality effects for bats are unproven, and further research is required before widespread implementation

At least 10 and perhaps many more 'wire gantries' have been built on UK and European roads with the specific aim of making them more permeable and safer to cross for bats. However, the most widely used design in the United Kingdom (Fig. 34.6 A), even when on the line of pre-construction flyways, does not work. Recent studies have shown that only a very small proportion of bats that approached the gantry used them and most crossed the road below at traffic height" (Berthinussen & Altringham 2012b). Many of these gantries were built over the last decade without adequate assessment. This wasteful and ineffective use of resources highlights the importance of making comprehensive assessment an integral part of mitigation practice (Chapters 10 and 16).

Conventional un-vegetated overpasses carrying minor roads or footpaths (Fig. 34.6 B) also do not appear to be effective crossing points for bats, relative to routes under roads (Bach et al. 2004; Abbott et al. 2012a), possibly because they are too exposed. Wildlife overpasses such as land bridges (Fig. 34.6 C, Chapter 21), if planted with tall

vegetation and linked with existing bat flyways, have considerable potential as bat crossing structures. Bats do use vegetated overpasses, but a recent study found that only a small proportion of Bechstein's bats crossed a busy motorway using a new land bridge; most crossed the road at locations lacking crossing structures (Stephan & Bettendorf 2011). Further research is required before conclusions can be drawn, but several features of the design and connectivity to surroundings are likely to encourage the use of land bridges by bats: strategic location on known flight lines, connectivity to treelines, mature vegetation on the bridge and increased bridge width.

'Hop-overs' (Limpens et al. 2005) have been suggested as a measure to encourage bats to cross road gaps at safe heights. These consist of close planting of tall vegetation up to the road edge on both sides of the road, with tall vegetation in the central median of wide roads. Branches should overhang the carriageway, ideally giving continuous canopy cover over the road (Fig. 40.1 A). Safety concerns arising from overhanging branches have led to reluctance to adopt hop-overs, but many roads already have overhanging trees along their margins (Chapters 40 and 49). The effectiveness of hop-overs has yet to be assessed. The height of bats flying above a 20 m road gap was higher with taller roadside vegetation (Russell et al. 2009), and there is a positive correlation between road-crossing height and height of roadside embankment (Berthinussen & Altringham 2012b).

CONCLUSIONS

Roads have the potential to vastly reduce bat populations. The best approach (and a legal requirement in the European Union) is to first try to avoid impacts on bats by conducting effective pre-construction surveys to locate roosts, flight paths and foraging habitats and to determine the likely effects of road developments. Where impacts cannot be avoided, bat specialists should be brought in as early as possible in the planning and design stages to ensure mitigation plans are both functional and cost-effective (Chapter 9). Post-construction surveys and monitoring, in conjunction with pre-construction surveys, are needed to assess the effectiveness of mitigation (Chapters 10, 15 and 16). Large underpasses built on original flight paths appear to be the most effective way to increase road permeability and reduce roadkill. Wildlife overpasses may be more practical where roads are in a cutting but require further assessment. Flight paths and crossing structures should be unlit. All crossing points should be on the line of pre-construction flyways and tree/hedge planting should link them effectively to undisturbed flyways in the surrounding habitat. Plants should be as large as practicably possible and given the maximum time to establish. Mitigation must be proven to be effective in protecting bat populations through objective and rigorous monitoring. The effects of roads on bat populations are likely to reveal themselves slowly, so this monitoring should be long term.

ACKNOWLEDGEMENTS

IA's PhD research was funded by the National Roads Authority of Ireland and AB's by the University of Leeds.

FURTHER READING

Altringham (2008): A critical assessment of bat mitigation measures for a major UK road.

Bat Conservation International website (www.batcon.org) and Bat Conservation Trust UK website (www.bats.org.uk): Both websites provide information about bats and bat research and also provide advice on roost creation, but as yet, few methods have been objectively evaluated. Long-term monitoring to assess success of roost measures should be integrated in future work.

Hundt (2012): Detailed methods to conduct bat surveys, with specific information for the United Kingdom but relevant for all regions.

Limpens et al. (2005): Accessible background ecological information and advice on mitigating the effects of roads on bats, but it has a non-quantitative evidence base.

Mitchell-Jones (2004): Guidelines focused mainly on roost mitigation measures, with specific information for UK species but relevant for all regions.

REFERENCES

Abbott, I. M. 2012. Assessment of the effectiveness of mitigation measures employed on Irish national road schemes for the conservation of bats. PhD thesis, University College Cork, Ireland.

Abbott, I. M., Butler, F., Harrison, S. 2012a. When flyways meet highways–The relative permeability of different motorway crossing sites to functionally diverse bat species. Landscape and Urban Planning **106**:293–302.

Abbott, I. M., Harrison, S., Butler, F. 2012b. Clutter-adaptation of bat species predicts their use of under-motorway passageways of contrasting sizes – a natural experiment. Journal of Zoology (London) **287**:124–132.

Allen, L. C., Turmelle, A. S., Widmaier, E. P., Hristov, N. I., McCracken, G. F., Kunz, T. H. 2011. Variation in physiological stress between bridge- and cave-roosting Brazilian free-tailed bats. Conservation Biology **25**:374–381.

Altringham, J. D. 2008. Bat Ecology and Mitigation; Proof of Evidence; Public enquiry into the A350 Westbury bypass. White Horse Alliance, Neston, UK.

Bach, L., Burkhardt, P., Limpens, H. J. G. A. 2004. Tunnels as a possibility to connect bat habitats. Mammalia **68**:411–420.

Bennett, V. J., Zurcher, A. A. 2013. When corridors collide: road-related disturbance in commuting bats. The Journal of Wildlife Management **77**:93–101.

Berthinussen, A., Altringham, J. 2012a. The effect of a major road on bat activity and diversity. Journal of Applied Ecology **49**:82–89.

Berthinussen, A., Altringham, J. 2012b. Do bat gantries and underpasses help bats cross roads safely? PLoS ONE **7**:e38,775.

Berthinussen, A., Richardson, O. C., Altringham J. D. 2014. Bat conservation: Global evidence for the effects of interventions. Conservation Evidence and Pelagic Publishing, Exeter, UK. pp. 110. Available from www.conservationevidence.com, Accessed on 22 September 2014.

Boonman, M. 2011. Factors determining the use of culverts underneath highways and railway tracks by bats in lowland areas. Lutra **54**:3–16.

Hundt, L., Editor 2012. Bat Surveys: Good Practice Guidelines, 2nd edn., Bat Conservation Trust. London, England.

Kerth, G., Melber, M. 2009. Species-specific barrier effects of a motorway on the habitat use of two threatened forest-living bat species. Biological Conservation **142**:270–279.

Lesiński, G., Sikora, A., Olszewski, A. 2010. Bat casualties on a road crossing a mosaic landscape. European Journal of Wildlife Research **57**:217–223.

Limpens et al. (2005). Bats and road construction. Dutch Ministry of Transport, Public Works and Water Management. Directorate-General for Public Works and Water Management, Road and Hydraulic Engineering Institute, Delft, the Netherlands and the Association for the Study and Conservation of Mammals, Arnhem, the Netherlands.

Medinas, D., Marques, J. T., Mira, A. 2013. Assessing road effects on bats: the role of landscape, road features, and bat activity on road-kills. Ecological Research **28**:227–237.

Mitchell-Jones, A. 2004. Bat Mitigation Guidelines. English Nature, UK. Available from http://publications.natural england.org.uk/publication/69046?category=3100, Accessed on 22 September 2014.

Russell, A. L., Butchkoski, C. M., Saidak, L., McCracken, G. F. 2009. Road-killed bats, highway design, and the commuting ecology of bats. Endangered Species Research **8**:49–60.

Rydell, J. 1992. Exploitation of insects around streetlamps by bats in Sweden. Functional Ecology **6**:744–750.

Santos, S. M., Carvalho, F., Mira, A. 2011. How long do the dead survive on the road? Carcass persistence probability and implications for road-kill monitoring surveys. PLoS ONE **6**:e25383.

Siemers, B. M., Schaub, A. 2011. Hunting at the highway: traffic noise reduces foraging efficiency in acoustic predators. Proceedings of the Royal Society B: Biological Sciences **278**:1646–1652.

Stephan, S., Bettendorf, J. 2011. Home ranges of Bechstein's bats overlapping a motorway. European Bat Research Symposium, Vilnius, Lithuania (Poster presentation).

Stone, E. L., Jones, G., Harris, S. 2009. Street lighting disturbs commuting bats. Current Biology **19**:1123–1127.

Stone, E. L., Jones, G., Harris, S. 2012. Conserving energy at a cost to biodiversity? Impacts of LED lighting on bats. Global Change Biology **19**:2458–2465.

Stone, E. L., Jones, G., Harris, S. 2013. Mitigating the effect of development on bats in England with derogation licensing. Conservation Biology **27**:1324–1334.

Chapter 35

CARNIVORES: STRUGGLING FOR SURVIVAL IN ROADED LANDSCAPES

Clara Grilo[1], Daniel J. Smith[2] and Nina Klar[3]

[1]Departamento de Biologia & CESAM, Universidade de Aveiro, Aveiro, Portugal
[2]Department of Biology, University of Central Florida, Orlando, FL, USA
[3]OEKO-LOG field research, Hamburg, Germany

SUMMARY

Carnivores are a diverse group of wildlife that occur in most environments around the world. Large, wide-ranging carnivores play key ecological roles in natural systems. They regulate population sizes of herbivores and other small- and medium-sized carnivores that in turn affect the growth, structure and composition of plant communities and habitats and the health of the small-animal populations that live in these habitats. Carnivores are particularly susceptible to the impacts of roads because many species require large areas to sustain their populations, have low reproductive output and occur in low densities.

35.1 Carnivores with large home ranges, long dispersal distances or inability to tolerate human disturbance are particularly vulnerable to the effects of roads and traffic.

35.2 Threats from roads and traffic such as wildlife-vehicle collisions barriers to movement, habitat disturbance and road avoidance jeopardise the persistence of certain carnivore populations.

35.3 Road and landscape-related features influence behavioural responses of carnivores to roads, mortality risk and barrier effects.

35.4 Different types of crossing structures are needed to increase habitat connectivity for the wide diversity of carnivore species.

35.5 Fencing, when paired with crossing structures, is critical to reducing the negative effects of roads on carnivores.

Handbook of Road Ecology, First Edition. Edited by Rodney van der Ree, Daniel J. Smith and Clara Grilo.
© 2015 John Wiley & Sons, Ltd. Published 2015 by John Wiley & Sons, Ltd.
Companion website: www.wiley.com\go\vanderree\roadecology

The effects of roads and traffic on carnivores are well understood and vary significantly because of the diversity in their body size, movement ecology, prey selection and habitat preferences. Consequently, carnivores require a diverse suite of mitigation options, many of which have been well studied. Further research is needed to evaluate effects of roads and mitigation success in maintaining genetic integrity that supports long-term viable populations of carnivores.

INTRODUCTION

Carnivores are a diverse group of predatory mammals that consume animal tissue as part of their diet. There are terrestrial and aquatic representatives adapted to nearly every continental environment and climate on earth. Terrestrial species range from very small to quite large sizes, such as the least weasel (37–50 g) and polar bear (420–500 kg). Their foraging strategies are diverse and include hunting (e.g. stone marten), scavenging (e.g. hyena) and omnivory (e.g. bears), and social structures range from relatively solitary individuals (e.g. jaguar) to complex interacting family groups (e.g. wolf). Carnivores play a key role in maintaining ecosystem integrity and preserving biodiversity in a number of ways. Many animal and plant species are protected when large areas of habitat are set aside for carnivore conservation because their needs are also addressed. The removal of carnivores from the top of the food chain will negatively impact the abundance of prey and other species (e.g. Palomares et al. 1996). These effects can cascade through the food chain, altering the interactions among species as well as the structure and function of ecological communities and ecosystem processes (Ripple et al. 2014).

Although carnivores are often major conservation icons today, such as tigers, wolves and jaguar (Chapters 36 and 37), they have historically been subjected to many anthropogenic threats. These include habitat loss and degradation, depletion of their prey and direct human persecution for the fur trade, trophy hunting and extermination because of fear, ignorance and perceived threats to livestock and human life, which in combination have resulted in massive population declines and range contractions (Ripple et al. 2014). Today, persecution and loss of prey are the immediate threats, but continued loss of habitat and the additional mortality and barrier effect of roads and traffic are the greatest long-term threats to their persistence (Burkey & Reed 2006). The aims of this chapter are to (i) highlight the ecological and biological traits that make carnivores susceptible to roads; (ii) summarise the impacts of roads and traffic on this group of wildlife; and (iii) review the mitigation strategies necessary to conserve viable carnivore populations.

LESSONS

35.1 Carnivores with large home ranges, long dispersal distances or inability to tolerate human disturbance are particularly vulnerable to the effects of roads and traffic

Roads are a significant direct cause of habitat loss, fragmentation and disturbance and indirectly lead to widespread land transformation for agriculture and urban development (Liu et al. 2014; Chapter 2). Many carnivore species are vulnerable to the effects of road-network expansion (Cardillo et al. 2004), such as increased human disturbance, mortality due to wildlife-vehicle collision (WVC), reduction of sufficient space for home ranges and isolation of populations because of their large spatial needs and other biological and ecological traits. Due to past and present human persecution, some species do not tolerate areas of high human activity (e.g. jaguar, gray wolf), further reducing the amount of suitable habitat. As a result, many carnivores now occupy only small portions of their former geographic range (e.g. gray wolf, Florida panther).

Those species most sensitive to fragmentation and human disturbance typically have large home ranges, low population densities, low reproductive outputs and display territorial behaviour (e.g. grizzly bear, Iberian lynx). These biological characteristics translate into the need for large undisturbed areas to support viable populations and the inability to sustain high levels of mortality. Many of the most imperilled carnivores require large home ranges, such as 99–241 km^2 for wolves (Okarma et al. 1998), 18–324 km^2 for grizzly bears (Craighead 1976), 195–520 km^2 for Florida panther (FFWCC, 2014) and about 30 km² for leopards (Simcharoen et al. 2008). In addition, carnivores with large home ranges tend to occur in low population densities. For example, the primary habitat zone (9190 km²) in south Florida for the Florida panther was estimated to support a stable population of only 71–84 individuals or at most one per 109 km² (Kautz et al. 2006). When the population in an area reaches its carrying capacity, young adults must disperse to new areas to find suitable habitat. Many carnivores disperse long distances, for example, 13–219 km for black bear (Rogers 1987) and an average of 123 km for puma in the United States (Maehr et al. 2002), further increasing the probability of encountering roads. When carnivores increase their movement range in search of food to feed their young or to find mates, they tend to search for new areas that may include suboptimal habitat, thus increasing the likelihood of encountering a road (Saeki & Macdonald 2004). This is especially significant for dispersers and other novice individuals exploring unknown areas that are also unfamiliar with roads or the danger they pose when attempting to cross. One study on black bears in Florida showed that 66 of 96 roadkills were inexperienced young males dispersing in search of mates in late spring/early summer or new food sources in autumn (Wooding & Brady 1987).

Although carnivores have a wide range of morphological, ecological and behavioural adaptations to coexist and adapt to diverse habitats (Gittleman et al. 2001), some are unable to compensate biologically for the increased mortality due to low reproductive output or overcome the barrier effect of roads (e.g. Iberian lynx, bears, African wild dogs – Chapter 38). In general, larger carnivores that require more space are more sensitive to the effects of habitat fragmentation and isolation than smaller carnivores that appear more able to adapt to human activities and land development (Crooks 2002).

35.2 Threats from roads and traffic such as wildlife-vehicle collisions barriers to movement, habitat disturbance and road avoidance jeopardise the persistence of certain carnivore populations

One of the major causes of mortality for carnivores is WVC (Fig. 35.1), which in certain cases is sufficient to threaten population viability. For example, in a high-traffic area of Ocala National Forest, Florida, United States, the rate of female black bear mortality due to WVC was 23%, which when combined with other sources of mortality was estimated to exceed the maximum sustainable annual mortality rate for populations of similar demographics and reproductive traits (Bunnell & Tait 1980; McCown et al. 2009; Textbox 35.1). Roadkills accounted for 35% of annual mortality of the federally endangered Florida panther (Taylor et al. 2002) and 17% of annual mortality of the most threatened felid species in southern Spain, the

Figure 35.1 Stone marten roadkill in southern Portugal. Source: Photograph by and reproduced with permission of Joaquim Pedro Ferreira.

Textbox 35.1 Landform, land cover, road alignment and traffic influence black bear movement and roadkill patterns.

Ocala National Forest (155,000 ha) in Florida, United States, contains the largest population of black bears (~825–1225 individuals) in the state (FFWCC 2012). WVC is a primary cause of mortality, with about 80 bears killed per year, including four hotspots of 5 or more roadkills annually on State Road 40 (SR40), which carries 5100 vehicles per day.

The occurrence and movement of bears around SR40 was studied from 1999 to 2003 using radiotracking and sand-tracking plots. In order to understand the factors influencing bear-vehicle collision hotspots, data on landform, road alignment and land cover was also collected and analysed. The number of bear tracks was consistent along a 19 km stretch of SR40 through the core of the forest; the analysis revealed no bias in bear movements or road crossings by land cover, habitat, road curvature, topography or presence of intersecting roads or trails (McCown et al. 2004). Rather, bear movements and crossings were in response to availability of food sources, and road geometry influenced the location of roadkills. Roadkills occurred on curves and hills (Fig. 35.2), strongly suggesting that reduced visibility was the ultimate cause. Relatively high vehicle speed (90 km per hour) along SR40 exacerbates the problem; combined with the hills and curves in the road, it reduces driver response times when encountering bears.

Subsequent research in a more fragmented and human-dominated area along SR40 with 15,700 vehicles per day revealed that (i) males crossed the road more frequently than females; (ii) the rate of road crossing by females was only slightly lower in the more fragmented area, despite having approximately three times more traffic volume; and (iii) more female WVC occurred at the higher traffic volume site (McCown et al. 2009). Males have larger home ranges than females and therefore encounter and cross roads more often. The higher rate of female mortality in the high-traffic volume site was somewhat unexpected because more traffic should increasingly act as a barrier to crossings. The most likely explanation is that bears increase their frequency of road crossings in fragmented areas because of a greater need to cover larger areas to access food and mates, attraction to human food sources, and a shift to more nocturnal movements to adapt to periods of minimal human activity and traffic levels. Vehicle collisions accounted for 23% of annual mortality of female bears in the high-traffic area.

Previous studies (Brody & Pelton 1989; Beringer et al. 1990) have documented road avoidance by bears when traffic volume is high; however, bears occurring in fragmented areas may need to cross busy roads (Fig. 35.3)

Figure 35.2 Typical topographic relief and curvature of State Road 40 through Ocala National Forest, Florida, United States, which resulted in high rates of bear-vehicle collision. Source: Photograph by D.J. Smith.

Figure 35.3 A black bear crossing State Road 40 in Ocala National Forest, Florida, Untied States. Source: Photograph by and reproduced with permission of Mark Cunningham.

to find mates, suitable den sites and food sources. The estimated annual mortality from all sources was 37.6% (McCown et al. 2004) in the fragmented area, making it unsustainable (Bunnell & Tait 1980), and any increase in mortality or habitat fragmentation will further imperil their existence (McCown et al. 2009). Current plans to widen the road to four lanes would exacerbate these effects; therefore, recommendations were made to include crossing structures and fencing in any future highway designs. This study demonstrates the importance of considering habitat use and movement patterns of the target species, landscape characteristics, traffic volume and road alignment in evaluating WVC locations and mitigation needs.

Iberian lynx (Ferreras et al. 1992). In Britain, more than 40% of the adult Eurasian badger population is killed annually by vehicles (Clarke et al. 1998). The estimated annual roadkill rates of carnivores along 314 km of national roads in southern Portugal was around 47 individuals per 100 km, with red fox and stone marten (Fig. 35.1) experiencing the highest rates (20 and 8 individuals per 100 km, respectively) (Grilo et al. 2009). While little is known about the implications of these mortality rates on the long-term viability of the populations, attention must also be paid to this added source of mortality on the potential reduction of genetic diversity (Jackson & Fahrig 2011).

Roads with high-traffic volumes and vehicle speeds can act as barriers to animal movement as well as disturb and displace carnivore populations due to road avoidance (Fig. 1.2; Riley et al. 2006). In some cases, populations of carnivores have declined due to increased hunting pressure as a result of improved access provided by the road (e.g. Van Dyke et al. 1986; Mech et al. 1988; Beldon & Hagedorn 1993; Chapter 37). Moreover, human disturbance while carnivores are hunting or feeding can reduce hunting efficiency and increase carcass abandonment, as shown for Amur tigers (Kerley et al. 2002). Similarly, the Asiatic leopard avoided habitat near a road bisecting a National Park in Thailand and also reduced their level of diurnal activity (Ngoprasert et al. 2007). Similarly, Eurasian lynx avoided areas with the highest road densities within their home ranges (Basille et al. 2013). The severity of these effects can vary between sexes, as shown by male jaguars that were more willing than females to use areas close to roads, and with higher levels of human occupation, even though the species generally avoids both land uses and preferentially moves in undisturbed forests (Colchero et al. 2011; Chapter 36).

35.3 Road and landscape-related features influence behavioural responses of carnivores to roads, mortality risk and barrier effects

The rate of WVC with carnivores is influenced by habitat suitability and landscape structure, as well as road and traffic characteristics (Gunson et al. 2010;

Textbox 35.1). Roads through high-quality habitat are especially problematic due to the high abundance and diversity of species (Barrientos & Miranda 2012) as well the often narrow and sinuous roads with low to medium traffic volumes. Roads in protected areas and parklands are often designed to blend into the landscape in order to improve aesthetics. However, the narrowness, curves and steep slopes in wilderness areas can severely limit driver visibility, decreasing their ability to detect and avoid animals on the road (Grilo et al. 2009). Vegetated roadsides can support populations of small mammals and other potential prey (Chapters 39 and 46), potentially attracting carnivores and increasing their risk of WVC. Scavenging carnivores are often attracted to roadkills as a source of food, which increases their probability of also being involved in WVC (Figs. 26.2B and 50.3). Roadkills are also more likely to occur where roadside fencing ends, such as where side roads join the fenced road (Clevenger et al. 2001; Cserkész et al. 2013; Chapter 20). Weather conditions also influence the rate of WVC. For example, the rate of mortality of river otters in England was higher during heavy rainfall periods when small culverts became impassable due to flooding or increased water velocity, forcing them to cross over the road (Philcox et al. 1999). Otters also travel over land more often during periods of drought in search of water, which increases their probability of WVC.

There are thresholds in traffic volume where the rate of WVC decreases and the barrier effect becomes intensified. For larger carnivores (e.g. coyote, wolf, puma), the threshold is approximately 2000–5000 vehicles per day (Alexander et al. 2005). In line with this study, stone martens seem to regularly cross a four-lane highway with nightly traffic volumes of 2000 vehicles (Grilo et al. 2012); consequently, WVC was the main threat to the population. On the other hand, significant genetic structuring was found in wildcat populations divided by a six-lane highway with 100,000 vehicles per day (Hartmann et al. 2013). Similarly, a 10–12 lane freeway in California with 150,000 vehicles per day was only permeable for dispersing bobcats and coyotes through the use of culverts or underpasses (Riley et al. 2006).

35.4 Different types of crossing structures are needed to increase habitat connectivity for the wide diversity of carnivore species

Many wildlife crossing structures (Chapter 21) built across roads in Europe and North America are used by carnivores (Beckmann et al. 2010). In addition to these,

some carnivore species will also use drainage culverts and other types of multi-use structures (e.g. Clevenger et al. 2001; Lesson 21.3) to cross roads. The suitability of both dedicated and multi-use crossing structures is influenced by the type and size of the structure itself and the characteristics of the surrounding vegetation and landscape, the road and the degree of human disturbance. In summary, carnivores use a diversity of structures to cross roads, and there are several design parameters that should be considered to make them attractive to a wide range of species (Chapter 21 and 59):

(i) *Structure type and size* – Larger structures generally have higher rates of use by carnivores than smaller ones (Kusak et al. 2009). Small-sized underpasses (1–1.5 m wide) are also used by small- to medium-sized species, such as marten, coyote and bobcat (e.g. Cain et al. 2003; Grilo et al. 2008). Carnivores that require cover or concealment, especially those which live in burrows or dens, appear to prefer or be more tolerant of more constricted underpasses, such as badgers (Fig. 35.4), black bears and cougars, while others that use open habitat, such as grizzly bears and wolves, appear to prefer overpasses or high, wide and short underpasses (Clevenger & Waltho 2005; Sawaya et al. 2014). The use of structures by subordinate individuals or species appears to be negatively affected by the presence of established, dominant individuals or species. For example, use of a specific structure by wolves may reduce or preclude use by coyotes, as well as potential prey species (Clevenger 2011; Chapter 23). Thus, multiple structures of a range of types and sizes are required to allow crossing by many different species (Mata et al. 2008). Wildlife crossing structures have been included in many road projects for large carnivores (e.g. bears, wolves and lynxes – Textbox 35.2), including overpasses ranging in width from 30–50 m to over 200 m (e.g. Wieren & Worm 2001). Wildlife overpasses are used by a wide diversity of species (e.g. Brodziewska 2005) and have numerous other advantages over underpasses including the provision of wider areas for crossing, exposure to natural rainfall, temperature and light conditions, and the provision of continuous, vegetated habitat corridors across the road (Glista et al. 2009).

(ii) *Landscape context and local features* – Crossing structures should be located in areas with ecological significance for carnivores, such as within highly used areas and connecting corridors of forested habitat, which may include waterways with riparian vegetation, and low levels of human disturbance (e.g. Clevenger & Waltho 2005).

(iii) *Structure enhancements* – Crossing structures should be as similar as possible to the preferred habitat of the

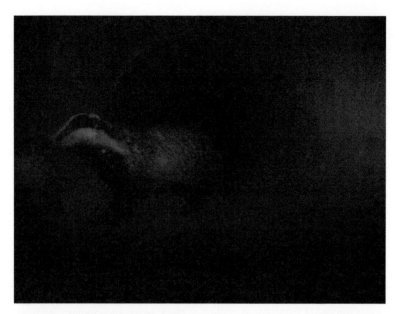

Figure 35.4 Badger using a tunnel. Source: Photograph by C. Grilo.

Figure 35.5 Logs and brush were included inside this underpass to provide cover for carnivores and small mammals. Source: Photograph by C. Grilo.

target species. Low rates of use and avoidance of structures by weasels and polecats was attributed to the unnatural characteristics of most underpasses (Grilo et al. 2008). Guiding carnivores towards structures with the help of linear strips of vegetation and placing logs, rocks and bushes inside and outside passages to provide cover is highly recommended to improve function (Fig. 35.5).

(iv) *Spacing intervals* – Road permeability can be improved by placing crossing structures at intervals that correspond to the movements of the target species and the goals for mitigation (see Chapter 21; Lesson 21.6).

(v) Enhancements *to drainage structures* – Many standard drainage structures, including culverts and bridge underpasses, are unsuitable for use by wildlife (Fig. 35.6A), especially when carrying water or during periods of flooding (e.g. stone martens and genets – Villalva et al. 2013). Options for dry passage should be provided (Serronha et al. 2013), and there are a number of strategies: (i) bridge underpasses should include a strip of dry land on one or preferably both banks (Figs. 35.6B and 45.4; Lesson 45.4); (ii) install multiple culverts with one or more culvert being elevated to remain dry most of the time (Figs. 21.4 and 45.5); and (iii) install shelves or ledges on the walls of underpasses above the water level, with ramps to provide access (Figs. 35.6C and 39.3). Dry ledges are also easily retrofitted to existing culverts (Fig. 35.7). Swimming ledges, which float on the water surface and therefore can adapt to changing water levels, are also a possibility. Such ledges or shelves can circumvent impassable dams and also enhance the attraction to culverts by swimming species, for example, otters have the opportunity to exit the water to mark territory.

However, behavioural differences among individuals play an important role regarding the efficiency of crossing structures. Adaptation to and acceptance of new structures may take an extended period of time (e.g. an average of 4–6 years for carnivores, Clevenger 2011). Following this acclimation period, certain resident individuals often become accustomed to many types of crossing structures within their home ranges and use them regularly, while other individuals may have preferences for a specific structure type or location (Klar et al. 2009). Dispersing animals are often reluctant to use crossing structures, particularly when moving through unfamiliar areas (e.g. Zimmermann 2004).

(A)   (B)   (C)

Figure 35.6 Examples of river otter crossings: (A) unsuitable culvert design with high risk of submersion from flooding, (B) superior design with riparian strips and (C) retrofitted culvert with dry ledge including aquatic entry points for otters. Source: Photographs by N. Klar.

Figure 35.7 Bolt-on wildlife shelf retrofitted to an existing drainage culvert for use by weasels. Source: Photograph by and reproduced with permission of Kerry Foresman, Critter-Crossing Technology L.L.C.

Textbox 35.2 Enhancing population connectivity and minimising road mortality for Iberian lynx in Spain.

Iberian lynx (Fig. 35.8) is a highly endangered felid species with just 400 breeding individuals occurring in two isolated populations: 75% in Sierra Morena and 25% in Doñana, a fragmented habitat area (Simón 2012). Mortality due to WVC is one of the primary threats to the viability of the Doñana population – in 2006, 12% of this population was killed in collisions. Consequently, a series of measures to reduce mortality and restore connectivity for this population was implemented along 150 km of roads in the Donãna–Aljarafe region in southern Spain, including: (i) reduction of the attractiveness of roadside vegetation to lynx; (ii) installation of traffic calming devices to reduce vehicle speed; (iii) installation of 40 km of fencing to funnel animals to underpasses (Fig. 35.9A); (iv) construction of 53 crossing structures (33 wildlife underpasses, 2 wildlife overpasses, 11 retrofitted culverts and 7 bridges or viaducts) (Figs. 35.9B, C and D); and (v) installation of roadside reflectors (but see Chapter 25) in areas used by dispersing lynx. By 2012, the rate of roadkill had decreased to 5% of the Doñana lynx population, and many crossing structures were being used regularly by lynx with demonstrated increases in gene flow (Simón 2012).

Figure 35.8 Iberian lynx is one of the world's most endangered carnivores, with just 400 breeding individuals remaining in the wild. Source: Photograph by and reproduced with permission of Joaquim Pedro Ferreira.

Figure 35.9 Mitigation measures to minimise road mortality of Iberian lynx and improve population connectivity: (A) fencing to prevent access to the roadway and funnel lynx to crossing structures, (B) earthen ramps as escape structures for Iberian lynx, (C) underpass with wooden ledge, (D) Iberian lynx using an underpass, (E) wildlife overpass specifically for lynx during construction and (F) the operational overpass. Source: Photographs by and reproduced with the permission of (A) Joaquim Pedro Ferreira, (B, C, E, F) Gema Ruiz, (D) Miguel Simón).

35.5 Fencing, when paired with crossing structures, is critical to reducing the negative effects of roads on carnivores

Fencing is an effective strategy to reduce WVC and is recommended for many species (Chapter 20), including carnivores (Klar et al. 2009). However, fences without crossing structures can exacerbate the barrier effect of the road and create genetically distinct populations, increasing the likelihood of local extinctions (e.g. Klar et al. 2006). Therefore, roadside fencing should always be combined with crossing structures to reduce fragmentation effects and provide the added benefit of directing animals to crossing structure entrances (Chapters 20 and 21). Standard livestock fences are typically ineffective at containing carnivores because most species are agile, capable climbers or persistent diggers (Grilo et al. 2009; Cserkész et al. 2013). Effective fences for species with a penchant for digging need to be buried, while those for

climbing species require sufficient height and an overhanging edge to contain them (Chapter 20, Textbox 35.3).

The length of fencing necessary to effectively funnel wildlife to crossing structures varies by species. For instance, fencing with a small mesh size extending for 100 m on each side of culvert entrances was not enough to prevent roadkills of small- and medium-sized carnivores with larger spatial requirements (Villalva et al. 2013). Similarly, a 100 m section of fence around culverts did not generally increase use by bobcats, but it may have contributed to increased use of culverts previously frequented by bobcats (Cain et al. 2003). Interestingly, a simulation study with a common species such as stone martens in areas of high rates of roadkill showed that partial wildlife fencing alone may be more effective than crossing structures at reducing genetic differentiation, given its ability to eliminate road mortality, which in turn increased genetic diversity (Ascensão et al. 2013).

Textbox 35.3 Designing fences to prevent road mortality of wildcats.

European wildcats, similar to many other species of carnivore, are able to climb and jump standard wildlife and livestock fences with relative ease. Consequently, a fence design that prevented wildcats from accessing the road and guided them towards crossing structures was urgently needed. A series of trials with European wildcat in different types of fence enclosures was conducted in Germany, resulting in a fence

that is 2 m high and has a mesh size of 5 cm², a 50 cm-wide overhanging metal sheet and a 30 cm-wide subterranean plastic board (Fig. 35.10; Klar et al. 2009). These are now installed as a standard measure along new motorways that traverse wildcat habitat. This fence is combined with a variety of crossing structures spaced a few kilometre apart and effectively reduces roadkill and restores permeability.

(A)

(B)

Figure 35.10 (A) The European wildcat is identified by the circular black rings on its tail. (B) Fences for wildcats in Germany are 2 m high with mesh size of 5 cm², a 50 cm-wide overhanging metal sheet and plastic board buried 30 cm deep. Source: (A) Photograph by and reproduced with permission of Heiko Müller-Stieß; and (B) Photograph by N. Klar.

CONCLUSIONS

There is an abundance of scientific literature identifying the direct impacts of roads on carnivores including quantifying road mortality and crossing rates (see also Chapter 28). However, little is known about the implications of those values on the viability of carnivore populations over time. For example, the typical recording of use of crossing structures by carnivores is not sufficient to fully assess their effectiveness; identifying the minimum number of breeding individuals required to cross the road barrier to ensure adequate gene flow and maintain sustainable populations must be determined (Corlatti et al. 2009). Research should be conducted to find the thresholds in road density that threaten the persistence of various carnivore populations and to identify minimum specifications for mitigation efficacy required to provide connectivity for individuals, genetic exchange and long-term population persistence.

Long-term monitoring programmes that incorporate pre- and post-construction evaluation (Chapter 10) should be more widely employed across the geographic ranges of different carnivore species. Importantly, these programmes should specifically address the variation in responses to roads by different carnivore species and individual mitigation preferences. This information can be used to assess the effectiveness of crossing structures and fencing to adequately reduce mortality rates and facilitate gene flow across roads and thereby maintain viable populations across large scales (Corlatti et al. 2009).

ACKNOWLEDGEMENTS

C. Grilo was supported by Fundação para a Ciência e a Tecnologia through a postdoctoral grant (SFRH/BPD/64205/2009). Special thanks to Gemma Ruiz and Miguel Simón for information on Iberian lynx road mitigation measures.

FURTHER READING

Basille et al. (2013): Emphasises the hierarchical nature of habitat selection at multiple spatial scales, in particular concerning road density, where one carnivore species can shift habitat selection to avoid areas with the highest road densities within their home range, using a compensatory mechanism at fine scales.

Gittleman et al. (2001): Summarises the problems, approaches and solutions for carnivore conservation and provides a conceptual framework for future research and management, especially in changing landscapes.

Ripple et al. (2014): Assesses how threats such as habitat loss, persecution by humans and loss of prey combined can promote declines of large carnivores which pose a global conservation problem.

Sawaya et al. (2014): Highlights the importance of wildlife crossing structures to provide for interactions between individuals and consequently promote gene flow restoring landscape connectivity for carnivores in roaded landscapes.

REFERENCES

Alexander, S.M., N.M. Waters and P.C. Paquet. 2005. Traffic volume and highway permeability for a mammalian community in the Canadian Rocky Mountains. Canadian Geographer **49**: 321–331.

Ascensão, F., A. Clevenger, M. Santos-Reis, P. Urbano and N. Jackson. 2013. Wildlife-vehicle collision mitigation: is partial fencing the answer? An agent-based model approach. Ecological Modelling **257**: 36–43.

Barrientos, R. and J.D. Miranda. 2012. Can we explain regional abundance and road-kill patterns with variables derived from local-scale road-kill models? Evaluating transferability with the European polecat. Diversity and Distributions **18**: 635–647.

Basille, M., B. Van Moorter, I. Herfindal, J. Martin, J.D.C. Linnell, J. Odden, R. Andersen and J.-M. Gaillard. 2013. Selecting habitat to survive: the impact of road density on survival in a large carnivore. PLoS ONE **8**: e65493. doi:10.1371/journal.pone.0065493.

Beckmann, J.P., A.P. Clevenger, M. Huijser and J.A. Hilty (Eds.). 2010. Safe passages: highways, wildlife, and habitat connectivity. Washington: Island Press. pp. 3–16.

Beldon, B.C. and B.W. Hagedorn. 1993. Feasibility of translocating panthers into northern Florida. Journal of Wildlife Management **57**: 388–397.

Beringer, J.J., S.G. Siebert and M.R. Pelton. 1990. Incidence of road crossing by black bears on Pisgah National Forest, North Carolina. International Conference on Bear Research and Management **8**: 85–92.

Brody, A.J. and M.R. Pelton. 1989. Effects of roads on black bear movements in western North Carolina. Wildlife Society Bulletin **17**: 5–10.

Brodziewska, J. 2005. Wildlife tunnels and fauna bridges In Poland: past, present and future, 1997–2013. In: International Conference on Ecology and Transportation. 2005 San Diego proceedings. Center for Transportation and the Environment: North Carolina State University, 2006. Raleigh, NC, pp. 448–460.

Bunnell, F.G. and D.E.N. Tait. 1980. Bears in models and in reality-implications to management. International Conference on Bear Research and Management **4**: 15–24.

Burkey, T.V., D.H. Reed. 2006. The effects of habitat fragmentation on extinction risk: mechanisms and synthesis. Songklanakarin Journal of Science and Technology **28**: 9–37.

Cain, A.T., V.R. Tuovila, D.G. Hewitt and M.E. Tewes. 2003. Effects of a highway and mitigation projects on bobcats in southern Texas. Biological Conservation **114**: 189–197.

Cardillo, M., A. Purvis, W. Sechrest, J.L. Gittleman, J. Bielby and G.M. Mace 2004. Human population density and extinction risk in the world's carnivores. PLoS Biology **2**: 909–914.

Clarke, G.P., P.C.L. White and S. Harris. 1998. Effects of roads on badger *Meles meles* populations in southwest England. Biological Conservation **86**: 117–124.

Clevenger, A.P. 2011. 15 years of Banff research: what we've learned and why it's important to transportation managers beyond the park boundary. In Long-Term Perspectives on Ecology and Sustainable Transportation. Proceedings of the 2011 International Conference on Ecology and Transportation, Seattle, WA, pp. 409–423.

Clevenger, A.P. and N. Waltho. 2005. Performance indices to identify attributes of highway crossing structures facilitating movement of large mammals. Biological Conservation **121**: 453–464.

Clevenger, A.P., B. Chruszcz and K. Gunson. 2001. Drainage culverts as habitat linkages and factors affecting passage by mammals. Journal of Applied Ecology **38**: 1340–1349.

Colchero, F., D.A. Conde, C. Manterola, C. Chávez, A. Rivera and G. Ceballos. 2011. Jaguars on the move: modeling movement to mitigate fragmentation from road expansion in the Mayan Forests. Animal Conservation **14**: 158–166.

Corlatti L., K. Hackländer and F. Frey-Roos. 2009. Ability of wildlife overpasses to provide connectivity and prevent genetic isolation. Conservation Biology **23**: 548–556.

Craighead, F.C., Jr. 1976. Grizzly bear ranges and movement as determined by radio tracking. Third International Conference on Bears **3**: 97–109.

Crooks, K.R. 2002. Relative sensitivities of mammalian carnivores to habitat fragmentation. Conservation Biology **16**: 488–502.

Cserkész T., B. Ottlecz, A. Cserkész-Nagy and J. Farkas. 2013. Interchange as the main factor determining wildlife-vehicle collision hotspots on the fenced highways: spatial analysis and applications. European Journal of Wildlife Research **59**: 587–597.

Ferreras, P., J.J. Aldama, J.F. Beltran and M. Delibes. 1992. Rates and causes of mortality in a fragmented population of Iberian lynx *Felis pardina* Temminck, 1824. Biological Conservation **61**: 197–202.

Florida Fish and Wildlife Conservation Commission (FFWCC). 2012. Florida black bear management plan. FFWCC, Tallahassee, FL. p. 215.

Florida Fish and Wildlife Conservation Commission (FFWCC). 2014. Florida panther net. FFWCC, Tallahassee, FL. http://www.floridapanthernet.org/index.php. Accessed 17 April 2014.

Gittleman, J.L., S.M. Funk, D. MacDonald and R.K. Wayne (Eds.). 2001. Carnivore conservation. Cambridge University Press, Cambridge.

Glista, D.J., T.L. DeVault and J.A. DeWoody. 2009. A review of mitigation measures for reducing wildlife mortality on roadways. Landscape and Urban Planning **91**: 1–7.

Grilo, C., J.A. Bissonette and M. Santos-Reis. 2008. Response of carnivores to existing highway culverts and underpasses: implications for road planning and mitigation. Biodiversity and Conservation **17**: 1685–1699.

Grilo, C., J.A. Bissonette and M. Santos-Reis. 2009. Spatial–temporal patterns in Mediterranean carnivore road casualties: consequences for mitigation. Biological Conservation **142**: 301–313.

Grilo, C., J. Sousa, F. Ascensão, H. Matos, I. Leitão, P. Pinheiro, M. Costa, J. Bernardo, D. Reto, R. Lourenço, M. Santos-Reis and E. Revilla. 2012. Individual spatial responses towards roads: implications for mortality risk. PLOS ONE. http://dx.plos.org/10.1371/journal.pone.0043811. Accessed 23 September 2014.

Gunson, K.E., G. Mountrakis and L.J. Quackenbush. 2010. Spatial wildlife-vehicle collision models: a review of current work and its application to transportation mitigation projects. Journal of Environmental Management **92**: 1074–1082.

Hartmann, S., K. Steyer, R. Kraus, G. Segelbacher and C. Nowak. 2013. Potential barriers to gene flow in the endangered European wildcat (*Felis silvestris*). Conservation Genetics **14**: 413–426.

Jackson, N.D. and L. Fahrig. 2011. Relative effects of road mortality and decreased connectivity on population genetic diversity. Biological Conservation **144**: 3143–3148.

Kautz, R., R. Kawula, T. Hoctor, J. Comiskey, D. Jansen, D. Jennings, J. Kasbohm, F. Mazzotti, R. McBride, L. Richardson and K. Root. 2006. How much is enough? Landscape-scale conservation for the Florida panther. Biological Conservation **130**: 118–133.

Kerley, L., J.M. Goodrich, D.G. Miquelle, E.N. Smirnov, H.B. Quigley and M.G. Hornocker. 2002. Effects of roads and human disturbance on Amur tigers. Conservation Biology **16**: 1–12.

Klar, N., M. Herrmann and S. Kramer-Schadt. 2006. Effects of roads on a founder population of lynx in the biosphere reserve "Pfälzerwald – Vosges du Nord" – a model as planning tool. Naturschutz und Landschaftsplanung **38**: 330–337.

Klar, N., M. Herrmann and S. Kramer-Schadt. 2009. Effects and mitigation of road impacts on individual movement behavior of wildcats. Journal of Wildlife Management **73**: 631–638.

Kusak J., D. Huber, T. Gomerčić, G. Schwaderer and G. Gužvica. 2009. The permeability of highway in Gorski Kotar (Croatia) for large mammals. European Journal of Wildlife Research **55**: 7–21.

Liu, S., Y. Dong, L. Deng, Q. Liu, H. Zhao and S. Dong. 2014. Forest fragmentation and landscape connectivity change associated with road network extension and city expansion: a case study in the Lancang River Valley. Ecological Indicators **36**: 160–168.

Mata, C., I. Hervás, J. Herranz, F Suárez and J.E. Malo. 2008. Are motorway wildlife passages worth building? Vertebrate use of road-crossing structures on a Spanish motorway. Journal of Environmental Management **88**: 407–415.

McCown, J.W., P. Kubilis, T. Eason and B. Scheick. 2004. Black bear movements and habitat use relative to roads in Ocala National Forest. Final Report. Florida Department of Transportation Contract BD-016. Florida Fish and Wildlife Conservation Commission, Tallahassee, FL. p. 118.

McCown, J.W., P. Kubilis, T.H. Eason and B.K. Scheick. 2009. Effect of traffic volume on American black bears in central Florida, USA. Ursus **20**: 39–46.

Okarma H., W. Jędrzejewski, K. Schmidt, S. Sniezko, A.N. Bunevich and B. Jedrzejewska. 1998. Home ranges of wolves in Białowieża Primeval Forest, Poland, compared with other Eurasian populations. Journal of Mammalogy **79**: 842–852.

Maehr, D.S., E.D. Land, D.B. Shindle, O.L. Bass and T.S. Hoctor. 2002. Florida panther dispersal and conservation. Biological Conservation **106**: 187–197.

Mech, L.D., S.H. Fritts, G.L. Radde and W.J. Paul. 1988. Wolf distribution and road density in Minnesota. Wildlife Society Bulletin **16**: 85–87.

Ngoprasert, D., A.J. Lynam and G.A. Gale. 2007. Human disturbance affects habitat use and behaviour of Asiatic leopard *Panthera pardus* in Kaeng Krachan National Park, Thailand. Oryx **41**: 343–351.

Palomares, F., P. Ferreras, J.M. Fedriani and M. Delibes. 1996. Spatial relationships between Iberian lynx and other carnivores in an area of south-western Spain. Journal of Applied Ecology **33**: 5–13.

Philcox, C.K., A.L. Grogan and D.W. MacDonald. 1999. Patterns of otter *Lutra lutra* road mortality in Britain. Journal of Applied Ecology **36**: 748–762.

Riley, S.P.D., J.P. Pollinger, R.M. Sauvajot, E.C. York, C. Bromley, T.K. Fuller and R.K. Wayne. 2006. A southern California freeway is a physical and social barrier to gene flow in carnivores. Molecular Ecology **15**: 1733–1741.

Ripple, W.J., J.A. Estes, R.L. Beschta, C.C. Wilmers, E.G. Ritchie, M. Hebblewhite, J. Berger, B. Elmhagen, M. Letnic, M.P. Nelson, O.J. Schmitz, D.W. Smith, A.D. Wallach and A.J. Wirsing 2014. Status and ecological effects of the world's largest carnivores. Science **343**(6167) 1241484. doi: 10.1126/science.1241484; doi: 10.1126/science.1241484.

Rogers, L.L. 1987. Factors influencing dispersal in the black bear. In Mammalian dispersal patterns: the effects of social structure on population genetics. B.D. Chepko-Sade and Z.T. Halpin (Eds.). University of Chicago Press, Chicago and London, pp. 75–84.

Saeki, M. and D.W. MacDonald 2004. The effects of traffic on the raccoon dog (*Nyctereutes procyonoides viverrinus*) and other mammals in Japan. Biological Conservation **118**: 559–571.

Sawaya, M.A, S.T. Kalinowski and A.P. Clevenger. 2014. Genetic connectivity for two bear species at wildlife crossing structures in Banff National Park. Proceedings of the Royal Society B **280**: 1780.

Serronha, A.M., R. Mateus, F. Eaton, M. Santos-Reis and C. Grilo. 2013. Towards effective culvert design: monitoring seasonal use and behavior by Mediterranean mesocarnivores. Environmental Monitoring and Assessment **185**: 6235–6245.

Simcharoen, S., A.C.D. Barlow, A. Simcharoen and J.L.D. Smith. 2008. Home range size and daytime habitat selection of leopards in Huai Kha Khaeng Wildlife Sanctuary, Thailand. Biological Conservation **141**: 2242–2250.

Simón, M. (Ed.) 2012. Ten years conserving the Iberian lynx. Consejeria de Agricultura y Medio Ambiente, Junta de Andalucia, Seville.

Taylor, S.K., C.D. Buergelt, M.E. Roelke-Parker, B.L. Homer and D.S. Rotstein. 2002. Causes of mortality of free-ranging Florida panthers. Journal of Wildlife Diseases **38**: 107–114.

Van Dyke, F.G., R.H. Brocke, H.G. Shaw, B.B. Ackerman, T.P. Hemker and F.G. Lindzey. 1986. Reactions of mountain lions to logging and human activity. Journal of Wildlife Management **550**: 95–102.

Villalva, P., D. Reto, M. Santos-Reis, E. Revilla and C. Grilo. 2013. Do dry ledges reduce the barrier effect of roads? Ecological Engineering **57**: 143–148.

Wieren, S.E. and P.B. Worm. 2001. The use of a motorway wildlife overpass by large mammals. Netherlands Journal of Zoology **51**: 97–105.

Wooding, J.B. and J.R. Brady. 1987. Black bear roadkills in Florida. Proceedings of the Annual Conference of the Southeastern Association of Fish and Wildlife Agencies **41**: 438–442.

Zimmermann, F. 2004. Conservation of the Eurasian Lynx (*Lynx lynx*) in a fragmented landscape-habitat models, dispersal and potential distribution. Faculté de Biologie et de Médecine, Université de Lausanne, Lausanne, p. 178.

Chapter 36

CASE STUDY: ROADS AND JAGUARS IN THE MAYAN FORESTS

Eugenia Pallares[1], Carlos Manterola[2], Dalia A. Conde[3] and Fernando Colchero[4]

[1]Jaguar Conservancy, Mexico City, Mexico
[2]Grupo Anima Efferus, A.C., Mexico City, Mexico
[3]Max-Planck Odense Center on the Biodemography of Aging and Institute of Biology, University of Southern Denmark, Odense, Denmark
[4]Max-Planck Odense Center on the Biodemography of Aging and Department of Mathematics and Computer Science, University of Southern Denmark, Odense, Denmark

Jaguars (Fig. 36.1) are one of the most elusive large carnivores on earth, and gathering information on their demography and behaviour is extremely challenging. Although their geographic range extends from northern Mexico to Argentina, it has shrunk to less than 54% of its original extent in the last few decades (Kinnaird et al. 2002). Jaguars are a key flagship species in the Americas, being the subject of Paseo Pantera, the continent-wide connectivity initiative (Sanderson et al. 2002; Rabinowitz & Zeller 2010).

Studies on jaguars have focused mostly on understanding foraging behaviour, dietary preferences and activity patterns (Rabinowitz 1986; Novack et al. 2005; Weckel et al. 2006), and only a few have attempted to understand habitat preferences and the impact of roads on their ecology (Ortega-Huerta & Medley 1999; Conde et al. 2010; Colchero et al. 2011). Recent studies in the Mayan Forest of Mexico, Belize and Guatemala have shed some light on the impact of roads on this charismatic carnivore. This region supports the major zone of tropical forest in North and Central America and is a key element of the Mesoamerican Hotspot (Myers et al. 2000).

The largest jaguar population in the northern hemisphere also occurs here, and this population is threatened by the expansion of road networks that have severely fragmented the habitat of jaguars and many other species (Conde et al. 2007).

A recent study on the habitat preferences of jaguars in the Mayan Forests showed that they occur with higher probability in well-preserved forest patches than on secondary growth or agricultural lands and that the probability of occurrence for jaguars declined with increasing proximity to roads (Conde et al. 2010). Male and female jaguars select different habitats and males show a higher tolerance to roads than females. From 1980 to 2000, 34% of female and 22% of male habitats were lost, while habitat fragmentation doubled (Conde et al. 2010). Mortality due to wildlife-vehicle collision (WVC) is higher in males than females.

In a subsequent study, radio-telemetry and GPS data were used to infer the movement of jaguars in response to vegetation, roads and human population density (Colchero et al. 2011). Analysis of jaguar movement patterns identified crossing sites on the Escárcega–Xpujil

Handbook of Road Ecology, First Edition. Edited by Rodney van der Ree, Daniel J. Smith and Clara Grilo.
© 2015 John Wiley & Sons, Ltd. Published 2015 by John Wiley & Sons, Ltd.
Companion website: www.wiley.com\go\vanderree\roadecology

Figure 36.1 Jaguar in Mexico being sedated and fitted with GPS. Source: Photograph by Carlos Manterola.

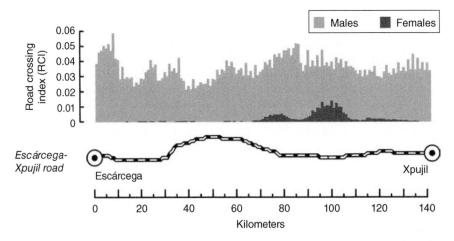

Figure 36.2 Road-crossing frequency from simulated female (dark bars) and male (light bars) jaguars along the Escárcega–Xpujil Road segment. Males cross the road along its entire length, but females will only cross around the 75 and 100 km sections of the road. Source: Colchero et al. (2011).

Road through the Calakmul Biosphere Reserve, one of the most important biological sanctuaries in the Mayan Forests of Mexico and Guatemala. The movement of male and female jaguars was affected, with females much more restricted than males by proximity to roads and even intermediate levels of human population density. Consequently, females have considerably fewer suitable locations to cross the Escárcega–Xpujil Road than males (Fig. 36.2). These crossing locations occurred in areas of high forest density and low density of humans. Still, this study identified a 1 km section of the road (~100 km from Escárcega; Fig. 36.3) where the likelihood of crossing was highest for both sexes and wildlife crossing structures should be built to re-establish connectivity. A comprehensive suite of mitigation measures were proposed to the Ministries of Communication and Environment in order to mitigate the impact of the existing road on Jaguars. As yet (February 2015), none

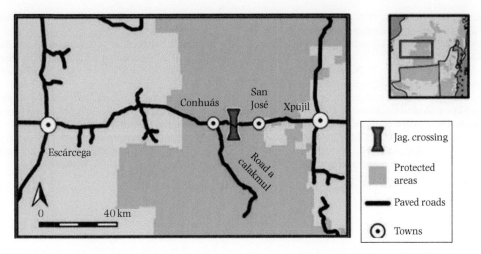

Figure 36.3 Location of proposed crossing structure for jaguar (Jag. crossing) along the Escárcega–Xpujil Road in southeast Mexico (see inset). The crossing site was identified as the location with the highest crossing index for male and female jaguars based on the simulations performed by Colchero et al. (2011). Source: Colchero et al. (2011).

have been installed along the 140 km length of road. As yet, no mitigation measures have been installed along the 140 km length of road.

Another negative effect is that roads improve access to the forest, leading to increased poaching of jaguars and hunting by people on prey populations important to large carnivores (Wilkie et al. 2000; Chapters 2, 37 and 56). Habitat loss and fragmentation also reduce the abundance and diversity of prey for jaguars, which increases the rate of depredation on cattle by jaguars (Polisar et al. 2003). Male jaguars are responsible for approximately 70% of cattle predation in the region, and this results in greater rates of male mortality by farmers and communal landowners who attempt to protect their stock (Rabinowitz 1986).

Consequently, it is imperative that the negative effects of roads on jaguars (e.g. mortality due to WVC, landscape fragmentation and reduction in high-quality habitat) play a key role in the decision-making process when planning and designing road projects. Importantly, priority habitat areas for the jaguar should be avoided when building roads, and where unavoidable, sufficient appropriate mitigation measures are required to maintain connectivity and reduce mortality due to WVC. Consideration of road impacts will reduce future conflicts between jaguars and people.

ACKNOWLEDGEMENTS

We thank the staff of Defensores de la Naturaleza and Fundación Arcas as well as Antonio Rivera, Heliot Zarza and Francisco Zavala for their invaluable work collecting jaguar data. Funding was provided by the Conservation International, Conservation Partnership Fund, Mesoamerican Biological Corridor–Mexico, CONACYT, Conservation Strategy Fund, National Fish and Wildlife Foundation and Safari Club International Foundation.

REFERENCES

Colchero, F., Conde, D.A., Manterola, C., Chávez, C., Rivera, A. & Ceballos, G. 2011. Jaguars on the move: modeling movement to mitigate fragmentation from road expansion in the Mayan Forest. Animal Conservation **14**: 158–166.

Conde, D.A., Burgués, I., Fleck, L., Manterola, C. & Reid, J. 2007. Análisis Ambiental y Económico de Proyectos Carreteros en la Selva Maya, un Estudio a Escala Regional. San Jose: Conservation Strategy Fund.

Conde, D.A., Colchero, F., Zarza, H., Christensen, N.L., Jr, Sexton, J.O., Manterola, C., Chávez, C., Rivera, A., Azuara, D. & Ceballos, G. 2010. Sex matters: modeling male and female habitat differences for jaguar conservation. Biological Conservation **143**: 1980–1988.

Kinnaird, M.F., Sanderson, E.W., O'Brien, T.G., Wibisono, H.T. & Woolmer, G. 2002. Deforestation trends in a tropical landscape and implications for endangered large mammals. Conservation Biology **17**: 245–257.

Myers, N., Mittermeier, R.A., Mittermeier, C.G., da Fonseca, G.A.B. & Kent, J. 2000. Biodiversity hotspots for conservation priorities. Nature **403**: 853–858.

Novack, A., Main, M., Sunquist, M. & Labisky, R. 2005. Foraging ecology of jaguar (*Panthera onca*) and puma (*Puma concolor*) in hunted and non-hunted sites within the Maya Biosphere Reserve, Guatemala. Journal of Zoology **267**: 167–178.

Ortega-Huerta, M. & Medley, K. 1999. Landscape analysis of jaguar (*Panthera onca*) habitat using sighting records in the Sierra de Tamaulipas, Mexico. Environmental Conservation **26**: 257–269.

Polisar, J., Maxit, I., Scognamillo, D., Farrell, L., Sunquist, M.E. & Eisenberg, J.F. 2003. Jaguars, pumas, their prey base, and cattle ranching: ecological interpretations of a management problem. Biological Conservation **109**: 297–310.

Rabinowitz, A. 1986. Jaguar predation on domestic livestock in Belize. Wildlife Society Bulletin **14**: 170–174.

Rabinowitz, A. & Zeller, K. 2010. A range-wide model of landscape connectivity and conservation for the jaguar, *Panthera onca*. Biological Conservation **143**: 939–945.

Sanderson, E., Redford, K., Chetkiewicz, C., Medellin, R., Rabinowitz, A., Robinson, J. & Taber, A. 2002. Planning to save a species: the jaguar as a model. Conservation Biology **16**: 58–72.

Weckel, M., Giuliano, W. & Silver, S. 2006. Jaguar (*Panthera onca*) feeding ecology: distribution of predator and prey through time and space. Journal of Zoology **270**: 25–30.

Wilkie, D., Shaw, E., Rotberg, F., Morelli, G. & Auzel, P. 2000. Roads, development, and conservation in the Congo Basin. Conservation Biology **14**: 1614–1622.

Chapter 37

CASE STUDY: FINDING THE MIDDLE ROAD – GROUNDED APPROACHES TO MITIGATE HIGHWAY IMPACTS IN TIGER RESERVES

Sanjay Gubbi[1,2] and H.C. Poornesha[2]

[1]Panthera, New York, USA
[2]Nature Conservation Foundation, Mysore, India

Tigers (Fig. 37.1) are one of the world's most endangered large carnivores with an estimated global population of approximately 3200 individuals. They currently occur in 13 countries, representing 7% of their former range (Dinerstein et al. 2007). The survival of tigers in the wild depends largely upon the willingness of the tiger-range countries to ensure adequate protection of sufficiently large areas from inappropriate development and activities such as roads and poaching. Tigers are threatened by roads and traffic. Research on Amur tigers in Russia suggests that direct mortality due to wildlife-vehicle collisions (WVC) can reduce survivorship and reproductive success of surviving animals (Kerley et al. 2002). Tigers have been affected in western Malaysia through construction of the North–South highway and another highway that bisected a bottleneck area in Taman Negara National Park (Kawanishi et al. 2010).

The current rate of mortality of tigers due to WVC in India appears to be relatively low, with approximately 20 documented tiger deaths in various reserves over the past 15 years (Prakash 2012), although this number is likely an underestimate due to non-detection or non-reporting. Furthermore, as the size of the tiger population declines and the road network expands, the direct and indirect effects of mortality due to WVC and fragmentation of tiger habitats will become of greater concern. In addition to direct mortality, the death of individual tigers results in social instability. The death of a territorial male can lead to infighting of transient males trying to establish territories and infanticide by the new territorial male, and it also affects tigresses due to unstable male ranges, possibly leading to depressed birth rates. Axis deer, a principal prey species for tigers in India, are also commonly killed by WVC, resulting in reduced food for tigers.

Furthermore, roads are used for illegal activities including hunting of tigers and their prey (see also Chapters 2, 36 and 56). In eastern Russia, at least six Amur tigers were poached over a 10-year period along one road (Kerley et al. 2002). In 2010, poachers apprehended in southern India confessed to the illegal hunting of axis and other deer species in Bandipur, Bhadra and Biligirirangaswamy Tiger Reserves by driving on roads at night.

Handbook of Road Ecology, First Edition. Edited by Rodney van der Ree, Daniel J. Smith and Clara Grilo.
© 2015 John Wiley & Sons, Ltd. Published 2015 by John Wiley & Sons, Ltd.
Companion website: www.wiley.com\go\vanderree\roadecology

Figure 37.1 A tiger crosses a road in Bandipur Tiger Reserve, southern India. Source: Photograph by and reproduced with permission of H. S. Basavanna.

In India, the National Wildlife Action Plan 2002–2016 (Government of India 2002) specifically prescribed regulation and mitigation measures for threats to wildlife posed by roads. However, there have been few serious attempts to implement the policy on the ground.

37.1 Night closures and alternative roads in nagarahole–bandipur tiger reserves

India has prioritised the conservation of tigers by establishing 47 tiger reserves, spread across the whole country, which get additional legal protection and funding. The Nagarahole (643 km^2) and Bandipur Tiger Reserves (990 km^2) are important conservation areas for tigers, with over 100 breeding tigers (Sanderson et al. 2010). Seven major roads pass through these two contiguous tiger reserves (Fig. 37.2) nestled in the Western Ghats in the Indian state of Karnataka.

A 27.3 km stretch of the Mysore–Mananthavadi Road (State Highway (SH-17D) passes through the southern part of Nagarahole. In July 2008, based on a proposal made by conservationists to government officials and a presentation to other stakeholders, the road was closed to all traffic at night (18.00-6.00 hrs) except emergency vehicles. However, as varied interest groups opposed the closure, upholding the closure needed persistent follow-up and support of well-intentioned

government officials and media. In a particular instance, conservationists forged alliances with animal rights groups who were opposed to the use of this road to illegally transport livestock to slaughter houses in violation of officially prescribed transportation standards. Finally, after various stages of turnovers, the closure of the road by the government was upheld in the court.

Under a World Bank-funded project, the Mysore–Mananthavadi Road was proposed to be upgraded to a high-speed highway. An alternative alignment was identified that would reduce the length of the road within the tiger reserve to 17.3 km and offer better access to 11 villages that were previously poorly connected. Combined with the night closure to traffic, realigning the road outside the reserve would further decrease the impacts of traffic during daylight hours when vehicles were allowed to pass through the tiger reserve. However, the alternative road was in poor condition and needed substantial investment from the government. In 2012, US$3.2 m was released by the state government to repair and upgrade the alternative road after the government was convinced of the conservation and public welfare merits. Now, a 10 km stretch of the road through the tiger reserve has been officially decommissioned and a corresponding alternate alignment authorised and developed outside the reserve.

Based on the experience of Mysore–Mananthavadi Road, in 2010, the government ordered nighttime

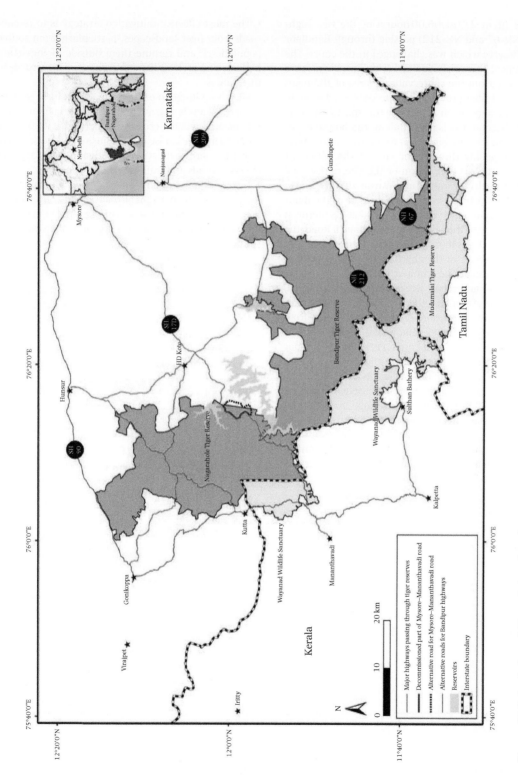

Figure 37.2 Roads through Nagarahole and Bandipur Tiger Reserves and alternate solutions provided.

closure (from 21.00–06.00 hours) of the two highways NH-67 and NH-212) passing through Bandipur Tiger Reserve which was challenged in the court. The government counsel, including key policymakers, were persuaded about the conservation merits of the night closure but also provided an alternative solution to reduce the impacts on night-time transportation. One of the alternate roads passed along the boundary of Nagarahole (SH-90, Fig. 37.2) which was less damaging than passing through the core of the reserve at night. Apart from this, 22 km of the alternative road passed through Nagarahole and Wayanad, while 29 km of the highway traversed through the core of Bandipur and the adjoining Wayanad Wildlife Sanctuary. However, as with the Mysore–Mananthavadi Road, the alternate road was in poor condition. Hence, the court ordered the repair of the alternate road for which the government made the required budgetary allocations (US\$8.03 m) and repair works have been completed. Led by these examples, other state governments have implemented night closure or diversion of roads from protected areas including Mudumalai Tiger Reserve in Tamil Nadu and Gir National Park and Velavadar Wildlife Sanctuary in the state of Gujarat.

Working closely with the media was an effective tool to increase public awareness and help create the atmosphere to build public acceptance and support. Educating drivers to slow down in protected areas is an important aspect of our campaign because commuters drive at high speeds in these areas. Improved scientific understanding of road impacts on tigers and other species in India is required.

37.2 Science to support conservation

We carried out a preliminary study to assess the relative use of roadside habitats by large mammals on the Mysore–Mananthavadi Road. Tiger, gaur, chital and elephant were more frequently detected along a closed section of the road than one with traffic (Gubbi et al. 2012). Our data also showed that existing culverts built for drainage were seldom used by wildlife, contrary to arguments advanced by developers seeking to expand this road (Hosmat & Gubbi 2010). Our results broadly vindicated the need for measures to mitigate the impact of roads and traffic in sensitive wildlife habitats.

37.3 Future course of action

These experiences offer key lessons on managing the impact of roads for tigers:

- The most effective mitigation strategy is to remove roads from tiger landscapes, particularly from source populations, and reroute them outside of important habitats and prevent the construction of new roads in those areas.
- Effective mitigation of road impacts requires engagement of all levels of government and the community.
- Dedicated wildlife crossing structures will likely be required in tiger landscapes because standard drainage structures are ineffective at mitigating the negative effects of roads and traffic on tigers.
- International funding agencies are financing the rapid rate of construction of roads in many tiger-range countries, and they must become involved in measures to ensure these developments do not further endanger the persistence of tigers.

ACKNOWLEDGEMENTS

I thank the Wildlife Conservation Society–India Program, Centre for Wildlife Studies and 21st Century Tiger for supporting part of this work and M.D. Madhusudan whose comments helped to greatly improve this chapter.

REFERENCES

Dinerstein, E., C. Loucks, E. Wikramanayake, J. Ginsberg, E. Sanderson, J. Seidensticker, J. Forrest, G. Bryja, A. Heydlauff, S. Klenzendorf, P. Leimgruber, J. Mills, T.G. O'Brien, M. Shrestha, R. Simons and M. Songer. 2007. The fate of wild tigers. BioScience **57**:508–514.

Government of India. 2002. National Wildlife Action Plan 2002–2016. Ministry of Environment and Forests, Government of India, New Delhi.

Gubbi, S., H. C. Poornesha and M. D. Madhusudan. 2012. Impact of vehicular traffic on the use of highway edges by large mammals in a South Indian wildlife reserve. Current Science **102**:1047–1051.

Hosmat, B. J. and S. Gubbi. 2010. Final report on the mitigation measures to reduce wildlife mortality on the Mysore–Mananthavadi highway, Nagarahole Tiger Reserve. Karnataka Forest Department, Wildlife Conservation Society-India Program, Centre for Wildlife Studies, Mysore and Bangalore.

Kawanishi, K., M. Gumal, L. A. Shepherd, G. Goldthorpe, C. R. Shepherd, K. Krishnasamy and A. K. A. Hashim. 2010. The Malayan tiger. Pages 367–376 in R. Tilson and P. J. Nyhus, editors. Tigers of the World: The Science, Politics and Conservation of *Panthera tigris*. Elsevier Inc., London.

Kerley, L. L., J. M. Goodrich, D. G. Miquelle, E. N. Smirnov, H. B. Quigley and M. G. Hornocker. 2002. Effects of roads and human disturbance on Amur tigers. Conservation Biology **16**:97–108.

Prakash, A. B. 2012. A study on the impacts of highways on wildlife in Bandipur Tiger Reserve. National Centre for Biological Sciences. Tata Institute of Fundamental Research, Bangalore.

Sanderson, E. W., J. Forrest, C. Loucks, J. Ginsberg, E. Dinerstein, J. Seidensticker, P. Leimgruber, M. Songer, A.

Heydlauff, T. O'Brien, G. Bryja, S. Klenzendorf and E. Wikramanayake. 2010. Setting Priorities for Tiger Conservation: 2005–2015. Pages 143–161 in R. Tilson and P. J. Nyhus, editors. Tigers of the World: The Science, Politics and Conservation of *Panthera tigris*. Elsevier Inc., London.

Chapter 38

CASE STUDY: AFRICAN WILD DOGS AND THE FRAGMENTATION MENACE

Brendan Whittington-Jones and Harriet Davies-Mostert

Endangered Wildlife Trust, Johannesburg, South Africa

The complex social dynamics, sleek cursorial design and semi-nomadic, fluid approach of African wild dogs (Fig. 38.1) to the broader landscape are in direct contrast to the human approach to infrastructure development (i.e. hardened, static, linear facilities and boundaries, such as those imposed by roads and fences). Wild dogs, like many wide-ranging carnivores, survive at comparatively low population densities, and their prey requirements often leave them vulnerable to persecution, reduced food availability and the fragmentation of prey populations and habitats. Such carnivores are increasingly exposed to anthropogenic threats such as wildlife-vehicle collisions (WVC), snares and direct persecution as they disperse through fragmented landscapes. Deterioration of the connectedness of landscapes can isolate sub-populations and reduce population viability, thereby increasing the risk of local extinctions and raising the management challenges for remaining populations.

Approximately 6600 free-ranging wild dogs remain in Africa (IUCN 2013), having disappeared from at least 25 countries over the past 50 years largely due to direct persecution and widespread habitat loss and fragmentation (Woodroffe & Ginsberg 1999). With fewer than 450 free-ranging African wild dogs left in South Africa, they are the country's rarest large carnivore (Lindsey & Davies-Mostert 2009). A managed metapopulation approach has been applied that comprises a collection of geographically isolated private and state reserves that are capable of sustaining wild dogs but that need continual, intensive, collaborative efforts and logistical support to manage these subpopulations as a single, collective population.

In South Africa, the familiar saying that 'good fences make good neighbours' underlies an approach to partitioning land into defined, intensively managed units, frequently disrupting and fragmenting habitats. Understanding the influences of this transformation on large carnivores is important for managing connectivity among core populations and for mitigating the artificially intense impact such restricted carnivores may have at a local scale.

Fences limit the movement of carnivores among populations, and they alter relationships between predators and prey (Chapter 20). Wild dogs alter their hunting strategies by using fences to their advantage, with the result that they can chase larger prey into fences that they wouldn't ordinarily be able to catch.

Handbook of Road Ecology, First Edition. Edited by Rodney van der Ree, Daniel J. Smith and Clara Grilo.
© 2015 John Wiley & Sons, Ltd. Published 2015 by John Wiley & Sons, Ltd.
Companion website: www.wiley.com\go\vanderree\roadecology

Figure 38.1 The African wild dog is endangered and is the subject of numerous studies to understand its ecology and aid in conservation. This individual has been fitted with a VHF collar to track its movements within a state game reserve. Source: Photograph by Brendan Whittington-Jones.

High perimeter-to-area ratios increase the chance of packs coming into contact with fences, and the extent of this will vary according to both reserve size and shape. In Venetia Limpopo Nature Reserve in northeastern South Africa, fence-hunting behaviour was found to influence the impact of wild dogs on prey populations by enabling the capture of different species, size classes, sexes, and conditions (Davies-Mostert et al. 2013). For example, fence-impeded kills comprised 40.5% of all kills ($n = 316$) over a 3-year period. When compared to kills made away from the fence, fence-impeded kills comprised larger prey species (kudu were twice as likely to be captured on the fence than off the fence) and enabled the capture of animals in better physical condition for one prey category (Davies-Mostert et al. 2013). Fence-impeded kills also provided greater catch per unit of hunting effort, resulting in longer inter-kill intervals (Davies-Mostert et al. 2013). Despite these potential benefits for wild dogs, certain fences may also act as barriers to dispersing individuals.

Dispersal direction may be influenced by artificial barriers such as fences, roads and railways, and this depends on whether such barriers interrupt dispersal or are semipermeable to a particular species. Selection of dispersal routes by wild dogs is a function of interacting ecological features and environmental and social pressures. In some landscapes, despite the inherent risks, wild dogs have been found to favour the use of rural roads for movement, a feature which may enable quicker movement than through dense vegetation. In a study of transient wild dogs outside of resident protected areas in eastern South Africa, many sightings were reported close to roads, and most were frequently found in areas with road densities of 0.6–0.7 km/km^2 (Whittington-Jones 2011).

Although few confirmed wild dog deaths have been recorded on roads when compared to other anthropogenic threats (e.g. poacher's wire snares), they do actually occur (Fig. 38.2). In Hluhluwe-iMfolozi Park in South Africa, four wild dogs (from a pack of 15 and a total population of 93 in the park) were killed in a single incident on the 20 km portion of the sealed R618 road which bisects the park. Such events can disrupt the intense social bond of the pack and even result in pack dissolution if key pack members are killed. Although there has been no formal quantification of the impact of the R618 on wildlife populations, traffic officials have attributed the frequent wildlife deaths on the road primarily to speeding vehicles (the speed limit varies between 60 and 80 km/h), particularly from dusk until dawn when no traffic enforcement takes place and visibility of wildlife on the winding road can be poor (L. Munro, personal communication).

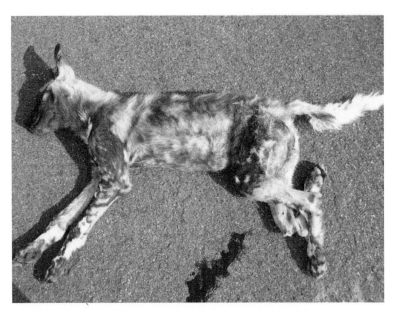

Figure 38.2 Wild dog mortality from collision with a vehicle. Source: Photograph by and reproduced with permission of Wendy J. Collinson.

To conserve wild dogs within heterogeneous landscapes, it is important to understand how their spatial movements contribute to dispersal risk. It is important to identify higher-risk linkages (e.g. those with high-speed roads, livestock farming or communal hunting) between subpopulations to develop tailored, pragmatic strategies for planning and management of contiguous landscapes that link core populations of large carnivores.

ACKNOWLEDGEMENTS

The Endangered Wildlife Trust's African Wild Dog projects which generated this information are, and were, supported by Jaguar Land Rover South Africa, Land Rover Centurion, Richard Bosman, Vaughan de la Harpe, RS Components, Knowsley Safari Park, Painted Wolf Wines, Investec, the Green Trust, the Stuart Bromfield Wild Dog Fund and De Beers Consolidated Mines.

REFERENCES

Davies-Mostert, H.T., M.G.L. Mills and D.W. Macdonald. 2013. Hard boundaries influence African wild dogs' diet and prey selection. Journal of Applied Ecology **50**: 1358–1366.

International Union for the Conservation of Nature (IUCN). 2013. The IUCN Red List of Threatened Species. Version 2013.1. http://www.iucnredlist.org. Accesssed on 23 September 2014.

Lindsey, P.A. & H.T. Davies-Mostert (Eds.) 2009. South African Action Plan for the Conservation of Cheetahs and African Wild Dogs. Report from a National Conservation Action Planning Workshop, Bela-Bela, South Africa, 17–19 June 2009.

Whittington-Jones, B.M. 2011. The dispersal of African Wild Dogs *Lycaon pictus* from protected areas in the northern KwaZulu-Natal province, South Africa. MSc Thesis, Rhodes University, Grahamstown.

Woodroffe, R. & J.R.. Ginsberg. 1999. Conserving the African wild dog *Lycaon pictus*. I. Diagnosing and treating causes of decline. Oryx **33**: 132–142.

Chapter 39

ROADS, TRAFFIC AND VERGES: BIG PROBLEMS AND BIG OPPORTUNITIES FOR SMALL MAMMALS

Fernando Ascensão[1], Scott LaPoint[2] and Rodney van der Ree[3]

[1]CE3C – Centre for Ecology, Evolution and Environmental Changes, Faculdade de Ciências, Universidade de Lisboa, Lisboa, Portugal
Centro Brasileiro de Estudos em Ecologia de Estradas, Universidade Federal de Lavras, Campus Universitário, Lavras, Brasi
[2]Department of Migration and Immuno-Ecology, Max Planck Institute for Ornithology, Radolfzell, Germany, Department of Biology, The University of Konstanz, Konstanz, Germany
[3]Australian Research Centre for Urban Ecology, Royal Botanic Gardens Melbourne, and School of BioSciences, The University of Melbourne, Melbourne, Victoria, Australia

SUMMARY

We consider small mammals as shrews and moles, small marsupials, hedgehogs, mice and rats, and other rodents, rabbits, and hares commonly designated as 'prey species' (ie, those weighing <5 kg). For these species, roads typically have negative, often detrimental impacts, but in a few circumstances roads may be beneficial, making road mitigation strategies for this group slightly more complex than for other taxa. For example, roads and traffic typically act as barriers to the movement of individuals, while road verges may provide suitable habitat within otherwise degraded areas. This chapter illustrates the complex relationship between roads and small mammals and suggests possible road management and mitigation solutions.

39.1 Road building may result in complete losses of habitat, territories and individuals.

39.2 Roads may filter small mammal movements and reduce genetic exchange.

39.3 The effect of roadkills on small mammal population persistence is often unknown.

39.4 Fenced verges may provide benefits for small mammals.

39.5 Crossing structures for small mammals must be optimally located and connected to surrounding habitat.

39.6 Road verges should be diverse and contain natural structure.

Handbook of Road Ecology, First Edition. Edited by Rodney van der Ree, Daniel J. Smith and Clara Grilo.
© 2015 John Wiley & Sons, Ltd. Published 2015 by John Wiley & Sons, Ltd.
Companion website: www.wiley.com\go\vanderree\roadecology

Despite numerous road ecology studies on small mammals, we lack information on several key points, particularly on the impact of roads on genetic exchange and the impact of roadkills for small mammal population persistence. Nevertheless, our current knowledge of small mammal ecology should allow road planners to avoid or mitigate the main negative impact of roads while encouraging the positive attributes of roads for small mammals.

INTRODUCTION

Small mammals play an important role within most ecosystems (Golley 1960), being key prey items for many predators (Moleón & Gil-Sánchez 2003) and voracious predators of invertebrates (Montgomery & Montgomery 1990). They are also important seed dispersers (Gómez et al. 2008) and ecosystem engineers capable of altering entire landscapes, such as the European rabbit within Mediterranean landscapes (Gálvez Bravo et al. 2008) or the prairie dogs of the North American plains (Baker et al. 2012). Thus, changes in small mammal populations can destabilise entire ecosystems. For example, the decline of the once abundant black-tailed prairie dog across the grasslands of North America resulted in an increased establishment of woody plants and an eventual transition to woodland ecosystems (Weltzin et al. 1997). Therefore, efforts to understand small mammal–road interactions and how to mitigate the negative impacts of roads and traffic will benefit the overall ecosystem health.

Small mammals typically respond negatively to roads but in some situations can also respond positively or both (Chapter 28), varying among species and landscapes, making the impacts of roads difficult to predict. As with other taxa, roads and traffic negatively impact small mammals mainly by reducing and fragmenting habitat, by restricting individual movements and by causing direct mortality through wildlife–vehicle collision (WVC) (Ruiz-Capillas et al 2015). Conversely, roadside verges can benefit some small mammals because access to the right of way is often limited by exclusion fences that reduce habitat disturbances caused by livestock and humans (Ascensão et al. 2012; Chapter 46) and may even prevent some carnivore species from hunting within the road verge (assuming the mesh size is small enough).

Because roads chiefly have negative effects on small mammal populations, yet are occasionally positive, general mitigation strategies for these species are sometimes difficult to design. Road agencies must strive for multifaceted and dynamic mitigation approaches that maintain landscape connectivity by reducing the barrier effect and roadkills while managing the road verges to better support small mammal populations. This chapter aims to detail the most significant impacts of roads on small mammals, highlight the importance of road verges as habitat and recommend some best practices for road mitigation targeting small mammals.

LESSONS

39.1   Road building may result in complete losses of habitat, territories and individuals

Road-construction activities and the completed road itself pose unique and direct threats to small mammals. Road construction involves major landscape modifications that often change the hydrology, connectivity and habitat suitability and availability within the landscape. As one may expect given their small size, many species typically have small territories, often less than 1 ha (Lindstedt et al. 1986), and thus, a small mammal can have its entire territory destroyed during road construction. Further, small mammals are likely to be directly killed during road-construction activities due to their limited mobility, and therefore, the construction of even a small road can have disproportionate impacts on small mammals (Andrén 1994). This is especially true for species whose forested habitats are removed along roads or species that require specific or scarce habitat that is replaced by roads, thereby extending the habitat gap and limiting future mitigation options (Lessons 39.5 and 39.6).

Where large expanses of contiguous habitat exist, road construction may be a relatively small threat to the persistence of most small mammal species. However, road construction and improvements are more frequent in landscapes already dominated by people and where habitat for small mammals is often fragmented or scarce. In these areas especially, efforts should be made to mitigate all reductions in landscape connectivity and habitat availability. Reducing the amount of landscape modification during road construction can immediately lessen habitat loss and mortality. Although these habitat disturbances cannot be completely avoided during road construction, it may be possible to limit its ecological significance by setting strict limits on the areas to be cleared (Chapter 8).

39.2 Roads may filter small mammal movements and reduce genetic exchange

Small mammal species vary in their willingness and ability to cross gaps in their habitat (e.g. Goosem 2001). Species that naturally forage in or live within open, grassy areas (e.g. meadow voles or European rabbits) may also do so within road verges. However, many small mammal species require overhead or complex ground cover and avoid open areas, decreasing their likelihood of attempting to cross a road (e.g. woodland jumping mouse; see McLaren et al. 2011). For some species, this road avoidance has been attributed to the road surface and less to traffic disturbance (McGregor et al. 2008). In fact, even narrow, unpaved or seldom-used roads can still pose a significant threat to small mammals (Rico et al. 2007).

For species that avoid roads or for individuals that are killed when attempting to cross a road (Lesson 39.3), the road is a barrier to movement, thereby 'filtering' out individuals that are either killed or otherwise fail to cross the road (Fig. 1.2). If individuals from opposite sides of a road are prevented from mating and producing offspring, the road creates a genetic subdivision between populations (Jackson & Fahrig 2011), reducing the genetic diversity of populations and increasing the probability of local extinction (Reed et al. 2007; Chapter 14). Only a handful of studies have focused on the genetic effects of roads on small mammal populations, and these have found only a weak genetic subdivision effect from the road (Holderegger & Di Giulio 2010), possibly due to the large population sizes of the study species (Gauffre et al. 2008). Further, multiple studies on the same species produced different results, probably due to different road attributes such as width or traffic volume. For example, one study suggested that a highway (>28,000 vehicles per day) was a genetic barrier for bank voles (Gerlach & Musolf 2000), whereas another did not find any effect on a less travelled road (Redeker et al. 2006). These contrasting results of the same species at different locations highlight the complexity of road–small mammal relationships.

39.3 The effect of roadkills on small mammal population persistence is often unknown

Small mammals that move slowly or have stationary defence responses to threats, such as hedgehogs who react to oncoming traffic by 'freezing' in place, often have high rates of mortality due to WVC (Huijser & Bergers 2000; Brockie et al. 2009; Lesson 32.2). However, our current understanding of the population-level effect of road mortality on many small mammal species is limited, partly because roadkill rates are usually underestimated due to their low detectability (Santos et al. 2011), rapid removal by scavengers (Antworth et al. 2005) or deterioration by passing vehicles (Santos et al. 2011). Thus, the actual impact of road mortality on population persistence for most species is unknown, limiting our ability to infer the significance of the loss of individuals to the population.

Nevertheless, small mammal populations are typically highly productive (Golley 1960), so most populations are able to sustain a certain rate of mortality before the overall population begins to decline. For example, the annual, sustainable mortality rate for 17 small (<1 kg) mammal species of conservation concern in Australia ranges from 94% to 29% (Hone et al. 2010). However, this study assumed that the species live in 'optimal conditions' (i.e. unlimited food and no predation, parasites or competitors; conditions that are not found in the wild) and the list of considered species was limited to those for which there was sufficient data, (i.e. many other species were not considered) (Hone et al. 2010). We are not suggesting that a certain rate of mortality is justifiable, but rather that the population-level effect of road mortality will vary considerably among species, habitat quality and population size, and that further research focusing on the importance of road mortality for small mammal population persistence is urgently required. This knowledge is even more crucial for endangered species. There are approximately 400 small mammal species currently classified by the International Union for the Conservation of Nature as 'conservation concern' or worse (IUCN 2012), and some of these species are severely impacted by roads, such as the mountain pygmy possum (Textbox 39.1).

39.4 Fenced verges may provide benefits for small mammals

Exclusion fencing can reduce livestock trampling and overgrazing to the benefit of many small mammal species (Bilotta et al. 2007; Torre et al. 2007). These fenced areas usually have taller and more complex vegetation than in adjacent areas, supporting a higher density of small mammals that prefer these habitats (Torre et al. 2007). Fenced road verges may create similar opportunities for small mammals, provided that the verge contains suitable habitat (Ascensão et al. 2012), with the additional advantages of limited human disturbances within verges (Chapter 20). As a result, road verges and medians can provide suitable habitat for some species, at times even supporting higher densities than in surrounding areas (e.g. Ascensão et al. 2012; Ruiz-Capillas et al. 2013b). Road verges may also aid the conservation of endangered species by providing habitat (e.g. Cabrera vole, Pita et al. 2006).

Appropriately designed and maintained fences can also prevent some predators of small mammals from accessing the road verge (Chapter 20). For example, white-footed mice were more abundant in areas with roads, as their predators may avoid the roads or their access to the road verges was limited by fencing (Rytwinski & Fahrig 2007). However, the presence of European rabbit within a road verge seemed to attract polecats, leading to a higher roadkill rate for the carnivore (Barrientos & Bolonio 2009), yet this relationship is not universal, even within the same species (Planillo & Malo 2012). In addition, high rates of roadkill of predators may lead to relaxed predation of small mammals (Chapter 28). Thus, the ability of road verge fencing to provide a refuge from carnivore predation requires more research.

Road verges can also serve as corridors for animals moving parallel to the road, allowing them to move between habitat patches and ultimately improving the viability of their population, particularly in fragmented landscapes (Huijser & Clevenger 2006; Chapter 46). However, the benefit of verges as habitat and corridors could also have negative impacts as some invasive species may also use verges as corridors, accelerating their access to new areas (Brown et al. 2006). To our knowledge, no studies addressing this issue for small mammals exist, which, given the high number of species classified as pests, constitutes a major lack of information.

39.5 Crossing structures for small mammals must be optimally located and connected to surrounding habitat

Maintaining connectivity across the road is vital, particularly for small mammals whose limited mobility increases their risk for local extinctions. Connectivity can be facilitated by installing new wildlife crossing structures or improving existing structures, including wildlife underpasses and overpasses as well as multi-use structures, such as drainage culverts (Chapter 22). Determining the location and type of crossing structures should be decided early during the road planning process (Chapters 4 and 9) at least because the presence of suitable habitat surrounding the crossing

structure entrances is critically important for facilitating its use by small mammals (e.g. Goosem et al. 2001, Lesson 39.1). As with other species, crossing structure placement and design should be tailored to the needs of the target species, rather than a generic 'small mammal' strategy. In fact, crossing structure designs for species that prefer open or grassy habitats may not be suitable for species that prefer dense habitat, for example, forests and heathland. Some small mammals prefer small tunnels, while others may prefer stream channels, and others may prefer large structures (van der Grift et al. 2013). Placing items such as logs, rocks or other debris inside crossing structures can encourage use by multiple species, particularly those species that avoid open spaces (Mata et al. 2009) (Figs 35.5, 39.1 and 39.2). For passages with standing or moving water, raised ledges or shelves that remain above the average water depth can provide dry walking surfaces for small mammals (Figs 35.6, 35.7 and 39.3). These design considerations are relatively inexpensive to incorporate into new structures and can be easily retrofitted to existing crossing structures (Meaney et al. 2007).

39.6 Road verges should be diverse and contain natural structure

Where possible, the verge outside the clear zone should be managed to provide suitable habitat for as many species of small mammal as possible, especially when adjacent land has been cleared. This can be achieved by managing the verge for a diversity of cover types (i.e. grasses, shrubs and trees) and by retaining as much of the pre-existing structural features (e.g. rocks and logs) as possible. This diversity and structural complexity provides more opportunities for foraging and escape routes from predators. The road verge should be a continuum that incorporates multiple structural elements, rather than a uniform and continuous verge characterised by an abrupt habitat edge and a strip of mowed grass adjacent to the pavement. In general, verges of roads through natural areas should be as narrow as possible to reduce the amount of habitat cleared or modified.

Ideally, the verge can be enhanced by planting short and ubiquitous vegetation (like grasses) adjacent to the pavement and increasing plant diversity (including height and complexity) with increased distance from the road (Fig. 39.4). Such a gradient can support a higher diversity of plant communities that can in turn support a higher diversity and density of resident

Figure 39.1 Logs and rocks in this underpass offer cover and protection to small animals. Ideally, the logs and rocks would provide continuous cover through the underpass. Source: Photograph by R. van der Ree.

Figure 39.2 A line of tree stumps as shelter for small mammals, reptiles, amphibians and invertebrates on a small multi-use bridge overpass in the Netherlands. Source: Photograph by R. van der Ree.

small mammals. Alternatively, management can create patches of different habitats adjacent to the road, for example, patches of shrubs or dense patches of tall grasses. This strategy however could hinder individuals moving parallel to the road but would still be more favourable than a tightly manicured swath of grass. Management plans should consider potential adverse effects, such as increased mortality rates of the target and non-target species, including other small mammals, as well as predators that may hunt small mammals within road verges or as they cross the road (Lesson 39.5, Chapter 33).

Figure 39.3 Shelves for small mammals on both walls of a culvert near Eindhoven, the Netherlands. Source: Photograph by R. van der Ree.

CONCLUSIONS

A small mammal that crosses a road does not represent the significant hazard to a motorist that a larger animal might, so there are practically no economic or human safety motivations for understanding or mitigating the impacts of roads on small mammals. Nevertheless, these species are ecologically important, and many are threatened with extinction, so mitigation efforts for small mammals deserve more attention. Most mitigation efforts towards small mammals are easy to implement and relatively inexpensive as they can often be incorporated into mitigation efforts already planned for larger species. Simple mitigation measures include reducing landscape modification during road construction or upgrades, managing road verges to allow a continuum of increasing vegetation complexity from the road, providing natural cover leading to crossing structures and installing new or retrofitting existing crossing structures to facilitate small mammal movements.

Avenues for future research on small mammals and roads include the significance of roadkills on population persistence, the role of road characteristics in genetic differentiation between populations on opposite sides of the road and the role of road verges as habitat and corridors for small mammals, including invasive species. Small mammals are ecologically important and require greater attention in both road planning and road mitigation strategies, despite the challenges associated with studying them.

Figure 39.4 Schematic of road verge to provide habitat for small mammals, moving from longer grass within the clear zone through to shrubs and trees outside the clear zone. The size of each zone will depend on local safety standards and size of the right of way, and the exact mix of plant species and other structural elements (e.g. logs, rocks) will depend on the needs of the target species and maintenance requirements. Source: Reproduced with permission of Zoe Metherell.

Textbox 39.1 Species with unusual breeding patterns require special attention.

A few small mammal species have unusual breeding strategies that depend on seasonal movements whose interruption can have drastic consequences for the population. For example, the mountain pygmy possum (Fig. 39.5) is a small marsupial restricted to three small populations in the alpine and subalpine areas of south-east Australia (Heinze et al. 2004). Females are sedentary and occur at higher altitudes than males. Each year, the males migrate upslope to mate and then return downslope ensuring that females have access to sufficient food to wean their young before winter. The construction of the Alpine Way road disrupted the possums' annual migration at Mt.

Higginbotham, preventing the males from migrating downslope, resulting in increased densities and reduced overwinter survival rates. To remedy this barrier effect, two small underpasses were installed under the road in 1985, allowing individuals to safely cross the road and ultimately improved survival rates (van der Ree et al. 2009). Unfortunately, subsequent analyses suggest that while the underpass did restore natural movement patterns, they did not completely mitigate all of the negative effects of the road, as the population still suffered a significant population reduction due to alteration in population dynamics (van der Ree et al. 2009).

Figure 39.5 The mountain pygmy possum is one of the most threatened mammals in the world, with approximately 2000 individuals remaining in the wild. Source: Photograph by and reproduced with permission of Glen Johnson.

FURTHER READING

Ascensão et al. (2012) and Ruiz-Capillas et al. (2013b): These studies suggest fenced highway verges support higher abundance of small mammals by promoting suitable vegetative conditions for the study species.

Ruiz-Capillas et al. (2013a): This study suggests small mammal activity patterns near a road are more strongly affected by the activity patterns of their predators (i.e. red fox that are tolerant of the road) than differences in the microhabitat.

Mader (1984): This study quantified the barrier effect of different road classes on the movements of two species of forest mice, showing that the barrier effect varied between species and that even small forest roads present an obstacle to mouse movements.

McGregor et al. (2008): By translocating individuals, this study suggests that small mammals respond more strongly to the road surface rather than traffic volume and its associated noise and emissions.

van der Ree et al. (2009): This study evaluated the population-level effectiveness of under-road tunnels for the mountain pygmy possum and concluded that the tunnel reduced, but did not eliminate, the effect of the road.

REFERENCES

Andrén, H. 1994. Effects of habitat fragmentation on birds and mammals in landscapes with different proportions of suitable habitat: a review. *Oikos* **71**:355–366.

Antworth, R. L., D. A. Pike and E. E. Stevens. 2005. Hit and run: effects of scavenging on estimates of roadkilled vertebrates. *Southeastern Naturalist* **4**:647–656.

Ascensão, F., A. P. Clevenger, C. Grilo, J. Filipe and M. Santos-Reis. 2012. Highway verges as habitat providers for small mammals in agrosilvopastoral environments. *Biodiversity and Conservation* **21**:3681–3697.

Baker, B. W., D. J. Augustine, J. A. Sedgwick and B. C. Lubow. 2012. Ecosystem engineering varies spatially: a test of the vegetation modification paradigm for prairie dogs. *Ecography* **36**:230–239.

Barrientos, R. and L. Bolonio. 2009. The presence of rabbits adjacent to roads increases polecat road mortality. *Biodiversity and Conservation* **18**:405–418.

Bilotta, G. S., R. E. Brazier and P. M. Haygarth. 2007. The impacts of grazing animals on the quality of soils, vegetation, and surface waters in intensively managed grasslands. *Advances in Agronomy* **94**:237–280.

Brockie, R. E., R. M. F. S. Sadleir and W. L. Linklater. 2009. Long-term wildlife road-kill counts in New Zealand. *New Zealand Journal of Zoology* **36**:123–134.

Brown, G. P., B. L. Phillips, J. K. Webb and R. Shine. 2006. Toad on the road: use of roads as dispersal corridors by cane toads (*Bufo marinus*) at an invasion front in tropical Australia. *Biological Conservation* **133**:88–94.

Gálvez Bravo, L., J. Belliure and S. Rebollo. 2008. European rabbits as ecosystem engineers: warrens increase lizard density and diversity. *Biodiversity and Conservation* **18**:869–885.

Gauffre, B., A. Estoup, V. Bretagnolle and J. F. Cosson. 2008. Spatial genetic structure of a small rodent in a heterogeneous landscape. *Molecular Ecology* **17**:4619–4629.

Gerlach, G. and K. Musolf. 2000. Fragmentation of landscape as a cause for genetic subdivision in bank voles. *Conservation Biology* **14**:1066–1074.

Golley, F. B. 1960. Energy dynamics of a food chain of an old-field community. *Ecological Monographs* **30**:187–206.

Gómez, J. M., C. Puerta-Piñero and E. W. Schupp. 2008. Effectiveness of rodents as local seed dispersers of Holm oaks. *Oecologia* **155**:529–537.

Goosem, M. 2001. Effects of tropical rainforest roads on small mammals: inhibition of crossing movements. *Wildlife Research* **28**:351–364.

Goosem, M., Y. Izumi and S. Turton. 2001. Efforts to restore habitat connectivity for an upland tropical rainforest fauna: a trial of underpasses below roads. *Ecological Management & Restoration* **2**:196–202.

Heinze, D., L. S. Broome and I. M. Mansergh. 2004. A review of the ecology and conservation of the mountain pygmy-possum *Burramys parvus*. In R. Goldingay and S. Jackson, editors. The Biology of Australian Possums and Gliders. Surrey Beatty & Sons, Chipping Norton, pp. 254–267.

Holderegger, R. and M. Di Giulio. 2010. The genetic effects of roads: a review of empirical evidence. *Basic and Applied Ecology* **11**:522–531.

Hone, J., R. P. Duncan and D. M. Forsyth. 2010. Estimates of maximum annual population growth rates (rm) of mammals and their application in wildlife management. *Journal of Applied Ecology* **47**:507–514.

Huijser, M. P. and P. J. M. Bergers. 2000. The effect of roads and traffic on hedgehog (*Erinaceus europaeus*) populations. *Biological Conservation* **95**:111–116.

Huijser, M. P. and A. Clevenger. 2006. Habitat and corridor function of rights-of-way. In J. Davenport and J. L. Davenport, editors. The Ecology of Transportation: Managing Mobility for the Environment, Springer, Dordrecht. pp. 233–254.

IUCN. 2012. IUCN Red List of Threatened Species. Version 2012.2. www.iucnredlist.org. Accessed on 19 April 2013.

Jackson, N. D. and L. Fahrig. 2011. Relative effects of road mortality and decreased connectivity on population genetic diversity. *Biological Conservation* **144**:3143–3148.

Lindstedt, S. L., B. J. Miller and S. W. Buskirk. 1986. Home range, time, and body size in mammals. *Ecology* **67**: 413–418.

Mader, H. J. 1984. Animal habitat isolation by roads and agricultural fields. *Biological Conservation* **29**:81–96.

Mata, C., I. Hervas, J. Herranz, J. E. Malo and F. Suarez. 2009. Seasonal changes in wildlife use of motorway crossing structures and their implication for monitoring programmes. *Transportation Research Part D* **14**:447–452.

McGregor, R. L., D. J. Bender and L. Fahrig. 2008. Do small mammals avoid roads because of the traffic? *Journal of Applied Ecology* **45**:117–123.

McLaren, A. A. D., L. Fahrig and N. Waltho. 2011. Movement of small mammals across divided highways with vegetated medians. *Canadian Journal of Zoology* **89**:1214–1222.

Meaney, C. A., M. Bakeman, M. Reed-Eckert and E. Wostl. 2007. Effectiveness of ledges in culverts for small mammal passage. Final Report No. CDOT-2007-9. Colorado Department of Transportation, Research Branch.

Moleón, M. and J. M. Gil-Sánchez. 2003. Food habits of the wildcat (*Felis silvestris*) in a peculiar habitat: the Mediterranean high mountain. *Journal of Zoology* **260**:17–22.

Montgomery, S. S. J. and W. I. Montgomery. 1990. Intra-population variation in the diet of the wood mouse *Apodemus sylvaticus*. *Journal of Zoology* **222**:641–651.

Pita, R., A. Mira and P. Beja. 2006. Conserving the Cabrera vole, *Microtus cabrerae*, in intensively used Mediterranean landscapes. *Agriculture Ecosystems & Environment* **115**: 1–5.

Planillo, A. and J. E. Malo. 2012. Motorway verges: paradise for prey species? A case study with the European rabbit. *Mammalian Biology—Zeitschrift für Säugetierkunde* **78**: 187–192.

Redeker, S., L. W. Andersen, C. Pertoldi, A. B. Madsen, T. S. Jensen and J. M. Jørgensen. 2006. Genetic structure, habitat fragmentation and bottlenecks in Danish bank voles (*Clethrionomys glareolus*). *Mammalian Biology—Zeitschrift für Säugetierkunde* **71**:144–158.

Reed, D. H., A. C. Nicholas and G. E. Stratton. 2007. Genetic quality of individuals impacts population dynamics. *Animal Conservation* **10**:275–283.

Rico, A., P. Kindlmann and F. Sedlacek. 2007. Barrier effects of roads on movements of small mammals. *Folia Zoologica* **56**:1–12.

Ruiz-Capillas, P., C. Mata and J. Malo. 2013a. Community response of mammalian predators and their prey to motorways: implications for predator–prey dynamics. *Ecosystems* **16**:617–626.

Ruiz-Capillas, P., C. Mata and J. E. Malo. 2013b. Road verges are refuges for small mammal populations in extensively managed Mediterranean landscapes. *Biological Conservation* **158**:223–229.

Ruiz-Capillas, P., C. Mata, and J. E. Malo. 2015. How many rodents die on the road? Biological and methodological implications from a small mammals' roadkill assessment on a Spanish motorway. Ecological Research:1-11. DOI 10.1007/s11284-014-1235-1

Rytwinski, T. and L. Fahrig. 2007. Effect of road density on abundance of white-footed mice. *Landscape Ecology* **22**:1501–1512.

Santos, S. M., F. Carvalho and A. Mira. 2011. How long do the dead survive on the road? Carcass persistence probability and implications for road-kill monitoring surveys. *PLoS One* **6** doi:10.1371/journal.pone.0025383.

Torre, I., M. Díaz, J. Martínez-Padilla, R. Bonal, J. Viñuela and J. A. Fargallo. 2007. Cattle grazing, raptor abundance and small mammal communities in Mediterranean grasslands. *Basic and Applied Ecology* **8**:565–575.

van der Grift, E. A., R. van der Ree, L. Fahrig, S. Findlay, J. Houlahan, J. A. G. Jaeger, N. Klar, L. F. Madriñan and L. Olson. 2013. Evaluating the effectiveness of road mitigation measures. *Biodiversity and Conservation* **22**: 425–448.

van der Ree, R., M. A. McCarthy, D. Heinze and I. M. Mansergh. 2009. Wildlife tunnel enhances population viability. *Ecology and Society* **14**:7. http://www.ecologyandsociety.org/vol14/iss12/art17/ Accessed on 19 April 2013.

Weltzin, J. F., S. Archer and R. K. Heitschmidt. 1997. Small-mammal regulation of vegetation structure in a temperate savanna. *Ecology* **78**:751–763.

Chapter 40

REDUCING ROAD IMPACTS ON TREE-DWELLING ANIMALS

Kylie Soanes and Rodney van der Ree

Australian Research Centre for Urban Ecology, Royal Botanic Gardens Melbourne, and School of BioSciences, The University of Melbourne, Melbourne, Victoria, Australia

SUMMARY

Arboreal animals need trees for some or all of their shelter, food and movement. This diverse group of wildlife includes mammals, amphibians and reptiles that climb, crawl and glide in trees. Since trees are a critical resource, arboreal animals are directly affected by habitat loss from road construction. The susceptibility of arboreal animals to barrier effects and wildlife-vehicle collisions (WVC) will depend on their willingness, opportunity and ability to cross gaps. Methods to mitigate the impacts of roads and traffic are often unique and specific to this group of wildlife.

40.1 Always avoid clearing trees where possible.

40.2 Canopy connectivity is important for most arboreal animals.

40.3 Not all arboreal animals need arboreal crossing structures.

40.4 Further research on impacts and mitigation for arboreal species is needed.

Recent studies have quantified the impacts of roads on some arboreal species, primarily mammals, and successful mitigation techniques are available. However, further research on the use and effectiveness of mitigation strategies for this group is urgently required, particularly for arboreal amphibians and reptiles.

INTRODUCTION

Wherever you find trees, you're likely to find arboreal animals, a group that includes reptiles, mammals and amphibians. Many species of bird and bat also require trees, and more specific information about them is given in Chapters 33 and 34. Arboreal animals may spend all, or only part, of their lives feeding, nesting or moving through trees. All depend on trees to some extent, but this dependence varies from species that never willingly descend to the ground, to those that do so frequently and also to those that glide or leap across the gaps between trees. Similar to other groups of wildlife, the major impacts of roads and traffic on arboreal animals include habitat loss, creating a barrier to movement and increased mortality through WVC. The consequences of these impacts are highlighted in Chapter 1 and can lead to local population declines, decreased genetic diversity and increased risk of extinction.

Handbook of Road Ecology, First Edition. Edited by Rodney van der Ree, Daniel J. Smith and Clara Grilo.

© 2015 John Wiley & Sons, Ltd. Published 2015 by John Wiley & Sons, Ltd.

Companion website: www.wiley.com\go\vanderree\roadecology

Most arboreal species are too small to cause significant damage to vehicles when struck by traffic (with the exception of some larger primates), so mitigation is primarily focused on conservation and animal welfare goals, rather than human safety concerns. Many species are endemic, specialised and highly sensitive to disturbance from roads and traffic (e.g. Laurance et al. 2008; McCall et al. 2010). Furthermore, arboreal species are often under pressure due to existing habitat loss and fragmentation from agriculture, forestry, urbanisation, hunting and poaching. Mitigating the impacts of roads is critical to conserving these species, especially in landscapes already modified by humans.

In this chapter, we describe the impacts of roads on arboreal animals and discuss the key approaches to mitigation: minimising habitat loss, maintaining tree canopy connectivity and reducing roadkill.

LESSONS

40.1   Avoid clearing trees where possible

Arboreal animals depend on trees but not just any tree will do. Many species are particular about the type or age of trees that are useful. For example, tree hollows, which some arboreal animals use for nest sites, often only develop in trees which are greater than 100 years old (Gibbons & Lindenmayer 2002). Furthermore, different tree species form hollows at different rates and provide resources (e.g. food) that are preferred by different species of arboreal animals. Because these relationships are complex and often poorly understood, it is best to avoid removing trees in the first place, especially old trees.

Detailed mapping of important trees early in the planning and design stages of a project can be used to select routes that require the fewest trees to be cleared. This 'tree-saving' approach should continue through to the end of the construction phase, by not clearing mature trees for temporary infrastructure (e.g. construction-site offices) that will be decommissioned post-construction or for infrastructure that can be easily located elsewhere (e.g. utility easements).

Retaining trees near roads can endanger motorists, particularly with species known to drop large branches. However, rather than removing the entire tree, consider removing the dangerous overhanging branches and installing guard rails (Figs 17.1, 20.2B) to address the safety risk. Important trees that are susceptible to

windfall during storms can be anchored and secured with cables.

Where tree removal is unavoidable, habitat loss can be mitigated through restoration. Revegetation should replace what was lost during clearing, using the same species and replicating the original habitat structure. However, if the primary goal is to restore connectivity, consider the use of other species on roadside verges if they grow taller and faster or are less prone to windfall or dropping branches than the original species. Replanted trees can take decades to mature, and additional measures, such as providing nest boxes to replace hollows or planting fast-growing food species, can quickly provide resources and prevent local population declines. Ideally, these measures would be implemented prior to habitat clearing, so that animals have alternative resources immediately available. The amount of habitat restoration should at least equal that expected to be lost during road construction (Chapter 7).

40.2   Canopy connectivity is important for most arboreal animals

Many arboreal species are so well adapted to treetop life that they rarely descend to the ground. These animals are 'gap limited', meaning they will not or cannot cross gaps in tree cover beyond a certain distance. The limit depends on the species. For example, the squirrel glider in south-eastern Australia regularly glides 30–40 m and, depending on tree height, has a maximum glide of about 70 m (van der Ree et al. 2003). Some species choose to never move along the ground and require connected canopy (Fig. 40.1A). In addition, arboreal animals are often slow or awkward when moving along the ground and are poor at avoiding traffic and predators. Fences that prevent arboreal animals from accessing the road and/or funnell them towards crossing structures are challenging because these species are good climbers and many can jump or glide above the fence (Chapter 20). Therefore, maintaining canopy connectivity directly over the road (Fig. 40.1) or on land bridges is probably the most effective approach to allow movement across the road, even for species that can move across the ground.

Retaining tall trees along roadsides and within the centre median ('vegetated median') during construction and maintenance can maintain the existing, natural canopy connectivity used by arboreal animals (Fig. 40.1A). This method is only useful for non-gliding species when tree canopy branches overlap and may

(A)

(B)

(C)

Figure 40.1 (A) Natural canopy connectivity, (B) glider pole and (C) canopy bridge. Source: (A) Photograph by and reproduced with permission of Tom Langen; and (B and C) Photographs by Kylie Soanes.

not be feasible across wider roads without a sufficiently wide centre median. For species that glide or jump, trees can provide a set of 'stepping stones' if they are within gliding or jumping range. Connectivity can also be created by planting trees to fill the gaps; however, it may take decades before they are tall enough to be effective crossing points. A similar approach (termed 'hop-overs') has been recommended for bats (Lesson 34.7), although further research is required.

Crossing structures specific to arboreal animals include glider poles and canopy bridges. Glider poles are 'surrogate trees'; tall poles, usually timber, with branch-like beams at the top (Fig. 40.1B). These can be placed in the centre median and/or the verge and provide launching and landing sites for gliding species. Glide paths and angles must be carefully calculated to ensure that animals stay above the traffic (Textbox 40.1). Canopy bridges (Fig. 40.1C) are commonly made of rope, steel or wood, and are suspended above the road by timber poles or roadside trees (Fig. 40.1C). These bridges should always be linked to trees on each side of the road with feeder ropes. The stability of canopy bridges should be considered, with single strands of rope or wire more likely than ladder designs to twist and potentially cause animals to fall, especially if spans are long. Canopy bridges can be incorporated into existing road infrastructure such as overhanging road signs, and both canopy bridges and glider poles can be installed on land bridges.

Do these structures work? A range of species have been detected using canopy bridges, glider poles and natural canopy to cross roads and other linear infrastructure ranging from 6 to 80 m wide. Canopy bridges of various designs have been used by arboreal and gliding mammals including possums (Fig. 40.2), gliders,

Figure 40.2 A common brushtail possum using a 70 m canopy bridge to cross a four-lane divided highway in south-east Australia. Source: Photograph by R. van der Ree.

primates, squirrels, martens, lemurs and the dormouse (e.g. Valladares-Padua et al. 1995; Bekker 2005; Mass et al. 2011; Weston et al. 2011; Goldingay et al. 2013; Soanes et al. 2013; Teixeira et al. 2013; Chapter 41). Gliding mammals have also been detected using glider poles and natural connectivity to cross four-lane divided highways and over land bridges (van der Ree et al. 2010; Goldingay et al. 2011; Taylor & Goldingay 2012; Kelly et al. 2013; Soanes et al. 2013). These structures are being used all over the world, however much of the monitoring work remains unpublished. The effects on population size, mortality rates and gene flow have yet to be assessed. While there is one published account of a lace monitor using a canopy bridge to cross a road (Soanes & van der Ree 2009), no studies have focused specifically on arboreal amphibians or reptiles.

Textbox 40.1 Getting the glide right.

To design effective crossing structures for gliding animals, we need a good understanding of their physical ability. This means knowing not only how far they can glide but also the glide ratio (i.e. how much height is lost for every unit of horizontal distance they travel). This tells us how far apart our glider poles (or trees, for natural connectivity) need to be and the necessary launch and landing heights to keep animals above traffic. Poorly designed poles could increase roadkill if a large number of animals collide with vehicles because they can't make the distance or miss poles that are too narrow. It's helpful when planning to use schematic diagrams to ensure that measurements are correct and poles are positioned effectively. For example, Figure 40.3 shows a set of glider poles over a four-lane divided freeway in south-east Australia. These were installed for the squirrel glider, a small, gliding marsupial with a glide ratio of 2.5:1 (distance/height) (Jackson, 2000). Based on the height of the existing roadside trees, poles were required in the centre median and roadsides. Note that minimum glide trajectories were to be at least three metres above traffic height.

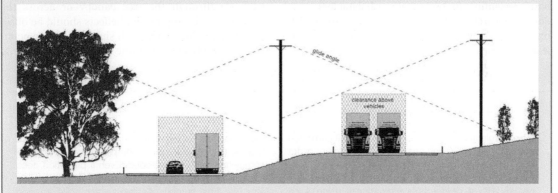

Figure 40.3 The spacing and height of glider poles are dependent on the glide capability of the target species, and glide angles must be calculated for each installation to ensure glides are well above the height of traffic. Source: Reproduced with permission of Scott Watson.

40.3 Not all arboreal animals need arboreal crossing structures

Arboreal-specific crossing structures may not work equally well for all species of arboreal animals (e.g. Chapter 41). Heavier animals, such as large primates, may avoid crossing long, narrow canopy bridges, while other species, such as some species of squirrel or possum, readily descend to the ground. However, if these species do not avoid roads or traffic disturbance and there is no exclusion fencing to prevent them from crossing the road, they risk being killed by traffic. A major challenge is the design of fencing that will exclude good climbers, and options include the use of smooth surfaces and floppy or overhanging tops (Fig. 20.4). Effective exclusion fencing combined with a wildlife underpass will keep some arboreal species off the road and provide connectivity. In these cases, it is best to fit underpasses with branch-like 'furniture', providing a preferred substrate for movement and opportunities to escape from terrestrial predators. The adjacent roadside should also have adequate cover to allow arboreal animals to safely access the crossing structure.

40.4 Further research on impacts and mitigation for arboreal species is needed

While the broad impacts of roads on arboreal mammals are fairly well understood, there is still much to learn. Key areas for further research include:
• Quantifying the specific impacts of roads and traffic on arboreal mammals and all impacts on arboreal reptiles and amphibians.

• Factors influencing the rate of crossing structure use (e.g. design features, adjacent habitat, traffic volume).
• The effectiveness of arboreal crossing structures for arboreal amphibians and reptiles.
• Species preference for different crossing structure types (e.g. canopy bridges, glider poles, natural canopy or underpasses).
• The maximum road width and traffic volumes that can be mitigated using crossing structures.
• Population-level effects of crossing structures (e.g. population size, survival rates, gene flow and predation (Textbox 40.2, Fig. 40.5)).
• The ability of habitat restoration actions (e.g. nest boxes, replanting) to compensate for habitat loss.
Finally, researchers should be encouraged to publish results from monitoring studies, so that others can learn from their work.

CONCLUSION

The best approach to prevent road impacts on arboreal animals is to avoid placing roads through wooded habitats. Where that is not possible, priorities should include strategies to minimise the width of the road and number of tall trees removed, thereby maintaining natural canopy connectivity, and where feasible, install barrier fencing to reduce roadkill. Retaining natural canopy connectivity or installing crossing structures such as canopy bridges, glider poles or underpasses can allow arboreal animals to safely cross roads. The best approach to mitigation depends on whether animals prefer to move through the tree canopy or across the ground. Lastly, any residual effects should be offset (Chapter 7). Future research should evaluate population-level effects, determine design limitations (e.g.

Textbox 40.2 Airborne predators.

If a particularly wise owl or eagle learns that it can catch an easy meal by waiting near a canopy bridge or glider pole, then crossing structures may not be such a safe place for arboreal animals. To date, there have been no records of predators systematically using such crossing structures to get an easy meal, but it's still a possibility (Chapters 23 and 35). Providing shelter on the structures, for example, refuge tubes (Fig. 40.4), or making the structure more complex can give animals somewhere to hide if needed. Even if predation does occur, it will only reduce the effectiveness of a crossing structure if the number of animals eaten while crossing is higher than the number of animals that would have been killed by vehicles if the structure was not installed.

Figure 40.4 PVC tubes on glider poles provide protection for arboreal animals, in this case squirrel gliders, from aerial predators. Camera for monitoring and antenna to transmit images are also shown. Source: Photograph by R. van der Ree.

Figure 40.5 Logs and ropes provide refuge for tree kangaroos from terrestrial predators in this underpass on the Atherton Tablelands, Australia. Source: Photograph by R. van der Ree.

what is the maximum length of an effective canopy bridge?) and include studies of arboreal amphibians and reptiles.

FURTHER READING

Goosem (2007) and Laurance et al. (2009): Both provide comprehensive reviews of the effects of roads and associated habitat fragmentation on rainforest communities, with specific reference to arboreal species and potential mitigation methods. While these reviews focus on rainforest habitats, the lessons can be applied to other forest types.

Soanes et al. (2013): This study shows that a gliding marsupial will use canopy bridges, glider poles and natural canopy connectivity to cross a four-lane divided freeway that was previously a barrier to movement. This research also demonstrates the importance of longer-term before–after–control–impact research when evaluating the effectiveness of wildlife crossing structures.

Taylor and Goldingay (2009): This research used population viability analysis to show that crossing structures which facilitate even small amounts of movement over roads can significantly reduce the risk of extinction in gliding mammal populations.

REFERENCES

Bekker, H. 2005. Taking the high road: Treetop bridges for arboreal mammals. International Conference on Ecology and Transportation, San Diego, CA.

Gibbons, P. and D. Lindenmayer. 2002. Tree Hollows and Wildlife Conservation in Australia. CSIRO Publishing, Melbourne, Australia.

Goldingay, R. L., B. D. Taylor and T. Ball. 2011. Wooden poles can provide habitat connectivity for a gliding mammal. Australian Mammalogy **33**: 36–43.

Goldingay, R. L., D. Rohweder and B. D. Taylor. 2013. Will arboreal mammals use rope-bridges across a highway in eastern Australia? Australian Mammalogy **35**: 30–38.

Goosem, M. 2007. Fragmentation impacts caused by roads through rainforests. Current Science **93**: 1587–1595.

Jackson, S. M. 2000. Glide angle in the genus *Petaurus* and a review of gliding in mammals. Mammal Review **30**: 9–30.

Kelly, C. A., C. A. Diggins and A. J. Lawrence. 2013. Crossing structures reconnect federally endangered flying squirrel populations divided for 20 years by a road barrier. Wildlife Society Bulletin **37**: 375–379. doi:10.1002/wsb.249.

Laurance, W. F., B. M. Croes, N. Guissouegou, R. Buij, M. Dethier and A. Alonso. 2008. Impacts of roads, hunting, and habitat alteration on nocturnal mammals in African rainforests. Conservation Biology **22**: 721–732.

Laurance, W. F., M. Goosem and S. G. W. Laurance. 2009. Impacts of roads and linear clearings on tropical forests. Trends in Ecology and Evolution **24**: 659–669.

Mass, V., B. Rakotomanga, G. Rakotondratsimba, S. Razafindramisa, P. Andrianaivomahefa, S. Dickinson, P. O. Berner and A. Cooke. 2011. Lemur bridges provide crossing structures over roads within a forested mining concession near Moramanga, Toamasina Province, Madagascar. Conservation Evidence **8**: 11–18.

McCall, S. C., M. A. McCarthy, R. van der Ree, M. J. Harper, S. Cesarini and K. Soanes. 2010. Evidence that a highway reduces apparent survival rates of squirrel gliders. Ecology and Society **15**: 27.

Soanes, K. and R. van der Ree. 2009. Highway impacts on arboreal mammals and the use and effectiveness of novel mitigation measures. International Conference on Ecology and Transportation, Duluth, MN.

Soanes, K., M. C. Lobo, P. A. Vesk, M. A. McCarthy, J. L. Moore and R. van der Ree. 2013. Movement re-established but not restored: Inferring the effectiveness of road-crossing mitigation for a gliding mammal by monitoring use. Biological Conservation **159**: 434–441.

Taylor, B. D. and R. L. Goldingay. 2009. Can road-crossing structures improve population viability of an urban gliding mammal?. Ecology and Society **14**: 13.

Taylor, B. D. and R. L. Goldingay. 2012. Restoring connectivity in landscapes fragmented by major roads: A case study using wooden poles as 'stepping stones' for gliding mammals. Restoration Ecology **20**: 671–678.

Teixeira, F. Z., R. C. Printes, J. C. G. Fagundes, A. C. Alonso, A. Kindel. 2013. Canopy bridges as road overpasses for wildlife in urban fragmented landscapes. Biota Neotropica **13**.117–123

Valladares-Padua, C., L. J. Cullen and S. Padua. 1995. A pole bridge to avoid primate road kills. Neotropical Primates **3**: 13–15.

van der Ree, R., A. F. Bennett and D. C. Gilmore. 2003. Gap-crossing by gliding marsupials: Thresholds for use of isolated woodland patches in an agricultural landscape. Biological Conservation **115**: 241–249.

van der Ree, R., S. Cesarini, P. Sunnucks, J. L. Moore and A. C. Taylor. 2010. Large gaps in canopy reduce road crossing by a gliding mammal. Ecology and Society **15**: 35. http://www.ecologyandsociety.org/vol15/iss34/art35/, Accessed on 25 September 2014.

Weston, N., M. Goosem, H. Marsh, M. Cohen and R. Wilson. 2011. Using canopy bridges to link habitat for arboreal mammals: Successful trials in the Wet Tropics of Queensland. Australian Mammology **33**: 93–105.

Chapter 41

CASE STUDY: CANOPY BRIDGES FOR PRIMATE CONSERVATION

Andrea Donaldson[1,2] and Pamela Cunneyworth[2]

[1]Department of Anthropology, Durham University, Durham, UK.
[2]Colobus Conservation, Diani Beach, Kenya.

Diani is an international tourist destination located on the south coast of Kenya. It is known for its beautiful white sand beaches, world-class beach hotels and the coral rag forest. The forest forms part of the Coastal Forests of eastern Africa ecosystem, noted as one of the top 25 global biodiversity hotspots by Conservation International. Remarkably, four monkey species (colobus, Sykes', vervet and baboon) – with a combined population of just over 1400 individuals – live within the 7 km^2 suburban setting.

In 1971, the Diani Beach Road was built and sealed to 10 m wide, bisecting the pristine forest. For the first time, the public had vehicle access to the area, and by 1996, noticeable numbers of monkeys were being injured or killed on the road due to wildlife-vehicle collision (WVC) (Fig. 41.1). As a response to a public outcry on this issue, Colobus Conservation (formerly Wakuluzu: Friends of the Colobus Trust), a primate and forest conservation organisation, was established.

To address the issue of primate injury and death caused by WVC, the Colobus Conservation installed the first canopy bridge in 1997 – locally known as a 'colobridge' (Fig. 41.2). By 2013, 28 bridges had been erected at mortality hotspots along the 10 km stretch of road, funded by the local community and international donors (35 and 65%, respectively).

A preliminary study in 2011 assessed traffic volume along the Diani Beach Road and rate of use of the canopy bridges (Colobus Conservation, unpublished data). Each of the 28 bridges was monitored continuously from 06:00 to 18:00 h for 2 days over several months. Daily traffic volume was approximately 2600 vehicles with an average of 800 primate crossings per day across the 28 bridges. Sykes' monkeys used the bridges most often (673 crossings per day), followed by the typically ground-foraging vervets (91 crossings per day) and the arboreal colobus (35 crossings per day); baboons were not observed using the bridges (Fig. 41.3). The rate of use of each bridge by different species was also highly variable (Fig. 41.4), with 7, 10 and 25 bridges used over the 2-day survey period by vervets, colobus and Sykes' monkeys, respectively.

We speculate that bridge location (e.g. adjacent to good habitat or at traditional crossing locations) is critical for attaining high rates of use. Specifically, a route through the tree canopy that funnels animals towards the bridge is important as monkeys do not change direction or climb trees in order to use a bridge. It is important to note that canopy bridges may not be used by larger primates or those of a terrestrial nature.

Handbook of Road Ecology, First Edition. Edited by Rodney van der Ree, Daniel J. Smith and Clara Grilo.
© 2015 John Wiley & Sons, Ltd. Published 2015 by John Wiley & Sons, Ltd.
Companion website: www.wiley.com\go\vanderree\roadecology

Figure 41.1 Adult Sykes' female with infant killed by a car on the Diani Beach Road. Source: Photograph by Andrea Donaldson.

Figure 41.2 Diani Beach Road with Sykes' monkeys on a colobridge. Source: Photograph by Andrea Donaldson.

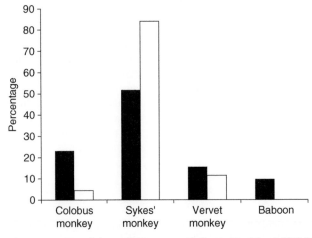

Figure 41.3 Population size (as a proportion of the total primate population in Diani ($n = 1421$), black columns) and rate of use of 28 colobridges (as a proportion of total daily crossings ($n = 800$), clear columns), recorded during two days in 2011. Sykes' monkeys used the bridges more than expected, while baboons and colobus used them less than expected ($X^2 = 292.4$, 2df, $p < 0.001$).

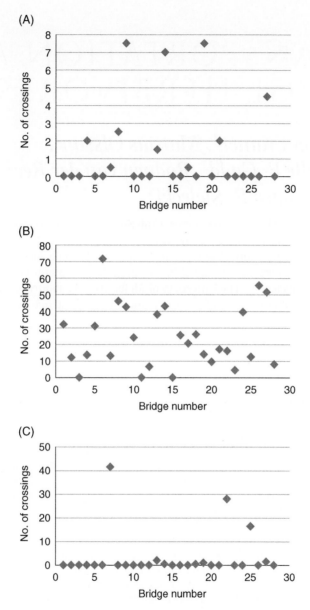

Figure 41.4 Average daily crossings on colobridges in 2011 for (A) colobus, (B) Sykes' and (C) vervet. Baboons were not observed using bridges. Note the *y*-axis scale differs among graphs.

Between 1999 and 2012, 518 monkeys were injured or killed by vehicles on the Diani Beach Road. This is approximately a 3% loss in the local primate population annually. Though no data is available on primate road injury or mortality prior to our colobridge installations, we believe that bridge use represents a significant reduction in risk of mortality for each crossing event.

For more information on considerations for building a colobridge, please see http://www.colobusconservation.org/index.php/conservation/151-colobridges.

For a guide on how to build a colobridge, please see http://www.colobusconservation.org/index.php/conservation/159-how-to-build-a-colobridge.

Chapter 42

TRANSPORTATION AND LARGE HERBIVORES

Patricia Cramer[1], Mattias Olsson[2], Michelle E. Gadd[3], Rodney van der Ree[4] and Leonard E. Sielecki[5]

[1]Department of Wildland Resources, Utah State University, Logan, UT, USA
[2]EnviroPlanning AB, Göteborg, Sweden
[3]Division of International Conservation, US Fish & Wildlife Service, Falls Church, VA, USA
[4]Australian Research Centre for Urban Ecology, Royal Botanic Gardens Melbourne, and School of BioSciences, The University of Melbourne, Melbourne, Victoria, Australia
[5]British Columbia Ministry of Transportation and Infrastructure, Victoria, British Columbia, Canada

SUMMARY

Large herbivores occur around the world and are often in conflict with roads and vehicles. Large herbivores are plant eaters and are generally hooved (e.g. deer, moose, elephant and buffalo) but also include kangaroos. All play important roles in ecosystem functioning. These are the animals typically involved in most reported wildlife-vehicle collisions (WVC), which have significant costs to human society and wildlife populations. It is important that transportation planners seek to avoid, minimise or mitigate these collisions in order to protect motorists and large herbivore populations.

42.1 Large herbivores need connectivity across the landscape because restriction of their movements by roads and vehicles will impact wildlife populations, ecosystems and humans.

42.2 Wildlife-vehicle collisions often involve large herbivores and cause large-scale costs to human societies; documenting WVC and their costs is the first step in preventing them.

42.3 Mitigating vehicle and road effects on large herbivores requires long-term transportation planning.

42.4 The type and size of wildlife crossing structures for large herbivores can be partly based on the North American and European experiences.

42.5 The placement of wildlife crossing structures for large herbivores should be based on animal size, their movement patterns and transportation constraints.

42.6 Additional mitigation measures include driver warning systems and vegetation management.

Handbook of Road Ecology, First Edition. Edited by Rodney van der Ree, Daniel J. Smith and Clara Grilo.
© 2015 John Wiley & Sons, Ltd. Published 2015 by John Wiley & Sons, Ltd.
Companion website: www.wiley.com\go\vanderree\roadecology

> It is important to consider large herbivores when planning and managing road networks because of their need for frequent and often large-scale movements, their large size which makes WVC dangerous to motorists and their important role in ecosystem function.

INTRODUCTION

Large herbivores need to move across the landscape to access food, water, shelter and mates as they become seasonally available. These movements can vary from several to thousands of kilometres and invariably bring animals into conflict with roads, railways, utility easements and vehicles. These interactions can restrict movement and cause mortality due to wildlife-vehicle collision (WVC) and endanger local populations and the entire species (Chapter 56), which in a cascade of effects can jeopardise ecosystems. In this chapter, large herbivores are defined as large mammals that eat plants. Most are hooved (i.e. ungulates) and include deer, antelope, horses, swine, elephants and llama, and those without hooves include kangaroos. Large herbivores are drivers of ecosystem processes (Hobbs 1996); they help recycle nutrients, promote grassland plant diversity and act as prey for carnivores. They are valuable to humans as harvestable food sources and income. They also serve as potential flagship species for landscape-level conservation, becoming the symbols of large-scale ecological areas and processes in need of protection (Thirgood et al. 2004).

Large herbivores are the animals most often involved in reported WVC. Collisions with these animals pose a danger to motorists; more than 500 humans died in Europe in 1996 in collisions with ungulates (Groot-Bruinderink & Hazebroek 1996). As a result, large herbivores in many parts of the world are the focus of mitigation to avoid WVC, thus protecting wildlife and humans. For example, there are over 700 wildlife crossing structures across roads for deer, moose and elk in North America (Bissonette & Cramer 2008). The goals of this chapter are to explain why large herbivores need to move, demonstrate that the effects of roads and WVC on herbivore populations and society can be costly and provide solutions to safely maintain the movement of large herbivores across transportation corridors.

LESSONS

42.1 Large herbivores need connectivity across the landscape because restriction of their movements by roads and vehicles will impact wildlife populations, ecosystems and humans

The daily and seasonal travels of large herbivores were once ubiquitous across continents, but their populations and movements have declined as human activity has expanded. Landscape connectivity is particularly important for migratory species, where all individuals within a population must undertake movements across landscapes. Roads, other linear features and vehicles can threaten connectivity by causing mortality through WVC, reducing the amount and quality of habitat, fragmenting habitat, restricting or preventing movement and causing animals to avoid areas near roads (Jaeger & Fahrig 2004). In turn, the sizes of these affected populations can decrease to levels that influence population persistence and ecosystem integrity and affect the human populations that rely on them for food and other resources.

Large herbivores that migrate long distances or remain as local residents can be affected by roads and railways. Roads can hinder animal movements, or the animals can be attracted to roads to feed on roadside vegetation or de-icing salts, thus making them more susceptible to WVC. Certain large herbivores migrate thousands of kilometres annually and may do so in herds numbering hundreds to thousands of individuals (e.g. Sawyer et al. 2005; Chapter 56). Large herbivores that do not migrate may also be subject to road effects. White-tailed deer in North America typically reside within several hectares, often living among humans and roads, thereby increasing the risk of WVC. There were an estimated 1.2 million deer–vehicle collisions reported to insurance companies in the United States during the

12 months from July 2012, the majority being with white-tailed deer (State Farm Insurance 2013). Eastern grey kangaroos are common in parts of Australia, and thousands are involved in WVC annually (e.g. Coulson 1982; Klocker et al. 2006), although reliable estimates are lacking.

Providing connectivity across roads for large herbivores is important because individual animals, populations of animals and ecosystems suffer if wildlife are confined to one side of a road. For example, migratory ungulates stabilise ecosystems by acting as keystone species (Kie & Lehmkuhl 2001). They are important to maintain plant and animal diversity (Chapter 56) and, if confined to an area, may overgraze the vegetation and cause crashes in the animal populations (Christianson & Creel 2009). Therefore, maintaining connectivity across infrastructure is important for large herbivore populations and the ecosystems they rely upon, which in turn are dependent on them. This connectivity is also crucial to humans that rely on large herbivores for food and income from hunting and wildlife viewing.

42.2 Wildlife-vehicle collision often involve large herbivores and cause large-scale costs to human societies; documenting WVC and their costs is the first step in preventing them

The cost of WVC with large herbivores is high and includes human injury and death, vehicle repair costs and the loss of wildlife populations (Table 42.1). In many cases, it is cost-effective to install mitigation (i.e. fencing and crossing structures) to prevent WVC with large herbivores (e.g. Huijser et al. 2009). While the construction costs of mitigation may be high, the reduction in the direct and indirect costs of WVC soon outweighs the initial investment. It is therefore important to document the number of collisions with large herbivores and the associated costs and use a cost–benefit analysis (see Huijser et al. 2009) to demonstrate the potential benefits of future mitigation.

Documenting the location and rate of WVC is a critical first step to develop a strategy to prevent them. Unfortunately, more than half of the WVC with large herbivores in Sweden are not reported (Seiler 2004); and counts of large herbivore carcasses from WVC are often more accurate than reported accidents. For example, in Virginia, United States, transportation agency staff found over nine times more white-tailed deer carcasses along roads than documented by crash reports (Donaldson & Lafon 2010). WVC and their costs are also documented by insurance agencies. In 2008, approximately 60,000 deer–vehicle collisions in Canada were reported to insurance companies (Sielecki 2013). The total annual cost of these collisions in Canada was CAD$400 million (L-P Tardif & Associates 2006). The 1.2 million deer–vehicle collisions in the United States reported to insurance companies were estimated to cost over US$4 billion in the 12 months from July 2012 (State Farm Insurance 2013). In Europe in 1996, Groot-Bruinderink and Hazebroek (1996) estimated that 30,000 people were injured in over 500,000 ungulate–vehicle collisions, which cost over one billion Euros in material damage. In Japan, the number of WVC in 1998 was estimated to be as high as 370,000 (Saeki & MacDonald 2004). With such high rates of WVC, the added costs of avoiding high-risk areas or installing mitigation – either on new roads or to retrofit on existing roads – may be a cost-effective solution.

42.3 Mitigating vehicle and road effects on large herbivores requires long-term transportation planning

A key to reducing WVC with large herbivores is to consider their movement needs early (5–20 years) in the transportation planning process (Chapter 9). This is necessary for all projects, including new roads as well as maintenance and upgrades. Planners should consult species experts (Chapter 9), data and maps of the occurrence and movement patterns of large herbivores, as well as analyses of WVC data to identify potential

Table 42.1 Estimates of average financial costs of WVC with large herbivores.

Species	Monetary cost	Location	Reference
Moose	US$30,773	United States	Huijser et al. (2009)
Moose	€34,426	Sweden	Swedish National Road Administration (2013)[a]
Mule deer	US$3,085	Utah, United States	Bissonette et al. (2008)
Mule deer	US$8,388	United States	Huijser et al. (2009)
Roe deer	€4,360	Sweden	Swedish National Road Administration (2013)[a]

[a] Average cost for a WVC when travelling at 100 km/h, in 2010.

collision and mortality hotspots. Where collision and mortality data are absent, it should be collected or models built using data from similar landscapes or locations elsewhere and applied to the road in question.

The mitigation hierarchy is to (i) avoid development in ecologically sensitive areas and areas with no roads or roads with low traffic volumes, (ii) minimise impacts, (iii) mitigate impacts and (iv) offset or compensate any residual impacts (Chapter 7). Every opportunity to improve existing infrastructure for large herbivores should be considered, especially when existing structures are being replaced or the road upgraded (Cramer et al. 2011; Kintsch & Cramer 2011). Predicting areas crucial for large herbivores and where they may cross transportation corridors involves the use of maps of wildlife critical habitats and linkage areas. These maps are created through a collaborative effort of multiple agencies and jurisdictions because of the often large-scale movements of large herbivores (Chapters 4, 9 and 13). These maps will help to identify areas that are to be avoided for ecological reasons, especially if mitigation is expensive, logistically impossible or likely to be ineffective, such as the likely ineffectiveness of crossing structures to facilitate the migration of 100,000 wildebeest across the Serengeti (Chapter 56). Documenting WVC carcasses on existing roads can help identify areas where large herbivores need to move across roads. These may not be the only locations important to large herbivore movements; there are areas where animals move across roads with low rates of WVC or where they avoid roads all together but need to cross. For example, pronghorn antelope in North America are typically impeded by roads and fences, and their need to move is underrepresented by WVC (Dodd et al. 2011; Theimer et al. 2012); thus, knowledge of their movement needs is not gained from WVC data. Comprehensive research and monitoring of the wildlife in the area of concern should be an integral part of the planning, construction, and post-construction monitoring of wildlife mitigation.

42.4 The type and size of wildlife crossing structures for large herbivores can be partly based on the North American and European experiences

Appropriately designed wildlife crossing structures with fencing are the most effective way to keep many species of wildlife off the road and to facilitate connectivity (Hedlund et al. 2004). An important consideration for large herbivores is their role as prey; they need to avoid predators when using crossing structures.

Structures which are more open than confined work best for these species.

Most crossing structures for large herbivores are underpasses; however, overpasses are increasingly being used (Chapter 21). The type and size of underpasses vary (Chapter 21) and, if large enough, are suitable for many species (Fig. 42.1A–E), including elephants (Fig. 43.2). Open-span bridge underpasses have higher success rates than culverts for mule deer and elk in Utah, United States (Cramer 2012, 2013; Fig. 42.1D), and may be the only option for more cautious large herbivores. The more wary species that need specific landscape features such as desert bighorn sheep in Arizona, United States, and pronghorn antelope in Wyoming, United States, have the highest successful crossing rates on wildlife overpasses (Gagnon et al. 2013; Sawyer & LeBeau 2013; Fig. 43.1F, G and H). While several individuals of a wary population may use underpasses, it is overpasses that appear to allow entire populations to cross the road (Fig. 42.1H). While these broad recommendations are helpful, rigorous research is necessary to determine a species' preference for the type and size of crossing structures (Chapter 10).

The dimensions of crossing structures influence use by large herbivores. Besides physical size constraints, herbivores are prey species and unlikely to use structures that they perceive may leave them vulnerable to predators. The most basic recommendation to maximise successful crossing rates by large herbivores is to keep the structure as short and as tall and wide as possible. Cramer (2013) and Schwender (2013) found that shorter underpasses in Utah, United States, had higher success rates for mule deer than longer ones. They recommended that underpasses be less than 43 m in length to ensure mule deer success rates of 70% or more and that height of structures was less important than width. Bridges were also more successful in passing large herbivores than culverts in Utah, United States (Cramer 2013), which may hold true for other large herbivores as well. Interestingly, the width and height of underpasses were more important than length in predicting the number of crossings by moose through conventional bridges and culverts in Sweden (Olsson & Seiler 2012). They estimated that underpasses wider than 23 m were likely to be used by moose at the same frequency with which they were detected on reference track beds close to the underpass. Wider crossing structures provide large herbivores with escape routes to evade predators (Chapter 36). In areas where native predators have been extirpated, the prey species may tolerate more constrained openings (Cramer, personal observations), but further research

Figure 42.1 (A) White-tailed deer use underpass in Montana, United States; (B) moose use underpass in British Columbia, Canada; (C) eastern grey kangaroos use underpass in Australia; (D) mule deer using bridge underpass in Utah, United States; (E) underpass for Sitka deer, Hokkaido, Japan; (F) elk use overpass in Utah, United States; (G) desert bighorn sheep on overpass in Arizona, United States; (H) pronghorn antelope use overpass in Wyoming, United States. Source: (A) Photograph by P. Cramer and Montana DOT; (B) Photograph by and reproduced with permission of AECOM Ministère des Transports du Québec, Québec, Canada; (C) Photograph by R. van der Ree; (D) Photograph by P. Cramer and Utah DOT; (E) Photograph by L. Sielecki; (F) Photograph by P. Cramer and Utah Division of Wildlife Resources; (G) Photograph by and reproduced with permission of J. Gagnon, Arizona Game and Fish; and (H) Photograph by and reproduced with permission of Wyoming Dept. of Transportation and West, Inc.

is required. Importantly, efforts to reintroduce top-order predators to restore ecosystem functions mean that crossing structures in these landscapes should be designed as if the predators still occur. Height may be the least important dimension of an underpass (Cramer 2013), but minimum thresholds will apply.

There has been a significant amount of mitigation and research in North America and Europe to quantify the preferences of large herbivores for wildlife crossing structures (Cramer et al. 2011). Transportation agencies should adopt a more experimental approach to mitigation generally (Chapter 10), ensuring sufficient resources are available for evaluating the success of mitigation and adopt an adaptive management approach to mitigation. In addition, research is urgently required to determine preferences outside of North America and Europe and to refine recommendations for crossing structures for common species.

42.5 The placement of wildlife crossing structures for large herbivores should be based on animal size, their movement patterns and transportation constraints

The placement and spacing of crossing structures for large herbivores should be based on the species' size, the distance and location of their daily and seasonal movements and transportation-related constraints. The placement of crossing structures should be determined collaboratively with species experts and the planners, designers and engineers (Chapter 9). Once wildlife habitat maps have been referenced for general areas where wildlife movements are bisected by roads, pinpointing the optimal locations of crossing structures and their spacing requires data on wildlife, existing structures and potential locations for new structures. Wildlife crossing structures and modified existing structures should be placed where large herbivores are known to move and if possible in conjunction with waterways and other natural features. Drainage structures can often be cost-effectively adapted (for new structures) and modified (for existing structures) for large herbivore passage. These actions may be as simple as removing large rocks to provide clear pathways (Fig. 42.2) or including dry banks under bridges that span waterways. Other changes to existing and future structures can be made to increase the likelihood of large herbivores finding and using them. If the structures are intended to be multi-use structures, they still must be sufficiently effective for wildlife so that the effectiveness is not compromised (Chapters 21 and 22). The spacing will depend on whether movement is required on a daily or seasonal basis and is discussed further in Lesson 21.6.

Wildlife exclusion fencing is typically required to prevent wildlife from accessing the road and to funnel them towards crossing structures (Chapter 20). Fences need to be designed for the target species, and ungulate fences in North America are typically 2.4 m tall. Fencing has been shown to significantly increase the

Figure 42.2 White-tailed deer move through an underpass in Montana, United States, with a pathway cleared through large rocks Source: Photograph by P. Cramer and Montana DOT.

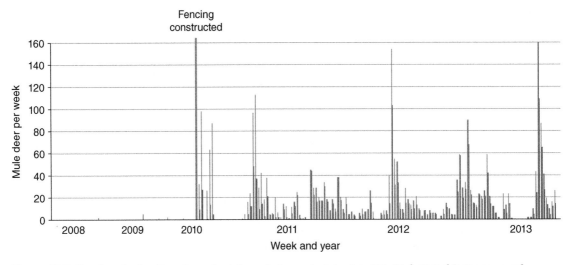

Figure 42.3 Use of a pair of multi-use box culverts by mule deer under Interstate 70 in Utah, United States, increased significantly after the installation of wildlife fencing. Red columns represent approaches by deer, and blue columns complete crossings. Source: Cramer (2014).

rate of crossing by large herbivores in the western United States (Fig 42.3; Dodd et al. 2007; Cramer 2013; Chapter 20).

42.6   Additional mitigation measures include driver warning systems and vegetation management

Wildlife crossing structures and fencing can effectively prevent WVC and maintain connectivity for large herbivores. In situations where crossing structures and fences are not feasible, additional options such as warning signs, wildlife detection systems and vegetation management may reduce the risk of WVC. Signs and detection systems aim to educate motorists of the risks and presence of wildlife on or adjacent to the road, but the effectiveness is variable (Chapter 24). On-board detections systems are an emerging infrared system that detects body heat of animals (and pedestrians) on the road and alerts drivers to the danger (Autoliv 2014). These systems, currently only available in luxury vehicles, require a contrast between the ambient heat and that of the animal and therefore work best at night or in cold climates. The goals of vegetation management are to make roadsides less attractive to large herbivores and increase visibility for both drivers and wildlife. Furthermore, vegetation along the verge and median should be unpalatable to discourage grazing (e.g. Textbox 46.1). Other approaches,

such as whistles and reflectors, which are purported to disturb and repel wildlife from the roadway, are ineffective (Chapter 25).

CONCLUSION

Large herbivores are important to natural systems and human populations. Accommodating the needs of large herbivores in transportation systems is critical to the healthy functioning of natural ecosystems and human safety. The most effective mitigation option is to avoid building new roads or widening existing roads in places that still support populations of large herbivores. Where avoidance and minimisation is not possible, the installation of crossings structures with wildlife fencing is the next most effective approach, which is often also cost-effective. Retrofits of existing culverts and bridges can also help move large herbivores beneath roads. Less costly but also often less effective mitigation options include signs, wildlife detection systems and vegetation management. Failures in mitigation efforts for large herbivores are very obvious to the public and can be deadly to motorists. The design of mitigation measures for areas outside of North America and Europe can initially be based on the results of monitoring from these two regions. Ultimately, the success of mitigation for large herbivores will depend on quality research and adaptive management of infrastructure.

FURTHER READING

Allison (2012): The film 'Highway Wilding' is an excellent entry-level introduction to the issue of WVC, wildlife mortality and options for mitigation.

Bissonette and Cramer (2008): A national study and review of the use and effectiveness of wildlife crossing structures in the United States.

Beckman et al. (2010): An accessible and wide-ranging book with a North American focus on the impacts of roads and traffic and the options for mitigation.

Groot-Bruinderink and Hazebroek (1996): An early publication quantifying the rates of WVC with ungulate in Europe.

US Fish and Wildlife Service (2010): This movie, Innovative Approaches to Wildlife Highway Interactions, targeting transportation agency staff, provides a range of state-of-the-art examples of mitigation projects from the United States. It includes an online instructional manual.

REFERENCES

Allison, L. 2012. Highway Wilding. A film. http://highway wilding.org/hw_movie.php (Accessed on 28 January 2014).

Autoliv. 2014. Do you have night vision? http://autolivnightvision. com/features/animal-detection/ (Accessed 29 January 2015).

Beckman, J. P., A. P. Clevenger, M. P. Huijser and J. A. Hilty. 2010. Safe Passages: Highways, Wildlife, and Habitat Connectivity. Island Press, Washington, DC, 396 pp.

Bissonette, J. A. and P. C. Cramer. 2008. Evaluation of the use and effectiveness of wildlife crossings. Report 615 for National Academies', Transportation Research Board, National Cooperative Highway Research Program, Washington, DC. http://onlinepubs.trb.org/onlinepubs/nchrp/nchrp_rpt_615.pdf (Accessed on 22 September 2014).

Bissonette J. A., C. A. Kassar and L. J. Cook. 2008. Assessment of costs associated with deer-vehicle collisions: human death and injury, vehicle damage, and deer loss. Human-wildlife conflicts 2:17–27.

Coulson, G. M. 1982. Road-kills of macropods on a section of highway in central Victoria. Australian Wildlife Research 9:21–26.

Cramer, P. C. 2012. Determining wildlife use of wildlife crossing structures under different scenarios. Final Report to Utah Department of Transportation, Salt Lake City, UT. 181 pages. http://www.udot.utah.gov/main/uconowner.gf?n=10315521671291686 (Accessed on 18 September 2014).

Cramer, P. C. 2013. Design recommendations from five years of wildlife crossing research across Utah. In: Proceedings of the 2013 International Conference on Ecology and Transportation, 2013. Center for Transportation and the Environment, North Carolina State University, Raleigh, NC. http://www.icoet.net/ICOET_2013/documents/papers/ICOET2013_Paper402A_Cramer_Formatted.pdf (Accessed on 22 September 2014).

Cramer, P. C. 2014. Wildlife crossing structures in Utah: determining the best designs. Final report to Utah Division of Wildlife Resources, from Utah State University. 331 pages.

Christianson, D. and S. Creel. 2009. Effects of grass and browse consumption on the winter mass dynamics of elk. Physiological Ecology 158:603–613.

Dodd, N. L., W. Gagnon, S. Boe, and R. E. Schweinsburg. 2007. Role of fencing in promoting wildlife underpass use and highway permeability. In: 2007 International Conference on Ecology and Transportation 2007 Proceedings, pp. 475–487. http://www.icoet.net/ICOET_2007/ (Accessed on 22 September 2014).

Dodd, N. L., J. W. Gagnon, S. Sprague, S. Boe and R. E. Schweinsburg. 2011. Assessment of pronghorn movements and strategies to promote highway permeability: U.S. Highway 89. Final project report 619, Arizona Department of Transportation Research Center, Phoenix, AZ.

Donaldson, B. and N. Lafon. 2010. Personal digital assistants to collect data on animal carcass removal from roadways. Transportation Research Record: Journal of the Transportation Research Board 2147: 18–24.

Gagnon, J. W., C. D. Loberger, S. C. Sprague, M. Priest, S. Boe, K. Ogren, E. Kombe and R. E. Schweinsburg. 2013. Evaluation of Desert Bighorn Sheep Overpasses along US Highway 93 in Arizona, USA. In: 2013 International Conference on Ecology and Transportation 2013 Proceedings. http://www.icoet.net/ICOET_2013/ (Accessed on 22 September 2014).

Groot-Bruinderink G. W. T. A. and E. Hazebroek. 1996. Ungulate traffic collisions in Europe. Conservation Biology 10: 1059–1067.

Hedlund, J. H., P. C. Curtis, G. Curtis and A. F. Williams. 2004. Methods to reduce traffic crashes involving deer: what works and what does not. Traffic Injury Prevention 5: 122–131.

Hobbs, N. T. 1996. Modification of ecosystems by ungulates. Journal of Wildlife Management 60: 695–713.

Huijser, M. P. P., J. W. Duffield, A. P. Clevenger, R. J. Ament and P. T. McGowen. 2009. Cost-benefit analyses of mitigation measures aimed at reducing collisions with large ungulates in the United States and Canada: a decision support tool. Ecology and Society 14: 15. http://www.ecologyandsociety.org/vol14/iss2/art15/ (Accessed 29 January 2015).

Jaeger, J. A. and L. Fahrig. 2004. Effects of road fencing on population persistence. Conservation Biology 18: 1651–1657.

Kie, J. G. and J. F. Lehmkuhl. 2001. Herbivory by wild and domestic ungulates in the intermountain west. Northwest Science 75:55–61.

Kintsch, J. and P. C. Cramer. 2011. Permeability of existing structures for terrestrial wildlife: a passage assessment system. For Washington Department of Transportation, WA-RD 777.1. Olympia, Washington, 188 pages. http://www.wsdot.wa.gov/research/reports/fullreports/777.1.pdf (Accessed on 22 September 2014).

Klocker, U., D. B. Croft and D. Ramp. 2006. Frequency and causes of kangaroo-vehicle collisions on an Australian outback highway. Wildlife Research 33: 5–15.

L-P Tardif & Associates. 2006. Update of data sources on collisions involving motor vehicles and large animals in Canada. Transport Canada, Ottawa. http://www.tc.gc.ca/eng/roadsafety/tp-tp14798-menu-160.htm (Accessed on 22 September 2014).

Olsson, M. and A. Seiler. 2012. The use of a moose and roe deer permeability index to develop performance standards for conventional road bridges. In: IENE 2012 International Conference, 21–24 October 2012. Swedish Biodiversity Centre, Berlin-Potsdam, Germany, p. 195.

Saeki, M. and D. W. Macdonald. 2004. The effects of traffic on raccoon dog (*Nyctereutes procyonoides viverrinus*) and other mammals in Japan. Biological Conservation **118**: 559–571.

Sawyer, H. and C. LeBeau. 2013. Trapper's Point wildlife crossing study, 2012 progress report. Prepared for Wyoming Department of Transportation, by Western EcoSystems Technology, Inc. Laramie, Wyoming.

Sawyer, H., F. Lindzey and D. McWhirter. 2005. Mule deer and pronghorn migration in western Wyoming. Wildlife Society Bulletin **4**:1266–1273.

Schwender, M. 2013. Mule deer and wildlife crossings in Utah, USA. Paper 1465. Master's Thesis, Utah State University. http://digitalcommons.usu.edu/etd/1465/ (Accessed on 22 September 2014).

Seiler, A. 2004. Trends and spatial patterns in ungulate-vehicle collisions in Sweden. Wildlife Biology **10**: 301–313.

Sielecki, L. E. 2013. Communicating the risk of wildlife hazards to new drivers in the United States and Canada with state and provincial driver manuals and handbooks. In: Proceedings of the 2013 International Conference on Ecology and Transportation, Raleigh, NC. http://www. icoet.net/ICOET_2013/documents/papers/ICOET2013_ Paper206D_Sielecki.pdf (Accessed on 22 September 2014).

State Farm Insurance. 2013. U.S. Deer-Vehicle Collisions Decline: State Farm survey shows trend more pronounced in Nation's mid-section. PR Newswire, Jersey City, NJ. http://www.multivu.com/mnr/56800-state-farm-survey-show-u-s-deer-vehicle-collisions-decline (Accessed on 22 September 2014).

Swedish National Road Administration. 2013. Trafiksäkerhet (traffic safety). Bygg om eller bygg nytt – effektsamband för transportsystemet (in Swedish). Trafikverket, Sweden, p. 14.

Theimer T. C., S. Sprague, E. Eddy and R. Benford. 2012. Genetic variation of pronghorn across US Route 89 and State Route 64. Final project report 659. ADOT Research Center, Arizona Department of Transportation, Phoenix, AZ.

Thirgood, S., A. Mosser, S. Tham, G. Hopcraft, E. Mwangomo, T. Mlengeya, M. Kilewo, J. Fryxell, A. R. E. Sinclair and M. Borner, 2004. Can parks protect migratory ungulates? The case of the Serengeti wildebeest. Animal Conservation **7**: 113–120. doi: 10.1017/S1367943004001404.

US Fish and Wildlife Service. 2010. Online video: Innovative Approaches to Wildlife Highway Interactions. National Conservation Training Center, Shepherdstown, WV. http:// bcove.me/m37cfirs (Accessed on 22 September 2014).

Chapter 43

CASE STUDY: THE MOUNT KENYA ELEPHANT CORRIDOR AND UNDERPASS

Susie Weeks

Mount Kenya Trust, Nanyuki, Kenya

Mount Kenya is Africa's second highest peak and is an important source of water for agriculture and hydro-electricity. It is also important for tourism and biodiversity conservation, with unique Afro-Alpine forests and bamboo zones which give way to shrubs, giant lobelia, grasses and heather on the moorlands beneath glacier-clad peaks. Its biodiversity and cultural heritage are internationally recognised, and the mountain is a National Park, a Man and Biosphere Reserve and a World Heritage Site. The fertile lower slopes of Mount Kenya are important for agriculture.

Historically, elephants moved between Mount Kenya and the Ngare Ndare Forest and the lower drier country to the north. These daily and seasonal movements are now restricted by farms, fences and roads, leading to increased rates of human–wildlife conflict, particularly crop raiding by elephants. For example, 28 reports of crop damage by elephants were recorded between Ontulili and Kibirichia on the western side of the mountain in 2005 (Kenya Wildlife Service, unpublished data). Fences have been installed to protect crops and prevent the need to cull problem elephants. Consequently, the Laikipia/Samburu elephant populations (Fig. 43.1) which were once continuous with the Mount Kenya population are now isolated from each other.

Elephants still need to move across their original habitat, and so the idea of building a fenced corridor between the northern forests of Mount Kenya and the Ngare Ndare Forest Reserve was born. The two main obstacles in constructing this corridor were obtaining the land and crossing the A2, a national single-lane highway with approximately 1000 vehicles per day (2006 data). Two large farms (Mariana and Kisima) on either side of the A2 donated the land for the corridor and the local community set about solving the challenge of allowing elephants to safely access and cross the A2.

A key to the project's success was that it was led by a group of strong local champions. The Mount Kenya Trust spearheaded the planning of the project, with assistance from the farms adjoining the corridor, the Lewa Wildlife Conservancy, the Ngare Ndare Forest Trust and a number of other non-government organisations (NGOs). Using data from the 'Save the Elephants' organisation, the movement of one particular elephant was used to identify the location of the corridor and A2 crossing. Furthermore, a detailed feasibility study, an EIA and a survey of road-crossing behaviour of elephants were commissioned by the Mount Kenya Trust between 2005 and 2007.

The members of the planning team understood elephant movement behaviour and effective elephant fences, but had no experience of wildlife crossing structures for elephants. Overpasses and underpasses were both initially considered, but the topography of

Handbook of Road Ecology, First Edition. Edited by Rodney van der Ree, Daniel J. Smith and Clara Grilo.
© 2015 John Wiley & Sons, Ltd. Published 2015 by John Wiley & Sons, Ltd.
Companion website: www.wiley.com\go\vanderree\roadecology

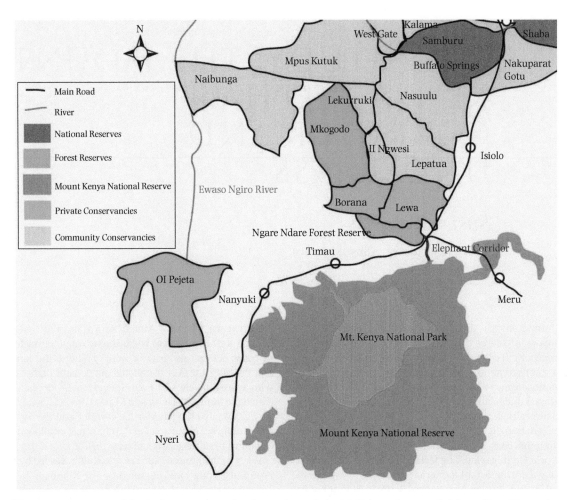

Figure 43.1 Location of the elephant corridor and underpass beneath the A2 Highway between the Mount Kenya National Reserve and the Ngare Ndare Forest Reserve, Kenya. Source: Map by Susie Weeks, Mount Kenya Trust.

the area best facilitated an underpass. The choice to construct an underpass was confirmed when elephants were observed using a culvert in KwaZulu-Natal, South Africa.

The Kenyan government was unable to provide any funds for the corridor or underpass so the Mount Kenya Trust began fundraising after obtaining permission to build the underpass in 2008. Tenders were released, and the total project cost was estimated at approximately US$1 million (2011), with one third for the underpass and the remaining two thirds for 28 km of full 'game-proof' fencing and housing for the corridor maintenance team. Ongoing costs to maintain the corridor fences are raised each year.

The underpass was opened in December 2010 and it was used by an elephant (now known as Tony) on

the first night. Data from remote cameras and direct inspections indicate over 300 crossings by elephants within the first year of use (Mount Kenya Trust, unpublished data) (Fig. 43.2). While this number includes multiple crossings by the same individuals, it exceeds the 15 crossings made at grade, recorded in the year prior to construction.

A range of strategies to encourage the use of the corridor and underpass by elephants was suggested during the planning phase, including the spreading of dung and urine of cows in oestrus at the entrances, as well as installing temporary pools of water, placing of crop waste as a food source and planting palatable plant species. Dung was placed in the area the day the fences were connected and underpass opened, and Tony used the underpass the following day. The extent

Figure 43.2 Elephants using the underpass beneath the A2 Highway, Kenya. Source: Photograph by Susie Weeks.

(A)   (B)

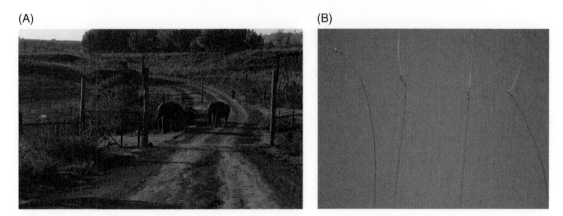

Figure 43.3 (A) The elephant corridor (running left to right) crossing a rural road and (B) a close-up of electrified wires that are suspended above the road to act as a curtain to discourage elephants from leaving the corridor and walk along the road. Source: Photographs by Susie Weeks.

to which the dung encouraged Tony to enter the underpass is unknown, but it probably helped. The culvert under the A2 (12 m long, 4.5 m high and 6 m wide) proved sufficiently large for elephant. However, evidence from India suggests that elephants are sensitive to the size of underpasses. In Rajaji National Park, tunnels (111 m long × 5 m wide × 5 m tall) on the Dugadda drainage line were too small, and there is no evidence of elephants having used them in 30 years from what local experts define as 'tunnel syndrome' (Singh et al. 2011).

Re-establishing migration for large species like elephants has improved the biodiversity and conservation status of the region, and in June 2013, the Ngare Ndare Forest Reserve and the Lewa Wildlife Conservancy were added to the Mount Kenya UNESCO World Heritage Site. In addition, the Marania and Kisima Farms that donated land for the elephant corridor

have also been included within the World Heritage Site, and elephant damage is now minimal. The elephant corridor crosses a dirt road (Fig. 43.3A), and elephants are discouraged from leaving the corridor via the road by electrified wires suspended above the road (Fig. 43.3B). The Mount Kenya Trust and partners are currently raising funds to build a similar underpass under this rural road.

ACKNOWLEDGEMENTS

The key partners in the project are the Kisima and Marania Farms, the Mount Kenya Trust, the Lewa Wildlife Conservancy, the Ngare Ndare Forest Trust, Trinnie Cartland, the Kenya Wildlife Service and the Kenya Forest Service. Major funding was sourced by the Mount Kenya Trust, Laikipia Wildlife Forum,

Borana Ranch and Lewa Wildlife Conservancy. Major donors include the Royal Netherlands Government (via the Laikipia Wildlife Forum), Virgin Atlantic, The Nature Conservancy, Rex Dobie, Zurich Zoo and friends of the Lewa Conservancy.

REFERENCE

Singh A., Negi A.S., Rasaily S.S., Mishra D.K. and Singh A. 2011. Eco Engineering Report. Uttarakhand Forest Department, the Wildlife Savers Society and the Tiger Protection Group, Uttarakhand, India.

Chapter 44

FORM AND FUNCTION: A MORE NATURAL APPROACH TO INFRASTRUCTURE, FISH AND STREAM HABITATS

Paul J. Wagner

Washington State Department of Transportation, Environmental Services Office, Olympia, WA, USA

SUMMARY

Rivers and streams are dynamic and their changing nature needs to be accommodated in planning, expanding or operating transportation infrastructure where roads and other linear infrastructure cross water or occur in a floodplain. This is necessary not only to minimise direct ecological effects to habitat, fish and other aquatic species but also to help reduce the potential damage to infrastructure from flooding, erosion and channel movement. Damage to infrastructure can often lead to additional environmental impacts.

44.1 Streams need special consideration when planning roads.

44.2 Protect natural stream processes: avoid constricting natural stream channels.

44.3 Roads often interfere with natural stream processes, impacting aquatic species.

44.4 Roads along shorelines: avoid simplifying stream channels and use dynamic natural systems as the model.

44.5 Reduce impacts at stream crossings by using designs that simulate natural channel conditions.

Fish and aquatic organisms, such as amphibians, reptiles and invertebrate species can be directly affected by roads, especially at water crossings. If these are not designed correctly, crossings result in the direct loss of natural habitat and can become barriers to movement caused by high velocity, inadequate water depth or excessive drops, especially at the outfall. To help minimise these problems, it is best to accommodate the natural conditions of streams, allow for a wide range of natural stream flow channels, avoid constricting flows and avoid reducing channel and shoreline complexity.

Handbook of Road Ecology, First Edition. Edited by Rodney van der Ree, Daniel J. Smith and Clara Grilo.
© 2015 John Wiley & Sons, Ltd. Published 2015 by John Wiley & Sons, Ltd.
Companion website: www.wiley.com\go\vanderree\roadecology

INTRODUCTION

Streams and rivers are dynamic systems that change seasonally as the water flow rises and falls. The constant motion of water, transporting sediment, nutrients and woody material creates and maintains complex habitats which support a diversity of aquatic organisms. Innumerable aquatic and terrestrial species depend on these habitats for all or part of their life cycle. Steps typically taken to create and protect roads near streams include straightening channels or placing very large rocks or concrete reinforcement to stabilise banks. These generally result in loss of habitat and potential impacts to fish and wildlife populations. The intent of this chapter is to discuss the importance of structural complexity in streams, the consequences of simplifying natural stream channels and to describe an approach for designing and managing roads along waterways that accommodate natural stream channel processes.

LESSONS

44.1   Streams need special consideration when planning roads

Transportation systems and river systems both exist as networks across landscapes, and their intersection is inevitable. Aquatic habitats support the most biologically diverse and ecologically productive places on earth and deserve a high level of protection and consideration when planning for roads and other linear infrastructure. Water courses are constantly in motion and shaped by dynamic forces, especially by periodic high flow events when tremendous energy moves bed material and reshapes channels. These forces help create and maintain the physical attributes of habitat and distribute nutrients. The dynamic physical processes sustain life in aquatic systems, but also pose a threat to static infrastructure like culverts, bridge abutments and roadway fill.

Streams naturally tend to have complex forms that lead to complex habitat associations. The deep portion of the channel shifts from one side of the stream to the other at bends and meanders. Deep pools on the bends provide cool temperatures and protective cover for fish. Shallower areas transition to riffles where turbulence introduces oxygen to the water and moves fine sediment out of the spaces between gravel in the bed. This creates optimal conditions for organisms living below the surface of the gravel: insects and the incubating eggs of fish such as trout and salmon. Streamside vegetation provides shade that helps protect stream temperatures and also provides nutrient input as vegetative matter and insects fall into the water (Groot & Margolis 1991). Fish have evolved to rely on the complexity of aquatic habitats (Fig. 44.1). When habitat elements are lost or altered, the system becomes more simplified and less capable of supporting ecological diversity (Fig. 44.2).

(A)   (B)

Figure 44.1 Natural stream channels which support aquatic organisms exhibit natural complexity with variation in substrate type, vegetation along the water's edge and a meandering course. Examples here from Washington State, USA, include (A) Ravensdale Creek, a small natural stream in a temperate forested area and (B) Dry Creek, which is located in a more arid climate. Source: Photographs by Paul J. Wagner.

(A)   (B)

Figure 44.2 Many waterways have been modified to move water quickly and prevent flooding. These simplified streams in (A) western Washington, USA, and (B) Putah South Canal, Solano County, California, USA, show gradations in the extent of modification. Source: Photographs by Paul J. Wagner.

44.2 Protect natural stream processes: avoid constricting natural stream channels

Careful planning is needed to provide reliable infrastructure that also protects aquatic habitats and the species on which they depend. The guiding general principle here is to *allow streams to occupy their naturally defined channels and seek to restore similar conditions where these have been altered or constricted.* Simply put, let streams be streams.

When considering the location of new roads, it is best to avoid locating transportation infrastructure where it may impede waterways. This means that roads should be located away from active channels and outside the floodplain wherever possible. Choosing a road alignment that avoids these areas to begin with can eliminate future challenges, many of which are complex and costly to repair or retrofit. The number of stream-crossing locations should be minimised to the extent possible. Crossings should be sized not simply for hydraulic capacity, but to span the natural channel width at high flow events along with some natural shoreline above the waterline. This exposed natural shoreline in turn provides potential connectivity for a host of terrestrial species (Chapter 21).

The location and configuration of the existing road system generally owes more to historic travel routes and transportation needs than a regard for ecological and stream flow processes. Historic travel routes were frequently along waterways, which means that the current roads are often located within floodplains where they are subject to periodic damage from high flows. This pressure and damage can be a significant challenge in regions that experience high seasonal precipitation (i.e. monsoons – Chapters 49 and 51) or severe storms (including some arid areas – Chapter 47). The same general ideas of minimising the number of stream crossings and reducing the degree to which these crossings and road fill constrict flows apply to existing roads as they do when considering the placement of a new road. Existing roads often present many established problems that have resulted from inattention to these principles. To reduce the impacts to aquatic habitats, crossings may need to be retrofitted, and existing undersized culverts and bridges may need to be replaced and made larger.

44.3 Roads often interfere with natural stream processes, impacting aquatic species

Roads that are too close to streambanks tend to confine or divert the stream flow, which generally leads to increased stream velocity and increased risk of erosion. This is compounded when the stream has been 'straightened' and the streamside (riparian) vegetation has been reduced or removed.

Erosion is a natural process, but human activity often causes it to accelerate above the natural rate. This leads to loss of streambed or bank habitat as substrate is washed away. This also leads to degradation of water quality when sediment is suspended in the water, which can injure fish and other aquatic organisms whose gills or filter feeding organs can become clogged. When eroded sediment is eventually deposited at high levels, it can smother organisms living in the stream bed habitat and reduce availability of food sources.

Many methods for protecting roads from erosion involve armouring shorelines with large rock, concrete or steel, effectively hardening these surfaces in an attempt to protect them from the forces of moving water (Figs. 44.3 and 47.3). These methods are widely used, but they often exacerbate the problem and also impact habitat and alter hydrology. Simplified channels and protective approaches, like bank armouring, work by deflecting the kinetic energy of water. This generally just moves the energy downstream where it exerts more erosive force elsewhere. Complex natural channels, on the other hand, tend to diffuse energy and as a result are not only more stable, but more productive from a habitat standpoint.

44.4 Roads along shorelines: avoid simplifying stream channels and use dynamic natural systems as the model

When additional protection is needed for roads along shorelines, it is preferable to use a more natural approach (e.g. Textbox 44.1), rather than hardening and simplifying shorelines. The basic principle is that by using methods that mimic natural processes of energy diffusion, greater success is likely not only in protecting road infrastructure but also in reducing the environmental impacts on aquatic habitat.

Consciously mimicking natural processes along shorelines also holds benefits with regard to permitting and mitigation requirements. Erosion repair which relies largely on replacing natural shorelines with hardened elements like rock or concrete (Fig. 44.3) means a loss of habitat and can require some form of compensatory mitigation to offset the impacts of the project. When complex elements such as large woody debris and plantings can be incorporated into the protective structure (known as 'soft armouring'), then the structure can provide habitat for aquatic organisms while also stabilising the stream bank. An example of this is the engineered log jam (Fig. 44.4) where large logs and rootwads are installed to provide physical protection to banks while also improving natural habitat.

44.5 Reduce impacts at stream crossings by using designs that simulate natural channel conditions

Traditionally, stream crossings have been designed from a hydraulic perspective, with the size of the drainage structures based on the ability to convey the volume and velocity of predicted flows. The least

---

Textbox 44.1 Integrated streambank protection guidelines in Washington State, USA.

In Washington State, USA, an approach called the Integrated Streambank Protection Guidelines (ISPG; Cramer et al. 2002) has been developed through coordination between natural resource and transportation agencies. The general principles are instructive and can be broadly applied. This guidance emphasizes the importance of allowing for natural stream processes to occur unimpeded as part of the planning and operation of infrastructure near streams.

The guidelines address geomorphic and biological processes. *Geomorphic processes* relate to the physical interaction of the landform, water, sediment and woody material that create channel and shoreline structure. Geomorphic processes include bank and bed erosion, channel migration, influence of debris, sedimentation and sediment transport. *Biological processes* include the ways plants and animals shape the physical habitat in streams such as nutrient cycling, species interactions, riparian and upland vegetation dynamics, and

species-mediated, habitat-forming processes such as beaver activity.

Guiding principles identified in Integrated Streambank Protection are as follows:
(i) Natural erosion processes and rates are essential for ecological health of the aquatic system.
(ii) Human-caused erosion that exceeds natural rates and amounts is usually detrimental to ecological functions.
(iii) Natural processes of erosion are expected to occur throughout the channel-migration zone. Project considerations should include the channel-migration zone and potential upstream and downstream effects.
(iv) Preservation of natural channel processes will sustain opportunities for continued habitat formation and maintenance.

(Cramer et al. 2002)

For more information on ISPG, see http://wdfw. wa.gov/publications/pub.php?id=00046

Figure 44.3 The use of large rock (or rip rap) is a common approach to protecting road infrastructure from damaging flows, here used as an emergency response along Highway 20 on the Skagit River, Washington, USA. Source: Photograph by Paul J. Wagner.

Figure 44.4 A change in approach is needed to address the habitat loss by continued use of rip-rap. The Sauk River wood cribwall is an example of a more natural approach, Washington, USA. Source: Photograph by Paul J. Wagner.

expensive solution is often sought, which generally translates to the smallest culvert or shortest bridge span that can meet the flow volume criteria. When crossings are built to minimal hydraulic criteria, they often constrict the higher flows. When flows are constricted by a narrow structure, the result is increased velocity, which leads to increased erosion. At culverts, this erosion scours the stream bed as it exits the culvert often causing significant drops at the outfall (Fig. 44.5). These outfall drops can become barriers to the upstream movement of fish and other species (Chapter 45). Constricting flows also reduces the natural transport of sediment and woody debris which is an essential process for maintaining good quality stream habitat for fish. Culverts that constrict flow can have maintenance problems if woody debris or sediment gets trapped, blocks flow or creates a backwater condition (see also Chapter 17).

Stream crossings that constrict the range of flows and do not span the ordinary high water level can lead to damaging flow rates, high volumes of water and downstream erosion, and can impede the movement of fish and other aquatic organisms upstream and downstream. To avoid these problems, crossing structures should be designed to mimic natural conditions and span widths sufficient to allow development of natural stream channel and bed characteristics within the structure. Stream simulation is a useful method to achieve this (Barnard et al. 2013), and it attempts to mimic the channel conditions (width, slope and bed material) within the crossing structure as are found in the upstream and downstream channel (Fig. 44.6). This results in crossing structures that span the natural stream channel as well as some amount of adjacent terrestrial area, even at typical high flow levels. These crossings also include bed material that is comparable to and at a slope that is consistent with the upstream and downstream channel. Specific details on road crossings for fish are given in Chapter 45.

CONCLUSIONS

Streams are dynamic systems which support complex habitat for fish and other aquatic species. The best approach to protecting these important resources is to plan and create road and structure designs that avoid roadway fill in floodplains, keep the number of stream crossings to a minimum and design culvert and bridge spans that accommodate dynamic natural streams and stream bank conditions to the greatest extent possible. At locations in existing road networks where stream erosion threatens roadways and realignment is not feasible, using soft armouring techniques for stream bank protection can provide a more natural approach that protects infrastructure while providing habitat.

(A)   (B)

Figure 44.5 These culverts have constricted the flow of water, which increases velocity and leads to erosion at the outfall. When originally installed, these culverts were at the same elevation as the streambed but over time, erosion has led to a significant outfall drop, creating a vertical barrier to the upstream movement of fish and other aquatic organisms. As this process continues, the integrity of the culvert and the road fill is at risk. Source: Photography by Paul J. Wagner - Washington State Department of Transportation.

Figure 44.6 Example of a stream crossing designed using the stream simulation concept. This bridge spans the channel even at high flows and under normal flows has a significant area which is available for passage by terrestrial species. This photo was taken when construction was just finishing up at Butler Creek in Washington State, USA, and deer had already begun to use it to cross under this rural highway. Source: Photograph by Paul J. Wagner.

FURTHER READING

Barnard et al. (2013): This document describes ecologically-minded design principles for stream crossings including stream simulation developed for Washington State, USA.

Forest Service Stream-Simulation Working Group (2008): This is guidance on stream simulation design developed by the US Forest Service, primarily for smaller roads in forest settings.

Furniss et al. (2006): A useful software and reference tool for designing culverts for the passage of fish and aquatic organisms.

McEnroe (2009): This review covers bridge and culvert designs over time and documents the evolution of the practice including some of the lessons learned.

Viessman et al. (1989): A good general description of hydrology principles.

Williams (1978): A helpful description of the importance of planning for high flow events in the design of crossing structures.

REFERENCES

Barnard, R. J., J. Johnson, P. Brooks, K. M. Bates, B. Heiner, J. P. Klavas, D. C. Ponder, P. D. Smith and P. D. Powers. 2013. Water Crossings Design Guidelines. Washington Department of Fish and Wildlife, Olympia, WA.

Cramer, M., K. Bates, D. Miller, K. Boyd, L. Fortherby, P. Skidmore and T. Hoitsma. 2002. Integrated Streambank Protection Guidelines. Washington State Aquatic Habitat Guideline Program. Washington Department of Fish and Wildlife, Olympia, WA.

Forest Service Stream-Simulation Working Group, Ed. 2008. Stream Simulation: An Ecological Approach to Providing Passage for Aquatic Organisms at Road-Stream Crossings. National Technology and Development Program, San Dimas, CA. U.S. Department of Agriculture, Forest Service National Technology and Development Program.

Furniss, M., M. Love, S. Firor, R. Gubernick, D. Dunklin and R. Quarles. 2006.FishXing, USDA Forest Service, San Dimas Technology and Developement Center, San Dimas, CA. Available from http://www.stream.fs.fed.us/fishxing/index.html (accessed 29 January 2015).

Groot, C. and L. Margolis. 1991. Pacific Salmon Life Histories. UBC Press, Vancouver, BC.

McEnroe, B. M. 2009. Hydrologic design of bridges and culverts: a historical review. Great Rivers History, pp. 83–90. doi:10.1061/41032(344)10.

Viessman, W. V., G. L. Lewis and J. W. Knapp. 1989. An Introduction to Hydrology. Harper Collins, New York.

Williams, G. P. 1978. Bankfull discharge of rivers. Water Resources Research **14**: 1141–1154.

Chapter 45

SOLUTIONS TO THE IMPACTS OF ROADS AND OTHER BARRIERS ON FISH AND FISH HABITAT

Fabrice Ottburg[1] and Matt Blank[2]

[1]Alterra, Wageningen UR, Environmental Science Group, Wageningen, Netherlands
[2]Western Transportation Institute (WTI), Montana State University, Bozeman, MT, USA

SUMMARY

As with all wildlife, fish need to move throughout their range in order to complete their life cycles. Unlike other animals, fish cannot leave the stream or river that they are living in or migrating through to bypass a barrier. Structures under roads that facilitate the flow of water, particularly during flood events, are critical to protect the infrastructure and if well designed can provide passage for fish and other aquatic species. However, improper design, construction or maintenance of road-stream crossings can limit or completely prevent fish passage. In addition, roads and traffic can also impact fish and fish habitat by degrading the quality of the streambed, adjacent riparian habitat and water quality, as well as changing patterns in the flow of ground and surface water.

45.1 Roads and the vehicles that travel on them can negatively affect fish habitat and water quality.

45.2 Roads and other in-stream structures can be barriers to the movement of fish.

45.3 Well-informed planning and design of roads can limit the impacts of roads on fish and fish habitat.

45.4 The most effective road-stream crossings for fish, when long-span 'floodplain' bridges are not an option, are culverts or shorter-span bridges that simulate the natural channel.

45.5 Specialty culverts and other technical solutions are possible but require careful design and do not provide all the qualities of uninterrupted natural waterways.

Roads, railways and other linear infrastructure inevitably intersect waterways, often restricting the movement of fish. New infrastructure should avoid waterways where possible and any crossings that are needed should be designed to allow the natural flow and function of the waterway. Existing road crossings that are barriers to the movement of fish should also be modified to be more natural and improve connectivity for fish. Better designed road-stream crossings also have the added benefit of accommodating flood events and ensuring static infrastructure is stronger and less prone to failure, thereby requiring less maintenance and repair and saving money.

Handbook of Road Ecology, First Edition. Edited by Rodney van der Ree, Daniel J. Smith and Clara Grilo.
© 2015 John Wiley & Sons, Ltd. Published 2015 by John Wiley & Sons, Ltd.
Companion website: www.wiley.com\go\vanderree\roadecology

INTRODUCTION

There are over 14,000 species of freshwater fish globally, and nearly all of them need to move in order to complete their life cycles (Eschmeyer 1998). Fish movements occur over a wide range of distances and timescales in order to access food resources, reach spawning grounds and use cover and security habitat when avoiding predators. The migration of fish is a well-known phenomenon worldwide, with some species travelling thousands of kilometres annually, including Atlantic salmon which migrate from salt water to fresh water and European eel from fresh to salt water. Many species migrate shorter distances, such as 10's to 100's of kilometre within a watershed or section of waterway, including pike and ide (Miller et al. 2001; Winter & Fredrich 2003). This chapter focuses on fish that spend at least a portion of their life in freshwater systems and provides solutions to minimise the effects of roads and road-stream crossings on fish and fish habitat. Other in-stream barriers to fish, such as low-head dams, are also discussed.

LESSONS

45.1   Roads and the vehicles that travel on them can negatively affect fish habitat and water quality

Roads and railways are often parallel or in close proximity to waterways (i.e. rivers, streams and creeks), and other waterbodies such as lakes or estuaries and in many cases cross them. This occurs in part because linear infrastructure often follows historic pathways along waterways. These pathways followed the easiest and most efficient routes between settlements. For example, it is easier to navigate through steep and mountainous terrain by following the paths carved out by rivers. Similarly, many towns were built along waterways to facilitate transportation of goods and services. However, this proximity can facilitate the transfer of pollutants from the road surface and vehicles into the aquatic system, with negative consequences. Storm-water run-off from roads and urban areas can change water quality, for example, pH and turbidity, and in some cases can be toxic (Kayhanian et al. 2008). The level of impact to water quality and potentially fish habitat is affected by the length of the dry period prior to a storm event and the intensity of the storm itself, both of which play a large part in the concentration of toxicants (Pitt et al. 1995).

A common practice in cold regions is to apply sand or some form of anti-icing or de-icing compound to the road surface in order to increase traction and remove ice and snow. When washed into waterways in sufficient quantities, small particles can clog the interstitial spaces between gravel in the riverbed, reducing the transfer of oxygen to buried fish eggs and thus affecting their development and survival (Williams 2006). Road run-off can also affect aquatic insect (macro-invertebrates) populations, which are an important food source for fish and an integral part of the aquatic ecosystem (Medeiros et al. 1983). Accidental chemical spills on roads do occur, and if the chemicals enter a waterbody in high enough concentrations, they can have deleterious effects on fish habitat and water quality, although few studies have documented the impact of individual spill events. Recently, there has been heightened awareness about the effect of dust particles from non-paved roads on aquatic and terrestrial habitat and organisms (Trombulak & Frissel 2000).

Roadways can alter surface water flow, groundwater flow and the interaction between them (Trombulak & Frissel 2000). Many waterways have been straightened or riparian habitats removed during road projects in order to simplify the construction process and reduce costs. In some cases, waterways have been converted to concrete-lined drains (Fig. 44.2B) or enclosed in pipes. Consequently, the functioning of the waterway is often compromised, with a concomitant reduction in quality of habitat for fish and other aquatic species (Trombulak & Frissel 2000).

45.2   Roads and other in-stream structures can be barriers to the movement of fish

Freshwater fish migrate longitudinally (upstream or downstream; e.g. salmon) and/or laterally from waterways to connected tributaries, lakes or wetlands (e.g. pike and bream; Fig. 45.1) over spatial scales ranging from metres to many thousands of kilometres depending on the species. Unfortunately, most waterways have man-made structures across them, such as culverts (Fig. 45.2) or low-head dams (Fig. 45.3A) that can restrict or prevent fish movement. Other structures that fragment fish populations include large dams for flood control, water storage or hydropower, as well as tidal barrages, shipping locks and sluices (Fig. 45.3B, C). By isolating habitats, barriers can decrease fish populations and in some cases contribute to the total loss of a species. For example, white-spotted char were absent upstream of dams at

(A)

(B)

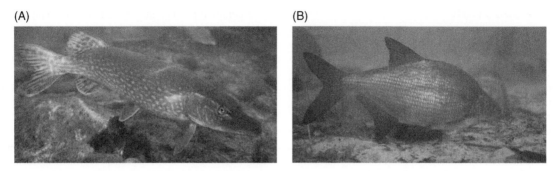

Figure 45.1 Adult pike (A) can travel 26 km/day as part of their annual migration, and adult bream (B) can migrate longitudinally and laterally. Source: Photographs by Fabrice Ottburg.

(A)

(B)

(C)

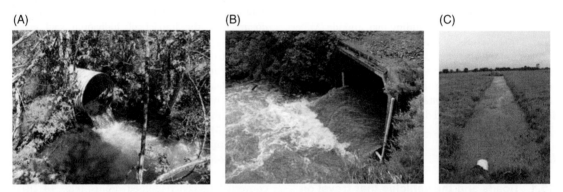

Figure 45.2 Culverts are barriers to fish movement when they are either (A) perched; (B) undersized, resulting in extreme water velocity; or (C) physically too small for large fish to pass. Source: (A and B) Photographs by Matt Blank; and (C) Photograph by Fabrice Ottburg.

(A)

(B)

(C)

Figure 45.3 Other examples of structures that are barriers to fish movement include (A) this low-head dam in Montana USA, (B) this large dam for hydropower in Canada and (C) this sluice to control water levels in North Brabant, the Netherlands. Source: (A and B) Photographs by Matt Blank; and (C) Photograph by Fabrice Ottburg.

17 of 52 study sites in Japan (Morita & Yamamoto 2002), and 4 minnow species were extirpated upstream of a dam in a prairie stream in the United States (Winston et al. 1991).

The swimming ability of a fish is influenced by many factors including species-specific ability, body size and condition, and the temperature, turbulence and dissolved oxygen content of the water (Liefferinge et al., 2005). Swimming ability is commonly categorised as either (i) 'sustained' where swim speeds are maintained for extended periods of time, usually greater than approximately 200 minutes; (ii) 'prolonged' where swim speeds are maintained for moderate periods, typically 15 seconds to 200 minutes; or (iii) 'burst' or 'sprint', where swim speeds are fast and maintained for short durations, typically less than 15 seconds (Hoar & Randall 1978). Culverts are a barrier to fish movement when they (i) are undersized and create excessive water velocity (Fig. 45.2B); (ii) have insufficient water depth; (iii) are physically too small for the species of fish (Fig. 45.2C); (iv) have large outlet drops, also called 'perched culverts' (Fig. 45.2A); and/or (v) are blocked by debris jams. In some cases, it is the combination of these conditions that make ordinarily benign conditions problematic. For example, a structure may create relatively fast water velocity that, by itself, may not be a barrier; however, the combination of fast and shallow water makes it more difficult for a fish to swim upstream. In addition to perched outlets, inlets can also restrict passage by causing the water level to drop (e.g. Ottburg & Th. de Jong 2006). Inlet drops occur when the culvert is undersized relative to the upstream channel and flows are severely restricted at the entrance.

45.3   Well-informed planning and design of roads can limit the impacts of roads on fish and fish habitat

The best solution to prevent pollution of aquatic systems from road-borne pollutants is to ensure roads are sufficiently distant from them. If this is not feasible, the next best approach to limit the impact of pollution from roads and traffic on water quality, habitat and fish is to provide a sufficient natural or semi-natural buffer between the two. A properly functioning buffer should capture, store and process the pollutants, rather than allowing them to enter the waterway. The size of the buffer (distance between waterway and road) will vary depending on the volume and condition of the stormwater run-off to be treated, vegetation, slope of the landscape, soil infiltration rates and numerous other site-specific factors. Engineering solutions, such as sediment traps, may be required in order to limit the effect of road pollutants. Other methods include modifying the type of chemicals used (e.g. non-polluting de-icing treatments) and their application rates in environmentally sensitive areas, as well as changes to vehicle designs and fuel to reduce the type and quantity of chemicals they produce. For example, some regions have policies which require the application rate of sand and salts to roads to be reduced or halted completely on and in close vicinity to bridges or culverts over waterways.

The optimal approach to maintain connectivity for fish and other aquatic species in waterways is to avoid constructing roads and other potential barriers across them. Where this cannot be avoided, road-stream crossings should be designed and constructed in a manner that allows long-term function of the waterway and its floodplain (Textbox 45.1, Lesson 45.4, Chapter 44). Long-span 'floodplain' bridges are the ideal but most costly solution in this regard and allow for natural stream dynamics, including the ability of the stream to migrate throughout the floodplain over time (Fig. 45.4). Other large structures that fully span the watercourse and banks provide the next best solution to ensure long-term passage of fish and are less expensive. In retrofit cases, it may be more feasible to remove the existing infrastructure and reinstate the natural system or replace it with a crossing structure designed using the stream-simulation approach (Lesson 45.4).

45.4   The most effective road–stream crossings for fish, when long-span 'floodplain' bridges are not an option, are culverts or shorter-span bridges that simulate the natural channel

The 'stream-simulation' approach is the preferred strategy to designing culverts or shorter-span bridges at road-stream crossings (see also Chapter 44) when long-span 'floodplain' bridges are not an option. Stream simulation allows aquatic and riparian processes to function without interruption through a road-stream crossing interface. Importantly, it provides passage for all life stages of fish and other aquatic species present in the system because the waterway channel is continuous through the crossing structure, and there are natural substrate, streambanks and, in some cases, vegetation. Bridges or large culverts (Figs 44.6, 45.4, 45.5 and 45.6) are used to either span the waterway and some bank and floodplain areas or allow placement of streambed, streambank and floodplain replicating materials within the structure. The US Forest Service has been instrumental in developing, applying and promoting this method, and many new and retrofitted road-stream crossings in the United States are now being designed using this approach (USDA 2008).

The stream-simulation approach relies upon the following basic principle:

> …designing crossing structures (usually culverts), that creates a structure that is as similar as possible to the natural channel.

Textbox 45.1 Steps in the planning, design and construction of road-stream crossings for fish.

There are many pieces of important information to gather and consider when designing a fish passage that effectively removes barriers and restores fish migrations. Some of this information includes:

Step 1. Carefully define the key planning parameters:

• Define the features and conditions of the river, including hydrology, hydraulic factors, geology, geomorphology, section profiles, substrate, type and amounts of debris.
• Describe the characteristics of the existing or proposed roadway.
• Identify all financial and legal requirements, such as permits and licences for construction.
• Identify the target species and other non-target species that may be affected, and river zone.
• Identify the ecological aims of the project, such as upstream and/or downstream connectivity for migration or other movements.

Step 2. Design the solution:

• The design must take into account the biological parameters of the target species as well as hydraulic and geomorphic criteria.
• Clearly specify the technical design criteria of the structure, such as its width, length, water flow, spacing and construction material.

Step 3. Construct and maintain the solution:

• Review the final design before construction commences, with a focus on relevant biological and hydraulic criteria (see also Chapter 9).
• Construction must include oversight by both the design engineer and consulting biologist.

• Develop and implement maintenance protocols, including a general description of the structure and its intended operation, as well as the frequency and methods for maintenance and any relevant health and safety issues (see also Chapter 17).

The following guiding principles will assist in deciding on the optimal approach to restore or maintain fish movement:

• The solution should allow sufficient movement by the target species during the specific life stage(s) where connectivity is required to maintain a viable population.
• The hydraulic function of the waterway should not be compromised by the solution.
• Crossings should be designed for all species that occur or could be expected to occur within the waterway.
• Avoid designing a single-species solution. Technical solutions are expensive and multi-species solutions are often more cost-effective.
• Where possible, the solution should also provide some biological functions (e.g. residence or nursery area for juvenile fish) for the target species.
• Waterways are part of a system and barriers upstream and downstream of the project site may also need improvement.
• The presence of native and non-native species can complicate the development of effective solutions. Consider if barriers should remain to prevent the movement of invasive species or if crossing structures should be designed to act as a barrier to protect native species.
• Ensure the solution does not negatively affect fish stock. For example, can fish reach their spawning grounds or migrate back to deeper water for the winter?

When channel dimensions, slope and streambed structure are similar, water velocities and depths also will be similar. Thus, the simulated channel should present no more of an obstacle to aquatic animals than the natural channel. (USDA 2008, Introduction, page xxiii)

An additional benefit of this approach is much greater flood conveyance because of its size (area of the opening) and span (width of the opening), resulting in a longer lifespan for the infrastructure with less risk of failure and need for costly replacement.

45.5 Specialty culverts and other technical solutions are possible but require careful design and do not provide all the qualities of uninterrupted natural waterways

Culverts are the traditional approach to road-stream crossings; however, many existing culverts are barriers to the movement of fish and other aquatic species (Lesson 45.2). If culverts are to be installed on new projects or existing culverts replaced, engineers and fish specialists must collaborate to ensure that the five major causes of the barrier effect of culverts for fish (Lesson 45.2) are avoided. Important culvert design parameters include

Figure 45.4 This large bridge across Merced River and floodplains, Hwy 49, at Old Bagby, Lake McClure Recreation Area, California, allows the waterway to function 'naturally', benefiting fish and other aquatic organisms as well as terrestrial wildlife. Source: Photograph by and reproduced with permission of Marcel Huijser.

Figure 45.5 This recently constructed crossing of Fleshman Creek, Montana, was designed using the stream-simulation approach. The culvert maintains the stream channel and bank lines through the entire structure, thus ensuring fish passage for most flows. The structure is designed to provide flood conveyance up to the 100 year event. Source: Photograph by Matt Blank.

water speed and depth through the culvert for the anticipated range of stream flows. In many cases, three flow rates, and associated passage parameters (such as water depth and velocity) based upon the species in the system, will be specified for the culvert design:

(i) A low-flow fish passage, typically the base flow of the stream or river. For the low-flow condition, the design should specify a minimum water depth based upon the requirements of the species of fish living in the stream, thus ensuring a depth barrier is not created.

(ii) A high-flow fish passage, typically similar to a bank-full flow rate or a 2-year recurrence interval flow (a 2-year flow is a peak flow that has a 50% chance

(A)

(B)

Figure 45.6 Pipe culverts (A) on the Clearwater River, Montana, were replaced with a bridge (B) to provide natural channel function and improve passage for all fish species and life stages. Source: Photographs by and reproduced with permission of Shane Hendrickson.

of occurring in any year). For the high-flow condition, where velocity is the main consideration, the design should specify a maximum water velocity based upon the weakest swimming fish species in the stream, ensuring a velocity barrier is not created.

(iii) A flood-flow fish passage typically has a 10–100-year recurrence interval flow. The type and size of the structure is influenced by many factors, including the risk to society if the structure fails, the cost to build the structure and the importance of the crossing to the ecosystem. The flood flow is evaluated against engineering criteria, primarily the height of the road deck relative to the water level during the flood event. Conditions in the culvert during these extreme flood flows are not evaluated relative to fish passage because they rarely occur.

(iv) If the culvert is intended to pass all fish, then the design should ensure that there is no outlet drop at any flow condition. If the culvert is intended to pass some fish species, but purposely limit others, as in the case of selective barriers for conservation of native species, an outlet drop can be used to achieve this. However, this approach requires a substantial amount of knowledge about the species of fish in the system, long-term effects of purposely creating a partial barrier and other factors.

(v) If the waterway transports large amounts of sediment, woody debris or ice, then a bridge or stream-simulation approach should be implemented to reduce the chances of the structure becoming blocked, causing the structure to fail.

Unfortunately, avoiding waterways, removing structures or using large open-span bridges is not always feasible, and other technical solutions are required.

While many technical and semi-natural solutions have been employed around the world (Gough et al. 2012), they are usually less effective than avoidance or use of large crossing structures. One of the most common approaches is that of the fish ladder or fishway (Clay 1995), which are structures designed to allow the upstream (and sometimes downstream) movement of fish over or through a barrier by modifying the velocity and depth of water to provide conditions suitable for fish movement (Clay 1995).

There are numerous fish ladder designs (Fig. 45.7), with each installation designed specifically for the location, target species, water conditions and available space. The most common types are (i) pool weir or step-pool, both of which have a series of steps, sometimes with an orifice in the base of each baffle, which are effective for jumping species; (ii) vertical slot, which uses a series of baffles to create narrow slots that control water velocity and depth of pools between slots; (iii) Denil, often described as roughened ramps which create a series of small rapids that allow many fish species to pass; (iv) fish lock, where fish accumulate in a holding area at the base of a weir which is periodically sealed and filled with water, allowing fish to swim through the lock; and (v) bypass, where a stream channel that resembles natural conditions is constructed around the barrier. Other approaches include (i) fish lifts or 'trap and transport' techniques, which mechanically lift fish up and over the barrier; and (ii) modifications to infrastructure, such as pumps that injure or kill fish, to be less damaging. While these strategies are typically installed to bypass dams and weirs, some can be installed within culverts to reduce water velocity, such

(A)

(B)

(C)

Figure 45.7 A series of step-pools or pool weirs (A), De Wit fish ladder inside the rectangular steel box (B) and semi-natural bypass (C) are used on rivers in the Netherlands to bypass barriers to fish movement. The water velocity through fish ladders, culverts and other structures must not exceed the swimming speed of any life stage or species living within the waterway. In lowland streams common to flat landscapes, water velocities < 1.0 m/s are typical, and most adult fish species are capable of moving through passages with such conditions. For subadults or juvenile fish, water velocities should be < 0.8 m/s or even < 0.4 m/s for some species (Jens 1982; Liefferinge et al. 2005; Gough et al. 2012). Source: Photographs by Fabrice Ottburg.

as a series of baffles or ropes to restore fish movement (David et al. 2014). Baffles come in a wide range of configurations and sizes (Ead et al. 2002), and some disadvantages include maintenance difficulties and potentially restricted passage for some fish (Liefferinge et al. 2005).

CONCLUSIONS

Roads intersect waterways in nearly every terrestrial environment on the planet and have the potential to restrict the movement of fish, often resulting in population declines and local extinctions. Where possible, avoiding water bodies altogether or at least providing a sufficient natural or semi-natural buffer between the road and the waterbody is the best approach to reduce impacts on fish and fish habitat. Where unavoidable, road-stream crossings should be properly designed, constructed and maintained to reduce the barrier to movement effect on fish and other aquatic species. The best approach for road-stream crossings is to design and build a bridge that fully spans the entire floodplain, allowing the waterway to function 'naturally'. If this is not possible, one proven and effective strategy to maintain fish passage at all types of road-stream crossings is to use a stream-simulation approach. This approach has the added benefit of accommodating larger flow events with reduced impact on the static infrastructure, such as the roadway. A system-wide approach that addresses aquatic species needs, riparian function and large flood events will produce road-stream crossings that will last longer and require less maintenance and replacement.

FURTHER READING

Clay (1995): A comprehensive book that presents the full range of engineering solutions for fish passage past barriers ranging from large hydropower dams to roadway culverts.

IECA (2014): The International Erosion Control Association is a non-profit, member organisation that provides education, resource information and business opportunities in the erosion and sediment control industry. The EICA has over 2500 members from 30 countries aiming to solve a broad range of problems caused by soil erosion and its by-product – sediment.

USDA (2008): The US Forest Service presents a road–stream crossing design approach that mimics natural channel and floodplain function through the crossing. The report has excellent examples, drawings and figures.

REFERENCES

Clay, C. 1995. Design of Fishways and Other Fish Facilities, 2nd ed. CRC Press. Boca Raton, FL.

David, B.O., J.D. Tonkin, K.W.T. Taipeti and H.T. Hokianga. 2014. Learning the ropes: Mussel spat ropes improve fish and shrimp passage through culverts. Journal of Applied Ecology **51**:214–223.

Ead, S.A., N. Rajaratnam and C. Katopodis. 2002. Generalized study of hydraulics of culvert fishways. Journal of Hydraulic Engineering **128**:1018–1022.

Eschmeyer, W.N. 1998. Catalog of Fishes, Vol. I and 2. California Academy of Sciences. San Francisco, CA.

Gough, P., P. Philipsen, P.P. Schollema and H. Wanningen. 2012. From Sea to Source; International Guidance for the Restoration of Fish Migration Highways. Regional Water Authority Hunze en Aa's, Veendam, the Netherlands.

Hoar, W.S. and D.J. Randall eds. 1978. Fish Physiology, Vol. 7: Locomotion. Academic Press, New York.

IECA. 2014. International erosion control association. www. ieca.org. Accessed 24 April 2014.

Jens, G. 1982. Der Bau von fischwegen. Fischtreppen, Aalleiturn und Fischschleusen. Verlag Paul Parey, Hamburg und Berlin. 96 p.

Kayhanian, M., C. Stransky, S. Bay, S.L. Lau and M.K. Stenstrom. 2008. Toxicity to urban highway runoff with respect to storm duration. Science of the Total Environment **389**:386–406.

Miller, L.M., L. Kallemeyn and W. Senanan. 2001. Spawning-site and natal-site fidelity by northern pike in a large lake: mark-recapture and genetic evidence. Transactions of the American Fisheries Society **130**:307–316.

Liefferinge, C., P. Meire, B. Jacobs, D. van Erdeghem, J.H. Kemper and F.T. Vrieze. 2005. [in Dutch]. Vismigratie. Een handboek voor herstel in Vlaanderen en Nederland. Uitgave van de Vlaamse Gemeenschap. ANIMAL, Afdeling Water. Brussel.

Medeiros, C., R. LeBlanc and R.A. Coler. 1983. An in situ assessment of the acute toxicity of urban runoff to benthic macroinvertebrates. Environmental Toxicology **2**:119–126.

Morita, K. and S. Yamamoto. 2002. Effects of habitat fragmentation by damming on the persistence of stream-dwelling charr populations. Conservation Biology **16**:1318–1323.

Ottburg, F.G.W.A. and Th. de Jong. 2006. [in Dutch]. Fishes in polder ditches; the influence of dredging in 'isolated' and open ditches on fresh water fish and amphibians. Alterra-report 1349. Wageningen, Alterra.

Pitt, R., R. Field, M. Lalor and M. Brown. 1995. Urban storm-water toxic pollutants: assessment, sources, and treatability. Water Environment Research **67**:260–275.

Trombulak, S.C. and C.A. Frissel. 2000. Review of ecological effects of roads on terrestrial and aquatic communities. Conservation Biology **14**:18–30.

United States Department of Agriculture (USDA). 2008. Stream Simulation: An Ecological Approach to Providing Passage for Aquatic Organisms at Road-Stream Crossings. USDA, National Technology Development Program, San Dimas, CA

Williams, R.N. ed. 2006. Return to the River: Restoring Salmon to the Columbia River. Elsevier Academic Press, Burlington, MA.

Winston, M.D., C.M. Taylor and J. Pigg. 1991. Upstream extirpation of four minnow species due to damming of a prairie stream. Transactions of the American Fisheries Society **120**: 98–105.

Winter, H.V. and F. Fredrich. 2003. Migratory behaviour of ide: A comparison between the lowland rivers Elbe, Germany, and Vecht, the Netherlands. Journal of Fish Biology **63**: 871–880.

Chapter 46

THE FUNCTION AND MANAGEMENT OF ROADSIDE VEGETATION

Suzanne J. Milton[1], W. Richard J. Dean[1], Leonard E. Sielecki[2] and Rodney van der Ree[3]

[1]DST/NRF Centre of Excellence at the Percy FitzPatrick Institute of African Ornithology, University of Cape Town, South Africa, and Renu-Karoo Veld Restoration cc, Prince Albert, South Africa

[2]British Columbia Ministry of Transportation and Infrastructure, Victoria, British Columbia, Canada

[3]Australian Research Centre for Urban Ecology, Royal Botanic Gardens Melbourne, and School of BioSciences, The University of Melbourne, Melbourne, Victoria, Australia

SUMMARY

The structure and composition of roadside vegetation vary from frequently mown grass to shrubs and trees and from artificial landscaping to natural plant communities. Roadside vegetation can perform many important functions, including the provision of habitat for rare plants and animals, a source of seeds for adjacent landscapes, a buffer to reduce the penetration of traffic noise and light, carbon sinks and enhanced aesthetics for road users. In certain situations, roadside vegetation can have negative effects, such as attracting wildlife and increasing rates of wildlife-vehicle collisions (WVC), creating movement corridors for weeds and invasive species, obscuring road signs and damaging road surfaces.

46.1 Roadsides can support rare and threatened species of plants and animals, and these should be managed for conservation.

46.2 Vegetation that reduces visibility or poses a traffic hazard should be managed to achieve a compromise between safety and biodiversity conservation.

46.3 Roadside habitats may act as ecological traps: It is preferable to recreate offset or compensation habitats away from roadsides.

46.4 The drainage of roads and roadsides must be designed to minimise impacts on adjacent vegetation and habitats.

46.5 Never plant invasive species (environmental weeds) along roads: Use plants native to the region for roadside soil stabilisation, shade, ornamental planting and control of noise and light pollution.

46.6 Perennial vegetation cover and ongoing management of roadside vegetation are required to control the continuous threat of weed invasion.

Handbook of Road Ecology, First Edition. Edited by Rodney van der Ree, Daniel J. Smith and Clara Grilo.
© 2015 John Wiley & Sons, Ltd. Published 2015 by John Wiley & Sons, Ltd.
Companion website: www.wiley.com\go\vanderree\roadecology

46.7 The reduction of fuel loads on roadsides should be compatible with biodiversity management objectives.

The challenge for management is to comprehensively quantify and understand the role and values of roadside vegetation and manage roadsides to enhance their positive impacts and reduce their negative effects.

INTRODUCTION

Most roads, especially those designed and built in recent years, include a strip of land on one or both sides of the road that remains undeveloped. These strips of land (i.e. roadsides or verges) are usually owned and managed by the government or department of transportation. The width of the roadside depends on the tenure, land use and competing demands (and hence purchase price) of the land at the time it was acquired for the road, the predicted growth in traffic volume (and hence need for future expansion) and the terrain (mountainous vs. flat). The uses and functions of roadsides are diverse, and they usually support natural or planted vegetation, footpaths or bicycle paths, utility infrastructure (e.g. power lines and pipelines) and safe places for vehicles to pull off the road. Roadside vegetation may play numerous other roles, including the provision of habitat for plants and animals; intercepting and buffering adjacent landscapes from noise, dust, light and other pollutants; controlling soil erosion; and improving aesthetics.

Roadside management is challenging because of the diversity of uses and often competing goals. For example, the provision of habitat for biodiversity along roads by maximising the amount of habitat (e.g. large trees) is often at odds with driver safety due to the risk of collision. Consequently, vegetation that remains, re-establishes itself or is planted on roadsides will always be influenced by the primary function of the road – that is, safe and rapid traffic flow. The aim of this chapter is to highlight ways in which roadside vegetation may influence the integrity of the road; the safety and comfort of motorists and adjacent settlements; the conservation of plants, wildlife and genetic resources; and the spread of fires and invasive weeds. We present general guidelines for roadside vegetation management that consider issues of road integrity, road user safety, biodiversity conservation and aesthetics.

LESSONS

46.1   Roadsides can support rare and threatened species of plants and animals, and these should be managed for conservation

Large areas of natural and semi-natural vegetation and wetlands occur in road reserves, potentially providing habitat or movement corridors for a wide range of plants and animals (Saunders & Hobbs 1991; Lamont & Atkins 2000). In highly cleared landscapes, roadsides can provide the majority of available habitat (e.g. van der Ree 2002). Fragments of natural vegetation along roadsides also have the potential to protect rare species, habitats and ecosystem functions (e.g. pollinators) that have been lost from crop monocultures (Figs. 46.1, 46.2 and 46.3) or that are threatened by urban development, mining, overgrazing and other forms of land transformation. In some situations, the quality of the habitat in roadside fragments may exceed that which is remaining in larger patches (e.g. van der Ree & Bennett 2001).

Roadside conservation requires identification and mapping of plant populations or communities of significant conservation value and integration of such maps with the infrastructure maps used for road planning and maintenance (Connor & Ralph 2006; Johnson 2008; Chapter 17). Depending on the vulnerability of the target plants to theft or vandalism, the sensitive section of the roadside could be demarcated with signs that provide information to the road user or

Figure 46.1 Contrast between density of flowering plants in the grazed rangeland (to the left of the fence) and ungrazed roadside (to the right of the fence) in the South African Karoo. Source: Photograph by S. J. Milton.

Figure 46.2 Fragment of threatened Renosterveld vegetation lies between cropland and the weedy mowed shoulder of the road near Wellington, South Africa. Source: Photograph by and reproduced with permission of Clement Cupido.

Figure 46.3 Woodland along roads and streams in this agricultural area in south-east Australia accounts for over 85% of the remaining woodland cover in this district. Source: Photograph by R. van der Ree.

marked only on maps used by road authorities. Conservation of rare plants and habitat on roadsides may require special management, for example, plant species that require fire for regeneration may need to be burned periodically (Johnson 2008).

46.2   Vegetation that reduces visibility or poses a traffic hazard should be managed to achieve a compromise between safety and biodiversity conservation

Tall or dense vegetation adjacent to roads can pose a traffic hazard by obscuring traffic, road signs, wildlife and pedestrians. This vegetation may need to be pruned or mowed to improve driver visibility (Forman & Alexander 1998; Johnson 2008) or to give drivers more time to observe and respond to animals that cross the road (see also Lesson 53.3). Non-frangible vegetation may be a safety hazard for out-of-control vehicles and can either be removed or safety barriers (e.g. guard rail or wire rope fencing) installed to prevent collisions. Guard rail is effective at preventing vehicle–tree collisions; however, it is expensive and doesn't solve the issue of driver visibility. Unfortunately, cutting of shrubs and herbaceous vegetation can result in a flush of new growth that may attract herbivores (e.g. bears; Textbox 46.1; Chapter 42) which increases the probability of WVC. Although the provision of diversionary feeding, forage repellents, establishment of unpalatable species or vegetation clearing along roads could reduce this problem (Rea 2003), these interventions are costly, with unproven effectiveness, and may have negative effects on rare plant communities or adjacent vegetation (e.g. facilitating the spread of invasive species). Vegetation in the median can improve safety by reducing headlight glare (Forman & Alexander 1998), but if vegetation is non-frangible, the median will need to be sufficiently wide and include safety barriers to protect motorists on high-speed roads.

Roadside vegetation must be managed to achieve both safety and conservation requirements; therefore, we recommend the use of parallel management zones. This means the priority for management of vegetation immediately adjacent to the road should be on driver safety, while areas outside the clear zone should focus on conservation (e.g. Fig. 39.4). Furthermore, where high-conservation-value vegetation exists close to the road, the use of safety barriers should be used to protect motorists and conserve plants. The amount of effort invested in conserving and managing rare or threatened species in roadside habitats should increase in proportion to its rarity and value.

46.3   Roadside habitats may act as ecological traps: It is preferable to recreate offset or compensation habitats away from roadsides

Roadside habitats may function as ecological traps that attract animals and increase rates of WVC and mortality. Wetlands are often built adjacent to new and existing roads to provide fill for the road construction and to receive and treat storm-water run-off before it enters waterways or groundwater recharge areas. In some instances, road agencies intentionally build wetlands immediately adjacent to the road as an offset for wetlands that were destroyed during construction (Textbox 31.2). These attract amphibians and birds and are likely to contribute to the millions of such animals killed on roads (Forman & Alexander 1998). As a rule of thumb, it is probably better to provide offset habitats away from the road to reduce the rate of mortality and avoid the road-effect zone (Chapter 7). However, it is essential to clearly understand the specific impact of the road on the species of concern and not just assume that roadside habitats are either good or bad for all species. For example, some species of wildlife that occupy roadside habitats may suffer low rates of WVC and mortality because they avoid the road surface or avoid traffic.

46.4   The drainage of roads and roadsides must be designed to minimise impacts on adjacent vegetation and habitats

Roads alter the flow of water by bisecting and damming waterways and by collecting and discharging surface water via channels or pipes (Chapters 44, 45 and 47). Roads that intersect streams and wetlands are often elevated on bridges or contain culverts and pipes to protect the road from flooding and damage. These roads have the potential to alter rates and patterns of water flow, depths and chemistry (Coffin 2007; Chapter 44), and adequate provision for water flow is essential to ensure that downstream wetlands do not dry out, resulting in habitat degradation or loss.

Culverts that drain water away from the road surface should discharge run-off into natural waterways or wetlands but only via treatment or retention ponds to filter out pollutants and sediment loads. Foreign liquids spilt on the road (e.g. oil, fuel, milk, sewerage), and seeds transported on vehicles find their way via natural waterways or overland flow to rivers, wetlands and groundwater. Drainage infrastructure can cause soil erosion or development of artificial wetlands where natural waterways are absent or unable to cope with peak water flows (Coffin 2007). Erosion can be reduced by installing drainage systems that mimic natural flows and facilitate absorption of water into the ground and by the strategic planting of vegetation to bind the soil (Chapter 44). Modifications of the physical and chemical properties of soil and hydrology caused by roads and drainage works can change the structure, composition and nutritional value of natural vegetation and promote weed growth, particularly in arid and semi-arid regions (Martinez & Wood 2006; Chapter 47).

46.5   Never plant invasive species (environmental weeds) along roads: Use plants native to the region for roadside soil stabilisation, shade, ornamental planting and control of noise and light pollution

Roadside rest areas, often providing shade and shelter near scenic lookouts, are a feature of national and international highways around the world. Native trees and shrubs adapted to the local conditions should be used for plantings in these areas to reduce maintenance costs and as an opportunity to educate motorists. Species that become environmental weeds when planted outside their natural ranges (e.g. Peruvian pepper, mesquite, black locust, black wattle, sugar gum) should be avoided because of the risk of invasion into adjacent habitats (Forman & Alexander 1998; The University of Queensland 2008; Milton & Dean 2010; Figs. 46.4 and 46.5). Most regions have lists of declared weeds, and these should be consulted to identify which floral species should be avoided.

Near-continuous noise produced by major roads has a negative effect on the quality of life of people, reduces the value of adjacent properties and affects vocal communication in many species of wildlife (Chapters 19, 33 and 34). Trees, hedges or vegetation-covered barriers can reduce high-frequency sound by 40% (Kalansuriya et al. 2009), but this approach may not be practical or ecologically acceptable if the species to be planted is invasive. Similarly, soil embankments of roads in steep topography or shoulders of roads through sand dunes are often stabilised with grasses (e.g. fountain grass, marram), shrubs (bramble) or ground covers (highway iceplant) which must be carefully selected to minimise the risk of invasion into adjacent natural and agricultural ecosystems.

46.6   Perennial vegetation cover and ongoing management of roadside vegetation are required to control the continuous threat of weed invasion

Roadside management that maintains an intact community of native perennial vegetation reduces the risk of weed establishment and slows the rate of its spread. In contrast, the clearing of natural vegetation and repeated grading to bare soil promote the establishment of weeds due to increased disturbance, creation of open areas for colonisation and availability of water and nutrients via run-off from the road surface (USDA 2003; Figs 46.4 and 46.5). Furthermore, pollutants including salt applied to road surfaces to reduce ice in cold climates contribute to the development of weedy vegetation adjacent to the road surface (Truscott et al. 2005). The constant supply of weed seeds carried by vehicles (Taylor

Figure 46.4 Blue morning glory is an aggressive climber and has smothered this broadleaf–podocarp forest along a roadside near Auckland, New Zealand. Source: Photograph by and reproduced with permission of Margaret Stanley.

Figure 46.5 Pampas grass, originally from South America, has invaded this roadside in the Hunua Ranges near Auckland, New Zealand. Source: Photograph by and reproduced with permission of Margaret Stanley.

Textbox 46.1 Roadside vegetation management to protect black bears in British Columbia, Canada.

The Sea to Sky Highway (Highway 99) in British Columbia, which connects Vancouver to Whistler, passes through the Pacific Ranges of the Coast Mountains of North America. Bears emerging from hibernation in spring 2010 were unable to find sufficient food at higher elevations because the melting of the mountain snowpack was delayed due to an unseasonably late and cold spring. Consequently, a large number of black bears grazed on vegetation in the highway verge where vegetation planted the previous year was emerging (Fig. 46.6). Highway verges typically green up earlier in the spring than adjacent areas because they are exposed to more sunlight.

Over the spring and summer, intense grazing by the bears altered the planned succession of rights-of-way plant species. Clover, one component of the seed mix, was intended to quickly stabilise soil, fix nitrogen, hinder the establishment of invasive plant species and then be outcompeted by the other species in the mix. The clover successfully withstood overgrazing by the bears and flourished, providing a high-quality food source for the bears. A number of bears were injured and killed due to collision with vehicles, and 'bear jams' began occurring on the highway in the District of Squamish when motorists slowed down and stopped near the bears. Motorists unfamiliar with wildlife risked their safety by leaving their vehicles to photograph the bears.

The British Columbia Ministry of Transportation and Infrastructure worked with its maintenance contractors, the Get Bear Smart Society and the District of Squamish to reduce the potential for human/bear conflict along the Sea to Sky Highway. An extensive programme was developed which included new seasonal warning signs, changeable message signs, access restriction and innovative vegetation management. Native plant species known to be unattractive to bears were planted to shade out and displace clover over time and/or provide visual barriers to reduce the number of bear jams.

Figure 46.6 Motorists along the Sea to Sky Highway (Highway 99) in British Columbia, Canada, stop to view and photograph black bears grazing on clover on the highway verge. Source: Photograph by and reproduced with permission of Sylvia Dolson.

---

et al. 2012) or dispersed from adjacent disturbed landscapes (Sullivan et al. 2009) exacerbates the risk of establishment and spread. Generalist birds, such as omnivorous corvids scavenging roadkills, will also disperse seeds of fleshy-fruited plants to roadsides under their roosts on fences, poles and signs. Once established on roadsides, invasive alien plants and agricultural weeds are further spread via overland water flows and culverts to rivers that intersect roads (Rahlao 2010). Through regular monitoring and the use of mechanical and chemical methods, road management authorities should better control weeds of environmental and agricultural significance along roadsides (Johnson 2008).

46.7 The reduction of fuel loads on roadsides should be compatible with biodiversity management objectives

Reduction in woody vegetation cover to improve visibility along roads creates suitable conditions for development of tall grass that can increase the risk of fire spread and exacerbate fragmentation in forested environments (Coffin 2007). Fires can be caused by ignition of dry roadside vegetation by vehicle accidents, sparks from engines or exhausts, discarded cigarettes and cooking fires at roadside pull-offs, potentially posing a hazard to passing traffic. Managers may plough or poison roadside vegetation to prevent roadside fires from destroying adjacent crops, plantations, rangelands, conservation areas or suburbs. Such practices are often in conflict with roadside conservation goals, especially for threatened but fire-prone vegetation, such as Chaparral, Fynbos and Kwongan.

Best-practice guidelines for roadside fuel reduction include the removal of dead trees and mowing, controlled burning or cutting of grass and brush in areas where fire risk is high; however, environmental authorisation should be required for use of herbicides and grading or for any intervention in high-conservation-value or specially protected vegetation or where it may impact threatened wildlife (Lamont and Atkins 2000; Johnson 2008).

CONCLUSIONS

Roadside vegetation can perform many roles, including pollution reduction, erosion control and aesthetics. Vegetation can pose risks to drivers by obscuring oncoming vehicles, signs and large animals approaching the road. Trees and other non-frangible vegetation can also be a hazard for out of control vehicles that leave the road. Roadsides can also aid in the conservation of plant and animal species and ecological communities in highly cleared or modified landscapes. While human safety is critical, there are viable and cost-effective alternatives to the destruction of vegetation. In many cases, vegetation can be maintained, while road safety is improved through reductions in traffic speed and the installation of crash barriers. In the case of rare or threatened plant communities, the rarity and value of that plant community should trigger a more detailed analysis of management options and a willingness to do more to protect it. Minimising damage to existing roadside vegetation also reduces the effort required to manage the weeds that colonise disturbed ground.

ACKNOWLEDGEMENTS

The WWF Table Mountain Fund through the Wildlife and Environment Society of South Africa and the Rufford Small Grants for Nature Conservation (United Kingdom) supported research by Suzanne Milton and Richard Dean on roadside vegetation management. Rodney van der Ree was supported by The Baker Foundation.

FURTHER READING

Dean and Milton (2000): Provides evidence that roadside furniture leads to predictable patterns in the dispersal and distribution of animal-dispersed plant species.
DECWA (2009): Useful guidelines for roadside vegetation management.
Harper-Lore and Wilson (2000): A useful resource for North America and possible model for other regions on roadside use of native plants.
Milton and Dean (2010): Discusses role of roads in arid areas in facilitating the spread of invasive alien plant species.

REFERENCES

Coffin, A.W. 2007. From roadkill to road ecology: a review of the ecological effects of roads. Journal of Transport Geography **15**:396–406.
Connor, P. and M. Ralph. 2006. Community Roadside Management Handbook. Yarriambiack Shire Council, Victoria, Australia, 20 pp. Available from http://www.yarriambiack.vic.gov.au/media/uploads/Community Handbook.pdf (Accessed on 26 September 2014).
Dean, W.R.J. and S.J. Milton. 2000. Directed dispersal of *Opuntia* species in the Karoo, South Africa: are crows the responsible agents? Journal of Arid Environments **45**:305–314.
Department of Environment and Conservation of Western Australia (DECWA), 2009. Declared Rare Flora and Road Maintenance. DECWA. Available from http://www.dpaw.wa.gov.au/images/documents/conservation-management/off-road-conservation/rcc/marking_roadside_declared_rare_flora.pdf (Accessed on 18 January 2015).
Forman, R.T.T. and L.E. Alexander. 1998. Roads and their major ecological effects. Annual Review of Ecology and Systematics **29**:207–231.
Harper-Lore, B.L. and M. Wilson (eds.). 2000. Roadside Use of Native Plants. Island Press, Washington, DC. 665 p.
Johnson, A. 2008. Best Practices Handbook for Roadside Vegetation Management. Minnesota Department of Transportation, Office of Research Services, St. Paul, MN. Available from http://www.ttap.mtu.edu/publications/2009/BestPracticesHandbookonRoadsideVegetation Management2008.pdf (Accessed on 26 September 2014).
Kalansuriya, C.M., A.S., Pannila and D.U.J. Sonnadara. 2009. Effect of roadside vegetation on the reduction of traffic noise levels. Proceedings of the Technical Sessions **25**:1–6.
Lamont, D.A. and K.J. Atkins. 2000. Guidelines for Managing Special Environmental Areas in Transport Corridors.

Roadside Conservation Committee, Kensington, Western Australia

Martinez, J.J. and D. Wood. 2006. Sampling bias in roadsides: the case of galling aphids on Pistacia trees. Biodiversity and Conservation **15**:2109–2121.

Milton, S.J. and W.R.J. Dean. 2010. Plant invasions in arid areas: special problems and solutions – A South African perspective. Biological Invasions **12**:3935–3948.

Rahlao, S.J. 2010. The distribution of invasive *Pennisetum setaceum* along roadsides in western South Africa, the role of corridor interchanges. Weed Research **50**:537–543.

Rea, R.V. 2003. Modifying roadside vegetation management practices to reduce vehicular collisions with moose *Alces alces*. Wildlife Biology **9**:81–91.

Saunders, D.A. and R.J. Hobbs (eds.). 1991. Nature Conservation 2: the Role of Corridors. Surrey-Beattie, Chipping Norton, Australia.

Sullivan, J.J., P.A. Williams, S.M. Timmins and M.C. Smale. 2009. Distribution and spread of environmental weeds along New Zealand roadsides. New Zealand Journal of Ecology **33**:190–204.

Taylor, K., T. Brummer, M.L. Taper, A. Wing and L.J. Rew. 2012. Human-mediated long-distance dispersal: an empirical evaluation of seed dispersal by vehicles. Diversity and Distributions **18**:942–951.

Truscott, A.M., S.C.F. Palmer, G.M. McGowan, J.N. Cape and S. Smart. 2005. Vegetation composition of roadside verges in Scotland: the effects of nitrogen deposition, disturbance and management. Environmental Pollution **136**:109–18.

The University of Queensland. 2008. Environmental Weeds of Australia. Special edition for Biosecurity Queensland. Available from http://keyserver.lucidcentral.org/weeds/data/03030800-0b07-490a-8d04-0605030c0f01/media/Html/search.html?zoom_query=Environmental&zoom_per_page=10&zoom_and=1&zoom_sort=0 (Accessed on 18 January 2015).

USDA. 2003. Backcountry Road Maintenance and Weed Management. Report 0371-2811-MTDC. USDA Forest Service, West Missoula, MT.

van der Ree, R. 2002. The population ecology of the Squirrel Glider *Petaurus norfolcensis*, within a network of remnant linear habitats. Wildlife Research **29**:329–340.

van der Ree, R. and A.F. Bennett. 2001. Woodland remnants along roadsides - a reflection of pre-European structure in temperate woodlands? Ecological Management and Restoration **2**:226–228.

Chapter 47

ROADS IN THE ARID LANDS: ISSUES, CHALLENGES AND POTENTIAL SOLUTIONS

Enhua Lee[1], David B. Croft[2] and Tamar Achiron-Frumkin[3]

[1]Eco Logical Australia, Sutherland, New South Wales, Australia
[2]School of Biological, Earth and Environmental Sciences, University of New South Wales, Care of Post Office, Adelaide River, Northern Territory, Australia
[3]Ecological and Environmental Consultant, Mevasseret Zion, Israel

SUMMARY

Roads in arid ecosystems present unique challenges for road planners and managers, as these ecosystems are highly sensitive to changes in water flows. Water from rainfall is the major force shaping these ecosystems and drives productivity. Roads concentrate water from rain or condensation at their edges as water runs off their surfaces. In arid ecosystems, this water-shedding generally results in an attraction for a number of animal groups, and as a consequence, wildlife-vehicle collisions (WVC) and roadkill are frequent. Arid landscapes are naturally open, and so roads generally do not fragment habitat and/or act as a barrier to animal movement to the same extent as in forested or other more densely vegetated ecosystems. Simple and effective solutions can be deployed to reduce the rate of roadkill, but care must be taken to avoid introducing other road impacts not currently issues for the majority of arid-zone roads. Highly engineered multi-lane roads in arid lands have additional barrier effects and influence water flows at significantly greater scales than the typical roads servicing the small populations of these regions.

47.1 The majority of roads in arid ecosystems act as an attractant for animals, which may increase animal densities around roads.

47.2 The attraction to arid-zone roads can lead to an increased incidence of WVC.

47.3 Roads in arid zones that are highly engineered, multi-lane and/or built on embankments cause additional effects, including habitat fragmentation.

47.4 Minimising the scale of microhabitat differences at the edges of arid-zone roads, or lengthening the distance over which water is shed, may reduce their attractiveness and lessen their impact.

Handbook of Road Ecology, First Edition. Edited by Rodney van der Ree, Daniel J. Smith and Clara Grilo.
© 2015 John Wiley & Sons, Ltd. Published 2015 by John Wiley & Sons, Ltd.
Companion website: www.wiley.com\go\vanderree\roadecology

47.5 New roads and upgrades to existing roads in arid environments must consider landscape features, topography and animal species likely to be present and be designed to maintain landscape function which is easily disturbed in this environment.

The arid lands represent a significant challenge to road builders as impacts to ecosystems and wildlife are easily manifested, but costs of mitigation are high relative to the number of road users. Roads are typically traversed at high speed between distant population centres and are best threaded through zones of prior heavy human use (e.g. stock or caravan routes) rather than pristine habitat.

INTRODUCTION

Arid (0–300 mm annual rainfall) and semi-arid (300–600 mm annual rainfall) lands are present on every continent and cover approximately one third of the earth's land surface (Kinlaw 1999). They are characterised by low rainfall and the presence of drought-adapted plants and unique wildlife that are unevenly distributed in the landscape and highly sensitive to disturbance. Given their extent and sensitivity, appropriate and informed management of arid-zone roads and other linear infrastructure is of major importance. The many roads running through arid and semi-arid areas have the potential to negatively impact specialised plant communities and unique wildlife which have low resilience to man-made disturbances and landscape degradation. However, strategic management of the impacts of arid-zone roads is hampered by a lack of detailed information on the interaction between the ecosystems traversed by roads and road effects. The majority of road ecology investigations have been undertaken in temperate regions where impacts are likely to be different.

The scale of road effects is likely to vary among ecosystems due to the different ways that roads interact with their ecological processes (Brooks & Lair 2005). For example, water run-off from road surfaces is likely to have greater effects on plant growth near roads in arid ecosystems as compared with more mesic (wetter) environments because water limits primary productivity in arid areas (Westoby 1980). Effects of enhanced plant growth along roads in arid ecosystems may subsequently affect animal distributions and abundance more so than in mesic environments, due largely to the patchier distributions and more limited amounts of food resources in arid regions (Stafford Smith & Morton 1990). Given the open nature of arid landscapes, typical single-lane roads in arid environments are less likely to act as barriers to animal movement compared to similar roads in forested environments. Animals that live in arid areas are adapted to traversing open

areas lacking cover to access resources present in patches unevenly distributed in the landscape. However, highly engineered, multi-lane arid-zone roads may nonetheless be barriers.

This chapter outlines the unique challenges for low-formation, single-lane, sealed roads typical in many arid landscapes. These are generally less than 10 m wide and lack features such as bridges, median strips and crash barriers. This chapter discusses simple and effective solutions, based on sound ecological principles, which can be easily deployed to manage arid-zone roads and minimise impacts on animals. We also outline the impacts and challenges for other types of arid-zone roads, such as those that are multi-lane, elevated on embankments or are highly engineered.

LESSONS

47.1 The majority of roads in arid ecosystems act as an attractant for animals, which may increase animal densities around roads

The condition of plant species and the response of animal species to arid-zone roads vary according to their specific habitat preferences and, in the case of animal species, also their behaviour (Brooks & Lair 2005; Lee & Croft 2009). However, roads in arid ecosystems generally increase the cover, quality and productivity of plants in the road verge, particularly in the area directly adjacent to the road where water runs off the road surface (Johnson et al. 1975; Lightfoot & Whitford 1991; Norton & Stafford Smith 1999; Brooks & Lair 2005), and therefore act as an attractant rather than deterrent for most species of wildlife. This is because water limits productivity in arid ecosystems and, where present, drives plant growth and animal distributions (Textbox 47.1; Fig. 47.1).

A range of animal groups are attracted to arid-zone roads and their verges for the increased

Textbox 47.1 A study of vegetation quality in relation to an arid-zone road and influencing factors.

Lee (2006) investigated the effects of an arid-zone road in inland Australia on vegetation quality and the factors that influenced vegetation quality. Five sites along the road verge and into the hinterland were investigated for 2 years in each of a hilly and floodplain landscape. Data on 'greenness' (a measure of quality) of vegetation classified into five categories (grasses, forbs, copper burrs, flat-leafed chenopods and round-leafed chenopods) were collected at survey sites. Soil moisture, chemistry and compaction data were also collected.

The study found that vegetation 'greenness' was higher at the road edge (Fig. 47.1) than into the

hinterland, with the effect more pronounced in the hilly than the floodplain landscape. Water availability governed vegetation quality, and the road's effect of redistributing water towards its edges contributed to enhanced vegetation quality at the roadside, especially in the hills. Soil nutrients had a lesser influence on vegetation quality relative to water availability.

The study demonstrated that the mechanism of vegetation enhancement by roads in arid ecosystems differed from those identified in other ecosystems, as water additions enhanced vegetation rather than nutrient additions as found in other ecosystems.

Figure 47.1 Increased cover, quality and productivity of plants along an arid-zone road in inland Australia. This arid-zone road is a typical low-formation, single-lane, sealed arid-zone road less than 10 m wide, lacking features such as bridges, median strips and crash barriers. Source: Photograph by E. Lee.

concentration of resources in these areas relative to surrounding areas, resulting in higher animal densities. Many animal species are attracted by the availability of green pick (soft, nutrient-rich grasses/ herbs), seeds and insects present in arid-zone road verges, as well as free-standing water on these roads (Garland & Bradley 1984; Boarman et al. 1997; Starr 2001; Lee et al. 2004). Sealed roads hold heat, so

animal species, such as basking reptiles, are attracted to this heat source during cool conditions (Rosen & Lowe 1994; Tanner & Perry 2007; Chapter 32). Predators are in turn attracted to the higher densities of their prey near arid-zone roads (Knight & Kawashima 1993; Dean et al. 2005; Dean et al. 2006; see also Chapters 33 and 35). In effect, roads in arid ecosystems create a man-made, linear, rich patch in which vegetation is highly productive and animals concentrate.

47.2 The attraction to arid-zone roads can lead to an increased incidence of WVC

The attraction of animals to resources concentrated along arid-zone roads ensures a relatively constant presence of animals that may stray into the path of vehicles. Arid-zone roads do not generally fragment the landscape and/or act as a barrier to animal movement to the same extent as roads in most other ecosystems. Arid ecosystems are naturally open and characteristically patterned, with productive vegetated areas scattered among a larger area of unproductive open spaces (Stafford Smith & Morton 1990). Mobile animals, including smaller animals such as small mammals, regularly traverse these open spaces to access resources. Thus, the open passageways of arid-zone roads, which are typically less than 10 m wide, are not generally perceived by such mobile animals as hostile environments that cannot be traversed.

The attraction and presence of animals can lead to an increased incidence of WVC and roadkill. For example, the attraction of the road as a heat source for snakes on cool evenings was likely a factor influencing snake mortality on a highway through the Sonoran Desert in Southern Arizona, United States (Rosen & Lowe 1994). Similarly, basking opportunities were a factor influencing lava lizard mortality on a road traversing arid environments in the Galapagos Islands (Tanner & Perry 2007). The attraction of desert tortoises in the Mojave Desert, United States, to preferred forage species along the road placed them at risk of mortality from vehicles (Boarman et al. 1997). Predators and scavengers attracted to roadkill, such as crows, ravens, owls, weasels, cats and dogs, often form secondary roadkill (i.e. killed in collisions with vehicles while preying upon roadkilled carcasses) (Dean et al. 2005, 2006; Bullock et al. 2011; Chapter 33).

It is possible that roadkills along arid-zone roads may impact on higher-order ecosystem responses, that is, populations and communities. In relation to animal communities, Lee and Croft (2009) examined roadkill patterns of three kangaroo species in arid central Australia and found that two kangaroo species were killed on roads more often than expected by chance alone, while the third was killed less often than expected. Snakes and frogs in the Sonoran Desert in Arizona, United States, are particularly susceptible to being killed on roads compared to other animal groups due to their particular behavioural characteristics and limited mobility (Grandmaison 2012). Members of the crow/raven family may also be killed more often than other birds of prey along roads in arid environments (e.g. the Karoo, South Africa; Dean & Milton 2003).

47.3 Roads in arid zones that are highly engineered, multi-lane and/or built on embankments cause additional effects, including habitat fragmentation

Low-formation, low-traffic-volume, single-lane roads are typical in many desert ecosystems where human population densities are low. However, some highly engineered, multi-lane and/or more elevated roads are present in desert environments supporting higher human population densities (e.g. south-western United States and Israel). Such roads have a host of different effects (Textbox 47.2; Figs 47.2, 47.3 and 47.4).

Elevated roads and railways (i.e. built on fill or embankments) in arid ecosystems can have significant impacts on the natural hydrology beyond the concentration of water along their edges, leading to more widespread impacts on surrounding biodiversity (see also Chapter 44; Fig. 58.3). Without adequate drainage structures to redistribute water, embankments can act as barriers to natural patterns of overland water flow. This can result in ponding on the upstream side of roads and the creation of downstream 'drainage shadows' which can cause moisture stress to both upstream and downstream vegetation communities. In arid Australia, trees such as mulga are dependent on overland sheetflow of water for survival (Dunkerley 2002). Mulga can show signs of moisture stress due to either drought (too little water) or drowning through saturated soils/long periods of inundation (too much water) and, if stressed for prolonged periods, can result in large-scale mulga death. Elevated arid-zone roads can also result in higher levels of scouring at floodways following high-intensity rain or flood events, degrading the surrounding arid environment.

Habitat fragmentation is also more of an issue along multi-lane and elevated arid-zone roads. Garland and Bradley (1984) found that seven out of eight species of arid rodents generally avoided crossing a four-lane highway in the Mojave Desert, and only a single individual of the species that did cross was recorded. The disturbance associated with the road environment (traffic and mowing of the verge) or the road surface was thought to be the cause of the avoidance as there were no major physical barriers preventing their

Multi-lane roads have additional impacts to the more typical low-formation, low-traffic-volume, single-lane arid-zone roads, including (i) habitat fragmentation, (ii) altered water regimes and (iii) the creation of new and unintended habitats. Large roads cut through the landscape and dissect habitat patches, thereby creating new patches. The effects of fragmentation depend on an animal's body size, the size of its home range and level of mobility and its willingness to cross gaps. Highly engineered roads with their concrete drainage systems, wide verges or solid crash barriers form additional barriers to wildlife movement, especially to smaller animals (Fig. 47.2).

While water is the most limiting resource in arid areas, the occasional floods are major forces that can reshape the landscape. The types of rock formation and of soil have a large impact on water flow behaviour. Highly engineered roads usually change natural water flows, along with energy, soil and nutrient fluxes, playing a major role in determining the distribution and abundance of animals in the landscape. Their impacts may vary greatly between years due to different precipitation levels. Where roads cross dry river beds with occasional strong floods, some engineering solutions disrupt landscape connectivity and may create additional problems. Such solutions include netting the surface below a bridge into gabions (Fig. 47.3) where animal movement and the natural flow of water, soil and organic material are negatively affected.

Changes in relief, water availability and shading create new local conditions and microhabitats that did not previously exist (Fig. 47.4), sometimes with unintended consequences.

Figure 47.2 A large, multi-lane and engineered arid-zone road may fragment habitat. Source: Photograph by T. Achiron-Frumkin.

Figure 47.3 Netting the surface below a bridge into gabions may affect the natural flow of water, soil and organic material and may also affect animal movement. Source: Photograph by and reproduced with permission of Dr. Ron Frumkin.

Figure 47.4 Cuts in hills create a new local conditions and microhabitats. Source: Photograph by T. Achiron-Frumkin.

movements across the road, the road was not elevated, and the species investigated regularly traversed distances equivalent to the road width. Other small animals such as reptiles could also avoid crossing multi-lane and elevated arid-zone roads (Chapter 32). Habitat fragmentation would also be expected to increase on multi-lane roads with artificial lighting that would deter nocturnal animals and expose them to increased predation (Chapter 18).

47.4 Minimising the scale of microhabitat differences at the edges of arid-zone roads, or lengthening the distance over which water is shed, may reduce their attractiveness and lessen their impact

Given the increased risk of WVC potentially impacting on animals at the population or community level, arid-zone roads should be managed to reduce their attractiveness to animals. However, this is difficult due to the nature of the attraction. Water will always be limiting in an arid environment, and arid-zone roads, including unsealed ones, will always be designed to shed water from rain or condensation to their edges and into the road verge, resulting in a concentration of water, green pick or other limiting resources in this area. There may

also be significant costs associated with implementing some engineering solutions to prevent roadkill, including fencing (given the large area traversed by arid-zone roads), distance between human population centres and the relatively low number of human road users in arid regions compared with more populated temperate zones. Cheaper management options include minimising rather than increasing the scale of microhabitat differences at the edges of arid-zone roads (between where the road surface ends and the verge starts) and attempting to draw animals away from the traffic zone of the road or prevent them from accessing the road.

There are a number of methods that could be deployed to minimise the scale of microhabitat differences at the road edge. One method is to change mowing and grading practices, which are common forms of roadside maintenance undertaken to improve driver visibility and allow safe movement of vehicles onto the road shoulder (Chapter 46). However, mowing promotes the growth and diversity of short palatable grasses and herbs that are attractive to many herbivores and often remove more unpalatable species and overshadowing shrubs (Chapter 42). Grading increases erosion and temporarily raises dust. Both mowing and grading may transport invasive plant materials and promote suitable conditions for their growth (Figs 46.4 and 46.5). Thus, mowing and grading practices usually increase rather than decrease microhabitat differences at the edges of arid-zone roads relative to surrounding areas. Managers should consider the costs and benefits of these practices and their transformative effects on the road verge (increasing microhabitat differences). They should also consider the effect of the timing of these practices. Arid-zone vegetation is typically slow growing or erupts briefly after rain when there is a burst of plant growth across the landscape. Thus, these practices may only be necessary, with an appropriate growth lag, after rare heavy and sustained rainfalls in arid zones.

Another method to minimise the scale of micro-habitat differences along arid-zone roads is the use of different road surfaces that do not retain as much heat as a dark, bitumen surface, for example, a white concreted surface.

In terms of measures that draw animals away from the traffic zone of the road, the distance over which water is shed from the traffic zone to the road verge can be lengthened, with road shoulders used to act as a buffer between the traffic zone and the road verge

(Fig. 47.5). Increasing the width of the road shoulder will also provide a larger buffer between animals and vehicles and give drivers more opportunity to detect and avoid animals.

Fencing may be appropriate in some circumstances to prevent animals from accessing the road (Boarman et al. 1997; Grandmaison 2012; Chapter 20). However, this option would be impractical along large stretches of arid-zone roads and should only be used on a case-by-case basis in areas specifically demonstrated to be roadkill hotspots. Effective fencing without sufficient and adequate crossing structures will also result in a complete barrier to movement of wildlife.

It should be remembered that management actions to reduce the attractiveness of arid-zone roads and lower the incidence of WVC and roadkill should produce a net benefit. We have suggested widening road shoulders to increase the distance between the traffic zone and the road verge. However, some rodent species in the Mojave Desert avoided crossing arid-zone roads that had a total clearance of 69 m (Garland & Bradley 1984), which is wider than the clearance of a typical,

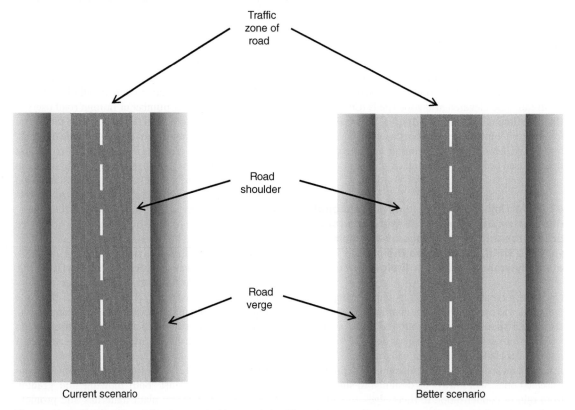

Figure 47.5 Increase the widths of water-shedding road shoulders (shown in yellow) to draw animals away from the traffic zone of the road and improve peripheral vision for drivers. Source: Illustration by E. Lee.

single-lane, arid-zone road. Thus, there will be a threshold in width where arid-zone roads are a barrier to the movement of wildlife. Once the threshold is reached, populations on either side of the road could become fragmented – an undesirable management outcome. The installation of fences to prevent roadkill could also prevent movement across arid-zone roads, fragmenting and preventing gene flow between populations on either side. Thus, in formulating and implementing management actions, we must consider all potential ecological impacts and conflicts that may exist among the possible solutions.

47.5 New roads and upgrades to existing roads in arid environments must consider landscape features, topography and animal species likely to be present and be designed to maintain landscape function which is easily disturbed in this environment

There are numerous factors that should be considered when planning and constructing new roads and upgrading existing roads in arid ecosystems. First, topography and drainage patterns must be considered. For example, roads that run straight up hill slopes in arid ecosystems will promote soil erosion and the development of rills and gullies. Roads built on embankments that run parallel to elevation contours will alter the natural run-off patterns in arid ecosystems by blocking the flow of surface water or redirecting water along roadside ditches to low points or culverts along the road. This results in ponding upslope and drainage shadows downslope or concentrated discharges of water. These potential impacts should be avoided, either through changes in the alignment of the road or the implementation of drainage or water retention/stabilisation measures (Chapter 44). Drainage measures include those that capture and redistribute surface flow in a manner that mimics natural flow. Water retention and stabilisation measures include those that have the capacity and stability to allow for the large volumes of water and sediment that flow through arid ecosystems during high-intensity rain and flood events.

Second, the use and response of animals to landscape features which increases vulnerability to WVC or habitat fragmentation must be considered. Arid-zone animals often cross roads to reach specific resources (e.g. water) and will often repeatedly use the same route. Some threatened desert species, such as the desert tortoise in the Mojave Desert, United States, are particularly vulnerable to becoming roadkill due to

their wide-ranging but slow movement (Boarman & Sazaki 2006). New roads should aim to avoid bisecting areas that animals are known to traverse or where threatened species are present and vulnerable to road impacts. If this is unavoidable, then careful design should be used to minimise and mitigate unavoidable impacts.

CONCLUSIONS

The collection of water at arid-zone road edges, both on the road shoulder and in the road verge, drives many of the ecological effects of roads in arid ecosystems, the most obvious of which is roadkill. To reduce the rate of WVC and roadkill and manage any potential impacts at the population and community level, management actions must seek to reduce the microhabitat differences at the road edge or implement measures to draw animals away from the traffic zone. The construction of new roads through arid landscapes must consider factors such as landscape positioning, animal use of the landscape and the presence of threatened animal species.

FURTHER READING

Boarman and Sazaki (2006): Study conducted in the Mojave Desert in California, United States, which found that the abundance of desert tortoise was depressed in a zone at least up to 400 m from the edge of a highway.

Morton et al. (2011): Provides a revised set of propositions about ecology in arid Australia and argues that most features of Australian deserts are explicable in terms of two dominant physical and climatic elements: rainfall variability, leading to extended droughts and occasional flooding rains, and widespread nutrient poverty. Discusses the extent to which the ecology of Australian deserts is 'different', 'accentuated' or 'universal' in comparison with other deserts of the world.

Stafford Smith and Morton (1990): Discusses the ecology of arid Australia and describes how arid ecosystems are naturally open and characteristically patterned, with productive vegetated areas (relatively fertile and moist) scattered among larger areas of unproductive (infertile and arid) open spaces.

Tanner and Perry (2007): A study of road impacts on lava lizards in arid areas of Santa Cruz, Galapagos, which showed lower abundances near roads and also greater rates of prior tail loss closer to roads, suggesting an impact on fitness for those living close to the road.

Westoby (1980): Discusses vegetation dynamics in arid rangelands to explain patterns and processes of vegetation succession.

REFERENCES

Boarman, W. I. and M. Sazaki. 2006. A highway's road effect zone for desert tortoises (*Gopherus agassizii*). Journal of Arid Environments **65**: 94–101.

Boarman, W. I., M. Sazaki and W. B. Jennings. 1997. The effects of roads, barrier fences and culverts on desert tortoise populations in California, USA. In Proceedings of the international conference on conservation, restoration and management of tortoises and turtles. New York Turtle and Tortoise Society, pp. 54–58.

Brooks, M. L. and B. Lair. 2005. Ecological effects of vehicular routes in a desert ecosystem. Report prepared for the United States Geological Survey, Recoverability and vulnerability of desert ecosystems program. US Department of the Interior, US Geological Survey.

Bullock, K. L., G. Malan and M. D. Pretorius. 2011. Mammal and bird road mortalities on the Upington to Twee Rivieren main road in the southern Kalahari, South Africa. African Zoology **46**: 60–71.

Dean, W. R. J. and S. J. Milton. 2003. The importance of roads and road verges for raptors and crows in the Succulent and Nama-Karoo, South Africa. Ostrich **74**: 181–186.

Dean, W. R. J., S. J. Milton and M. D. Anderson. 2005. Use of road-kills and roadside vegetation by Pied and Cape Crows in semi-arid South Africa. Ostrich **77**: 102–104.

Dean, W. R. J, R. I. Yeaton and S. J. Milton. 2006. Foraging sites of Turkey Vultures *Cathartes aura* and Common Ravens *Corvus corax* in central Mexico. Vulture News **54**: 30–33.

Dunkerley, D. 2002. Infiltration rates and soil moisture in a groved mulga community near Alice Springs, arid central Australia: Evidence for complex internal rainwater redistribution in a runoff–runon landscape. Journal of Arid Environments **51**: 199–219.

Garland, J. T. and W. G. Bradley. 1984. Effects of a highway on Mojave Desert rodent populations. The American Midland Naturalist **111**: 47–56.

Grandmaison, D. 2012. Wildlife linkage research study in Pima County: Crossing structures and fencing to reduce wildlife mortality. Report prepared for the Arizona Game and Fish Department.

Johnson, H. B., F. C. Vasek and T. Yonkers. 1975. Productivity, diversity and stability relationships in Mojave Desert roadside vegetation. Bulletin of the Torrey Botanical Club **102**: 106–115.

Kinlaw, A. 1999. A review of burrowing by semi-fossorial vertebrates in arid environments. Journal of Arid Environments **41**: 127–145.

Knight, R. L. and J. Y. Kawashima. 1993. Responses of raven and red-tailed hawk populations to linear right-of-ways. Journal of Wildlife Management. **57**: 266–271.

Lee, E. 2006. The ecological effects of sealed roads in arid ecosystems. PhD Thesis, University of NSW.

Lee, E. and D. B. Croft. 2009. The effects of an arid-zone road on vertebrates: Priorities for management? In D. Lunney, A. Munn and W. Meikle, editors. Too Close for Comfort: Contentious Issues in Human-Wildlife Encounters. The Royal Zoological Society of New South Wales, Mosman, pp. 105–117.

Lee, E., U. Klocker, D. B. Croft and D. Ramp. 2004. Kangaroo-vehicle collisions in Australia's sheep rangelands, during and following drought periods. Australian Mammalogy **26**: 215–226.

Lightfoot, D. C. and W. G. Whitford. 1991. Productivity of creosote foliage and associated canopy arthropods along a desert roadside. American Midland Naturalist **125**: 310–322.

Morton, S. R., D. M. Stafford Smith, C. R. Dickman, D. L. Dunkerley, M. H. Friedel, R. R. J. McAllister, J. R. W. Reid, D. A. Roshier, M. A. Smith, F. J. Walsh, G. M. Wardle, I. W. Watson and M. Westoby. 2011. A fresh framework for the ecology of arid Australia. Journal of Arid Environments **75**: 313–329.

Norton, D. A. and M. Stafford Smith. 1999. Why might roadside mulgas be better mistletoe hosts? Australian Journal of Ecology **24**: 193–198.

Rosen, P. C. and C. H. Lowe. 1994. Highway mortality of snakes in the Sonoran Desert of Southern Arizona. Biological Conservation **68**: 143–148.

Stafford Smith, D. M. and S. R. Morton. 1990. A framework for the ecology of arid Australia. Journal of Arid Environments **18**: 255–278.

Starr, M. J. 2001. Assessing the effects of roads on desert ground squirrels. Papers and Proceedings of Applied Geography Conferences **24**: 35–40.

Tanner, D. and J. Perry. 2007. Road effects on abundance and fitness of Galápagos lava lizards (*Microlophus albemarlensis*). Journal of Environmental Management **85**: 270–278.

Westoby, M. 1980. Elements of a theory of vegetation dynamics in arid rangelands. Israel Journal of Botany **28**: 169–194.

Chapter 48

ROAD ECOLOGY IN AN URBANISING WORLD

Darryl Jones[1], Hans Bekker[2] and Rodney van der Ree[3]

[1]Environmental Futures Research Institute, Griffith University, Nathan, Queensland, Australia
[2]Centre for Transport and Navigation, Rijkswaterstaat, Delft, Netherlands
[3]Australian Research Centre for Urban Ecology, Royal Botanic Gardens Melbourne, and School of BioSciences, The University of Melbourne, Melbourne, Victoria, Australia

SUMMARY

Urban areas are unique ecosystems with many distinctly human-centric features not found in other environments; both humans and animals of urban environments must be able to habituate and adapt to many novel influences. Such adaptation includes learning to cope with roads and traffic, which are found in high densities in cities and towns. Perhaps unexpectedly, urban environments often support high levels of biodiversity. Habitats provided by gardens, parks, reserves and roadsides are essential for the survival of wildlife, and these areas are often greatly valued by human residents.

48.1 Urban environments support biodiversity.
48.2 The ecological effects of roads are more intense in urban areas.
48.3 Roads are often directly connected to important riparian habitats.
48.4 Road impacts and mitigation measures are more conspicuous in urban areas.
48.5 Urban roads affect people as well as wildlife.
48.6 Better urban road planning requires community involvement.
48.7 Raise the profile of road ecology research in the urban landscape.

The presence of people provides both a major obstacle and potentially the key element to successful application of urban and road ecology principles to urban road issues. We contend that people – the defining characteristic of cities – must be included at all levels of road design and implementation.

Handbook of Road Ecology, First Edition. Edited by Rodney van der Ree, Daniel J. Smith and Clara Grilo.
© 2015 John Wiley & Sons, Ltd. Published 2015 by John Wiley & Sons, Ltd.
Companion website: www.wiley.com\go\vanderree\roadecology

INTRODUCTION

For the first time in human history, most people on Earth now live in cities. While the definition of what constitutes 'urban' varies greatly, we use the term *broadly* to refer to areas where people live and work in large numbers or high density. As such areas expand, so too do road networks and the number of vehicles. These developments pose significant challenges for urban planners seeking to balance demands for the safe and efficient movement of people and goods via private motor vehicles, public transport, cycling and walking, along with the provision of green space and the protection of biodiversity. The relatively new fields of urban ecology and road ecology offer important opportunities to address these challenges. Much of the focus of road ecology has, however, been on rural areas and wildlands; urbanised areas, the cities and towns where most people live and work (Miller 2006), have largely been ignored by road ecologists. This is unfortunate because many urban areas support diverse and, in some cases, rare and endangered species, especially in the rapidly urbanising fringes. The presence of the dense road networks typical of cities may be particularly detrimental to local biodiversity, green space, safety and quality of life in urban areas. In many cases, the expansion of road networks associated with urbanisation includes little consideration for biodiversity or, indeed, the people either. While it may seem obvious that roads are constructed to aid human movement, the associated impacts of poorly planned roads on local residents may be significant. People care about the places they live in and are often willing to fight for the conservation of local wildlife sites more than for places further away (Rhode & Kendell 1994). As many road planning and design committees have experienced, well-organised and motivated community groups contesting local issues can be either formidable opponents or valuable allies (Chapter 59).

In this chapter, we outline some key issues associated with the concepts and implementation of road ecology in cities and towns, emphasising the centrality of people – as concerned residents, members of community groups and road users – in discussions on road planning and design.

LESSONS

48.1 Urban environments support biodiversity

It is often suggested that cities are effectively devoid of nature, due to the overwhelming changes associated with human development. However, a common finding by urban ecologists has been the presence of a remarkable diversity of wildlife species and natural habitats within many urban areas (McDonnell et al. 2009). Historically, people have tended to settle in areas with fertile soils, reliable water supplies and other natural resources that can sustain human settlements. As a result, many urban areas occur on biodiversity hotspots, and evidence of this former biodiversity is still present in many cities and towns (e.g. van der Ree & McCarthy 2005). Urbanisation is now recognised globally as a major threatening process and has led to dramatic and often catastrophic impacts on local biodiversity, with many more extinctions likely within the present century (Marzluff 2001).

Most species that are sensitive to habitat loss and fragmentation or that require specific ecological processes (e.g. fire or flooding) do not persist following urbanisation. With existing cities, urbanisation primarily occurs at the interface with rural lands (McDonnell et al. 2009). While such urban sprawl is often an intensification of land use (typically rural to suburban), there may be vastly different levels of biodiversity present prior to development. The total area of suburban habitat available to wildlife is often substantial: in the United Kingdom, for example, gardens in total occupy a much greater area than all conservation reserves combined (Chamberlain et al. 2009). Thus, the matrix of vegetated house yards, urban parklands and remnants of natural ecosystems that make up many suburban areas are major components of the landscape, and often support unexpectedly rich faunas. In addition, structures found in cities may be utilised by large numbers of native species; the immense roosts of bats under certain bridges being an obvious example (Fig. 34.3).

One of the most obvious outcomes of urbanisation is the loss of many species (Marzluff 2001) along with the proliferation of a few resilient and adaptable species. White-tailed deer of North America, roe deer in many parts of Europe and eastern grey kangaroos in Australia are large mammals now abundant in the suburbs or on the urban–rural fringe of some cities (Huijser & McGowen 2010, Ramp 2010; Chapter 42). While each of these species is associated with serious motorist safety issues (Clevenger & Ford 2010), there are many other common animals that concern the community because of high rates of roadkill.

Many cities contain green spaces that have been protected from urbanisation because of their benefits to humans (e.g. recreation and aesthetics) or the biodiversity they protect (Marzluff 2001). These urban green spaces are often threatened by proposals to develop them for housing, roads or other infrastructure. Because of their value to society, community opposition to such development can be especially pronounced. If carefully handled, however, such controversies can also lead to innovative solutions and long-term community support (Bond & Jones 2008; Textbox 48.1, Fig. 48.1).

48.2 The ecological effects of roads are more intense in urban areas

It is in urban areas that the most dense and concentrated networks of roads are found, as well as the highest volumes of traffic (Beckman et al. 2010). Since the size of the 'road-effect zone' (Chapter 1)

increases with road density, road width and volume of traffic, the scale and intensity of related impacts would be expected to be greater in urban areas. These impacts are exacerbated by the fact that urban landscapes have already been dramatically altered. Despite these impacts, many species and habitats continue to exist in cities and towns throughout the world. For biodiversity to persist, it is critical that issues such as habitat connectivity and measures to reduce the rate of roadkill are integrated into new road and upgrade projects. This is particularly relevant where urban areas are expanding into rural or natural landscapes with moderate to high levels of biodiversity. Without careful planning, such areas are likely to become degraded and dominated by urban-adapted and invasive species.

Textbox 48.1 Compton Road, Australia: Successful mitigation through dialogue.

When the Brisbane City Council in Queensland, Australia, proposed to double the width of a major arterial road that traversed a nationally significant nature reserve, many predicted a bitter and long-running dispute with the well-organised local community group. Instead, the city initiated a planning committee consisting of all stakeholders and road ecology experts. This resulted in the highest density and variety of wildlife crossing structures found anywhere in the world, with two underpasses, three canopy rope bridges, a set of glider poles and a wildlife overpass along 1.3 km section of the four-lane road (Fig. 48.1). The council also committed to a 10-year monitoring programme to evaluate use and effectiveness of the mitigation.

Figure 48.1 Wildlife overpass at Compton Road, in the suburbs of Brisbane, Australia. Due to successful negotiations between local government, community groups, ecologists and engineers, this site has a wide variety of mitigation measures to allow wildlife to cross the road. Source: Photograph by Darryl Jones.

48.3   Roads are often directly connected to important riparian habitats

In many cities, major roads are often built adjacent to existing waterways, typically because these places provide space for such developments. Rivers and creeks are almost always important habitats and movement corridors for wildlife, and are vulnerable to the impacts associated with excessive run-off (Iuell et al. 2003; Chapters 44 and 45). Roads frequently intersect with urban stormwater systems, which often feed directly into local riparian networks. Roads and drainage systems are designed to move the water off the road as quickly as possible to improve driver safety and minimise damage to the road infrastructure. Typically, this means channelling run-off directly into natural waterways adjacent to the road through curbs, drains and pipes. The result is often large volumes of water entering the waterway, moving at considerable velocities and carrying high quantities of pollutants. The combination of periodic major flushes and contaminants represent serious threats to the health of species that live in and move along the riparian systems (Chapter 45). Changing the design of the water management systems associated with roads in many urban areas is among the most important yet achievable changes to the design of roads. Small mammals and amphibians often explore and become trapped in the drainage infrastructure (e.g. curbs and drains), and exclusion grids or escape mechanisms should be included on all future projects that dissect habitat for these species.

48.4   Road impacts and mitigation measures are more conspicuous in urban areas

Roadkill and other impacts of roads and traffic often appear more apparent in urban areas because of the high density of humans. This also means that wildlife crossing structures are more conspicuous and accessible to people. By contrast, most wildlife crossing structures are located far from urban centres and access by the public is limited or prevented entirely. This means that local groups and residents – often including those who were involved in campaigns for their construction – have enhanced 'ownership' of such structures in urban areas and may remain frequent visitors and even willing participants in monitoring of fauna use (e.g. the *Groene Woud* wildlife overpass in the Netherlands, van der Grift et al. 2010).

Wildlife crossing structures are often also utilised by people. Depending on the intention of the planners of the structure, human usage may be allowed, encouraged or actively discouraged, although complete exclusion of access to determined trespassers is often difficult to ensure. The extent to which crossing structures may be used beneficially by both humans and wildlife animals remains a vigorous debate (Chapter 22; van der Grift et al. 2010), especially in Europe where the density of human settlements has meant that many such structures are within easy access to local recreationists. Chapter 22 summarises the extent of knowledge on recreational co-use of wildlife crossing structures and presents some guiding principles to inform their design and to evaluate effects co-use on wildlife.

48.5   Urban roads affect people as well as wildlife

The main focus of road ecology is on wildlife and natural habitats, but it is important to appreciate that roads and traffic also affect people. In some cases, roads may divide communities when safe crossing passages are not provided. This is especially important in regions where many people walk and cycle (Forman 1999). Traffic noise, light, particulates and chemical pollution associated with roads and traffic also affect the quality of life of people living near major roads (Forman et al. 2003). Poorly designed roads impact motorists and the community through traffic congestion, increased commuter times, travel costs, vehicle maintenance and associated issues. By contrast, road design that incorporates ecological principles also benefits humans, especially when other modes of transport, specifically bicycle paths and public transport, are provided and promoted. Issues associated with railway lines and industrial linear infrastructure are addressed elsewhere (Chapters 26 and 28), but paths and trails for cyclists, pedestrians and horse riders can also affect wildlife (Fig. 32.3), and must be designed accordingly.

48.6   Better urban road planning requires community involvement

Construction of a new road or an upgrade to an existing road is a complex and challenging process (Fig. 4.1, Table 4.1). These challenges are accentuated in urban settings because of competing demands for space, high costs of land and the involvement of a wide range of stakeholders. Addressing these common dilemmas in an urban setting requires two fundamental considerations: (i) the necessity for landscape-scale planning; and (ii) the genuine engagement of all key stakeholder groups. Both must be included throughout the planning and design process since redressing a neglected issue is often impossible or expensive once the road has been completed (Chapter 9).

It is almost always the case that the least expensive route for a new arterial road in urban areas is through existing greenspace such as parks and conservation reserves. However, the cheapest route is not always the optimal because such areas are typically the source of much of the local biodiversity and of significant value as open space to the community, with health and recreational benefits. Plans to dissect or destroy these green spaces are commonly subject to vehement opposition from the public and subsequently low levels of support for a planned road project (Chapter 59). A transparent dialogue with a wide variety of stakeholder groups and relevant experts is essential to achieving an outcome that benefits people and biodiversity. Achieving a successful and mutually productive outcome requires trust and respect between the parties, which takes time and effort.

Experience from committees (from many countries) formed to encourage dialogue between the road design team and the public has demonstrated some common themes, with perhaps the most fundamental being the extent to which one side is ignorant of the key concerns of the others (Iuell et al. 2003). For example, road engineers are often unprepared for the level of preparation and personal commitment shown by many community members. On the other hand, private individuals are frequently focused on specific issues or locations, while consulting ecologists may have unrealistic expectations of the flexibility of the process. A key element of success is the willingness and ability of each party to explain their concerns and expectations clearly and to listen to alternative views. Of course, such a process depends on effective facilitation, preferably by a neutral party, or at least a person respected by all sides.

48.7   Raise the profile of road ecology research in the urban landscape

Road ecology research and practice has occurred mainly in rural areas and wildlands, typically far from the scrutiny and demands of the urban public. As we have described in this chapter, the impacts and influences of roads in urban areas are often severe and more concentrated than elsewhere. Road ecology research that focuses on urban landscapes is critically needed (van der Ree 2009). Important research questions include the following: (i) What is the optimal design of multi-use crossing structures that allow effective use by people and wildlife? (Chapter 22); (ii) How to evaluate if a crossing structure is likely to be effective at preventing an endangered species from going extinct, given other urban pressures?; (iii) What are the relative cost and benefits of the apparently competing values ascribed by local communities to features such as transportation convenience, preservation of local biodiversity and protection of local greenspace?

CONCLUSIONS

Cities have become the destination for an increasingly large proportion of humanity. Human activities and infrastructure, including roads and traffic, are concentrated in these areas. The road networks in urban areas have expanded to cope with increasing human populations, often encroaching into the remaining greenspace. These areas are vital for the survival of local biodiversity and often greatly valued by residents who fear that roads and developments will eliminate or diminish the quality of their communities. Furthermore, many communities around the world are beginning to encourage the local and regional governments to reconsider the overwhelming priority given to private vehicle transportation. While modern developed economies will always need road systems, serious discussions about alternative transportation options within urban landscapes are now needed.

Road ecology research has led to several technical and practical solutions to many of these challenges, with certain mitigation structures and designs well suited to application in urban areas. A close and effective working relationship among all parties is essential to ensure a successful balance between the demands of road users and the communities they serve.

FURTHER READING

Marzluff et al. (2001): An edited volume about the impacts of urbanization on birds.

McDonnell et al. (2009): A published volume of 35 chapters focusing on the impacts of urbanization on biodiversity and recommendations on directions for sustainable urban developments.

Miller (2006): An influential and provocative article drawing attention to the increasing disconnection between people in cities and nature.

van der Ree (2009): A review of road ecology research in urban areas.

REFERENCES

Beckman, J. P., A. P. Clevenger, M. P. Huijser and J. A. Hiffy. 2010. Safe Passages: Highways, Wildlife and Habitat Connectivity. Island Press, Washington, DC.

Bond, A. R. and D. N. Jones. 2008. Temporal trends in use of fauna-friendly underpasses and overpasses. Wildlife Research **35**: 105–112.

Chamberlain, D., A. Cannon, M. Toms, D. Leech, B. Hatchwell and K. Gaston. 2009. Avian productivity in urban landscapes: A review and meta-analysis. Ibis **151**: 1–18.

Clevenger, A. P. and A. T. Ford. 2010. Wildlife crossing structures, fencing and other highway design considerations. Pages 17–49 in: J. P. Beckmann, A. P. Clevenger, M. P. Huijser and J. A. Hilty, editors. Safe Passages: Highways, Wildlife and Habitat Connectivity. Island Press, Washington, DC.

Forman, R. T. T. 1999. Horizontal processes, roads, suburbs, societal objectives and landscape ecology. Pages 35–53 in: J. M. Klopatek and R. H. Gardner, editors. Landscape Ecological Analysis: Issues and Applications. Springer, New York.

Forman, R. T. T., D. Sperling, J. A. Bissonette, A. P. Clevenger, C. D. Cutshall, V. H. Dale, L. Fahrig, R. France, C. R. Goldman, K. Heanue, J. A. Jones, F. J. Swanson and T. C Turrentine. 2003. Road Ecology. Science and Solutions. Island Press, Washington, DC.

Huijser, M. P. and P. T. McGowen. 2010. Reducing wildlife-vehicle collisions. Pages 51–74 in: J. P. Beckmann, A. P. Clevenger, M. P. Huijser and J. A. Hilty, editors. Safe Passages: Highways, Wildlife and Habitat Connectivity. Island Press, Washington, DC.

Iuell, B., C. J. Bekker, R. Cuperus, J. Dufek, G. Fry and C. Hicks. 2003. Wildlife and Traffic: A European Handbook for Identifying Conflicts and Designing Solutions. Office for Official Publications of the European Communities, Luxembourg.

Marzluff, J. M. 2001. Worldwide urbanization and its effects on birds. Pages 1–16 in: J. Marzluff, R. Bowman and R. Donnelly, editors. Avian Ecology and Conservation in an Urbanizing World. Kluwer, Norwell, MA.

Marzluff, J. M., R. Bowman and R. Donelly, editors. 2001. Avian Ecology and Conservation in an Urbanizing World. Kluwer Academic Publishers, Norwell, MA.

McDonnell, M. J., A. K. Hahs and J. H. Breuste. 2009. Comparative Ecology of Cities and Towns. Cambridge University Press, Cambridge.

Miller, J. R. 2006. Biodiversity conservation and the extinction of experience. Trends in Ecology and Evolution **20**: 430–434.

Ramp, D. 2010. Roads as drivers of change for macropodids. Pages 279–290 in: G. Coulson and E. Eldridge, editors. The Biology of Kangaroos, Wallabies and Rat-Kangaroos. CSIRO Publishing, Melbourne.

Rhode, C. L. and A. D. Kendell. 1994. Human Well-Being, Natural Landscapes and Wildlife in Urban Areas: A Review. English Nature, London.

van der Grift, E. A., J. Dirksen, F. G. W. A. Ottburg and R. Pouwels. 2010. Recreatief medegebruik van ecoducten; effecten op het functioneren als faunapassage. Alterra 2097, Wageningen, the Netherlands.

van der Ree, R. 2009. The ecology of roads in urban and urbanising landscapes. Pages 187–194 in: M. J. McDonnell, A. Hahs and J. Breuste, editors. Comparative Ecology of Cities and Towns. Cambridge University Press, Cambridge.

van der Ree, R. and McCarthy, M. A. 2005. Inferring persistence of indigenous mammals in response to urbanisation. Animal Conservation **8**: 309–319.

Chapter 49

TROPICAL ECOSYSTEM VULNERABILITY AND CLIMATIC CONDITIONS: PARTICULAR CHALLENGES FOR ROAD PLANNING, CONSTRUCTION AND MAINTENANCE

Miriam Goosem

Centre for Tropical and Environmental Sustainability Sciences, College of Marine and Environmental Sciences, James Cook University, Cairns, Queensland, Australia

SUMMARY

The tropics cover 40% of the Earth but support most of the world's biodiversity in terms of numbers of species, endemism and global hotspots. More than 49% of the Earth's terrestrial species are contained within the 7% of the Earth's land surface covered by tropical rainforests. However tropical drier forests, savannahs and wetlands also rank comparatively high in biodiversity. The construction and operation of roads in tropical areas can result in particularly severe ecological impacts, nevertheless many of the roads currently planned or under construction are found in remote tropical regions.

49.1 The tropical rainforest biota is especially vulnerable to environmental impacts of roads due to specialisations and diversity.

49.2 Edge effects, barrier effects and weed growth are particularly severe in tropical rainforests because the unique stable forest microclimate contrasts with the extreme conditions in cleared roadways.

49.3 Tropical animals are extremely vulnerable to hunting and roadkill when roads create access into remote areas.

Handbook of Road Ecology, First Edition. Edited by Rodney van der Ree, Daniel J. Smith and Clara Grilo.
© 2015 John Wiley & Sons, Ltd. Published 2015 by John Wiley & Sons, Ltd.
Companion website: www.wiley.com\go\vanderree\roadecology

49.4 Natural catastrophes and environmental conditions associated with tropical ecosystems exacerbate road impacts.

49.5 Narrow and unpaved roads with continuous canopy cover are preferred over the greater impact caused by wide and paved roads in tropical rainforest.

49.6 Tropical roads often are located in developing nations aiming to improve their economies and standard of living by opening frontiers of previously undeveloped lands.

49.7 Where it is impossible to avoid undisturbed regions and sensitive habitats, sensitive tropical areas should either be protected before road construction or access should be restricted.

49.8 Mitigation of tropical road impacts at the roadway scale can be achieved by maintaining canopy connectivity, minimising road width and restricting logging roads to low-impact areas.

Due to the extreme environmental conditions that often occur in tropical areas and the sensitivity of many tropical ecosystems, avoiding sensitive and undisturbed areas is the best strategy for preventing road impacts. Due to the engineering challenges involved, avoiding such areas may also prove the least expensive option.

INTRODUCTION

The tropics cover 40% of the Earth's surface between latitude 23.438°N and 23.438°S and include the greatest amount of forest cover in any climatic zone (Goosem & Tucker 2013). About 60% of the tropical forests remaining are highly diverse rainforests, many of which are included in 18 of the 35 global biodiversity hotspots (Zachos & Habel 2011). These generally receive more than 2000 mm of precipitation a year which can occur as periods of extremely heavy rainfall (Stork et al. 2011). Tropical rainforests are believed to house more than half of the world's biodiversity because they support high levels of endemism and large numbers of rare or threatened species (Stork et al. 2011). These areas also often support indigenous people and remote minority groups with ancient cultures. Globally important tropical ecosystems also include dry forests, woodlands and wetlands, all of which rank highly in terms of diversity of biota (Zachos & Habel 2011). However, this chapter will mainly concentrate on the great biodiversity and particular vulnerability of tropical rainforests to the ecological impacts of roads. Rainforest vulnerability is a result of the specialised biota, complex structure and challenging environmental conditions found in this ecosystem.

The construction and operation of roads and highways in tropical areas can result in particularly severe impacts when compared with many other regions. Despite the severity of impacts, many major new roads through road-free areas are planned or are already under construction in tropical regions (Chapters 2, 50, 51, 52, 53, 55 and 56). Additionally, tropical forest roads are often located in developing countries which have reduced financial means and capability to design and implement low-impact roads.

This chapter aims to provide insights regarding the impacts of roads in tropical forests and explain why they are particularly severe. Avoidance of these sensitive areas is preferred, but mitigation measures that are currently considered best practice for rainforest roads are also discussed.

LESSONS

49.1 The tropical rainforest biota is especially vulnerable to environmental impacts of roads due to specialisations and diversity

Although roads are known to cause serious environmental impacts in natural habitats all around the world, tropical habitats and especially rainforests are particularly vulnerable (Laurance et al. 2009). This is partly due to specialisation to the extremely complex architecture and unique humid, dark, stable microclimate found within rainforests. Additionally, the great biodiversity of tropical rainforests means that many more species are likely to be affected than in other terrestrial habitats. Many of these species are rare and/or restricted to very small areas, and a large number are now also endangered. In comparison with other habitats, these species are severely impacted by edge and barrier effects, and remote tropical areas suffer from human invasion and hunting when roads are constructed.

49.2 Edge effects, barrier effects and weed growth are particularly severe in tropical rainforests because the unique stable forest microclimate contrasts with the extreme conditions in cleared roadways

The majority of tropical rainforest animal and plant species are specialised for living in the forest interior and do not survive in the modified conditions found in roadway clearings or forest edges along roads. A clearing through rainforest for a road results in brighter, warmer, windier and drier conditions with greater fluctuation in light levels, temperature and humidity through day and night. This occurs not only in the roadway clearing but also at the edge of adjacent rainforest (Pohlman et al. 2009). Increased moisture stress and higher wind speeds along the edge cause death or damage to trees, while greater light levels assist weeds and light-loving vines and pioneer trees to proliferate, thus changing the structure and floristics of the forest adjacent to the road (Laurance et al. 2009).

Because plant growth is prolific in the tropics, grasses and weeds which can endure microclimate extremes quickly invade along roadsides, assisted by vehicles and animals transporting seeds (Fig. 49.1) (Laurance & Goosem 2008; Chapter 46). Weed species transform wide roadsides through a self-perpetuating cycle that alters fire regimes or prevents recruitment of other species. The modified vegetation structure in the roadside provides corridors that allow rapid dispersal of non-forest fauna, feral generalists (e.g. fire ants, feral pigs and mice, and cane toads), feral predators (e.g. domestic cats and dogs) and diseases. These invaders may prey on, poison, destroy or simply out-compete native plants and animals (Laurance et al. 2009).

Changes in the clearing and rainforest edge severely impact rainforest plant and animal species, because so many of the diverse rainforest species are specialised for the cool, humid, stable conditions of the rainforest understory. Rainforest animals ranging from ants (Dejean & Gibernau 2000), through understory birds (Laurance 2004), bats (Delaval & Charles-Dominique 2006), amphibians (Hoskin & Goosem 2010) up to mammals as large as forest elephants (Blake et al. 2008; Gubbi et al. 2012) avoid not only the hot, dry road surface, traffic noise and human presence in the road clearing but also the altered microclimate and vegetation within the roadside and forest edge (Hoskin & Goosem 2010). Such edge effects can penetrate the forest for distances of 100 m or more (Laurance et al. 2009).

Avoidance of noise, traffic and humans associated with roads and altered forest edges means that many rainforest understory animals rarely or never cross even narrow roads. Certain rainforest arboreal species, including monkeys, kinkajou, small primates such as pottos and some possums seldom or never venture to ground level and therefore have no viable means of crossing roads (Wilson et al. 2007; Laurance et al. 2009). Therefore, clearings for larger roads and highways may form a complete barrier to many rainforest specialists, threatening population viability in the short term and potentially leading to genetic isolation and decay in the long-term (Laurance et al. 2004; Goosem 2007; Laurance et al. 2009).

(A)

(B)

Figure 49.1 (A) Weeds, particularly grasses, grow prolifically along wide road clearings in the tropics but (B) can be controlled by allowing canopy to extend over the road. Source: (A) Photograph by and reproduced with permission of S. Goosem; and (B) Photograph by M. Goosem.

(A)   (B)

Figure 49.2 Hunting and roadkill are serious problems for the many rare and endangered species of tropical forests. (A) Malayan tapir roadkill, Perak, Malaysia. (B) Poachers captured by camera trap in underpass crossing structure, Malaysia. Source: (A) Photograph by and reproduced with permission of © WWF Malaysia/Sara Sukor; and (B) Photograph by and reproduced with permission of Rimba/Reuben Clements.

Because tropical open forest, woodland and savannah habitats are drier and more open, the contrast in microclimate and habitat structure with the cleared roadway is less extreme than with rainforest. However, changes in vegetation structure and floristic composition and in particular, weed invasions, still occur along roadsides (Hoffman et al. 2004; Goosem & Pohlman 2014). Similar to rainforests, road avoidance and difficulties in crossing by specialised fauna have been observed (Asari et al. 2010), although these problems are generally less pronounced in these habitats.

49.3 Tropical animals are extremely vulnerable to hunting and roadkill when roads create access into remote areas

Many tropical animals are vulnerable to roadkill (Goosem 1997; Vijayakumar et al. 2001) and hunting as well as predation along roads (Lee et al. 2005; Laurance et al. 2009). New roads allow easy access for people to hunt in rainforest areas that were previously impenetrable, resulting in greater mortality and reduced abundance of large mammals at distances of 5–10 km from the road (Blake et al. 2008; Chapters 55 and 56). Roads also facilitate the trade in bush meat and wildlife products by allowing rapid transport to markets (Lee et al. 2005; Fig. 56.5). Natural predators and predatory invaders are also able to use roads as a base for hunting and scavenging (Laurance et al. 2009).

Mortality rates that are artificially elevated through roadkill can be serious for rare species, of which many occur in rainforests (Fig. 49.2A). For example, in the Wet Tropics bioregion in north-east Australia, 13% of rainforest vertebrate animal species (106 species) are considered rare or threatened under Queensland State legislation, even though most of the remaining rainforest is conserved in national parks or as World Heritage areas (WTMA 2009). High rates of road mortality are also a problem for wide-ranging species with low reproductive rates that need to cross roads multiple times during their normal movements, as these are often specialised keystone species. An example is the endangered southern cassowary, which is the only disperser of large-seeded rainforest trees in the Wet Tropics of Australia. High road mortality in this species also places the viability of many rainforest plant species in danger (Moore 2007).

49.4 Natural catastrophes and environmental conditions associated with tropical ecosystems exacerbate road impacts

Tropical environments are particularly vulnerable to road impacts because of the extreme rainfall events that commonly occur. Providing sufficient culverts to carry large amounts of stormwater is extremely difficult and expensive. During tropical downpours culverts often become overloaded or blocked, which results in wide-scale flooding upstream of roads and impeded flow downstream of roads (Fig. 49.3, Chapters 44 and 45). This causes death of vegetation through inundation upstream and lack of normal flow downstream (Laurance et al. 2009). Long bridges (e.g. Fig. 45.4) may be the only viable solution to this problem.

(A) (B)

Figure 49.3 Insufficient or undersized culverts result in (A) blockages upstream and flooding over the road causing (B) washouts of the road surface, landslides and scouring of downstream habitats. Source: Photograph by and reproduced with permission of W.F. Laurance.

During extremely heavy rainfall events, concentrated stream volumes with increased velocity are generated when culverts are undersized, which degrades streambed quality downstream of the road by scouring, channelizing and simplifying aquatic habitats (Iwata et al. 2003, Chapter 44). The erosion of road cuttings and embankments during downpours (Fig. 49.3B) can result in landslides (Sidle et al. 2006) and greater loads of sediment which accumulate downstream, also changing stream habitats (Bruijnzeel 2004). Major loads of pollutants and nutrients that accumulate on highways during the dry season wash into the streams in a sudden pulse with the first downpours of the tropical wet season. These contaminants can travel long distances (Pratt & Lottermoser 2007), polluting streams and estuaries and adversely affecting the health and persistence of flora and fauna. Toxins can biomagnify up the food chain, whereas sudden pulses of nutrients can deplete oxygen levels in streams and rivers by promoting rapid algal and plant growth. Both can potentially lead to widespread death of aquatic biota (e.g. fish kills).

Although roads do not appear to increase the impacts of extremely severe tropical cyclones and hurricanes on forests (Pohlman et al. 2008), increased treefall does occur along rainforest edges during smaller cyclones and seasonal windstorms (Laurance

et al. 2009). Tree damage from lesser storms is similarly elevated along rainforest road edges.

In savannahs, woodlands and open forests of the tropics, the impact of common but severe droughts can be exacerbated by the potential for roads to facilitate access for humans who ignite catastrophic fires accidentally or intentionally (Hoffman et al. 2004; Roman-Cuesta & Martinez-Vilalta 2006). Alternatively, roads acting as firebreaks in these drier tropical habitats can alter natural fire regimes, and thereby change the vegetation structure and floristics (Harrington & Sanderson 1994; Vigilante et al. 2004).

49.5 Narrow and unpaved roads with continuous canopy cover are preferred over the greater impact caused by wide and paved roads in tropical rainforest

In tropical rainforests, edge and barrier effects of roads are particularly important in comparison with other habitats (Lesson 49.2). These changes become more severe as the road becomes wider. More light penetrates into the forest floor and temperature and humidity fluctuate through greater extremes, consequently affecting a larger area of forest and greater numbers of species. By contrast, maintaining a substantially intact forest canopy above a narrow road limits the amount of light reaching the road surface and forest edge and maintains more stable humidity and temperature levels. This will reduce the much larger and more extensive alterations to vegetation structure and floristics and thus barrier effects incurred with a wider road (Laurance et al. 2004, 2009).

A narrow road with connected rainforest canopy extending above it also has several more general advantages (Goosem 2007). First, weed invasion is almost completely prevented (Fig. 49.1B). Second, traffic travels more slowly, so disturbance to fauna from noise, vibrations, headlights and vehicular movements is less. Unpaved roads may be noisier when traversed, but they generally carry fewer vehicles than paved roads so that on average, disturbance is less. Third, slower speeds allow greater reaction time for both driver and animal, which may mean lower levels of animal mortality. Narrow, unsurfaced roads also often carry minimal traffic at night, reducing roadkill of nocturnal animals, which are very common in the diverse rainforest fauna. Finally, the canopy intercepts some of the precipitation in tropical downpours reducing road runoff, erosion and stream modifications.

At the global scale, larger all-weather roads in wet tropical environments are often the drivers of large-scale clearing of forest for agriculture, ranching, forestry and mining (Chapters 2 and 51). Without large all-weather roads to transport products and supplies such investments would not be profitable. Narrow, unpaved roads are far less likely to support large-scale agricultural clearing or year-round resource extraction as they often need to be closed during extremely wet periods. Larger paved roads are also more likely to encourage the illegal colonisation of undisturbed areas by people who wish to settle in the new frontier (Chapters 2 and 51). However it should be remembered that the creation of a small track or road often acts as a catalyst for expansion. Although every tropical road cannot be retained in a narrow and unpaved state, expansion of any such road requires careful consideration of the actual need, design, ecosystem and cultural sensitivity as well as the environmental conditions that might exacerbate their impact (Laurance et al. 2014).

49.6 Tropical roads often are located in developing nations aiming to improve their economies and standard of living by opening frontiers of previously undeveloped lands

Many recently constructed tropical roads and those proposed for construction are concentrated in developing countries (Laurance et al. 2009; Clements et al. 2014; Chapters 51, 52, 54 and 57). These nations often have high population growth rates and are seeking to expand their economies and improve their standard of living through intense exploitation of natural resources such as timber, oil or minerals, or through creation of new agricultural industries. Such new frontiers are common in the tropics. From the perspective of many economists and regional planners the opening up of regions for development of natural resources is an advantage. By contrast, environmental scientists generally believe this to be a serious problem for biodiversity conservation (Butler & Laurance 2008, Chapters 2 and 51). Impacts on biota and the culture of indigenous and minority groups are severe when roads and highways open up large areas of previously undeveloped lands.

The construction of roads provides the impetus for a raft of environmental impacts. These range from hunting of bush meat along roads (Blake et al. 2008) and illegal logging, to clearing for small-scale human settlements that gradually expand, eventually resulting in huge cleared swaths of previously forested land (see Chapter 51). Although these latter impacts are far more severe and occur more frequently in unprotected forests, hunting and clearing are still likely to a lesser extent in protected areas (Roman-Cuesta & Martinez-Vilalta 2006). Roads are also required by wealthy investors and corporations currently involved in clearing of enormous areas of tropical forests for large-scale industrial agriculture (e.g. palm oil, corn and soybean plantations) and for large cattle ranches (Laurance & Balmford 2013). Developing nations that need income to provide for large, expanding populations often welcome such investment. However, there may be alternatives whereby they could retain biodiversity and increase economic benefits of current cleared land (Laurance et al. 2014; Chapter 2).

49.7 Where it is impossible to avoid undisturbed regions and sensitive habitats, sensitive tropical areas should either be protected before road construction or access should be restricted

Avoidance of undisturbed wilderness regions, sensitive habitats such as rainforest and wetlands, and areas subject to physical constraints (e.g. erosion and landslips) will always be the most successful strategy to prevent road impacts (Chapter 3). By avoiding remote frontier regions, the problems caused by human access to and colonisation of such areas are also avoided, as are the usually consequent increases in habitat clearing, extraction of timber and other plants and hunting of animals. Where economic growth is needed, it may be possible to improve agricultural efficiency by upgrading roads through regions that are already cleared, avoiding the need to open and clear new areas (Laurance & Balmford 2013; Laurance et al. 2014; Chapter 2). Although this mitigation strategy applies globally, the sensitive nature of tropical habitats and biota and the potential for severe natural catastrophes means that it is particularly important in the tropics.

If road access must be created to alleviate poverty through resource extraction or improved transport, several actions can mitigate against impacts of colonisation, the spawning of networks of secondary roads (Fig. 2.1, Lessons 2.3 and 2.4) and the illegal extraction of other resources (Laurance et al. 2009):

• For a large road or highway, creating a protected area around the road prior to construction reduces potential colonisation by legal and illegal squatters. However, adequate resources must be provided in perpetuity to ensure that protected area staff can adequately combat illegal activities such as poaching, timber removal and forest fires.

• For smaller mining or logging roads, secured gates can assist with preventing large-scale ingress of people. Contracts with staff that preclude subsistence removal of bushmeat and other resources also reduce depletion of forest areas adjacent to gated roads.

• Logging road networks need to be managed to avoid subsequent deforestation using careful pre-harvest planning to minimise unnecessary roads. Additionally, roads should be closed after harvest by destroying bridges and rendering long sections of road impassable.

• Environmental impact assessments (EIAs) must be comprehensive and thorough, taking into account potential secondary and cumulative impacts (Chapter 5). In many developing nations, EIAs for highways focus on a narrow strip of land along the proposed road, and ignore the potential for forest invasions, secondary road expansion and hunting (Chapter 53). Similarly, EIAs for mining, hydroelectric dams and oil and gas projects often ignore the impacts of the roads required to build the projects (see also Lesson 53.5).

49.8 Mitigation of tropical road impacts at the roadway scale can be achieved by maintaining canopy connectivity, minimising road width and restricting logging roads to low-impact areas

For rainforest roads, retaining connected canopy above the road (Fig. 40.1A) will mitigate against many road impacts (Goosem 2007; Lesson 40.2),

including reducing the severity of edge effects and invasion by weeds and feral animals. Connected canopies provide a crossing route for arboreal animals and reduce barrier effects for terrestrial fauna by providing cover from aerial predators and moderating the microclimate contrast between forest and road clearing. Road maintenance costs are reduced due to lowered requirement for mowing, spraying or grading of roadsides, and less pesticide reaches waterways. Maintaining canopy also increases the attractiveness of roads to tourists, creating a 'green tunnel' effect without tall weeds that obscure the view. This strategy is particularly appropriate for narrow roads with low traffic levels. For roads that carry higher traffic levels, maintaining continuous canopy above the road may not be feasible due to safety concerns or road width. However, keeping the road and roadside as narrow as possible will minimise edge effects, invasions and barrier effects for animals. Steep road embankments and cuttings will help to minimise clearing width, and these can be stabilised using the roots of native tree species such as figs which also provide a spreading canopy that is not subject to safety issues with falling limbs. Understory removal below such species can provide safe visibility for drivers. Gabions can be included for extremely unstable embankments but should also be planted with suitable vegetation (Goosem et al. 2010).

Restricting logging roads to ridgelines and gentle slopes, minimising the number of stream crossings and the inclusion of well-designed and sufficient culverts or bridges will reduce erosion, landslides and impacts to streams as well as barriers to terrestrial and aquatic connectivity (Laurance et al. 2009; Chapters 44 and 45). Additionally, if road construction is prohibited during wet seasons/periods, the consequent damage to aquatic habitats up- and downstream will be further limited.

Many tropical terrestrial animals will cross under roads via culverts, bridges and viaducts (Fig. 49.4; Goosem 2008; Clements 2013). Rope canopy bridges are used by tropical arboreal species including specialised rainforest possums in Australia and primates in other tropical countries (Weston et al. 2011; Chapters 40 and 41). Suitable existing road culverts can be retrofitted at low cost to increase permeability for some fauna. Unfortunately, not all species are prepared to use such structures, which is compounded by the difficulty of maintaining effective fences year-round (Goosem et al. 2010; Clements 2013; Chapter 20). In such areas, reducing and enforcing speed limits and including traffic calming will provide

(A)

(B)

(C)

(D)

Figure 49.4 Open span bridges and viaducts can be an effective approach to restoring connectivity for large tropical rainforest animals because of their size and distance from vehicles. This bridge in (A) Perak, Malaysia, is used by generalist rainforest species, including (C) Asian elephant and (D) barking deer. The use of viaducts by rainforest specialists which rarely or never use them, such as leopard and (B) Malayan tiger (Clements 2013), should be encouraged by planting rainforest trees and shrubs adjacent to and under the viaduct while crossing at grade around viaducts can be discouraged by keeping roadsides at each end of viaducts cleared of rainforest vegetation. Source: Photograph by and reproduced with permission of Rimba/Reuben Clements.

more time for reaction by both driver and animal to avoid collision.

Existing roads that are no longer needed should be closed with culverts and bridges removed and road surfaces rehabilitated (Chapter 3). Alternative routes that avoid sensitive areas should be investigated so that existing roads proven to create high levels of environmental damage can be closed.

CONCLUSIONS

Roads in the tropics present a unique set of challenges from broad landscape scales to the fine scale of an individual road. Avoiding sensitive tropical habitats is by far the best strategy to reduce impacts. This approach will minimise secondary and cumulative road impacts at the landscape scale, including forest colonisation and clearing, secondary road expansion, hunting and illegal extraction of resources. At the scale of the individual road, alternative routes that avoid sensitive regions may be less expensive than dealing with the greater engineering challenges resulting from extreme rainfall and vulnerable and diverse flora and fauna.

ACKNOWLEDGEMENTS

I acknowledge all the colleagues, students and volunteers who have contributed to rainforest road ecology research in the tropics over the past 25 years.

FURTHER READING

Laurance et al. (2009): An easily understood but comprehensive review of the impacts of roads in tropical rainforest at the scale of landscapes and individual road. It concludes with a section describing ways to prevent and mitigate these impacts.

Goosem et al. (2010): A book, freely available online. Part A provides an easily interpreted set of principles and guidelines for road agency personnel to implement best practice planning, design and management of ecologically sustainable roads within tropical rainforest. Part B summarises scientific findings which support the principles and guidelines in Part A and includes a comprehensive list of references and further readings.

Goosem (2008): This book chapter summarises many of the numerous road ecology studies conducted in the Australian Wet Tropics, which established the basis for the design of successful mitigation strategies. Available from: http://research.jcu.edu.au/portfolio/miriam.goosem1/

Laurance and Balmford (2013): Describes impacts of roads in undeveloped regions and principles for avoiding those impacts whilst maximising financial benefits, and is particularly appropriate in developing nations with wilderness areas.

REFERENCES

Asari, Y., C.N. Johnson, M. Parsons and J. Larson. 2010. Gap-crossing in fragmented habitats by mahogany gliders (*Petaurus gracilis*). Do they cross roads and powerline corridors? Australian Mammalogy **3**: 10–15.

Blake, S., S.L. Deem, S. Strindberg, F. Maisels, L. Momont, I.-B. Isia, I. Douglas-Hamilton, W.B. Karesh and M.D. Kock. 2008. Roadless wilderness area determines forest elephant movements in the Congo Basin, PLoS One **3**: e3546. doi:10.1371/journal.pone.0003546

Bruijnzeel, L.A. 2004. Hydrological functions of tropical forests: not seeing the soil for the trees? Agriculture, Ecosystems and Environment **104**: 185–228.

Butler, R.A. and W.F. Laurance. 2008. New strategies for conserving tropical forests. Trends in Ecology and Evolution **23**: 469–472.

Clements, G.R., A.J. Lynam, D. Gaveau, W.L. Yap, S. Lhota, M. Goosem, S. Laurance and W.F. Laurance. 2014. Where and how are roads endangering mammals in southeast Asia's forests? PLoS ONE **9**(12): e115376. doi:10.1371/journal.pone.0115376

Clements, G.R. 2013. The environmental and social impacts of roads in Southeast Asia. PhD thesis, James Cook University.

Dejean, A. and M. Gibernau. 2000. A rainforest ant mosaic: the edge effect (Hymenoptera: Formicidae). Sociobiology **35**: 385–401.

Delaval, M. and P. Charles-Dominique. 2006. Edge effects of frugivorous and nectarivous bat communities in a neotropical primary forest in French Guiana. Revue D Ecologie-la Terre Et La Vie **61**: 343–352.

Goosem, M. 1997. Internal fragmentation: the effects of roads, highways and powerline clearings on movements and mortality of rainforest vertebrates. Pages 241–255, in W.F. Laurance and R.O. Bierregaard, Jr., editors. Tropical forest remnants: ecology, management and conservation of fragmented communities. University of Chicago Press, Chicago, IL.

Goosem, M. 2007. Fragmentation impacts caused by roads through rainforests. Current Science **93**: 1587–1595.

Goosem, M. 2008. Rethinking road ecology. Pages 445–459, in N. Stork and S. Turton, editors. Living in a dynamic forest landscape. Blackwell, Australia.

Goosem, M., E.K. Harding, G. Chester, N. Tucker, C. Harriss and K. Oakley. 2010. Roads in rainforest: Best practice guidelines for planning, design and management. Roads in rainforest: the science behind the guidelines. Guidelines prepared for the Queensland Department of Transport and Main Roads and the Australian Government's Marine and Tropical Sciences Research Facility. Reef and Rainforest Research Centre, Cairns, Australia. Available from: http://www.rrrc.org.au/wp-content/uploads/2014/06/guidelines.pdf, http://www.rrrc.org.au/wp-content/uploads/2014/06/science-background.pdf (accessed 9 October 2014).

Goosem, M. and C. Pohlman. 2014. Rapid assessment of environmental monitoring sites. Report to Powerlink, Queensland. James Cook University, Cairns.

Goosem, S. and N.I.J. Tucker. 2013. Repairing the rainforest (2nd ed.) Wet Tropics Management Authority and Biotropica Australia Pty. Ltd., Cairns. 158 pp.

Gubbi, S., H.C. Poornesha and M.D. Madhusudan. 2012. Impact of vehicular traffic on the use of highway edges by large mammals in a South Indian wildlife reserve. Current Science **102**: 1047–1051.

Harrington, G.N. and K.D. Sanderson. 1994. Recent contraction of wet sclerophyll forest in the wet tropics of Queensland due to invasion by rainforest. Pacific Conservation Biology **1**: 319–327.

Hoffman, W.A., V.M.P.C. Lucatelli, F.J. Silva, I.N.C. Azeuedo, M. da S. Marinho, A.M.S. Albuquerque, A de O. Lopes and S.P. Moreira. 2004. Impact of the invasive alien grass *Melinis minutiflora* at the savanna-forest ecotone in the Brazilian Cerrado. Diversity and Distributions **10**: 99–103.

Hoskin, C. and M. Goosem. 2010. Road impacts on abundance, call traits and body size of rainforest frogs in north-east Australia. Ecology and Society **15**: 15. http://www.ecologyandsociety.org/vol15/iss3/art15/ (accessed 9 October 2014).

Iwata, T., S. Nakano and M. Inoue. 2003. Impacts of past riparian deforestation on stream communities in a tropical rainforest in Borneo. Ecological Applications **13**: 461–473.

Laurance, S. G. 2004. Responses of understory rain forest birds to road edges in central Amazonia. Ecological Applications **14**: 1344–1357.

Laurance, S.G., P. Stouffer and W. Laurance. 2004. Effects of road clearings on movement patterns of understory rainforest birds in Central Amazonia. Conservation Biology **18**: 1099–1109.

Laurance, W.F. and A. Balmford. 2013. A global map for road building. Nature **495**: 308–309.

Laurance, W.F. and M. Goosem. 2008. Impacts of habitat fragmentation and linear clearings on Australian rainforest biota. Pages 295–306, in N. Stork and S. Turton, editors. Living in a dynamic forest landscape. Blackwell, Australia.

Laurance, W.F., M. Goosem and S.G. Laurance. 2009. Impacts of roads and linear clearings on tropical forests. Trends in Ecology and Evolution **24**: 659–669.

Laurance, W.F., G.R. Clements, S. Sloan, C. O'Connell, N.D. Mueller, M. Goosem, O. Venter, D.P. Edwards, B. Phalan, A. Balmford, R. van der Ree and I.B. Arrea. 2014. A global strategy for road building. Nature, **513**: 229–232.

Lee, R.J., A.J. Gorog, A. Dwiyahreni, S. Siwu, J. Riley, H. Alexander, G.D. Paoli and W.D. Ramono. 2005. Wildlife trade and implications for law enforcement in Indonesia: a case study from North Sulawesi. Biological Conservation **123**: 477–488.

Moore, L.A. 2007. Population ecology of the southern cassowary, *Casuarius casuarius johnsonii*, Mission Beach, north Queensland. Journal of Ornithology **148**: 357–366.

Pohlman, C., M. Goosem and S. Turton. 2008. Effects of severe tropical cyclone Larry on rainforest vegetation and understorey microclimate near a road, powerline and stream. Austral Ecology **33**: 493–515.

Pohlman, C., S. Turton and M. Goosem. 2009. Temporal variation in microclimate edge effects near powerlines, highways and streams in tropical rainforest. Agricultural and Forest Meteorology **149**: 84–95.

Pratt, C. and B.G. Lottermoser. 2007. Mobilisation of traffic derived trace metals from road corridors into coastal stream and estuarine sediments, Cairns, northern Australia. Environmental Geology **52**: 437–448.

Roman-Cuesta, R.M. and J. Martinez-Vilalta. 2006. Effectiveness of protected areas in mitigating fire within their boundaries: case study of Chiapas, Mexico. Conservation Biology **20**: 1074–1086.

Sidle, R.C., A.D. Ziegler , J.N. Negishi, A.R. Nik, R. Siew and F. Turkelboom. 2006. Erosion processes in steep terrain – truths, myths and uncertainties related to forest management in Southeast Asia. Forest Ecology and Management **224**: 199–225.

Stork, N.E., S. Goosem and S.M. Turton. 2011. Status and threats in the dynamic landscapes of Northern Australia's tropical rainforest biodiversity hotspot: the Wet Tropics. Pages 311–332, in F.E. Zachos and J.C. Habel, editors. Biodiversity hotspots: distribution and protection of conservation priority areas. Springer, Berlin.

Vigilante, T., D.M.J.S. Bowman, R. Fisher, J. Russell-Smith and C. Yates. 2004. Contemporary landscape burning patterns in the far North Kimberley region of north-west Australia: human influences and environmental determinants. Journal of Biogeography **31**: 1317–1333.

Vijayakumar, S.P., K. Vasudevan and N.M. Ishwar. 2001. Herpetofaunal mortality on roads in the Anamalai Hills, southern Western Ghats. Hamadryad **26**: 265–272.

Weston, N., M. Goosem, H. Marsh, M. Cohen and R. Wilson. 2011. Using canopy bridges to link habitat for arboreal mammals: successful trials in the Wet Tropics of Queensland. Australian Mammalogy **33**: 93–105.

Wilson, R.F., H. Marsh and J. Winter. 2007. Importance of canopy connectivity for home range and movements of the rainforest arboreal ringtail possum (*Hemibelideus lemuroides*). Wildlife Research **34**: 177–184.

Wet Tropics Management Authority (WTMA). 2009. State of the Wet Tropics report 2008-2009. WTMA, Cairns, Australia.

Zachos, F.E. and J.C. Habel. 2011. Biodiversity hotspots: distribution and protection of conservation priority areas. Springer, Berlin.

Chapter 50

THE INFLUENCE OF ECONOMICS, POLITICS AND ENVIRONMENT ON ROAD ECOLOGY IN SOUTH AMERICA

Alex Bager[1], Carlos E. Borghi[2] and Helio Secco[1]

[1]Brazilian Center for Road Ecology Research, Department of Biology, Federal University of Lavras, Lavras, Brazil
[2]CIGEOBIO (UNSJ-CONICET), CUIM, Facultad de Ciencias Exactas, Físicas y Naturales, San Juan, Argentina

SUMMARY

Rapid economic growth in several South American countries combined with high species diversity in tropical regions has raised great concern among ecologists on the future of wildlife in those areas. One of the consequences and drivers of economic growth is widespread infrastructure development. The economic and social development of most countries in South America is a higher priority than biodiversity conservation, especially when compared with transport infrastructure. Over the past decade, several research groups have focussed on the impacts of roads on wildlife in South America.

50.1 The economic development of some countries in South America has strongly influenced the expansion of road networks.

50.2 Although there are numerous organisations involved in planning, development and administration of highways, only few of them evaluate the impacts on biodiversity.

50.3 Numerous protected areas in South America are directly and indirectly affected by roads.

50.4 Road ecology is an emerging discipline in South America, and Brazil and Argentina are leading the field.

There are enormous challenges to effectively incorporate ecological considerations into the planning, design, construction and operation of roads in South America. While much of the current practise has been adapted from international experience, the time has come to invest in local experts and improve the quality of the scientific knowledge generated from within South America. Government policies must also support the development of an ecologically sustainable transportation network.

Handbook of Road Ecology, First Edition. Edited by Rodney van der Ree, Daniel J. Smith and Clara Grilo.
© 2015 John Wiley & Sons, Ltd. Published 2015 by John Wiley & Sons, Ltd.
Companion website: www.wiley.com\go\vanderree\roadecology

INTRODUCTION

South America comprises 12 countries across 18.4 million km², each with diverse geographic, economic and social characteristics. South America is largely a developing continent, with its countries ranked between the 44th and 117th positions (out of 187 countries) by the Human Development Index (HDI), a statistic that measures life expectancy, education and income. The inequality in HDI amongst South American countries is reflected in the level of the development of national transportation networks, with road densities ranging from 0.015 km/km² in Bolivia to 0.049 km/km² in Uruguay, and a railroad density of 0.00009 km/km² in Paraguay and of 0.014 km/km² in Argentina (IIRSA 2010).

South America is physiographically, biologically and culturally diverse, including extreme environments, such as the permanent snow-covered mountains in the high Andes, dry deserts (e.g. the Atacama) and the wettest tropical forests (e.g. the Amazon). Biologically, South America includes 5 of the 17 countries considered megadiverse (Mittermeier et al. 1997). South American countries also vary culturally, economically and socially and, consequently, exhibit different environmental protection perspectives. The planning and construction of infrastructure, including road, rail and air transport, is a priority because it contributes to economic development. Measures to protect the environment or support sustainable development are often neglected and usually inadequate.

LESSONS

50.1 The economic development of some countries in South America has strongly influenced the expansion of road networks

The development of the road network is heterogeneous in South America. Brazil, the largest country and one of the most developed in South America, has been described as an emerging economy and is currently the seventh largest economy in the world. The gross domestic product (GDP) in Brazil was worth US$2.3 trillion in 2012 (4% of the global economy), a value similar to the GDP of the United Kingdom. To support and drive this economic growth, the Brazilian Federal Government has prioritised the construction of highways and railroads. Between 2007 and 2010, an additional 4731 km of highways and 356 km of railroads was added to the network, and by 2014, approximately

7000 km of highway and 3000 km of railroads will be built, duplicated or repaired. In addition to these works, the Brazilian Government will allocate US$39 billion to highways and railroads over the next 5 years and US$26 billion in the following 25 years. This will result in 7,500 km of new highways and 10,000 km of new railroads, an increase of nearly 20% of paved roads. To achieve this, the Brazilian Government has systematically reduced the environmental licensing requirements for the construction of highways and railways, preferring to fast-track construction at the cost of adequate environmental impact assessments (EIA). While other countries are also planning to expand their transport networks, their growth rate is slower than in Brazil. Bager (unpublished data) estimated that over 400 million vertebrates are killed on Brazilian highways each year. If the road network increased 20%, Brazil will cause the loss of half a billion vertebrates annually.

Another country that is rapidly expanding its road network is Colombia; paving and building new roads and planning to increase its network of paved roads by nearly 8000 km in the next 5 years. Furthermore, the government of Venezuela has a National Railway Development Plan, aiming to expand the country's railway network to 13,665 km of railroads by the year 2030, more than a 10-fold increase over its current system. Economic development requires adequate connections among countries and between coasts to ensure efficient regional and intercontinental trading.

In this context, the Initiative for the Integration of Regional Infrastructure of South America (IIRSA, a joint program of governments of 12 countries) was established in 2000 with the goal of promoting the sustainable and equitable development of transport, energy and telecommunication infrastructure through the integration of all South American countries. The IIRSA projects are organized around 10 hubs spread around South America, which were established according to their productive economic activities and potential for development. In the first 10 years (2000–2010), 474 integration projects were executed, of which 225 were highways (valued at US$49.28 billion), and 61 were railroads (US$14.16 billion) (IIRSA 2010). The program is continuing and transportation projects remain a priority (see www.iirsa.org).

The approach to highway construction across South America is changing, with several governments awarding highway concession contracts to private companies, with varying consequences for biodiversity. Brazilian private companies with concessions are controlled by regulatory agencies that establish more rigid environmental controls than those imposed on highways built and managed by the state or federal governments. These companies usually charge tolls for the use of the highway and in return must commit to making infrastructure improvements and maintaining the highway for a period of time, including environmental requirements. However, this is not consistent among all South American countries, for example private companies in Argentina have lower environmental standards than that of government road agencies.

50.2 Although there are numerous organisations involved in planning, development and administration of highways, only few of them evaluate the impacts on biodiversity

Poor quality EIA are a serious problem in South America (see also Chapters 5, 51, 53, 54 and 56). Minimal controls during the road planning and construction phase and poor quality environmental monitoring programmes produce inefficient mitigation measures. The only organisation that exclusively addresses ecological impacts of roads in South America is the Brazilian Center for Road Ecology Research (CBEE) (http://cbee.ufla.br). CBEE was established in 2011 and is affiliated with the Federal University of Lavras. Its main goal is to undertake research, train specialists, develop technology related to the mitigation of road effects, and contribute

to the development of public policies. Currently, the CBEE produces a monthly newsletter and organises the biannual Road Ecology Brazil Congress, which started in 2010 with more than 200 participants. Because most road ecology studies being conducted in Brazil include monitoring wildlife road-kills (either by highway concessionaires or researchers), a primary task of the CBEE is to develop a standardised national protocol to collect road-kill data and a unified database (Textbox 50.1).

In addition to the CBEE, there are a range of other research groups in South America working on road ecology. In 2008, INTERBIODES, an Argentine research group was formed; associated with the National Council for Scientific Investigation and the National University of San Juan. This group conducts research and also advises the Federal Highway Administration on road ecology issues. Additionally, in the province of Misiones, an NGO (Conservation Argentina) and the highway planning body have a research group collaborating on road planning. Currently in Brazil, there are at least four research groups focussed on road ecology and many others that informally study the topic. These groups collaborate with research and other activities that contribute to the growth of the road ecology discipline in the country. Although we are not aware of the existence of formal research groups in Ecuador, some studies have been undertaken on different groups of fauna (Tanner & Perry 2007; Gottdenker et al. 2008; Carpio et al. 2009) and road ecology topics (Byg et al. 2007; Suárez et al. 2009; Suárez et al. 2012).

Textbox 50.1 Brazilian National Wildlife Roadkill Database.

The CBEE, together with the federal Brazilian Institute for Biodiversity Conservation (*Instituto Chico Mendes de Conservação da Biodiversidade*), are developing a wildlife roadkill database that will compile data from a wide array of sources, such as scientific research projects, drivers, and surveys conducted during the planning of highways and railroads http://cbee.ufla.br/portal/sistema_urubu/. This database will integrate roadkill data with a web-based geographical platform, modules of data analysis and reports. Mobile phone applications (Fig. 50.1) have been developed to collect WVC data in the field, including for the general public to record opportunistic sightings; and scientists, with capacity for systematic sampling, control of the monitored distance and images of the highway and

surrounding landscape (http://goo.gl/zy9VCF and http://goo.gl/x2h6Sx). The database will be connected to the Brazilian Biodiversity Information System (SIB-Br), thereby allowing the use of the data in the development of state and federal public policies, both for biodiversity protection and the planning of highways and railroads. This database is linked to the development of the standardized national protocol to collect roadkill data (Lesson 50.2). We believe that if a collection tool is available to store and analyse data, various research bodies and institutions will adopt it as their protocol, and Brazil will have spatially and temporally comparable data from every region of the country. This system can also be adapted for other regions in South America.

Figure 50.1 The welcome page of the Urubu System to collect and analyse wildlife road-kill data in Brazil. Source: Alex Bager.

50.3   Numerous protected areas in South America are directly and indirectly affected by roads

More than half (62%) of the federally protected conservation areas in Brazil are intersected by highways and 72% are indirectly affected by highways, accounting for 5.6% of the total park area (Botelho et al. 2012). Many of the most affected areas are in the Atlantic Rainforest (*Mata Atlântica*) biome, which is close to the coastal region and the major urban centers. No similar studies for other South American countries have been published.

In Brazil, the managers of 300 state and federal conservation reserves were interviewed and most identified conflicts between roads and wildlife as a significant issue in their reserve (A. Bager, unpublished data). The managers also reported that 10% of the protected areas contained roads constructed within the previous 5 years, and 25% will have new roads within the next 5 years. Sixty percent of the protected areas had public roads, including major highways, running through them. Managers of 74 % of the protected areas reported wildlife-vehicle collisions (WVC) in their parks, with 41% reporting occasional collisions, 36% constant but at a low frequency, 14% constant and

daily and 1% had high roadkill rates; 8% didn't quantify the rate of WVC. This study also demonstrated that mortality from WVC was a significant cause of mortality in the protected areas for 23 of the 29 species of medium and large mammals that are officially threatened with extinction in Brazil. Park managers identified native feline, canine and arboreal species as most affected.

Recent surveys within 11 conservation reserves of Argentina found 12 species under threat (including the jaguar and the Andean condor), which are directly impacted by WVC (C. Borghi, unpublished data; Speziale et al. 2008; Chapter 36).

50.4   Road ecology is an emerging discipline in South America, and Brazil and Argentina are leading the field

Despite the rapid growth of road ecology in Brazil since 2004, most studies have focused on rates of roadkill (e.g. Bager et al. 2007; Dornas et al. 2012) (Figs 50.2 and 50.3) and were published in regional portuguese language journals (e.g. Cherem et al. 2007; Kunz & Ghizoni Jr. 2009; Zaleski et al. 2009). However, sampling effort was usually not documented

Figure 50.2 Capybara, the largest rodent in the world (weighing up to 50–60 kg), are common in most countries in South America and suffer high rates of mortality from WVC, especially where roads cross wetlands. Source: Photograph by Alex Bager.

Figure 50.3 Secondary mortality, where scavengers feeding on carcasses become roadkill themselves, is a major problem for raptors in South America as the rate of WVC continues to increase. Source: Photograph by Alex Bager.

in these studies, making it difficult to evaluate the effect of mortality on population persistence. There was a significant increase in publications in international journals after 2010 (e.g. Bager & Rosa 2010, 2011; Caceres 2011; Hartmann et al. 2011; Oliveira Jr. et al. 2011; Coelho et al. 2012; Freitas et al. 2012; Rosa & Bager 2012; Bager & Fontoura 2013; Teixeira et al. 2013; D'Anunciação; Ratton et al. 2014; Secco et al. 2014). In 2012, the first book on the road ecology

of South America (Bager 2012) was published, which addresses political and methodological topics. From this perspective, it was possible to identify major knowledge gaps in road ecology after the analysis of 41 articles published by Brazilian researchers (Table 50.1).

In 2008, Argentina was the first country in South America to install overpasses for wildlife, near Iguaçu National Park in the northern part of the country. In Brazil, wildlife underpasses and fences are routinely

Table 50.1 Summary of Brazilian studies in road ecology.

Knowledge level	Road ecology topic	Number of publications
No information, anecdotal at best	Physical disturbance	0
	Chemical pollution	0
	Exotic species invasions	0
Incipient level	Barrier effects	1
	Edge effects	2
Low level	Human invasions	3
	Fragmentation	3
	Spatial planning and mitigation	5
Medium level	Wildlife roadkill	27

used to mitigate the fragmentation caused by roads, although no rigorous monitoring programs have assessed effectiveness. The inclusion of fencing and underpasses in the design of new roads is positive; however, the use of standard-size underpasses for drainage (both culverts and bridges) and calling them wildlife crossing structures appears to be a justification to approve all new road projects. More detailed evaluation of effectiveness is required to ensure that new roads do not further endanger wildlife and ensure that damaging roads avoid high quality conservation areas.

CONCLUSIONS

To advance the study of road ecology in South America, we need stronger government legislation, expanded road ecology research programmes to increase the number and quality of investigative studies, and the creation of new course curriculums and training opportunities at the academic and technical levels. Our main challenge is to integrate these sectors and normalise the mitigation hierarchy: avoid impacts first, minimise second, mitigate third and compensate or offset any remaining impacts. The road and rail network across South America is rapidly expanding and there is an urgent need to better understand their effects on biodiversity to guide the growth of the transport sector.

ACKNOWLEDGEMENTS

We are grateful for financial support provided by Fapemig (Process CRA – PPM-00139/14; 453 and CRA – APQ-03868-10), CNPq (Process 303509/2012-0), Fundação O Boticário de Proteção à Natureza (Process 0945-20122), Tropical Forest Conservation Act – TFCA (through of Fundo Brasileiro para Biodiversidade – FUNBIO).

FURTHER READING

Bager (2011 and 2014): Annals of second and third Brazilian conference of road ecology, available at http://cbee.ufla.br.

Bager (2012): First book about road ecology in Portuguese, with three sections: politics, methods and study cases.

Bager and Rosa (2011): This article proposes different sampling efforts to measure rates of road kill, depending on the objectives of the study target group of wildlife.

Laurance et al. (2009): This article summarizes all types of negative effects caused by roads in tropical areas.

REFERENCES

Bager, A. 2011. Anais do Road Ecology Brazil 2011. Conference. Lavras, MG, Brasil, 247 p.

Bager, A. 2012. Ecologia de estradas: Tendências e perspectivas. UFLA, Lavras.

Bager, A. and C. A. Rosa. 2010. Priority ranking of road sites for mitigating wildlife roadkill. Biota Neotropica **10**: 149–143.

Bager, A. and C. A. Rosa. 2011. Influence of sampling effort on the estimated richness of road-killed vertebrate wildlife. Environmental Management **47**:851–858.

Bager, A., S. R. N. Piedras, T. S. M. Pereira and Q. Hobus. 2007. Fauna selvagem e atropelamento – diagnóstico do conhecimento científico brasileiro. Pages 49–62, in A. Bager, editor. Áreas Protegidas: Repensando as escalas de atuação. Armazém Digital, Porto Alegre.

Bager, A. and V. Fontoura. 2013. Evaluation of the effectiveness of a wildlife roadkill mitigation system in wetland habitat. Ecological Engineering **53**:31–38.

Botelho, R. G. M., J. C. L. Morelli, C. Bueno and A. Bager. 2012. O impacto das rodovias em unidades de conservação no Brasil. Page 232, in D. Boscolo, editor. Anais do II Congresso Brasileiro de Ecologia de Paisagens e II Simpósio SCGIS-BR, Salvador, Bahia.

Byg, A., J. Vormisto and H. Balslev. 2007. Influence of diversity and road access on palm extraction at landscape scale in SE Ecuador. Biodiversity and Conservation **16**:631–642.

Caceres, N. C. 2011. Biological characteristics influence mammal road kill in an Atlantic Forest-Cerrado interface in south-western Brazil. Italian Journal of Zoology **78**:379–389.

Carpio, C., D. A. Donoso and G. Ramón. 2009. Short term response of dung beetle communities to disturbance by road construction in the Ecuadorian Amazon. Annales de la Société Entomologique de France **45**:455–469.

Cherem, J. J., M. Kammers, I. R. Guizoni Jr. and A. Martins. 2007. Mamíferos de médio e grande porte atropelados em rodovias do Estado de Santa Catarina, sul do Brasil. Biotemas **20**:81–96.

Coelho, I. P., F. Z. Teixeira, P. Colombo, A. V. P. Coelho and A. Kindel. 2012. Anuran road-kills neighboring a peri-urban reserve in the Atlantic Forest, Brazil. Journal of Environmental Management **112**:17–26.

D'Anunciação, P. E. R., P. S. Lucas, V. X. Silva and A. Bager. 2013. Road Ecology and Neotropical amphibians: contributions for future studies. Acta Herpetologica **8**:129–140.

Dornas, R., A. Kindel, S. R. Freitas and A. Bager. 2012. Avaliação da mortalidade de vertebrados em rodovias no Brasil. Pages 139–152, in A. Bager, editor. Ecologia de estradas: Tendências e perspectivas. UFLA, Lavras.

Freitas, S. R., M. M. Alexandrino, R. Pardini and J. P. Metzger. 2012. A model of road effect using line integrals and a test of the performance of two new road indices using the distribution of small mammals in an Atlantic Forest landscape. Ecological Modelling **247**:64–70.

Gottdenker, N. L., T. Walsh, G. Jimenez-Uzcategui, F. Betancourt, M. Cruz, C. Soos, E. R. Miller and P. G. Parker. 2008. Causes of mortality of wild birds submitted to the Charles Darwin Research Station, Santa Cruz, Galapagos, Ecuador from 2002-2004. Journal of Wildlife Diseases **44**:1024–1031.

Hartmann, P. A., M. T. Hartmann and M. Martins. 2011. Snake road mortality in a protected area in the Atlantic forest of Southeastern Brazil. South American Journal Herpetology **6**: 35–42.

INICIATIVA PARA LAINTEGRACIÓN DE LA INFRAE-STRUCTURA REGIONAL SURAMERICANA (IIRSA). 2010. Capítulo II – Infraestructura Vial. Pages 1–38, in Facilitación del Transporte en los Pasos de Frontera de Sudamérica. IIRSA, Montevideo.

Kunz, T. S. and I. R. Ghizoni Jr.. 2009. Serpentes encontradas mortas em rodovias do Estado de Santa Catarina, Brasil. Biotemas **22**: 91–103.

Laurance, W. F., Goosem, M. and Laurance, S. G. W. 2009. Impacts of roads and linear clearings on tropical forests. Trends in Ecology and Evolution **24**: 659–669.

Mittermeier, R. A., Robles Gil, P. and Mittermeier, C. G. 1997. Megadiversity: earth's biologically wealthiest nations. CEMEX, Conservation International, Agrupación Sierra Madre, Cidade do México.

Oliveira Jr., P. R. R., C. C. Alberts and M. R. Francisco. 2011. Impacts of road clearings on the movements of three understory insectivorous bird species in the Brazilian Atlantic Forest. Biotropica **43**: 628–632.

Ratton, P., H. Secco and C. A. Rosa. 2014. Carcass permanency time and its implications to the roadkill data. European Journal of Wildlife Research. DOI 10.1007/s10344-014-0798-z.

Rosa, C. A. and A. Bager. 2012. Seasonality and habitat types affect roadkill of neotropical birds. Journal of Environmental Management **97**: 1–5.

Secco, H., P. Ratton, E. Castro, P. S. Lucas and A. Bager. 2014. Intentional snake road-kill: a case study using fake snakes on a Brazilian road. Tropical Conservation Science **7**:561–571.

Speziale, K. L., S. A. Lambertucci and O. Olsson. 2008. Disturbance from roads negatively affects Andean condor habitat use. Biological Conservation **141**: 1765–1772.

Suárez, E., M. Morales, R. Cueva, V. Utreras Bucheli, G. Zapata-Rios, E. Toral, J. Torres, W. Prado and J. Vargas Olalla. 2009. Oil industry, wild meat trade and roads: indirect effects of oil extraction activities in a protected area in north-eastern Ecuador. Animal Conservation **12**: 364–373.

Suárez, E., G. Zapata-Rios, V. Utreras, S. Strindberg and J. Vargas. 2012. Controlling access to oil roads protects forest cover, but not wildlife communities: a case study from the rainforest of Yasuní Biosphere Reserve (Ecuador). Animal Conservation **16**: 265–274

Tanner, D. and J. Perry. 2007. Road effects on abundance and fitness of Galápagos lava lizards (*Microlophus albemarlensis*). Journal of Environmental Management **85**: 270–278.

Teixeira, F. Z., A. V. P. Coelho, I. B. Esperandio and A. Kindel. 2013. Vertebrate road mortality estimates: effects of sampling methods and carcass removal. Biological Conservation **157**: 317–323.

Zaleski, T., D. Rocha, S. A. Filipaki and E. L. A. Monteiro-Filho. 2009. Atropelamentos de mamíferos silvestres na região do município de Telêmaco Borba, Paraná, Brasil. Natureza e Conservação **7**: 81–94.

Chapter 51

HIGHWAY CONSTRUCTION AS A FORCE IN THE DESTRUCTION OF THE AMAZON FOREST

Philip M. Fearnside

Department of Ecology, National Institute for Research in the Amazon (INPA), Manaus, Brazil

SUMMARY

Roads act as drivers of deforestation by drawing migrant workers and investment to previously inaccessible areas of forest. In Amazonia, deforestation is stimulated not only by roads that increase profitability of agriculture and ranching, but also by the effect of roads on land speculation and clearing for establishing and defending land tenure. Major highways are accompanied by networks of side roads built by loggers, miners and others. Deforestation spreads outwards from highways and their associated access roads. Highways also provide avenues for migration of landless farmers and others, thereby driving deforestation into adjacent areas.

51.1 Roads are important forces influencing the rate of deforestation in Amazonia.

51.2 Major roads stimulate deforestation by facilitating the construction of smaller side roads and human settlements in remote areas.

51.3 The alleged benefits of roads to the Amazon forest are illusory.

51.4 Roads must be included in deforestation models.

51.5 No amount of mitigation will prevent deforestation from occurring after a road is built.

51.6 Deforestation in Brazil is unregulated and future road projects will accelerate clearing.

51.7 'Governance scenarios' serve to justify approval of damaging roads.

51.8 Environmental safeguards are needed for approval of international financing of road development.

The consequences of the pattern of development associated with previously constructed Amazonian highways need to be recognised and lessons learned quickly, as plans for additional highways are rapidly moving forwards that would provide deforesters with access to much of the remaining area of Amazonian forest.

Handbook of Road Ecology, First Edition. Edited by Rodney van der Ree, Daniel J. Smith and Clara Grilo.
© 2015 John Wiley & Sons, Ltd. Published 2015 by John Wiley & Sons, Ltd.
Companion website: www.wiley.com\go\vanderree\roadecology

INTRODUCTION

The Amazon forest is by far the largest area of tropical rainforest in the world. About two thirds of the forest is in Brazil, the remainder being divided between Bolivia, Peru, Ecuador and Colombia. The Brazilian portion of the forest originally covered an area roughly the size of Western Europe. By 1995, an area the size of France had been cleared, and by 2012, areas equivalent to Portugal, Belgium and the Netherlands had been added to this. Approximately 80% of Brazil's portion of the Amazon forest remains, although logging and fire have disturbed a significant part of this. The size of the remaining forest is misleading; its vastness creates the illusion that there are no limits. However, the process of deforestation is cumulative and can rapidly advance through immense areas. The remaining blocks of relatively undisturbed forest owe their current state of preservation mainly to lack of access, especially the lack of roads. The Amazon forest provides environmental services such as conserving biodiversity, water cycling that provides rainfall to central and southern Brazil and to neighbouring countries and maintaining carbon stocks in biomass and soil that avoid greenhouse gas emissions. These services are lost when forests are cleared. Decisions on road construction and improvement have consequences for deforestation that last for decades and extend far beyond the roads themselves. Understanding the consequences of these decisions and the process of deforestation is essential to better decision-making about roads in the Amazon Basin and in many other tropical forest areas.

LESSONS

51.1 Roads are important forces influencing the rate of deforestation in Amazonia

The presence and quality of roads have been shown to be major factors in predicting deforestation throughout the Amazon Basin, including Brazil (e.g. Nepstad et al. 2001; Laurance et al. 2002; Soares-Filho et al. 2006; Pfaff et al. 2007), Bolivia (Kaimowitz 1997), Peru (Imbernon 1999) and Ecuador (Southgate et al. 1991; Mena et al. 2006). Understanding why roads lead to more deforestation is essential information both for designing ways to control clearing and for better decision-making about road construction.

Roads play multiple roles in faci litating deforestation. Roads improve the profitability of ranching and agriculture, thereby attracting investment in clearing. Roads also improve access to timber, unleashing a chain of events that spreads deforestation by providing money for clearing and facilitating entry of migrant workers along logging tracks. These clandestine roads, built primarily for logging, amount to tens of thousands of kilometres. Roads in forest areas attract migrants, forming part of a positive feedback loop justifying still more roads (Fearnside 1987a). Roads also provoke population turnover, replacing small individual farms with more capitalised landholders who deforest more. Roads increase land values in the areas near them, which in turn increases speculation and deforestation in order to establish and maintain land tenure (Fearnside 1987a). These effects evolve over time, with the ultimate result of a largely deforested landscape, usually dominated by cattle pasture in medium- or larger-sized properties.

The effect of roads is embedded in the social context of the country in which they are built. In the case of Brazilian Amazonia, they open frontiers to squatters (Fearnside 2001). The time when land tenure was established by squatting has faded from human memory in most areas of the world, but it is still common practice in Brazilian Amazonia (Fig. 51.1). In other tropical regions, such as Southeast Asia and Africa, this is an uncommon expectation. However, public land in Brazil is often illegally seized by individual squatters, organised groups of landless people and wealthy *grileiros* (landgrabbers who obtain land titles through various forms of fraud) (Fig. 51.2) (Fearnside 2008). A 2009 Brazilian law allowing illegal claims of up to 1500 ha to be legalised represents a setback in a transition to a land tenure system based on the expectation that illegal invasions will not be rewarded.

51.2 Major roads stimulate deforestation by facilitating the construction of smaller side roads and human settlements in remote areas

Major roads stimulate the construction of side roads that provide access to land far from the main highway route. An important example is the planned reconstruction of the BR-319 (Manaus–Porto Velho) Highway (Fig. 51.3; Textbox 51.1). Side roads would open the large block of intact forest in the western

Figure 51.1 Squatters near the BR-319 (Manaus–Porto Velho) Highway. Squatters, including organised landless farmers (*sem terras*), are one of several groups of actors that deforest when roads are built. Source: Photograph by P. Fearnside.

Figure 51.2 Cattle pasture in large and medium properties is the dominant land use in deforested areas. Here, *grileiros* (illegal landgrabbers) have established large ranches in the municipality of Lábrea (in the state of Amazonas) in an area with access by a privately constructed side road connected to the BR-364 Highway in Rondônia. Source: Photograph by P. Fearnside.

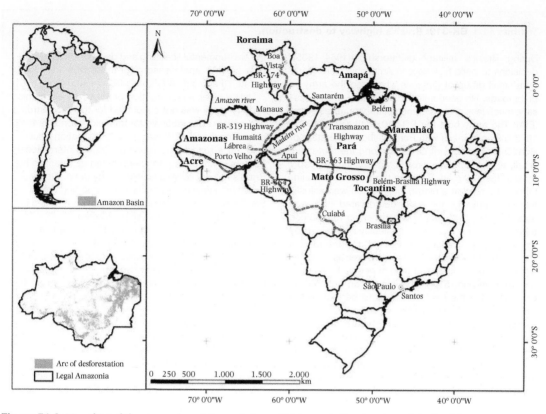

Figure 51.3 Map of Brazil showing major existing and planned highways in the nine states of the Legal Amazon region: Amazonas, Roraima, Pará, Amapá, Maranhão (western half), Tocantins, Mato Grosso, Rondônia and Acre. Source: Map by P. Fearnside.

part of the state of Amazonas that includes vast areas of public land – the category most vulnerable to invasion by *grileiros* and squatters (Fearnside & Graça 2006).

Highways also impact the forest by promoting migration of people. The proposed reconstruction of the BR-319 Highway would link Rondônia with areas in central and northern Amazonia that already have road access from Manaus. Rondônia is in the 'arc of deforestation': the crescent-shaped strip along the southern and eastern edges of the Amazon forest where approximately 80% of past deforestation is concentrated. Migration of people to Roraima along the BR-319 and BR-174 Highways is expected to stimulate rapid deforestation of new areas to the north and recolonisation of the largely abandoned agriculture and ranching district of the Manaus Free Trade Zone in the state of Amazonas (Fig. 51.3).

51.3   The alleged benefits of roads to the Amazon forest are illusory

Roads allegedly have positive effects on Amazonian forests, although these have often been contested. One claim is that roads bring governance, providing access for inspection and management by environmental authorities which prevents illegal deforestation (e.g. Nepstad et al. 2002a, b; Câmara et al. 2005; but see: Laurance & Fearnside 2002; Laurance et al. 2005). A persistent idea is that economic development, to which roads contribute, leads to a forest transition where the area of forest increases after an initial phase of deforestation. Unfortunately, in parts of the world where this has occurred, it has been largely based on planting non-native tree farms (such as eucalypts), as opposed to maintaining original forest. This is part of the theory in which increased wealth eventually leads to an improved environment through a shift in the

Textbox 51.1 BR-319: Brazil's highway to destruction.

During Brazil's military dictatorship (1964–1985), decisions to build highways in Amazonia were made by a small group of generals who used the army to build roads. No economic viability study was necessary, much less an environmental impact assessment (EIA). The army built the BR-319 Highway in 1972–1973 linking Manaus with Porto Velho, in the state of Rondônia. The amount of traffic along the route was small, since freight could reach Manaus more cheaply by ship than by road. Because the expense of maintaining the road in a high-rainfall region was unjustified economically, the road deteriorated and was abandoned by the highway department in 1988. Since then, the bridges have been minimally maintained by the telecommunications department, but the road is impassable to normal vehicles and serves as an effective barrier preventing migration of people to the central Amazon (Fig. 51.4). This may soon change, as reconstructing the road continues to be featured in the Federal Government's development plans. Together with existing and planned roads connected to this highway, the BR-319 would provide deforesters with access to approximately half of what remains of Brazil's Amazon forest. The tremendous potential impacts of the road and the multiple deficiencies in

the environmental licensing and decision-making process provide ample lessons for those who are willing to listen (Fearnside & Graça 2006).

Unlike other reconstruction projects, an economic viability study was not conducted for this reconstruction project. The rationale for this exception was that reopening the BR-319 was a matter of 'national security'. Yet a road far from any of Brazil's international borders is not among the items listed as priorities by Brazil's military. Were a viability study done, the result would be unfavourable because transporting freight to São Paulo from the Manaus Free Trade Zone would be significantly cheaper by ship (Teixeira 2007). The renewed interest in rebuilding the BR-319 appears to be primarily its value for electoral politics in Manaus.

The EIA for the project fails to compare the proposed road with the alternative options for transporting freight to São Paulo and confines itself to comparing transport modes between Manaus and Porto Velho (UFAM 2009). Porto Velho is not the destination of the freight, but rather a stopping place on the way to São Paulo. The EIA even admits that industry in Manaus does not consider the highway to be a priority. Furthermore, the EIA does not consider the highway's major impacts, namely, increased deforestation

Figure 51.4 The BR-319 (Manaus–Porto Velho) Highway was abandoned in 1988 and is a barrier to human migration. Its planned reconstruction would connect central and northern Amazonia with the 'arc of deforestation' where clearing has been concentrated along the southern and eastern edges of the Amazon forest (see Fig. 51.3). Source: Photograph by P. Fearnside.

and migration of people. Instead, it presents a scenario of 'strong environmental governance' and offers Yellowstone National Park, United States, as an example (Fig. 51.5). Yellowstone includes a network of roads yet no deforestation occurs. Unfortunately, the chaos of the Amazonian frontier is a completely different setting from Yellowstone National Park. It would be hard to exaggerate the unreality of the expectation that those who gain access to the forest by means of the new road would behave like visitors to Yellowstone.

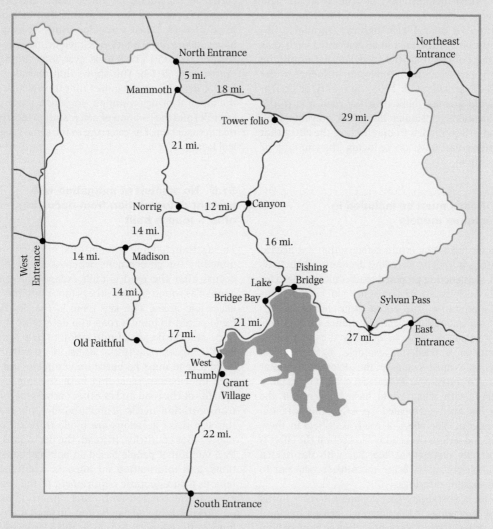

Figure 51.5 The EIA for the BR-319 Highway presents this map of Yellowstone National Park illustrating the presence of roads without the threat of deforestation (UFAM 2009). The National Department of Transport Infrastructure (DNIT) logo in the corner of the map represents the road-building branch of Brazil's Ministry of Transportation. Unrealistic governance scenarios incite approval of highway projects despite major impacts in the real world. Source: UFAM 2009.

equilibrium between destruction and restoration (known as the environmental Kuznets curve). Despite being generally discredited, this theory is a major influence on Brazilian planning, especially with regard to Amazonia (Hecht 2011).

Another supposed benefit of roads is that they act as 'magnets' that attract deforestation out of the interior and focus it in areas along roads (Aguiar 2006; Câmara 2007; contested by Fearnside et al. 2009). Unfortunately, roads stimulate clearing at the roadside as well as further away. The claim that roads attract deforestation to the roadside and thereby protect more distant areas was based on a simulation for the BR-319 Highway (Aguiar 2006) where the size of the area to be deforested each year was fixed externally by the modeller, with only the location of clearings being subject to influence by the road (see Fearnside et al. 2009). In reality, deforestation would increase both near the road and in the other locations. A similar limitation prevents such demand-driven models from reflecting the effect that protected areas have on reducing the amount of deforestation.

51.4   Roads must be included in deforestation models

Models for predicting future deforestation are needed for various purposes, including decisions on the wisdom of constructing proposed roads. Another need for modelling deforestation is for use in Reducing Emissions from Deforestation and Forest Degradation (REDD) projects for mitigation of climate change through forest protection. In order to quantify the amount of deforestation avoided, for example, by creating a reserve, one must compare the deforestation that occurs after the reserve is created (i.e. what takes place in reality) with what would have happened if the reserve had not been created (i.e. a hypothetical baseline scenario). The effect of roads is critical in these models, and serious distortions can occur if the effect is not correctly represented, together with the related effect of deforestation being more likely adjacent to already-existing clearings.

The Juma Sustainable Development Reserve in the state of Amazonas is the first such project in Brazil and provides an example of how choices in modelling of road effects can cause greatly distorted deforestation predictions and calculated carbon benefits. Roads are a key factor in the baseline scenario, based on Soares-Filho et al. (2006), which indicates 80.7% of

the reserve being deforested by 2050 in the absence of the project. The simulation calculated the total amount of deforestation each year for the subregion that includes the Juma Reserve by multiplying a proportion times the area of forest remaining in the subregion, which covers all of the state of Amazonas and parts of Mato Grosso and Pará. The calculated area to be cleared is therefore considerable even with a low proportion being cleared each year. The location where this clearing occurs is determined based on weights of evidence, which direct the clearing to the corner of the subregion where roads and previous clearings exist, namely, in the area that includes the Juma Reserve. When simulated using clearing rates based on local deforestation behaviour, clearing in the reserve by 2050 (18.9%) is over four times lower (Yanai et al. 2012). This shows that realistic modelling of deforestation requires that the calculation of the amount of deforestation (in which roads play a key role) and the location of deforestation (also closely tied to roads) must be referring to the same geographical location.

51.5   No amount of mitigation will prevent deforestation from occurring after a road is built

The decision to build a major road in Brazil is determined by the government, whereas much of what occurs after the road is built escapes government control as squatters and others move into the newly accessible areas. The key issue is the decision to build or not to build a road and determine its location – not the details of mitigation to help lower the road's impact. The initial decision on whether to build a road must be based on complete and unbiased information regarding both the impacts and benefits of the road and of other options for meeting transportation needs. Unfortunately, this is rarely the way these decisions are made in Brazil and in many other countries. Instead, the decision is made by a handful of people based on political considerations, and information on impacts is only sought later as a bureaucratic requirement of the licensing process. Comparison with other options is likewise an afterthought and can be done in such a way as to ignore the main options (see BR-319 example in Textbox 51.1). The decision-making process in Brazil (as in many other countries), including the EIA system, is in obvious need of reform (Fearnside 2012; Chapters 5 and 53).

51.6   Deforestation in Brazil is unregulated and future road projects will accelerate clearing

One of the barriers to instilling greater caution in road-building decisions is the belief that deforestation is under control and a thing of the past. This has become commonplace as a result of misinterpretation of the declining rate of deforestation in Brazil over 2004–2012 (e.g. Fearnside 2009), although planning officials made such claims before any decline in deforestation occurred (e.g. Silveira 2001; contested by Laurance et al. 2001). The belief that deforestation is under control is pervasive throughout Brazilian planning circles and affects the current Plan for the Acceleration of Growth (PAC-2). However, from 2004 to 2008, when most of the decline in the rate of deforestation took place, the main explanations were decreases in international prices of soy and beef and a worsening exchange rate between the Brazilian Real and other currencies for exporters (Assunção et al. 2012). Only from 2009 onwards did deforestation actually follow a trend that would best be explained by governance, not economic instability. Whether increases in governance mean that Amazonia can be criss-crossed by new highways without increasing deforestation is open to question.

51.7   'Governance scenarios' serve to justify approval of damaging roads

Planning of new highways is often done on the strength of 'governance scenarios' that portray a future with highways that bring benefits but minimal impacts. For example, the effect of the proposed reconstruction of the BR-163 (Cuiabá–Santarém) Highway was simulated, with a governance scenario indicating much less deforestation than a 'business-as-usual' scenario (Soares-Filho et al. 2004). The decision-maker is presented with choices as if he or she were in an all-you-can-eat buffet restaurant where one is free to choose anything with essentially no consequences. Any planner will choose the governance scenario over one indicating environmental destruction. However, the governance scenario implies tremendous change in government action and in individual behaviour, which is unrealistic. Consider the BR-163, a road intended to transport soybeans from Mato Grosso to a port on the Amazon River, in which the crash in global soy prices from 2003 led to successive postponements of the project (but the road is now expected to be completed in 2015) (Fig. 51.6). Even without the reconstructed highway, deforestation in the area has exceeded even that projected in the business-as-usual scenario, and much of the

Figure 51.6 The BR-163 (Santarém–Cuiabá) Highway was built in 1973 and is now planned for reconstruction – essentially the building of a new highway on the route of the old one. In its present condition, it is inadequate for transporting soybeans from Mato Grosso to an Amazon River port in Santarém. Source: Photograph by and reproduced with permission from M. Torres.

forest expected to be saved in the governance scenario is already gone (Fearnside 2007). The BR-319 Highway offers another example of a completely unrealistic governance scenario being used to justify approval of the project, in this case using Yellowstone National Park in the United States as a representation of the government control proposed for the area (Textbox 51.1).

51.8   Environmental safeguards are needed for approval of international financing of road development

Road development in Amazonia has often been funded and influenced by international entities. Funding from multilateral development banks for reconstruction of the BR-364 Highway that opened up Rondônia and Acre to migration and deforestation is the classic example (Fearnside 1986, 1987b). The creation of the World Bank's Environment Department in March 1987, together with a system of EIA within the Bank, was a direct reaction to the dramatic surge of rainforest destruction in Rondônia as a result of the highway. The announcement of the Environment Department occurred less than 48 hours after an exposé of the highway project aired on the 60 Minutes television programme in the United States.

Brazil is a major source of financing and construction resources for road projects in Peru, Bolivia, Ecuador and Guyana, including the Transoceanic Highway in Peru and the Highway to the Pacific in Brazil. The primary purpose of this road is to transport commodities from Brazilian Amazonia to Pacific ports in Peru. It is being built by Brazilian construction firms with funding from Brazil's National Bank for Economic and Social Development (BNDES). The road, although officially completed in 2010 but still undergoing upgrading (as at early 2015), is already causing an upsurge in deforestation in the biodiversity hotspot in the Madre de Dios region at the base of the Andes in the Amazonian portion of Peru. The highway is part of the Initiative for the Integration of the Regional Infrastructure of South America (better known as 'IIRSA'), a massive programme to integrate transportation infrastructure in South America (Killeen 2007). Despite the deficiencies of Brazil's environmental review and licensing system, the neighbouring countries in Amazonia have even less protections against road impacts. Therefore, Brazil's road-building activities are a major force in Amazonian deforestation.

CONCLUSIONS

Roads built in Amazonia expose areas of rainforest that previously remained intact largely due to inaccessibility to deforesters. Decisions to build new highways and upgrade or reopen marginal or abandoned highways have consequences for forest loss that are far-reaching, both in space and time. Rational decisions require realistic modelling of future deforestation, including the critical effect of roads. Unfortunately, the widespread belief that deforestation in Amazonia is under control and that highways can therefore be built at will without increasing deforestation is erroneous.

ACKNOWLEDGEMENTS

The author's research is supported exclusively by academic sources: Conselho Nacional de Desenvolvimento Científico e Tecnológico (CNPq: Proc. 305880/2007-1; 304020/2010-9; 573810/2008-7; 575853/2008-5) and Instituto Nacional de Pesquisas da Amazônia (INPA: PRJ15.125).

FURTHER READING

Fearnside (2002): Discusses the roads and other infrastructure planned for Brazil under a massive development plan. A decade later, many of the projects remain undone, but Brazil is moving forwards under the current 'Plan for the Acceleration of Growth' (PAC-2).

Fearnside (2006): Discusses decision-making on roads in Brazil using the BR-163 and BR-319 as examples.

REFERENCES

Aguiar, A.P.D. 2006. Modelagem de mudança do uso da terra na Amazônia: explorando a heterogeneidade intra-regional. Doctoral thesis in remote sensing, Instituto Nacional de Pesquisas Espaciais, São José dos Campos, São Paulo, Brazil, 204 pp. (Available at: http://urlib.net/sid.inpe.br/MTC-m13@80/2006/08.10.18.21, Accessed on 25 September 2014).

Assunção, J., C.C. Gandour & R. Rocha. 2012. Deforestation slowdown in the legal Amazon: Prices or policies? Climate policy initiative (CPI) Working Paper, Rio de Janeiro, RJ, Brazil: Pontifícia Universidade Católica (PUC), 37 pp. (Available at: http://climatepolicyinitiative.org/publication/deforestation-slowdown-in-the-legal-amazon-prices-or-policie/, Accessed on 25 September 2014).

Câmara, G. 2007. Developments in land change modelling in Amazonia: Governance and public policies. Global Land

Project, Scientific Steering Committee Meeting, Copenhagen, October 2007. Powerpoint presentation, 38 pp. (Available at: http://www.dpi.inpe.br/gilberto/present/camara_glp_oct_2007.ppt, Accessed on 25 September 2014).

Câmara, G., A.P.D. Aguiar, M.I. Escada, S. Amaral, T. Carneiro, A.M.V. Monteiro, R. Araújo, I. Vieira & B. Becker. 2005. Amazonian deforestation models. Science **307**: 1043–1044.

Fearnside, P.M. 1986. Spatial concentration of deforestation in the Brazilian Amazon. Ambio **15**: 72-79.

Fearnside, P.M. 1987a. Causes of deforestation in the Brazilian Amazon. In: R.F. Dickinson (ed.) The Geophysiology of Amazonia: Vegetation and Climate Interactions. John Wiley & Sons, Inc., New York. pp. 37-61. 526 pp.

Fearnside, P.M. 1987b. Deforestation and international economic development projects in Brazilian Amazonia. Conservation Biology **1**: 214-221.

Fearnside, P.M. 2001. Land-tenure issues as factors in environmental destruction in Brazilian Amazonia: The case of southern Pará. World Development **29**: 1361–1372.

Fearnside, P.M. 2002. Avança Brasil: Environmental and social consequences of Brazil's planned infrastructure in Amazonia. Environmental Management **30**: 748–763.

Fearnside, P.M. 2006. Containing destruction from Brazil's Amazon highways: Now is the time to give weight to the environment in decision-making. Environmental Conservation **33**: 181–183.

Fearnside, P.M. 2007. Brazil's Cuiabá-Santarém (BR-163) Highway: The environmental cost of paving a soybean corridor through the Amazon. Environmental Management **39**: 601–614.

Fearnside, P.M. 2008. The roles and movements of actors in the deforestation of Brazilian Amazonia. Ecology and Society **13**: 23. (Available at: http://www.ecologyandsociety.org/vol13/iss1/art23/, Accessed on 25 September 2014).

Fearnside, P.M. 2009. Brazil's evolving proposal to control deforestation: Amazon still at risk. Environmental Conservation **36**: 176–179.

Fearnside, P.M. 2012. A tomada de decisão sobre grandes estradas amazônicas. In: A. Bager (ed.) Ecologia de Estradas: Tendências e Pesquisas. Editora da Universidade Federal de Lavras, Lavras, Brazil. pp. 59–76. 314 pp.

Fearnside, P.M. & P.M.L.A. Graça. 2006. BR-319: Brazil's Manaus-Porto Velho Highway and the potential impact of linking the arc of deforestation to central Amazonia. Environmental Management **38**: 705–716.

Fearnside, P.M., P.M.L.A. Graça, E.W.H. Keizer, F.D. Maldonado, R.I. Barbosa & E.M. Nogueira. 2009. Modelagem de desmatamento e emissões de gases de efeito estufa na região sob influência da Rodovia Manaus-Porto Velho (BR-319). Revista Brasileira de Meteorologia **24**: 208–233. (English translation available at: http://philip.inpa.gov.br/publ_livres/mss%20and%20in%20press/RBMET-BR-319_-engl.pdf, Accessed on 25 September 2014).

Hecht, S.B. 2011. From eco-catastrophe to zero deforestation? Interdisciplinarities, politics, environmentalisms and reduced clearing in Amazonia. Environmental Conservation **39**: 4–19.

Imbernon, J. 1999. A comparison of the driving forces behind deforestation in the Peruvian and the Brazilian Amazon. Ambio **28**: 509–513.

Kaimowitz, D. 1997. Factors determining low deforestation in the Bolivian Amazon. Ambio **26**: 537–540.

Killeen, T.J. 2007. A Perfect Storm in the Amazon Wilderness: Development and Conservation in the Context of the Initiative for the Integration of the Regional Infrastructure of South America (IIRSA). Conservation International, Arlington, VA, 98 pp. (Available at: http://www.conservation.org/publications/pages/perfect_storm.aspx, Accessed on 25 September 2014).

Laurance, W.F., A.K.M. Albernaz, G. Schroth, P.M. Fearnside, S. Bergen, E.M. Venticinque & C. da Costa. 2002. Predictors of deforestation in the Brazilian Amazon. Journal of Biogeography **29**: 737–748.

Laurance, W.F., M.A. Cochrane, P.M. Fearnside, S. Bergen, P. Delamonica, S. D'Angelo, T. Fernandes & C. Barber. 2001. Response. Science **292**: 1652–1654.

Laurance, W.F. & P.M. Fearnside. 2002. Issues in Amazonian development. Science **295**: 1643. doi:10.1126/science.295.5560.1643b

Laurance, W.F., P.M. Fearnside, A.K.M. Albernaz, H.L. Vasconcelos & L.V. Ferreira. 2005. Response. Science **307**: 1044.

Mena, C.F., R.E. Bilsborrow & M.E. McClain. 2006. Socioeconomic drivers of deforestation in the northern Ecuadorian Amazon. Environmental Management **37**: 802–815. doi:10.1007/s00267-003-0230-z

Nepstad, D.C., G. Carvalho, A.C. Barros, A. Alencar, J.P. Capobianco, J. Bishop, P. Moutinho, P. Lefebvre, U.L. Silva, Jr. & E. Prins. 2001. Road paving, fire regime feedbacks, and the future of Amazon forests. Forest Ecology and Management **154**: 395–407.

Nepstad, D.C., D. McGrath, A. Alencar, A.C. Barros, G. Carvalho, M. Santilli & M. del C. Vera Diaz. 2002a. Frontier governance in Amazonia. Science **295**: 629.

Nepstad, D.C., D. Mcgrath, A. Alencar, A.C. Barros, G. Carvalho, M. Santilli & M. del C. Vera Diaz. 2002b. Response [to Laurance et al.]. Science **295**: 1643–1644.

Pfaff, A., J. Robalino, R. Walker, S. Aldrich, M. Caldas, E. Reis, S. Perz, C. Bohrer, E. Arima, W. Laurance & K. Kirby. 2007. Road investments, spatial spillovers, and deforestation in the Brazilian Amazon. Journal of Regional Science **47**: 109–123.

Silveira, J.P. 2001. Development of the Brazilian Amazon. Science **292**: 1651–1652.

Soares-Filho, B.S., A. Alencar, D.C. Nepstad, G. Cerqueira, M. del C. Vera Diaz, S. Rivero, L. Solórzano & E. Voll. 2004. Simulating the response of land-cover changes to road paving and governance along a major Amazon highway: The Santarém-Cuiabá corridor. Global Change Biology **10**: 745–764.

Soares-Filho, B.S., D.C. Nepstad, L.M. Curran, G.C. Cerqueira, R.A. Garcia, C.A. Ramos, E. Voll, A. Mcdonald, P. Lefebvre &

P. Schlesinger. 2006. Modelling conservation in the Amazon Basin. Nature **440**: 520–523.

Southgate, D., R. Sierra & L. Brown. 1991. The causes of tropical deforestation in Ecuador: A statistical analysis. World Development **19**: 1145–1151.

Teixeira, K.M. 2007. Investigação de Opções de Transporte de Carga Geral em Conteineres nas Conexões com a Região Amazônica. Doctoral thesis in transport engineering, Universidade de São Paulo, Escola de Engenharia de São Carlos, São Carlos, São Paulo, Brazil, 235 pp. (Available at http://philip.inpa.gov.br, Accessed on 25 September 2014).

UFAM. 2009. Estudo de Impacto Ambiental – EIA: Obras de reconstrução/pavimentação da rodovia BR-319/AM, no segmento entre os km 250,0 e km 655,7. Universidade Federal do Amazonas (UFAM), Manaus, Amazonas, Brazil. 6 Vols. + Annexes. (Available at http://philip.inpa.gov.br, Accessed on 25 September 2014).

Yanai, A.M., P.M. Fearnside, P.M.L.A. Graça & E.M. Nogueira. 2012. Avoided Deforestation in Brazilian Amazonia: Simulating the effect of the Juma Sustainable Development Reserve. Forest Ecology and Management **282**: 78–91.

Chapter 52

ROAD ECOLOGY IN SOUTH INDIA: ISSUES AND MITIGATION OPPORTUNITIES

K. S. Seshadri[1] and T. Ganesh[2]

[1]Department of Biological Sciences, National University of Singapore, Singapore
[2]Ashoka Trust for Research in Ecology and the Environment (ATREE), Royal Enclave, Bangalore, India

SUMMARY

India has one of the highest rates of economic and population growth of all the developing nations as well as being a biologically and culturally rich country. Therefore, finding the delicate balance between development and nature conservation is very important. Unfortunately, more attention and resources are currently committed to India's economic development than to conservation of biodiversity, and the resulting rapid increase in roads and vehicles is causing numerous ecological problems.

52.1 The extensive road networks within and around protected areas are a major challenge for conserving biodiversity.

52.2 Roads to religious enclaves inside forests are an increasing threat to wildlife.

52.3 India needs ecological principles in its road-construction policies, stronger political will and simple engineering solutions to effectively avoid and mitigate road impacts.

India and its citizens have historically shown tremendous respect to wildlife, and a lot can be achieved based on this understanding. The lessons in this chapter demonstrate that solutions to many of the problems associated with roads and traffic can and have been found through dialogue between the stakeholders.

Handbook of Road Ecology, First Edition. Edited by Rodney van der Ree, Daniel J. Smith and Clara Grilo.
© 2015 John Wiley & Sons, Ltd. Published 2015 by John Wiley & Sons, Ltd.
Companion website: www.wiley.com\go\vanderree\roadecology

INTRODUCTION

India is a rapidly growing developing country and plays a substantial role in the global economy. To foster this growth, India has an ambitious and aggressive development programme (Sengupta 2012). Roads are important for economic growth, and by March 2011, India had 4.7 million kilometres of roads (ranked fourth in the world in terms of total road length) with an average road density of 1.4 km per km² (MoRTH 2011). The number of vehicles on India's roads is set to quadruple from just over 100 million in March 2011 to 450 million by 2020 (India Transport Portal 2012).

India is the second most populated country in the world and is home to four of the world's 25 biodiversity hotspots (Myers et al. 2000). It has a large network of protected areas (PA) as well as important areas for conservation outside PA, all of which are traversed by a network of roads. Studies on the impacts of roads on wildlife in India has primarily focused on documenting the rates of roadkill, with a few quantifying the barrier effects of roads on large fauna (Prakash 2012) and the effects of night-time traffic on smaller fauna (Vijaykumar et al. 2001; Seshadri & Ganesh 2011).

The Indian government has only recently officially recognised the impacts of roads on wildlife. The National Board for Wildlife commissioned a report (Raman 2011) to highlight the problem and developed a framework for mitigation. Raman (2011) identified poor enforcement of existing legislation and disregard for standard procedures as major issues and emphasised the need for a national policy for implementing ecologically and socially sound infrastructure. Despite the recognition of the seriousness of road-related problems, numerous religious, social and development pressures also hinder ecologically sound road construction and management. In this chapter, we describe examples, mostly from the biodiversity hotspot of the Western Ghats, where attempts were made to understand and mitigate the wildlife conservation problems caused by roads.

LESSONS

52.1 The extensive road networks within and around protected areas are a major challenge for conserving biodiversity

Less than 4% of India's forests are protected (MoEF 2012), and roads traverse many of them, including those set aside for tiger conservation (e.g. Kudremukh Tiger Reserve, Nagarahole National Park, Bandipur

Tiger Reserve and Anamalai Tiger Reserve). There is continual pressure to construct new roads through PA and widen existing roads, as well as to remove existing restrictions on traffic. This persistent pressure to upgrade roads is demonstrated at Nagarahole National Park, one of India's premier tiger habitats (Karanth et al. 2011). The Mysore-Mananthavadi Road is a state highway that traverses the park and separates it from the adjoining Bandipur Tiger Reserve. In 2004, a proposal to upgrade this road to national highway status faced opposition, and two alternative alignments were proposed (see Chapter 37).

A considerable diversity and abundance of wildlife also exist in forests and other habitats outside PA (Das et al. 2006), most of which are also bisected by roads. In several cases, the roads pass through forest corridors or migration paths resulting in high rates of wildlife-vehicle collisions (WVC) and wildlife mortality (Fig. 52.1, Seshadri et al. 2009). Historically, most roads connecting cities and towns were single lane and lined with century-old trees, which provided shade and fodder for cattle that pulled carts. Coincidentally, they also served as corridors for the movement of birds and arboreal animals. The use of cattle-drawn carts in rural areas has declined dramatically in recent times, being replaced with cars and trucks as they have become affordable. Consequently, roads designed for carts are now congested with vehicles and are being widened, resulting in the loss of the old roadside trees. Expansion and/or construction of new roads outside PA is readily approved by the Ministry of Environment because there is no legislative protection for forests or roadside trees outside of PA and development projects take priority.

52.2 Roads to religious enclaves inside forests are an increasing threat to wildlife

Religion is a fundamental part of Indian life and culture, with greater than 93% of the population associating themselves with a religion. Many places of worship are located within forests (including within PA), and people visit them to attend festivals, consequently impacting the vegetation and wildlife. These impacts are exacerbated when large numbers of pilgrims arrive in private vehicles and set up temporary camps. This is particularly evident in the Periyar Tiger Reserve where approximately 50 million pilgrims visit the temple throughout the year, with a peak during the annual Sabarimala Festival in January. Similarly, about half a million people congregate for a week at Sorimuthian

Figure 52.1 Leopard roadkill on the MM Hills–Ponnachi Road inside the Cauvery Wildlife Sanctuary. Unregulated vehicle movement at night on this state highway causes such mortalities. Implementing a ban on night-time traffic is thus important. Source: Photograph by and reproduced with permission from Anjali Anantharam.

Temple inside the Kalakad Mundanthurai Tiger Reserve during the Aadi Amavasai Festival. While pilgrim movement occurs throughout the year at numerous religious enclaves, annual pilgrimages are associated with high traffic densities and high rates of wildlife mortality. For example, pilgrims to the annual Sorimuthian Festival must travel along a 14 km road passing through three forest types to reach the temple, often leading to major traffic jams (Fig. 52.2). Night-time traffic is ordinarily banned in the reserve, but during the festival, the restriction is removed and vehicles are allowed to enter at night. Three surveys conducted during the day on the access road in 2008 and 2009 found 1399 dead animals from > 54 species during the festival compared to 230 deaths of 16 species before the festival (Seshadri & Ganesh 2011).

Figure 52.2 Traffic congestion inside the Kalakad Mundanthurai Tiger Reserve becomes a physical barrier for wildlife. Vehicles travelling in batches with 10–15 min gaps between batches would help ease the barrier effect. Source: Photograph by Seshadri K.S.

Apart from roadkill during the festival, poaching, fuel-wood collection, garbage dumping and deliberate burning of the forest also occur.

Potential strategies to reduce wildlife mortality associated with pilgrimages include: (i) applying and enforcing bans on night-time vehicle movement through the forest (Chapter 37); (ii) reducing the number of vehicles by encouraging the use of public transport; (iii) implementing traffic calming measures like batching vehicle movements with half-hour gaps in between; (iv) lowering vehicle speed by installing speed bumps; and (v) managing roadside vegetation to slow traffic and maintain canopy connectivity (Chapter 40). The implementation of these solutions, however, varies because the stakeholders of places of worship typically oppose any measure that may restrict or inconvenience pilgrims, consequently reducing income. Any hindrance to the movement of pilgrims becomes a volatile issue, and political and public support usually favours the place of worship. Consequently, park managers and conservation advocacy groups must work hard to obtain community support to implement mitigation strategies.

52.3   India needs ecological principles in its road-construction policies, stronger political will and simple engineering solutions to effectively avoid and mitigate road impacts

The judiciary has been used to prevent and mitigate some of the ecological impacts of roads and traffic. However, using the courts to solve conflicts with road projects is problematic because it is a lengthy process, delays are likely and in most cases the outcome is uncertain since environmental matters are often a low priority. Furthermore, the person who petitions the court may face persecution and even death threats from vested interests.

Most conservation battles in India have been won by dedicated individuals and groups. A massive increase in community support for conservation is essential to increase political support for reform in road ecology. But the contrary seems to occur more often. For example, the Supreme Court of India passed a ruling in March 2013 that removed the need for approval of road projects from both the central Ministry of the Environment and Forests and the State Forest Department, allowing numerous major projects to commence (Das 2013). This ruling means that projects can commence without adequate protection for wildlife and forests, effectively undermining the Wildlife (Protection) Act of 1972 and the Forest (Conservation) Act of 1980. It is important for all conservationists and community members to vehemently protest this move as the PA will be permanently transformed, resulting in severe adverse effects to wildlife populations, increasing their risk of extinction. International pressure against this policy ruling is necessary because it violates the Convention on Biological Diversity and contravenes the sustainability policies of many of the international banks that finance road projects.

Many options to mitigate the ecological impacts of roads on wildlife are inexpensive. Importantly, planners and engineers should use all construction and maintenance projects as opportunities to modify existing structures (e.g. drainage culverts) when suitable, to provide safe passage for wildlife. Despite the massive benefits of road-free areas (Chapter 3) and existing laws that prevent new roads through PA and the use of roads at night, developmental pressures lead to relaxing these laws. Hence, there is a need to incorporate mitigation measures like wildlife crossing structures to facilitate wildlife movement and fencing to prevent WVC in areas where vehicular movement cannot be avoided.

India may be a relatively poor country, but adopting these minimum standards represents a small proportion of the overall road budget, and international funding agencies, such as the World Bank, should require greater standards as a condition of funding.

CONCLUSIONS

It is evident that roads are a persistent feature of developing countries like India (see also Chapters 2, 36–38, 41, 43, 50, 51, 53–58). The negative ecological impacts of roads passing through biologically and culturally diverse areas of India are numerous and widespread. With existing road networks and vehicle usage poised for rapid expansion, it is important to reconcile development with that of nature conservation. There are many challenges to achieving a sustainable road network in India, but the important areas include: (i) overcoming the lack of knowledge regarding the ecological impacts of roads; (ii) ensuring the avoidance and mitigation of impacts is mandatory for all road projects; (iii) ensuring road users and road authorities understand each other and work collaboratively; (iv) controlling unplanned growth and vested interests; and (v) ensuring a productive and collaborative dialogue between road authorities and ecologists/biologists.

REFERENCES

Das, A., Krishnaswamy, J., Bawa, K. S., Kiran, M. C., Srinivas, V., Samba Kumar, N. and Karanth, K. U. 2006. Prioritisation of conservation areas in the Western Ghats, India. Biological Conservation, **133**:16–31.

Das, M. 2013. Highway projects: Apex court delinks forest from environment clearance. The Hindu Business Line, India. Available from http://www.thehindubusinessline.com/industry and economy/logistics/highway-projects-apex-court-delinks-forest-from-environment-clearance/article4504512.ece (Accessed on 25 September 2014).

India Transport Portal. 2012. How Many Vehicles on India Roads? India Transport Portal, India. Available from http://indiatransportportal.com/vehicles-in-india-20292 (Accessed on 12 January 2015).

Karanth, K. U., Nichols, J. D., Kumar, N. S. and Devcharan, J. 2011. Estimation of demographic parameters in a tiger population from long-term camera trap data. In A. F. O'Connell, J. D. Nichols and K. U. Karanth, editors. Camera Traps in Animal Ecology. Springer, Tokyo. pp. 145–161.

Ministry of Environment and Forests (MoEF). 2012. Protected Area Network in India. MoEF, New Delhi. Available from http://envfor.nic.in/public-information/protected-area-network (Accessed on 12 January 2015).

Ministry of Road Transportation and Highways (MoRTH). 2011. Basic Road Statistics of India. MoRTH, New Delhi. Available from http://morth.nic.in/showfile.asp?lid=839 (Accessed on 12 January 2015).

Myers, N., Mittermeier, R. A., Mittermeier, C. G., Da Fonseca, G. A. and Kent. J. 2000. Biodiversity hotspots for conservation priorities. Nature, **403**:853–858.

Prakash, A. 2012. A study on the impacts of highways on wildlife in Bandipur Tiger Reserve. MSc Thesis. National Centre for Biological Sciences, Bangalore, India.

Raman, T. R. S. 2011. Framing ecologically sound policy on linear intrusions affecting wildlife habitats. Background paper for the National Board for Wildlife. Ministry of Environment and Forests, New Delhi, India. Available from http://envfor.nic.in/assets/Linear%20intrusions%20background%20paper.pdf (Accessed on 25 September 2014).

Sengupta, S. 2012. India: the next superpower?: Managing the environment: A growing problem for a growing power. IDEAS reports – special reports, Kitchen, Nicholas (ed.) SR010. LSE IDEAS, London School of Economics and Political Science, London.

Seshadri, K. S. and Ganesh, T. 2011. Faunal mortality on roads due to religious tourism across time and space in protected areas: A case study from south India. Forest Ecology and Management, **262**:1713–1721.

Seshadri, K. S., Yadav, A. and Gururaja, K. V. 2009. Road kills of amphibians in different land use areas from Sharavathi River Basin, central Western Ghats, India. Journal of Threatened Taxa, **1**:549–552.

Vijaykumar, S. P., Vasudevan, K. and Ishwar, N. M. 2001. Herpetofaunal mortality on roads in the Anamalai Hills, Southern Western Ghats. Hamadryad, **26**:265–272.

Chapter 53

PLANNING ROADS THROUGH SENSITIVE ASIAN LANDSCAPES: REGULATORY ISSUES, ECOLOGICAL IMPLICATIONS AND CHALLENGES FOR DECISION-MAKING

Asha Rajvanshi and Vinod B. Mathur

Wildlife Institute of India, Dehradun, India

SUMMARY

Adequate protection of biodiversity in the planning and development of road projects is typically rare in India and other countries in Asia. This issue creates conflict among planners, decision-makers and the conservation community and causes significant delays in land transfers and decision-making. This situation warrants the need for regulatory reforms to ensure environmental impact assessments (EIAs) are adequate, comprehensive and focused on assessing relevant impacts.

53.1 The EIA of road projects that cross jurisdictional boundaries or different land categories should be a single report, not multiple stand-alone reports.

53.2 Regulatory reforms are needed to define the width of the road corridor for EIAs based on the likely extent and severity of the road-effect zone.

53.3 Greater collaboration and coordination among transportation planners, land managers and wildlife experts is essential to ensure that the ecological requirements of wildlife are included in the design of roads and mitigation measures.

Handbook of Road Ecology, First Edition. Edited by Rodney van der Ree, Daniel J. Smith and Clara Grilo.
© 2015 John Wiley & Sons, Ltd. Published 2015 by John Wiley & Sons, Ltd.
Companion website: www.wiley.com/go/vanderree/roadecology

53.4 Reliable information on biodiversity issues facilitates decision-making and can improve cost-effectiveness.

53.5 The potential impact of non-road linear infrastructure associated with major developments is often overlooked and should be considered through cumulative environmental assessment.

Many of these lessons are applicable to other developing countries in South and Southeast Asia (see case studies in Rajvanshi et al. (2001)), and improvements in the EIA and planning of new road projects are urgently required.

INTRODUCTION

The importance of road and rail transportation for improving access and economic development is rarely questioned in most developing countries in Asia (Raghuram & Babu 2001). The National Highways Development Project (NHDP) currently being implemented by the National Highways Authority of India (NHAI) is the largest cross-country highway project in India. It aims to add new roads, upgrade, rehabilitate or widen nearly 14,000 km of existing highways to four or six lanes in three successive 5-year plans (2005–2020). The 143,000 km Asian Highway criss-crossing 32 countries with links to Europe (ADBI 2013) is another major transport system promoting the development of the region.

Regulatory frameworks for environmental approval of road projects exist in most Asian countries. Despite this, the neglect of biodiversity considerations in road planning has exacerbated challenges to conserving wildlife in many sensitive landscapes. In India, the Vision 2021 (IRS 2001) prioritises road projects that upgrade existing highways to multiple lanes, construction of bypasses around congested urban areas and connection of rural areas with all-weather roads. Unfortunately, this vision overlooks the need to incorporate biodiversity concerns in proposed projects. Likewise, in Peninsular Malaysia, the Highway Network Development Plans that were prepared in accordance with guidelines for EIA (DOE 2011) and implemented by the Ministry of Works failed to consider the impacts of the tens of thousands of km of existing roads criss-crossing the habitat of the threatened Malayan tiger. In Indonesia, decisions on the location of forestry roads are often made 'on the bulldozer' (Hüttche 1999), despite the existence of the Indonesian EIA process referred to as Analisis Mengenai Dampak Lingkungan (AMDAL). A significant failing of a recent road development through forest in north-west Laos was the opening of formerly inaccessible forests to further logging, hunting and development (Stidbig et al. 2007) (see also Chapters 2, 3 and 51).

This chapter highlights lessons learned from the shortcomings in current planning procedures and regulatory processes and their implications for biodiversity conservation for roads through ecologically sensitive landscapes in South and Southeast Asia.

LESSONS

53.1 The EIA of road projects that cross jurisdictional boundaries or different land categories should be a single report, not multiple stand-alone reports

The compartmentalisation of a single highway project into separate projects on the basis of jurisdictional boundaries, land use or landownership is common practice in Southeast Asia (e.g. India and Laos). Generally, the rationale for splitting a single project is to facilitate the approval of land take in different areas. Adoption of such an approach, based on seemingly rational criteria, is also often intended to 'force' the approval for land clearing in conservation areas. Approvals for land diversion and clearing are often obtained first for sections of road that pass through biologically poor areas (e.g. privately owned or non-forested lands) because they are relatively easy to acquire. This action can then be used to justify the construction or widening of the section of road through the protected area because it is the final section to complete the project (Textbox 53.1). Splitting projects into sub-projects can sometimes result in additional delays in project implementation due to disputes among contractors and insufficient supply of equipment, partly offsetting the cost savings in other areas.

From a planning perspective, several sub-projects can make impact assessment and decision-making complex and challenging. Dividing a single project into multiple sections may not halt or prevent the widening of a road through a conservation area if the road at each end of the conservation area has already been widened. Planning documents must be based on single projects, and the EIA must accordingly focus on evaluating the impacts of the entire road on all ecological units within the road corridor. The road should only be divided into multiple sections for detailed design to facilitate logistics during construction and to minimise disruption to traffic.

Textbox 53.1 Piecemeal planning and EIA of a road project leads to poor outcomes for biodiversity.

The National Highways Authority of India (NHAI) proposed widening of the National Highway (NH-)37 from two to four lanes in the states of Madhya Pradesh and Maharashtra in India. The 92 km project was divided into three sections: (i) private lands in Madhya Pradesh and Maharashtra, (ii) forested areas of South Seoni Forest Division in Madhya Pradesh and (iii) Pench Tiger Reserve in both states (Fig. 53.1).

Numerous wildlife species, including the endangered tiger, currently cross the two-lane highway (with about 3000 vehicles per day) where it passes through Pench Tiger Reserve. Wildlife–vehicle collisions (WVC) and subsequent wildlife mortality (1035 total roadkills in 430 days) are a significant impact of the existing two-lane highway (Rajvanshi et al. 2013). The risk of WVC becomes greater as traffic volume and speed increase until a threshold is reached and wildlife begin to avoid the road and traffic. While fragmentation is a significant effect of major roads, WVC still occurs as some animals attempt to cross the road.

The preferred approach would be to avoid and/or remove or not widen existing roads through ecologically significant and biologically diverse habitats. In this specific case, road widening outside the reserve had been approved, and the fast-moving vehicles would be funnelled into the narrow section through the reserve, creating a bottleneck and traffic congestion (Fig. 53.2). The higher density of slow-moving vehicles in the bottleneck is likely to pose a significant barrier to the movement of wildlife, including tigers.

The NHAI prepared a stand-alone EIA for each of the three road sections to obtain the necessary approvals. This did not provide a 'big picture' perspective of the ecological impacts and also led to decisions that justified adopting piecemeal mitigation. Presently (as at early 2014), the widening of the NH-7 has been halted because the sections through the forest and Tiger Reserve have not been approved, despite approval having been given for the sections through private land. Since this is an existing NH, its closure through the reserve is unlikely. Approval to widen the road through the reserve appears inevitable because the adjacent sections have already been widened. The challenge for managers is to design the road such that WVC and the barrier effect of the widened road do not further endanger wildlife, especially the endangered tiger. An optimal approach is to fence the highway to prevent WVC and provide adequate numbers of effective crossing structures to maintain connectivity.

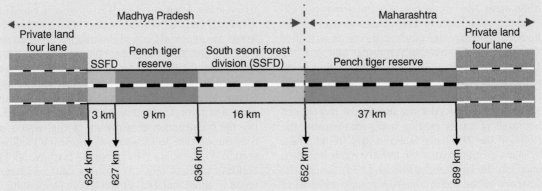

Figure 53.1 Status of the widening of NH-37 through Pench Tiger Reserve in Madhya Pradesh and Maharashtra, India. Source: Asha Rajvanshi.

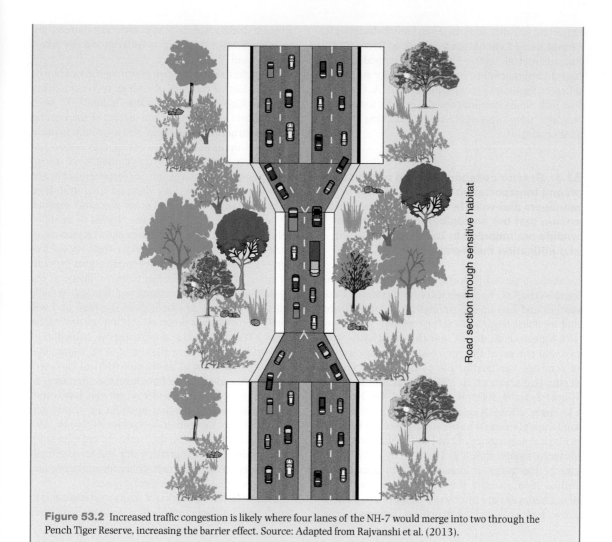

Road section through sensitive habitat

Figure 53.2 Increased traffic congestion is likely where four lanes of the NH-7 would merge into two through the Pench Tiger Reserve, increasing the barrier effect. Source: Adapted from Rajvanshi et al. (2013).

53.2   Regulatory reforms are needed to define the width of the road corridor for EIAs based on the likely extent and severity of the road-effect zone

Many Asian countries have legislation to guide the process and scope of EIAs. For example, legislation in India stipulates that EIAs in ecologically sensitive areas, including for road projects, should extend 15 km from proposed developments. However, EIA practitioners often conduct inadequate assessments because survey effort is either thinly spread over a large area or they focus effort on a narrow corridor and miss the wider-ranging impacts. Ideally, the spatial limit for

assessments would not be universally applied to all developments because the zone of influence will vary. While the physical footprint of the project is simply the width of the road plus the right of way, the size and severity of the road-effect zone (Chapter 1) will vary depending on road width, traffic volume, topography, vegetation type and species of wildlife (Fig. 1.1).

We recommend that the spatial extent of the investigation area be based on the likely extent and severity of the road-effect zone, the species present therein and their conservation importance. Once a suite of alternative potential corridors have been identified, they should be examined using geographic information system (GIS) data and field visits to identify ecologically important features and species and to set the limits for investigation. Existing guidelines (e.g. Byron 2000; Rajvanshi et al. 2007; NRA 2009) provide a solid basis to begin defining corridor widths for surveys of different taxonomic groups during an EIA. Countries that lack these guidelines should consider using those available from other regions and adapting them to their conditions.

53.3 Greater collaboration and coordination among transportation planners, land managers and wildlife experts is essential to ensure that the ecological requirements of wildlife are included in the design of roads and mitigation measures

Roadside verges are important habitats and corridors for movement of many species of invertebrates, birds, reptiles and mammals, especially in highly cleared and modified landscapes (Chapters 29, 39, 46 and 48). Numerous studies around the world have identified and reiterated the potential conservation value of roadside habitats for the conservation of biodiversity (e.g. Oxley et al. 1974; Bellamy et al. 2000; Hlavac & Andel 2002; Brock & Kelt 2004; Huijser & Clevenger 2006; Noordijk et al. 2009). Road and landscape planners who are unaware of the potential ecological importance of roadside habitats may inadvertently compromise this function by reducing the size of the verge or managing it inappropriately. Unlike other parts of the world where road verges along highways are intensively managed (e.g. mown grass and potential obstructions removed), road

verges in India, especially in conservation areas, are typically a natural transition between interior forest and the road (Fig. 53.3).

The preferred option when planning the location of new or widened roads is, of course, to avoid placing them in high-conservation-value habitats. If such areas cannot be avoided, a combination of fencing and crossing structures is the next best approach to avoid WVC and maintain connectivity. If this is also not feasible (e.g. due to costs), verges consisting of transitional vegetation through interior forest should not be made narrower to reduce demand on forest land without taking into account other potential impacts. While minimising the size of the cleared or modified right of way in conservation areas and tropical rainforest (Chapter 49) should always be a priority, the loss of the verge can have negative implications for conservation of some species and highway safety. In the absence of crossing structures and fencing, wildlife will attempt at-grade crossings of the road and will use the verge to make decisions about when and where to cross (Fig. 53.4). Therefore, excessive reduction of verge width can increase the risk of WVC and animal mortality because wildlife are more difficult to detect by motorists and animals have less space to retreat to avoid vehicles. However, wider verges may have other negative impacts, including invasion by weeds and avoidance by forest-interior species (Chapter 49). Careful planning and further research are necessary to determine the appropriate width and management of the verge that maximises safety for motorists and minimises the suite of ecological impacts.

A common misconception of many road engineers in South and Southeast Asia is that drainage structures

(A)

(B)

Figure 53.3 The natural features of the edge habitat are generally retained on highways in Asia as is visible along NH-37 in India through Kaziranga National Park (A) and barking deer feeding in edge habitat along NH-37 (B). Source: Photographs by A. Pragatheesh © WII 2013. Reproduced with permission from Wildlife Institute of India.

under roads are suitable and effective crossing structures for all wildlife. While this may be true for some species (e.g. wild pig, civets, porcupine and mongoose), there are others (e.g. spotted deer, leopards and tigers) that appear unwilling to use drainage culverts (Rajvanshi et al. 2013). Coordination among road planners and ecologists is essential to integrate the critical ecological requirements of wildlife into the design and location of multifunctional drainage structures for wildlife passage (Textbox 53.2).

53.4 Reliable information on biodiversity issues facilitates decision-making and can improve cost-effectiveness

A 24 km section of the 108 km-long Nandyal–Giddalur–Thokapalli Road in Andhra Pradesh state of India was proposed for widening in the mid-1990s. A review of the EIA report by government found it deficient on several counts. For example, it did not consider the potential barrier effect of the road on the

Figure 53.4 Elephant from the Kaziranga National Park in India approaches NH-37. In situations like this, a cleared verge provides wildlife and drivers with time and space to see each other and potentially avoid WVC. Source: Photograph by A. Pragatheesh © WII 2013. Reproduced with permission from Wildlife Institute of India.

Textbox 53.2 More collaboration between road planners and ecologists is needed in South and Southeast Asia to ensure wildlife crossing structures are optimally designed and located.

The proposed upgrade of the Gujarat Highway (NH-14) from two lanes to four through the Balaram-Ambaji Wildlife Sanctuary in India (Fig. 53.5) shows how inadequate collaboration between planners and ecologists resulted in suboptimal outcomes for wildlife. This sanctuary supports the most western population of the endangered sloth bear (Fig. 53.6) in India, and the highway severs a movement corridor for bears between the Balaram-Ambaji and Jessore Wildlife Sanctuaries (Singh 2000).

The NHAI proposed a number of crossing structures to facilitate the movement of wildlife including leopard, sloth bear, hyena, nilgai and mongoose.

Before approving the project, the National Board for Wildlife called for the evaluation of the efficacy of the proposed mitigation measures. It was concluded that planners who had no understanding of the ecological requirements of the species proposed far too many crossing structures in inappropriate locations in the hope of expediting approval for the project (WII 2007). For example, one culvert was located near high-tension power lines where animal movements had not been detected, and another was located in a highly disturbed agricultural area. Wildlife crossing structures must be appropriately designed and located to be effective.

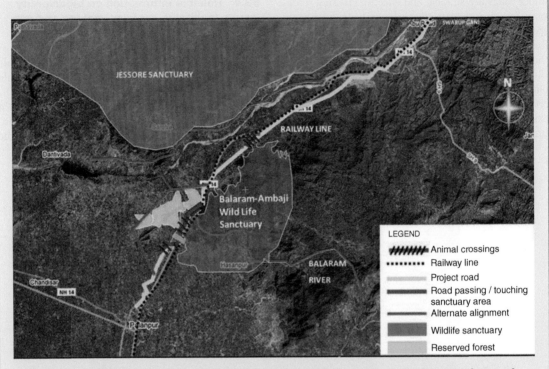

Figure 53.5 Proposed widening of NH-14 in India between two wildlife sanctuaries showing locations of proposed animal crossing structures. Source: Wildlife Institute of India (WII) (2007).

Figure 53.6 The endangered sloth bear in India is threatened by roads. Highway projects, including the NH-14 upgrade, would further impact the species. Source: Photograph by and reproduced with permission from Saurabh Desai.

endangered giant squirrel, a gliding mammal (see also Chapter 40). The EIA also incorrectly reported the location of the road as being outside the southern border of Gundla-Brahmeswaram Sanctuary. This necessitated the involvement of a new team of biodiversity experts, who concluded that the road would not restrict the movements of the giant squirrel because the 5 m-wide carriageway allowed overhanging trees to form a connected canopy and thus allow movement (Chapters 40 and 49). Since the impacts were likely to be modest, no costly mitigation measures were proposed by the biodiversity experts. This experience demonstrates the key lesson that including specialists in the field, even for a quick inspection, is vastly superior to desk-based analyses and on this occasion resulted in cost savings (WII & CEC 1998).

53.5 The potential impact of non-road linear infrastructure associated with major developments is often overlooked and should be considered through cumulative environmental assessment

Roads often form secondary or supporting infrastructure for major development projects such as mining, energy generation and airports (Chapter 27). The ecological impacts of such roads rarely receive adequate attention in the EIA because the focus is usually to assess the primary impacts of the major development.

Many examples highlight that the ecological implications of secondary developments such as roads can sometimes be more significant than those associated with the primary development. For example, the access road to the Bodai-Daldali Bauxite Mine in Chattisgarh state of India required widening to 8 m to accommodate trucks. An EIA for the project concluded that the widened road would increase the patchiness of critical tiger habitat in central India to an unacceptable level (Rajvanshi 2011). The mine project was recommended on the condition that a 5.5 km-long conveyer belt be used to transport the ore from the mine to the stocking yard instead of widening a 50 km-long section of road through sensitive areas for truck transport of ore. However, this could not be achieved as the cost of the conveyer system was not considered in the economic analysis of the project.

This lesson highlights the need to develop more stringent scoping requirements for EIA of projects involving secondary developments with potentially significant impacts. Also, regulatory reforms are necessary to ensure projects with a large footprint from subsidiary developments use cumulative environmental assessment (CEA; Chapter 5) to capture all direct, indirect, larger, multiple or combined impacts associated with major developments. This lesson also highlights the need to conduct a thorough and fair cost–benefit analysis of the project that incorporates the cost of effective mitigation, not just the cheapest or simplest.

CONCLUSIONS

The road network across South and Southeast Asia will continue to grow in length, width and sophistication of design and engineering. The ecological impacts of roads and traffic must receive greater attention to ensure the future road network is functional and safe for people as well as not being an impediment to maintaining healthy, sustainable ecosystems and wildlife populations. Importantly, this requires looking beyond the horizons of individual stakeholders and drawing from a larger perspective and synthesis of knowledge and experience of planners, engineers and biodiversity experts. The benefits of encouraging landscape-level impact assessment of road projects are huge and many. Regulatory reforms and development policy must encourage comprehensive EIA and CEA as a planning support tool to resolve the often complex and competing issues around road projects.

ACKNOWLEDGEMENTS

We thank the director of the Wildlife Institute of India for his encouragement and support for conducting the research on the impacts of NH-7 in Pench Tiger Reserve, Madhya Pradesh, and Shri A. Pragatheesh for his inputs in the field surveys under this project.

FURTHER READING

Davenport and Davenport (2006): Share perspectives of ecological effects and their causes and ameliorative approaches through future design options for transport systems.

Goosem et al. (2010): Presents a framework for understanding the primary ecological issues to be addressed in the planning, design and management of roads in rainforest environments that is also highly relevant in the context of Asian region.

Hilty et al. (2006): Guiding principles and cautionary notes for achieving habitat connectivity while implementing corridor projects.

Kintsch and Cramer (2011): Provides guidance on improving wildlife permeability needs in existing wildlife structures and to plan and design effective passage for wildlife in new road corridors.

Rajvanshi et al. (2007): A best-practice manual to provide 'start to end' procedures for identifying entry points for mainstreaming biodiversity in impact assessment of developments in different sectors including the road sector.

REFERENCES

Asian Development Bank Institute (ADBI). 2013. Connecting South Asia and Southeast Asia. ADBI, Tokyo, Japan.

Bellamy, P. E., R. F. Shore, D. Ardeshir, J. R. Treweek, and T. H. Sparks. 2000. Road verges as habitat for small mammals in Britain. Mammal Review **30**: 131–139.

Brock, R. E. and D. A. Kelt. 2004. Influence of roads on the endangered Stephens' kangaroo rat (*Dipodomys stephensi*): are dirt and gravel roads different? Biological Conservation **118**: 633–640.

Byron, H. 2000. Biodiversity and Environmental Impact Assessment: A Good Practice Guide for Road Schemes. The RSPB, WWF-UK, English Nature and the Wildlife Trusts, Sandy, UT.

Clements, R. 2013.How will Road Expansion Affect Tiger Landscapes in Peninsular Malaysia? Spot Image, Toulouse, France. Available from http://www.planet-action.org/web/85-project-detail.php?projectID=10268 (accessed on September 2013).

Davenport, J. and J. L. Davenport, editors. 2006. The Ecology of Transportation: Managing Mobility for the Environment. Springer, Dordrecht, the Netherlands.

Department of Environment (DOE). 2011. Handbook of Environmental Impact Assessment Guidelines. Department of Environment, Ministry of Science, Technology and Environment (MSTE), Kuala Lumpur, Malaysia.

Goosem, M., E. K. Harding, G. Chester, N. Tucker, C. Harriss, and K. Oakley. 2010. Roads in Rainforest: Best Practice Guidelines for Planning, Design and Management. Guidelines prepared for the Queensland Department of Transport and Main Roads and the Australian Government's Marine and Tropical Sciences Research Facility. Reef and Rainforest Research Centre Limited, Cairns, Australia.

Hilty, J. A., Z. W. Lidicker Jr., and A. M. Merenlender. 2006. Corridor Ecology: The Science and Practice of Linking Landscapes for Biodiversity Conservation. Island Press, Washington, DC.

Hlavac, V. and P. Andel. 2002. On the Permeability of Roads for Wildlife: A Handbook. Agency for Nature Conservation and Landscape Protection of the Czech Republic, Prague, Czech Republic.

Huijser, M. P. and A. P. Clevenger. 2006. Habitat and corridor function of rights-of-way. Pages 233–254. In: J. Davenport and J. L. Davenport (eds.). The Ecology of Transportation: Managing Mobility for the Environment. Springer, Dordrecht, the Netherlands.

Hüttche, C. 1999. Making AMDAL a Tool for Road Planning in Forests in Indonesia. Discussion Paper. Natural Resource Management Programme, USAID, Indonesia.

Indian Road Congress (IRS). 2001. Road Development Plan Vision, 2021. IRS, Ministry of Shipping and Transport, New Delhi, India.

Kintsch, J. and P. C. Cramer. 2011. Permeability of Existing Structures for Terrestrial Wildlife: A Passage Assessment System. Research Report No. WA-RD 777.1. Washington State Department of Transportation, Olympia, WA.

National Roads Authority (NRA). 2009. Ecological Surveying Techniques for Protected Flora and Fauna during the Planning of National Road Schemes. NRA, Dublin, Ireland.

Noordijk, J., I. P. Raemakers, A. P. Schaffers, and K. V. Sýkora. 2009. Arthropod richness in roadside verges in the Netherlands. Terrestrial Arthropod Review **2**: 63–76.

Oxley, D. J., M. B. Fenton, and G. R. Carmody. 1974. The effects of roads on populations of small mammals. Journal of Applied Ecology **11**: 51–59.

Raghuram, G. and R. Babu. 2001. Alternative means of financing railways. In: G. Raghuram, R. Jain, S. Sinha, P. Pangotra, and S. Morris (eds.). Infrastructure Development and Financing: Towards a Public–Private Partnership. Macmillan, New Delhi, India.

Rajvanshi, A. 2011. Site Appraisal Report of Bodai-Daldali Open Cast Bauxite Mine of Bharat Aluminum Company Limited. Submitted to the Ministry of Environment and Forests. Government of India, India.

Rajvanshi, A., V. B. Mathur, G. Teleki, and S. K. Mukerjee. 2001. Roads, Sensitive Habitats and Wildlife: Environmental Guideline for India and South Asia. Wildlife Institute of India, Dehradun and Canadian Environmental Collaborative Ltd., Toronto, ON, Canada.

Rajvanshi, A., V. B. Mathur, and U. A. Iftikhar. 2007. Best Practice Guidance for Biodiversity-Inclusive Impact Assessment: A Manual for Practitioners and Reviewers in South Asia. International Association for Impact Assessment (IAIA), North Dakota.

Rajvanshi, A., V. B. Mathur, and A. Pragatheesh. 2013. Ecological Effects of Roads through Sensitive Habitats: Implications for Wildlife Conservation. Research Report. Wildlife Institute of India, Dehradun, Uttarakhand, India.

Singh, R. 2000. Management Plan for Balaram-Ambaji Wildlife Sanctuary. Banaskantha Forest Division, Gujarat, India.

Stidbig, H. J., F. Stolle, R. Dennis, and C. Feldkotter. 2007. Forest Cover Change in Southeast Asia – The Regional Pattern. JRC Scientific and Technical Report, EUR 22896 EN – 2007. European Commission Joint Research Centre, Ispra, Italy.

Wildlife Institute of India (WII). 2007. Review of the Mitigation Measures Proposed for the Widening of NH-14 through Balaram–Ambaji Wildlife Sanctuary, Gujarat State. WII–EIA Technical Report 33. Wildlife Institute of India, Dehradun, Uttarakhand, India.

Wildlife Institute of India & Canadian Environmental Collaborative Ltd. (WII & CEC). 1998. The Andhra Pradesh State Highway Rehabilitation and Maintenance Project: The Nandyal–Giddalur–Thokapalli Road. Technical Report. Wildlife Institute of India, Dehradun, Uttarakhand, India.

Chapter 54

SETJHABA SA, SOUTH AFRIKA: A SOUTH AFRICAN PERSPECTIVE OF AN EMERGING TRANSPORT INFRASTRUCTURE

Wendy Collinson[1], Dan Parker[2], Claire Patterson-Abrolat[1], Graham Alexander[3] and Harriet Davies-Mostert[1]

[1]Endangered Wildlife Trust, Johannesburg, South Africa
[2]Department of Zoology and Entomology, Wildlife and Reserve Management Research Group, Rhodes University, Grahamstown, South Africa
[3]School of Animal, Plant, and Environmental Sciences, University of the Witwatersrand, Johannesburg, South Africa

SUMMARY

Roads are integral to the financial development and prosperity of the South African economy. However, a multitude of species and habitats are under increasing pressure from human development, and the need for a quick method of recognising the latent threat caused by roads is becoming more urgent. It is therefore important to reach a compromise between conserving the country's wildlife from the impacts of roads and providing road networks which enable South Africa's economy to function effectively.

54.1 Wildlife-vehicle collisions in South Africa are costly and likely to be under-reported.
54.2 Knowledge of the impacts of roads and traffic in South Africa is limited because most studies have focused solely on roadkill.
54.3 South Africa needs a standardised protocol to detect and record roadkill.
54.4 Fencing may be a successful mitigation measure in South Africa, but it can exacerbate fragmentation.

The national anthem of South Africa. The lyrics employ the five most widely spoken languages of South Africa's eleven official languages – Xhosa, Zulu, Sesotho, Afrikaans and English. 'Setjhaba sa, South Afrika' means 'The nation of South Africa' in Sesotho.

Handbook of Road Ecology, First Edition. Edited by Rodney van der Ree, Daniel J. Smith and Clara Grilo.
© 2015 John Wiley & Sons, Ltd. Published 2015 by John Wiley & Sons, Ltd.
Companion website: www.wiley.com\go\vanderree\roadecology

The threat of roads and traffic to wildlife in South Africa is generally poorly understood. Consequently, there is a need to define the extent of the impact of roads on the country's wildlife and develop national guidelines that not only contribute towards effective conservation planning but also to the sustainable development of infrastructure in South Africa.

INTRODUCTION

South Africa is the 25th largest country in the world, ranked 18th globally in terms of total road length (65,600 km paved and 689,000 km unpaved out of 1.2 million km^2 of the country's surface) and 74th for the number of cars (123 cars per 1000 people) (CIA 2012). Almost half (47%) of South Africa's roads are classified as state or provincial roads, followed by 29% rural, with 23% metropolitan or municipal and 1% national (Karani 2008). Recent annual budgets of US$140 million (National Treasury 2013) have been allocated for building, upgrading and maintaining roads over the next 3 years, but no offset is mentioned for the indirect and direct effects of roads or their cumulative effects on wildlife (Karani 2008). Furthermore, the South African population is ranked 26th in the world (51 million people or 42 people per km^2; CIA 2012) and with a positive economic growth of 4%, pressure is anticipated on all modes of transport through greater demand and increased use (Karani 2008; Statistics South Africa 2012). In South Africa, approximately 75% of freight is transported by road (Karani 2008).

There is a perennial conflict between development and biodiversity objectives in South Africa. South Africa's future economic development requires infrastructure and the construction of new transport routes is inevitable. South Africa is estimated to have the world's fifth largest mining sector in terms of GDP (18%) and mining accounts for 50% of transportation volume (Statistics South Africa 2012). In addition, tourism is an important revenue earner, currently accounting for 7.9% of GDP and supporting one in every 12 jobs in South Africa. South Africa contains eight World Heritage Sites (from 166 globally; Dudley 2008) and 19 national parks. Tourism is expected to generate an annual contribution of US$56 billion by 2020, placing the country's transport network under increasing pressure to meet these demands (Statistics South Africa 2012).

South Africa is the most affluent country in Africa, and yet 60% of the population lives below the poverty line (<US$67 per month). Roads are continually being constructed to facilitate effective transport of goods and services and to improve access to resources and employment opportunities. Innovative strategies are required to minimise the ecological impact of roads as well improve people's livelihoods. Although South Africa has a legislative framework that necessitates environmental impact assessments for development, these tools are not always used optimally. This is in part due to the lack of capacity to ensure compliance and enforcement, and also partly due to a lack of understanding of the real impacts of development. This chapter aims to illustrate the challenges and some solutions facing transportation planners and ecologists in South Africa.

LESSONS

54.1 Wildlife-vehicle collisions in South Africa are costly and likely to be under-reported

In 2011, there were 13,932 human fatalities from road accidents in South Africa (RTMC 2011; PMG 2013). Wildlife-vehicle collision (WVC) did not rate as a category for describing the type of collision, but came under the heading of 'other' or 'unknown', of which there were 714. Annually, around US$150 million is spent on accident insurance claims, with US$7.7 million attributed to possible WVCs (RTMC 2011). It is difficult to state why little WVC data is available; it may be because some vehicles are uninsured, belief of 'getting in trouble' with the law for killing an animal, or simply because many people drive four-wheel drive vehicles in South Africa and are therefore less likely to be involved in a serious collision (RTMC 2011). This may be particularly so for smaller species that cause little or no vehicle damage, and may not be considered a priority. And while vehicle owners are financially compensated through their insurance, the biodiversity costs of these collisions are never calculated (Eloff & van Niekerk 2005). A report on the progress of achieving road safety, road safety education and data collection goals presented at the African Road Safety Conference held in Ghana in 2007 (PMG 2013) concluded that the main challenge in South Africa was a lack of resources.

54.2 Knowledge of the impacts of roads and traffic in South Africa is limited because most studies have focused solely on roadkill

Research on the ecological impacts of roads has been slower in Africa than elsewhere in the world, and mitigation of such impacts is rarely considered during road projects (Chapters 55 and 56). The few studies that have been conducted in southern Africa (e.g. Jackson 2003; Mkanda & Chansa 2010; Bullock et al. 2011) have

focused primarily on the rate and location of wildlife mortality (Textboxes 54.1 and 54.2). However, simple counting of the number of mortalities does not estimate the extent to which roads and vehicles are endangering wildlife populations or species. This reflects the need for a greater understanding of other potential impacts of roads, including habitat fragmentation and pollution.

One of the earliest studies in South Africa recorded bird roadkill in the Northern Cape Province (Siegfried 1966). Later studies included surveys in the Eastern Cape (Eloff & van Niekerk 2008), Nama-Karoo (Dean & Milton 2009), the southern Kalahari (Bullock et al. 2011), northern Limpopo (Collinson 2013) as well as unpublished data from various locations (I. McDonald, Percy FitzPatrick Institute of African Ornithology, University of Cape Town,

---

Textbox 54.1 Roads and grain spill threaten owls in South Africa.

In 2004, owls (including the African grass owl, barn owl, marsh owl and spotted eagle owl; Fig. 54.1) were being killed in significant numbers ($n = 554$) on a 41 km stretch of the N17 toll road between Springs and Devon, South Africa (Ansara 2004) (Fig. 54.2). Research indicated that rodent populations were flourishing due to the availability of good habitat and an abundant food supply from trucks spilling grain onto the road. Owls hunting the overabundant rodents on the road at night were consequently being hit by vehicles and killed. A survey in the Free State Province identified a

similar problem for the threatened African grass owl.

South Africa produces greater than 10 million tons of maize and other grain annually, and trucks are critical for the distribution of these stocks. Primary agriculture provides about 7% of formal employment in South Africa and contributes about 3% to the country's GDP, with the agro-industrial sector comprising about 12% of GDP (Statistics South Africa 2012). Mitigation of problems caused by grain spillage includes ongoing education of transport companies to cover their loads and installation of tighter seals on trucks (Ansara 2004).

Figure 54.1 Spotted eagle owl roadkill, South Africa. Source: Photograph by W. Collinson.

Textbox 54.2 Reptiles and amphibians are vulnerable to road mortality in South Africa.

With 480 species, southern Africa has the highest reptile diversity on the continent (Branch 1998). In 2009, 13 surveys along a 30 km stretch of the R523 road north of the western Soutpansberg, Limpopo Province (Fig. 54.2), revealed that reptiles comprised 60% of kills, while mammals and birds made up only 19 and 21%, respectively (G. Alexander, unpublished data). In this study, amphibians were excluded from surveys because the high number of kills and the difficulty in identifying the much-damaged carcasses made quantification of amphibians impractical. Under the conservative assumption that reptiles are active for 4 months of the year, these measures translate to an annual mortality rate of 33 ± 19 (s.d.) individuals per kilometre. Snakes made up the majority of reptile roadkill recorded (89 ± 15%), with mortality rates higher than those reported in a similar study in Arizona, United States (Rosen & Lowe 1994). The impacts of roads and traffic on amphibians and reptiles in South Africa are potentially of great conservation concern.

Roadkill may also have important demographic consequences (Steen et al. 2006). For example, male puff adders travel long distances during the mating season (March to May) and accordingly cross roads more often. At these times, males may encounter roads and be killed greater than 10 times more frequently than females (G.J. Alexander & B. Maritz 2011, unpublished data). Such differential mortality rates have the potential to skew sex ratios, leading to depressed reproductive output (Aresco & Gunzburger 2004).

---

personal communication; C. Vernon, East London Museum, personal communication).

While data are yet to be used in road planning, the Endangered Wildlife Trust formed the Wildlife and Transport Programme (WTP) in 2012 to address identified concerns. Related concerns included undertaking risk assessments; identifying, developing and implementing relevant mitigation strategies; and raising the awareness of transportation and planning agencies.

54.3 South Africa needs a standardised protocol to detect and record roadkill

In 2013, South Africa's first standardised protocol to quantify roadkill rates was developed (Collinson 2013). A study reviewed methods used globally for the study of roadkill and incorporated them into the design of a standardised detection protocol (Table 54.1). The protocol guides users in collecting mortality data for multiple vertebrate species and was used to collect baseline estimates of roadkill in the Greater Mapungubwe Transfrontier Conservation Area (GMTFCA), Limpopo Province (Fig. 54.2). Fabricated roadkill of two sizes, replicating a large bird (e.g. spurfowl) and a small rodent (e.g. bushveld gerbil) were used to assess detection rates by observers travelling at different vehicle speeds. A 96% detection rate was achieved for the large simulated roadkill at speeds of 80 km h^{-1}, but a lower detection rate was noted for the smaller simulated roadkill

Table 54.1 Recommended protocols to maximise detection rates of vertebrates in South Africa.

Trial	Recommendation
Speed	40–50 km h^{-1}
Start time	1.5 h after sunrise
Stop time	1.5 h before sunset
Number of observers	1
Observer skill level	Trained
Distance to be driven	100 km
Number of days to be sampled	40

Source: Adapted from Collinson et al. (2014).

(92%). This decreased slightly to 91% at 50 km h^{-1} (but remained at 96% for the larger roadkill).

This study also identified the biophysical, environmental and physical factors affecting roadkill rates. The study area was selected because it is home to a wide range of vertebrates, some of which are endangered (e.g. African wild dog and Pel's fishing owl). It has a high species richness for reptiles (120 species; Branch 1998), birds (at least 429 species; Hockey et al. 2005) and mammals (about 100 species; Skinner & Chimimba 2005) and a low species richness for amphibians (about 12 species; Braack 2009). Traffic volumes are expected to increase in the near future with the recent development of a nearby coal mine and increased tourism.

Over a 120-day period, 1121 roadkill carcasses were detected and 166 species identified. Birds were the most commonly detected species (52%), and mammals,

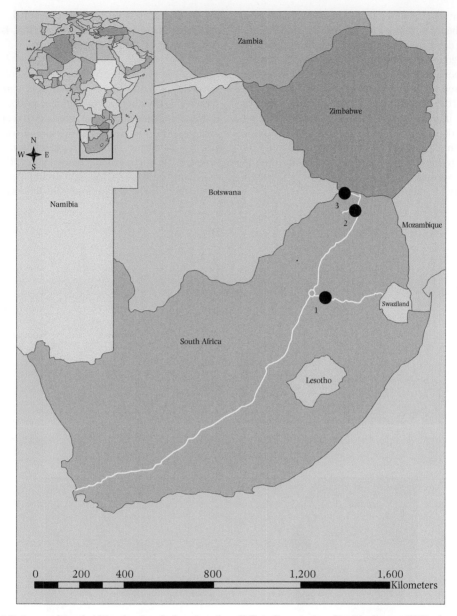

Figure 54.2 A map of South Africa showing the location of roadkill studies mentioned in this chapter: (1) Textbox 54.1, (2) Textbox 54.2 and (3) Lesson 54.3. Source: Map by and reproduced with permission from Armand Kok, Rhodes University.

reptiles and amphibians comprised 26, 20 and 2%, respectively. Additional records indicated that five cheetahs were killed on roads adjacent to the GMTFCA between January 2006 and June 2009 (no comparative data available on road mortality rates for elsewhere in South Africa) and nine African wild dog road fatalities in a 3-month period in 2012. With only 450

African wild dogs left in South Africa, roads may have serious impacts on this species (Chapter 38).

An alternate method to survey roadkill (Guinard et al. 2012; Chapter 12) used slow driving combined with walking transects in order to subsample sections of road by monitoring carcass persistence and removal (both on and off the road). While likely to be more time-consuming, resulting in shorter overall sampling distances being covered, foot surveys are likely more effective at detecting carcasses on verges that may have otherwise gone undetected.

It is important that future research on roadkill be standardised to enable comparisons among studies (Chapter 10). Representative road stretches should be chosen in a study, and each stretch should have a set length to allow for easier analysis and later comparison. The conservation implications of this protocol are far-reaching because roads are important for economic development but also threaten biodiversity.

54.4 Fencing may be a successful mitigation measure in South Africa, but it can exacerbate fragmentation

South Africa is a country with a 'fence culture' with hundreds of thousands of kilometres of game fencing (Fig. 54.3) dividing farms, national parks and other properties. While an effective method to reduce WVCs (Seiler 2005; Chapter 20), widespread use of game fencing increases habitat fragmentation (see also Chapter 38). Furthermore, many South African fences are electrified (Fig. 54.3B), and animals can be killed trying to move through or under the fences (Fig. 54.4).

(A)   (B)

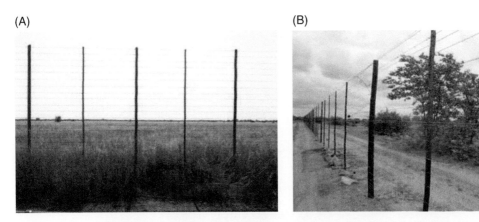

Figure 54.3 (A) An example of a game fence and (B) and electric game fence in South Africa. Source: Photographs by W. Collinson.

Figure 54.4 Rock monitor killed by an electric fence. The electric strands on the ground deter wildlife from crossing but allow vehicles to pass, reducing the need for a gate. Source: Photograph by W. Collinson.

There are currently no formal national guidelines pertaining to the design of electrified game fences in South Africa (Beck 2010).

A study on the presence and absence of game fencing in the Eastern Cape, South Africa, showed that fewer roadkills were detected where there was full fencing on both sides of the road (Eloff & van Niekerk 2005). More roadkill were detected when there was partial fencing or when only one side of the road had a game fence, with 80% of roadkill detected when there was no fencing. In addition, where there was no fencing, there was often dense bush that ungulates could feed on, increasing the likelihood of WVC (Eloff & van Niekerk 2005). Furthermore, almost 40% of roadkills were detected near fence ends (Eloff & van Niekerk 2005, see also Chapter 20).

More roadkills were detected in the GMTFCA in areas without electric or game fences (Collinson 2013), suggesting that the taller and more robust electric and game fences are more effective at preventing wildlife from accessing roads. However, fencing does not stop all animals from accessing a road. Many South African antelope can easily jump over 2.4 m fences (Eloff & van Niekerk 2005) and other species dig under or push through fences, providing an opening for other animals (Beck 2010).

'To fence or not to fence?' That is the question surrounding solutions to prevent WVCs in South Africa. Consideration needs to be given to whether a species is more at risk of becoming roadkill or whether fragmentation by fencing will increase population isolation. Many species with long-distance dispersal (e.g. African wild dog, Chapter 38) are currently managed as metapopulations within fenced nature reserves and would therefore benefit from more effective fencing that prevents their dispersal and lowers the risk of WVC. The same is true for most African antelope that are often translocated among protected areas.

However, many species hold 'less value' to landowners (e.g. African civet) and often do not form part of a reserve's management programme. Other species have different values and uses, even as roadkill (Textbox 54.3). The loss of an African civet will not impact a game farmer's livelihood, whereas a roadkill kudu that can no longer be sold to a trophy hunter will reduce their income. Consequently, species such as the African civet may be affected by both population isolation through the presence of fences and WVCs in areas without effective fencing and therefore require a combination of wildlife crossing structures and fencing.

When solving a 'barrier versus roadkill' problem, the solution is likely to vary among species, as there does not appear to be a 'one-size-fits-all' remedy. Decisions are likely to be based on the willingness of adjacent landowners to connect their properties and ultimately their wildlife. A combination of fencing and crossing structures could be an effective solution to barriers and roadkill in South Africa. However, until each landowner is prepared to engage in developing large conservancies, this solution is unlikely to take effect. Instead, we should examine alternatives and create crossing structures for species of 'lesser monetary value' (and highlight their ecological value) that fall outside of reserve management programmes and still rely on natural dispersal and movement, often across roads.

Textbox 54.3 Some roadkill in South Africa is likely intentional, but there is little supporting data.

Wildlife road mortality has a unique and valuable use in South Africa. In addition to natural scavenging by wildlife, carcasses may be removed from roads by humans for use as food, trophy and *muti* (traditional medicine/folklore). African traditional medicine makes use of various natural products, many derived from trees and other plants as well as animals, which is prescribed by a *sangoma* (a traditional healer). For example, a brown hyena roadkill was found in the northern Limpopo Province with the animal's tail removed (Fig. 54.5; W. Collinson, unpublished data). Reports in Namibia show that hyena body parts are used for *muti*: if the tail is burnt and the smoke blown under doors, it is believed that people inside fall asleep, making it possible for burglars to steal property undetected. Other purposes for hyena body parts include medicine (e.g. treatment for asthma) and cosmetics (e.g. perfume) (Low 2008). It is likely that other wildlife species may have similar traditional uses and roadkill is an obvious source.

Many western cultures frown upon the idea of consuming roadkill carcasses, despite it being a potentially abundant source of protein (Klein 2011). In South Africa, scrub hare roadkill may be removed by people as a food source or for the treatment of earache. *Muti* comprises knowledge systems that have developed over generations and are often culturally embedded beliefs that may be difficult to change. This challenges wildlife conservationists to be sensitive to these beliefs, as well as to address the 60% who live below the poverty line (Statistics South Africa 2012) and may not have ready access to meat.

Figure 54.5 Brown hyena roadkill with the tail removed. Source: Photograph by W. Collinson.

CONCLUSIONS

There is a lack of information about the effects of roads on wildlife in South Africa and consequently little mitigation of these impacts. South Africa is fundamentally different to Europe and North America in its faunal diversity, landscapes and geography, density of roads and humans, and funding and support for road ecology research and mitigation. However, the information and lessons learned in developed countries can be implemented and adapted in order to develop unique African solutions.

A balance between the need for an efficient transport network and a sustainable environment is a challenge facing most developing nations (Chapters 36–38, 41, 43 and 50–58). This commitment will require financial resources and a legislative framework to support the implementation of mitigation measures.

The current review of the IUCN (2012) Red List (Mammals of South Africa) by the Endangered Wildlife Trust is an opportunity to summarise the threat of roads to one group of taxa as well as raise awareness. As a logical starting point, such information will be used to develop guidelines for environmental impact assessments and produce a handbook on mitigation measures for reducing the incidence of roadkill. Over time, similar reviews will be conducted for birds, amphibians and reptiles. An advocacy programme will then be established to ensure that mitigation measures are mainstreamed and that the environmental sustainability of transport is adequately considered in the early stages of development.

ACKNOWLEDGEMENTS

We thank Bridgestone SA and the De Beers Group of Companies for funding and the Endangered Wildlife Trust, Rhodes University and University of the Witwatersrand for support.

FURTHER READING

Collinson (2013): An outline of the standardised protocol (as currently being implemented in South Africa) to examine roadkill rates.

Endangered Wildlife Trust (EWT) website: Insights into the current initiatives and work being conducted by the EWT's Wildlife and Transport Programme (https://www.ewt.org.za/WTP/wtp.html and blog http://endan-geredwildlifetrust.wordpress.com/).

EWT et al. (2012): Minutes of two international road ecology workshops coordinated by EWT in South Africa in 2012, which outline a 5-year action plan to be undertaken by the EWT, including a national network to raise awareness, quantify road impacts through a range of techniques and develop best-practice guidelines.

Guinard et al. (2012): An alternate method to assess roadkill that examines detection rates by taking into account carcass persistence and removal rates.

REFERENCES

Ansara, T.M. 2004. Determining the ecological status and possible anthropogenic impacts on the grass owl, *Tyto capensis* population in the East Rand Highveld, Gauteng. MSc dissertation. Rand Afrikaans University, Johannesburg, South Africa.

Aresco, M.J., and M.S. Gunzburger. 2004. Effects of large-scale sediment removal on herpetofauna in Florida wetlands. Journal of Herpetology **38**: 275–279.

Beck, A. 2010. Electric fence induced mortality in South Africa. Msc thesis. University of the Witwatersrand, Johannesburg, South Africa.

Braack, H.H. 2009. Report on a survey of Herpetofauna on the proposed mining site: Vele Colliery Project. In: Nel, G.P. and E.J. Nel. Biodiversity Impact Assessment of the Planned Vele Colliery. Prepared for Coal of Africa Limited. Dubel Integrated Environmental Services, Polokwane, South Africa.

Branch, B. 1998. Field Guide to Snakes and other Reptiles of Southern Africa, 3rd edition. P. Barker and H. Reid (eds.). Struik Publishers, Cape Town, South Africa.

Bullock, K.L., G. Malan, and M.D. Pretorius. 2011. Mammal and bird road mortalities on the upington to twee rivieren main road in the southern Kalahari, South Africa. African Zoology **461**: 60–71.

Central Intelligence Agency (CIA). 2012. CIA World Factbook. Available from https://www.cia.gov/library/publications/the-world-factbook/index.html (accessed on August 2012).

Collinson, W.J. 2013. A standardised protocol for roadkill detection and the determinants of roadkill in the greater mapungubwe transfrontier conservation area, Limpopo Province, South Africa. MSc dissertation. Rhodes University, Grahamstown, South Africa.

Collinson, W.J., D.M. Parker, R.T.F. Bernard, B.K. Reilly and H.T. Davies-Mostert. 2014. Wildlife road traffic accidents: a new technique for counting flattened fauna. Ecology and Evolution **4**(15): 3060–3071.

Dean, W.R.J., and S.J. Milton. 2009. The importance of roads and road verges for raptors and crows in the Succulent and Nama-Karoo, South Africa. Ostrich **74**: 181–186.

Dudley, N. (ed.). 2008. Guidelines for Applying Protected Area Management Categories. IUCN, Gland, Switzerland.

Eloff, P.J., and A. van Niekerk. 2005. Game, fences and motor vehicle accidents: spatial patterns in the Eastern Cape. South African Journal of Wildlife Research **35**: 125–130.

Eloff, P. and A. van Niekerk. 2008. Temporal patterns of animal-related traffic accidents in the Eastern Cape, South Africa. South African Journal of Wildlife Research, **38**(2): 153–162.

Endangered Wildlife Trust (EWT), D.J. Smith, and R. Van Der Ree. 2012. Road ecology workshops minutes. Arrive Alive, South Africa. Available from http://www.arrivealive.co.za/documents/Road%20Ecology%20Workshop%20Minutes%20July%202012%20Final.pdf (accessed on January 2013).

Guinard, É., R. Julliard, and C. Barbraud. 2012. Motorways and bird traffic casualties: carcasses surveys and scavenging bias. Biological Conservation **147**: 40–51.

Hockey, P.A.R., W.R.J. Dean, and P.G. Ryan (eds.). 2005. Roberts – Birds of Southern Africa, VII edition. The Trustees of the John Voelcker Bird Book Fund, Cape Town, South Africa.

International Union for Conservation of Nature (IUCN). 2012. The IUCN Red List of Threatened Species. Version 2012.2. IUCN, Gland, Switzerland. Available from http://www.iucnredlist.org (accessed on October 2012).

Jackson, H.D. 2003. A field survey to investigate why nightjars frequent roads at night. Ostrich **74**: 97–101.

Karani, P. 2008. Impacts of roads on the environment in South Africa. DBSA – Development Bank of Southern Africa, Midrand, South Africa.

Klein, D. 2011. The Perennial Plate Trailer. Vimeo, New York. Available from http://vimeo.com/17957265 (accessed on January 2013).

Low, C. 2008. Animals in Bushman Medicine: full research report ESRC end of award report, RES-000-23-1326. ESRC, Swindon.

Mkanda, F.X., and W. Chansa. 2010. Changes in temporal and spatial pattern of road kills along the Lusaka-Mongu (M9) highway, Kafue National Park, Zambia. South African Journal of Wildlife Research **411**: 68–78.

National Treasury. 2013. Budget Review 2013. National Treasury Republic of South Africa, South Africa. Available from http://www.treasury.gov.za/documents/national%20budget/2013/review/Prelims%202013.pdf (accessed on July 2013).

PMG (Parliamentary Monitoring Group). 2013. Road Accident Rates: Road Traffic Management Corporation Intervention. Parliamentary Monitoring Group, South Africa. Available from http://www.pmg.org.za/report/20130219-towardssafer-roads-decade-action-road-safety-2011-2020-strate-gic-int (accessed on July 2013).

RTMC (Road Traffic Management Corporation). 2011. Road Traffic Report 31 March 2011. RTMC, Pretoria, South Africa. Available from http://www.arrivealive.co.za/documents/March%202011%20Road%20Traffic%20Report.pdf (accessed on August 2013).

Rosen, P.C., and C.H. Lowe. 1994. Highway mortality of snakes in the Sonoran desert of southern Arizona. Biological Conservation **68**: 143–148.

Seiler, A. 2005. Predicting locations of moose–vehicle collisions in Sweden. Journal of Applied Ecology **42**: 371–382.

Siegfried, W.R. 1966. Casualties among birds along a selected road in Stellenbosch. Ostrich: Journal of African Ornithology **37**: 146–148.

Skinner, J.D., and C.T. Chimimba. 2005. The Mammals of the Southern African Subregion, 3rd edition. D. Van Der Horst (ed.). Cambridge University Press, Cape Town, South Africa.

Statistics South Africa. 2012. Census 2011: Census in Brief. Statistics South Africa. Pretoria, South Africa. Available from http://www.statssa.gov.za/ (accessed October 2012).

Steen, D.A., M.J. Aresco, S.G. Beilke, B.W. Compton, E.P. Condon, C.K. Dodd Jr., H. Forrester, J.W. Gibbons, J.L. Greene, G. Johnson, T.A. Langen, M.J. Oldham, D.N. Oxier, R.A. Saumure, F.W. Schueler, J.M. Sleeman, L.L. Smith, J.K. Tucker, and J.P. Gibbs. 2006. Relative vulnerability of female turtles to road mortality. Animal Conservation **9**: 269–273.

Chapter 55

UNFENCED RESERVES, UNPARALLELED BIODIVERSITY AND A RAPIDLY CHANGING LANDSCAPE: ROADWAYS AND WILDLIFE IN EAST AFRICA

Clinton W. Epps[1], Katarzyna Nowak[2,3] and Benezeth Mutayoba[4]

[1]Department of Fisheries and Wildlife, Oregon State University, Corvallis, OR, USA
[2]Department of Zoology & Entomology, University of the Free State, Phuthaditjhaba, South Africa
[3]Department of Anthropology, Durham University, Durham, UK
[4]Department of Veterinary Physiology, Biochemistry, Pharmacology and Toxicology, Sokoine University of Agriculture, Faculty of Veterinary Medicine, Morogoro, Tanzania

SUMMARY

Roads in East Africa present unique challenges to wildlife conservation because reserves are largely unfenced and biodiversity is high. Few roads in the region carry high-speed, high-volume traffic, but this is changing rapidly. This chapter reviews existing research on the impact of roads on East African wildlife, research needs and mitigation strategies.

55.1 Road type and protected area status affect the direct impacts of roads and traffic on wildlife in East Africa.

55.2 Indirect impacts of East African roads may be even more important than the direct impacts.

55.3 Documenting remaining movements and migrations in these rapidly changing landscapes is critical.

55.4 Mitigation of road impacts on East African wildlife has mostly relied on low-cost approaches.

Handbook of Road Ecology, First Edition. Edited by Rodney van der Ree, Daniel J. Smith and Clara Grilo.
© 2015 John Wiley & Sons, Ltd. Published 2015 by John Wiley & Sons, Ltd.
Companion website: www.wiley.com\go\vanderree\roadecology

55.5 New highway construction should avoid protected areas and requires timely conservation action.
55.6 Early outreach to planners, designers and funders of East African road projects is needed.

We recommend avoiding construction of high-speed roads through protected areas in East Africa. Outside of protected areas, identifying remaining wildlife movement pathways across current or proposed high-speed roads is a critical and time-sensitive research need. Because relatively few roads have been expanded and paved for high-speed traffic, engaging with road planners early during the proposed expansion of such roads could allow more cost-efficient design modifications to facilitate safe movement of wildlife across roads.

INTRODUCTION

As human populations have rapidly increased, East Africa has seen tremendous changes in land use and patterns of human settlement. East African nations (here, defined as the member states of the East African Community (EAC): Tanzania, Kenya, Uganda, Rwanda and Burundi) have some of the highest human population growth rates in the world. Although paved roads in Africa have a lower impact on poverty reduction than expected (Gachassin et al. 2009), road infrastructure is often considered a key to socio-economic development, and there is continual pressure to enlarge and improve existing roads and highways under the East African Road Network Project (EARNP). As seen worldwide, roads have direct and indirect negative effects on species and ecosystems. Given the importance of wildlife to economies of countries in the EAC (e.g. Kweka et al. 2001), road development will likely have negative economic impacts as well.

Despite commonalities with other regions, road ecology in East Africa has unique aspects. Most importantly, unlike much of southern Africa (Chapter 54) and Europe, East African reserves are often unfenced; thus, animals cross roads in larger numbers and more regularly than in many other parts of the world (Fig. 55.1). Conservation planning in this region is also complicated because East Africa has a heterogeneous environment, with habitats including savannah, woodland, plains, swamps, mountains, semi-desert and wet tropical forests. Vertebrate biodiversity, particularly of large mammals, is extremely high (Chapter 56).

(A)

(B)

Figure 55.1 Impala (A) and elephants (B) crossing the Mikumi Highway. Where there is road infrastructure, wildlife species may adopt a siege or skirmish strategy (Blake et al. 2008). Siege strategists (e.g. elephants) reduce home range size and restrict movements in order to avoid roads. Skirmish strategists (likely including impala) continue to range widely, crossing roads despite risk of mortality. Source: Photographs by K. Nowak.

Many of the larger mammals pose potential risk to humans, either directly or in wildlife-vehicle collisions (WVC), and human–wildlife conflict (i.e. interaction resulting in negative effects on people, wildlife species or the environment) along roads is common (e.g. Sitati et al. 2003). Finally, East African roads are comparatively undeveloped. For instance, the two largest nations in the region, Tanzania and Kenya, have approximately 86,000 km (TANROADS 2012) and 63,000 km (KRB 2012), respectively, of officially recognised roads, but most of those are unpaved and not suitable for high-speed traffic. However, many new road-building or existing road improvement projects have recently been initiated. For example, in Tanzania, extensive roadway development is planned, including increasing the density of all paved roads from 6.86 km/1000 km² as of 2009 to 9.98 km/1000 km² by 2015 and increasing the percentage of paved roads from 39% as of 2009 to 45% by 2015. Similar levels of expansion are planned for elsewhere in the EAC (2011).

In this chapter, we review the direct and indirect effects of roads on wildlife, identify research needs and discuss key aspects of studying and mitigating ecological impacts of roads in East Africa.

LESSONS

55.1 Road type and protected area status affect the direct impacts of roads and traffic on wildlife in East Africa

Relatively few studies have addressed the direct impacts of roads on wildlife and ecosystems in East Africa; although, studies elsewhere in Africa (Chapter 54) or those involving wide-ranging or edge species may be informative. Most research has focused on large mammals, particularly elephants and primates. Direct impacts of roads on wildlife mortality and movement have been shown to vary widely with road type, location and species. Most roads in East Africa are currently unpaved, have low-speed traffic and have few vehicle strikes of larger mammals; however, paved, high-speed and high-traffic roads have a much greater impact on species, especially wide-ranging carnivores such as wild dogs (Woodroffe & Ginsberg 1997; Chapter 38). Some of the best documentation of how high-speed paved highways affect local movements and mortality of East African wildlife has occurred in Mikumi National Park, Tanzania (Fig. 55.2), which is bisected by the Tanzania–Zambia Highway. Since the 1970s, thousands of road mortalities of dozens of species of mammals, birds and reptiles have been recorded in Mikumi (Drews 1995; Rugaimukamu 2009).

Many wildlife species avoid roads, including elephants (Blake et al. 2008) and chimpanzees (Olupot & Sheil 2011; Hicks et al. 2012). In Mikumi National Park, Newmark et al. (1996) documented avoidance of the highway by elephants (Fig. 55.1B), four species of ungulates and black-backed jackals. In Central Africa, nocturnal animals occurred at lower densities in areas near roads even where hunting was not implicated (Laurance et al. 2008). However, some species may be attracted to roads. Yellow baboons pick up garbage thrown from passing vehicles on the Tanzania–Zambia Highway between Mikumi National Park and Iringa and feed on rice that falls off passing vehicles in the Kilombero Valley (Fig. 55.2; Gupfinger 2012). Baboons and other opportunistic species may therefore be disproportionately affected by roadkill (Drews 1995).

Road location is also important. Outside of protected areas, mortality has mostly been documented for small nocturnal carnivores, monkeys, reptiles, amphibians and invertebrates (Kamau et al. 2010; Senzota 2012). High-speed roads within or near protected areas are likely to have the greatest impact on large mammals and primates (e.g. Drews 1995; Holdo et al. 2011), such as the endangered Zanzibar red colobus monkey moving between protected forest and farms (Fig. 55.2). Protected areas often have extensive road networks because tourists viewing wildlife are largely restricted to vehicles. However, animals may not be greatly affected by roads utilised for tourism inside protected areas because commercial traffic is minimal or absent, night driving is often restricted, and vehicle speeds are low. Elephants in Congo routinely cross roads located in protected areas; elsewhere, they avoid villages and increase speed when crossing (Blake et al. 2008).

Even greater consequences for species may occur if roads affect long-distance movements, dispersal or seasonal migration (Chapter 56). Elephants still appear to make long-distance movements across south-central Tanzania between the Ruaha, Udzungwa and Selous–Mikumi ecosystems but apparently cross the paved Tanzania–Zambia Highway in only one or two locations (Jones et al. 2009; Epps et al. 2011. Development associated with the highway in that location could sever the only known route for dispersal and gene flow across this large region. The famous migration of wildebeest and other species in the Serengeti could be threatened by construction of a new highway there (Chapter 56; Dobson et al. 2010; Holdo et al. 2011).

Figure 55.2 Road ecology study sites in Kenya and Tanzania. (A) dead African rock python on the recently re-paved Mikumi–Iringa Highway; (B) baboon on the highway which bisects Mikumi NP; (C) colobus crossing sign alongside speed bumps in Jozani, Zanzibar; (D) baboons feeding on rice on the unpaved road next to Udzungwa Mts. NP, Kilombero Valley (there are plans underway to pave this road up to Ifakara). Existing roads (red) are from the Africover database; protected areas (green) are from the Protected Planet database. Source: Map and photographs (A, B, C) by K. Nowak; (D) Photograph by and reproduced with permission of Christina Gupfinger.

55.2 Indirect impacts of East African roads may be even more important than the direct impacts

As in many developing countries, the effects of increased human access following road construction or improvement may be highly significant (Chapters 2, 3 and 51). After road construction, human settlement rates typically increase, and roads facilitate extraction of bushmeat, charcoal, timber and other resources (Laurance et al. 2006; Brugiere & Magassouba 2009) by people from both local and outside communities. Human–wildlife conflict increases along roads: for example, in Kenya, elephant poaching (Maingi et al. 2012), wildlife snaring (Wato et al. 2006) and instances of elephants injuring people (Sitati et al. 2003) are more common along roads.

55.3 Documenting remaining movements and migrations in these rapidly changing landscapes is critical

Much remains to be done to document how roads affect East African ecosystems and develop suitable mitigation measures. Given the rapid pace of recent landscape change in East Africa, it is most critical to document and protect remaining movement corridors and migration routes. Conservation planning in the region is complicated by the diversity of wildlife species with different dispersal abilities and habitat requirements. Even for well-studied species, long-distance movements are poorly understood in East Africa (Caro et al. 2009). High-tech methods such as radio, satellite, or cell phone-based telemetry or population genetics (e.g. Epps et al. 2013; Chapter 14) are providing new insights on movements of large animals but are expensive. Low-tech methods such as interviewing local people (perhaps the only way to identify past migration or movement routes), tracking, visual counts, roadkill documentation and remote cameras have been used to describe road crossings, habitat use and landscape connectivity over larger scales (Caro et al. 2009; Jones et al. 2009; Epps et al. 2011). Such methods can be rapidly implemented on proposed road projects. Likely, future road projects should be identified early using analyses of settlement patterns and gaps in the existing transportation network; such analyses could be used to guide low-cost studies.

55.4   Mitigation of road impacts on East African wildlife has mostly relied on low-cost approaches

Current practices to limit WVC and roadkill on paved highways in East Africa have mostly been limited to low-cost options such as signage, reduced speed limits and speed bumps to slow drivers in key locations (e.g. Mikumi National Park). Because of lack of enforcement in rural areas, physical barriers such as speed bumps are probably most effective, but do not eliminate roadkill (e.g. Rugaimukamu 2009). There are only two known instances of wildlife crossing structures installed in East Africa: the underpass recently constructed to allow elephant movements between Ngare Ndare Forest Reserve and Mt. Kenya National Park (Chapter 43) and 'colobridges' for black-and-white colobus monkeys to cross safely above the busy Diani Beach road in Kenya (Chapter 41). Near Jozani-Chwaka Bay National Park, Zanzibar (Fig. 55.2), colobridges for Zanzibar red colobus were tested but failed because that population prefers terrestrial movement. Subsequently, speed bumps were installed that have reduced colobus deaths by greater than 80% (Tom Struhsaker, personal communication, 2012).

55.5   New highway construction should avoid protected areas and requires timely conservation action

New highway construction projects can affect connectivity of wildlife populations that are already threatened (Dobson et al. 2010; Chapter 56). However, the relatively undeveloped state of many East African roads provides opportunities to design wildlife-friendly crossings that can be implemented most cost-efficiently at the time of construction and to influence future road-siting decisions. Input from countries with a long history of road ecology research and mitigation will be required to make this process cost-efficient and effective. Given that wildlife corridors are a national priority in Tanzania, we recommend (i) avoiding or minimising construction of roads across major migration routes or through protected areas; (ii) mitigating by attempting to make existing and new road corridors permeable for the movement of wildlife, particularly in key locations for connectivity; and (iii) establishing a road ecology unit to coordinate efforts between the Ministries of Transport and the Ministries of Natural Resources and Tourism. Even where purpose-built wildlife crossing structures are too costly, bridges and large culverts for drainage may suffice if the approaches and structures themselves are accessible and wildlife friendly (e.g. van der Hoeven et al. 2010; Senzota 2012). The complex communities of large mammals (including large-scale migrations), ground-dwelling birds and other vertebrates could make designing appropriate crossings for all species challenging, but taxa-specific designs from elsewhere (see Chapter 59 for a list of best-practice guidelines) will provide excellent starting points.

Best-practice guidelines for road design and mitigation for African ecosystems must be developed because all the negative effects of roads on wildlife cannot be eliminated (Chapter 59). Basic recommendations for unavoidable roads include avoiding construction of roads with steep embankments unless crossing structures are also installed. In African forests, van der Hoeven et al. (2010) recommended maintaining tree cover above and adjacent to forest roads to facilitate crossing by forest animals, but such cover may increase the risk of WVC if traffic is high speed. We reiterate that the identification and documentation of existing wildlife crossing points are critical to planning any aspect of future road construction. Important crossing points must be protected, such as the only known location where elephants move between Ruaha, Udzungwa and Mikumi National Parks in Tanzania (Epps et al. 2011; Fig. 55.2). Because so many roads are as yet unpaved, modifications to facilitate wildlife crossings could be implemented relatively efficiently when roads are paved or enlarged.

55.6   Early outreach to planners, designers and funders of East African road projects is needed

Clearly, a primary planning need is to identify potential road projects early and communicate with funders (e.g. African Development Bank) and project leaders about ways to minimise and mitigate ecological impacts of road projects, perhaps using a road zoning approach (Chapter 2; Laurance et al. 2014). In East Africa, this likely requires communication with national entities that approve and manage those projects (Ministries of Transport and Infrastructure), but also outside entities that provide funding for road projects (i.e. development banks). Where possible, outside funders should be convinced to require ecologically friendly road planning as a condition for funding, as has recently been put forth by the German government with regard to alternatives to the Serengeti road (Chapter 56). If developed nations value East Africa's wildlife, then they must consider the impact of aid money on wildlife.

CONCLUSIONS

In East Africa, high human population growth has led to rates of landscape change unprecedented in that region and increasing road development. As seen in other parts of the world, indirect effects of road construction such as increased rates of human settlement and access may have the biggest impact of all road impacts (Chapter 51). Wildlife is a key resource to many East African economies, and conservation planning is complicated by high species diversity and heterogeneous environments. Where animals regularly cross roads, the large size of many East African species increases human safety concerns. Unlike southern Africa, most reserves in East Africa are unfenced, many species still occur outside of protected areas (Epps et al. 2011), and there are many seasonally migratory populations of wildlife. However, the concentration of large mammals and other species in protected areas means that avoiding construction of high-speed roads in or near protected areas is of paramount importance (Blake et al. 2008). Given the rate of landscape change, documenting remaining corridors and migration routes for wildlife by any means is critical, as is anticipating and planning where new road development may occur. Finally, conservation planners must communicate with government entities and external donors early in the planning phases so that new roads avoid, minimise and mitigate ecological impacts.

ACKNOWLEDGEMENTS

We thank Trevor Jones, Keith Lindsay, Tom Struhsaker and Christina Gupfinger for their assistance in identifying and documenting studies of wildlife interacting with roads.

FURTHER READING

Caro et al. (2009): Describes an initiative to document known wildlife movement corridors across Tanzania, based on all available information ranging from anecdotal evidence to carefully designed research. The initiative involved many scientists and resulted in a report to the Tanzanian Wildlife Institute.

EAC (2011): This web page describes planned changes to main travel corridors in the East Africa Community.

Epps et al. (2011): Describes a recent assessment of the distribution and potential connectivity of many large mammal populations between three protected areas in central Tanzania. Thus, it depicts how the large and complex large mammal community in East Africa interacts with many types of land management, human activities and habitats.

Laurance et al. (2006): This paper, although focused on Central Africa, details some of the indirect impacts of roads that can be expected in East Africa, particularly in forested areas.

Newmark et al. (1996). One of the few attempts to directly assess the effects of major roads on protected areas. It describes the impact of a large high-speed paved highway that bisects one of Tanzania's National Parks on the movement and distribution of several large mammal species.

REFERENCES

Blake, S., S. L. Deem, S. Strindberg, F. Maisels, L. Momont, I.-B. Isia, I. Douglas-Hamilton, W. B. Karesh, and M. D. Kock. 2008. Roadless wilderness area determines forest elephant movements in the Congo Basin. PLoS One **3**: e3546.

Brugiere, D. and B. Magassouba. 2009. Pattern and sustainability of the bushmeat trade in the Haut Niger National Park, Republic of Guinea. African Journal of Ecology **47**: 630–639.

Caro, T., T. Jones, and T. R. B. Davenport. 2009. Realities of documenting wildlife corridors in tropical countries. Biological Conservation **142**: 2807–2811.

Dobson, A., M. Borner, and T. Sinclair. 2010. Road will ruin Serengeti. Nature **467**: 272–273.

Drews, C. 1995. Road kills of animals by public traffic in Mikumi National Park, Tanzania, with notes on baboon mortality. African Journal of Ecology **33**: 89–100.

East African Community (EAC). 2011. Road transport in East Africa. EAC, Arusha, Tanzania. Available from http://www.infrastructure.eac.int/index.php?option=com_content&view=article&id=109&Itemid=129 (accessed 15 November 2012).

Epps, C. W., B. M. Mutayoba, L. Gwin, and J. S. Brashares. 2011. An empirical evaluation of the African elephant as a focal species for connectivity planning in East Africa. Diversity and Distributions **17**: 603–612.

Epps, C. W., J. L. Keim, S. Wasser, B. M. Mutayoba, and J. S. Brashares. 2013. Quantifying past and present connectivity illuminates a rapidly changing landscape for the African elephant. Molecular Ecology **22**: 1574–1588.

Gachassin, M., B. Najman, and G. Raballand. 2009. The impact of roads on poverty reduction: a case study of Cameroon. Policy Research Working Paper Series, Volume **5209**. World Bank, Washington, DC, 39 pp.

Gupfinger, C. 2012. A behavioral analysis of primate poly-specific associations in relation to the park boundary of Udzungwa Mountains National Park, Tanzania. Senior thesis. Princeton University, Princeton, NJ.

Hicks, T. C., P. Roessingh, and S. B. J. Menken. 2012. Reactions of Bili-Uele chimpanzees to humans in relation to their distance from roads and villages. American Journal of Primatology **74**: 721–733.

Holdo, R. M., J. M. Fryxell, A. R. E. Sinclair, A. Dobson, and R. D. Holt. 2011. Predicted impact of barriers to migration on the Serengeti wildebeest population. PLoS One **6**: e16370.

Jones, T., T. Caro, and T. R. B. Davenport. 2009. Wildlife Corridors in Tanzania. Unpublished Report, Tanzania Wildlife Research Institute (TAWIRI), Arusha, Tanzania.

Kamau, J. G., W. O. Ogara, J. J. McDermott, and P. M. Kitala. 2010. Community- and road-kill rabies surveillance in Kibwezi, Kenya. Journal of Commonwealth Veterinary Association **26**: 10–16.

Kenya Roads Board (KRB). 2012. Kenya Road Network. KRB, Nairobi, Kenya. Available from http://www.krb.go.ke/road-network/road-network.html (accessed 14 October 2013).

Kweka, J., O. Morrissey, and A. Blake. 2001. Is tourism a key sector in Tanzania? Input–output analysis of income, output, employment and tax revenue. Discussion Paper 2001–1. Tourism and Travel Research Institute, University of Nottingham, Nottingham.

Laurance, W. F., B. M. Croes, L. Tchignoumba, S. A. Lahm, A. Alonso, M. E. Lee, P. Campbell, and C. Ondzeano. 2006. Impacts of roads and hunting on central African rainforest mammals. Conservation Biology **20**: 1251–1261.

Laurance, W. F., B. M. Croes, N. Guissouegou, R. Buij, M. Dethier, and A. Alonso. 2008. Impacts of roads, hunting, and habitat alteration on nocturnal mammals in African rainforests. Conservation Biology **22**: 721–732.

Laurance, W. F., G. R. Clements, S. Sloan, C. O'Connell, N. D. Mueller, M. Goosem, O. Venter, D. P. Edwards, B. Phalan, A. Balmford, R. van der Ree, and I. B. Arrea. 2014. A global strategy for road building. Nature **513**: 229–232.

Maingi, J. K., J. M. Mukeka, D. M. Kyale, and R. M. Muasya. 2012. Spatiotemporal patterns of elephant poaching in south-eastern Kenya. Wildlife Research **39**: 234–249.

Newmark, W. D., J. I. Boshe, H. I. Sariko, and G. K. Makumbule. 1996. Effects of a highway on large mammals in Mikumi National Park, Tanzania. African Journal of Ecology **34**: 15–31.

Olupot, W. and D. Sheil. 2011. A preliminary assessment of large mammal and bird use of different habitats in Bwindi Impenetrable National Park. African Journal of Ecology **49**: 21–30.

Rugaimukamu, E. 2009. Studies on Road Kill Dynamics and Its Potential for Wild Animal Health Investigations: A Case Study of Mikumi National Park. Sokoine University of Agriculture, Morogoro, Tanzania.

Senzota, R. 2012. Wildlife mortality on foot paths of the University of Dar es Salaam, Tanzania. Tropical Ecology **53**: 81–92.

Sitati, N. W., M. J. Walpole, R. J. Smith, and N. Leader-Williams. 2003. Predicting spatial aspects of human–elephant conflict. Journal of Applied Ecology **40**: 667–677.

Tanzania National Roads Agency (TANROADS). 2012. Welcome to TANROADS: Classified Road Network. TANROADS, Dodoma, Tanzania. Available from http://www.tanroads.org (accessed 14 October 2013).

van der Hoeven, C. A., W. F. de Boer, and H. H. T. Prins. 2010. Roadside conditions as predictor for wildlife crossing probability in a Central African rainforest. African Journal of Ecology **48**: 368–377.

Wato, Y. A., G. M. Wahungu, and M. M. Okello. 2006. Correlates of wildlife snaring patterns in Tsavo West National Park, Kenya. Biological Conservation **132**: 500–509.

Woodroffe, R. and J. R. Ginsberg. 1997. Past and future causes of wild dogs' population decline. Pages 58–74. In: R. Woodroffe, J. R. Ginsberg, and D. Macdonald, editors. The African Wild Dog: Status Survey and Conservation Action Plan. International Union for the Conservation of Nature (IUCN), Gland, Switzerland.

Chapter 56

EXPECTED EFFECTS OF A ROAD ACROSS THE SERENGETI

Michelle E. Gadd

US Fish & Wildlife Service, Division of International Conservation, Falls Church, VA, USA

SUMMARY

Every year, more than 1.5 million wildebeest, zebras and gazelles migrate between the Serengeti National Park in Tanzania and the Masai Mara National Reserve in Kenya and back. In early 2010, plans were announced to construct a commercial road across the Serengeti National Park, bisecting this migration route. In addition to the direct impact of the road and traffic on migratory wildlife, a road would also increase access to a previously remote area, making it highly susceptible to human activities that are difficult to regulate. This region of the Serengeti is also home to endangered species targeted by poachers, and increased public access would make protection of these imperilled species even more challenging.

56.1 The Serengeti–Mara ecosystem is home to one of the last great mammal migrations on earth.

56.2 The migration is essential to ecosystem health of the Serengeti.

56.3 The direct and indirect effects of a commercial road would irreversibly damage the Serengeti.

56.4 Increased public access will increase poaching in the Serengeti.

56.5 Damage to the Serengeti would be impossible to mitigate or offset.

   The Serengeti–Mara ecosystem is one of the most productive ecosystems on earth and remains relatively intact. Constructing a highway across the Serengeti would result in a significant crash in the migrating populations of large wildlife due to illegal hunting, increased mortality due to wildlife vehicle collisions and barrier effects. Wildlife crossing structures are unlikely to be feasible or effective because of the large number of migrating animals and the chaotic nature of the migration. Whether the road is paved or unpaved is irrelevant – it will have a profound and far-reaching effect on this fragile and irreplaceable globally significant resource.

Handbook of Road Ecology, First Edition. Edited by Rodney van der Ree, Daniel J. Smith and Clara Grilo.
© 2015 John Wiley & Sons, Ltd. Published 2015 by John Wiley & Sons, Ltd.
Companion website: www.wiley.com\go\vanderree\roadecology

INTRODUCTION

The Serengeti–Mara ecosystem is home to one of the last great mammal migrations on earth; twice each year, more than 1 million wildebeest and half a million zebras and gazelles make the 300 km trek between the Serengeti National Park in Tanzania and the Masai Mara National Reserve in Kenya. Serengeti is one of the most spectacular and economically valuable destinations in Africa, and the migration is the primary attraction. More than 1 million tourists visit Tanzania per year, primarily for photographic and wildlife safaris, generating more than $1.5 billion USD in revenue (Mtweve 2013).

In early 2010, the president of Tanzania, Jakaya Mrisho Kikwete, announced plans to build a new east–west commercial road through the Serengeti National Park (Fig. 56.1A). A road on the same alignment had been rejected for unacceptable ecological impacts by the World Bank more than 20 years earlier yet continues to resurface in Tanzanian politics. Although the proposed road is only 54 km long within the protected area, it would serve as a link to Mwanza in the west and Arusha in the east, thus opening a new connection from Lake Victoria to the Tanzanian coast, and it can be expected to have significant commercial traffic. A commercial road already crosses southern Serengeti National Park and is not without environmental impact, but it is south of the core wildebeest dry season habitat and migration (Fig. 56.1B and C). By contrast, the proposed road would cross through the northern, most remote section of Serengeti, bisecting the north–south migration.

The announcement provoked an outpouring of international concern, highlighting the catastrophic impact such a road would have on wildlife, ecosystem health and tourism revenue in Tanzania and Kenya. Even proponents of infrastructure development questioned the purpose of such a road, as the government of Tanzania has not provided a credible explanation of the anticipated benefits of this costly undertaking. Concerned citizens, scientists, economists, diplomats and infrastructure development agencies acknowledged that people residing to the west and to the east of the Serengeti National Park need improved connectivity to other towns and cities in Tanzania but were unified in questioning the rationale behind building a road through such a fragile and valuable wilderness area when alternative routes outside the park would reach more settlements, serve more rural people and be less destructive (see Hopcraft et al. in press a for analysis of the socio-'economic and demographic advantages of

alternative routes). In response to international pressure, Tanzania issued a statement at a UNESCO meeting in Paris in 2011 indicating that the road would not be paved and that the government would consider alternative routes south of the protected area boundaries. In 2014, the northern route has been marked with surveyors' beacons. Some Tanzanian politicians and government officials continue to insist that the road will be built through the park. Pressure from local residents for a road will continue until those isolated communities are adequately connected to urban hubs and receive adequate benefits from tourism.

LESSONS

56.1 The Serengeti–Mara ecosystem is home to one of the last great mammal migrations on earth

Each year as the southern grasslands in the Serengeti National Park in Tanzania become drier, more than 1 million wildebeest and half a million zebras and gazelles (Figs 56.2 and 56.3) leave their wet season range (Fig. 56.1B) and head north, in search of greener pastures in the Masai Mara (Fig. 56.1C), on the Kenyan side of the border. While the general location of the start and finish of the migration has remained the same for millennia, the pathway followed by the animals is unpredictable and dynamic. The ungulates track grass suitability, moving across a patchy mosaic, seeking the fresh green flush that arises after rain. Rainfall patterns, herd dynamics, predator presence, river levels and human disturbance are among the myriad of factors affecting which way the herds go and how much time they spend in a given area (Hopcraft 2012).

Along the way, wildebeest males spar, trying to mate with and monopolise females on their way north (Estes 1991). At times, the migration is an impressive stampede; at others, the herds mill about chaotically. The migration can have tremendous momentum but it can also be fickle. A splash or a snort can trigger mass panic (Fig. 56.3). Lines of wildebeest double back on themselves, mindlessly going back the way they came, until for unknown reasons they realign and make another push forwards.

When the great herds reach the higher rainfall areas in Kenya, they disperse across the Masai Mara National Reserve and surrounding grasslands, grazing for several months. When rain returns to the southern grasslands, the wildebeest, zebra and gazelles make the

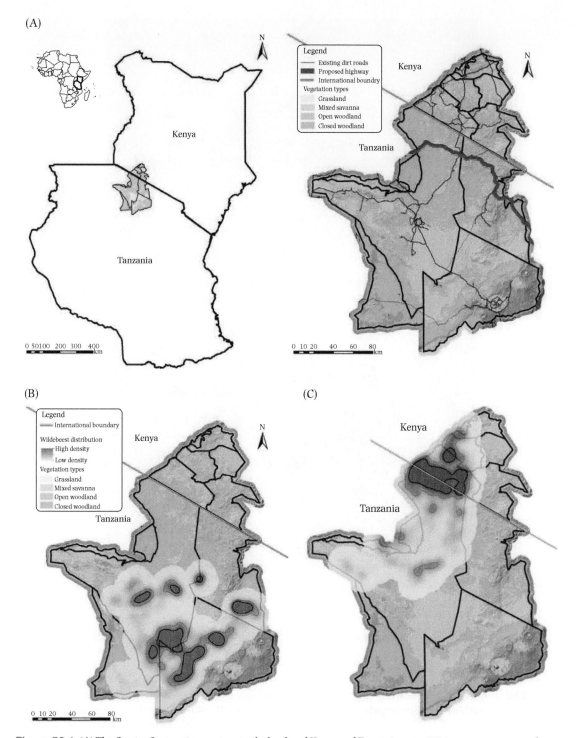

Figure 56.1 (A) The Greater Serengeti ecosystem on the border of Kenya and Tanzania, comprising numerous protected areas (delineated by black lines), the proposed commercial road (red line) and existing roads (brown lines). Distribution of wildebeest in the (B) wet and (C) dry seasons, respectively. Darker shadings represent higher densities of wildebeest in the landscape. Source: Reproduced with permission of J.G.C. Hopcraft.

Figure 56.2 Wildebeest in Serengeti beginning the trek north. Source: Photograph by M.E. Gadd.

Figure 56.3 Chaos of the migration. Source: Photograph by M.E. Gadd.

trek back, reaching the Serengeti plains in time for the females to calve and repeat the cycle.

56.2   The migration is essential to ecosystem health of the Serengeti

The Serengeti–Mara ecosystem is a mosaic of grasslands and wooded savannahs. Topography, geology, rainfall, herbivory and fire create and reinforce the boundaries between grass and woody plants. On the vast open plains, the sequence of rainfall, rapid grass growth, intense grazing by selective grazers (wildebeest) and bulk grazers (zebra) and return of nutrients through defecation of extremely high densities of mammals (Dobson et al. 2010) has created one of the most fertile and productive ecosystems on earth in terms of mammal biomass production. The herbivores in turn support an incredible diversity and quantity of carnivores.

Suppressing the movement of the great herds would disrupt the system irreversibly. Constraining the seasonal movement of the megaherds would prevent them from following optimal pasture, consuming and redistributing nutrients as they go. The feeding behaviour, social structure and ecological impact of migratory ungulates on the landscape are altered when they are forced to remain in one place. If they are forced to continually graze on the same area, rather than grazing on a patch and moving on, that area would become depleted and exhausted. Areas would become overgrazed and, without time to recover and replenish nutrients, would become less productive. Migratory ungulate populations collapse to low densities when forced to be sedentary (Textbox 56.1; Gadd 2012; Harris et al. 2009); numbers of wildebeest in the Karoo, Kruger and Kalahari ecosystems decreased by 100% (i.e. local extinction), 88% and 99%, respectively, after their migrations were impeded and range was restricted (Table 56.1).

Lower primary production would affect the whole food chain, resulting in lower secondary production (herbivore biomass) and the carnivores that feed upon them. Wildebeest and zebra numbers would eventually decline to the lowest density that could be supported at the leanest times of year. In the case of the Serengeti where positive feedback loops between grass growth, grazing and nutrient deposition have created an incredibly productive ecosystem (Hopcraft 2012), the overall carrying capacity would plummet.

> **Textbox 56.1 What do we know about susceptibility of migrations to infrastructure development?**
>
> Worldwide, 24 ungulate species once migrated (Harris et al. 2009) and six of these no longer migrate anywhere. Africa was home to 14 of the 24 migrations, and five are now entirely sedentary (springbok, black wildebeest, blesbok and scimitar horned oryx) or are extinct (quagga). In 2009, nine migratory species remained in six locations in Africa, but wildlife populations were stable or increasing in only two of these six locations (Boma-Jonglei in South Sudan and Serengeti–Mara Ecosystem in Tanzania and Kenya), while the other four (Tarangire in Tanzania, Liuwa in Zambia and Angola, Chobe in Botswana and Kalahari in Botswana) were in decline. Two factors were common in driving the disappearance of these migrations: (i) death of a significant number of individuals and (ii) restricted movement or restricted access to water or grazing areas (see also Chapter 42). Transportation infrastructure was a causal factor in the crash of four migrations: the pronghorn in North America and Mongolian gazelle, saiga antelope and kulan in Eurasia (Harris et al. 2009; Chapter 58).

Serengeti's primary tourist (and economic) draw, the migration, would disappear, and overall wildlife densities would decline.

Table 56.1 Decline of migratory wildebeest populations across Africa.

Locality	Historic population size	Population after movement impeded	Die-off (%)	Causes
Karoo, South Africa	100,000	0	100.0	Fencing, overhunting, habitat conversion
Etosha, Namibia	30,000	2,000	93.3	Fencing, disease
Kruger, South Africa	6,000	750	87.5	Fencing, restricted habitat
Kalahari, Botswana	262,000	260	99.9	Fencing, overhunting, competition with livestock
Amboseli, Kenya	Unknown	Unknown		Human settlement
North Mara–Loita Plains, Kenya	100,000	25,000	75.0	Habitat conversion
Tarangire, Tanzania	50,000	5,000	90.0	Habitat conversion, overhunting

Primary sources cited in Estes (1991); Gadd (2012); Harris et al. (2009); Tambling & duToit (2005).

56.3 The direct and indirect effects of a commercial road would irreversibly damage the Serengeti

The proposed road through Tanzania would cross one of the most remote, roadless wilderness areas left in the region (see also Chapter 3). Based on the pattern unfolding across Africa, one can anticipate the sequence of events expected to occur. Non-transparent awarding of road contracts, weak legislation regulating environmental impact and lack of oversight on construction sites elsewhere in Africa have led to the influx of poorly paid, poorly equipped construction crews, who make no effort to minimise the construction footprint (Fig. 56.4). Reports of foreign construction crews soliciting bushmeat (illegally hunted wild animals), ivory and rhino horn are widespread (northern Kenya, northern Republic of Congo, western Gabon; personal observations). Wherever paths are cleared, an influx of vehicles, passengers, pedestrians, livestock and eventually settlements (even where settlement is not legally allowed; Chapters 2 and 51) soon follows. In the case of a wilderness area like the Serengeti, a road would open up previously inaccessible wilderness areas, bisect the pathway of the migration and allow uncontrolled ingress and egress of people.

This route is likely to be used by heavy freight haulers looking to shorten their commute from the Congo Basin to the Indian Ocean travelling at high speeds, not by tourists on a relaxing game drive. If the road is constructed, numerous wildlife-vehicle collisions (WVC) are anticipated based on high wildlife densities, vehicle speed, the naiveté of wildlife to vehicles and the innate 'need to move' displayed via mass migrations. In addition to the migrants, animals that disperse long distances or require extremely large areas are likely to attempt to traverse the road and therefore be at risk of WVC. In this system, elephants and African wild dogs (Chapters 38, 43 and 55) (both critically endangered) are among the wide-ranging species likely to be affected.

For species that are hunted from roads or actively avoid human activity, roads can fragment habitat and isolate sub-populations. Transect data from Central Africa demonstrated that elephant densities are lower near roads and are often absent from blocks nearest roads or settlements (Yackulic et al. 2011). Tracking data from 27 collared elephants revealed that 26 of them avoided crossing roads in poorly protected areas, even when intact habitat was present on the other side. In 28.5 years of cumulative tracking, only one elephant crossed a public road while outside a protected area, and it did so at the point furthest from any two settlements, indicating that the elephant may have searched for the area of lowest human activity to make the crossing (Blake et al. 2008).

For rare species which require interbreeding, dispersal and recolonisation by conspecifics in the Masai Mara to remain viable (especially black rhinos, cheetah, lion and African wild dog), isolation into separate sub-populations would be detrimental.

Figure 56.4 Footprint of construction crews building the road north of Isiolo in Kenya. Source: Photograph by M.E. Gadd.

Figure 56.5 Movement of wildlife and wildlife products out of wilderness areas, in this case dwarf crocodiles in the Republic of Congo. Source: Photograph by M.E. Gadd.

56.4 Increased public access will increase poaching in the Serengeti

In Tanzania, poaching and bushmeat hunting of both common and rare species have reached unsustainable levels. Protected area personnel are losing the battle to protect wildlife and to prevent the flow of wildlife contraband onto the international black market; public roads into wilderness areas will make their jobs even tougher (Fig. 56.5).

Poaching of common game for meat is prevalent in the Serengeti ecosystem. Estimates range from 60,000 wildebeest (Mduma et al. 1998 in Loibooki et al. 2002) to 160,000 large mammals (Hofer et al. 1996 in Hofer et al. 2000) killed illegally each year. The meat is consumed locally and traded commercially as far away as the Democratic Republic of Congo. Bushmeat poaching is usually conducted within walking distance of the nearest road or settlement (Hofer et al. 2000).

Although it is illegal to trade in elephant ivory or rhino horn, black market prices have soared to record highs, putting tremendous poaching pressure on elephants and rhinos. Tanzania was previously home to the second largest elephant population in Africa but in 2009–2011 was the largest exporter of illegal ivory in the world (Burnett 2012). By its own estimates, Tanzania is losing 30–60 elephants per day to poachers, or 10,000 elephants per year (Member of Parliament James Lembeli, quoted by Burnett 2012). Nearly half of the country's surviving elephants have been killed in the 5 years to 2013. Eastern black rhinos in the Serengeti–Mara ecosystem have dwindled from approximately 450 to fewer than 100 (Metzger et al. 2007).

56.5 Damage to the Serengeti would be impossible to mitigate or offset

The government of Tanzania and the team contracted to write the Environmental and Social Impact Assessment (ESIA) assert that speed bumps or seasonal/night-time road closures would be sufficient to mitigate the impact of a commercial road (Intercontinental Consultants and Technocrats Pvt. Ltd. (India) for the United Republic of Tanzania Ministry of Infrastructure Development Tanzania National Roads Agency (TANROADS) 2010). The mitigation efforts described in the ESIA include the following: (i) avoid bisection of habitats; (ii) establish and implement procedures (by laws) to limit speeding; and (iii) conduct seminars on road safety for communities, but the total budget for these activities was less than US$2000. No methods have been described, and no budget has been allocated to identify, designate and preserve critical wildlife movement pathways through research, baseline data collection or road design sensitive to the species' habits. The design does not include any wildlife crossing structures.

The only proposed intervention with regard to WVC was to instal speed signs, which are largely ineffective (Chapter 24). The ESIA notes that Mikumi National Park in Tanzania, a small park with a fraction of Serengeti's wildlife density, had an average of three mammal roadkills per day many years after road construction there. In Serengeti, with much higher wildlife densities and more migratory animals, we should expect this number to be significantly higher, particularly in the initial years when the area is still wildlife rich. No mention was made of fencing or barriers that are normally used to prevent WVC and that undoubtedly will be called for in future years (see Textbox 56.2 on the impact of such barriers).

The government tried to deflect criticism by saying the road would be unpaved, but gravel is not necessarily better for wildlife. Vehicle speed, traffic volume and animal densities are better predictors of WVC than whether the road is paved or unpaved. Furthermore, the ESIA authors calculated minimum initial volume to be 300 vehicles per day based on data collected on an isolated road west of Serengeti in 1999 when the local human population was a fraction of its current size and determined that this is 'more than two times the commonly adopted upper limit for gravel roads in Tanzania. Therefore gravel pavement for this section is not appropriate'.

Other well-meaning suggestions include speed limits, temporary road closures, overpasses, underpasses and a 54 km elevated span. However, all of these mitigation suggestions are fundamentally flawed. Speed controls are rarely heeded in Africa. Closures are unlikely to be implemented because industry with financial interests in a speedy route across Northern Tanzania would likely push for unlimited access later and closures would be unpopular. The migration moves throughout the day so night-time closures would be insufficient to preserve mass movement. Expensive overpasses, underpasses or suspended spans were not included in the proposed road budget, nor are they ecologically realistic. The various species in the Serengeti ecosystem have very different habits and behaviours that would require different crossing structures. Given the unpredictable direction and fluidity of the migration, it would be impossible to design (much less enforce) crossings that would successfully convey 1.5 million unruly wildebeest and zebras from one side of the road to the other, twice per year. Elephants and rhinos, targeted for poaching, would require passages that cater to their aversion to traffic and avert bottlenecks or ambush sites for poachers.

The impact assessment correctly states that if the road is built, 'the ecosystem value of the Serengeti would be at stake' and mentions the Serengeti's unique tourism and natural heritage value. The consultants accurately concluded that 'given the complex nature of the dynamics of the migration it is estimated that the replacement costs would far exceed the prevention costs'. The migration is irreplaceable – no amount of money would be able to recreate it. The ESIA does not calculate the economic value of the Serengeti, including the economic cost of loss in tourism if the road is built, World Heritage status is revoked, poaching increases, rhinos go extinct and wildebeest numbers decline. The report does not address the value of 'roadless wilderness' itself (Chapter 3). Tourists are drawn to the Serengeti as a wilderness destination. Having a road through the area will eliminate this wilderness appeal, and one must forecast a significant drop in tourism when this is lost.

Textbox 56.2 Roads are semipermeable barriers to African wildlife.

There is increasing evidence that roads act as barriers to wildlife movement in Africa: animals fail to cross roads because they avoid roads (Blake et al. 2008) or are heavily hunted near roads or due to WVC (Chapter 55). While there are few longitudinal studies of road effects on wildlife in Africa, other manmade structures provide insight into how barriers like roads, settlements and fences affect wildlife. Thousands of kilometres of fences have been built across southern Africa since the 1900s for livestock management (see also Chapter 54). The ecological cost of these fences is often overlooked, but the evidence from 34 reports amounts to significant ongoing damage ranging from wildlife mortality to the disappearance of entire migrations (Gadd 2012). Fences disrupt daily and seasonal movements and may lead to death by starvation, dehydration or entanglement (e.g. Fig. 58.2). There is also evidence that fences influence hunting strategies and success by African wild dogs (Chapter 38). Fencing can divide populations, prevent recolonisation and render sub-populations prone to the increased risks of extinction faced by small populations. Large-bodied, migratory ungulates and elephants have been the most severely affected (Gadd 2012). In at least seven locations in East and southern Africa, wildebeest populations crashed to a fraction of their previous size after their movement was impeded (Table 56.1).

The ESIA mentions alternative routes could be built outside the protected area, but does not assess costs or benefits or provide a comparison between the options. Not surprisingly, a route through inhabited areas south of the Serengeti would serve more people and connect more communities than a route through the uninhabited protected area (Hopcraft et al. in press b).

CONCLUSIONS

The Serengeti–Mara ecosystem is still relatively intact and has maintained its unique natural history and exceptional biological productivity. It is one of the most productive ecosystems on earth both in terms of biomass and income generated for a developing country. Opening a road in the pristine north would cause a significant crash in the migrating populations and further endanger rare species by making them even more susceptible to illegal hunting, increased mortality due to WVC and loss of connectivity among individuals and interbreeding populations.

Due to the chaotic and unpredictable nature of the migration and the varying needs of other species, crossing structures are unlikely to be feasible or effective. Other species will be threatened by proximity to a road: the extremely high poaching pressure on some species, particularly elephants and rhinos, makes them especially unlikely to survive near roads or human presence.

Whether the road is paved or unpaved is irrelevant: opening a road through northern Serengeti would significantly disrupt the mass migration, depress productivity of the ecosystem and open up a wilderness area to illegal activities. As roadless wilderness becomes increasingly rare, this gem will become even more valuable for Tanzania and Kenya.

FURTHER READING

Berger (2004): Reviews similarities among terrestrial mammal populations that migrate, outlining the common causes of cessation of migration and population decline. Recommends applied actions that can be taken to retain long-distance migrations, using the Greater Yellowstone region and migratory elk, bison and pronghorn as an example.

Harris et al. (2009): Meta-analysis of the state of migratory land mammals worldwide, highlighting the disappearance of the majority of the planet's migrations and the outstanding threats to the remaining migrants.

Hopcraft et al. (in press): A discussion of the 'human' factors that should be taken into account to measure the economic

and social impacts and benefits of alternative routes in contrast to a route directly through the Serengeti National Park.

REFERENCES

Berger, J. 2004. The last mile: how to sustain long-distance migration in mammals. Conservation Biology **18**: 320–331.

Blake, S., S.L. Deem, S. Strindberg, F. Maisels, L. Momont, I.-B. Isia, I. Douglas-Hamilton, W.B. Karesh, and M.D. Kock. 2008. Roadless wilderness area determines forest elephant movements in the Congo Basin. PLoS One **3**: 1–9.

Burnett, J. 2012. Poachers Decimate Tanzania's Elephant Herds. National Public Radio Broadcast, Washington, DC. http://www.npr.org/2012/10/25/163563426/poachers-decimate-tanzanias-elephant-herds (accessed on 25 October 2012).

Dobson, A.P. et al. 2010. Road will ruin Serengeti. Nature **467**: 272–273. doi:10.1038/467272a.

Estes, R.D.E. 1991. The Behavior Guide to African Mammals. University of California Press, Berkeley, CA.

Gadd, M.E. 2012. Barriers, the beef industry and unnatural selection: a review of the impacts of veterinary fencing on mammals in southern Africa, pages 153–186. In: M. Hayward and M. Somers, editors. Fencing for Conservation. Springer Science, New York.

Harris, G., S. Thirgood, J.G.C. Hopcraft, J.P.G.M. Cromsigt, and J. Berger. 2009. Global decline in aggregated migrations of large terrestrial mammals. Endangered Species Research **7**: 55–76.

Hofer, H., K.L.I. Campbell, M.L. East and S.A. Huish. 1996. The impact of game meat hunting on target and non-target species in the Serengeti, in The Exploitation of Mammal Populations. V.J. Taylor and N. Dunstone, eds. Chapman and Hall, London, UK, pp. 117–146.

Hofer, H., K.L.I. Campbell, M.L. East, and S.A. Huish. 2000. Modeling the spatial distribution of the economic costs and benefits of illegal game meat hunting in the Serengeti. Natural Resource Modelling **13**: 151–177.

Hopcraft, J.G.C. 2012. Ecological implications of food and predation risk for herbivores in the Serengeti. Ph.D. thesis. University of Groningen, Groningen, the Netherlands.

Hopcraft, J.G.C., G. Bigurube, J.D. Lembeli, and M. Borner. In press a. Alternatives to the Serengeti highway: Finding solutions that benefit both socio-economic development and conservation. PLoS One.

Hopcraft, J.G.C., S.A.R. Mduma, M. Borner, G. Bigurube, A. Kijazi, D.T. Haydon, W. Wakilema, D. Rentsch, A.R.E. Sinclair, A.P. Dobson and J.D. Lembeli. In press b. Road around Serengeti provides greater economic development opportunities than a road through it. Conservation Biology.

Intercontinental Consultants and Technocrats Pvt. Ltd. (India) for the United Republic of Tanzania Ministry of Infrastructure Development Tanzania National Roads Agency (TANROADS). 2010. Draft report: consultancy services for engineering design, environmental and social impact assessment, and preparation of tender documents for upgrading of Natta-Mugumu-Loliondo (171.9 km) road to bitumen standard. Tanzania draft report submitted to the National Environmental Management Council of Tanzania, Dar es Salaam, Tanzania. Intercontinental Consultants and Technocrats Pvt. Ltd., New Delhi, India.

Loibooki, M., H. Hofer, K.L.I. Campbell, and M.L. East. 2002. Bushmeat hunting by communities adjacent to the Serengeti National Park, Tanzania: the importance of livestock ownership and alternative sources of protein and income. Environmental Conservation **29**: 391–398. doi:10.1017/S0376892902000279.

Mduma, S.A.R., R. Hilborn and A.R.E. Sinclair 1998. Limits to exploitation of Serengeti wildebeest and implications for its management, in Dynamics of Tropical Communities. D.M. Newbury, H.H.T. Prins and N. Brown, eds. Blackwell Science, Oxford, UK, pp. 243–265.

Metzger, K.L., A.R.E. Sinclair, K.L.I. Campbell, R. Hilborn, J.G.C. Hopcraft, S.A.R. Mduma, and R. M. Reich. 2007. Using historical data to establish baselines for conservation: the black rhinoceros (*Diceros bicornis*) of the Serengeti as a case study. Biological Conservation **139**: 358–374.

Mtweve, S. 2013. Tanzania tops latest list of best tourist destinations in the world. *The Citizen*. The Nation Media Group, Dar es Salaam, Tanzania. http://www.thecitizen.co.tz/News/Tanzania-tops-latest-list-of-best-tourist-destinations/-/1840392/2020442/-/10p1cisz/-/index.html (accessed 31 January 2015).

Tambling, C. and J. DuToit. 2005. Modelling wildebeest population dynamics: implications of predation and harvesting in a closed system. Journal of Applied Ecology **42**: 431–441.

Yackulic, C.B., S. Strindberg, F. Maisels, and S. Blake. 2011. The spatial structure of hunter access determines the local abundance of forest elephants. Ecological Applications **21**: 1296–1307.

Chapter 57

CHINA: BUILDING AND MANAGING A MASSIVE ROAD AND RAIL NETWORK AND PROTECTING OUR RICH BIODIVERSITY

Yun Wang, Yaping Kong and Jiding Chen

Research Centre for Environmental Protection and Transportation Safety, China Academy of Transportation Sciences, Ministry of Transport of the People's Republic of China, Beijing, China

SUMMARY

China is experiencing rapid growth in its economy, human population and transportation network. Environmental protection (e.g. slope stabilisation, vegetation protection and establishment, storm water collection and treatment) during road construction has been a priority over the past few decades. More recently, China has begun to protect its biodiversity when planning, designing and constructing new roads and railway lines. However, China still lags behind many developed countries in some areas of road ecology. A concerted and sustained effort is required to achieve an ecologically sustainable transportation network for the future.

57.1 The rate of growth in the Chinese road and railway network is rapid and will continue to be so into the foreseeable future.

57.2 China has recently adopted road ecology principles and concepts to be used in future road projects.

57.3 Wildlife-sensitive road designs based on recent road ecology research are being implemented on new roads in China.

57.4 Further research and dissemination of findings is essential to improve the ecological sustainability of China's roads.

New roads are continuing to be built and existing roads widened across much of China to accommodate an increasingly mobile human population. To protect China's unique biodiversity, we need to (i) better understand the impact of the road network on plants, animals and ecosystem processes; (ii) initiate and complete long-term studies of the effectiveness of mitigation measures on populations; and (iii) develop systems and processes to ensure experts from relevant disciplines are involved in the planning, design, construction, operation and evaluation of the road and rail network.

Handbook of Road Ecology, First Edition. Edited by Rodney van der Ree, Daniel J. Smith and Clara Grilo.
© 2015 John Wiley & Sons, Ltd. Published 2015 by John Wiley & Sons, Ltd.
Companion website: www.wiley.com\go\vanderree\roadecology

INTRODUCTION

China has the largest human population density on Earth, who drive the most vehicles along an extensive road network that includes the second largest expressway network on the planet (MOT 2011). China's economy is growing rapidly, and gross domestic product is expected to increase by approximately 7% annually over the next few years, one of the highest in the world (MOT 2011). An important component of China's growing economy is the construction of transportation infrastructure, including roads and railway. China covers almost 9,600,000 km^2 and includes tropical and subtropical forest, coniferous forest, arctic and alpine habitats, deserts and grasslands and savannah. An enormous diversity of species persists within these diverse habitats and wilderness areas, including numerous rare and threatened species.

Chinese road agencies have traditionally focused on achieving high levels of environmental protection along its major roads. The protection and restoration of vegetation is a priority, and the recently completed Qinghai–Tibet Highway and Ring Changbai Mountain Scenic Highway projects exemplify this focus (Chen et al. 2004; Wang et al. 2013b). The impact of roads on water quality and hydrology has been extensively studied, and protective measures are routinely included during highway construction (Kong & Liu 2013). Protecting landscape aesthetics and scenery has recently become a priority because of the rapid growth in China's economy and standard of living. The China Academy of Transportation Sciences (CATS) carried out the development of the first provincial-level scenic highway network with the planning of the 'Scenic Highway Network Development of Hainan Province Project' in 2011. Nationally, scenic qualities and landscape aesthetics were important in the design of the China–Pakistan Karakoram Highway, the Jilin–Yanji Expressway and the Ring Changbai Mountain Highway (Lu et al. 2010). Protecting farmland is also a very high priority in China, and road design focuses on reducing the land take for highways, thereby maximising the amount of land for agriculture (Tao et al. 2010).

While environmental protection has dominated road planning and design in China for many years, it is only relatively recently (in the last decade or so) that ecological issues have even been considered. The challenge for China in the years ahead is to identify the most cost-effective solutions from elsewhere and integrate these approaches into Chinese practice. The aims of this chapter are to highlight the rate of growth in China's surface transportation network, summarise the key achievements in ecologically sensitive road design and prioritise areas for future research and policy development.

LESSONS

57.1 The rate of growth in the Chinese road and railway network is rapid and will continue to be so into the foreseeable future

The expansion of the road and rail network in China is a high priority to facilitate economic growth and improve the standard of living for its people. By the end of 2012, the total length of roads in China had reached 4.24 million km (from 3.73 million km in 2008), including 96,200 km of expressway (up from 60,300 km in 2008) (DOCP 2012). According to China's latest transportation strategy (MOT 2011), the rapid expansion of its road and rail network is set to continue, reaching 4.5 million km of road and 108,000 km of expressway by the end of 2015.

In 2010, China had 91,000 km of railway lines, which was expected to increase to 98,000 km by the end of 2012, making it the second longest network for a single country in the world (Zhu 2013). By 2015, the length of railway is expected to reach 120,000 km (MOR 2011).

57.2 China has recently adopted road ecology principles and concepts to be used in future road projects

In 2002, the Chinese Ministry of Transport (MOT) incorporated road ecology principles and concepts into the first demonstration project, the Chuanzhusi to Jiuzhaigou Scenic Highway in Sichuan Province. This was the first scenic highway in China, and since then, numerous other projects have incorporated ecological aspects into their design. The importance of road ecology in China was further acknowledged when CATS undertook numerous road ecology research projects and implementation throughout China. The MOT continues to fund CATS to do research and provide input into the planning and design of roads and the evaluation of the use and effectiveness of mitigation measures. In 2008, the Forman (2003) classic *Road Ecology: Science and Solutions* was translated into Chinese and published by Dr. Taian Li from Lanzhou University and Dr. Yun Wang from CATS. In 2009, CATS published their own version: *Road Ecology in China* (Mao et al. 2009; Forman et al. 2011). International exposure of Chinese road ecology research is encouraged, and researchers from CATS regularly attend and present their research findings at the International Conference on Ecology and Transportation (Chapter 60).

57.3 Wildlife-sensitive road designs based on recent road ecology research are being implemented on new roads in China

Road ecology research and mitigation began in China approximately 15 years ago and has primarily focused on quantifying the rates of roadkill, the barrier effect and the size of the road-effect zone at a number of locations across China (Fig. 57.1). The results of these

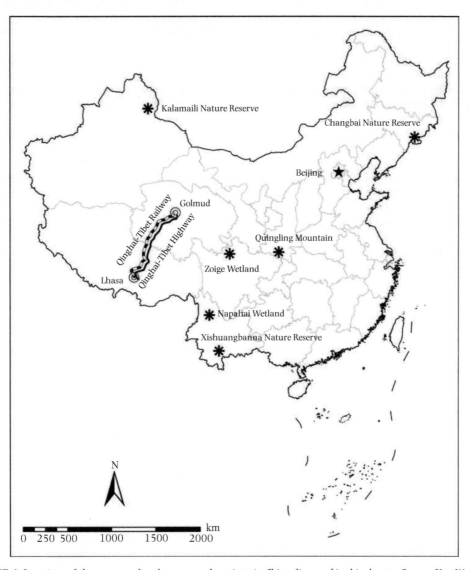

Figure 57.1 Locations of the seven road ecology research projects in China discussed in this chapter. Source: Yun Wang.

studies have shown that Chinese wildlife suffer similar effects to those in other countries. For example, mortality of wildlife has been identified as a serious problem for amphibians where a highway dissects the Zoige wetland (Gu et al. 2011) and for mammals, birds, amphibians and reptiles on the Ring Changbai Mountain Scenic Highway (Wang et al. 2013a). From August to October in 2007, five Przewalski's wild horses (out of a population of 27) were killed by vehicles on roads through the Kalamaili Nature Reserve (Zhang et al. 2008). Research has also demonstrated that many species appear to avoid highways or have lower population densities near to the highway. This avoidance zone appears to be evident for up to 5 km for the giant panda (F. Wang, Peking University, personal communication), 1 km for wild yak, 600 m for kiang, 300 m for Tibetan antelope, 200 m for Tibetan gazelle (Lian et al. 2012) and just under 150 m for black-necked crane (Wang et al. 2011). Roads have also affected the movement of wildlife, and the number of locations that Asian elephants are willing to cross the Simao to Xiaomengyang Expressway has decreased from 28 to 23 following its construction in 2006 (Pan et al. 2009).

The Chinese MOT has built multipurpose crossing structures across new roads that typically allow for the movement of wildlife and water, local residents or domestic animals. These structures include open-span bridges, culverts and tunnels, and fencing that is always included along expressways to keep animals off the road and direct them towards crossing structures. A wide range of species use these structures (e.g. Table 57.1). The Asian elephant used multipurpose crossing structures (16 open-span bridges, 2 tunnels) along the Simao to Xiaomengyang Expressway cutting through Xishuangbanna Nature Reserve, and 44% of individuals that approached or entered the structures passed through (Pan et al. 2009). More than 10 species of wildlife crossed the Ring Changbai Mountain Scenic Highway, also using multi-use crossing structures (Textbox 57.1). Dedicated wildlife crossing structures on the Qinghai–Tibet Railway were regularly used by Tibetan antelope (Textbox 57.2).

57.4 Further research and dissemination of findings is essential to improve the ecological sustainability of China's roads

In an effort to reduce its carbon footprint and protect the environment, the Chinese MOT issued a policy to accelerate the development of a 'green' and low-carbon transport system by 2020. Road ecology is recognised as a critical component of this strategy. Chinese road ecology research has mostly focused on small areas or single roads, which is out of step with the rate of growth of the overall network (Mao et al. 2009). Long-term studies are needed to properly identify the ecological impacts of roads and traffic as well as to quantify the effectiveness of mitigation measures, vegetation succession and water quality. The right of way of roads in China is typically narrower than in other countries because of the imperative to protect valuable agricultural land to feed the growing human population. Consequently, the

Table 57.1 Species observed using 84 culverts and 22 bridges under the Ring Changbai Mountain Scenic Highway, November 2008–February 2013.

Chinese species name	English species name	Conservation status in China[a]
紫貂	Sable	First class
青鼬	Yellow-throated marten	Second class
花尾榛鸡	Hazel grouse	Second class
环颈稚	Common pheasant	*
野猪	Wild boar	*
狍子	Siberian roe deer	*
黄鼬	Siberian weasel	*
松鼠	Eurasian red squirrel	*
东北兔	Manchurian hare	*
伶鼬	Least weasel	*

Source: List of beneficial or important terrestrial wildlife species in economy and science in China (http://baike.baidu.com/view/1496360.htm).

[a]The first class is the most endangered, the second class is the next most endangered, and * are threatened but also beneficial or important to the economy and/or science in China.

Textbox 57.1 Effectiveness of mitigation along the Ring Changbai Mountain Scenic Highway.

The Ring Changbai Mountain Scenic Highway (85 km in length) was constructed from 2007 to 2009, with extensive design input from wildlife ecologists to reduce the impact on biodiversity. Mitigation measures included 190 culverts and 25 extended bridges, to allow passage of wildlife, people and drainage. In 2010, traffic volume was about 200 vehicles per day. Studies on the use of the mitigation structures by wildlife have shown a high diversity of species using 84 of the culverts and 22 of the bridges, including some species listed as rare and endangered under Chinese legislation (Table 57.1 and Fig. 57.2).

Figure 57.2 Tracks of Siberian weasel through a culvert under the Ring Changbai Mountain Scenic Highway. Source: Photograph by Yun Wang.

quantification of the road-effect zone is critical in China, and multidisciplinary planning teams must cooperate to plan, design, construct and manage China's road network.

CONCLUSIONS

The rate of expansion of the road and railway network in China will continue to be rapid until at least 2015 and almost certainly for many years after that. Road ecology has recently become an important topic to the Chinese MOT, and future planning and research should focus on the improved design of multipurpose and dedicated wildlife crossing structures and fencing, the impacts of roads on endangered species and the development of national standards for the design of wildlife-friendly roads and cost-effective crossing structures.

ACKNOWLEDGEMENTS

We thank Richard Forman from Harvard University and Marcel Huijser from the Western Transportation Institute at Montana State University, United States, for advice and support. Our research is funded by the Western China Communications Construction and Technology Project (Grant No. 2011 318 670 1290) and International S&T Cooperation Program of China (Grant No. 2012DFA20980) and National Key Technology Research and Development Program of the Ministry of Science and Technology of China (Grant No. 2014BAG05B06).

Textbox 57.2 Tibetan antelope use crossing structures under the Qinghai–Tibet Railway.

The Qinghai–Tibet Railway is 1142 km in length and is the highest elevation railway in the world, connecting Qinghai to Tibet (Fig. 57.1). Constructed from 2001 to 2006, the railway line includes 7 at-grade crossings, 1 tunnel, 25 bridges (similar to the Hoh Xil bridge in Fig. 57.3) and many more smaller bridges, all purpose built for wildlife. Studies conducted from 2004 to 2007 have demonstrated that the migration of the Tibetan antelope has increased over time. Of animals that approached the train line, 100% crossed underneath it in 2007, compared to only 60% during the construction phase in 2004. Furthermore, the animals that do use it have become accustomed to the noise and other disturbances, and the length of time that individuals waited before passing through has gone from 1 to 2 weeks in 2004 to several minutes in 2007 (Fig. 57.3; Li et al. 2008).

Figure 57.3 Tibetan antelopes using the Hoh Xil wildlife crossing structure in August 2006, shortly after the railway was opened. Source: Reproduced with permission of Hongfeng Zhang, Northwest Institute of Endangered Zoological Species.

FURTHER READING

Forman et al. (2011): A short review describing the state of road ecology in China for an international audience.

Li et al. (2008): Evaluated the rate of use of wildlife crossing structures by Tibetan Antelope along Qinghai–Tibet Railway.

Mao et al. (2009): The first book to summarise the state of road ecology in China, describing numerous Chinese case studies and examples.

Wang et al. (2013b): The first published comprehensive investigation of the impacts of roads (in this case the Ring Changbai Mountain Scenic Highway) and traffic in China and the effectiveness of mitigation measures.

REFERENCES

Chen, J. D., Z. W. He, H. Fang and Q. J. Li. 2004. Study on techniques for slope protection along Qinghai–Tibet highway. Journal of Glaciology and Geocryology **26** (supplement): 291–295.

Department of Comprehensive Planning, Ministry of Transport (DOCP). 2012. Statistical bulletin of highway and waterway transport sector development. Ministry of Transport of the People's Republic of China, Beijing, China. Available from http://www.moc.gov.cn/zhuzhan/zhengwugonggao/jiaotongbu/guihuatongji/201304/t20130426_1403039.html (accessed 25 September 2014).

Forman, R. T. T., D. Sperling, J. A. Bissonette, A. P. Clevenger, C. D. Cutshall, V. H. Dale, L. Fahrig, R. France, C. R. Goldman, K. Heanue, J. A. Jones, F. J. Swanson, T. Turrentine, and T. C. Winter. 2003. Road Ecology. Science and Solutions. Island Press, Washington, DC.

Forman, R. T. T., M. P. Huijser and A. P. Clevenger. 2011. Emergence of road ecology in China. Committee on Ecology and Transportation Newsletter Summer: 2–6.

Gu, H. J., Q. Dai, Q. Wang and Y. Z. Wang. 2011. Factors contributing to amphibian road mortality in a wetland. Current Zoology **57**: 768–774.

Kong, Y. P. and X. X. Liu. 2013. Analysis of water resource protection technology during highway operation period. Journal of Highway and Transportation Research and Development **30**: 146–151.

Li, Y. Z., T. J. Zhou and H. B. Jiang. 2008. Utilization effect of wildlife passages in Golmud–Lhasa section of Qinghai–Tibet railway. China Railway Science **29**: 127–131.

Lian, X. M., X. X. Li and T. Xu. 2012. Avoidance distances of four ungulates from roads in Kekexili and related protection suggestions. Chinese Journal of Ecology **31**: 81–86.

Lu, X. D., J. Z. Dong, W. Xu, Z. Xue and T. Wang. 2010. View point construction along Ring Changbai Mountain Scenic Highway. Journal of Highway and Transportation Research and Development **10**: 405–408.

Mao, W. B., C. Q. Duan, H. F. Li, X. Y. Zhen, B. Chen, Y. Wang, P. Tao, Z. C. Zhang, J. Wang, X. C. Qin, Y. B. Xu and J. Y. Wu. 2009. Road Ecology in China. China Communications Press, Beijing, China.

Ministry of Railways (MOR). 2011. The 12 Five-Year Development Planning of Railway. Baidu Library, Beijing, China. Available from http://wenku.baidu.com/view/ 2850a5d626fff705cc170a7c.html (accessed 25 September 2014).

Ministry of Transport (MOT). 2011. The 12th Five-Year Plan for Transportation. Ministry of Transport of the People's Republic of China, Beijing, China. Available from http:// www.mot.gov.cn/zhuantizhuanlan/jiaotongguihua/shier wujiaotongyunshufazhanguihua/ (accessed 25 September 2014).

Pan, W. J., L. Lin, A. D. Luo and L. Zhang. 2009. Corridor use by Asian elephants. Integrative Zoology **4**: 220–231.

Tao, S. C., B. Sun, Y. Liu and Y. P. Kong. 2010. Saving land resource during expressway construction in plain area. Journal of China & Foreign Highway **30**: 347–349.

Wang, Y., Q. L. Li, L. Guan, R. Fang and R. Jiang. 2011. Effect of traffic noise around Napahai wetland highway on birds. Chinese Journal of Zoology **46**: 65–72.

Wang, Y., Z. J. Piao, L. Guan, X. Y. Wang and Y. P. Kong. 2013a. Road mortalities of vertebrate species on Ring Changbai Mountain Scenic Highway, Jilin Province, China. Northwestern Journal of Zoology **9**: 399–409.

Wang, Y., Z. J. Piao, L. Guan and Y. P. Kong. 2013b. Influence of Ring Changbai Mountain Scenic Highway on wildlife. Chinese Journal of Ecology **32**: 425–435.

Zhang, F., D. F. Hu, J. L. Chen and T. T. Cao. 2008. Build safe passage for *Equus ferus przewalskii*. Chinese Nature **5**: 58–59.

Zhu, J. 2013. The length of Chinese railway has arrived at 98,000 km and ranks second in the world. China News Agency, Beijing, China. Available from http://finance. chinanews.com/cj/2013/01-17/4496731.shtml (accessed 25 September 2014).

Chapter 58

RAILWAYS, ROADS AND FENCES ACROSS KAZAKHSTAN AND MONGOLIA THREATEN THE SURVIVAL OF WIDE-RANGING WILDLIFE

Kirk A. Olson[1] and Rodney van der Ree[2]

[1]Smithsonian Conservation Biology Institute, National Zoological Park, Front Royal, VA, USA
[2]Australian Research Centre for Urban Ecology, Royal Botanic Gardens Melbourne, and School of BioSciences, The University of Melbourne, Melbourne, Victoria, Australia

SUMMARY

The temperate grasslands of Central Asia are habitat for a number of wide-ranging and endangered species such as Mongolian gazelle, saiga antelope, black-tailed gazelle and Asiatic wild ass. These species' habitat covers hundreds of thousands of square kilometres of largely ecologically intact grassland. Unless carefully planned and managed, the development of railways, highways and fences will be the catalyst for population decline and loss of important wild natural resources.

58.1 The temperate grasslands of Kazakhstan and Mongolia are the largest in the world and are critically important to the survival of Mongolian gazelle, Asiatic wild ass and saiga antelope.

58.2 Kazakhstan and Mongolia are rapidly expanding their overland transportation network to support increased transcontinental trade and resource extraction.

58.3 Railways, highways and fences prevent access to important seasonal resources for various species and cause the decline of wildlife populations.

58.4 Future roads and railways must avoid further fragmentation, existing roads and railways should be modified to restore wildlife movements, and fences should be modified or removed to increase connectivity.

58.5 GPS tracking of long-distance migratory species that identifies preferred movement paths and existing barriers to movement is essential to properly plan infrastructure projects.

Achieving a balance between healthy ecosystems and economic development is a significant and critical challenge for developing nations such as Kazakhstan and Mongolia. Experience from other regions has led to an established hierarchy of measures to meet this challenge – avoid, minimise, mitigate and offset. Kazakhstan and Mongolia should adopt these approaches now to ensure that the imminent massive expansion of their road and railway networks has a minimal effect on biodiversity and may even result in a positive effect.

INTRODUCTION

Kazakhstan and Mongolia are situated between China and Russia in Central Asia. With a traditionally rural way of life, Kazakhstan and Mongolia have embarked upon an economic transformation fuelled by the extraction of extensive and valuable fossil fuel (natural gas and oil) and mineral (copper, gold and coal) deposits, respectively (Batsaikhan et al. 2014). Increased trade between Asia and Europe via overland routes will also benefit Mongolia and Kazakhstan because much trade will pass through them. Both countries are at crossroads in their development, and decisions and plans made today will influence the extent to which the future growth and expansion in Central Asia is ecologically sustainable, even though some of this progress occurs at the cost of losing biodiversity.

LESSONS

58.1   The temperate grasslands of Kazakhstan and Mongolia are the largest in the world and are critically important to the survival of Mongolian gazelle, Asiatic wild ass and saiga antelope

The finest examples of ecologically intact temperate grasslands in the world are found in west and central Kazakhstan and eastern Mongolia (Batsaikhan et al. 2014). These grasslands are situated within a nearly 7000 km long band of arid rangelands and grasslands that stretch from the Hungarian plains to the eastern steppes of Mongolia. These grasslands are habitat for one of the largest populations of Mongolian gazelles in the world as well as other endangered and critically endangered wildlife, such as Asiatic wild ass, saiga antelope (Fig. 58.1), black-tailed gazelle and wild Bactrian camels (Mallon & Jiang 2009). These species require access to large parts of their range over the course of a single year and over their lifetime in order to survive and successfully raise offspring (Chapter 42). The availability and quality

of forage within these grasslands are constantly changing, and often unpredictably, due to drought, snow (e.g. Fig. 58.3), burning or occupation by herdsmen. The global population size of saiga antelope is approximately 208,600, with 183,000 in Kazakhstan, 12,500 in Russia/Kalmykia, and 13,000 in Mongolia. As many as 1 million Mongolian gazelles and up to 20,000 Asiatic wild ass survive within Mongolia (Reading et al. 2001; Olson et al. 2011). Together, these latter two species occupy an area greater than 800,000 km², approximately 32 times the size of Serengeti National Park. While the total area is large, the region is subdivided into numerous smaller areas by roads, railways and fences (Lesson 58.3).

58.2   Kazakhstan and Mongolia are rapidly expanding their overland transportation network to support increased transcontinental trade and resource extraction

The Asian Highway Network consists of over 141,000 km of roads which link 32 countries in Asia as well as to Europe. This transport network has been in existence since 1959 to facilitate global trade and to meet the needs of emerging markets and national development goals. Administered by the United Nations Economic and Social Commission for Asia and the Pacific, the goal is to develop transport within the region using a network of standardised roads (UNESCAP 2014). The Central Asia Regional Economic Cooperation (CAREC) programme, administered by the Asian Development Bank (ADB 2014), consists of six major rail routes that connect Asia and Europe. Five of the six corridors pass through Kazakhstan, and the sixth route passes through Mongolia. The CAREC partnership consists of 10 countries which aim to improve economic development and poverty reduction through cooperation on transport issues.

As trade between Asia and Europe increases, the CAREC corridors and the Central Asian Highway

(A)

(B)

Figure 58.1 (A) Mongolian gazelles can be found throughout the gobi–steppe ecosystem of Mongolia and have been observed in herds of up to 250,000. (B) The Asiatic wild ass, or khulan in Mongolian, has lost more than 50% of its range in Mongolia in the past 70 years due to poaching and competition from grazing livestock. Source: (A) Photograph by and reproduced with permission of J. Kerby; and (B) Photograph by K. Olson.

Network are gaining in importance. Less than 5% of the total volume of trade between Asia and Europe currently goes overland; the vast majority is sent by sea and a small fraction of time-sensitive products go by air. Road transport is faster than rail or sea, but more expensive. Rail transport is almost twice as fast as by sea, but also more expensive. Irrespective of price, air cargo will maintain its status as a special needs option. Improvements to transit times by road and rail will make overland transit more competitive against sea routes and increase the volume of traffic along the major routes.

The government of Kazakhstan is eager to benefit from the revenue that is generated from increased traffic volume and is supporting more trade. This includes creating new routes, increased customs

capacity, faster container transfer times between different gauge railways, improvements to existing roads and railways to handle more traffic and higher travel speeds. Kazakhstan hopes to increase the volume of freight from its current 2.5 million containers per year (2013) to 7.5 million by 2020, necessitating extensive expansion of its current 8700 km of railways. One example of this expansion is a newly constructed east–west railway to Europe from Asia via the Caspian Sea port of Aktau. This route allows trains to change tracks if one is congested and reduces the reliance on Russian transit routes. Mongolia has similarly ambitious plans to develop their road and rail network and plan to construct more than 6000 km of paved roads by 2030 and add 5684 km of railway in three phases (tied to the development of mining projects). The private sector, primarily mining companies, is also promoting linear infrastructure projects that are specific to their needs, often independently of regional development goals.

58.3   Railways, highways and fences prevent access to important seasonal resources for various species and cause the decline of wildlife populations

The barrier effect of the existing transport corridors in Kazakhstan and Mongolia for wildlife will increase with increased traffic volume (Ito et al. 2013). For highly mobile species, a single barrier can have wide-reaching long-term consequences on their ability to persist in what is otherwise suitable habitat. Although they remain as the largest tracts of temperate grasslands in the world, barrier effects exist from fences along national borders and railways, as well as high-traffic-volume transcontinental roads and railways (Ito et al. 2013). Saiga populations (with the exception of within Mongolia) appear defined by railway corridors which have been in use for over a century and which have recently become major intercontinental rail routes. Genetic differences can be detected in Asiatic wild ass populations that have been separated by a combination of natural landscape features and anthropogenic disturbance (Kaczensky et al. 2011). In Kazakhstan, the addition of extra tracks along the existing railway through habitat of the Ustyurt and the Betpak-Dala populations of saiga antelope may be enough to cause local extinctions (Olson 2013). The fenced Trans-Mongolian Railway, which connects Ulan-Ude in Russia with Erenhot in China through Ulaanbaatar in Mongolia, prevents Asiatic wild asses from repopulating their former range in the east. This same barrier entangles many Mongolian gazelle each year (Fig. 58.2), and many more are denied access to important foraging resources. It would be more cost-effective to remove the fence and compensate herders for the occasional loss of livestock from a train collision than pay for its continual maintenance. Fence removal would also have the added benefit of reducing gazelle mortality from fence entanglements and restore landscape-scale movement of wildlife. Train volumes are sufficiently low at present that wildlife mortality is unlikely to have a significant impact; however, this needs to be monitored, especially if train volumes increase (see also Chapter 26).

58.4   Future roads and railways must avoid further fragmentation, existing roads and railways should be modified to restore wildlife movements, and fences should be modified or removed to increase connectivity

As for elsewhere (Chapter 9), the planning and design of road and railway projects in Central Asia occurs years in advance of their implementation. Organisations in development assistance and international lenders have a responsibility to ensure that their actions do not degrade the environment and natural heritage of the regions that development projects aim to improve. Stakeholders interested in better environmental planning must be involved throughout the planning and design stages so that planners and designers are informed of the related issues and appropriate mitigation and budgets can reflect the additional costs. Implementing such a process would help to avoid costly redesigns of plans and retrofitting of mitigation measures which are often resisted. Importantly, conservation scientists and practitioners have an obligation to make their findings available in a timely manner to ensure they can be integrated into project planning and design. The international development community has a responsibility to ensure that projects do indeed lead to a better world by improving livelihoods and environmental standards.

Minimising the barrier effect of existing railways and highways on saiga antelope in Kazakhstan might be achieved by limiting disturbance and human activity within areas where saiga are known to occur or where potential for reconnecting two populations exists. Unfenced railways are passable except where embankments are high or steep (Fig. 58.3) – mitigation here may be as simple as decreasing the slope of the embankments to facilitate crossings.

In places where the movements and distribution of wildlife have already been altered but the habitat remains in good condition, railways, highways and fences can be retrofitted or mitigated to improve habitat connectivity. There are long segments of the Trans-Mongolian Railway where there are no people; it is here that the fence can simply be removed to allow animals to cross at will. Where a fence is necessary for safety or to avoid wildlife–train collision, design changes can be incorporated to allow smaller animals, such as a gazelle, to crawl under the fence but prevent larger livestock (e.g. horses, camels and cows) from entering the tracks. Crossing structures for open plain ungulates along high-traffic roads and railways with fencing that funnels wildlife without entangling them are

Figure 58.2 Many Mongolian gazelles become entangled and die in fences each year, in this case along the Trans-Mongolian Railway. Source: Photograph by K. Olson.

Figure 58.3 High and steep road and railway embankments can be a barrier for movement of saiga. Source: Photograph by K. Olson.

being developed with preliminary success. Overpasses are likely to be more effective than underpasses for ungulates of the vast open plains of Central Asia. Wildlife and cattle guards (Fig. 20.9) placed across fenced roads and railway tracks will discourage animals from accessing the roadway or railway when they cross at grade. However, in general, there is still much uncertainty surrounding effective mitigation for plain ungulates, and much more work is required (Chapter 56).

58.5 GPS tracking of long-distance migratory species that identifies preferred movement paths and existing barriers to movement is essential to properly plan infrastructure projects

The vast expanses of Central Asia combined with the low density of humans and long-distance nomadic movements of many large ungulate species make the identification of movement patterns challenging. The use of GPS technology overcomes many of these issues (Chapter 11). It is important to collect data on animal movements within these arid grasslands over multiple years because the vegetation dynamics are highly variable and animal movements typically differ from year to year. Carefully designed research and monitoring programmes (Chapter 10) of animal movements are imperative to (i) avoid placing new infrastructure in important areas; (ii) identify locations on new or existing infrastructure where barriers or mortality occur that require mitigation; and (iii) evaluate the effectiveness of mitigation. GPS tracking of ungulate migrations has helped identify problem areas for mule deer and pronghorn antelope in North America (Sawyer et al. 2009; Cohn 2010) and zebra in Botswana (Bartlam-Brooks et al. 2013). This research and monitoring should be integrated into road and rail projects and be supported through the overall funding package for the project. Similarly, the added costs of avoiding critically important habitats or movement corridors and mitigating the barrier and mortality effects along new and existing infrastructure through appropriately designed fencing and crossing structures should be a routine and accepted part of infrastructure development. This means that development banks, private companies and governments have a responsibility to ensure that the current and future transportation infrastructure in Central Asia enhances, rather than degrades, the survival prospects for large ungulates.

CONCLUSIONS

The people of Kazakhstan and Mongolia are rightly proud of their wild heritage, and development agencies would be wise to take this into account when planning and creating a more integrated and globalised economy. Society is demanding that development models incorporate legitimate discussions and take necessary actions to ensure that the effects on ecosystems and biodiversity are minimised or even have a positive effect. Global trade patterns and demand for natural resources are driving many of the development projects and threats to conservation of wildlife in Kazakhstan and Mongolia, particularly animals that move long distances. Institutions whose goal is to promote development have an obligation to promote growth and development that do not represent a step backwards with respect to responsible ecosystem management and wildlife stewardship.

ACKNOWLEDGEMENTS

We thank the UNDP/GEF/Govt. of Mongolia Eastern Steppes Biodiversity Project, Wildlife Conservation Society, Disney Worldwide Conservation Fund, University of Massachusetts, Smithsonian Conservation Biology Institute, World Bank, Association for the Conservation of Biodiversity of Kazakhstan, Frankfurt Zoological Society, Convention on the Conservation of Migratory Species of Wild Animals and Fauna & Flora International.

FURTHER READING

ADB (2014): This website provides a continuously updated reference to development activities related to roads and railways in Central Asia.

Olson (2013): An overview of the current threats to habitat connectivity for saiga antelope and options for minimising their effects.

REFERENCES

Asian Development Bank (ADB). 2014. Central Asia Regional Economic Cooperation (CAREC) program. ADB, Manila, Philippines. http://www.adb.org/countries/subregional-programs/carec (accessed 27 January 2015).

Bartlam-Brooks, H.L.A., P.S.A. Beck, G. Bohrer and S. Harris. 2013. In search of greener pastures: Using satellite images to predict the effects of environmental change on zebra migration. Journal of Geophysical Research: Biogeosciences **118**:1427–1437.

Batsaikhan, N., et al. 2014. Conserving the world's finest grassland amidst ambitious national development. Conservation Biology **48**: 35–46.

Cohn, J.P. 2010. A narrow path for pronghorns. Bioscience **60**: 480.

Ito T.Y., B. Lhagvasuren, A. Tsunekawa, M. Shinoda, S. Takatsuki, B. Buuveibaatar and B. Chimeddorj. 2013. Fragmentation of the habitat of wild ungulates by anthropogenic barriers in Mongolia. PLoS ONE **8**: e56995. doi:10.1371/journal.pone.0056995.

Kaczensky, P., R. Kuehn, B. Lhagvasuren, S. Pietsch, W. Yang, and C. Walzer. 2011. Connectivity of the Asiatic wild ass population in the Mongolian Gobi. Biological Conservation **144**: 920–929.

Mallon, D.P. and Z. Jiang. 2009. Grazers on the plains: challenges and prospects for large herbivores in central Asia. Journal of Applied Ecology **46**:516–519.

Olson, K.A. 2013. Saiga crossing options: guidelines and recommendations to mitigate barrier effects of border fencing and railroad corridors on saiga antelope in Kazakhstan. Smithsonian Conservation Biology Institute. Available from http://www.cms.int/sites/default/files/publication/Kirk_Olson_Saiga_Crossing_Options_English.pdf (accessed 27 January 2015).

Olson, K.A., T. Mueller, J.T. Kerby, S. Bolortsetseg, P. Leimgruber, C. Nicolson and T.K. Fuller. 2011. Death by a thousand huts? Effects of household presence on density and distribution of Mongolian gazelles. Conservation Letters **4**:304–312.

Reading, R.P., H.M. Mix, B. Lhagvasuren, C. Feh, D.P. Kane, S. Dulumtseren and S. Enkhbold. 2001. The status and distribution of khulan (*Equus hemionus*) in Mongolia. Journal of Zoology **254**:381–389.

Sawyer, H., Kauffman, M.J., Nielson, R.M. and Horne, J.S. 2009. Identifying and prioritizing ungulate migration routes for landscape-level conservation. Ecological Applications **19**: 2016–2025.

United Nations Economic and Social Commission for Asia and the Pacific (UNESCAP). 2014. About the Asian Highway. United Nations ESCAP, Bangkok, Thailand. http://www.unescap.org/our-work/transport/asian-highway/about (accessed 27 January 2015).

Chapter 59

BEST-PRACTICE GUIDELINES AND MANUALS

Marguerite Trocmé

Environmental Technology, Road Network Division, Swiss Federal Road Office, Bern, Switzerland

SUMMARY

Best-practice manuals in road ecology provide state-of-the-art information on how to avoid, minimise, mitigate and compensate for impacts of roads and traffic and other linear infrastructure on animals, plants and natural habitats. The first relevant guidelines appeared in the late 1960s and were comparatively simple documents, oriented at avoiding wildlife-vehicle collisions. Today's manuals are more detailed (see Sections 'References' and Appendix 59.1) and reflect the diversity and complexity of the potential impacts and the range of possible mitigation measures. For maximum effect, best-practice guidelines and manuals should address or contain the following:

59.1 Define the purpose of the guidelines and identify the audience.
59.2 Guidelines should be practical, implementable and based on state-of-the-art knowledge.
59.3 Describe the negative effects of roads and traffic on wildlife, ecosystems and human safety.
59.4 Bring together a multidisciplinary team to write the guidelines and to test the feasibility of the proposed mitigation measures.
59.5 Structure the manual to follow the typical stages of road projects.
59.6 Describe the range of mitigation measures and identify the preferred option.
59.7 Clearly identify mitigation measures that don't work.
59.8 Address transport infrastructure in the context of spatial planning.
59.9 Describe monitoring techniques and schemes.
59.10 Schedule a time frame for revising and updating the manual.

Best-practice guidelines are critically important as a resource for people who plan, design and manage roads and other linear infrastructure. The preparation and updating of these documents should be seen as an opportunity for collaboration among a diversity of professions to improve the ecological sustainability of roads as well as develop a shared responsibility.

Handbook of Road Ecology, First Edition. Edited by Rodney van der Ree, Daniel J. Smith and Clara Grilo.
© 2015 John Wiley & Sons, Ltd. Published 2015 by John Wiley & Sons, Ltd.
Companion website: www.wiley.com\go\vanderree\roadecology

INTRODUCTION

The rapid growth of motorways and vehicle ownership in Europe and North America in the 1950s and 1960s brought about a significant increase in wildlife-vehicle collisions (WVC), typically with deer. This presented engineers with a new road safety issue, and in 1968, the first guidelines on wildlife fencing along motorways were written in Switzerland (VSS 1968). This set of guidelines provided species-specific information on fence design based on a concise overview of the fieldwork and collision-reduction tests of the time. In 1969, Reed published his research on deer motorway crossings and on the effectiveness of illuminated signs in the United States (Reed 1969). Since then, numerous guidelines have been published, and their scope has evolved as the global understanding of the problems and solutions in road ecology has grown (see Sections 'References' and Appendix 59.1).

Guidelines began to focus on ecological issues and solutions in the 1980s when environmental impact assessments (EIAs) became common practice. These guidelines and manuals were typically written for ecologists and wildlife specialists who were contributing to the EIAs (e.g. SETRA 1985; Ryser 1988; Carsignol 2006). Although the guidelines contained detailed information on potential impacts and mitigation measures, this information failed to reach the road planners and engineers, resulting in rare or poor implementation of mitigation. A collaborative effort among 15 countries, the *European Handbook for Identifying Conflicts and Designing Solutions* (Iuell et al. 2003), was the first attempt to gather and define best-practice data and methods at a continental scale. It set the standard for the current generation of guidelines that aim to reach beyond wildlife specialists to include transport policy-makers, planners and designers. The 2011 *North American Wildlife Crossing Structure Handbook* (Clevenger & Huijser 2011) goes a step further. It includes detailed 'hands-on' information that describes field-data sampling techniques, habitat models to identify where wildlife need to cross roads and the principles underlying the design, number and spacing of wildlife crossing structures. Regional best-practice manuals have also been produced (e.g. QDTMR 2011), where the state of the art is adapted to local policy and conditions.

Best-practice manuals currently address a broad array of impacts of roads and traffic on the environment. State-of-the-art manuals should guide users through the entire planning process, from the selection of the road alignment to the evaluation of mitigation effectiveness following construction. They should describe the most effective approaches to avoid, minimise, mitigate and compensate for impacts while seeking integrated solutions. These manuals should clearly describe the range of potential mitigation measures, including types of crossing structures, as well as solutions to other impacts (e.g. noise and light pollution). Guidelines can be very broad and cover a range of issues, or they can be very targeted and specific. For example, the Swiss Association of Road and Transportation Experts (VSS 2011) has written a series of guidelines, with each covering a specific issue such as planning, wildlife crossing structures and protective measures to avoid trapping small animals on the road. Written by a team of wildlife specialists and civil engineers, these guidelines use engineering terms and have extensive flow charts and decision trees to help guide the engineer towards the most appropriate mitigation strategy. Other specialised guidelines focus on habitat fragmentation issues (e.g. Anděl et al. 2005) and reducing the rate and severity of WVC using animal detection systems and enhanced signs (Huijser et al. 2006).

This chapter aims to highlight important steps and components when commissioning, writing, reviewing and updating best-practice guidelines and manuals.

LESSONS

59.1   Define the purpose of the guidelines and identify the audience

Each guideline or manual has a different focus and target audience, and these should be carefully considered and defined before the writing commences. Is the manual describing (i) approaches and methods to identify the ecological impacts of roads and traffic; (ii) the range of potential solutions; (iii) methods and approaches to evaluate mitigation success; or (iv) a combination of topics? What geographic region will the guidelines apply to? Will the manual present guidelines to consider or will it specify minimum standards that must be met? Avoid trying to address all potential stakeholders in one manual as different audiences will likely require different information or the same information presented in different ways. Policy-makers may need higher-level strategic reports to enhance their awareness of certain issues, while road planners and designers need practical handbooks with detailed information. Text must be concise and drawings and photos should be used to highlight important design aspects. Decision flow charts can be helpful to guide users through the planning, design and construction process. Most manuals target a specific region, and the recommendations and guidelines must be adapted to suit particular species, legislation and environment.

Consider whether the publication will be printed, made available online or both. Understanding the needs of the user group and where and how they access information will influence how the guidelines are published.

59.2 Guidelines should be practical, implementable and based on state-of-the-art knowledge

Guidelines and manuals in road ecology should be based on rigorous scientific studies or on a consensus among practitioners. Numerous best-practice manuals have been written for continents, countries and regions (see Sections 'References' and Appendix 59.1). These manuals are a good foundation to begin writing new or revising existing manuals, but the efficacy of all strategies should be reviewed before inclusion. There are often multiple solutions to a problem, and manuals should provide enough information to allow a cost–benefit analysis of the different possible solutions. Uncertainty and risks associated with particular strategies should be acknowledged so these can be incorporated into the cost–benefit analysis.

Guidelines should aim to provide consistency across a region in adopting procedures but also to improve standard practice. A pragmatic approach should be favoured when describing and recommending preferred strategies rather than overly ambitious goals that are difficult to achieve.

59.3 Describe the negative effects of roads and traffic on wildlife, ecosystems and human safety

The negative effects of roads and traffic are numerous and varied (Chapter 1), and best-practice manuals should give enough information for the target audience (Lesson 59.1) to adequately identify and assess the potential impacts of proposed projects. In many situations, additional fieldwork will be necessary for a comprehensive EIA (Chapters 5 and 6), and guidelines are required to ensure these are completed to a satisfactory standard. Importantly, manuals should identify impacts and provide solutions for rare and endangered species as well as common and widespread species.

Manuals should clearly explain the importance of thorough surveys at different road planning stages, the seasonal nature of fieldwork and the potential need for multi-year surveys to detect long-term trends and account for year-to-year variation.

59.4 Bring together a multidisciplinary team to write the guidelines and to test the feasibility of the proposed mitigation measures

A multidisciplinary team should collaborate to write the guidelines, or, at the very least, review them, to ensure the content is understandable by the user group and feasible in the field and policy context. External review will help to identify solutions that are inconsistent with other goals, processes or strategies (Chapter 9). Use language appropriate to the target audience and promote ownership of the guidelines by inviting user groups to participate in the process.

59.5 Structure the manual to follow the typical stages of road projects

Most users of manuals do not 'read' the document from cover to cover, but will access information as and when they need it to solve a specific problem. Therefore, manuals must be logical, concise and easy to follow. The use of standard formatting, indexes and tables of contents allows easy navigation around the document.

Most road projects follow a broadly similar process that begins with planning and design and concludes with construction, maintenance and evaluation (Chapters 4 and 9). Manuals could be similarly structured, or at the very least, each of these stages should be clearly identified, enabling different disciplines to readily identify the relevant sections. Where appropriate, manuals should commence with a discussion of the importance of taking a landscape-scale approach to planning (Lesson 59.8) and following the mitigation hierarchy: avoiding, minimising, mitigating and offsetting impacts (Chapter 7).

59.6 Describe the range of mitigation measures and identify the preferred option

There are a wide range of strategies and measures to avoid, minimise, mitigate and offset the ecological impacts of transportation infrastructure. While all of the possible solutions should be described, the optimal or preferred approach(es) should be clearly identified and justified. Manuals should include information on strategies and designs for new roads, as well as how to retrofit mitigation measures to existing roads or modify existing structures to be more wildlife friendly.

The section of the manual that describes the mitigation solutions will likely be the most read and scrutinised. Be as practical and descriptive as possible and provide minimum standards or sizes for engineers to work from. Design sketches, final plans, photos and actual locations (to allow field visits) of installed mitigation measures are useful, especially to convince senior managers that others have adopted this mitigation before, thereby minimising the risk of failure. Where possible, use construction terms familiar to engineers and provide a detailed glossary for terms that may be uncommon. For example, do not assume that the reader understands the nuances of terms used to describe the type of crossing structures (Lesson 21.2). Identify potential unintended consequences of mitigation measures, such as increased rates of mortality of birds if they collide with glass noise walls or small animals trapped on the road surface by guard rail or kerbing. Include a table which summarises and ranks the mitigation options for specific issues, species and habitats. Information on the pros and cons and indicative cost of each mitigation measure should also be provided to allow planners and designers to undertake their own cost–benefit analysis.

59.7 Clearly identify mitigation measures that don't work

Ineffective mitigation measures should be declared as not meeting standards for best practice and therefore not recommended. It is also essential to note situations or conditions when otherwise effective strategies are ineffective or less effective. For example, numerous guidelines still promote the use of mitigation measures for which there is very little evidence of effectiveness, such as roadside reflectors, wildlife warning whistles and some types of signs (Chapters 24 and 25).

59.8 Address transport infrastructure in the context of spatial planning

New and improved roads encourage and facilitate development in adjacent areas, which, depending on the local context, may or may not be appropriate. In some situations, the indirect or secondary effects of the road may have greater impacts than the road itself (Chapters 2, 3, 51 and 56). Best-practice manuals must highlight the importance of landscape-scale planning and include indirect effects when assessing potential impacts (Chapter 5).

59.9 Describe monitoring techniques and schemes

The manual should explain the importance of thorough surveys at different stages of the road project and why the method and timing of the field surveys matter (Chapter 11). Thoroughly evaluating the implementation of the mitigation measure is essential to ensuring that it has been built as intended and designed (Chapter 9). This evaluation also serves to gather further knowledge on the method's efficacy that can be used to improve its design and implementation and to provide policy-makers with results that enhance future adoption of mitigation measures.

59.10 Schedule a time frame for revising and updating the manual

Research on the impacts of roads and traffic and effectiveness of solutions is ongoing, and best-practice manuals must undergo a regular review to incorporate new knowledge and changes to legislation and safety standards. Based on the rate of current research, this should occur at intervals of 5–10 years. Guidelines published on the web or a series of stand-alone volumes on different topics can be more rapidly reviewed and updated than printed and/or single-volume manuals.

CONCLUSIONS

Every road agency needs access to relevant, current and comprehensive guidelines in order to plan, design, construct and maintain ecologically sustainable transportation systems. The expertise of wildlife specialists and civil engineers must be combined to successfully avoid and minimise the negative impacts of transportation infrastructure on natural habitats and implement effective mitigation and compensation solutions. Writing guidelines should be an opportunity for collaboration and a means to develop a common language and understanding as well as a sense of shared responsibility.

ACKNOWLEDGEMENTS

I thank Jürg Bärlocher, former head engineer in the Canton of Thurgau, who invited me in 2004 to join the VSS to develop an environmental approach to road planning and thereby opened a new, productive dialogue between wildlife specialists and engineers.

APPENDIX 59.1 EXAMPLES OF ADDITIONAL BEST-PRACTICE MANUALS AND GUIDELINES NOT ALREADY CITED.

Achiron-Frumkin, T. 2012. Israeli handbook on habitat fragmentation by transportation infrastructure. Commissioned by Israel's National Road Company. 260 p. Available from http://www.iroads.co.il/sites/default/files/handbook-israel2012-light.pdf (accessed 12 January 2015).

Anděl P., Hlaváč V., Lenner R. et al. 2006. Migracni objekty pro zajistěnì pruchodnosti dàlnic a silnic pro volnê zijlìcì zivocichy; Technicke Podminky, EVERNIA. 92 p. Available from http://www.evernia.cz/publikace/Migacni_objekty_pro_zajisteni_pruchodnosti_dalnic.pdf (accessed 12 January 2015).

DFT (UK Department for Transport). 2014. Design manual for roads and bridges. Volume 10 – Environmental design. Available from http://www.standardsforhighways.co.uk/ha/standards/dmrb/vol10/index.htm (accessed 12 January 2015).

Dinetti, M. 2000. Infrastrutture ecologiche, Il Verde Editoriale, Milano, Italy, 214 p.

Dinetti, M. (ed.). 2005. Atti del Convegno 'Infrastrutture viarie e biodiversità. Impatti ambientali e soluzioni di mitigazione'. Pisa, 25 Novembre 2004. Provincia di Pisa e Lipu. Stylgrafica Cascinese, Cascina PI, Italy.

Dinetti, M. 2008. Infrastrutture di trasporto e biodiversità: lo Stato dell'Arte in Italia. Il problema della frammentazione degli habitat causata da autostrade, strade, ferrovie e canali navigabili. IENE Infra-Eco-Network-Europe, Sezione Italia, 160 p.

Dinetti, M., C. Sangiorgi, and F. Irali, 2012. Progettazione ecologica delle infrastrutture di trasporto. Felici Editore, Pisa, Italy, 150 p.

Hlavac, V., and P. Andel. 2002. On the Permeability of Roads for Wildlife: a Handbook. Agency for Nature Conservation and Landscape Protection of the Czech Republic, Prague, Czech Republic.

Jędrzejewski, W., S. Nowak, R. Kurek, R.W. Mysłajek, K. Stachura, B. Zawadzka, M. Pchałek. 2009. Animals and Roads: Methods of mitigating the negative impact of roads on wildlife. Mammal Research Institute, Polish Academy of Sciences, Bialowieza, Poland.

Lundin, U., Sjölund, A., Skoog, J., Eriksson, O., Jakobi, M., Olsson, M., Seiler, A. & Andrén, C. 2005. Wildlife and infrastructure – a handbook for mitigation. Swedish Transport Administration, 72, in Swedish, 124 pp. Available from http://www.rms.

nsw.gov.au/documents/about/environment/biodiversity_guidelines.pdf (accessed 12. January 2015).

Ministerio de Medio Ambiente. 2006. *Prescripciones Técnicas para el diseño de pasos de fauna y vallados perimetrales.* Documentos para la reducción de la fragmentación de hábitats causada por infraestructuras de transporte, número 1. O.A. Parques Nacionales. Ministerio de Medio Ambiente, Madrid, Spain, 108 pp.

Ministerio de Medio Ambiente y Medio Rural y Marino. 2008. *Prescripciones técnicas para el seguimiento y evaluación de la efectividad de las medidas correctoras del efecto barrera de las infraestructuras de transporte.* Documentos para la reducción de la fragmentación de hábitats causada por infraestructuras de transportes, número 2. O.A. Parques Nacionales. Ministerio de Medio Ambiente y Medio Rural y Marino. Madrid, Spain, 138 pp.

Ministerio de Medio Ambiente y Medio Rural y Marino. 2010. *Prescripciones técnicas para la reducción de la fragmentación de hábitats en las fases de planificación y trazado.* Documentos para la reducción de la fragmentación de hábitats causada por infraestructuras de transportes, número 3. O.A. Parques Nacionales. Ministerio de Medio Ambiente y Medio Rural y Marino. Madrid, Spain, 145 pp.

Ministerio de Agricultura, Alimentación y Medio Ambiente. 2013. *Desfragmentación de hábitats. Orientaciones para reducir los efectos de las infraestructuras de transporte en funcionamiento.* Documentos para la reducción de la fragmentación de hábitats causada por infraestructuras de transporte, número 5. O.A. Parques Nacionales. Ministerio de Agricultura, Alimentación y Medio Ambiente. Madrid, Spain, 159 pp.

Ministry of Environment, Korea. 2010. Guidelines for design and management of wildlife crossing structures in Korea. Ministry of Environment, Seoul, Korea. Available from http://webbook.me.go.kr/DLi-File/077/203004.pdf (accessed 12. January 2015).

Pepper, H. W., M. Holland, and R. Trout. 2006. Wildlife fencing design guide. CIRIA, London, UK.

Piepers, A. A. G. editor. 2001. Infrastructure and nature; fragmentation and defragmentation. Dutch State of the Art Report for COST Activity 341. Road and Hydraulic Engineering Division Defragmentation Series, part 39A. Road and Hydraulic Engineering Division, Delft, the Netherlands.

Rajvanshi, A., V. B. Mathur, G. C. Teleki, and S. K. Mukherjee. 2001. Roads, sensitive habitats and wildlife: Environmental guidelines for India and South Asia, Wildlife Institute of India, Dehradun, India and Canadian Environmental Collaborative Ltd., Toronto, Canada.

Rosell, C. & Velasco Rivas, J.M. 1999. *Manual de prevenció i correcció dels impactes de les infraestructures viàries sobre la fauna.* Documents dels Quaderns de Medi Ambient, 4. Departament de Medi Ambient, Barcelona, Spain, 95pp. (Incluye traducción al castellano e inglés).

RTA (Roads and Traffic Authority of New South Wales). 2011. Biodiversity guidelines: Protecting and managing biodiversity on RTA projects. Roads and Traffic Authority of NSW, Sydney, Australia.

SETRA (Service d'Etudes Techniques des Routes et Autoroutes). 1993. Passage pour la grande faune, guide technique. SETRA, France. 124 pp.

SETRA (Service d'Etudes Techniques des Routes et Autoroutes). 2005. Facilities for small fauna, technical guide. SETRA, France. 264 pp.

SETRA (Service d'Etudes Techniques des Routes et Autoroutes), CETE EST, CETE NP. 2009. Bats and road transport infrastructure, threats and preservation measures, information note. SETRA, France. 22 pp.

CETE EST, ONEMA. 2013, Petits ouvrages hydrauliques et continuités écologiques: cas de la faune piscicole, note d'information – SETRA – 25 p.

SETRA (Service d'Etudes Techniques des Routes et Autoroutes). 2006. Bilan d'expériences, Routes et passages à faune, 40 ans d'évolution, rapport. SETRA, France. 55 pp.

US Department of Transportation, Federal Highway Administration, Keeping it simple-easy ways to help wildlife along roads, PUB. No FHWA-EP-03-066, 58 pp. Available from http://www.fhwa.dot.gov/environment/critter_crossings/intro.cfm (accessed 18 January 2015).

van der Grift, E., Biserkov, V., and Simeonova, V. 2008. Restoring ecological networks across transport corridors in Bulgaria: Identification of bottleneck locations and practical solutions. Alterra, Wageningen, The Netherlands.

White, P. A. and Ernst, M. 2005. Second nature: Improving transportation without putting nature second. Defenders of Wildlife, Washington, DC, USA.

FURTHER READING

Australasian Network for Ecology and Transportation (ANET): Network, similar to ICOET and IENE, for the Australasian region (http://www.ecoltrans.net).

Beckman et al. (2010): A well-written book that summarises the impacts of roads and traffic on wildlife, with a focus on the United States. It includes specific recommendations on siting, design and mitigation, as well as case studies on successful mitigation programmes.

Forman et al. (2003): The first road ecology textbook, with an extensive and detailed discussion of impacts and solutions.

Infra Eco Network Europe (IENE): The European network for the interaction between green infrastructure and the surface transportation network (http://www.iene.info/; Chapter 60).

International Conferences on Ecology and Transport (ICOET): The international conference held biannually in the United States for ecology and transportation (http://www.icoet.net/; see Chapter 60).

Tsunokawa and Hoban (1997): A publication that promotes best practice in EIA and road design for World Bank-funded projects – with relevance to all linear infrastructure projects, regardless of funding source.

REFERENCES

Anděl, P., I. Gorčicová, V. Hlaváč, L. Miko and H. Andělová. 2005. Assessment of landscape fragmentation caused by traffic. Agency for Nature Conservation and Landscape Protection of the Czech Republic, Liberec, Czech Republic.

Beckman, J. P., A. P. Clevenger, M. P. Huijser and J. A. Hilty, editors. 2010. Safe passages: highways, wildlife and habitat connectivity. Island Press, Washington, DC.

Carsignol, J. 2006. Routes et passages à faune – 40 ans d'évolution. Sétra Référence 0641W, 57 pp.

Clevenger, A. P. and Huijser, M. P. 2011. Wildlife crossing structure handbook design and evaluation in North America. Federal Highway Administration, Lakewood CO, 211 pp.

Forman, R. T. T., D. Sperling, J. A. Bissonette, A. P. Clevenger, C. D. Cutshall, V. H. Dale, L. Fahrig, R. France, C. R. Goldman, K. Heanue, J. A. Jones, F. J. Swanson, T. Turrentine and T. C. Winter. 2003. Road ecology. Science and solutions. Island Press, Washington, DC.

Huijser, M. P., P. McGowen, W. Camel, A. Hardy, P. Wright and A. Clevenger. 2006. Animal vehicle crash mitigation using advanced technology. Phase 1: review design and implementation. Western Transportation Institute-Montana State University, Bozeman, Montana.

Iuell, B., et al. 2003. Wildlife and traffic: a European handbook for identifying conflicts and designing solutions. KNNV Publishers, Brussels, Belgium, Europe.

Queensland Department of Transport and Main Roads (QDTMR). 2011. Fauna sensitive road design manual. Volume 2: Preferred practises. Queensland Department of Main Roads, Brisbane, Australia.

Reed, D. F. 1969. Techniques for determining potentially critical deer highway crossings. Outdoor Facts 73. Colorado Department of Natural Resources, Denver, CO.

Ryser, J. 1988. Amphibien und Verkehr, Teil2. Amphibien-rettungsmassnahmen an Strassen in der Schweiz. Koordinationsstelle für Amphibien und Reptilienschutz in der Schweiz (KARCH), Bern, Switzerland, 10 pp.

Service d'Etudes Techniques des Routes et Autoroutes (SETRA). 1985. Routes et Faune Sauvage. Actes du colloque. Strasbourg, Conseil de l'Europe. Référence Sétra: B8764. 406 pp.

Swiss Association of Road and Transportation Experts (VSS). 1968. Swiss standards 640 693 "Clotures à faune". www.vss.ch

Swiss Association of Road and Transportation Experts (VSS), SN640690-640699, 9 norms on fauna and traffic. ww.vss.ch. 2011.

Tsunokawa, K. and C. Hoban, editors. 1997. Roads and the environment: a handbook. World Bank, Washington, DC.

Chapter 60

CASE STUDY: THE ROLE OF NON-GOVERNMENTAL ORGANISATIONS (NGOS) AND ADVOCATES IN REDUCING THE IMPACTS OF ROADS ON WILDLIFE

Patricia White

National Wildlife Federation, Washington, DC, USA

Today, a growing cadre of committed professionals is making progress towards raising the ecological standards of our transportation infrastructure. One important contributor to this progress has been conservation non-governmental organisations (NGOs). Through their special skill set, including media and public outreach, bringing volunteer labour and a capacity to influence policy and funding changes, NGOs play an important role in making positive changes for wildlife.

WHY NGOS?

Protecting and restoring core habitat and corridors is a daunting task, requiring collaboration and cooperation among many stakeholders including local officials, regional and national authorities, transportation and natural resource agencies, elected officials, community leaders, landowners, academia, conservation organisations and concerned citizens. No single sector can do it alone. For a wildlife and transportation project to be planned, designed, funded, built and maintained and safely move wildlife, many factors must be in place, including science, technology, funding, policy/law, best practices, public support and political will.

A project – such as installing a wildlife crossing structure – might have many of these factors, but if it is missing even one, it may fail. For example, biologists can conduct the necessary research on wildlife movement to determine the location and design for the structure, but without funding and agency support, it may not be built. Or that same project may have funding and agency support, but if an unenlightened public does not support it, the political will can disappear, taking the funding with it. Many good, promising projects have been derailed before they were finished because just one of these factors was missing.

Fortunately, many conservation NGOs possess the skills and the spirit to meet these challenges. Indeed, conservation NGOs are uniquely positioned to address the various challenges and bring together the stakeholders, policy, implementation and public support to make ecologically sensitive roads a reality. Specifically,

Handbook of Road Ecology, First Edition. Edited by Rodney van der Ree, Daniel J. Smith and Clara Grilo.

conservationists can contribute by increasing public awareness and support, persuading lawmakers and recruiting volunteers.

CHANGING HEARTS AND MINDS

While millions of people are involved in wildlife-vehicle collisions (WVC) globally, the general public remains woefully uneducated about the conflict between wildlife and transportation. And only a small proportion of those who notice their roadsides littered with roadkill understand the full scope of ecological effects of roads upon wildlife beyond the pavement. Even fewer are aware of solutions to reduce these impacts or understand how they can participate in the process.

A 2006 study by the University of Denver (Archerd-Bingham 2006) found four major barriers to effective citizen participation in wildlife-sensitive transportation projects: (i) insufficient awareness of wildlife and transportation issues; (ii) public apathy or lack of citizen interest in wildlife and transportation issues; (iii) ineffective citizen participation techniques and processes; and (iv) transportation agencies' poor communication with local citizens. Before road ecology can achieve and sustain progress, it will need to penetrate the mindset of the general driving and taxpaying public and normalise not only the problem but the solutions as well.

Conservation NGOs are in a prime position to educate the public on the conflict between wildlife and transportation. Conservation NGOs often employ communication staff, skilled at crafting persuasive messages for the public ear. Utilising public outreach and media, conservation NGOs can teach people about the impacts of transportation on wildlife and the variety of solutions. In doing so, citizens will be convinced that reducing WVC and improving habitat connectivity across roads are in the best interest of the motoring public and a valid use of taxpayer dollars. Conservation NGOs can also help spread the message to educate drivers on the need to drive with caution and reduce their speed when driving in areas containing wildlife.

CHILDREN AND THE I-90 WILDLIFE BRIDGES

To build support for wildlife crossings in Washington, United States, the I-90 Wildlife Bridges Coalition visited schools in the area to educate students about wildlife and roads. The children then created colourful posters to demonstrate their own designs for crossings that would benefit both animals and people. The coalition partnered with the transportation agency to select the winning posters and award prizes to the children. The winning poster was then proudly displayed on a billboard where passing motorists could see it (Fig. 60.1), thereby increasing public support for the project.

CHANGING POLICY

In addition to transportation and natural resource agency professionals, elected officials – from mayors to regional and national lawmakers – can influence public policies regarding transportation infrastructure and how it will impact wildlife and habitat. Elected

Figure 60.1 Road sign created for the I-90 Highway Upgrade, by school children, the Washington State Department of Transportation and the I-90 Wildlife Bridges Coalition. Source: Reproduced with permission of Charlie Raines, I-90 Wildlife Bridges Coalition.

leaders at every level may be involved with allocating transportation funds, setting development priorities and even having a hand in project selection. They often control funding and priorities for natural resource and environmental protection agencies as well.

Many different sectors of society have a stake in transportation infrastructure, and all of them seek out elected leaders, attempting to influence how transportation funds are allocated, which projects are built and where and when they are constructed. In the United States, government agency staff are prohibited from communicating with lawmakers or their staff to influence the passage of laws, appropriation of funds or other measures. They are also forbidden to communicate their support or opposition to any laws or appropriation of funds to the general public. Under these circumstances, conservation NGOs can carry the message of support for wildlife-friendly policies and allocation of funding for habitat connectivity programmes and projects to elected leaders. Conservation NGOs can also encourage their expansive support networks to write letters or emails and call their lawmakers to pressure them to support wildlife-friendly policies and funding.

BEING THE CHANGE

Wildlife crossing structures vary greatly in size and variety, from small underpasses under a rural road to wide, vegetated wildlife overpasses over major highways. Many effective measures require little or no construction at all. And even with large projects like wildlife overpasses, there are several pre- and post-construction activities that can be done by a novice. While the primary responsibility lies with road agencies, unpaid and enthusiastic volunteers can help with a wide range of tasks, thereby saving money and building ownership in projects by engaging with the community.

Conservation NGOs often have support networks ranging from a dozen in smaller local groups to several hundred thousand in national and international NGOs. They can use their relationships with supporters and reach out to the general public to recruit volunteers for small projects including:
• Clearing invasive vegetation near selected crossing locations;
• Roadside vegetation management;
• Collecting pre- and post-construction wildlife mortality and/or crossing data;
• Repairing or removing fencing;
• Clearing debris from drainage culverts and wildlife underpasses.

THE MIISTAKIS INSTITUTE AND THE 'ROAD WATCH IN THE PASS' PROJECT

The Miistakis Institute, a non-profit organisation affiliated with the University of Calgary, British Columbia, Canada, launched the 'Road Watch in the Pass' project, asking drivers using Highway 3 through Crowsnest Pass to report sightings of wildlife on a special website (www.rockies.ca/roadwatch). If a participant observes an animal (alive or dead) while driving through the pass, they are asked to log on to the website and fill out a simple report on the species, location and status of each sighting. Data collected from the website are tallied and provided to transportation and natural resource officials to inform decisions regarding safety upgrades like wildlife crossings.

CONCLUSIONS

Together, all sectors involved have made great strides in advancing the new science of road ecology. There is no shortage of available data, technology, policy, best practices and expertise to protect and restore core habitats and corridors. However, much work remains to be done.

Protecting and restoring essential wildlife habitat and corridors will take considerable investments from all sectors from transportation and natural resource agencies, academia, lawmakers, landowners and communities. At this time, the greatest contribution conservation NGOs can offer is in bringing all partners together.

FURTHER READING

White (2007): An accessible book that seeks to crack the code on transportation, demystifying the policies and practices to provide conservationists with the necessary foundation to become informed, more effective stakeholders in transportation debates.

REFERENCES

Archerd-Bingham, L.E. 2006. The role of citizen participation in wildlife-sensitive transportation projects. Capstone project. University of Denver, Denver, CO.

White, P. 2007. Getting Up to Speed: A Conservationist's Guide to Wildlife and Highways. Defenders of Wildlife, Washington, DC. Available at http://bit.ly/14xqYyU (accessed 10 October 2014).

Chapter 61

CASE STUDY: BUILDING A COMMUNITY OF PRACTICE FOR ROAD ECOLOGY

Paul J. Wagner[1] and Andreas Seiler[2]

[1]Washington State Department of Transportation, Environmental Services Office, Olympia, WA, USA
[2]Department of Ecology, Swedish University of Agricultural Sciences, Uppsala, Sweden

CONGREGATING AND COMMUNICATING IS ESSENTIAL

The practice of evaluating the ecological effects of roads and railroads and applying this knowledge to the design and operation of transportation infrastructure has come into being largely in the last two decades. Its origins stem from the pursuit of local solutions to address specific ecological concerns, often related to individual transportation projects or facilities. As practical experience was gained, lessons learned and knowledge shared, the network of practitioners and body of knowledge grew. This knowledge is now applied at broader scales and higher levels. A significant catalyst in the development of this field has been the opportunities for people from various disciplines, organisations and regions to share knowledge and to network as part of various conferences, workshops and seminars. Two well-established forums are the Infra Eco Network Europe (IENE) and the International Conference on Ecology and Transportation (ICOET) in the United States, and these are the primary focus of

this chapter. More recently, similar networks have been established in Brazil (REB – Chapter 50) and Australasia (ANET 2013).

A community of practice (CoP) is a group of people who share a craft or profession (Wenger et al. 2002). The group can evolve naturally because of the members' common interest in a particular domain, or it can be created specifically to gain knowledge. A CoP is particularly important for questions where different domains interact and if resources and knowledge is limited. The IENE, ICOET, ANET and REB are CoP for those who seek to better understand and address the ecological consequences of traffic and infrastructure. The distinguishing qualities of this work are (i) the typically pressing need for direct application of the best available science to on-the-ground project decisions and (ii) the frequent need for cross-disciplinary coordination to develop effective solutions. The cross-disciplinary nature is important because individuals or disciplines (i.e. biology, ecology, engineering, hydrology, planning, traffic management and maintenance) do not possess all the information needed, and they can

Handbook of Road Ecology, First Edition. Edited by Rodney van der Ree, Daniel J. Smith and Clara Grilo.
© 2015 John Wiley & Sons, Ltd. Published 2015 by John Wiley & Sons, Ltd.
Companion website: www.wiley.com\go\vanderree\roadecology

use others within the CoP to analyse the problem and implement an effective solution.

THE INFRA ECO NETWORK OF EUROPE

In Europe, professional interest in wildlife and traffic arose during the 1970s as traffic safety concerns grew due to a rising number of wildlife-vehicle collisions (WVC) and pollution from de-icing salt and lead was increasingly acknowledged. Habitat fragmentation due to roads and noise disturbance by traffic gained attention in the following years (e.g. Mader 1981; Bernard et al. 1987) leading to an international conference arranged by the Dutch Ministry of Transport, Public Works and Water Management in the Netherlands in 1995 (Canters et al. 1997). The conference highlighted similarities in the problems and solutions addressed by the participating countries and outlined the potential benefits from closer collaboration and exchange of experience. This gathering produced a declaration from over 25 countries for collaboration, which initiated the official formation of the IENE (IENE 2013).

Originally, the IENE was a group of designated national representatives for the sectors of transport or environment who met annually. This group developed and proposed the Cooperation in Science and Technology (COST) Action 341 on habitat fragmentation due to transport infrastructure, an action of the European (EU) framework for COST. This action gathered experience and knowledge from within the IENE network and produced national and EU-wide state-of-the-art reports on habitat fragmentation (Trocmé et al. 2003) and a seminal EU handbook on wildlife and traffic (Iuell et al. 2003). The handbook has been adopted and translated for use by many countries since then. The COST Action 341 closed in 2003 with an international conference arranged by the IENE. Thereafter, a shortage in central funding for network coordination put the IENE into a dormant state until activities resumed in 2008. The IENE has developed into a network with free individual membership, an elected steering committee and an approved obligation to serve as a forum for cross-border and interdisciplinary exchange in the fields of transportation and ecology.

The IENE arranges regional annual meetings and workshops across Europe and organises biannual international conferences on alternate years to ICOET conferences. In addition, the IENE initiates research activities and applications for EU-wide communication projects. It is through these activities that the network survives and grows. They provide the necessary incentives for personal contacts, funding opportunities and practical collaboration. Although most communication is managed online, it is the personal contact among the members that keeps the network alive.

At present (in 2014), the IENE has greater than 250 members from 45 countries and even more subscribe to the newsletter or participate at conferences. IENE networks have been established in some EU countries, and they produce newsletters or reports, arrange seminars or provide advice in construction projects. All members and national networks benefit from the broader international CoP of the IENE, either directly through knowledge transfer or indirectly when the reputation of the network grows.

As the network grows and becomes increasingly recognised as a resource for agencies, governments or individual researchers, there is also a growing demand for a stable node or secretariat that responds to official requests and maintains communication activities. Securing this central service, however, requires reliable and sufficient funding, which, in the international setting of the IENE, is extremely difficult to come by. Despite all attempts to establish a multinational funding consortium, only the Swedish Transport Administration currently funds the IENE secretariat.

THE INTERNATIONAL CONFERENCE ON ECOLOGY AND TRANSPORTATION

In the early 1990s, biologists seeking solutions for WVC in the United States drew inspiration from work undertaken in Europe with wildlife crossing structures. They sought to address situations where WVC represented serious safety concerns as well as where losses to wildlife constituted a significant threat to populations of at-risk species (e.g. Florida panther). Various efforts up to that point to address WVC and other ecological issues had been largely experimental, based on the professional judgement of the people involved.

In 1996, a national conference was hosted in Orlando, Florida, by the Florida Department of Transportation (FDOT) and Federal Highway Administration (FHWA), called the 'Transportation-Related Wildlife Mortality Seminar'. About 100 people attended this conference which was the first national forum in the United States for sharing practical experience, discussing the state of knowledge and identifying gaps and needs in understanding the problems and solutions related to roadside ecology. In September 1996, another national conference was hosted in Tacoma, Washington, by the Washington State Department of Transportation entitled 'Connections: Transportation, Wetlands and the Natural Environment', which expanded the discussion to a broader range of environmental topics beyond WVC. This discussion continued through a series of conferences. The name 'International Conference on Wildlife Ecology and Transportation (ICOWET)' was employed by the FDOT for the meeting jointly hosted in February 1998 with the FHWA, the US Forest Service (USFS) and the Defenders of Wildlife in Fort Myers, Florida. This meeting shared examples of best practices from Canada, the Netherlands and Australia and from around the United States. This meeting was next hosted in 1999, by the Montana Department of Transportation in Missoula, Montana.

The next step in evolution dropped the word 'wildlife' from the name, embracing a broader range of ecological issues. The organisers drafted the ICOET mission: *to identify and share quality research applications and best management practices that address wildlife, habitat, and ecosystem issues related to the delivery of surface transportation systems.* The meetings of the ICOET have been held every other year by transportation agencies from different regions of the United States (ICOET 2013). From the start, ICOET organisers recognised that this subject relates to a number of different perspectives, disciplines and stakeholder groups; therefore, productive discussion and effective solutions would need to embody input from a number of sources. Both the content of the conferences and the make-up of the steering committee represent partnerships among transportation agencies and regulatory/resource agencies, planning, engineering, academia and non-governmental advocacy organisations.

BUILDING A CoP: LESSONS LEARNED

Both IENE and ICOET help to support a group of people who are mutually engaged in a joint enterprise who utilise a shared repertoire of information; as such, they are successful examples of CoP (Wenger 1998). Although there are obvious differences between them (e.g. national vs. international organisation, members who speak different languages), there are also parallels in their origin, development and international participation. From these experiences, the following recommendations derive for current and future CoP in road ecology, such as ANET and REB:

(i) Design the community to evolve naturally – Because the nature of a CoP is dynamic, in that the interests, goals and members are subject to change, the CoP should be encouraged to refine their focus over time as the needs of the members change.

(ii) Create opportunities for open dialogue with internal and external perspectives – While the members and their knowledge are the CoP's most valuable resource, it is also beneficial to look outside the CoP to identify different potential solutions. This may come naturally to a CoP that operates at the intersection between different domains, such as between transportation and ecology.

(iii) Welcome and allow different levels of participation – Wenger (1998) identified three main levels of participation: (a) the core group who participate intensely in the community through discussions and projects and typically take on leadership roles; (b) active members who attend and participate regularly, but not to the level of the leaders; and (c) the peripheral and typically largest group who, while they are passive participants in the community, still learn from their level of involvement.

(iv) Develop both public and private community spaces – While CoPs typically operate in public spaces where all members share, discuss and explore ideas, they should also offer private exchanges. Different members of the CoP may cultivate relationships and share resources in an individualised approach based on specific needs.

(v) Create opportunities for personal encounters – Although most communication within the community will likely occur online or through email, the personal face-to-face meetings are a critical part of human relationships. Conferences, workshops and other gatherings must form an important part of a CoP.

(vi) Focus on the value of the community – CoP should create opportunities for participants to explicitly discuss the value and productivity of their participation in the group.

(vii) Combine familiarity and excitement – CoP should offer routine learning opportunities as well as opportunities for members to shape their learning experience together by brainstorming and examining the conventional and radical wisdom related to their topic.

(viii) Find and nurture a regular rhythm for the community – CoP should coordinate a thriving cycle of activities and events that allow members to regularly meet, reflect and evolve. The rhythm, or pace, should maintain an anticipated level of engagement to sustain the vibrancy of the community, yet not be so fast-paced that it becomes unwieldy and overwhelming in its intensity (Wenger et al. 2002).

(ix) Secure central funding for administration – A CoP, just like any other organisation, requires funding for coordination and common services. This cost should be shared among the community, but achieving this may be difficult, especially if the network and its benefits are intended to be available at no expense.

(x) Find a host for the central communication node – Hosting a network or CoP provides several benefits and may put the host in a strategic and powerful position. The host must ensure that the CoP remains rooted in the shared interests of its members and does not unduly influence the direction of the CoP for its own benefit.

Road ecology needs effective CoPs to plan, design, build and manage an ecologically sustainable transportation network. The potential ecological impacts of roads and traffic are so great that the combined knowledge and expertise of the CoP is essential to achieving cost-effective solutions. The future of the ICOET and IENE, and the other fledgling road ecology CoPs outside of North America and Europe, relies on the engagement and support of many different disciplines, organisations and government agencies.

REFERENCES

Australasian Network for Ecology and Transportation (ANET). 2013. Australasian Network for Ecology and Transportation, Melbourne, Australia. Available from www.ecoltrans.net (accessed 10 October 2014).

Bernard, J. M., M. Lansiart, C. Kempf and M. Tille. 1987. Actes du colloques 'Route et fauna sauvage'. Strasbourg, Conseil de l'Europe, 5–7 Juin 1985; p. 403. In: J. M. Bernard, M. Lansiart, C. Kempf and M. Tille, editors. Ministère de l'Équipement, du Longement, de l'Aménagement du Territoire et des Transports, Colmar, France.

Canters, K. A., A. A. G. Piepers and D. Hendriks-Heersma, editors. 1997. Habitat fragmentation and infrastructure. Proceedings of the international conference on habitat fragmentation, infrastructure and the role of ecological engineering, 17–21 September 1995, Maastricht and The Hague, the Netherlands. Directorate-General for Public Works and Water Management, Road and Hydraulic Engineering Division, Delft, the Netherlands.

Infra Eco Network Europe (IENE). 2013. Infra Eco Network Europe, Linköping, Sweden. Available from www.iene.info (accessed 10 October 2014).

International Conference on Ecology and Transportation (ICOET). 2013. Center for Transportation and the Environment, Raleigh, NC. Available from www.icoet.net (accessed 10 October 2014).

Iuell, B., H. Bekker, R. Cuperus, J. Dufek, G. L. Fry, C. Hicks, V. Hlavac, J. Keller, B. Le Maire Wandall, C. Rosell Pagès, T. Sangwine and N. Torslov, editors. 2003. Wildlife and Traffic – A European Handbook for Identifying Conflicts and Designing Solutions. Prepared by COST 341 – Habitat Fragmentation due to Transportation Infrastructure. Ministry of Transport, Public Works and Water Management, Road and Hydraulic Engineering Division, Delft, the Netherlands.

Mader, H. J. 1981. Der Konflikt Strasse-Tierwelt aus ökologischer Sicht. Schriftenreihe für Landschaftspflege und Naturschutz, Heft 22. Bundesforschungsanstalt fuer Naturschutz und Landschaftoekologie, Bonn – Bad Godesberg, Germany.

Trocmé, M., S. Cahill, J. G. De Vries, H. Farall, L. Folkeson, G. L. Fry, C. Hicks and J. Peymen 2003. COST 341 – Habitat Fragmentation Due to Transportation Infrastructure: The European Review. Office for Official Publications of the European Communities, Luxembourg, Europe.

Wenger, E. 1998. Communities of Practice: Learning, Meaning, and Identity. Cambridge University Press, Cambridge, UK.

Wenger, E., R. McDermott and W. M. Snyder. 2002. Cultivating Communities of Practice. Harvard Business Press, Boston, MA.

Chapter 62

WILDLIFE/ROADKILL OBSERVATION AND REPORTING SYSTEMS

Fraser Shilling[1], Sarah E. Perkins[2] and Wendy Collinson[3]

[1]University of California, Davis, United States
[2]Cardiff University, Cardiff, UK
[3]Endangered Wildlife Trust, Johannesburg, South Africa

SUMMARY

Wherever wildlife habitat and roadways overlap, roadkill seems inevitable. Observing and recording carcasses resulting from wildlife-vehicle collisions (WVC) provides data critical for sustainable transportation planning and species distribution mapping. Across the world, systems have been created to record WVC observations by researchers, highway maintenance workers, law officers, wildlife agency staff, insurers and volunteers. These wildlife/roadkill observation systems (WROS) can include mobile recording devices for data collection, a website for data management and visualisation and social media to reinforce reporting activity.

62.1 The specific purpose and goals of the WROS may vary among systems but should always be clearly defined.

62.2 Extensive social networks are needed for comprehensive observation systems.

62.3 Adopt a methodical approach to developing a WROS.

62.4 Analysis and visualisation of data collected within a WROS should correspond to the goals of the system.

62.5 Address issues in reporter bias by using standardised data collection methods or post hoc analyses.

62.6 The advantages and disadvantages of opportunistic and targeted data collection must be carefully considered when developing a WROS.

Volunteer science and web-based information tools have advanced to the point where transportation or wildlife agencies and their allies can develop, support or implement WROS to improve the sustainability of transportation systems. However, while numerous WROS have been developed and implemented around the world, the full potential of many systems has not been realised because they were not developed or maintained according to the basic principles outlined in this chapter. We provide suggestions and guidance useful for updating existing systems and developing new ones.

Handbook of Road Ecology, First Edition. Edited by Rodney van der Ree, Daniel J. Smith and Clara Grilo.
© 2015 John Wiley & Sons, Ltd. Published 2015 by John Wiley & Sons, Ltd.
Companion website: www.wiley.com\go\vanderree\roadecology

INTRODUCTION

Reporting the occurrence of wildlife on roads, whether alive or as carcasses resulting from wildlife-vehicle collisions (WVC), has recently exploded as an area of road ecology research and practice and has spawned a new type of volunteer involvement. Globally, there are dozens of web-based systems for reporting WVC (Table 62.1). Many have appeared over the last 5 years, and they vary in their specific purpose, taxonomic breadth and use of social networks for collecting data and outreach. A few use smartphone-based applications to facilitate data entry from the field, and some use social media and communication tools to receive observations (Table 62.1).

Web-based reporting of wildlife observations, including WVC, is a rapidly growing source of data for understanding the impacts of roads on wildlife and, in some cases, mitigation effectiveness. The largest system in the world conducted by government agencies is that of Sweden's national police, which collects and reports accidents involving 12 species of wildlife (Table 62.1). The largest, longest-running system that relies on volunteer observers reporting any species is the 'California Roadkill Observation System' (CROS), run by the Road Ecology Center at the University of California, Davis (UC Davis, Table 62.1). In the latter case, data is collected from all roads as well as on targeted 'transect' roads, which have been selected for regular surveys.

There are four main ways that observations are recorded (Table 62.1): (i) inclusion of historical records of accidents or carcasses that preceded the web-based system; (ii) form-based reporting on a website, including drop-down menus; (iii) smartphone application-assisted web systems; and (iv) social media sites, such as Twitter and Facebook. There are two main types of data collection strategies: opportunistic/random observations and transect/targeted route observations (Lesson 62.6). In the first case, observers report carcasses wherever and whenever they are seen (e.g. Endangered Wildlife Trust, South Africa). In the second case, observers regularly drive, walk or cycle routes and report carcasses and, less usually, absence of carcasses ('null' observations, e.g. Road Ecology Center for Maine, United States).

Existing WROS can consist of tens of thousands of data points (Table 62.1) and represent a potential source of 'big data' for road ecology, community ecology, biodiversity mapping and other scientific/engineering disciplines. Big data refers to data sets that are large and usually geographically extensive and so require novel solutions for storage, analysis, processing

and visualisation. At a global level, WROS provide the largest known, continuous source of data on animal occurrence and distribution while also providing opportunities for tissue sampling of genetics, disease and other testing (Textbox 62.1). It is important to carefully structure the informatics (i.e. collection, management and sharing) systems for these observations to facilitate analyses and other uses of the data.

Textbox 62.1 The importance of sampling roadkill otters for evaluating pollution and parasites in the United Kingdom

The Cardiff University Otter Project has collected otter carcasses (95% from roadkill) in the United Kingdom for 20 years, across a period of population expansion, and from 2010 has examined approximately 200 carcasses per year. At the top of the aquatic food chain, and a wide-ranging predator, otters form an excellent 'sentinel' for environmental health, enabling researchers to determine spatial and temporal variation in contaminants (Chadwick et al. 2011) and parasites (Chadwick et al. 2013) of relevance to human as well as animal health. Making use of roadkill is particularly important when studying elusive species, such as the otter, that are otherwise difficult to monitor and can offer insights into population structure and behaviour (Hobbs et al. 2011) as well as basic biology (Sherrard-Smith & Chadwick 2010).

The aims of this chapter are to highlight key issues that should be considered during the planning, design and implementation phases of WROS to ensure the data collected are accurate, reliable and useful to multiple end users. Many WROS systems around the world have been unsuccessful because the lessons we outlined in this chapter were not adequately considered.

LESSONS

62.1   The specific purpose and goals of the WROS may vary among systems but should always be clearly defined

The rationales for creating a WROS include informing transportation mitigation planning (Gunson et al. 2011), improving driver safety (Bissonette et al. 2008) and contributing to biodiversity observations (Elmeros et al.

Table 62.1 Examples of wildlife/roadkill observation systems from around the world.

System name (country)	URL	Description	Data collection and feedback methods
National Wildlife Accident Council (Sweden)	http://www.viltolycka.se/hem/	Operated by the Swedish National police, it is the largest of its kind in the world, with over 200,000 records of WVC in the last 5 years. Online reporting and data display have been in place since 2010, but data are available back to 1985	Collection: website, iOS and android apps Feedback: real-time map of WVC on web of specific species, plus data downloaded to app to alert drivers of proximity to a recent WVC
California Roadkill Observation System and Maine Audubon Wildlife Road Watch (United States)	http://wildlifecrossing.net	Operated by the UC Davis Road Ecology Center, and beginning in 2009, this was one of the first in the world to use web reporting of roadkill by any organisation or individual. It allows entries for any species in California or Maine using websites for each of these states and as of October 2014 had greater than 30,000 entries on transect and non-transect routes	Collection: website Feedback: regular updates (last 30, 90 days) on a map of WVC, searchable by species. Lists and photo gallery of observations. News page on website
Idaho Fish and Wildlife Information System (IFWIS) (United States)	https://fishandgame.idaho.gov/species/roadkill	Operated by the Idaho Department of Fish and Game, for all wildlife species in Idaho since 2011. As of October 2014, there were greater than 22,000 entries	Collection: website by registered users or phone number Feedback: none available
Animals under the Wheels (Belgium)	http://waarnemingen.be	Operated by Natuurpunt of Belgium and since 2013 has collected greater than 14,000 roadkill observations (as of October 2014). Volunteers can report observations for any Belgian wildlife species on transect or other routes	Collection: website by registered users, iOS and android app Feedback: data on most observed species and top observers, species identification offered
Project Splatter (United Kingdom)	http://projectsplatter.co.uk/	Operated by scientists at Cardiff University and since starting in 2013 has collected greater than 10,000 entries of roadkill (as of October 2014). Volunteers can report ad hoc observations for any UK wildlife vertebrate species	Collection: website via form, Facebook, Twitter (@ProjectSplatter), iOS and android app and email to projectsplatter@gmail.com Feedback: weekly report on most observed species and top observers via social media, https://www.facebook.com/SplatterProject13. Map of WVC, updated maps every 30–90 days.

Name	URL	Description	Collection / Feedback
The Endangered Wildlife Trust (EWT) (South Africa)	http://endangeredwildlifetrust.wordpress.com/category/wildlife-and-transport-programme/	Operated by the South African non-government organisation (NGO), the Endangered Wildlife Trust (EWT) collects roadkill observations via phone, email, smartphone apps and social media tools from volunteer observers or organisations. It allows entries for any species in southern Africa and has greater than 4000 entries as of October 2014 dating back to 2011	Collection: website via form https://www.ewt.org.za/WTP/DataForm.html, iOS and android app and email to roads@ewt.org.za. Feedback: monthly blog, http://endangeredwildlifetrust.wordpress.com; newsletters, https://www.ewt.org.za/WTP/news.html; Twitter feed, @EWtRoads; Facebook groups, https://www.facebook.com/EndangeredWildlifeTrust https://www.facebook.com/groups/roadecology/ A number of other South African wildlife observation reporting platforms include roadkill as a category (e.g. Africa Live, iSpot, Latest Sightings Facebook)
Animal–Vehicle Collisions (Czech Republic)	http://www.srazenazver.cz/en/	Operated by the Czech Republic Transport Research Centre and began with entries from a police database of traffic accidents going back to 2007. It allows entries for 12 species by any organisation or individual and has greater than 7000 entries as of October 2014	Collection: website by registered users Feedback: map of location of WVC
Road Kill Survey (Ireland)	http://www.biology.ie/home.php?m=npws	Operated by Biology that has used web and map-based tools since 2005. Volunteers and organisation can report roadkill from a list of 22 species and species groups (e.g. 'bats'), and it has 2700 entries dating back to 2002	Collection: website by non-registered users Feedback: map of location of WVC, searchable by species and date

2006). The purpose of a system often drives its methods for data collection and determines the types of data collected. This not only makes development of the purpose very important, but potentially can constrain uses of the data for other functions.

There is a clear need to develop a goal or purpose statement for a WROS. This can begin with a fairly broad goal for the system and include a series of objectives that clearly link to types and modes of data collection. For example, one broad purpose statement that reflects the goals of most WROS is: *This system is designed to monitor the occurrences of roadkill in order to improve safety for drivers, reduce impacts to wildlife populations, and contribute to the understanding of regional biodiversity.* The statement has four main objectives, each of which is important to different stakeholders and requires different emphases in data collection, analysis and reporting (Fig. 62.1).

62.2   Extensive social networks are needed for comprehensive observation systems

Broad and inclusive networks are required for a WROS to grow and persist. Also called 'crowd-sourced science', volunteer science networks (sometimes called 'citizen science') consist of managers and scientists from transportation and wildlife agencies, NGOs (Chapter 60), colleges and universities and the general community. Volunteer science provides a large and robust pool of enthusiastic people interested in problem solving and data collection. Furthermore, volunteer science has facilitated analysis of ecological processes operating at broad spatial and temporal scales, far beyond the limit of traditional field studies (Wilson et al. 2013). Some of the largest systems in the world rely primarily on volunteer effort to develop reliable, verified wildlife data (Schmeller et al. 2009; Ryder

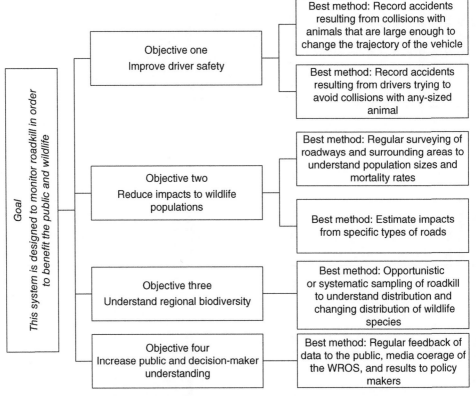

Figure 62.1 Linking goals, objectives and data collection methods to reporting activities in a WROS. The goal and the corresponding objectives and methods in this example are typical of many WROS. The 'best method' is most appropriate for the corresponding objective but may also contribute to other objectives.

et al. 2010; Cooper et al. 2014). These volunteers are often professional biologists making wildlife observations 'on the side' and contributing these observations to various WROS (e.g. CROS, United States). One perception of volunteer science-gathered data is that it may suffer from observer bias and identification error (Cooper et al. 2014). However, this has not often been the case and inaccuracies may be outweighed by the size of data sets available from volunteers (Schmeller et al. 2009; Ryder et al. 2010). As the volunteer science movement becomes an industry, it is anticipated that data collection will become more streamlined and standardised, with the volunteer scientist benefitting from the knowledge that they have helped advance in a scientific field they are passionate about.

Social media platforms (e.g. Facebook, Twitter) have revealed public concern about the rate of animals killed on roads and have become valuable volunteer science tools (e.g. Project Splatter from the U.K., Table 62.1). In addition to the actual collection of WVC data, social media can raise concern and awareness over WVC and their impacts on biodiversity, thereby encouraging more individuals across broader geographic areas to collect WVC data.

When it comes to submitting WVC data, public preferences will influence the researcher's choice of platform, of which there are many (Table 62.2). Ideally, photographs of the animal(s) should accompany the submission of an observation, with the location (preferably the GPS coordinates) and date/time of the observation. Photos do assist the WROS with species identification and allow the scheme to quantify the accuracy of submissions. While many new technologies are available for data collection, we recognise that some data collection may still rely on analogue devices such as paper and pen.

To allow for maximum public participation, we recommend a combination of platforms (e.g. smartphone application, social media and email) be adopted for collecting the data. The most robust data for examining long-term abundance trends and identifying hotspots are those that record observations of both the presence and absence of roadkill on set transects (Chapter 13). As such, contributors should be encouraged to submit null observations on defined journeys. This approach has been adopted at certain times of the year, for example, by the Belgian 'Animals under the Wheels' programme (Table 62.1) who run a 'report your commute' campaign to gain high-quality standardised survey data but also to re-inject enthusiasm into the volunteer base.

62.3 Adopt a methodical approach to developing a WROS

Developing a successful WROS depends on a wide range of activities and skills. This lesson lists the five critical features of a WROS and can be thought of as a checklist for existing schemes as well as guidelines for new WROS.

Table 62.2 Digital technologies currently available for WROS observation reporting by the public.

Platform	Hardware/technology required	Advantages	Disadvantages
Smartphone application	Smartphone/tablet	Data easily submitted by participant, usually providing an immediate and accurate data point	Not all cell phones and tablets support applications, and these applications must be frequently updated
Instant messaging/SMS message	Cell phone/tablet	Relatively accessible to all in possession of a cell phone, or tablet report usually immediate	Not all cell phones and tablets support photographic submissions or GPS locations
Social media (e.g. Facebook/Twitter)	Web-based (phone, tablet or PC)	Data easily submitted by participant	Does not always provide a GPS point and not always immediate
Email	Web-based (phone, tablet or PC)	Data easily submitted by participant	Does not always provide a GPS point and not always immediate

Communicate complex ideas simply and completely to everyone from politicians to scientists. If the purpose of the WROS is to understand and resolve the impacts of roads and traffic on wildlife to meet safety and conservation goals, then effective messaging to many types of audiences through social and traditional media is important to support the WROS itself as well as the conclusions generated from the data.

Continuous inclusion of broad participant types. People with many skills and education levels and types are required to make a WROS succeed. Social networks of participants (Lesson 62.2) are necessary to provide a stream of data. Web programmers and app designers are essential to design efficient data collection tools and to update them regularly. Transportation and wildlife agency staff should provide important feedback on what kinds of data and analysis are needed for decision support. Statisticians and GIS experts are necessary to ensure the records collected will be sufficient for rigorous data analysis.

Understand and implement principles of scientific data collection. Expectations are growing for WROS to include rigorous methods for data collection to test hypotheses, discover previously unknown relationships and increase understanding of the impacts of transportation systems. Design the system to encourage the collection of high-quality data, allow for verification of data quality and include essential information regarding sampling effort and observer skill, such that a scientific user is confident that the data and subsequent analysis is robust.

Use web systems, smartphones and social media to improve data collection. The metadata that can automatically accompany every roadkill observation in a web database means that many tools can be used to enter or retrieve the data (Olson et al. 2014). For example, it is theoretically possible to use the metadata attached to an image file sent from a smartphone to automatically create a roadkill record associated with a known user, geolocation and time stamp and potentially other information (such as observation method). What would remain is for an expert user to examine the photograph and update the record to include the animal's identity. Social media tools, such as Twitter or texting, could be used to collect such observations (Tables 62.1 and 62.2).

Data collection is a critical input of WROS. Highway maintenance staff cleaning up WVC carcasses cannot be expected to have the same diligence for taxonomy as a dedicated professional biologist volunteering their time observing WVC. A trade-off exists in the data gathered using different schemes and each poses challenges in terms of data analysis (see Lesson 62.5; and see Bird et al. 2014 for review).

62.4 Analysis and visualisation of data collected within a WROS should correspond to the goals of the system

The data collected in WROS are fundamentally spatial (i.e. the location of animals along a road), and spatial statistics (Chapter 13) are well suited to analyse and interpret the data. For example, spatial statistics are clearly relevant to (i) mapping the distribution and abundance of species (George et al. 2011) and impacts to species; (ii) identifying landscape or other factors, such as vehicle speed or time of day that are related to WVC (Langen et al. 2009); and (iii) statistically determining if WVC are clustered in space and/or time, otherwise known as identifying 'hotspots' (e.g. Barthelmess 2014; Chapter 13).

There are many tools to measure impacts to species from WVC, to determine causes and correlations with WVC and for finding places where transportation agencies can focus remedial action to reduce impacts to wildlife and improve driver safety. Analyses to identify non-random clusters of WVC's (hotspots) have utilised Geographic Information Systems (GIS), a promising tool where statistics have been used to identify spatial clusters (Chapter 13). Examples of analytical approaches and methods include the Nearest Neighbor Index (e.g. Matos et al. 2012); 'SaTScan', borrowed from epidemiological studies, which looks for non-random clusters of events (i.e. disease outbreaks); the Getis-Ord Gi statistic for spatial autocorrelation; and the Kernel density estimator plus method for estimating locations of high densities of events.

Maps can be both informative and evocative and thus useful in public relations, in scientific reporting and in supporting decision-making. Maps should be produced regularly, and a GIS is an efficient tool to generate and visually display the data (Fig. 62.2). Maps should be displayed on the WROS website, as well as via other mediums, such as scientific reports and social media. These maps typically display the locations and rates of WVC for specific species or groups of species thereby addressing many of the primary motivators for setting up the WROS (Textbox 62.2).

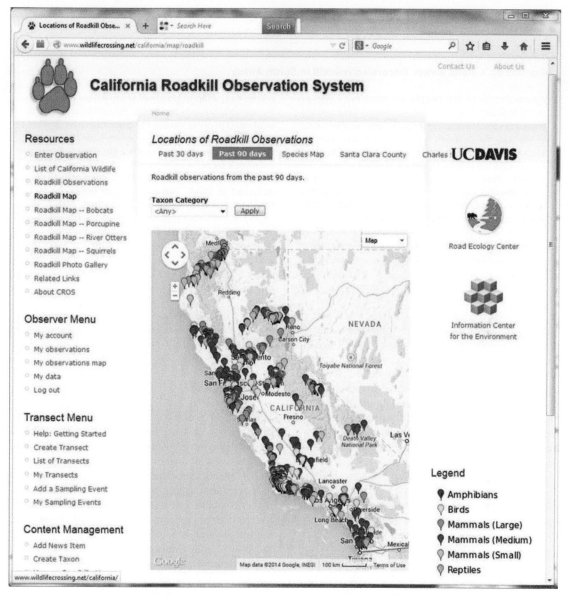

Figure 62.2 Website graphic of the California Roadkill Observation System (CROS, http://wildlifecrossing.net/california) demonstrating visual display of roadkill data submissions. Source: http://www.wildlifecrossing.net/california/map/roadkill.

62.5 Address issues in reporter bias by using standardised data collection methods or post hoc analyses

Considerable investment in both time and money is often needed to initially set up and maintain a WVC data collection system, although free/open-source tools such as *EpiCollect* (an app) and *cartoDB* (mapping) are available, which will significantly decrease the upfront cost of systems. Recruiting and retaining volunteers in a WROS can take considerable time investment. The trade-off in this time allocation, however, is small when considering that volunteers have been shown to collect high-quality, usable data (Schmeller et al. 2009) and can provide extensive geographical coverage of the type that would typically be prohibitively expensive if carried out without volunteers. As such, once established, the system has the potential to be cost- and time-effective as big data are obtained.

A standardised and systematic approach to data collection that is user-friendly (and potentially incorporating a number of platforms) is the ideal, with clear

---

Textbox 62.2 Road watch: Recording roadkill in South Africa

Approximately 14,000 people are killed each year on roads in South Africa, of which collisions between animals and vehicles account for many injuries and human mortalities, as well as extensive vehicle damage. Insurance claims in South Africa suggest that approximately US$150 million is paid each year to drivers due to WVC, though the biodiversity costs of these collisions are never calculated. To address the threat to biodiversity from WVC, the Endangered Wildlife Trust (EWT) launched the Wildlife and Roads Project in 2012.

In the beginning of 2014, a smartphone app, *Road Watch*, was launched to enable the public to submit roadkill sightings. This allowed the project to develop a sensitivity map of potential areas where wildlife are threatened by roads and traffic. *Road Watch* has been publicised in press releases and social media, and the Roadkill Research LinkedIn site has over 1000 members from around the globe, the Road Ecology Facebook page has almost 800 regularly contributing members, and an EWT Roads Twitter account was activated in 2014. The project also linked up with other reporting systems such as iSpot and Africa Live. The volunteer database has almost doubled in size from 2013 to 2014, with over 150 volunteers collecting roadkill data, and over 4000 data points for the country.

---

designs of how the data will be analysed and reported. This will assist with data collected from diverse sources that may be biased by taxonomy or location. Once the data have been obtained, quality control and assurance steps (e.g. use of photo verification) are needed to reassure users of data quality. While photographs submitted in addition to data will help eliminate identification error, one has to consider the safety of the people reporting the roadkill and other road users. It is therefore important that all projects provide safety information and issue a liability disclaimer.

62.6   The advantages and disadvantages of opportunistic and targeted data collection must be carefully considered when developing a WROS

Opportunistic observations of WVC provide 'presence-only' data, which identifies locations where WVC occur, but not locations where they do not. Opportunistic data should be treated cautiously and either used in 'presence-only' statistical analyses or as a tool to warrant further in-depth data collection. In contrast, targeted data collection on set transects can provide records of where wildlife are not getting killed (e.g. they are safely crossing or do not cross, or roadside fencing or other mitigations are effective), thereby allowing more robust identification of hotspots and the factors influencing them. However, targeted data collection will often be more costly and time consuming than opportunistic data collection, and we therefore recommend both data types be collected. The WROS system developed by the Road Ecology Center for Maine, United States

(http://wildlifecrossing.net/maine), includes both targeted and opportunistic observations by volunteers and allows the reporting of 'no-animal' observations.

CONCLUSION

The flood of options available for reporting wildlife sightings is a growing field, and it is easy to become bogged down by the availability of so many possibilities and examples of implementation. It is therefore important to ensure that clear goals, objectives and desired outcomes are in place before implementation occurs. A WROS should start with a targeted understanding of the methods or components required – that is, what is the 'supply and demand', in short, who will be using the WROS, and what will the data be used for? The combination of goals, objectives and methods should provide the framework for an implementable system that satisfies the users and participants.

ACKNOWLEDGEMENTS

F. Shilling thanks David Waetjen (UC Davis) and Barbara Charry (Maine Audubon) for their assistance of the http://wildlifecrossing.net websites. W. Collinson thanks Harriet Davies-Mostert, Claire Patterson-Abrolat and Shelley Lizzio of the Endangered Wildlife Trust and Bridgestone SA for funding. SE Perkins thanks Rita Harris and Darren Westcott of Project Splatter for their assistance with http://www.projectsplatter.co.uk and the Universities Federation for Animal Welfare for funding.

FURTHER READING

Bissonette et al. (2008): Estimated the cost to the public and drivers from deer–vehicle collisions on state highways, which in combination with WVC occurrence data has been very useful in proposing driver safety projects to reduce WVC.

Olson et al. (2014): Outlined methods for collecting WVC data using smartphone technology. Code supplied as an appendix.

Paul (2007): Determined that for a highway in Canada there was no statistically-significant difference between hotspots identified using volunteer-collected data or data collected by professionals.

Schmeller et al. (2009): Surveyed hundreds of volunteer science programmes across Europe and found that the data collected by the thousands of volunteers in these programmes represented millions of Euros of effort and resulted in large, reliable data sets.

REFERENCES

Barthelmess, E. L. 2014. Spatial distribution of road-kills and factors influencing road mortality for mammals in Northern New York State. Biodiversity and Conservation **23**: 2491–2514.

Bird, T. J., A. E. Bates, J. S. Lefcheck, N. A. Hill, R. J. Thomson, G. J. Edgar, R. D. Stuart-Smith, S. Wotherspoon, M. Krkosek, J. F. Stuart-Smith, G. T. Pecl, N. Barrett, and S. Frusher. 2014. Statistical solutions for error and bias in global citizen science datasets. Biological Conservation **173**: 144–154.

Bissonette, J. A., C. A. Kassar, and J. C. Lawrence. 2008. Assessment of costs associated with deer-vehicle collisions: human death and injury, vehicle damage, and deer loss. Human-Wildlife Conflicts **2**: 17–27.

Chadwick, E. A., V. R. Simpson, A. Nicholls, and F. M. Slater. 2011. Lead levels in Eurasian otters decline with time and reveal interactions between sources, prevailing weather, and stream chemistry. Environmental Science and Technology **45**: 1911–1916.

Chadwick, E. A., J. Cable, A. Chinchen, J. Francis, E. Guy, E. F. Kean, S. C. Paul, S. E. Perkins, E. Sherrard-Smith, C. Wilkinson, and D. W. Forman. 2013. Seroprevalence of *Toxoplasma gondii* in the Eurasian otter (*Lutra lutra*) in England and Wales. Parasites and Vectors **6**: 75.

Cooper, C. B., J. Shirk, and B. Zuckerberg. 2014. The invisible prevalence of citizen science in global research: migratory birds and climate change. PloS one **9**: e106508.

Elmeros, M., M. Hammershoj, A. Madsen, and B. Sogaard. 2006. Recovery of the otter *Lutra lutra* in Denmark monitored by field surveys and collection of carcasses. Hystrix: Italian Journal of Mammalogy **17**: 17–281.

George, L., J. L. Macpherson, Z. Balmforth, and P. W. Bright. 2011. Using the dead to monitor the living: can road kill counts detect trends in mammal abundance? Applied Ecology and Environmental Research **9**: 27–41.

Gunson, K. E., M. Giorgos, and L. J. Quackenbush. 2011. Spatial wildlife-vehicle collision models: a review of current work and its application to transportation mitigation projects. Journal of Environmental Management **92**: 1074–1082.

Hobbs, G. I., E. A. Chadwick, M. W. Bruford, and F. M. Slater. 2011. Bayesian clustering techniques and progressive partitioning to identify population structuring within a recovering otter population in the UK. Journal of Applied Ecology **48**: 1206–1217.

Langen, T. A., K. M. Ogden, and L. L. Schwarting. 2009. Predicting hot spots of herpetofauna road mortality along highway networks. Journal of Wildlife Management **73**: 104–114.

Matos, C., N. Sillero, and E. Argaña. 2012. Spatial analysis of amphibian road mortality levels in northern Portugal country roads. Amphibia-Reptilia **33**: 469–483.

Olson, D. D., J. A. Bissonette, P. C. Cramer, A. D. Green, S. T. Davis, P. J. Jackson, and D. C. Coster. 2014. Monitoring wildlife-vehicle collisions in the information age: how smartphones can improve data collection. PLoS One **9**: e98613.

Paul, K.J.S. (2007). Auditing a monitoring program: can citizen science document wildlife activity along highways. Thesis, University of Montana, Missoula. p. 67.

Ryder, T. B., R. Reitsma, B. Evans, and P. P. Marra. 2010. Quantifying avian nest survival along an urbanization gradient using citizen- and scientist-generated data. Ecological Applications **20**: 419–426.

Schmeller, D. S., P. Y. Henry, R. Julliard, B. Gruber, J. Clobert, F. Dziock, S. Lengyel, P. Nowicki, E. Deri, E. Budrys, T. Kull, K. Tali, B. Bauch, J. Settele, C. Van Swaay, A. Kobler, V. Babij, E. Papastergiadou, and K. Henle. 2009. Advantages of volunteer-based biodiversity monitoring in Europe. Conservation Biology **23**: 307–316.

Sherrard-Smith, E. and E. A. Chadwick. 2010. Age structure of the otter (*Lutra lutra*) population in England and Wales, and problems with cementum ageing. IUCN Otter Specialist Bulletin **27**: 42–49.

Wilson, S., E. M. Anderson, A. S. Wilson, D. F. Bertram, and P. Arcese. 2013. Citizen science reveals an extensive shift in the winter distribution of migratory western grebes. PloS One **8**: e65408.

GLOSSARY

Acoustic interference Occurs when background noise reduces the distance over which a sound can be detected

Adaptive genetic variability Genetic variation that contributes to the encoding of a property of an organism, and is thus potentially adaptive and under natural selection

Aestivation A period of dormancy during a hot or dry period, physiologically similar to the warm season version of hibernation when an animal shelters from harsh conditions

Amplitude The volume or loudness of a sound, which can be measured in pressure or intensity, both of which are expressed in decibels (dB). A-weighted decibels, abbreviated to dB(A) or dBA, describe the relative loudness of sounds in air as perceived by humans

Arboreal Meaning tree; related to animals that depend on trees to provide all or part of their life requirements

Arboreal crossing structures Wildlife crossing structures for arboreal species (e.g. glider pole, rope ladder)

Assignment test Statistical approach to ascribing individuals to their most probable natal populations on the basis of multiple DNA markers

At-grade crossing A section of road where animals are allowed or encouraged to cross the road by crossing the road surface directly

BACI Before-After-Control-Impact study design, in which data are gathered before and after roads or road-mitigation measures are constructed. Data from areas with roads or mitigation measures (impact) are compared with data obtained from areas without roads or mitigation measures (control)

Badger pipe A pipe specifically designed to allow badgers to pass beneath roads, often with funnel fencing

Barotrauma Internal injuries to animals caused by high-intensity, impulsive underwater sound

Barrier Any structure that restricts or prevents the movement of flora or fauna

Barrier effect The extent to which roads or other linear features prevent, or filter animal movement. The barrier effect can be quantified by species, populations and so on

Barrier fencing Wildlife fencing that forms a barrier to the movement of wildlife

Biodiversity The variety of life at any given spatial scale including all the levels contained within, including genes, species, communities and ecosystems and their complex interactions

Canopy bridge (also called rope bridge) An aerial crossing structure (often made of rope, steel or wood) providing a link between tree canopies over roads, allowing arboreal wildlife to cross safely

---

Handbook of Road Ecology, First Edition. Edited by Rodney van der Ree, Daniel J. Smith and Clara Grilo.
© 2015 John Wiley & Sons, Ltd. Published 2015 by John Wiley & Sons, Ltd.
Companion website: www.wiley.com\go\vanderree\roadecology

Canopy connectivity A method of maintaining habitat connectivity across a road for arboreal animals, where tall trees are present close to roadsides and/or in the centre median and where the adjacent canopies are connected or nearly so

Capture–mark–recapture Method used in ecology to estimate population size, where animals are initially captured, marked in some way (e.g. tattoo, leg band) and then resighted or recaptured at a later date. The ratio of marked versus unmarked individuals is used to estimate size of the population

Centre median The strip of land separating the lanes of a divided road. Often vegetated with grass, shrubs and/or trees

Chytrid fungus A globally distributed water-borne fungal pathogen that can injure and kill amphibians. Evidence suggests it can be transferred between species and it is partially responsible for the widespread decline and disappearance of several amphibian species, particularly anurans

Circadian rhythm Light-driven functions within animals and plants that affect growth, development, disease resistance, diurnal and seasonal functions, and reproduction, as well as animal foraging behaviour and avoidance of predation

Clear zone A strip of land outside the travel lanes of a road that improves driver visibility and serves as an area for drivers to recover control of vehicles. The clear zone is absent of large obstacles (e.g. trees), and may be paved, have a gravel surface and/or mowed grass

Coalescent analysis Describes mathematically the properties of samples of genes from their mutational processes and genealogical relationships

Co-evolution Interactions between species that impact how both evolve (e.g. bees and pollination of flowering plants)

Connectivity A measure of connectivity between locations, based on actual movement of individuals or genes

Connectivity modelling Analyses to identify how environmental features affect genetic or demographic connectivity

Conspecific Of or belonging to the same species

Contagious development Human developments following the construction and upgrading of roads. When roads provide access to previously remote areas, they open the area up for more roads and developments,

thus triggering land-use changes, resource extraction and human disturbance

Coordinated distributed experiment Experiments that are conducted collaboratively in different locations, using standardised and controlled protocols

Corridors Linear strips of vegetation or habitat that differ from the adjacent areas that allow for the movement of individuals or genes between discrete habitat patches. Corridors may also refer to linear landscape elements, such as roads, railways, utility easements, that may faciliate or impede movement across the landscape.

Cumulative effects The increasing impacts resulting from the combination of effects from several projects or activities over a period of time. Their assessment is called cumulative effect assessment (CEA)

Cut and fill balance The balance between how much earth is extracted and how much fill is required during road construction. Ideally, the cut and fill balance is zero – where any earth removed during construction is used elsewhere on the same project

Decibel (dB) The unit of measurement for the amplitude of a sound, expressed on a logarithmic scale

Demogenetic modelling Population modelling that recognizes the interdependence of genetic and demographic factors and integrates them

Demographic Adjective describing the operation and outcomes of four processes in populations – births, deaths, immigration and emigration

Detailed design stage The stage of a road project where detailed plans are drawn up showing the exact locations and dimensions of every feature of the road and having detailed calculations of excavations, fill and all other materials needed

Direct effects The effects of a project or action that are a direct and immediate consequence of the project or action, without any intervening steps (compare to indirect effects). Habitat loss due to clearing for road construction is a direct effect

Dispersal Ecological process that involves the movement of an individual or multiple individuals away from the population in which they were born to another location, or population, where they will settle and reproduce

DNA Deoxyribonucleic acid, the substance comprising the genetic material passed from parents to offspring. DNA has major influences on how individuals develop and differ from each other

DOT   Abbreviation for Department of Transportation

Earth berm   A constructed mound of earth, usually along the road, to provide a visual screen or absorb sound

Echo-location   The ability to perceive surroundings through the echoes of self-generated high frequency sounds

Ecosystem   The community of living organisms (plants, animals, microbes) and nonliving components of their environment (e.g. air, water, soil) interacting as a system

Ecosystem functioning   Involves the ecological and evolutionary processes, including gene flow, disturbance, pollination and nutrient cycling

Ecosystem services   Benefits such as goods and services, provided to society by ecosystems, for example production of food and water, control of climate and disease, nutrient cycles and crop pollination, and spiritual/recreational benefits

Edge effect   Changes in population or community structures that occur at the boundary of two landscape elements or types

E-DNA   Environmental DNA: DNA purified from samples such as soil and other ambient materials, typically containing complex mixtures derived from many organisms

Effective mesh size (m_{eff}), effective mesh density (s_{eff})   Metrics for quantifying the degree of landscape fragmentation, based on the probability that two randomly located points (or animals) in an area are connected and are not separated by a barrier (e.g. roads, urban area). The smaller the effective mesh size, the more fragmented the landscape. The effective mesh density gives the effective number of meshes per square kilometre, that is the density of the meshes. The effective mesh density value rises when fragmentation increases. The two measures contain the same information, but the effective mesh density is more suitable for quantifying trends. See Textbox 5.2

Effective population size   The number of interbreeding adults in a population (smaller than the total population because it excludes juveniles, non-reproductive and post-reproductive individuals)

Emission spectra   The distribution of wavelengths emitted by a light source

Environmental Impact Assessment (EIA)   A procedure for assessing the impacts on the environment likely to result from proposed projects or activities

Equator principles (EPs)   Equator principles (EPs) is a credit risk management framework for determining, assessing and managing environmental and social risk in project finance transactions. The EPs are primarily intended to provide a minimum standard for due diligence to support responsible risk decision-making

Equator principles financial institutions (EPFIs)   Equator principles financial institutions (EPFIs) commit to only providing loans to projects where the borrower will comply with their respective social and environmental policies and procedures. http://equator-principles.com/index.php/about-ep/about-ep (accessed 2 February 2015)

Experiment – manipulative   An experiment where the researcher has control over the study design and the manipulation (location, timing, extent, etc.), such that the location, number, intensity, and/or timing of treatments are included in the design of the experiment. It is also called 'true experiment'

Experiment – pseudo/natural/non-manipulative/observational   An observational study that is almost experimental, except that the researcher is not in control of the factors that influence the parameter of interest, that is the manipulation. For example, a study of the effects of road noise on wildlife is observational if the researcher measures road noise at a number of roads and relates it to the species of interest. By contrast, a manipulative experiment is where the noise levels are manipulated under the control of the researcher and the response of the species of interest is recorded

Experiment – scientific   The use of manipulation and testing under controlled conditions to understand the causal relationship between two or more variables while controlling for potential confounding factors. Essential components of experiments are the application of a treatment, randomization of the assignment of treatment and control sites and replication to account for uncontrollable variability among the treatment and control sites

External screening   Screening (e.g. walls, plantings, earth berms) along the sides of landbridges and at approach areas to wildlife underpasses to reduce the amount of noise, light and visual disturbance from traffic that can reduce or deter animal use of the structure

Extinction debt   Used to denote the number of existing populations that will go extinct in the very near future because of the changes that have already occurred

Fecundity Potential reproductive capacity of an individual or population

Fencing – fauna Fencing specifically to prevent animals from accessing the road and/or to funnel animals towards wildlife crossing structures

Fencing – game Fencing designed to contain game species (e.g. kudu, zebra, impala), common in southern Africa. Game fences are typically tall and often electrified

Fencing – stock Fencing designed to contain stock (e.g. sheep, cattle). Typically 1.2 m in height, and may be electrified

Flight diversion/diverter Mitigation measure intended to alter the flight path of birds and bats, such as poles, fences and walls which encourage higher flight

Frangible Breakable; often used to describe roadside objects/vegetation which are intended to break upon impact, rather than remain intact and damage a vehicle

Freeway, expressway, motorway, tollway, autobahn Major roads with more than two lanes in each direction with high traffic speeds, high traffic volume and very limited/restricted access points

Frequency The frequency or pitch of a sound is the number of pressure cycles it completes per second, measured in Hertz (Hz). Humans can hear across a frequency range of 20–20,000 Hz

F_{ST} A classic measure of population genetic differentiation based on differences in frequencies of genetic polymorphisms. It varies in theory between 0 (no differentiation) and 1 (completely different), although absolute values depend on details of the genetic assay.

Furniture, fauna furniture Logs, branches, rocks and other enrichment structures placed in wildlife crossing structures to provide shelter and/or protection from predators

Gabions A wire container or basket filled with rock or other material used in the construction of dams and retaining walls

Gene flow The transfer of alleles or genes from one population to another

Genetic differentiation The level of difference of genetic variation among samples

Genetic diversity The level of variability of genetic data within a sample or population, commonly measured through metrics such as heterozygosity and allelic richness

Genetic markers Variable, heritable characters that can be detected in a set of individual organisms and used for population genetic analysis

Genetic structure The patterns of genetic variation among samples

Genome The genetic 'blueprint' of an individual

Genomics Technical approaches to studying genomes, examining many genes

Genotypes States at genetic markers scored for individual organisms

Genotypic analyses Analyses based on combinations of genotypes across multiple loci, that is genotypic arrays

Genotypic clustering A class of analyses that find the number of genetic groupings in a set of genotypic data and place individuals in one or more of those groups

Glider pole A road-crossing structure for gliding animals where tall poles (usually timber) are placed in roadsides, centre medians and/or on landbridges in lieu of natural canopy connectivity. Glider poles act as stepping stones, providing connectivity for gliding animals across a habitat gap.

GPS telemetry Tracking technology for organisms, involving Global Positioning System technology

Green pick Soft, nutrient rich, palatable grasses/herbs

Habitat The area or environment where an organism or community normally lives or occurs

Hawking Catching flying prey in the air

Hertz (Hz) The unit of measurement for the frequency of a sound

Heterozygosity The proportion of genes in an individual in which the two copies (one inherited from the mother, one from the father) differ

Hibernacula A shelter used by animals to escape extreme seasonal climates (winter or dry season) by undergoing hibernation or torpor

Highway Major road, usually with more than two-lanes in each direction. Has many more access points than freeway/expressway/motorway/tollway/autobahn

Home range The area an individual accesses during its normal daily activities of food gathering, obtaining shelter, mating and caring for offspring

Hop-over A continuous or near-continuous 'bridge' formed by trees overhanging a road

Illuminance A measure of the brightness of a light source – defined as the total luminous flux incident on a surface per unit area, measured in units of lux = lumen per square metre

Indirect effect Impact of one organism, species or land use on another that is mediated or transmitted by a third. Examples include increased rates of predation on prey species, as a result of non-native predators using a road clearing to access new areas; or the avoidance by wildlife of a weed-infested area adjacent to a roadway, where the weed infestation is due to the presence of the road. Compare to direct effects

Inferential strength The ability of an experiment or analysis to adequately / fully answer the question posed, that is the validity of the inference that the hypothesis tested is true or false, given a set of data. Inferential strength relies on the study design, on the extent to which one must extrapolate from the context in which the study was conducted to the context of concern (the particular decision context), on the number of competing hypotheses tested, and on adequate statistical power

Insectivorous Description of the diet of a species, in this case the eating of insects or arthropods such as spiders

Interaction From an ecological perspective, the effects that organisms have on one another. Intraspecific interactions involve individuals of the same species; interspecific are individuals of different species

Internal screening Screening, such as earth berms, fencing or dense plantings installed on multi-use crossing structures to demarcate and/or provide screening between the wildlife zone and human use zone. See also external screening.

Isolation-by-resistance A pattern in which gene flow is restricted by differential ability of organisms to move through different types of habitat more than by distance alone

Land bridge, landscape bridge, ecoduct, green bridge Overpass, mostly wider than 50 m (some designs for specific species or situations approx 20 m wide), with natural vegetation and habitat elements

Landscape fragmentation The physical process where habitats become separated, usually through clearing and dividing of habitat

Landscape genetics A discipline combining landscape ecology and population genetics, used here as a catch-all for the broad application of genetics in ecological management. Landscape scale planning - Planning in a wider area than the local scale or infrastructure

Light intensity Number of photons per wavelength striking photoreceptors per second, measured as μmol m$^{-2}\cdot$s^{-1}

Low-formation road A road that is level with the surrounding ground, lacking significant fill and raised embankments

Lumen Luminous flux or power from a light source, measured as candela steradian

Mantel tests A classic non-parametric statistical approach for estimating associations among two or more matrixes. Mantel tests account for non-independence of pairwise data by permutations

Mass migration Seasonal and round-trip movement in aggregations that include hundreds to thousands of animals

Mesopredator Predator in the mid-trophic levels, such as opossums, raccoons and skunks

Meta-analysis The quantitative evaluation using a specific set of statistical methods to compare and synthesise the results of multiple studies

Metapopulation Set of local populations of a species within some larger area, where genetic diversity is maintained by the dispersal of individuals from one local population to another

Microchip An identifying integrated circuit placed under the skin that can be read with a scanner

Microhabitat Small-scale differences in habitat

Microsatellite Class of highly resolving DNA locus often applied in molecular population biology

Migration Journey undertaken by some species in response to changing seasons or climatic events, such as rainfall

Mitigation Methods used to eliminate or minimise the negative impacts of developments

Mitochondrial DNA (mtDNA)   DNA within the mitochondria found within cells, typically inherited through the maternal line in animals

Mitochondrial DNA sequencing markers   Genetic markers based on obtaining mtDNA sequences

Molecular ecology   A field of evolutionary biology that uses molecular population genetics, phylogenetics and genomics to address ecological questions.

Monitoring   A form of research where repeated measurements or observations are taken over time, usually to assess the change in a parameter over time or in response to a disturbance/intervention

Non-frangible   Non-breakable; often used to describe roadside objects/vegetation which do not break upon impact, and may damage a vehicle

Non-invasive sampling   Obtaining observations or genetic samples without the need for capturing individuals

Overpass   A road or wildlife crossing structure that facilitates movements over/above the road

Parentage analysis   Attribution of offspring to parents based on genotypic data

Paved road   Road surface made with asphalt, bitumen, concrete or tarmac; also called 'metalled' in Britain, or 'sealed' in Australia and New Zealand.

Pearson correlation coefficient   A measure of the degree to which two variables are correlated. Pearson correlation is the linear relationship between two variables, ranging from -1.00 (perfect negative correlation), through 0 (no linear relationship) to $+1.00$ (perfect positive correlation).

Photoperiod   Duration of daily exposure to light

Photoreceptor   Specialized cells (found in the retinae of an animal's eye) sensitive to light

Polymerase chain reaction (PCR)   A molecular biology reaction that produces many (millions) of copies of defined regions of DNA

Polymorphism   Polymorphic locus which has more than one allele

Precautionary principle   A principle to guide decision-making in the absence of scientific certainty which states that precautionary measures should be taken when an activity may harm human health or the environment and that the proponent for an activity must prove that the action will not cause harm

Resilience   The ability of an ecosystem to respond to a perturbation or disturbance by resisting damage and recovering quickly

Right of way   The entire width of the reserved strip of land on which the linear infrastructure is built.

Rip rap   Medium- to large-sized rocks and boulders, usually placed at the foot of embankments or bridge abutments and piers to prevent scouring and erosion during floods

Road avoidance   Avoidance by wildlife of the road due to a lack of cover and/or to the character of the road, roadside and pavement which is different from natural habitat

Road ecology   Science that seeks an understanding of the interactions between roads/railways/utility easements etc and the natural environment, including wildlife, natural resources, land use and climate change

Roadkill   Animals that have died as a result of collisions with vehicles on roads

Roadkill – hot moment   Periods of time with more wildlife–vehicle collisions (WVCs) than expected by chance, or an aggregation of road-kills in time

Roadkill – hotspot   Road stretches with more WVCs than expected by chance, that is, aggregation of road-kills in some road stretches

Roadless area   An area of land that is not dissected by a road, and are thus relatively undisturbed by humans

Road profile (at grade/below grade/above grade)   Position of the road or linear infrastructure relative to the surrounding land: roads above grade may be elevated by fill/soil, or built on bridges or culverts and roads below grade are built within a cut

Road reclamation   The physical treatment of a roadbed to restore the form and integrity of associated hillslopes, channels and flood plains and their related hydrologic, geomorphic, and ecological processes and properties. Also known as road removal or road decommissioning

Road shoulder   The section of the road between the traffic zone and the road edge which is generally kept clear of vehicular traffic. It is often used as an emergency or break down lane/area

Roadside   Area adjacent to travel lanes, generally includes the road shoulder and the road verge

Road-traffic noise   The noise generated by vehicles travelling on a road

Road verge   The vegetated area adjacent to roads; generally located outside the road shoulder

Secondary roadkill   Predators that have become road-kill themselves as a result of their presence at road-kill

Sedentary   Tends to stay in one place, not moving long distances

Spatial autocorrelation   A statistical technique that relates some kind of similarity in pairs of individuals (or samples) to the geographic distance between them

Spectroradiometer   Instrument designed to measure the spectral distribution, power and intensity of light from a source, reflected, transmitted or absorbed

Speed bump   Raised band or strip of paving material placed across the road to reduce vehicle speed

Statistical power   The ability of a statistical test or analysis to detect a relevant effect, should one exist

Strategic environmental assessment (SEA)   An assessment of a proposed plan, policy or program that incorporates social, economic and environmental sustainability. SEA exists above the single project and are of particular interest for assessing the landscape-scale effects of roads and road networks

Study class   With respect to research and monitoring, study class relates to whether the study is manipulative or non-manipulative. In manipulative studies, the researcher has control over the variable of interest and the response to manipulating it is measured. Non-manipulative studies (sometimes called observational or natural experiments) occur when the researcher takes advantage of changes that have happened (by using existing data) or are about to happen (by taking measurements, i.e. making observations) to understand its effect

Study type   With respect to research and monitoring, study type describes whether measurements are taken before, during and after manipulation or treatment and if it includes impact and/or control sites. Studies with the greatest inferential strength are BACI – measurements taken before and after at control and impact sites

Success rate   The number of individual animal movements through a crossing structure as a proportion of the number of approaches

Surface tunnel   Wildlife underpass designed to minimise tunnel length with a slotted top that forms a part of the wearing course of the road surface

Target species   The species or group of species for which the mitigation is intended

Taxa (singular - Taxon)   A classification or grouping of animals such as a class (e.g. mammal) or life-history-related characteristic (e.g. ground-dwelling birds)

Tracking surveys   Following the movement path of an animal by reading and recording locations of animal footprints

Thermoregulation   The ability of an organism to keep its body temperature within certain limits, even when the surrounding temperature varies

Traffic disturbance avoidance   Avoidance of roads from a distance due to traffic disturbance (e.g. lights, noise, chemical emissions)

Underpass   A road or wildlife crossing structure that facilitates the movements of things under the road

Unpaved road   A road without a bitumen, asphalt, concrete or cobble-stone surface

Vegetated median   Where the centre median of a divided road contains vegetation, in some cases suitable for use by some species of wildlife. Tall trees in the median can provide canopy connectivity across a divided road, potentially assisting the movement of birds, bats and arboreal animals

Water-shedding   Process where roads concentrate water at their edges, as water runs off the road surface from the centre of roads

Wildlife crossing structure   Any structure designed and purpose built to facilitate the safe movement of wildlife across roads

WVC   Wildlife-vehicle collisions - when wildlife are hit by moving vehicles.

SPECIES

African civet	*Civettictis civetta*	Black wattle	*Acacia mearnsii*
African grass owl	*Tyto capensis*	Black wildebeest	*Connochaetes gnou*
African rock python	*Python sebae*	Black-backed jackal	*Canis mesomelas*
African wild dog	*Lycaon pictus*	Blackcap	*Sylvia atricapilla*
American bison	*Bison bison*	Black-necked crane	*Grus nigricollis*
American black bear	*Ursus americanus*	Black-tailed gazelle	*Gazella subgutturosa*
Amphibians	*Amphibia*	Blackthroated blue warbler	*Dendroica caerulescens*
Amur tiger	*Panthera tigris altaica*	Blesbok	*Damaliscus dorcas*
Antelope	*Bovidae*	Blue morning glory	*Ipomoea indica*
Asiatic leopard	*Panthera pardus fusca*	Blue wildebeest	*Connochaetes taurinus*
Asiatic wild ass	*Equus hemionus*	Bluespotted salamander	*Ambystoma laterale*
Atlantic salmon	*Salmo salar*	Bobcat	*Lynx rufus*
Axis deer	*Axis axis*	Boneseed	*Chrysanthemoides*
Bactrian camel wild	*Camelus ferus*		*monilifera*
Badger-European	*Meles meles*	Bramble	*Rubus* spp.
Bank vole	*Clethrionomys glareolus*	Brazilian free-tailed bat	*Tadarida brasiliensis*
Barking deer	*Muntiacus muntjak*	Bream	*Abramis brama*
Barn owl	*Tyto alba*	Brown bear	*Ursus arctos*
Barn swallow	*Hirundo rustica*	Brown hare	*Lepus europaeus*
Bats	*Chiroptera*	Brown hyaena	*Parahyaena brunnea*
Bechstein's bat	*Myotis bechsteinii*	Brown pelican	*Pelecanus occidentalis*
Beecroft's scalytailed	*Anomalurus beecrofti*	Brown treecreeper	*Climacteris picumnus*
squirrel		Bullock's oriole	*Icterus bullockii*
Bighorn sheep	*Ovis canadensis*	Bushveld gerbil	*Tatera leucogaster*
Birds	*Aves*	Cabrera's vole	*Microtus cabrerae*
Black and white colobus	*Colobus angolensis*	California vole	*Microtus californicus*
Black kite	*Milvus migrans*	Cane toad	*Rhinella marina* or *Bufo*
Black locust	*Robinia pseudoacacia*		*marinus*
Black rat snake	*Pantherophis alleghaniensis*	Capybara	*Hydrochoerus hydrochaeris*
Black rhino	*Diceros bicornis*	Caribou	*Rangifer tarandus*

Carp-Common	*Cyprinus carpio*	Fallow deer	*Dama dama*
Cheetah	*Acinonyx jubatus*	House mouse	*Mus musculus*
Chinook salmon	*Oncorhynchus tshawytscha*	Finless porpoise	*Neophocaena phocaenoides*
Chital	*Axis axis*	Firethorn	*Pyracantha* spp.
Common brushtail possum	*Trichosurus vulpecula*	Fisher	*Martes pennanti*
Common chimpanzee	*Pan troglodytes*	Florida key deer	*Odocoileus virginianus clavium*
Common European adder	*Vipera berus*		
Common impala	*Aepcyros melampus*	Florida panther	*Puma concolor coryi*
Common lizard	*Zootoca vivipara*	Forest elephant	*Loxodonta cyclotis*
Common musk turtle	*Sternotherus odoratus*	Fountain grass	*Pennisetum Cenchrus setaceum*
Common pheasant	*Phasianus colchicus*		
Common raven	*Corvus corax*	Gambel's quail	*Callipepla gambelii*
Common redshank	*Tringa tetanus*	Gaur or Indian bison	*Bos gaurus*
Common snapping turtle	*Chelydra serpentina*	Gazelles	*Nangera granti* and *Gazella thomsoni*
Common spadefoot toad	*Pelobates fuscus*		
Common toad	*Bufo bufo*	Genet	*Genetta genetta*
Coyote	*Canis latrans*	Giant panda	*Ailuropoda melanoleuca*
Crustaceans	Form a large group of invertebrates which includes, for example crabs, crayfish, shrimp, copepods or barnacles, but also the terrestrial woodlice	Giant squirrel	*Ratufa indica*
		Gopher tortoise	*Gopherus polyphemus*
		Grain wart ground beetle	*Carabus cancellatus*
		Grass snake	*Natrix natrix*
		Gray/Grey wolf	*Canis lupus*
		Great crested newt	*Triturus cristatus*
		Great tit	*Parus major*
Desert bighorn sheep	*Ovis canadensis nelsonii*	Greater kudu	*Tragelaphus strepsiceros*
Desert kangaroo rat	*Dipodomys deserti*	Greater mouse-eared bat	*Myotis myotis*
Desert tortoise	*Gopherus agassizii*	Greater sage-grouse	*Centrocercus urophasianus*
Diamondback terrapin	*Malaclemys terrapin*	Green and golden bell frog	*Litoria aurea*
Domestic cat	*Felis catus*	Green bush-cricket	*Tettigonia viridissima*
Domestic dog	*Canis familiaris*	Green snake	*Phylodryas aestiva*
Dwarf crocodile	*Osteolaemus tetraspis*	Grey shrikethrush	*Colluricincla harmonica*
Eastern box turtle	*Terrapene carolina*	Grey squirrel	*Sciurus carolinensis*
Eastern chipmunk	*Tamias striatus*	Grizzly bear	*Ursus arctos horribilis*
Eastern diamondback rattlesnake	*Crotalus adamanteus*	Growling grass frog	*Litoria raniformis*
		Harbour porpoise	*Phocoena phocoena*
Eastern grey kangaroo	*Macropus giganteus*	Hawk-moth	*Macroglossum stellatarum*
Eastern hognosed snake	*Heterodon platirhinos*	Hazel grouse	*Bonasa bonasia*
Eastern massasauga rattlesnake	*Sistrurus catenatus catenatus*	Hedgehog	*Erinaceus europaeus*
		Hermit beetle	*Osmoderma eremita*
Eastern newt	*Notophthalmus viridescens*	Highway iceplant	*Carpobrotus* sp.
Elephant-African	*Loxodonta africana*	Hilgert's vervet monkey	*Chlorocebus pygerythrus hilgerti*
Elephant-Indian	*Elephas maximus indicus*		
Elk	*Cervus elaphus*	Hoary bat	*Lasiurus cinereus*
Eurasian lynx	*Lynx lynx*	Hyena	*Hyaena hyaena*
Eurasian red squirrel	*Sciurus vulgaris*	Ibean yellow baboon	*Papio cynocephalus ibeanus*
European eel	*Anguilla anguilla*	Iberian hare	*Lepus granatensis*
European rabbit	*Oryctolagus cuniculus*	Iberian lynx	*Lynx pardinus*
European robin	*Erithacus rubecula*	Ide	*Leuciscus idus*
European roe deer	*Capreolus capreolus*	Indo-Pacific humpback dolphin	*Sousa chinensis*
European tree frog	*Hyla arborea*		
European wildcat	*Felis silvestris*	Insects	*Insecta*

Jaguar	*Panthera onca*	Oleander	*Nerium oleander*
Kangaroos	*Macropus spp.*	Osprey	*Pandion haliaetus*
Kiang	*Equus kiang*	Ovenbird	*Seiurus aurocapillus*
Kinkajo	*Potos flavus*	Owls	*Strigiformes*
Kulan	*Equus hemionus*	Painted turtle	*Chrysemys picta*
Lace monitor	*Varunus varius*	Pallid bat	*Antrozous pallidus*
Lava lizard	*Microlophus albemarlensis*	Pampas grass	*Cortaderia selloana*
Least chipmunk	*Tamias minimus*	Passerines	*Passeriformes*
Least weasel	*Mustela nivalis*	Pels fishing owl	*Scotopelia peli*
Leopard	*Panthera pardus*	Peregrine falcon	*Falco peregrinus*
Lesser horseshoe bat	*Rhinolophus hipposideros*	Peruvian pepper	*Schinus molle*
Lesser long-nosed bat	*Leptonycteris yerbabuenae*	Peter's Angola colobus	*Colobus angolensis*
Lion	*Panthera leo*		*palliatus*
Little fire ant	*Wasmannia auropunctata*	Peter's duiker	*Cephalophus callipygus*
Little owl	*Athene noctua*	Phasianidae	Family of birds which
Lord Derby's scalytailed squirrel	*Anomalurus derbianus*		includes pheasants partridges and quail,
Macrozoobenthos	Consists of a taxonomic selection of the fauna larger than ca. 1 mm that lives at the bottom of a water body		junglefowls, chickens and peafowls
		Pike-Northern	*Esox lucius*
		Polar bear	*Ursus maritimus*
Malayan tapir	*Tapirus indicus*	Polecat	*Mustela putorius*
Malayan tiger	*Panthera tigris jacksoni*	Potto	*Perodicticus potto*
Manchurian hare	*Lepus mandschuricus*	Pronghorn/pronghorn antelope	*Antilocapra americana*
Marram	*Ammophila arenaria*	Przewalski's wild horses	*Equus ferus caballus*
Marsh owl	*Asio capensis*	Puff adder	*Bitis arietans*
Massasauga rattlesnake	*Sistrurus catenatus*	Quagga	*Equus quagga*
Meadow vole	*Microtus pennsylvanicus*	Rabbit-European	*Oryctolagus cuniculus*
Mesquite	*Prosopis glandulosa hybrids*	Rat-Black	*Rattus rattus*
Mitchell's hopping mouse	*Notomys mitchellii*	Rat-Brown	*Rattus norvegicus*
Mojave fringetoed lizard	*Uma scoparia*	Rattle grasshopper	*Psophus stridulus*
Mongolian gazelle	*Procapra gutturosa*	Red crab	*Gecarcoidea natalis*
Mongoose	*Herpestes edwardsii*	Red deer	*Cervus elaphus*
Moor frog	*Rana arvalis*	Red fox	*Vulpes vulpes*
Moose	*Alces alces*	Red kangaroo	*Macropus rufus*
Mountain pygmy possum	*Burramys parvus*	Red-backed salamander	*Plethodon cinereus*
Mule deer	*Odocoileus hemionus*	Red-necked wallaby	*Macropus rufogriseus*
Mulga	*Acacia aneura*	Red-sided garter snake	*Thamnophis sirtalis*
Musk beetle	*Aromia moschata*		*parietalis*
Nashville warbler	*Vermivora ruficapilla*	Reptiles	*Reptilia*
Nene	*Branta sandvicensis*	Otter-European	*Lutra lutra*
Nilgai	*Boselaphus tragocamelus*	Otter-North American River	*Lontra canadensis*
North American elk	*Cervus canadensis*		
Northern goshawk	*Accipiter gentilis*	Rock bunting	*Emberiza cia*
Northern hairynosed wombat	*Lasiorhinus krefftii*	Rock monitor	*Varanus albigularis*
		Rodents	*Rodentia*
Northern leopard frog	*Lithobates pipiens*	Roe deer	*Capreolus capreolus*
Northern two-lined salamander	*Eurycea bislineata*	Rotund disc	*Discus rotundatus*
		RoyaltTern	*Thalasseus maximus*
Northern wheatear	*Oenanthe oenanthe*	Sable	*Martes zibellina*

Sage-grouse	*Centrocercus* sp.
Saiga antelope	*Saiga tatarica tatarica* (mainly Kazakhstan and Russia) and *Saiga tatarica Mongolica* (in Mongolia)
Sand lizard	*Lacerta agilis*
Scimitar horned oryx	*Oryx dammah*
Scrub hare	*Lepus saxatilis*
Sea turtles	*Chelonioidea*
Seal salamander	*Desmognathus monticola*
Sedge wren	*Cistothorus platensis*
Siberian roe deer	*Capreolus pygargus*
Siberian weasel	*Mustela sibirica*
Sloth bear	*Melursus ursinus*
Slow worm	*Anguis fragilis*
Song sparrow	*Melospiza melodia*
Southern brown bandicoot	*Isoodon obesulus*
Southern cassowary	*Casuarius casuarius johnsonii*
Spotted eagle owl	*Bubo africanus*
Spotted hyena	*Crocuta crocuta*
Spotted turtle	*Clemmys guttata*
Spring peeper	*Pseudacris crucifer*
Springbok	*Antidorcas marsupialis*
Spurfowl	*Pternistes swainsonii*
Squirrel glider	*Petaurus norfolcensis*
Stag beetle	*Lucanus cervus*
Steppe grasshopper	*Chorthippus dorsatus*
Stoat	*Mustela erminea*
Stone marten	*Martes foina*
Straw-coloured fruit bat	*Eidolon helvum*
Sugar gum	*Eucalyptus cladocalyx*
Sun bear	*Helarctos malayanus*
The elk or wapiti	*Cervus canadensis*

Tibetan antelope	*Pantholops hodgsonii*
Tibetan gazelle	*Procapra picticaudata*
Tiger	*Panthera tigris*
Tiger salamander	*Ambystoma tigrinum tigrinum*
Timber rattlesnake	*Crotalus horridus*
Tree kangaroos	*Dendrolagus* spp.
Water vole	*Arvicola sapidus* and *Arvicola terrestris*
Weasels	*Mustelidae*
White fruit bat	*Ectophylla alba*
White-footed mouse	*Peromyscus leucopus*
White-spotted char	*Salvelinus leucomaenis*
White-tailed deer	*Odocoileus virginianus*
White-throated treecreeper	*Cormobates leucophaea*
Wild boar wild or feral pig	*Sus scrofa*
Wild turkey	*Meleagris gallopavo*
Wild yak	*Bos grunniens*
Wildebeest	*Connochaetes gnu*
Winter wren	*Troglodytes troglodytes*
Wolverine	*Gulo gulo*
Wood frog	*Rana sylvatica*
Wood turtle	*Glyptemys insculpta*
Woodland caribou	*Rangifer tarandus caribou*
Woodland jumping mouse	*Napaeozapus insignis*
Yellow baboon	*Papio cynocephalus*
Yellow-backed duiker	*Cephalophus silvicultor*
Yellow-bellied marmot	*Marmota flaviventris*
Yellow-throated marten	*Martes flavigula*
Zanzibar red colobus	*Procolobus kirkii*
Zanzibar Sykes' monkey	*Cercopithecus mitis albogularis*
Zebra	*Equus burchelli/Equus quagga/Equus zebra*

INDEX

Note: Page numbers in *italics* refer to illustrations; those in **bold** refer to tables

Handbook of Road Ecology, First Edition. Edited by Rodney van der Ree, Daniel J. Smith and Clara Grilo.
© 2015 John Wiley & Sons, Ltd. Published 2015 by John Wiley & Sons, Ltd.
Companion website: www.wiley.com\go\vanderree\roadecology

Printed and bound by CPI Group (UK) Ltd, Croydon, CR0 4YY

18/03/2024

14472084-0001